T0145226

Lecture Notes on Data Engineering and Communications Technologies

179

Series Editor

Fatos Xhafa, *Technical University of Catalonia, Barcelona, Spain*

The aim of the book series is to present cutting edge engineering approaches to data technologies and communications. It will publish latest advances on the engineering task of building and deploying distributed, scalable and reliable data infrastructures and communication systems.

The series will have a prominent applied focus on data technologies and communications with aim to promote the bridging from fundamental research on data science and networking to data engineering and communications that lead to industry products, business knowledge and standardisation.

Indexed by SCOPUS, INSPEC, EI Compendex.

All books published in the series are submitted for consideration in Web of Science.

Faisal Saeed · Fathey Mohammed ·
Errais Mohammed · Tawfik Al-Hadhrami ·
Mohammed Al-Sarem
Editors

Advances on Intelligent Computing and Data Science

Big Data Analytics, Intelligent Informatics, Smart Computing, Internet of Things

 Springer

Editors
Faisal Saeed
School of Computing and Digital Technology
Birmingham City University
Birmingham, UK

Errais Mohammed
Faculty of Sciences
University of Hassan II Casablanca
Casablanca, Morocco

Mohammed Al-Sarem
College of Computer Science
and Engineering
Taibah University
Medina, Saudi Arabia

Fathey Mohammed
Information Systems Department, School
of Computing
Universiti Utara Malaysia (UUM)
Sintok, Malaysia

Tawfik Al-Hadhrami
Department of Computing and Technology
Nottingham Trent University
Nottingham, UK

ISSN 2367-4512   ISSN 2367-4520 (electronic)
Lecture Notes on Data Engineering and Communications Technologies
ISBN 978-3-031-36257-6   ISBN 978-3-031-36258-3 (eBook)
https://doi.org/10.1007/978-3-031-36258-3

This Springer imprint is published by the registered company Springer Nature Switzerland AG
The registered company address is: Gewerbestrasse 11, 6330 Cham, Switzerland

Preface

We would like to thank all the authors for the participation in the 3rd International Conference of Advanced Computing and Informatics (ICACIN 2022) that was held in Hassan II University on October 15–16, 2022, and organized by Faculty of Sciences Ain Chock Casablanca, Hassan II University, and the Yemeni Scientists Research Group (YSRG) in collaboration with Data Analytics and Artificial Intelligence Research Group (DAAI) at Birmingham City University (UK), Information Service Systems and Innovation Research Group (ISSIRG) in Universiti Teknologi Malaysia (Malaysia), Higher Normal School of Casablanca (Morocco), and Association of Information Systems—Malaysian Chapter (MyAIS).

ICACIN 2022 is a forum for the presentation of technological advances in the field of computing and informatics. In this conference, 98 papers have been submitted by researchers from more than 20 countries. Of these 98 submissions, only 60 submissions have been selected to be included in this book.

The book presents several research topics which include data science, big data analytics, Internet of things (IoT), smart computing, artificial intelligence, machine learning, information security, intelligent communication systems, health informatics and information systems theories and applications.

We would like to express our appreciations to all the authors and keynote speakers for sharing their expertise with us. Special thanks go to the organizing committee for their great efforts in managing the conference. In addition, we would like to thank the technical committee for reviewing all the submitted papers; Prof. Dr. Fatos Xhafa, the series editor; Dr. Thomas Ditzinger and all editors from Springer.

Organization

ICACin'22 Organizing Committee

Honorary Chairs

Saddiqi Omar — Dean of the Faculty of Science, Ain Chock, Hassan II University Casablanca, Morocco

Noorminshah Iahad — Head of Association of Information Systems—Malaysian Section (MyAIS)

International Advisory Board

Naomie Salim	Universiti Teknologi Malaysia, Malaysia
Rose Alinda Alias	Universiti Teknologi Malaysia, Malaysia
Ahmad Lotfi	Nottingham Trent University, UK
Funminiyi Olajide	Nottingham Trent University, UK
Essa Hezzam	Taibah University, KSA
Mohammed Alshargabi	Najran University, KSA
Abdullah Ghareb	University of Saba Region, Yemen
Shadi Basurra	Birmingham City University, UK

Conference General Chairs

Errais Mohammed — Hassan II University, Casablanca, Morocco

Faisal Saeed (President) — Birmingham City University, UK; Yemeni Scientists Research Group (YSRG)

Program Committee Co-chairs

Raouyane Brahim — Hassan II University—Casablanca, Morocco

Mohammed Al-Sarem — Taibah University, KSA

Publications Committee

Fathey Mohammed — Universiti Utara Malaysia (UUM), Malaysia

Tawfik Al-Hadhrami — Nottingham Trent University, UK

Publicity Committee

Abdullah Aysh Dahawi	Universiti Teknologi Malaysia, Malaysia
Mandar Meriem	National School of Applied Science of Berrechid, Morocco
Yousef Fazea	Marshall University, West Virginia, USA

IT Committee

Moussaid Khalid	Hassan II University, Casablanca, Morocco
Chbihi Louhdi Mohammed Réda	Hassan II University, Casablanca, Morocco
Nahal Tarik	Hassan II University, Casablanca, Morocco

Secretary

Baddi Youssef	Chouaib Doukkali University, Morocco
Chiba Zouhair	Hassan II University, Morocco
Seddik El Kasmi	Hassan II University, Morocco
Chenyour Tarik	Hassan II University, Morocco

Finance Committee

Mohammed Errais	Hassan II University, Casablanca, Morocco
Abdullah Dahawi	Universiti Teknologi Malaysia, Malaysia

Logistic Committee

Rachdi Mohamed	ISGA–Casablanca, Morocco
Dehbi Rachid	Hassan II University, Casablanca, Morocco
Yassine Khazri	Hassan II University, Casablanca, Morocco

Registration Committee

Sameer Hasan Albakri	Universiti Teknologi Malaysia, Malaysia

Sponsorship Committee

Mandar Meriem	Hassan II University, Casablanca, Morocco
Fejtah Leila	Hassan II University, Casablanca, Morocco
Jai Andaloussi Said	Hassan II University, Casablanca, Morocco
Rachdi Mohammed	ISGA–Casablanca, Morocco

Technical Program Committee

Track Chairs

Wadii Boulila	Prince Sultan University, KSA
Maha Idriss	Prince Sultan University, KSA
Mohammed Al-Sarem	Taibah University, KSA
Faisal Saeed	Birmingham City University, UK
Jawad Ahmad	Edinburgh Napier University, UK
Mandar Meriem	ENS, CASABLANCA, Morocco
Raouyane Brahim	Hassan II University, Morocco
Mohamed Rachdi	ISGA Casablanca, Morocco
Tawfik Al-Hadhrami	Nottingham Trent University, UK

Committee Members

Aamre Khalil	Lorraine University, France
Abbas Dandache	Lorraine University, France
Abdallah AlSaeedi	Taibah University, KSA
Abdelaziz Ettaoufik	Hassan II University—Casablanca, Morocco
Abdelhamid Belmakki	INPT, Rabat, Morocco
Abdellah Kassem	UND, Lebanon
Abdelltif El Byed	Hassan II University—Casablanca, Morocco
Abderhmane Jarrou	Lorraine University, France
Abderrahim Maizat	Hassan II University—Casablanca, Morocco
Abderrahim Tragha	Hassan II University—Casablanca, Morocco
Abdullah Ghareb	University of Saba Region, Yemen
Abdulrahman Alsewari	Universiti Malaysia Pahang, Malaysia
Abdulrazak Alhababi	UNIMAS, Malaysia
Adil Sayouti	Ecole Royale Navale, Morocco
Ahmed Lakhssassi	University of Quebec in Outaouais, Canada
Ahmed Madian	NISC, Egypt
Ali Balaid	Aden University, Yemen
Ali Jwaid	Nottingham Trent University, UK
Arfat Ahnad Khan	Suranaree University of Technology, Thailand
Ashraf Osman	Alzaiem Alazhari University, Sudan
Asmaa Benghabrit	ENIM, Rabat, Morocco
Assia Bakali	Ecole Royale Navale, Morocco
Azouazi Mohamed	FS Ben M'sik, Casablanca, Morocco
Beverley Cook Essa Hezzam	Nottingham Trent University, UK
Bouragba Khalid	EST, Casablanca, Morocco

Brahim Hmedna	UIZ Agadir, Morocco
Brahim Raouyane	Hassan II University—Casablanca, Morocco
Chiba Zouhair	Hassan II University—Casablanca, Morocco
El Filali Sanae	FS Ben M'sik, Casablanca, Morocco
El Handri Kaoutar	ISGA, Casablanca, Morocco
El Houssine Ziyati	EST, Casablanca, Morocco
El Mehdi Bendriss	Vinci, Morocco
Elmoufidi Abdelal	FST, Beni Mellal, Morocco
El Moukhtar Zemmouri	Moulay Ismail University, Meknes, Morocco
Emmanuel Simeu	Grenoble-Alpes University, France
Er-Rafyg Aicha	Université Mohamed VI Rabat, Morocco
Fabrice Monteiro	Lorraine University, France
Faisal Saeed	Taibah University, KSA
Fathey Mohammed	Universiti Utara Malaysia (UUM), Malaysia
Georgina Cosma	Nottingham Trent University, UK
Hilal Imane	ESI, Rabat, Morocco
Israa Al_Barazanchi	University, Baghdad, Iraq
Issam Zahraoui	ISGA Casablanca, Morocco
Kamal Eddine El Khadri	Abdelmalek Essaadi University—Tangier, Morocco
Khalid Elfahssi	Sidi Mohamed Ben Abdellah University—Fez, Morocco
Khalid Moussaid	Hassan II University—Casablanca, Morocco
Lamiae Demraoui	EMSI, Morocco
Leila Fejtah	Hassan II University—Casablanca, Morocco
Meriem Mandar	Hassan I University, Settat, Morocco
Mohammed Alshargabi	Najran University, KSA
Mohammed Alsarem	Taibah University, KSA
Mohammed Baidada	ISGA Rabat, Morocco
Mohamemd El Youssfi	Hassan II University—Casablanca, Morocco
Mohammed H. Jabreel	Universitat Rovira i Virgili, Spain
Mohammed Hadwan	Qasim University, KSA
Mohammed Karim Guennoun	EHTP, Morocco
Mohammed Kolib	Taibah University, KSA
Mohammed Lahbi	Hassan II University—Casablanca, Morocco
Mohammed Laraqui	Ibnou Zohr University, Agadir, Morocco
Mohammed Rachdi	Hassan II University—Casablanca, Morocco
Mohammed Réda Chbihi Louhdi	Hassan II University—Casablanca, Morocco
Mohammed Rida	Hassan II University—Casablanca, Morocco
Mohammed Talea	Hassan II University—Casablanca, Morocco
Mostafa Bellafkih	INPT, Rabat, Morocco
Mostafa Belmakki	INPT, Rabat, Morocco

Mouhamad Chehaitly	Lorraine University, France
Mounia Abik	ENSIAS, Rabat, Morocco
Moussaid Laila	ENSEM, Casablanca, Morocco
Nabil Madrane	Hassan II University—Casablanca, Morocco
Nadhmi Gazem	Taibah University, KSA
Nadia Saber	EMSI, Morocco
Najib El Kamoun	Chouaib Doukkali University—El Jadida, Morocco
Neesrin Ali Kurdi	Taibah University, KSA
Noorminshah Iahad	Universiti Teknologi Malaysia, Malaysia
Noura Aknin	Abdelmalek Essaadi University—Tangier, Morocco
Noureddine Abghour	Hassan II University—Casablanca, Morocco
Ouail Ouchetto	Hassan II University—Casablanca, Morocco
Rabab Chakhmoune	Brams, Morocco
Rachida Aitabdelouahid	Faculté des sciences Ben M'sik, Morocco
Rachid Dehbi	Hassan II University—Casablanca, Morocco
Rachid Saadane	EHTP, Casablanca, Morocco
Rashiq Marie	Taibah University
Rashiq Rafiq	Taibah University, KSA
Rim Koulali	Hassan II University—Casablanca, Morocco
Said Jai Andaloussi	Hassan II University—Casablanca, Morocco
Said Harchi	ISGA Rabat, Morocco
Said Lotfi	Université Hassan II, Casablanca, Morocco
Said Ouatik El Alaoui	Ibn Tofail University—Kenitra, Morocco
Sanaa Elfilali	Hassan II university—Casablanca, Morocco
Siham Lamzabi	ISGA Rabat, Morocco
Slim Lammoun	ENSIT, Tunis University
Soufiane Jounaidi	ISGA Casablanca, Morocco
Stephen Clark	Nottingham Trent University, UK
Taha Hussein	Universiti Malaysia Pahang, Malaysia
Tariq Saeed Mian	Taibah University, KSA
Tarik Nahal	Hassan II University—Casablanca, Morocco
Tawfik Al-Hadhrami	Nottingham Trent University, UK
Wadii Boulila	University of Manouba, Tunisia
Wafaa Dachry	Hassan I University—FST—Settat, Morocco
Yassine Khazri	Hassan II University of Casablanca, Morocco
Yousef Fazea	Marshall University, West Virginia, USA
Youssef Baddi	Chouaib Doukkali University - El Jadida, Morocco
Zahour Omar	FS Ben M'sik, Casablanca, Morocco

Contents

Artificial Intelligence

Networking and IoT

Data Science

Information Security

Computational Informatics

Artificial Intelligence

Opinion Mining on Paul W. S. Anderson's *Monster Hunter* from Chinese Social Media Using Sentiment Analysis

Lan Zhenghua[1], Pantea Keikhosrokiani[1,2](✉) (iD), and Moussa Pourya Asl[3,4] (iD)

[1] School of Computer Sciences, Universiti Sains Malaysia, 11800 Penang, Malaysia
pantea@usm.my
[2] Faculty of Information Technology and Electrical Engineering, University of Oulu, Oulu, Finland
[3] School of Humanities, Universiti Sains Malaysia, Penang, Malaysia
moussa.pourya@usm.my
[4] Faculty of Humanities, University of Oulu, Oulu, Finland

Abstract. The growth of social media platforms has enabled people to publicly respond to matters such as racial and gender inequalities. On a recent case in China, individuals used online platforms to express their outrage at racial insults against the Chinese in Paul W. S. Anderson's *Monster Hunter* (2020). This article uses sentiment analysis and natural language processing to analyze Chinese users' views and responses on short-text social media platform called Weibo. More than 3,000 users' reviews are collected through a web crawler tool. An unsupervised way is chosen to carry out Sentiment Analysis. Findings reveal that the overall proportion of negative reviews about the movie is not high. Positive comments are relatively normal, and most of them are comments on the movie itself. There were negative comments before the movie was released but these comments did not contain content related to "辱华". The negative reviews after the release of the movie include different types of content: some are about the movie itself, some are because of "辱华", and some mention "腾讯". This study can benefit large-scale decision-makers on matters related censorship and filtering.

Keywords: Natural language processing · Sentiment analysis · Monster Hunter · Social Media

1 Introduction

Social media has provided a universal platform for people around the world to interact with each other, share information, and express their thoughts and emotions about a particular topic. Recent studies on social networking sites such as Twitter, Facebook, MySpace, and YouTube have sufficiently underlined the unparalleled significance of these platforms in providing the public with a free space to create and present their opinions. What remains under researched, however, is the way social media can help governments and companies with their large-scale decision-makings. Currently, there is

F. Saeed et al. (Eds.): ICACIn 2022, LNDECT 179, pp. 3–15, 2023.
https://doi.org/10.1007/978-3-031-36258-3_1

no public dataset available from the movie review of Paul W. S. Anderson's "Monster Hunter" (2020). In addition, existing studies lack a proper analytical model to monitor the public opinion presented in social media and to find the possible relationship between users' responses for the movie "Monster Hunter". In the movie, China is insulted in a dialogue with a racist playground. Therefore, we argue that a systematic analysis and thorough understanding of public response toward a product or a policy can help to control, limit, and stop the spread of negative reactions.

This study aims to develop a data-analytics model to monitor the public opinion presented in social media and to find the possible relationship between users' responses and the product or the topic in question. As a case study, we focus on a recent incident in China where movie theatres across the country had to stop screening of a Hollywood movie named *Monster Hunter* (2020), directed by Paul W. S. Anderson, due to increasing public outrage in social media against the racially insulting content of the movie. The initial spread of public opinion against the movie took place in China-based social media platforms such as Douban (similar to IMDB) and Weibo (similar to Twitter and Facebook) as well as applications like Tao Piaopiao that sells movie tickets and has user comment functions. The proposed analytical model first uses Opinion Mining to scrap public opinions expressed in Weibo about the movie trailer *Monster Hunter* (2020). Then Sentiment Analysis is used to determine the different levels of the extracted data.

This paper is structured as follows: First, the literature on Natural Language Processing (NLP) and Sentiment Analysis is thoroughly reviewed and a comparison between Chinese and English NLP technologies is provided. Then, the materials and methodology are fully described. Next is the results, the discussion on the experimental results as well as an evaluation of the model. And finally, the conclusion and future works are described in last section.

2 Literature Review

2.1 Natural Language Processing and Opinion Mining

NLP is an interdisciplinary subject that includes computer science, artificial intelligence, and linguistics. These subjects are both different and intersecting. The development of NLP can be divided into four stages: the budding period before 1956, the rapid development period from 1957 to 1970, the low development period from 1971 to 1993, and the recovery and integration period from 1994 to now. As an important branch of artificial intelligence, NLP also occupies an increasingly important position in the field of data processing and is applied for opinion mining and text mining. Applications of NLP include several fields of studies, such as machine translation, natural language text processing and summarization, user interfaces, multilingual and cross language information retrieval (CLIR), speech recognition, artificial intelligence, and expert systems, and so on [1–4]. Opinion mining is the systematic identification, extraction, quantification, and study of emotional states and subjective information through the use of natural language processing, text analysis, computational linguistics, and biometrics [5–7]. Natural Language Processing (NLP), and more especially sentiment analysis, may be used to determine how people truly feel about a topic.

2.2 Sentiment Analysis

Sentiment analysis and opinion mining is the study of people's views, feelings, assessments, attitudes, and emotions as expressed in written language. It is one of the most active study fields in natural language processing, and it is also extensively researched in data mining, Web mining, and text mining [5]. There are three main types of sentiment analysis algorithms. One is the Learn-based method, which trains the classifier based on the co-occurrence frequency of different words in the document [8–12]. The second is the Lexicon-based method, which uses a sentiment dictionary to determine the sentiment polarity of a sentence. The third is a hybrid method, which uses both the Lexicon-based method and the Learn-based method.

3 Materials and Methodology

According to the characteristics of this project, combined with the life cycle of the data science project, the project framework diagram shown in Fig. 1 is proposed. The first step, after defining the goal of study, is to collect the comment data form Weibo. Then, there is the work of text pre-processing, which is mainly to prepare for data modelling. The fourth step is the data modelling stage. The main modelling methods used in this project is Sentiment Analysis. The last step is to evaluate the model.

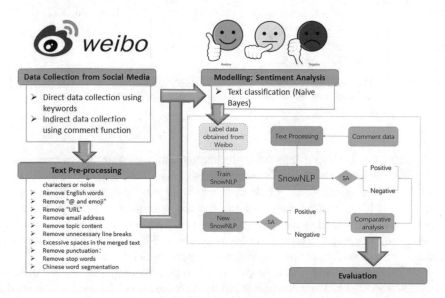

Fig. 1. The proposed data science lifecycle.

The public opinion event involved in this project is that the movie *Monster Hunter* (2020) was banned because of insulting the Chinese. There are many ways to obtain comments related to the event, such as obtaining related discussions via social platforms like Weibo, or obtaining comments related to the movie on a film review website like Douban or obtaining comments via other platforms like Zhihu. We chose to collect experimental

6   L. Zhenghua et al.

data from the Chinese social platform Weibo for the following three reasons: (1) The incident in which the movie *Monster Hunter* (2020) was banned for insulting China took place in China. The use of Chinese comments is conducive to establishing an effective model. (2) Weibo is currently the largest social network platform in China, with very high user activity and a wide range of information categories. Users can post their own views or comment on other users' views. (3) The relevant data of many platforms has been deleted, and currently only Weibo still retains some data.

In order to speed up the collection process, only the data used in the project is collected. The data description is shown in Table 1.

Table 1. Data description.

Feature Name	Description	Type	Example
user_id	The id of the user who posted the comment, used to identify the data, not a key feature	Text	C_4578444643933838 C_2351434912083141
comment	The comment data posted by users is the main data used for analysis	Text	"这真忽悠观众听不懂英语吗。(Does this really fool the audience that they don't understand English?)"
date	The time for users to post comments, in days	Category	2020-12-04, 2020-12-06
thumbs_up	The number of likes a user has made for a comment	Numeric	625, 206
source	Data acquired through direct collection or data acquired indirectly	Category	C(Direct collection), s (Indirect collection)

The text data obtained using crawler technology often contains some meaningless characters, which can also be called noise. These noises have negative impact on the model. In order to eliminate the influence of noise and make the data meet the requirements of modelling, some preprocessing of the text is required. In this paper, different steps were performed in data preprocessing step such as useless symbol removal, removing stop words, and Chinese word segmentation. We chose an open-source word segmentation tool with better python language support which is called jieba.

It can be seen from Fig. 1 that the core emotion polarity judgment method of semi-supervised learning used in this article is provided by the SnowNLP module. SnowNLP is a class library written by Python, which can easily handle Chinese text content. SnowNLP was written inspired by TextBlob, with the purpose of facilitating Chinese natural language processing. The SnowNLP library has the following characteristics: Chinese word segmentation, part of speech tagging, sentiment analysis (training data mainly uses review data of purchased products), text classification (Naive Bayes), convert to Pinyin (the maximum match achieved by the Trie tree), traditional Chinese to Simplified Chinese (Maximum matching achieved by Trie tree), extract text keywords, extract text summaries, tf, idf, tokenization (divide the article into sentences), text is similar (BM25), support python 3.

Although SnowNLP supports sentiment analysis, the data used by this model is mainly review data of purchased products. For the Weibo comment data, we used in this project, the effect may not be very good. However, SnowNLP provides a training framework, which means that we can train a more suitable sentiment analysis model through SnowNLP. In this experiment, we first directly use SnowNLP's sentiment analysis to analyze Weibo comment data and use the judgment result as the baseline. Then we use 100,000 marked Weibo comment data SnowNLP for training to get a new sentiment analysis model. Finally, the two sentiment analysis models are compared and analyzed, and the final sentiment analysis result is obtained.

4   Results and Discussion

4.1   Distribution of Sentiment Scores and the Changes in the Number of Negative and Positive Comments Over Time

First, using the data cleaning method, the final comment data that can be used to calculate sentiment scores is obtained. Then we used the sentiment analysis module provided in SnowNLP to calculate the sentiment score of each comment data. We drew the distribution of sentiment scores, as shown in Fig. 2. Take 0.5 as the median value of the sentiment score. Those with a score greater than 0.5 are considered positive comments, and those with a score less than 0.5 are considered negative comments. By observing Fig. 2, it can be found that the distribution of sentiment scores is relatively uniform in the interval of 0.1 to 0.9, and the total number of positive comments is greater than negative.

Before analyzing the relationship between negative and positive comments over time, it is important to note the relationship between the total number of comments over time. Figure 3 shows how the total number of comments changes over time. It can be seen from the image that the number of comments suddenly increased three times, on November 18th, November 25th, and December 4th. The subsequent analysis will explain why there has been a sudden increase in the number of comments three times.

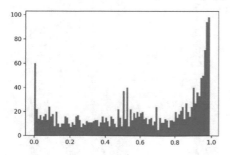

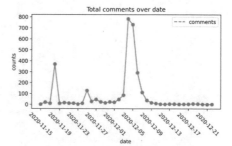

Fig. 2. Distribution map of sentiment score based on Original-SnowNLP.

Fig. 3. The relationship between the number of comments over time.

The relationship between the number of negative comments over time is shown in Fig. 4. Here, the number of negative comments increased sharply on December 4, 2020, and reached a peak on December 5, and then dropped rapidly. We know from the background knowledge of the event that the movie "Monster Hunter" was released on December 4, 2020. That is to say, the negative comments of the movie increased sharply as soon as it was released. The movie was subsequently banned from broadcasting on December 5. With the ban, negative reviews also declined. Thus, we can deduce that there is a strong correlation between negative reviews and the release of movies. The relationship between the number of positive comments over time is shown in Fig. 5. The change in the number of positive comments is very similar to the change in the total number of comments. There are three sudden increases in the number of comments.

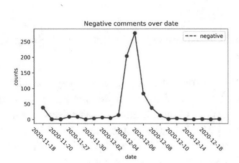 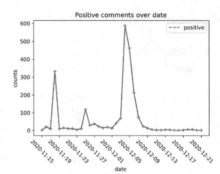

Fig. 4. The relationship between the number of negative comments over time.

Fig. 5. The relationship between the number of positive comments over time.

4.2 Results of Positive and Negative Reviews

The wordcloud for the positive and negative comments are shown in Fig. 6 and Fig. 7 respectively.

Fig. 6. Wordcloud analysis results of positive reviews.

Fig. 7. Wordcloud analysis results of negative reviews.

The analysis of these words shows that the words that appear frequently in positive comments can be roughly divided into three categories. The first category is related to names. Most of these names are actor names, such as: "米拉", "乔治维奇", and "保罗安德森". The second category is related to the movie, such as the location of the movie: "内地", when the movie will be released: "发布", "定档", movie trailer: "预告", movie type: "剧情". The third category is to express feelings about the movie, such as: "期待". Table 2 listed the 10 most frequent words in positive and negative reviews.

Table 2 shows the word frequency table corresponding to Fig. 7. Still take the top 10 words that appear more frequently. The analysis of these words showed that in addition to the more obvious words like "辱华", there are also two neutral words that appear more frequently, such as "腾讯" and "膝盖". The reason for knees can be known from background knowledge, because the key word in a nursery rhyme involving racial discrimination is knees. So why does "腾讯" appear in negative comments. By looking up relevant information, we learned that Tencent was the importer of this movie. In order to control negative reviews, Tencent deleted a large number of movie-related reviews on the Internet. This is why "腾讯" has appeared so many times. The second is some words that express anger and dissatisfaction, such as: "死", "恶心", "垃圾", "退票" and so on.

Table 2. The 10 most frequent words in positive and negative reviews.

Number	Frequency	Words	Translation	Number	Frequency	Words	Translation
Positive Reviews (Total words: 31956)				Negative Reviews (Total words: 4936)			
1	276	米拉	Milla	**1**	276	米拉	Milla
2	207	乔沃维奇	Jovovich	**2**	207	乔沃维奇	Jovovich
3	196	预告	Trailer	**3**	196	预告	Trailer
4	187	怪物	Monster	**4**	187	怪物	Monster
5	181	生化危机	Resident Evil	**5**	181	生化危机	Resident Evil
6	180	定档	Confirm the time	**6**	180	定档	Confirm the time
7	155	真人	Real person	**7**	155	真人	Real person
8	148	内地	Mainland	**8**	148	内地	Mainland
9	142	期待	look forward to	**9**	142	期待	look forward to
10	138	发布	Released	**10**	138	发布	Released

4.3 Analysis of Reviews Before and After the Movie Release Date

Here, the user reviews are divided by the date the movie was released. We wanted to see how the user reviews change before and after the movie is released. Figure 8 shows the wordcloud image generated by the reviews data before and after the movie is released,

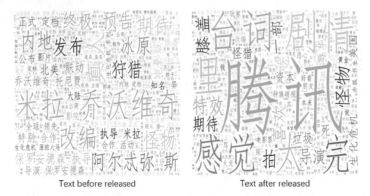

Text before released　　　　　　　　Text after released

Fig. 8. Wordcloud analysis results of reviews before and after the movie release date.

and Table 3 shows the word frequency table corresponding to the wordcloud image (take the top 10 most frequently).

The word frequency table shows that the reviews before the movie's release date are mostly positive, because many of the top 10 most frequent words overlap with the positive reviews (Table 3). Regarding the reviews after the movie was released, the frequency of the words "Tencent" and "knee" was also very high, indicating that the proportion of negative reviews began to increase after the movie was released.

Table 3. The 10 most frequent words in reviews before and after the movie release date.

Number	Before - Total words: 14072			After - Total words: 22820		
	Frequency	Words	Translation	Frequency	Words	Translation
1	222	米拉	Milla	95	腾讯	Tencent
2	182	定档	Confirm the time	92	剧情	Plot
3	173	乔沃维奇	Jovovich	81	台词	Actor's lines
4	167	预告	Trailer	71	怪物	Monster
5	143	内地	Mainland	70	感觉	feeling
6	137	真人	Real person	69	膝盖	Knee
7	124	发布	Released	68	米拉	Milla
8	122	怪物	Monster	68	特效	Special effects
9	116	生化危机	Resident Evil	67	生化危机	Resident Evil
10	112	保罗安德森	Paul Anderson	65	导演	Director

We know that the movie *Monster Hunter* (2020) was banned for insulting China, and the reviews that mentioned the film insulting China are often negative reviews, so the analysis of negative reviews is very necessary. Figure 9 and Table 4 show the wordcloud images before and after the movie is released and the corresponding word frequency table.

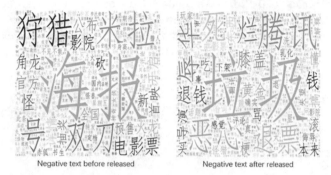

Negative text before released   Negative text after released

Fig. 9. Wordcloud analysis results of negative reviews before and after the movie release date.

In the negative reviews before the movie was released, there were no comments that contained "辱华" in the top 10 words, and even words that expressed emotions rarely appeared. However, when we checked the word cloud and word frequency list of comments after the movie was released, we found that it contained a large number of words expressing dissatisfaction, in addition to words related to "辱华". The negative reviews after the movie were released shows that there must be a plot in the movie that aroused the patriotism of some viewers, causing them to post a large number of negative reviews that have nothing to do with the movie.

Table 4. The 10 most frequent words in negative reviews before and after movie release date.

Number	Before - Total words: 706			After - Total words: 4204		
	Frequency	Words	Translation	Frequency	Words	Translation
1	9	海报	Poster	39	垃圾	Garbage
2	9	米拉	Milla	36	腾讯	Tencent
3	8	狩猎	Hunting	30	恶心	Nausea
4	7	双刀	Double knife	27	死	Dead
5	7	电影票	Movie ticket	23	退票	Return a ticket
6	5	角龙	Horned Dragon	22	辱华	Insult China
7	5	公布	Announce	21	钱	Money
8	5	影院	Cinema	18	烂	Very bad
9	5	书生	Student	18	退钱	Refund
10	4	火龙	Fire dragon	17	膝盖	Knee

4.4   Analysis of Extreme Points in Positive and Negative Reviews

The analysis of the changes in the number of positive reviews over time (Fig. 10) revealed that the number of reviews suddenly increased at three points in time, namely 2020-11-18, 2020-11-25, and 2020-12-04. We conducted a word cloud analysis on these three time points to find out why the number of reviews on these three dates suddenly increased through the analysis. First, we extracted the comment data related to these three days and plotted the corresponding word cloud (Fig. 10).

Fig. 10. Wordcloud analysis results of extreme points in positive reviews.

The corresponding top 10-word frequency table is shown in Table 5. By analyzing the word frequency table, we can see that on the day of 2020-11-18, the top two words mentioned in the reviews are "定档" and "内地", which means the release date of the movie "Monster Hunter" in China. On the day of 2020-11-25, the first two words that appeared the most in comments were "预告" and "终极". This is because the movie released the final trailer on this day, which led to an increase in the number of reviews.

Table 5. The 10 most frequent words in in the extreme points of positive reviews

Number	2020–11–18		2020–11–25		2020–12–04	
	Words	Translation	Words	Translation	Words	Translation
1	定档	Confirm the time	预告	Trailer	剧情	Plot
2	内地	Mainland	终极	Final	特效	Special effects
3	米拉	Milla	活动	Activity	生化危机	Resident Evil
4	乔沃维奇	Jovovich	发布	Released	完	Finish
5	预告	Trailer	特别	Special	米拉	Milla
6	怪物	Monster	米拉	Milla	怪物	Monster
7	正式	Official	真人	Real person	期待	look forward to
8	真人	Real person	联动	Linkage	玩家	Player
9	改编	Based on	冰原	Ice field	火龙	Fire dragon
10	期待	look forward to	合作	Cooperation	怪猎	Monster Hunter

Finally, regarding 2020-12-04, we all know that the movie was released in China on this day. Therefore, most of the positive reviews mentioned are about the movie's "剧情" and "特效" and so on.

For the extreme points of negative reviews, we chose the three days of 2020-11-18, 2020-12-04 and 2020-12-05. The related wordcloud analysis is shown in Fig. 11, and the word frequency table is shown in Table 6. The word frequency table shows that there are no obvious features in terms of the number and distribution of negative reviews on 2020-11-18. But the negative comments on 2020-12-04 and 2020-12-05 are more interesting. On 2020-12-04, the two most common words in negative reviews are "恶心" and "垃圾". These two words can be used to express dissatisfaction with the movie and can also be used to express dissatisfaction with the plot of insulting China. But on 2020-12-05, the word that appeared the most was "腾讯". This word is just a common term referring to Tencent. "腾讯" was mentioned the most in the negative reviews, which shows that the audience's dissatisfaction with the movie turned to Tencent. Of course, "辱华" was mentioned in the negative comments in the two days.

Fig. 11. Wordcloud analysis results of extreme points in negative reviews.

Table 6. The 10 most frequent words in in the extreme points of negative reviews.

Number	2020-11-18		2020-12-04		2020-12-05	
	Words	Translation	Words	Translation	Words	Translation
1	怪	Monster	恶心	Nausea	腾讯	Tencent
2	角龙	Horned dragon	垃圾	Garbage	垃圾	Garbage
3	双刀	Double knife	死	Dead	钱	Money
4	定档	Double knife	烂	Very bad	退票	Return a ticket
5	砍	Slash	火龙	Fire dragon	死	Dead
6	国内	Domestic	退票	Return a ticket	退钱	Refund
7	终于	Finally	辱华	Insult China	东西	Things
8	喷火	Spitfire	退钱	Refund	恶心	Nausea
9	走心	Go with heart	怪猎	Monster Hunter	辱华	Insult China
10	火	Fire	本来	At first	电影票	Movie ticket

5   Conclusion

The primary goal of this study was to obtain public opinions related to the movie trailer *Monster Hunter* (2020) from Weibo. More than 3,000 users' comments were collected through a web crawler tool and converted into a data set that can be used for analysis. The second goal was to perform sentiment analysis on the comment dataset and analyze the extracted negative comments. We chose an unsupervised way to carry out this stage. In order to ensure the validity of the sentiment analysis results, we used the data from Weibo to train a sentiment analysis model and used the analysis results of this model as a control group. The comparison found that the effect of the model we trained is not as good as that of the original model; besides, it also shows that the sentiment analysis results of the original model are reliable. After using the sentiment analysis results of the original model to conduct word cloud analysis, we found that although there were negative comments before the movie was released, these comments did not contain content related to "辱华". As for the negative reviews after the movie was released, there are many types of content: some are about the movie itself, some are because of "辱华", and some mention "腾讯". After searching for news related to the incident, we found that "腾讯" was mentioned because Tencent had deleted many comments about the movie on the internet in order to make the movie available for release, which aroused strong dissatisfaction among netizens. And this dissatisfaction is finally reflected in the negative reviews about the film. Since only a few people know about the plot of "辱华" in the movie, the overall proportion of negative reviews about the movie is not high. Positive comments are relatively normal, and most of them are comments on the movie itself. The main limitation of this study was related to data collection from only one social media platform called Webio as well as the exclusive focus on Chinese comments. Future studies can focus on opinion mining of English comments from other social media platforms using deep learning to get more accurate results.

References

1. Ying, S.Y., Keikhosrokiani, P., Asl, M.P.: Comparison of data analytic techniques for a spatial opinion mining in literary works: a review paper. In: Saeed, F., Mohammed, F., Al-Nahari, A. (eds.) Innovative Systems for Intelligent Health Informatics. IRICT 2020. LNDECT, vol. 72, pp. 523–535. Springer, Cham (2021). https://doi.org/10.1007/978-3-030-70713-2_49
2. Malik, E.F., Keikhosrokiani, P., Asl, M.P.: Text mining life cycle for a spatial reading of Viet Thanh Nguyen's The Refugees (2017). In: 2021 International Congress of Advanced Technology and Engineering (ICOTEN) (2021)
3. Ghosh, S., Gunning, D.: Natural Language Processing Fundamentals: Build Intelligent Applications that can Interpret the Human Language to Deliver Impactful Results. Packt Publishing Ltd. (2019)
4. Nadkarni, P.M., Ohno-Machado, L., Chapman, W.W.: Natural language processing: an introduction. J. Am. Med. Inform. Assoc. **18**(5), 544–551 (2011)
5. Keikhosrokiani, P., Asl, M.P. (eds.): Handbook of Research on Opinion Mining and Text Analytics on Literary Works and Social Media, pp. 1–462. IGI Global, Hershey, PA, USA (2022)

6. Yun Ying, S., Keikhosrokiani, P., Pourya Asl, M.: Opinion mining on Viet Thanh Nguyen's the sympathizer using topic modelling and sentiment analysis. J. Inf. Technol. Manage. (2022). 14(5th International Conference of Reliable Information and Communication Technology (IRICT 2020)), pp. 163–183

7. Paremeswaran, P.a., Keikhosrokiani, P., Asl, M.P.: Opinion mining of readers' responses to literary prize nominees on twitter: a case study of public reaction to the Booker Prize (2018–2020). In: Saeed, F., Mohammed, F., Ghaleb, F. (eds.) Advances on Intelligent Informatics and Computing. IRICT 2021. LNDECT, vol. 127. Springer, Cham (2022). https://doi.org/10.1007/978-3-030-98741-1_21

8. Sofian, N.B., Keikhosrokiani, P., Asl, M.P.: Opinion mining and text analytics of reader reviews of Yoko Ogawa's The Housekeeper and the Professor in Goodreads. In: Keikhosrokiani, P., Pourya Asl, M. (eds.) Handbook of Research on Opinion Mining and Text Analytics on Literary Works and Social Media, pp. pp. 240–262. IGI Global: Hershey, PA, USA (2022)

9. Al Mamun, M.H., et al.: Sentiment analysis of the Harry Potter series using a lexicon-based approach. In: Keikhosrokiani, P., Pourya Asl, M. (eds.) Handbook of Research on Opinion Mining and Text Analytics on Literary Works and Social Media, pp. 263–291. IGI Global, Hershey, PA, USA (2022)

10. Jafery, N.N., Keikhosrokiani, P., Asl, M.P.: Text analytics model to identify the connection between theme and sentiment in literary works: a case study of Iraqi life writings. In: Keikhosrokiani, P., Pourya Asl, M. (eds.) Handbook of Research on Opinion Mining and Text Analytics on Literary Works and Social Media, pp. 173–190. IGI Global, Hershey, PA, USA (2022)

11. Asri, M.A.Z.B.M., Keikhosrokiani, P., Asl, M.P.: Opinion mining using topic modeling: a case study of Firoozeh Dumas's *Funny in Farsi* in goodreads. In: Saeed, F., Mohammed, F., Ghaleb, F. (eds.) Advances on Intelligent Informatics and Computing, IRICT 2021. LNDECT, vol. 127. Springer, Cham (2022). https://doi.org/10.1007/978-3-030-98741-1_19

12. Suhendra, N.H.B., et al.: Opinion mining and text analytics of literary reader responses: a case study of reader responses to KL Noir volumes in Goodreads using sentiment analysis and topic. In: Keikhosrokiani, P., Pourya Asl, M. (eds.) Handbook of Research on Opinion Mining and Text Analytics on Literary Works and Social Media, pp. 191–239. IGI Global, Hershey, PA, USA (2022)

Kevin Kwan's *Crazy Rich Asians:* Opinion Mining and Emotion Detection on Fans' Comments on Social Media

Ong Mei Yee[1], Pantea Keikhosrokiani[1,2]([✉]) [iD], and Moussa Pourya Asl[3,4] [iD]

[1] School of Computer Sciences, Universiti Sains Malaysia, 11800 Penang, Malaysia
pantea@usm.my
[2] Faculty of Information Technology and Electrical Engineering, University of Oulu, Oulu, Finland
[3] School of Humanities, Universiti Sains Malaysia, Penang, Malaysia
moussa.pourya@usm.my
[4] Faculty of Humanities, University of Oulu, Oulu, Finland

Abstract. The Singapore-born American novelist Kevin Kwan's *Crazy Rich Asians* (2013) and its cinematic adaptation have evoked conflicting responses from the public in social media. Analyzing the huge amount of data posted online is an arduous task for literary scholars who use traditional methods to understand readers and viewers' responses to creative works of art. This study offers a data science approach for opinion mining and detecting embedded emotions in people's comments. To achieve this goal, a Machine Learning Model Classifier is developed to analyze embedded emotions in public responses to the movie adaptation of Kevin Kwan's *Crazy Rich Asians* on YouTube. A total of 41,161 comments posted before or on 27 December 2021 are collected using the YouTube Data API. An emotion classifier with Incremental Learning Approach is built to detect emotions associated with the comments. The built emotion classifier showed accuracy ranging from 84% to 93%. The findings reveal that most of the comments are associated with the emotions of curiosity, love, and surprise.

Keywords: Opinion Mining · Text Analytics · Emotion Detection · Machine Learning · Incremental Learning · *Crazy Rich Asians*

1 Introduction

With the advent of social media platforms, where users across the globe can express their thoughts and feelings about fictional writings, public and critical disputes over the production, dissemination, and reception of literary works have dramatically increased. One recent example that has attracted much public reaction in social media is the Singapore-born American novelist Kevin Kwan's *Crazy Rich Asians*. Published in 2013, the novel is a sarcastic romantic comedy that depicts Singapore's upper-crust society with their lavish culture. Following the commercial success of the novel, a major Hollywood studio produced a movie adaptation of the story by the same name in 2018. Although both the

F. Saeed et al. (Eds.): ICACIn 2022, LNDECT 179, pp. 16–28, 2023.
https://doi.org/10.1007/978-3-031-36258-3_2

novel and its cinematic version have won numerous accolades, they have evoked conflicting, yet largely negative, responses from the public. The huge amount of data—i.e., comments and criticisms—available on social media makes the task of literary scholars who use traditional methods to analyze readers and viewers' responses to the creative works arduous, flawed, and impossible.

Over the past two decades, developments in data science approaches and techniques have helped disciplines such as arts and humanities with objective and more comprehensive analysis of readers and audiences' responses to creative works of art. Opinion mining and emotion classification are used to classify readers' responses in social media using various data analytical techniques such as sentiment analysis, topic modelling, machine learning, deep learning, etc. [1–7]. This study aims to use a computerized method to analyze public reactions on social media to the movie adaptation of *Crazy Rich Asians*. The primary objective is to detect and classify the emotions associated with the opinions towards the movie with a classifier built with incremental learning approach. The data is collected from the movie's official YouTube channel that has received over sixteen thousand comments from viewers across the world. Machine learning techniques are used to create tow classifiers for the classification of the emotion categories of joy, love, optimism, annoyance, disappointment, sadness, surprise, curiosity, and confusion. The first machine learning model is called Classifier 1 which utilized Multinomial Naïve Bayes. It is created with the Binary Relevance approach, which treat each emotion as a single binary labelling task. The second classifier is built with Classifier Chains which is trained on the original data distribution with TF-IDF vectorizer as the features encoder. The algorithm used in the incremental learning is Logistic Regression.

This paper begins with an introduction where the objectives are stated. This is followed by literature review that focuses on text analytics, emotion classification, and incremental learning approaches. Next, materials and methodology are explained which elaborates on the techniques used to achieve the primary goal of the study. Results and findings are presented next, and finally, the paper ends with discussion and conclusion.

2 Literature Review

2.1 Social Media Text Analytics

Social media has become an inseparable part of our daily life. Besides serving as a platform to connect people, it plays an essential role as a platform for people to express their opinions, thoughts, and emotions about certain topics. With the increasing number of users of social media sites, these platforms contain a significant number of text data worth for opinion mining to identify people's reactions and needs.

Natural Language Processing (NLP) is a domain of machine-based processing of human communication [3, 8, 9]. With the available NLP techniques, machines could learn to process and understand human languages and process large-scale text data from social media. However, text data from social media are usually informal, associated with numerous accidental and deliberate errors and grammatical inconsistencies [1–3]. These text data could also be presented with emoticons, emojis, hashtags, abbreviations, and slangs, as they convey more emotional connection [10]. Noisy user-generated contents have become a challenge in applying NLP applications because of the difficulties for machines to derive the actual meaning from the text [5, 8, 9, 11]. To deal with the informal writing style on social media, it is recommended to pre-process text data based on the task to be performed. Studies [1–4, 12] have explored the impact of different pre-processing methods (URLs, negation, repeated letters, stemming and lemmatization) on the performance of Twitter sentiment classification. It is found that implementing URLs features reservation, negation transformation and repeated letters normalization increases the accuracy of sentiment classification. Conversely, the accuracy descends when stemming and lemmatization are used.

2.2 Emotion Classification

Developments in sentiment analysis have enhanced the formalization of emotion representation in digital media [13]. The techniques of detecting and classifying the emotion from text have become popular due to the essential role of emotions in human-machine interaction. Through identifying and classifying human emotions, machines are able to provide value-added services or generate more appropriate responses. A common approach to emotion analysis and modelling is characterization. For example, the basic emotions, mentioned by [14], are anger, disgust, fear, happiness, sadness, and surprise.

To extract textual features and create emotion dictionary, a practical technique should incorporate several statistical measures, such as the Pointwise Mutual Information parameter (PMI), information gain, and the term frequency-inverse document frequency (TF-IDF) to calculate the correlation score between each popular term in YouTube and each emotion type in lexica. TF-IDF is the most widely adopted method [13]. The underlying principle is that when a term appears frequently in one article but rarely in others, it is considered a highly representative term. TF refers to the frequency at which a term appears in an article, whereas IDF represents the inverse of articles containing this term. This method is usually adopted for building feature vectors.

Different methods were used in previous studies to build emotion detection and classification systems from text. The methods can be grouped into three categories, which are knowledge-based, machine learning based, and hybrid. Table 1 shows the summary of the comparison of three discussed approaches.

2.3 Incremental Learning Approaches

Incremental Learning (IL) refers to a learning system that can continuously learn new knowledge from new samples and can maintain most of the previously learnt knowledge [15]. An IL model can learn new knowledge and retain the one in lifelong time. IL is a special scenario of machine learning technology, which can deal with applications that are more consistent with human behavior and thinking. In recent years, it has played increasingly important role in the fields of automation.

Unstructured data are irregular information with no predefined data model. Comments that are obtained from social media are unstructured text data. Classification of unsupervised data is a challenging task due to their variability and the missing of labels. As the data keeps increasing, it becomes difficult to train and create a model from scratch each time. IL, a self-adaptive algorithm that utilizes the previously learnt model information, then learns and accommodates new information from newly arrived data providing

Table 1. Comparison between different approaches for Emotion Classification.

Method	Machine Learning Methods		Knowledge-based/Lexicon-based Method	Ensemble/Hybrid Method
Features	Supervised learning	Unsupervised learning	Domain knowledge and the semantic and syntactic characteristics of language	Combination of ML and lexicon-based approaches
Benefits	– Higher efficacy than UML	– No labelled data is required – Flexible to allow easy update of the data corpus	– High accessibility – Economic	Gain benefits from both ML and lexicon-based methods
Limitations	A training dataset with labels is required	Lower efficacy than supervised ML	– Rules are not comprehensive and need to be manually defined – Inability to handle complex linguistic rules	Computation complexity
Algorithms/Resources	– BOW – Naïve Bayes – SVM	– CNN – LSTM – ELM	– WordNet – SenticNet – ConceptNet – EmotiNet	– Combination of discussed methods

a new model, which avoids the retraining. The IL has been introduced recently as an alternative approach to classifying the unstructured data.

In addition, when applying IL approach, the adjustment of parameter weighs is made by a backpropagation algorithm according to loss on available sequential data. This will significantly lower model performance on knowledge that was learned previously. Madhusudhanan, S. et al. [16] have proposed a framework CUIL (Classification of Unstructured data using Incremental Learning) which clusters the metadata by training the data in batches. Others [17] have also implemented IL on the raw data permanently accepted from social networks while conducting the emotion analysis and classification. This IL approach has successfully reduced the memory resources and the training time significantly, compared to other methods.

3   Methodology

The research methodology of this project is based on the data science project lifecycle, which is the Cross-Industry Standard Process for Data-Mining (CRISP-DM) method. The lifecycle has been adopted and modified for this project as illustrated Fig. 1. This project focuses mainly on emotion classification. The lifecycle of emotion classification includes all stages in the lifecycle.

Due to the limitations of human factors, analyzing the data traditionally could lead to inefficient, inaccurate, and biased results. Hence, data science techniques are applied to

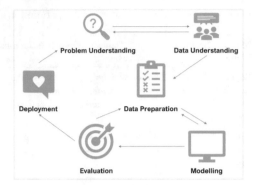

Fig. 1. Proposed data science lifecycle based on CRISP-DM method.

resolve the underlying problem with minimum requirement of traditional human reading and computation resources.

After recognizing the problem and setting up well-defined goals, we are now on the stage of data understanding. At this stage, the required data for this project has been identified. Two datasets are used in this project: (1) the GoEmotions dataset (published by Google) and (2) YouTube comments dataset (collected using available API).

GoEmotions dataset is used as the dataset to train the emotion classifier. This dataset is a human annotated dataset of 58k English-written Reddit comments with 27 emotion categories (12 positive, 11 negative, 4 ambiguous emotion categories). As shown in Fig. 2, these emotions can be clustered into three sentiment types which are positive, negative, and ambiguous. In this project, the emotion categories are reduced to 9 categories are joy, love, optimism, surprise, curiosity, confusion, disappointment, sadness, and anger.

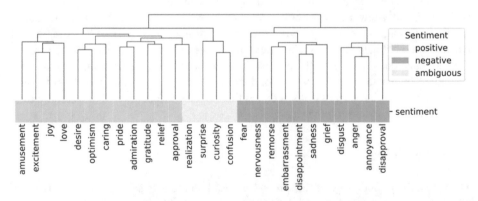

Fig. 2. Sentiment clustering of the emotion categories [18].

Each sentiment type has three emotion categories as shown in Table 2.

Table 2. Emotion categories for each sentiment type.

Sentiment	Emotion Category
Positive	Joy
	Optimism
	Love
Neutral	Confusion
	Curiosity
	Surprise
Negative	Disappointment
	Sadness
	Annoyance

The public opinion used in this project is the YouTube comments collected from several videos related to the movie. The 33 selected videos are those videos include only movie scene and the movie trailer. A total of 41,161 comments posted before or on 27 December 2021 are collected using the YouTube Data API. Table 3 shows the description of data collected from YouTube.

Table 3. Description of data collected from YouTube.

Variable	Description	Data Type
Author	Username of reviewer (author)	String
Comment	Actual content of the comment in string form	String
Time	Time of comment is published	String
Likes	Number of likes obtained by the comment	Integer
Reply Count	Number of replies to the comment	Integer
Comment/Reply	1 indicates the comment is a reply to the thread, 0 indicates the top-level comment	Boolean
VideoID	The ID number of the video where the comment is posted	String

As the datasets used for this project are the noisy text data from social media (Reddit and YouTube), they often include emojis, emoticons, hashtags, abbreviations, and slang. Therefore, text preparation to clean and process the text data is required to suit the use case of this project. The steps involved in this stage are Removal of duplicated posts, removal of noise, expanding of contractions, removing non-English comments, removal of punctuations, tokenization, conversion to lowercase, parts of speech tagging, lemmatization, and removal of stop words.

After preparing the text corpus, text mining experiments are done to build models for emotion classification. An emotion classifier is trained on the 211,225 rows of labelled data, GoEmotions. Since the task is a multi-label emotion classification, the problem is transformed into 3 approaches: Binary Relevance, Classifier Chains and Label Powerset. Binary Relevance treats each emotion label as a single classification problem while the Classifier Chains considers the correlation between the labels. Label Powerset transforms the multi-label classification task into a multi-class classification task. Naïve Bayes algorithm is implemented during the experiment as it is often used in supervised learning task and could perform with high accuracy.

To build a prototype, a sample of 60,000 data is used as this is the largest amount of data that could run without crashing the Google Colab after running out of RAM. TF-IDF is used to extract features from the texts for training. The RAM increased tremendously while encoding the texts into features. After the classifier models are built, the performance of the models will be evaluated with accuracy. The results are as shown in Table 4. The results from the first trial are extremely low for Naïve Bayes, as well as Decision Tree, Logistic Regression and Support Vector Machine. Although the data pre-processing steps are revised, the model built in second trial did not perform well.

Table 4. Results from two of the trials

No	Problem Transformation	Accuracy (%)	
		Trial 1	Trial 2
1	Binary Relevance	1.86	4.05
2	Classifier Chains	4.55	12.9
3	Label Powerset	26.80	28.36

Then, the next experiment is to implement incremental learning with all data for training the emotion classifier. The data are streamed into the pipeline of incremental learning model with TF-IDF as the features encoder. The extracted features will be input into the classifier to predict the class label. Then, the model will evaluate its performance by updating the metrics with the prediction and learn from the data. This process will loop until the model has learned from all training data. After the classifier model is trained, the performance of the models will be evaluated with the accuracy, precision, and recall metrics. The methodology used in this project to build the emotion classifier with incremental learning is as shown in Fig. 3.

Then, the model is implemented on the dataset of collected YouTube comments towards the movie *Crazy Rich Asians*. The built emotion classifier is used to identify the emotions present in the comments.

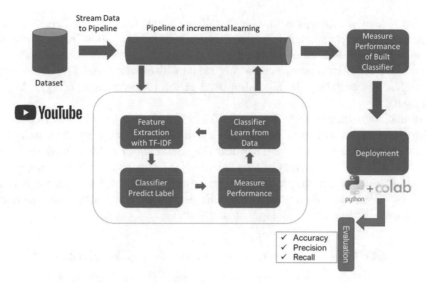

Fig. 3. Methodology of training emotion classifier with incremental learning

4 Analysis and Results

There are two classifiers built with the setting as shown in Table 5 to classify the emotion categories of joy, love, optimism, annoyance, disappointment, sadness, surprise, curiosity, and confusion. The first model is Classifier 1 built with the Binary Relevance approach, which treat each emotion as a single binary labelling task. TF-IDF Vectorizer is used in Classifier 1 to extract features from the text when the text data is streamed into the incremental learning model. Resampling is done in the incremental learning model as the real-world data do not always exist in a balanced distribution. The classifier is trained on an oversampled dataset with a data distribution of 50% labelled as True and 50% labelled as False. The selected algorithm for Classifier 1 is Multinomial Naïve Bayes.

Table 5. Settings of the classifier models.

Classifier	1	2
Problem Transformation	Binary Relevance	Classifier Chains
Resampling	Oversampling	No
Data Distribution	50%, 50%	Original
Feature Encoder	TF-IDF	TF-IDF
Incremental Learning	Yes	Yes
Resampling Before Data Streaming	No	No
Classifier Algorithm	Multinomial Naïve Bayes	Logistic Regression

The second classifier is built with Classifier Chains. Like Binary Relevance, each emotion is treated as a single binary classification task, but Classifier Chains also include the output label of other emotions. Hence, the correlation between the labels is also considered. The classifier is trained on the original data distribution with TF-IDF vectorizer as the features encoder. The algorithm used in the incremental learning is Logistic Regression.

Both classifiers are trained on Google Colab with the available RAM of 12GB. The performance of the model is measure during the training and after the training with the accuracy, precision, and recall metrics. The performance of the trained model is as shown in Table 6. The accuracy of the Classifier 1 is around 82% to 89% while training incrementally. However, the precision and recall are low. After the incremental learning, the classifier is run on the test data to test its performance on unseen data. Its overall performance is better after training with all the data.

Table 6. Performance of Classifier 1 with accuracy, precision, and recall.

Classifier 1	Training			Testing		
Performance (%)	Accuracy	Precision	Recall	Accuracy	Precision	Recall
Joy	85.23	11.31	42.50	90.82	20.33	48.97
Love	89.25	20.28	60.42	93.88	35.70	72.34
Optimism	89.81	17.08	38.12	90.21	20.73	48.59
Annoyance	84.79	14.81	28.58	84.23	18.59	42.77
Disappointment	87.38	9.58	25.48	88.69	13.31	33.06
Sadness	86.70	10.42	41.55	92.44	20.83	48.59
Surprise	87.23	8.48	39.74	93.41	19.16	47.33
Curiosity	82.30	11.51	42.70	90.29	25.04	55.99
Confusion	83.51	8.36	37.47	87.44	13.06	46.06

Table 7 shows the performance of Classifier 2 during training and testing with micro average accuracy, micro average precision and micro average recall. During training, the classifier has a micro average accuracy of 95.99%, micro average precision of 75.24% and micro average recall of 0.51%. After training, the built classifier is tested on the test data, which has obtained a micro average accuracy of 96.04%, micro average precision of 68.23% and micro average recall of 1.35%. Although the accuracy of the model during the training and testing are high, the recall rate is extremely low in both run.

The two trained emotion classifiers are deployed on 38,618 YouTube comments to identify the emotions of the opinions. The predicted results from the two classifiers are then compared with the human annotation to measure their performance. Their performance is as shown in Table 8 based on the comparison to 73 comments with human annotated emotion categories. Although Classifier 2 has more correct labels, it predicted almost all entries with False, which indicates there is no emotion detected in the comments. This can be because Classifier 2 is trained on an imbalanced dataset, so

Table 7. Performance of Classifier 2 with Micro Average Accuracy, Precision, and Recall

Classifier 2	Training			Testing		
Performance (%)	Accuracy	Precision	Recall	Accuracy	Precision	Recall
Micro Average	95.99	75.24	0.51	96.04	68.23	1.35

Table 8. Performance of Classifier 1 and Classifier 2 on YouTube comments

Emotions	Classifier 1		Classifier 2	
	Correct Labels	Incorrect Labels	Correct Labels	Incorrect Labels
Joy	51	22	63	10
Love	41	32	46	27
Optimism	55	18	58	15
Annoyance	49	24	57	16
Disappointment	53	20	66	7
Sadness	66	7	69	4
Surprise	46	27	64	9
Curiosity	42	31	38	35
Confusion	61	12	67	6

it predicted all outcome as the majority class to get higher accuracy. Classifier 1 did not get more correct labels, but it could detect the emotions in the comments and label them correctly. Hence in the following part, the results of Classifier 1 will be discussed.

4,373 (11.32%) out of 38,618 YouTube comments towards the videos related to the movie are labelled as "joy' by Classifier 1 as shown in Fig. 4. The most mentioned word for joy is "movie", which indicates that people talk a lot about the movie in their comments. When users talk about the movie, they usually refer to a certain scene, such as the wedding scene and the scene when Astrid and Michael are separated. Users are happy to see beautiful and romantic wedding scenes, and to see Astrid is finally freed from the relationship with her husband who has cheated on her. 37.27% of the comments are detected with the emotion of love as shown Fig. 4. The comments associated with the emotion of love talk about the character or the scenes which have well-presented the true Asian or Singaporean.

Furthermore, only 5.37% of the comments are identified with the emotion of optimism. People are optimistic towards this movie because it is different from the previous Hollywood movies. It is a non-martial art movie with Asian leads. Users are excited to see a Hollywood movie that is not a typical one. 47.97% of the comments are predicted to have emotion of curiosity. When looking at the 5 most liked comments related to "accent", users are curious about the accent of the characters in the movie. They also want to know if the characters should have spoken with British or Singaporean accent. 16.47%

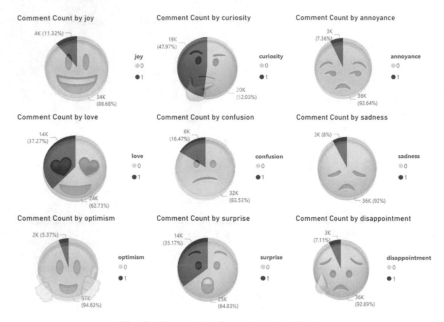

Fig. 4. The results of emotion classification.

of the comments are detected with the emotion of confusion. When a word cloud is generated on the 5 most liked comments, the emotion of confusion can be linked to several topics: (1) the Mahjong scene between Rachel and Eleanor Young, (2) the countries and people in Asia and (3) the English accent. 35.17% of the comments are associated with the emotion of surprise. People are surprised by the movie on how they convey Asian's perspective on the reputation and how Asian family gossip can spread quickly. 7.36% comments are detected with the emotion of annoyance. Users feel annoyed by the luxury living lifestyle of some rich people. Besides, this emotion can also be linked to the scene where Astrid's husband is cheating on her. 8% of the comments have the emotion of sadness. When looking at the comments related to Astrid, viewers are feeling sad for Astrid because of her situation in the marriage with Michael. They think that Astrid is a great person and deserves better. 7.11% of the comments are identified to associate with the emotion of disappointment. From the comments, people are expecting "Singlish" or Singaporean English to be used more frequently in the movie as its geographical setting is Singapore. There are also some users who think that the character of Astrid is more interesting than Rachel although Astrid is the second heroine.

5 Discussion and Conclusion

This study used machine learning model classifier to analyze embedded emotions in public responses to the movie adaptation of Kevin Kwan's *Crazy Rich Asians* on its official YouTube channel. The most detected emotions in the comments are curiosity, love, and surprise. The viewers' preoccupation with curiosity is due to the fact that this movie is the first contemporary movie by a major Western studio that showcase Asian culture with all Asian casts. The topics discussed in the comments can be summarized as follow:

Users are happy to see the beautiful and romantic wedding scene in the movie. As described in the comments, they have never seen a wedding that can be such beautiful.

The public is confused by the word "Asians" used in the name of the movie as the people in Asia are not just limited to Chinese. Hence, some of the users commented that the movie should be named as "Crazy Rich Southeast Asians". Moreover, some users commented that the movie failed to include other ethnic groups in Singapore.

Users are curious about the English accent used in the movie. As the story in the movie happens in Singapore, viewers expect to hear a lot of Singaporean English while watching the movie. However, some of them feel disappointed because only several actors or actresses use Singaporean English.

The Mahjong scene between Rachel and Eleanor could be one of the most important scenes in the movie. Some people feel confused while watching these scenes because they have little or no experience in playing Mahjong. Hence, there are a lot of conversations in the comments that people are discussing about the story behind Rachel's decision to let Nick's mother, Eleanor to win the game to show her love to Nick

Comments also show that users are more interested in Astrid's character, whose storyline is mainly on resolving her marriage with Michael. As Astrid is portrayed as a great loving person, users feel annoyed when her husband cheats on her and feel happy when Astrid finally decides to separate from Michael.

Machine learning classifiers were used in this study to analyze embedded emotions in public responses to the movie adaptation of Kevin Kwan's *Crazy Rich Asians*; however, the review comments from other social media platforms were not included. Therefore, the future direction of this study can focus on comparison of emotion classification for audiences' responses on other social media platforms. In addition, hybrid deep learning models can be used for emotion classification.

References

1. Asri, M.A.Z.B.M., Keikhosrokiani, P., Asl, M.P.: Opinion mining using topic modeling: a case study of Firoozeh Dumas's Funny in Farsi in goodreads. In: Saeed, F., Mohammed, F., Ghaleb, F. (eds.) IRICT 2021. LNDECT, vol. 127, pp. 219–230. Springer, Cham (2022). https://doi.org/10.1007/978-3-030-98741-1_19
2. Fasha, E.F.B.K., Keikhosrokiani, P., Asl, M.P.: Opinion mining using sentiment analysis: a case study of readers' response on long Litt Woon's the way through the woods in goodreads. In: Saeed, F., Mohammed, F., Ghaleb, F. (eds.) IRICT 2021. LNDECT, vol. 127, pp. 231–242. Springer, Cham (2022). https://doi.org/10.1007/978-3-030-98741-1_20
3. Paremeswaran, P.a.p., Keikhosrokiani, P., Asl, M.P.: Opinion mining of readers' responses to literary prize nominees on twitter: a case study of public reaction to the booker prize (2018–2020). In: Saeed, F., Mohammed, F., Ghaleb, F. (eds.) IRICT 2021. LNDECT, vol. 127, pp. 243–257. Springer, Cham (2022). https://doi.org/10.1007/978-3-030-98741-1_21
4. Sofian, N.B., Keikhosrokiani, P., Asl, M.P.: Opinion mining and text analytics of reader reviews of Yoko Ogawa's The Housekeeper and the Professor in Goodreads. In: Keikhosrokiani, P., Asl, M.P., (eds.) Handbook of Research on Opinion Mining and Text Analytics on Literary Works and Social Media, pp. 240–262. IGI Global, Hershey (2022)

5. Suhendra, N.H.B., et al.: Opinion mining and text analytics of literary reader responses: a case study of reader responses to KL Noir volumes in Goodreads using sentiment analysis and topic. In: Keikhosrokiani, P., Asl, M.P., (eds.) Handbook of Research on Opinion Mining and Text Analytics on Literary Works and Social Media, pp. 191–239. IGI Global: Hershey (2022)

6. Yun Ying, S., Keikhosrokiani, P., Asl, M.P.: Opinion mining on Viet Thanh Nguyen's the sympathizer using topic modelling and sentiment analysis. J. Inf. Technol. Manag. **14**, 163–183 (2022). 5th International Conference of Reliable Information and Communication Technology (IRICT 2020).

7. Hakak, N.M., et al.: Emotion analysis: a survey. In: 2017 International Conference on Computer, Communications and Electronics (Comptelix) (2017)

8. Keikhosrokiani, P., Asl, M.P. (eds.) Handbook of Research on Opinion Mining and Text Analytics on Literary Works and Social Media, pp. 1–462. IGI Global, Hershey (2022)

9. Al Mamun, M.H., et al.: Sentiment analysis of the Harry Potter series using a lexicon-based approach. In: Keikhosrokiani, P., Asl, M.P. (eds.) Handbook of Research on Opinion Mining and Text Analytics on Literary Works and Social Media, pp. 263–291. IGI Global, Hershey (2022)

10. Hutto, C., Gilbert, E.: Vader: a parsimonious rule-based model for sentiment analysis of social media text. In: Proceedings of the International AAAI Conference on Web and Social Media (2014)

11. Jafery, N.N., Keikhosrokiani, P., Asl, M.P.: Text analytics model to identify the connection between theme and sentiment in literary works: a case study of Iraqi life writings. In: Keikhosrokiani, P., Asl, M.P., (eds.) Handbook of Research on Opinion Mining and Text Analytics on Literary Works and Social Media, pp. 173–190. IGI Global, Hershey (2022)

12. Ying, S.Y., Keikhosrokiani, P., Asl, M.P.: Opinion mining on Viet Thanh Nguyen's the sympathizer using topic modelling and sentiment analysis. J. Inf. Technol. Manag. **14**, 163–183 (2022). 5th International Conference of Reliable Information and Communication Technology (IRICT 2020)

13. Chung, W., Zeng, D.: Dissecting emotion and user influence in social media communities: an interaction modeling approach. Inf. Manag. **57**(1), 103108 (2020)

14. Ekman, P.: Are there basic emotions? Psychol. Rev. **99**(3), 550–553 (1992)

15. Luo, Y., et al.: An appraisal of incremental learning methods. Entropy **22**(11), 1190 (2020)

16. Madhusudhanan, S., Jaganathan, S., Ls, J.: Incremental learning for classification of unstructured data using extreme learning machine. Algorithms **11**(10), 158 (2018)

17. Egorova, E., Tsarev, D., Surikov, A.: Emotion analysis based on incremental online learning in social networks. In: 2021 IEEE 15th International Conference on Application of Information and Communication Technologies (AICT). IEEE (2021)

18. Alon, D., Ko, J.: GoEmotions: a dataset for fine-grained emotion classification. In: Google AI Blog (2021)

Voltage Profile Improvement and Active Power Loss Reduction Through Network Reconfiguration Using Dingo Optimizer

Samson Oladayo Ayanlade[1(✉)], Abdulrasaq Jimoh[2], Sunday Olufemi Ezekiel[3], and Adedire Ayodeji Babatunde[3]

[1] Lead City University, Ibadan, Nigeria
samson.ayanlade@lcu.edu.ng
[2] Obafemi Awolowo University, Ile-Ife, Nigeria
[3] Olabisi Onabanjo University, Ago-Iwoye, Nigeria
ezekiel.sunday@oouagoiwoye.edu.ng

Abstract. One of the challenges faced by power system engineers all over the world is finding solutions to drastically reduce power system losses. Solving these problems has necessitated the invention of several methods, of which network reconfiguration is one. However, in the network reconfiguration technique, reconstructing the network optimally is a significant difficulty that must be solved. This research proposes the Dingo Optimizer (DOX), a nature-inspired optimization approach for reconfiguring distribution networks optimally. The network's objective function was established, and DOX was utilized to solve the optimization problems subjected to network constraints to obtain the network's ideal topology, which resulted in minimal total active power loss and an improved voltage profile. The concept was applied to the IEEE 33-bus network using the MATLAB software suite. The results of other approaches from the literature were used to make comparisons. The results showed that the minimum voltage magnitude was increased by 8.4%, from 0.913 to 0.990 p.u. Furthermore, the total real power loss was reduced by 36.76%, indicating significant network performance and operational improvements. In addition, DOX outperformed other optimization methods in the literature and thus is an effective optimization strategy for resolving network reconfiguration challenges.

Keywords: Dingo Optimizer · Voltage Magnitude · Voltage Profile · Real Power Loss

1 Introduction

A power system is a collection of electrical networks that work together to supply electricity to end customers [1]. It is a complicated system that is broken down into three key subsystems: generation, transmission, and distribution. On the distribution networks, the bulk of the electricity generated at power plants and delivered via transmission networks is squandered as a result of power losses in the distribution networks [2].

F. Saeed et al. (Eds.): ICACIn 2022, LNDECT 179, pp. 29–39, 2023.
https://doi.org/10.1007/978-3-031-36258-3_3

These power losses in the distribution networks are caused by several circumstances. The majority of distribution networks are structured radially, with electricity flowing radially from the substations to the loads, which are interconnected to the networks.

Because the phase conductors of distribution networks are not transposed like those of transmission networks, the impact of mutual inductance cannot be disregarded [3]. Furthermore, due to the high resistance to reactance ratios of radial distribution networks, massive power losses are unavoidable [4]. According to several studies [5–7], the distribution network is responsible for a larger percentage of the total power loss in the overall power system. As a result, strategizing on how to reduce power loss in this portion of the power system is a need that merits investigation. Similarly, radial distribution networks have unacceptably high bus voltage profiles, with most bus voltage magnitudes breaching the standard voltage limits, particularly those of the buses that are located far from the substation. This has a significant impact on the distribution networks' efficiency.

After substantial study, several approaches have been developed to alleviate the problem of power losses in distribution networks [8]. Shunt capacitor placement, network reconfiguration technique, FACTS controller placement, distributed generation (DG) placement, and so on are only a few of these strategies that have been reported in the literature [9–13]. The simplest, cheapest, and most straightforward strategy for addressing power loss issues in networks is network reconfiguration [14, 15].

The technique of restructuring a radial distribution network to minimize total active power loss while simultaneously improving the network voltage profile is known as network reconfiguration [16]. Two switches, the sectionalizing and tie switches, exist in radial distribution networks [17]. Tie switches are called "normally open switches," whereas sectionalizing switches are called "normally closed switches." Changing the state of these switches causes the network to be re-configured for power flow. However, considerable care must be taken while altering the states of these switches to guarantee that no loads are disconnected and that the network preserves its radiality following reconfiguration. Network reconfiguration is an optimization problem because it entails determining the network architecture that results in a better voltage profile and lower power losses. As a result, problems involving network reconfiguration optimization must be tackled using a precise optimization approach.

Several optimization approaches for tackling various sorts of optimization issues in power systems have been documented in the literature. Metaheuristic optimization algorithms are the most widely used of these techniques. These metaheuristic algorithms have been documented and used in the literature to solve network reconfiguration optimization problems. The firefly algorithm (FA) [18], particle swarm optimization (PSO) [19], grey wolf optimization (GWO) [20], bat algorithm (BA) [21], whale optimization algorithm (WOA) [22], improved genetic algorithm (IGA) [23] and others are examples of these algorithms.

In network reconfiguration optimization problems, these algorithms produce satisfactory results. However, their main shortcoming is their reliance on the initial adjusting parameters. In the case of distribution networks of varying sizes, this aspect increases the likelihood of premature convergence. The major purpose of this research, per the

preceding analysis, is to present a new strong optimization technique for extensively analyzing the network reconfiguration challenge. The Dingo Optimizer (DOX), a successful evolutionary-based optimization algorithm, was used in this regard. Various equality
and inequality constraints were fulfilled when addressing the network reconfiguration
problem throughout the optimization process. The proposed method's feasibility and
superiority were evaluated using the IEEE 33-bus standard distribution network, and the
results were compared to those of other optimization algorithms.

2 Network Reconfiguration Problem Formulation

The formulation of the network reconfiguration problem can be expressed as:

$$\text{Min} f = \sum_{i=1}^{NR} R_i \left| i^2 \right| \tag{1}$$

where, f = fitness function corresponding to the system's overall network power loss,
R = branch resistance.

Subjected to the following equality and inequality constraints.

2.1 The Voltage Magnitude

To maintain power quality, the bus voltage magnitude should be kept within allowable
limits.

$$V_{\min} \leq |V_i| \leq V_{\max} \forall_i \in N_b \tag{2}$$

where, V_i = voltage magnitude, V_{min} and V_{max} = minimum and maximum voltages,
respectively.

2.2 The Current Limit of Branches

The feeder's branch current (I_i) must not be more than the allowable current flowing through the branch (I_{imax}), thereby preventing insulation breakdowns. Provided the
thermal constraints are met,

$$|I_i| \leq I_{i\max} \forall_i \in N_R \tag{3}$$

2.3 Radial Topology

The distribution network must be radial and should not contain any meshes. In most
cases, all loads are served without interruption.

Closing all the ties results in the total number of main loops given as:

$$N_{\text{main loops}} = (N_b - N_R) + 1 \tag{4}$$

The number of sectionalising switches

$$N_R = N_b - 1 \qquad\qquad (5)$$

where, N_b = bus number, N_R = branch number.

The objective function is computed after solving the power flow equations with the backward/forward sweep method, which has proven to be very robust and effective in radial distribution system solutions. To assess the radiality restrictions for a particular configuration, a strategy based on the bus incident matrix A was utilized, where a graph may be defined in terms of an incidence matrix.

3 Dingo Optimizer

Dingoes are complex, clever, and extremely sociable creatures. Dingoes are skilled hunters who live in packs of an average size of 12–15. The social hierarchy is extremely organized, with the alpha being at the top and being either male or female. They may be distinguished depending on the responsibilities such as decision-making, sleeping quarters, and hunting. The Alpha is the pack's most dominating and powerful member and is regarded as the pack's leader. The pack is forced to follow the alpha's choice. In most cases, all pack members bow to the alpha by lowering their tails [24].

The second level of the hierarchy is the beta dingo, serving as a bridge between the alpha and the rest of the pack for the duties at hand. It serves as the alpha's counselor and keeps the pack's discipline in check. The beta in the group confirms the alpha's directives and interacts with him. The beta dingo follows the alpha dingo in the hierarchy. If alpha does not survive, for whatever reason, the beta will take over the management of the other lower-level dingoes.

A dingo is deemed subordinate if it is not an alpha or beta. Alphas and betas are followed by these subordinates. Scouts will be responsible for keeping an eye on the regions and alerting the group if any threat or condition arises. Hunters will assist the alphas and betas in catching prey and feeding the group.

Dingoes, according to research, have an excellent sense of communication. They communicate by sensing varied levels of sound intensity in the air. Dingoes in DOX make sound feedback in such a manner that they may share their knowledge and develop shared community details. As the dingo moves from one location to another, the strength of the person influences the amplitude of the vibration.

Group hunting is an intriguing social trait in dingoes that enables its application to social behavior to be even more intriguing. The following are the phases in which hunting strategies are classified: chasing and approaching, encircling and harassing, and attacking. Dingo's hunting behavior and social structures are also statistically studied to produce DOX, which does nature-inspired optimization. The two major components of DOX are exploration and exploitation.

3.1 Mathematical Models

The dingo optimization is performed using a mathematical description of the searching, encircling, and attacking of prey.

Encircling. Dingoes are intelligent enough to track out their prey. The pack, led by the alpha, encircles the prey after tracing its whereabouts. The target or aim prey strategy is supposed to be the best agent technique to describe Dingo's social hierarchy, which is comparable to the ideal since the quest region is unknown beforehand. Other search firms are still refining their techniques for the next probable approach in the meantime. The mathematical Eqs. (6)–(10) are used to model the dingoes' behavior.

$$\vec{D}_c = \left| \vec{A} \cdot \vec{P}_p(x) - \vec{P}(x) \right| \tag{6}$$

$$\vec{P}(i+1) = \vec{P}_p(i) - \vec{B} \cdot \vec{D}(d) \tag{7}$$

$$\vec{A} = 2 \cdot \vec{a}_1 \tag{8}$$

$$\vec{B} = 2\vec{b} \cdot \vec{a}_2 - \vec{b} \tag{9}$$

$$\vec{b} = 3 - \left(I * \left(\frac{3}{I_{max}} \right) \right) \tag{10}$$

where, $\vec{D}_d$ = separation between the prey and dingo, $\vec{P}_P$, $\vec{P}$ = prey and dingo position vectors, $\vec{A}$ = coefficient vector, $\vec{B}$ = coefficient vector, $\vec{a}_1$ = random vector in [0, 1], $\vec{a}_2$ = random vector in [0, 1], $\vec{b}$ = at each iteration, it decreases linearly from 3 to 0, I = 1,2,3,..., I_{max}.

Hunting. Agents do not generally have a computation of the position of the prey in the search space, according to the notion of optimum. It is assumed that all pack members have good knowledge of prospective prey locations while designing the dingoes' hunting strategy mathematically. The alpha dingo is always in charge of the hunt. However, beta and other dingoes may occasionally join in the hunt. As a result, the first two best values obtained thus far are selected. Other dingoes must update their positions following the location of the best search agent. Equations (11)–(16) are modeled in this regard, per the discussion.

$$\vec{D}_\alpha = \left| \vec{A}_1 \cdot \vec{P}_\alpha - \vec{P} \right| \tag{11}$$

$$\vec{D}_\beta = \left| \vec{A}_2 \cdot \vec{P}_\beta - \vec{P} \right| \tag{12}$$

$$\vec{D}_o = \left| \vec{A}_3 \cdot \vec{P}_o - \vec{P} \right| \tag{13}$$

$$\vec{P}_1 = \left| \vec{P}_\alpha - \vec{B} \cdot \vec{D}_\alpha \right| \tag{14}$$

$$\vec{P}_2 = \left| \vec{P}_\beta - \vec{B} \cdot \vec{D}_\beta \right| \tag{15}$$

$$\vec{P}_3 = \left| \vec{P}_o - \vec{B} \cdot \vec{D}_o \right| \tag{16}$$

Each dingo intensity is calculated using Eqs. (17)–(19):

$$\vec{I}_a = \log\left(\frac{1}{F_\alpha - (1E - 100)} + 1 \right) \tag{17}$$

$$\vec{I}_\beta = \log\left(\frac{1}{F_\beta - (1E - 100)} + 1 \right) \tag{18}$$

$$\vec{I}_o = \log\left(\frac{1}{F_o - (1E - 100)} + 1 \right) \tag{19}$$

where, F_α and F_β = α and β-dingo fitness values, respectively, and F_o = other dingo fitness value.

Attacking Prey. Dingo has completed the hunt by attacking the victim if there is no position update. The value of $\vec{b}$ is lowered linearly to mathematically construct the approach. It should be noted that the change range of $\vec{D}_\alpha$ is likewise reduced by $\vec{b}$. $\vec{D}_\alpha$ is a random variable in the $[-3b, 3b]$ interval where $\vec{b}$ is lowered from 3 to 0 between iterations. When $\vec{D}_\alpha$ has random values between [1, 1], a search agent's next position could be anywhere between its current and the prey's.

Searching. Dingoes seek prey based on the position of the group. They always go forward in search of predators and attack them. As a result, $\vec{B}$ is employed for random values, where a value smaller than -1 indicates that the prey is fleeing from the hunter, while a value bigger than 1 indicates that the pack is approaching the prey. This approach assists the DOX in scanning the targets globally. $\vec{A}$ is another component of DOX that increases the likelihood of exploration. For variable prey weights, the vector $\vec{A}$ in Eq. (8) can create any random integer between [0, 3]. DOX denotes a probabilistic variable, with vector ≤ 1 preceding vector ≥ 1 to analyse the influence of the gap described in Eq. (6).

3.2   Dingo Optimizer Algorithm

The following steps describe the dingo optimization algorithm process:
  Input: The population of dingoes D_n (n = 1, 2,..., n).
  Output: The best dingo.

1. Create the first search agents D_{in}
2. Set the values of $\vec{b}$, $\vec{A}$, and $\vec{B}$.
3. While termination conditions not reached do
4. Determine the fitness and intensity cost of each dingo.
5. D_α = Best-searching dingo

6. D_β = Second best-searching dingo
7. D_o = Dingoes search results afterwards
8. Iter 1
9. Repeat
10. For $i = 1{:}D_{in}$ do
11. Update the status of the most recent search agent.
12. End for
13. Calculate the cost of dingoes' fitness and intensity.
14. Keep track of the values of $\overrightarrow{b}$, $\overrightarrow{A}$, and $\overrightarrow{B}$.
15. Iter = Iter + 1
16. Check if, Iter ≥ criterion for stopping
17. Output
18. End while

The DOX algorithm is implemented in MATLAB software and applied to the IEEE 33-bus network, as shown in Fig. 1, with continuous lines representing sectionalizing switches and broken lines representing tie switches. The line and load data of this radial distribution network were obtained from [25].

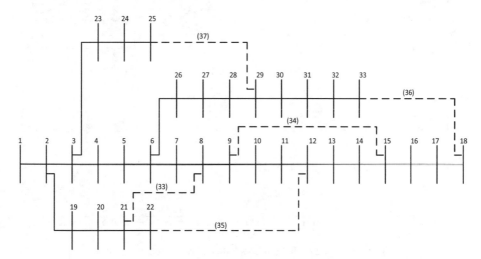

Fig. 1. IEEE 33-bus network

4 Results and Discussion

The simulations were run with 200 iterations and 50 searching agents, with a minimum and a maximum number of attacking dingoes of 2 and 25, respectively. Table 1 summarizes the simulation results for the IEEE 33-bus network. Figure 2 depicts the voltage profile before and after optimal network reconfiguration for visual inspection of the results and better understanding. The least voltage magnitude occurred at bus 18 before the network reconfiguration, and it was 0.913 p.u. Following network reconfiguration with switches 3, 7, 11, 12, and 24 now open, the least voltage magnitude increased to 0.990 p.u., which corresponds to 8.4%. After network reconfiguration, the new least voltage magnitude of 0.98237 p.u. was observed at bus 10. Similarly, the voltage magnitudes on all buses have improved overall, as seen in Fig. 2. As a result of optimal network reconfiguration with DOX, the network's overall performance greatly improved.

Table 1. Simulation results

	Before reconfiguration	After reconfiguration
Tie switches:	33, 34, 35, 36, 37	3, 7, 11, 12, 24
Power loss:	202.7052 kW	128.1941 kW
Power loss reduction:		36.7583%
Least voltage:	0.91308 pu	0.98237 pu

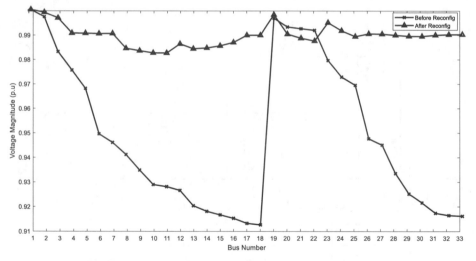

Fig. 2. Voltage profile before and after network reconfiguration

The total active loss was reduced by 36.76% for the optimal network reconfiguration, from 202.7052 to 128.1941 kW. Figure 3 shows the new network topology after optimal reconfiguration using DOX, with solid lines denoting closed sectionalizing switches and broken lines denoting open tie switches. The network's radiality was preserved, and all loads were connected to the supply. Furthermore, after optimum network reconfiguration using the proposed DOX algorithm, active power losses in all branches were considerably reduced, and thus the overall efficiency of the network was significantly increased.

The DOX results were compared to the FA [26], GA [27], ABC [28], GWO [29], and WAO [30] algorithms in the literature as presented in Table 2. The proposed DOX method reduced the total active power loss to 128.19 kW after network reconfiguration,

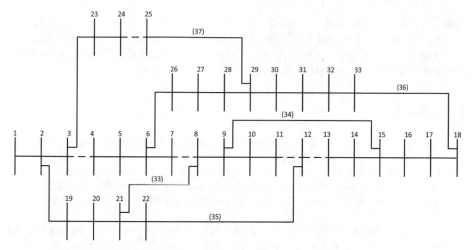

Fig. 3. The optimal network reconfiguration obtained by DOX

which was significantly less than the values obtained by FA, ABC, GA, GWO, and WAO, which were 136.14, 139.50, 140.60, 139.51, and 139.57, respectively. Table 2 has a wealth of additional information on the reconfiguration results for comparison with the other methods. These results show that the DOX algorithm is superior in optimally reconfiguring radial distribution networks to minimize active power loss while simultaneously enhancing the network voltage profile.

Table 2. Comparison of the results obtained by DOX with other algorithms

Algorithm	Ties switches	Overall active power loss (kW)	% Minimization
Base case	33, 34, 35, 36, 37	202.71	–
FA	7, 9, 14, 32, 37	136.14	32.57
ABC	8, 14, 28, 32, 33	139.50	31.20
GA	9, 28, 33, 34, 36	140.60	30.60
GWO	7, 14, 9, 32, 37	139.51	31.16
WAO	7, 9, 14, 32, 37	139.57	31.15
DOX	3, 7, 11, 12, 24	128.19	36.76

5   Conclusion

The DOX algorithm was utilized to determine the optimal network topology for the IEEE 33-bus system, to minimize total active power loss and improve the voltage profile. The optimal configuration obtained satisfies the requirements; it preserves the network's radial structure and ensures the supply of all network loads. In comparison to the results obtained by the FA, GA, and ABC algorithms in the literature, DOX's algorithm proved to be more effective. In a nutshell, the DOX algorithm outperforms the other algorithms in terms of finding the optimal solution.

References

1. Okelola, M.O., Ayanlade, S.O., Ogunwole, E.I.: Particle swarm optimization for optimal allocation of STATCOM on transmission network. J. Phys. Conf. Ser. Proc. 1–7 (2021)
2. Ayanlade, S.O., Komolafe, O.A., Adejumobi, I.O., Jimoh, A.: Distribution system power loss minimization based on network structural characteristics. In: 1st International Conference on Engineering and Environmental Sciences Proceedings, pp. 849–861. Osun State University (2019)
3. Jimoh, A., Ayanlade, S.O., Ariyo, F.K., Jimoh, A.B.: Variations in phase conductor size and spacing on power losses on the Nigerian distribution network. Bull. Electric. Eng. Inform. **11**(3), 1222–1233 (2022)
4. Ayanlade, S.O., Komolafe, O.A.: Distribution system voltage profile improvement based on network structural characteristics. In: Faculty of Technology Conference 2019 (OAUTEK-CONF 2019), pp. 75–80. OAU, Ile-Ife, Osun State, Nigeria (2019)
5. Xie, J., Chen, C., Long, H.: A loss reduction optimization method for distribution network based on combined power loss reduction strategy. Complexity **2021**, 1–13 (2021)
6. Diaz, S.: Electric power losses in distribution networks. Turk. J. Comput. Math. Educ. **12**(12), 581–591 (2021)
7. Adegboyega, G.A., Franklin, O.: Determination of electric power losses in distribution systems: Ekpoma, Edo state, Nigeria as a case study. Int. J. Eng. Sci. **3**(1), 66–72 (2014)
8. Salimon, S.A., Adepoju, G.A., Adebayo, I.G., Ayanlade, S.O.: Impact of shunt capacitor penetration level in radial distribution system considering techno-economic benefits. Niger. J. Technol. Dev. **19**(2), 101–109 (2022)
9. Magadum, R.B., Timsani, T.M.: Minimization of power loss in distribution networks by different techniques. Int. J. Sci. Eng. Res. **3**(5), 521–527 (2012)
10. Salau, A.O., Gebru, Y.W., Bitew, D.: Optimal network reconfiguration for power loss minimization and voltage profile enhancement in distribution systems. Heliyon **6**, e04233 (2020)
11. Okelola, M.O., Adebiyi, O.W., Salimon, S.A., Ayanlade, S.O., Amoo, A.L.: Optimal sizing and placement of shunt capacitors on the distribution system using whale optimization algorithm. Niger. J. Technol. Dev. **19**(1), 39–47 (2022)
12. Okelola, M.O., Salimon, S.A., Adegbola, O.A., Ogunwole, E.I., Ayanlade, S.O., Aderemi, B.A.: Optimal siting and sizing of D-STATCOM in distribution system using new voltage stability index and bat algorithm. In: 2021 International Congress of Advanced Technology and Engineering (ICOTEN), pp. 1–5 (2021)
13. Adepoju, G.A., Aderinko, H.A., Salimon, S.A., Ogunade, F.O., Ayanlade, S.O., Adepoju, T.M.: Optimal placement and sizing of distributed generation based on cost-savings using a two-stage method of sensitivity factor and firefly algorithm. In: 1st ICEECE & AMF Proceedings, pp. 52–58 (2021)
14. Syahputra, R., Soesanti, I., Ashari, M.: Performance enhancement of distribution network with DG integration using modified PSO algorithm. J. Electric. Syst. **12**(1), (2016)
15. Al Samman, M., Mokhlis, H., Mansor, N.N., Mohamad, H., Suyono, H., Sapari, N.M.: Fast optimal network reconfiguration with guided initialization based on a simplified network approach. IEEE Access **8**, 11948–11963 (2020)

16. Kashem, M.A., Ganapathy, V., Jasmon, G.B.: Network reconfiguration for load balancing in distribution networks. IEEE Proc. Gener. Transm. Distrib. **146**(6), 563–567 (1999)
17. Setif, A.: Optimum distribution network reconfiguration using firefly algorithm
18. LakshmiReddy, Y., Sathiyanarayanan, T., Sydulu, M.: Application of firefly algorithm for radial distribution network reconfiguration using different loads. IFAC Proc. **47**(1), 700–705 (2014)
19. Huang, W.T., et al.: A two-stage optimal network reconfiguration approach for minimizing energy loss of distribution networks using particle swarm optimization algorithm. Energies **8**(12), 13894–13910 (2015)
20. Reddy, A.S., Reddy, M.D., Reddy, M.S.K.: Network reconfiguration of distribution system for loss reduction using GWO algorithm. Int. J. Electric. Comput. Eng. **7**(6), 3226–3234 (2017)
21. Salman, N.: Optimal network reconfiguration of distribution systems for improving the performance in term of power quality using bat algorithm: electrical. Diyala J. Eng. Sci. **8**(4), 376–385 (2015)
22. Reddy, A.S., Reddy, M.D.: Application of whale optimization algorithm for distribution feeder reconfiguration. Imanager's J. Electric. Eng. **11**(3), 17–24 (2018)
23. Abubakar, A.S., Ekundayo, K.R., Olaniyan, A.A.: Optimal reconfiguration of radial distribution networks using improved genetic algorithm. Niger. J. Technol. Dev. **16**(1), 10–16 (2019)
24. Bairwa, A.K., Joshi, S., Singh, D.: Dingo optimizer: a nature-inspired metaheuristic approach for engineering problems. Math. Probl. Eng. **2021**, 1–12 (2021)
25. Dharageshwari, K., Nayanatara, C.: Multiobjective optimal placement of multiple distributed generations in IEEE 33 bus radial system using simulated annealing. In: 2015 International Conference on Circuits, Power and Computing Technologies (ICCPCT-2015), pp. 1–7 (2015)
26. Djabali, C., Boukaroura, A., Ketfi, N., Bouktir, T.: Optimum distribution network reconfiguration using firefly algorithm. In: International Conference on Recent Advances in Electrical Systems, Tunisia (2016)
27. Hong, Y.Y., Ho, S.Y.: Determination of network configuration considering multiobjective in distribution systems using genetic algorithms. IEEE Trans. Power Syst. **20**(2), 1062–1069 (2005)
28. Rao, R.S., Narasimham, S., Ramalingaraju, M.: Optimization of distribution network configuration for loss reduction using artificial bee colony algorithm. Int. J. Electric. Power Energy Syst. Eng. **1**, 116–122 (2008)
29. Mohammedi, R.D., Zine, R., Mosbah, M., Arif, S.: Optimum network reconfiguration using grey wolf optimizer. TELKOMNIKA Telecommun. Comput. Electron. Control **16**(5), 2428–2435 (2018)
30. Soliman, M., Abdelaziz, A.Y., El-Hassani, R.M.: Distribution power system reconfiguration using whale optimization algorithm. Int. J. Appl. Power Eng. **9**(1), 48–57 (2020)

Machine Learning and Marketing Campaign: Innovative Approaches and Creative Techniques for Increasing Efficiency and Profit

Nouri Hicham[✉] and Sabri Karim

Research Laboratory on New Economy and Development (LARNED), Faculty of Legal
Economic and Social Sciences, AIN SEBAA, Hassan II University of Casablanca,
Casablanca, Morocco
nhicham191@gmail.com

Abstract. Companies must use one-to-one marketing with their customers to
preserve a leading position in the current market competitiveness. Data mining
and machine learning have enabled companies to gain a competitive advantage
by forecasting client behavior. The capacity of machine learning algorithms to
effectively predict the future is mainly responsible for the explosion in popularity
of these programs over the past few years. When a customer is met with an unfore-
seen event, a customer service issue may arise; it is not conceivable to speculate
on how they will carry themselves in the coming years. Numerous algorithms are
created and put to the test in the same way.

In this study, we utilized five distinct machine learning algorithms to achieve
positive outcomes on consumer behavior prediction and to obtain a clear vision of
the marketing campaign's return on investment (ROI). LightGBM, Random Forest
(RF), Support Vector Machine (SVM), Naive Bayes (NB),and K-nearest neighbor
(KNN) are the five classification methods used in this study. The LightGBM model
provides a more accurate forecast. Comprehensive testing is used to determine the
overall effectiveness of the LightGBM at the level of precision, MCC, accuracy,
and Cohen's Kappa.

Keywords: Customer behavior · marketing campaign · machine learning ·
prediction

1 Introduction

Suppose the concept that the public should be at the center of all actions made is not
instilled in the organization. In that case, it isn't easy to realize how vital it is for an
organization, particularly the business manager, to understand the public thoroughly.
It involves expertise from various domains to comprehend a consumer dynamically,
altering their behavior correctly. However, the most realistic way to explain this phenomenon is to take a multidisciplinary approach to the problem. This is the opinion of
several experts who have looked into the matter. This growth demonstrates the importance of learning features in consumer habits. It is also the result of a variety of external

© The Author(s), under exclusive license to Springer Nature Switzerland AG 2023
F. Saeed et al. (Eds.): ICACIn 2022, LNDECT 179, pp. 40–52, 2023.
https://doi.org/10.1007/978-3-031-36258-3_4

and internal or cultural variables that influence the processes that are engaged constantly. In addition, this growth results from learning features in consumer habits are essential. As a result of this change, it is clear that it is crucial to know what customers do and how they do it [1].

Customers shop for a diverse array of products and services consistently, and they do so for various reasons. Because of this, it can be challenging to ascertain which service or product a client likes, given that customers' decisions to make purchases are influenced by a wide variety of other factors, one of which is pricing. Several different factors influence the decisions that customers make regarding their investments. For instance, every aspect of cultural, societal, and individual decision-making is accounted for here. Every one of them has had either a direct or indirect effect on the behavior of consumers [2, 3].

Consumer behavior research is a good illustration of how businesses can enhance the efficiency of their marketing to increase revenue generation by attracting new consumers, developing long-term relationships with existing ones, and retaining more of their current clients. Many different machine learning algorithms have been presented to analyze the behavior of customers [4]. Even though customers do not adhere to established guidelines when making a purchase decision, we can accurately forecast which of our services will be purchased the most frequently. To correctly anticipate the shopping behaviors of a new consumer, the first thing we need to do is understand the purchasing behaviors of past customers [2]. If a company can guess how a customer will buy something, they can market the services they think will most appeal to that customer. This will make the customer's experience better.

In this study, we utilized five distinct machine learning algorithms to achieve positive outcomes on consumer behavior prediction and to obtain a clear vision of the marketing campaign's return on investment (ROI). LightGBM, Random Forest (RF), Support Vector Machine (SVM), Naive Bayes (NB), and K-nearest neighbor (KNN) are the five classification methods used in this study. The following outline is the body of this paper: In the second section, a concise summary of contemporary research in consumer behavior is presented. Section 3, the research methodology is broken down, along with specifics on the machine learning algorithms and the evaluation criteria applied. The acquired results are broken down in depth in Sect. 4.

2 Related Work

For the last decade, machine learning (ML) has steadily increased prominence in the business and marketing sectors [5]. Companies quickly recognize the benefits of applying language that is unique to machine learning to make strategic decisions based on the analysis of massive volumes of data and knowledge in ML [6, 7]. In particular, machine learning (ML) has been quite successful in the many different endeavors that the organization has embarked on throughout its history [8, 9]. The vast bulk of this success can be credited to the recent emphasis on machine learning in studies into artificial intelligence [10, 11].

This accomplishment was reflected in the number of machine learning algorithms presented to analyze customers' behavior. When it comes to accurately predict the behaviors of clients, some examples of machine learning algorithms that are also easy to understand include Random Forests (RF), Support Vector Machines (SVM), and Decision Trees (DT) [12, 13]. The results acquired by the Random Forest classifier are more accurate than those provided by other machine learning approaches, as determined by the assessment metrics Accuracy, Precision, Recall, and F1-score, they constructed a hybrid prediction model for user purchasing behavior by merging SVM and logistic regression (LR) techniques. This model is used to predict user purchasing behavior. After that, they carry out an empirical investigation to determine whether or not the model that they developed is useful. The results show that the fusion model better predicts what will happen in the future than the single model.

In a similar vein, the authors of the paper titled [8] describe an innovative strategy for predicting the consumption patterns of users. By integrating LR with the XGBoost algorithm, this system can predict the purchasing decisions that customers will make. The researchers [14] use the cat boost model to investigate and predict the likelihood that individuals will purchase a particular product based on real unbalanced surfing data obtained from an online shopping platform. These investigations and predictions are based on the researchers' findings [14] gleaned from the data. When analyzing whether or not a prediction is reliable, many model characteristics, including accuracy and precision, are considered. Using this data set led to a better result: 88.51% accuracy in predicting how people would buy something.

The dataset utilized in the course of our investigation is made up of 2240 customers [3] and has a total of 28 characteristics associated with the iFood organization. The iFood firm is a multinational food company that serves several hundred thousand clients worldwide who are active participants in the food industry. It provide a wide variety of things, including wines, products, unique fruits, and fresh fish, which may be purchased from them. It is also possible to distinguish between standard objects and those that are gold. When buying products from the company, consumers can choose from one of three distinct sales channels available to them. These can be found on the company's website and in their physical stores and catalogs. Figures 1 and 2 shows depicts some descriptions of our dataset.

When analyzing the statistical summary, one of the essential things is plotting the distribution. It gives you a picture to help you understand the data set better.

Outliers are numbers that are significantly off from the rest of the values for a variable. In other words, this observation's value differs from other observations of the same variable. For example:

*Age: Ironically, the Age Boxplot appears to be an anomaly that was never cleaned. At least two consumers have been determined to be older than 100. Due to the rarity of this occurrence, it can be inferred that this was likely a data entry error, and we can easily remove it from the dataset. Aside from this, however, the Age variable appears somewhat regularly distributed, with a mean age of little less than 50 years.

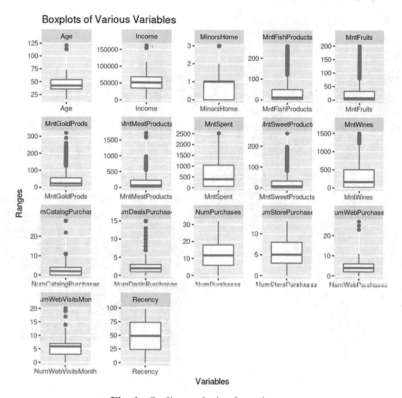

Fig. 1. Outlier analysis of our dataset

*Income: Having seen this plot previously in the preprocessing step, the only thing to note is that the 600k+ income outlier is no longer present, and the average pay can now be shown to be approximately 50k, which is comparable to the larger population of people Minors.

*MinorsHome: Most consumers have zero or one child at home; however, 2 and 3 children are also typical. Nobody has more than three children at home.

*MntFishProducts: This is the first extremely right-skewed distribution that we've discovered. While most customers purchase relatively few fish products, a few purchase more than 200 fish products, indicating either mass purchasing or continuous interest in the store.

LightGBM, Random Forest (RF), Support Vector Machine (SVM), Naive Bayes (NB), and K-nearest neighbor (KNN) were used in this study. Finally, each model's evaluation performance is assessed based on its precision, MCC, accuracy, and Cohen's Kappa.

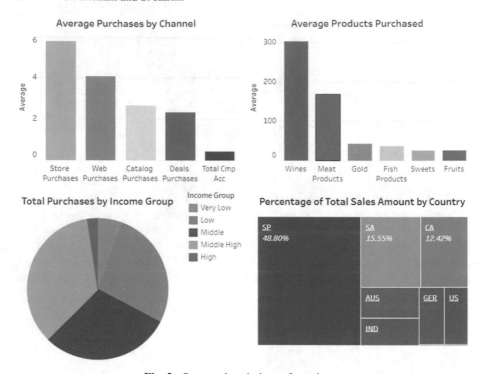

Fig. 2. Somme descriptions of our dataset

3   Methodology

In the parts that are to follow, we will talk about the several machine learning techniques that we used in our work. Before we talk about that, though, let's look at the overall structure on which our research will be based.

3.1   Overall Structure Architecture

The overall structure can be decomposed into four primary steps, as shown in Fig. 2.

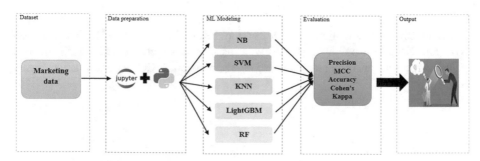

Fig. 3. Overall architecture

In terms of the overall architecture of Fig. 3, the following is a summary of what you should pay attention to:

First, get a database. It is based on data from the iFood firm, a worldwide food conglomerate with several hundred thousand existing consumers and a database of 2240 clients and 28 factors associated with its business. The iFood firm has several hundred thousand existing consumers. At this point, the dataset underwent the preliminary processing necessary to make it suitable for feature extraction. The data was preprocessed to prepare for being fed into the learned classifier. Using the evaluation criteria, we then choose the most successful model to predict what our customers will do.

3.2 Evaluation Metrics

To compare the performance of four algorithms, we used a variety of metrics, including precision, MCC, accuracy, and Cohen's Kappa.

3.2.1 Precision

The precision can be calculated by dividing the number of samples correctly identified as positive by the total number of pieces correctly classed as positive. Precision measures how accurate a model's classifications are and is based on whether or not a sample can be accurately identified as positive [15].

$$Precision = \frac{True\ Positive\ (TP)}{True\ Positive\ (TP) + False\ Positive\ (FP)} \tag{1}$$

3.2.2 MCC

In 1975, Brian Matthews developed the concept of Matthew's correlation coefficient, also known as MCC for short. An MCC is a statistical tool that is utilized in the process of analyzing the model's presumptions and is abbreviated as MCC. Between the predicted and actual values, the discrepancy can be measured with chi-square statistics using a contingency table that is two rows wide and two columns deep [16, 17].

$$MCC = \frac{TN * TP - FN * FP}{\sqrt{(FP + TP)(TP + FN)(FP + TN)(TN + FN)}} \tag{2}$$

3.2.3 Accuracy

The term "accuracy" refers to a measurement that summarizes the model's performance across all classes. It is good when all classes are of equal relevance to one another. Calculating it requires taking the proportion of accurate predictions and dividing that by the total number of predictions made [18].

$$Acc = \frac{Number\ of\ correct\ classification}{Total} \tag{3}$$

3.2.4 Cohen's Kappa

Cohen's Kappa is a metric of inter-rater agreement that is frequently utilized. Despite this, it is commonly used based on an opinion rather than a precise measurement. The Kappa coefficient is a measurement of the likelihood of agreement. It does this by contrasting the actual probability of understanding with the value that would be expected if the assessments were utterly unrelated to one another. The set of possible values is the range when it is situated inside the interval [0,1], where 1 represents total concordance, and 0 illustrates full autonomy [19, 20].

$$Kappa = \frac{p_x - p_y}{1 - p_y} \qquad (4)$$

where: The actual agreement amongst raters is denoted by P_x, whereas P_y indicates the theoretical probability of random agreement.

3.3 The Used ML Models

This study uses a machine learning algorithm broken down and explained in this part. As well as LightGBM, Random Forest (RF), Support Vector Machine (SVM), Naive Bayes (NB), and K-nearest neighbor (KNN).

3.3.1 Naive Bayes (NB)

NB learning is probabilistic learning based on the Bayesian theorem but uses a more straightforward method than traditional Bayesian learning [21]. The data classification process is built on learning from past mistakes and using that knowledge in future classifications. One example of data categorization is when math and probability gather information based on a hypothesis. The newly generated models will be used to categorize and modify data to either lessen the likelihood of something happening or raise the possibility of something active [10]. A new batch of data is collected, and the configurations already in place are adjusted to account for this additional information.

3.3.2 Support Vector Machines (SVM)

Due to the few available datasets, SVM was chosen as one of the most effective classifiers. It remains one of the most popular binary classification methods that utilize discrete attribute values. Support Vector Machines are linear classifiers. However, the kernel model can be modified to make them nonlinear. Support Vector Machines are used to ascertain the border or hyperplane that is most suitable for categorizing parameters into their respective two separate class datasets. The unique hyperplane can be considered a subspace with dimensions of (N-1), where N stands for hyperparameters [22]. SVM will be used to make as much difference as possible between the selection hyperplane and the support vectors.

3.3.3 K-nearest Neighbor (KNN)

Despite having the fewest total features, KNN has proven the most successful method. When classifying a new sample point, KNN uses a similarity measure such as the

Euclidean or Manhattan distance. The k-value indicates the number of categories drawn from the k-nearest neighbors that make up the voting group for a new sample data point. The most prevalent label that comes the closest to fitting the majority of the categories represented by the k sample data points will be the one applied to the new sample point. The k-value and the weights are the two most essential components of the KNN algorithm. The k-value has the most significant influence on how well KNN performs. [18] say that if the k-value is set too low, the method is more susceptible to being influenced by outliers, and if it is set too high, it has too many sample points.

3.3.4 LightGBM

One of the Gradient-Boosted Decision Trees (GBDT) frameworks developed by Microsoft Research Asia was named LightGBM and released in 2017. This project aims to enhance computer efficiency to solve significant data prediction issues timelily [23]. The goal of the GBDT, similar to other boosting techniques, is to merge multiple weak learners into a single strong learner. Because every tree in the GBDT learns the conclusions and residuals of the trees that came before it, the only type of decision tree used in this scenario is a regression tree. At this time, it has been suggested that the Light-GBM algorithm be used in applications. The accuracy of the forecast isn't changed, but both the speed of the prediction and the amount of memory it uses go down significantly [24].

3.3.5 Random Forest (RF)

RF is a form of supervised machine learning that can be used to solve problems involving regression and classification. The mean is used for regression, and the majority vote results are used to categorize the data. It generates DT based on several different data sources. The capacity of Random Forest to manage data sets containing both categorical and continuous variables, as is required for classification and regression, is one of its most essential and distinguishing features. It generates efficient solutions to the problems associated with a classification [25].

4   Result and Discussions

In marketing efforts, one frequently uses the prediction model to attempt to anticipate customers' behaviour. The performance of a forecast can be evaluated by looking at how well it predicted the results. To forecast the behavior of customers, numerous classification strategies are utilized. Also, the performance of the suggested system is tested using many different criteria, such as precision, MCC, accuracy, and Cohen's Kappa.

The accuracy of numerous classification schemes is seen in Fig. 4. The preceding graph demonstrates that when compared to the other methodologies, LightGBM offers a higher level of precision. The support vector machine (SVM) has lower accuracy than competing approaches. The most accurate version of the lightGBM model has a score of 88.39%.

The precision analysis of a wide variety of techniques is shown in Fig. 5. The accuracy of the Random Forest model is superior to that of the lightGBM, Support Vector Machines

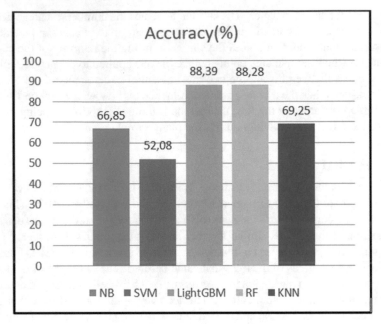

Fig. 4. Analytical Accuracy

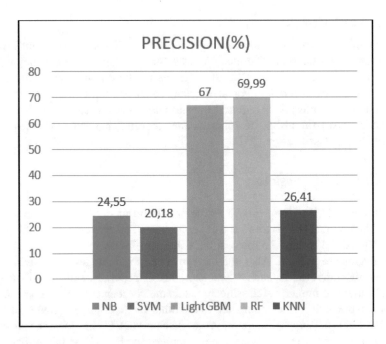

Fig. 5. Analytical Precision

(SVM), Naive Bayes (NB), and Linear K-nearest neighbor models (KNN). Random Forest has a precision value of 69.99%, but ligtGBM only has a precision value of 67%.

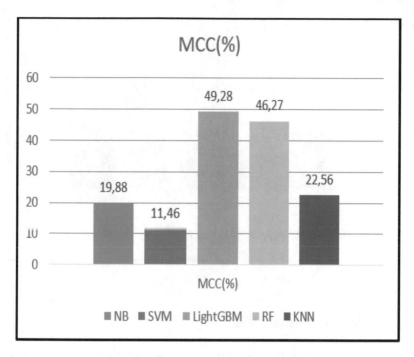

Fig. 6. Analytical MCC

Matthew's correlation coefficient is depicted in examining various methods in Fig. 6. In the MCC analysis, the performance of LightGBM is superior to that of Random Forest (RF), Support Vector Machines (SVM), Naive Bayes (NB), and K-nearest neighbor (KNN). In comparison, the MCC value for Random Forest (RF) comes in at 46.27%, while the MCC value for LightGBM is 49.28%.

The Cohen's Kappa analysis of various methods is illustrated in Fig. 7. LightGBM outperforms Random Forest, Support Vector Machines, Naive Bayes, and K-nearest neighbor regarding their Cohen's Kappa values (KNN). The Cohen's Kappa value for LightGBM is 46.27%, whereas the score for Random Forest (RF) is 38.80%.

In the evaluation metric that we used to know the performance of the mentioned models, we notice that the LightGBM model presents a considerable performance compared to the others and that this performance rises to interesting levels at the level of accuracy, precision, kappa, and MCC, the values obtained are 88.39%, 67%, 46.27%, and 49.28% respectively.

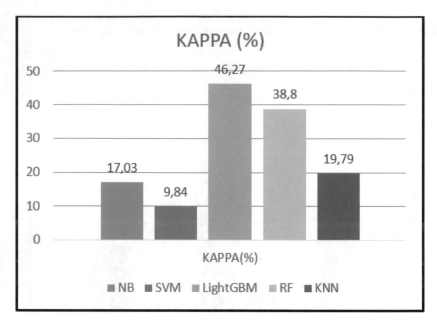

Fig. 7. Analytical Cohen's Kappa

5   Conclusion and Future Work

By using machine learning in their marketing strategies, companies can potentially learn more about their customers and improve the whole experience they have with those customers. It is also underlined that businesses can improve their marketing effectiveness to generate more revenue by increasing the percentage of customers they keep and recruiting new consumers. Predicting the behavior of consumers is a common practice in marketing, as it can assist in creating new service offers and help narrow the target of marketing campaigns. Knowledge is obtained by the utilization of the following five classification algorithms: LightGBM, Random Forest (RF), Support Vector Machine (SVM), Naive Bayes (NB), and K-nearest neighbor. As a direct consequence of this, LightGBM produces a more accurate forecast. A comprehensive investigation is utilized in order to assess LightGBM's level of functionality. The findings demonstrate that LightGBM is superior to other homogenous classification methods in terms of precision, MCC, accuracy, and Cohen's Kappa. In the future work, deep learning and hybrid models can be incorporated into the investigation to broaden its scope. There are a variety of performance measures that can be chosen to conduct an analysis of performance. The model can also be tested with a number of different datasets from different fields.

References

1. Saura, J.R.: Using data sciences in digital marketing: framework, methods, and performance metrics. J. Innov. Knowl. **6**(2), 92–102 (2021). https://doi.org/10.1016/j.jik.2020.08.001

2. Ghosh, S., Banerjee, C.: A predictive analysis model of customer purchase behavior using modified random forest algorithm in cloud environment. In:2020 IEEE 1st International Conference for Convergence in Engineering (ICCE), Kolkata, India, pp. 239–244 (2020). https://doi.org/10.1109/ICCE50343.2020.9290700
3. Hicham, N., Karim, S., Habbat, N.: Customer sentiment analysis for Arabic social media using a novel ensemble machine learning approach. IJECE **13**(4), 4504 (2023). https://doi.org/10.11591/ijece.v13i4.pp4504-4515
4. Paul, A., Mukherjee, D.P., Das, P., Gangopadhyay, A., Chintha, A.R., Kundu, S.: Improved random forest for classification. IEEE Trans. Image Process. **27**(8), 4012–4024 (2018). https://doi.org/10.1109/TIP.2018.2834830
5. van Giffen, B., Herhausen, D., Fahse, T.: Overcoming the pitfalls and perils of algorithms: a classification of machine learning biases and mitigation methods. J. Bus. Res. **144**, 93–106 (2022). https://doi.org/10.1016/j.jbusres.2022.01.076
6. Valecha, H., Varma, A., Khare, I., Sachdeva, A., Goyal, M.: Prediction of consumer behaviour using random forest algorithm. In: 2018 5th IEEE Uttar Pradesh Section International Conference on Electrical, Electronics and Computer Engineering (UPCON), Gorakhpur, pp. 1–6 (2018). https://doi.org/10.1109/UPCON.2018.8597070
7. Hicham, N., Karim, S.: Machine learning applications for consumer behavior prediction. Lecture Notes in Networks and Systems, vol. 629, LNNS, pp. 666–675 (2023). https://doi.org/10.1007/978-3-031-26852-6_62
8. Xing Fen, W., Xiangbin, Y., Yangchun, M.: Research on User Consumption Behavior Prediction Based on Improved XGBoost Algorithm. In: 2018 IEEE International Conference on Big Data (Big Data), Seattle, WA, USA, pp. 4169–4175 (2018). https://doi.org/10.1109/BigData.2018.8622235
9. Ravi, L., Subramaniyaswamy, V., Vijayakumar, V., Jhaveri, R.H., Shah, J.: Hybrid user clustering-based travel planning system for personalized point of interest recommendation. In: Sahni, M., Merigó, J.M., Jha, B.K., Verma, R. (eds.) Mathematical Modeling, Computational Intelligence Techniques and Renewable Energy. AISC, vol. 1287, pp. 311–321. Springer, Singapore (2021). https://doi.org/10.1007/978-981-15-9953-8_27
10. Assegie, T.A., Tulasi, R.L., Kumar, N.K.: Breast cancer prediction model with decision tree and adaptive boosting. In: IAES Int. J. Artif. Intell. IJ-AI **10**(1), 184 (2021). https://doi.org/10.11591/ijai.v10i1.pp184-190
11. Singh, S.P., Dhiman, G., Tiwari, P., Jhaveri, R.H.: A soft computing based multi-objective optimization approach for automatic prediction of software cost models. Appl. Soft Comput. **113**, 107981 (2021). https://doi.org/10.1016/j.asoc.2021.107981
12. Hu, X., Yang, Y., Chen, L., Zhu, S.: Research on a prediction model of online shopping behavior based on deep forest algorithm. In: 2020 3rd International Conference on Artificial Intelligence and Big Data (ICAIBD), Chengdu, China, pp. 137–141 (2020). https://doi.org/10.1109/ICAIBD49809.2020.9137436
13. Hu, X., Yang, Y., Zhu, S., Chen, L.: Research on a hybrid prediction model for purchase behavior based on logistic regression and support vector machine. In: 2020 3rd International Conference on Artificial Intelligence and Big Data (ICAIBD), Chengdu, China, pp. 200–204 (2020). https://doi.org/10.1109/ICAIBD49809.2020.9137484
14. Dou, X.: Online purchase behavior prediction and analysis using ensemble learning. In: 2020 IEEE 5th International Conference on Cloud Computing and Big Data Analytics (ICCCBDA), Chengdu, China, pp. 532–536 (2020). https://doi.org/10.1109/ICCCBDA49378.2020.9095554
15. Habbat, N., Anoun, H., Hassouni, L.: A Novel Hybrid Network for Arabic Sentiment Analysis Using Fine-Tuned AraBERT Model, p. 12

16. Chicco, D., Jurman, G.: The advantages of the Matthews correlation coefficient (MCC) over F1 score and accuracy in binary classification evaluation. BMC Genomics **21**(1), 6 (2020). https://doi.org/10.1186/s12864-019-6413-7

17. Chicco, D., Warrens, M.J., Jurman, G.: The Matthews Correlation Coefficient (MCC) is more informative than Cohen's Kappa and Brier score in binary classification assessment. IEEE Access **9**, 78368–78381 (2021). https://doi.org/10.1109/ACCESS.2021.3084050

18. Habbat, N., Anoun, H., Hassouni, L.: Sentiment analysis and topic modeling on arabic twitter data during Covid-19 pandemic. Indones. J. Innov. Appl. Sci. **2**(1), 60–67 (2022). https://doi.org/10.47540/ijias.v2i1.432

19. Warrens, M.J.: Five ways to look at Cohen's Kappa. J. Psychol. Psychother. **5**(4) (2015). https://doi.org/10.4172/2161-0487.1000197

20. Vergni, L., Todisco, F., Di Lena, B.: Evaluation of the similarity between drought indices by correlation analysis and Cohen's Kappa test in a Mediterranean area. Nat. Hazards **108**(2), 2187–2209 (2021). https://doi.org/10.1007/s11069-021-04775-w

21. Hicham, N., Karim, S., Habbat, N.: An efficient approach for improving customer Sentiment Analysis in the Arabic language using an Ensemble machine learning technique. In: 2022 5th International Conference on Advanced Communication Technologies and Networking (CommNet), pp. 1–6 (2022). https://doi.org/10.1109/CommNet56067.2022.9993924

22. Nik Hashim, N.N.W., Basri, N.A., Ahmad Ezzi, M.A.-E., Nik Hashim, N.M.H.: Comparison of classifiers using robust features for depression detection on Bahasa Malaysia speech. IAES Int. J. Artif. Intell. **11**(1), 238 (2022). https://doi.org/10.11591/ijai.v11.i1.pp238-253

23. Ju, Y., Sun, G., Chen, Q., Zhang, M., Zhu, H., Rehman, M.U.: A model combining convolutional neural network and LightGBM algorithm for ultra-short-term wind power forecasting. IEEE Access **7**, 28309–28318 (2019). https://doi.org/10.1109/ACCESS.2019.2901920

24. Liang, W., Luo, S., Zhao, G., Wu, H.: Predicting hard rock pillar stability using GBDT, XGBoost, and LightGBM algorithms. Mathematics **8**(5), 765 (2020). https://doi.org/10.3390/math8050765

25. Palmatier, R.W., Crecelius, A.T.: The "first principles" of marketing strategy. AMS Rev. **9**(1–2), 5–26 (2019). https://doi.org/10.1007/s13162-019-00134-y

A Bottom-Up 2-Stage Approach
for Constructing Arabic Knowledge Graph

Amani D. Alqarni[1] , Khaled M. G. Noaman[2] ,
Fatima N. AL-Aswadi[3,4(✉)] , and Hamood Alshalabi[5,6]

[1] Information Technology and Security Department, College of Computer
Science and Information Technology, Jazan University, Jazan, Saudi Arabia
aalqarni@jazanu.edu.sa
[2] Deanship of E-Learning and Information Technology, Jazan University, Jazan,
Saudi Arabia
knoaman@jazanu.edu.sa
[3] School of Computer Sciences, Universiti Sains Malaysia, 11800 Gelugor,
Pulau Pinang, Malaysia
fnsal5_com016@student.usm.my
[4] Faculty of Computer Science and Engineering, Hodeidah University, P.O.
Box 3114, Hodeidah, Yemen
[5] Faculty of Information Science and Technology, CAIT, Universiti Kebangsaan
Malaysia, Selangor, Malaysia
hmoud.shalabi@siswa.ukm.edu.my
[6] Sana'a University, Sana'a, Yemen

Abstract. Since Google has coined the term 'knowledge graph' (KG) in 2012, it has garnered much attention from the commercial and scientific domains. This term has been applied in numerous applications, such as automated fraud detection, supply chain management, semantic search, and intelligent question-answering (QA). The KG organises and represents strands of information as a semantic graph with higher usage efficiency, better modification, and clearer comprehension. However, KG has faced multiple challenges during the construction process particularly for Arabic language. As such, this study identified the Arabic KG (AKG) construction difficulties from Arabic text in order to propose a bottom-up 2-stage approach to construct AKG from Arabic texts. This proposed approach has dual stages: linguistic-based and machine learning-based (ML-based) stages. The linguistic-based stage is based on Natural Language Processing (NLP) techniques and three main tasks are deployed: corpus analysis, term extraction, and concept conceptualisation. As for the ML-based stage, it is based on semi-supervised ML techniques and three main tasks are applied: taxonomic classification, semantic mapping, and knowledge visualisation. The proposed methodology could successfully address the complex nature of the Arabic language in an effective manner to construct AKG.

Keywords: Knowledge Acquisition · Arabic Knowledge Graph · Knowledge base · Ontology

F. Saeed et al. (Eds.): ICACIn 2022, 179, pp. 53–63, 2023.
https://doi.org/10.1007/978-3-031-36258-3_5

1   Introduction

The revolution of big data across the Web is coupling with the rising demand for constructing KG, which is a key ingredient for many semantic Web and intelligent systems applications. Davies et al. [1] had recently defined KG as "a knowledge graph that acquires and integrates information into ontology and applies a reasoner to derive new knowledge." According to Ahmed et al. [2], "AKG is a collection of attributes, entities, relations, facts, and rules or other forms of knowledge for the Arabic language that present and define some kind of facts, relations or connections as a paradigm rather than a specific class of things." Simply put, AKG denotes the Arabic knowledge base (AKB) represented in a semantic graph and in machine-understandable format [2, 3]. Figure 1 illustrates an example of AKG (with English translation). Figure 1(a) displays the entities, relations, and attributes of AKG, while Fig. 1(b) presents instances of AKG facts and rules. Nevertheless, processing Arabic data and acquiring the simple information presented in Fig. 1 to construct AKG is not a simple task. More details pertaining to these difficulties are discussed in Sect. 2.

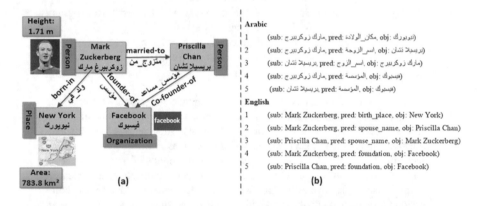

Fig. 1. An Example of Arabic Knowledge Graph

Many recent studies had constructed Arabic ontologies and knowledge databases, such as DBPedia [4–6], Arabic VerbNet (AVN) [7], Arabic PropBank [8, 9], Interactive Arabic Dictionary (IAD) [10], and Semantic Quran [11]. Although these studies highlight the good practices for constructing AKG, they tend to suffer from multiple issues and limitations. For example, AVN and IAD rely on manual construction and lack modern terms. Manual construction is time-consuming, costly, extremely laborious, and its progress is slower than data streaming across the Web [12, 13]. Besides, most studies on Arabic knowledge focused on constructing Arabic ontology than AKG, which is a huge ontology rich with instances and extensions that applies a reasoner to derive new knowledge [1, 2]. Since Google introduced the term 'KG' in 2012, no study has proposed an effective methodology for AKG construction. Hence, the significance of this study is amplified.

This study outlined the AKG construction difficulties and challenges from Arabic text. Next, an automatic bottom-up 2-stage approach is proposed to construct AKG from Arabic texts. This proposed approach is composed of the following stages: linguistic-based and ML-based stages.

The rest of this paper is organised as follows: Sect. 2 presents the AKG construction difficulties and challenges and discusses the related work. Section 3 elaborates on the proposed methodology for automatic AKG construction. Lastly, this study is concluded in Sect. 4.

2 Related Work

Recent scholarly work on AKG [2, 20] had reviewed the challenges, issues, and opportunities revolving around AKG. A variety of AKG has been built to involve globally linked data on the Web. Some examples of AKG are DBPedia [4–6], AVN [7], Arabic PropBank [8, 9], IAD [10], Makhtota+ [26], and Semantic Quran [11]. The DBPedia is a multilingual KG that supports many languages, inclusive of the Arabic language. The DBPedia extracts the Arabic chapter from the Arabic version of Wikipedia [2, 6, 20]. However, the Arabic chapter extracted from DBPedia is insufficient and is shallower than chapters of English and some other languages.

Qamus Almaeani[1], *Almuajam Alwasseet*[2], Arabic WordNet (AWN) [27], and AVN [7, 28] were built manually. Upon involving 1,000 nominal and 500 verbal Arabic synsets, AWN develops WordNet in order to turn it into a multilingual knowledge base (KB). As for AVN, which consists of 173 classes (4392 verbs and 498 frames); it develops WordNet by including semantics and syntactic frames for Arabic verb categories. The WordNet [29] is one of the most famous and oldest KB of the English language. Meanwhile, Almuajam Alwasseet refers to an old Arabic dictionary that is composed of more than 50,000 verbs and 75,000 nouns. The IAD is an electronic and open access version of Almuajam Alwasseet. Next, Qamus Almaeani or meanings dictionary is an open access dictionary that groups the Arabic lexical vocabularies meanings into literature, Islamic, quote, context, and rhetoric.

Both Arabic PropBank (APB) [8, 9] and Columbia Arabic Treebank (CATiB) [30] are initiated to construct AKB. The APB relies on Penn Arabic Treebank (PATB) [31] and Arabic Morphological Analyser[3] (Aramorph). On the other hand, PATB depends on the syntactic structure, excluding semantic dimension that can significantly affect its accuracy [28, 30].

While Aramorph is an Arabic morphological analyser that annotates Arabic text with POS tags, APB comprises of two parts: verb frameset database and annotated corpus. These annotations have many problems [28]. Despite slightly improving the AKB, CATiB; which depends on the syntactic structure with annotating dependencies, still suffers from the same problem as the APB.

Makhtota+ [26] is an expansion of the Makhtota (ancient manuscript) maintained by the King Saud University. It refers to a manual platform that applies crowd-resources of the Arabic cultural heritage for manual annotations. It has annotated more than 12,000 pages with linked data. Next, Semantic Quran [11] and Azhary [16] builds linked-data from Islamic resources. Semantic Quran is used to build a large dataset from two semi-structured sources, including an RDF scheme of all Quranic translations aligned with ontology. It has 43 distinctive under-represented languages in the Linking Open Data project and feeds 26,735 linkages to both Wiktionary and DBpedia. Meanwhile, Azhary is another Arabic lexical ontology based on Quran. It sets Arabic words into groups of synonyms known as synsets. Azhary includes 26,195 words, which are arranged in 13,328 synsets. Another study regarding AKB is the Arabic dependency treebank (ADTB) [32], which builds AKB for the travel domain. Some researchers have used Arabic news [33], the Holy Quran, and its translation [34] or Hadith [35] as the input data to build AKB.

Although many studies have attempted to develop the AKG, the AKG still suffers from some lacks and defects. For instance, most of the AKG systems are still cooperative or manual and demand human intervention in most construction tasks. Typically, manual construction is a time-consuming, intricate, and costly process [12, 13]. Notably, most of the AKG systems depend on two main sources: WordNet and Wikipedia. However, these AKG systems and sources lack plenty of modern Arabic terms and Arabic data when compared to English and some other languages [2, 13]. Besides, techniques that can address the complex nature of the Arabic language in an effective manner are in scarcity [2, 35].

2.1 Arabic Knowledge Graph Challenges

Many recently developed AKG systems appear to suffer from limitations and setbacks. Data insufficiency, incomplete and incorrect knowledge, inconsistent knowledge, knowledge explainability, and low knowledge integrity are some major concerns of AKG construction.

Referring to several prior studies [2, 3, 15, 20–25], one may conclude that the construction of AKG involves several complex challenges in many aspects pertaining to information processing, knowledge extraction, knowledge construction, knowledge visualisation, and many more. Figure 2 presents some of the identified AKG challenges.

2.2 Arabic Language Processing Difficulties

The Arabic language is a Semitic right-to-left language that includes 29 alphabetic characters. It is agglutinative, cursive, derivational, and highly inflectional [15]. Classical Arabic (CA) (or Ancient Arabic), Modern Standard Arabic (MSA), and Colloquial Arabic (CoA) are the three varieties of the Arabic language [2, 16, 17]. The CA is the Quran, Hadith, and Arabic cultural heritage language (e.g., Makhtota, pre-Islam literature, and poetry). Next, MSA is the official language of most Arabic States and is used in news bulletins, books, academic materials, publications, as well as solemn occasions. The CoA refers to the dialects of daily communication that vary from one country to another.

The Arabic sentence structure differs from that of the English. The sentence structure in English is as follows: *Subject+ Predicate*. On the contrary, the two types of sentences in the Arabic language are given as follows:

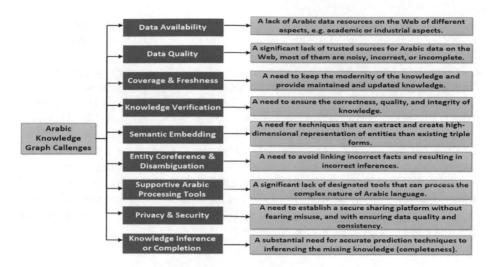

Fig. 2. Arabic Knowledge Graph Challenges

Nominal sentence: *Subject+ Predicate* and *Verbal sentence*: *Verb+ Predicate*

The predicate may include another subject (noun phrase (NP)), another verb phrase (VP), an adjective phrase (ADJP), a prepositional phrase (PP), an adverb phrase (ADVP) or another clause (CL).

Unlike English sentences, the verb is not necessary in the Arabic nominal sentence [2]. The Arabic nominal sentence may have two nouns or noun and any other part of speech (POS) without a verb. For instance, the sentence in English: "the teacher is tall" → in Arabic: "الأُستاذُ طَويل", consists of the subject "الأُستاذُ" and the adjective "طَويل". The subject in a verbal sentence could be implied and not mentioned in the sentence, but it can be understood from the context. For instance, the sentence in English: "I can" → in Arabic: "أَستَطِيع", consists of the verb "أَستَطِيع" and the subject "أنا" is covert subject that is understood from the context. It is considered as a single-word verbal sentence. The Arabic language is highly inflectional. A single word may present a complete sentence; the word may express different grammatical categories. For example, in English: "we ate it" → in Arabic: "أكلناها" is a single word sentence where "أَكَل" is the verb, "نا" is the subject, and "ها" is the object.

Each alphabetic character in Arabic could be written in one of the main four diacritics (تَشْكِيل, tashkīl) called ḥarakāt (حَرَكَات). The same word with different diacritic marks often possesses different meanings [15]. For example, the word "بر" has three different meanings based on diacritics marks: "بِر" (righteousness), "بَر" (land), and "بُر" (wheat). The existing Arabic analysis and parsing tools, such as Nooj[4] and Stanford-CoreNLP[5] (their last versions were released in 2022), can analyse Arabic sentences regardless of the diacritics in the words. As a result, incorrect outcomes were frequently generated due to the ambiguous issue in the Arabic language. In addition, these tools have failed to provide accurate analysis and yields (e.g., POS or Parse) for many Arabic sentences. For example, the Stanford CoreNLP did not analyse the simple sentence (أكلناها) correctly to detect the subject pronoun (نا). Besides, it failed to explicit the covert subject (أنا) from the simple sentence (أَسْتَطِيع). Figure 3 illustrates the Parse and POS outputs of CoreNLP 4.4.0 after analysing the instances of Arabic sentences, thus highlighting analytical problems.

Acquiring knowledge from Arabic data is more difficult when compared to the English language. For instance, the Arabic sentence ".²كم 783.8 نيويورك مَسَاحَة" has two translations in English: (i) "The area of New York is 783.8 km²" and (ii) "New York's area is 783.8 km²"). Extracting the area attribute of New York (see Fig. 1) is more difficult from Arabic data than from English data. In English data, the area attribute can be extracted based on the preposition "of" that appears in the first sentence or based on the apostrophe mark of the word as in the second sentence. Turning to the Arabic sentence, both words are noun and nothing syntactically indicates that the word "مَسَاحَة" (area) is an attribute for the word "نيويورك". The meaning is comprehensible through the context. It is worth noting that not all two nouns coming after each other signify that the first one is an attribute for the next one.

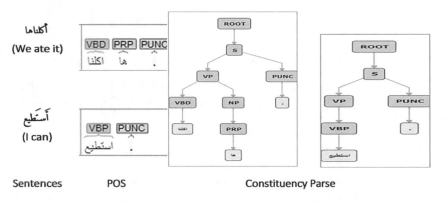

Fig. 3. Examples of analysis problems for Arabic Sentences

The existing Arabic NLP tools, such as stemmer and lemmatizer tools, have many imperfections [2, 18, 19] that could not extract the correct stem, lemm, and root of the Arabic words. At times, they could also generate words not available in the Arabic language. The Arabic language distinguishes feminine from masculine words of things (distinguishes between the gender), which could also affect the meaning.

In summary, the Arabic language has a very complex nature, apart from being a high inflectional and derivational language. Consequently, processing and analysing Arabic data to acquire knowledge are indeed intricate [15, 17].

3 Methodology

This section presents the bottom-up 2-stage approach proposed in this study to construct AKG from Arabic texts. This proposed approach is composed of two stages: linguistic-based and ML-based stages. The linguistic-based stage is based on NLP techniques and it includes three main tasks: corpus analysis, term extraction, and concept conceptualisation. Meanwhile, the ML-based stage is based on semi-supervised ML techniques and it includes three main tasks: taxonomic classification, semantic mapping, and knowledge visualisation. The methodology deployed in this study is briefly illustrated in Fig. 4.

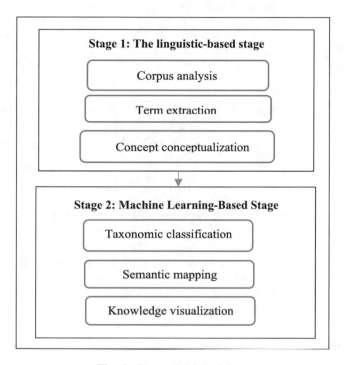

Fig. 4. Research Methodology

3.1 Linguistic-Based Stage

Corpus Analysis. Arabic patterns and roots are essential to comprehend the derivational aspects of the lexicon. This step involves the lexico-syntactic analysis of Arabic texts. First, a comprehensive dictionary was built by integrating the existing dictionaries, such as AWN, AVN, and IDA. Next, this dictionary is enriched by incorporating several extracted inflectional and derivational forms based on recent studies, such as [18, 19]. Lastly, a set of morphological grammar elements are modelled to detect the component morphemes of the Arabic agglutinative forms inspired by prior studies [15, 36]. This strategy is helped to solve a huge part of the ambiguity issue.

Term Extraction. After analysing the corpus, the term extraction technique is developed. This technique is based on patterns predefined by experts and previous studies, such as [37, 38]. Table 1 presents the proposed Arabic patterns for terms extraction. Finally, in order to retain the relevant terms and investigate the modern terms, the Domain Time Relevance (DTR) measure suggested in [14] is employed. This measure is aimed at updating and maintaining the extracted information.

Table 1. Arabic Term Extraction Patterns

Patterns	Examples
Noun + Noun*	بنت، ابن أخ، خارطة الوطن العربي
Noun + Noun* + Adj	تقنيات التعلم العميق، التعليم العالي، ذوي الدخل المحدود
Noun + Noun* + (adj or Prep) + Noun + Noun*	التعليم عن بعد، التعليم عالي الجودة، تقنيات التعلم عن بعد
Noun + Noun* + conj + Noun + Noun*	التربية والتعليم، وزارة التربية والتعليم، عمادة تطوير المناهج والبحث العلمي
Noun + Noun* + conj+ (adj or Prep) + Noun + Noun*	تقنيات و أساليب التعلم عن بعد، تقنيات و أساليب التعليم عالي الجودة

Concept Conceptualisation. After filtering the extracted terms and selecting relevant and modern concepts; the concepts, as well as their taxonomic and synonym relations, are derived from the extracted terms. This step identified the candidate concepts that share a number of syntactic contexts (e.g., sharing the same head or the same expansion). For example, التعليم العالي، التعليم عن بعد، التعليم عالي الجودة share the same head (التعليم), while التربية والتعليم، وزارة التربية والتعليم have similar expansion (التعليم). By using this method of conceptualisation, knowledge explainability issue is resolved.

3.2   Machine Learning-Based Stage

Taxonomic Classification. This step is aimed at enriching and enhancing the extracted concepts by applying the Skip-Gram model – an unsupervised learning technique [39, 40] that generates new concepts based on input (previous extracted terms). Next, the concepts are classified. This step is helped to solve the knowledge inference and embedding issues.

Semantic Mapping. In this step, the semantic relations between concepts are extracted. The method described in [13] is deployed for semantic relations extraction and classification by changing the training learning to be semi-supervised instead of supervised. In order to build a semi-supervised deep belief network (DBN) model, the results retrieved from the previous steps are used as pre-training data and the original data are applied as real input (training data) to the DBN model. This strategy is aimed at enhancing knowledge inference or completion and semantic embedding.

Knowledge Visualisation. This final step aims to store the extracted knowledge in a knowledge database format (e.g., RDF). Next, Neo4j[6] is used to visualise the knowledge in a graph. In addition, Cypher query language is deployed for visualising and representing the stored AKG.

4 Conclusion

This study has outlined the AKG construction difficulties and challenges from Arabic text. Next, an automatic bottom-up 2-stage approach is proposed to construct AKG from Arabic texts. This proposed approach is comprised of dual stages: linguistic-based and ML-based stages. The linguistic-based stage includes three main tasks: corpus analysis, term extraction, and concept conceptualisation, whereas the ML-based stage includes three main tasks: taxonomic classification, semantic mapping, and knowledge visualisation. This proposed approach is aimed at addressing the AKG construction problems, particularly the knowledge processing issues in the Arabic language. These issues include maintaining and updating the extracted information, knowledge explainability, knowledge ambiguity, and knowledge inference. In future work, the experimental results are discussed in detail regarding the performance of the proposed approach for building AKG.

Notes

1. قاموس المعاني https://www.almaany.com/
2. Mustafa I., Alzayat A. H., Abdel-kader h., Alnajar M. A, 3 المعجم الوسيط, rd edition, Alnouri Press, Damascus, 1960.
3. Arabic morphological analyser Version 1.0 (see https://catalog.ldc.upenn.edu/LDC2002L49).
4. http://www.nooj4nlp.org/index.html
5. Version 4.4.0, https://corenlp.run/
6. https://neo4j.com/

References

1. Davies, J., Studer, R., Warren, P.: Semantic Web Technologies: Trends and Research in Ontology-Based Systems: John Wiley & Sons (2006)
2. Ahmed, I.A., Al-Aswadi, F.N., Noaman, K.M.G., Alma'aitah, W.Z.A.: Arabic knowledge graph construction: a close look in the present and into the future. J. King Saud Univ. Comput. Inf. Sci. (2022)
3. Tiwari, S., Al-Aswadi, F.N., Gaurav, D.: Recent trends in knowledge graphs: theory and practice. Soft. Comput. **25**(13), 8337–8355 (2021). https://doi.org/10.1007/s00500-021-05756-8
4. Lehmann, J., et al.: DBpedia – a large-scale, multilingual knowledge base extracted from Wikipedia. Semantic Web. **6**(2), 167–195 (2015)
5. Bizer, C., et al.: DBpedia – a crystallization point for the Web of Data. J. Web Semant. **7**(3), 154–165 (2009)
6. Al-Feel, H.: The roadmap for the Arabic chapter of DBpedia. Mathematical and computational methods in electrical engineering. In: Proceedings of the 14th International Conference on Telecommunications and Informatics (TELE-INFO 2015), pp. 115–125 (2015)
7. Mousser, J.: A Large Coverage Verb Taxonomy for Arabic. LREC (2010)

8. Palmer, M., et al.: A Pilot Arabic Propbank. LREC (2008)
9. Zaghouani, W., Diab, M., Mansouri, A., Pradhan, S., Palmer, M.: The revised Arabic propbank. In: Proceedings of the fourth linguistic annotation workshop, pp. 222–226 (2010)
10. Rebdawi, G., Desouki, S., Ghneim, N.: The interactive Arabic dictionary: another collaboratively constructed language resource. J. Comput. Sci. Appl. **1**(2), 17–22 (2013)
11. Sherif, M.A., Ngonga Ngomo, A.-C.: Semantic Quran. Semant. Web **6**(4), 339–345 (2015)
12. Al-Aswadi, F.N., Chan, H.Y., Gan, K.H.: Automatic ontology construction from text: a review from shallow to deep learning trend. Artif. Intell. Rev. **53**(6), 3901–3928 (2019). https://doi.org/10.1007/s10462-019-09782-9
13. AL-Aswadi, F.N., Chan, H.Y., Gan, K.H.: Extracting semantic concepts and relations from scientific publications by using deep learning. In: Saeed, F., Mohammed, F., Al-Nahari, A. (eds.) IRICT 2020. LNDECT, vol. 72, pp. 374–383. Springer, Cham (2021). https://doi.org/10.1007/978-3-030-70713-2_35
14. AL-Aswadi, F.N., Chan, H.Y., Gan, K.H., Alma'aitah, W.Z.: Enhancing relevant concepts extraction for ontology learning using domain time relevance. Inf. Process. Manage. **60**(1), 103140 (2023). https://doi.org/10.1016/j.ipm.2022.103140
15. Bounhas, I., Soudani, N., Slimani, Y.: Building a morpho-semantic knowledge graph for Arabic information retrieval. Inf. Process. Manage. **57**(6), 102124 (2020)
16. Ishkewy, H., Harb, H., Farahat, H.: Azhary: an Arabic lexical ontology. Int. J. Web Semant. Technol. **5**(4), 71 (2014)
17. Ryding, K.C.: A Reference Grammar of Modern Standard Arabic. Cambridge University Press (2005)
18. Alshalabi, H., Tiun, S., Omar, N., Al-Aswadi, F.N., Ali Alezabi, K.: Arabic light-based stemmer using new rules. J. King Saud Univ. Comput. Inf. Sci. (2021)
19. Alshalabi, H., Tiun, S., Omar, N., Abdulwahab Anaam, E., Saif, Y.: BPR algorithm: new broken plural rules for an Arabic stemmer. Egypt. Inform. J. (2022)
20. Ktob, A., Li, Z.: The Arabic knowledge graph: opportunities and challenges. In: 2017 IEEE 11th International Conference on Semantic Computing (ICSC), pp. 48–52 (2017)
21. Albukhitan, S., Helmy, T., Alnazer, A.: Arabic ontology learning using deep learning. In: Proceedings of the International Conference on Web Intelligence, pp. 1138–1142. ACM, Leipzig, Germany (2017)
22. Stylianou, N., Vlahavas, I.: A neural entity coreference resolution review. Expert Syst. Appl. **168**, 114466 (2021)
23. Xie, K., Jia, Q., Jing, M., Yu, Q., Yang, T., Fan, R.: Data analysis based on knowledge graph. In: International Conference on Broadband and Wireless Computing, Communication and Applications, pp. 376–385. Springer (2020). https://doi.org/10.1007/978-3-030-61108-8_37
24. Alma'aitah, W.Z.A., Talib, A.Z., Osman, M.A.: Towards adaptive structured Dirichlet smoothing model for digital resource objects. Multimedia Tools Appl. **80**(8), 12175–12194 (2021)
25. Ji, S., Pan, S., Cambria, E., Marttinen, P., Yu, P.S.: A survey on knowledge graphs: representation, acquisition and applications. IEEE Trans. Neural Netw. Learn. Syst. **33**(2), 494–514 (2022)
26. Al-Rajebah, N.I., Al-Khalifa, H.S.: Makhtota+: enhancing old Arabic manuscripts with linked data. In: Proceedings of the 14th International Conference on Information Integration and Web-based Applications & Services, pp. 323–7. Association for Computing Machinery, Bali, Indonesia (2012)
27. Black, W., et al.: Introducing the Arabic wordnet project. In: Proceedings of the Third International WordNet Conference, pp. 295–300. Citeseer (2006)

28. Mousser, J.: A large coverage verb lexicon for Arabic. Fachbereich Sprachwissenschaft, p. 567. Universit¨at Konstanz, 78457 Konstanz, Germany (2013)
29. Miller, G.A.: WordNet: a lexical database for English. Commun. ACM. **38**(11), 39–41 (1995)
30. Habash, N., Roth, R.: Catib: the columbia arabic treebank. In: Proceedings of the ACL-IJCNLP 2009 Conference Short Papers, pp. 221–4 (2009)
31. Maamouri, M., Bies, A., Buckwalter, T., Mekki, W.: The Penn Arabic treebank: building a large-scale annotated Arabic corpus. In: NEMLAR Conference on Arabic Language Resources and Tools, pp. 466–467. Cairo (2004)
32. Taji, D., Gizuli, J.E., Habash, N.: An Arabic dependency treebank in the travel domain. arXiv preprint arXiv:190110188 (2019)
33. Al-kouz, A., Awajan, A., Jeet, M., Al-Zaqqa, A.: Extracting Arabic semantic graph from Aljazeera.net. In: 2013 IEEE Jordan Conference on Applied Electrical Engineering and Computing Technologies (AEECT), pp. 1–6 (2013)
34. Ismail, R., Rahman, N.A., Bakar, Z.A.: Ontology learning framework for Quran. Adv. Sci. Lett. **23**(5), 4175–4178 (2017)
35. Saloot, M.A., Idris, N., Mahmud, R., Ja'afar, S., Thorleuchter, D., Gani, A.: Hadith data mining and classification: a comparative analysis. Artif. Intell. Rev. **46**(1), 113–128 (2016)
36. Mesfar, S.: Named Entity Recognition for Arabic Using Syntactic Grammars, pp. 305–316. Springer Berlin Heidelberg, Berlin, Heidelberg (2007).https://doi.org/10.1007/978-3-540-73351-5_27
37. Belhoucine, K., Mourchid, M., Mbarki, S., Mouloudi, A.: A bottom-up approach for Moroccan legal ontology learning from Arabic texts. In: Bekavac, B., Kocijan, K., Silberztein, M., Šojat, K. (eds.) Formalising Natural Languages: Applications to Natural Language Processing and Digital Humanities. CCIS, vol. 1389, pp. 230–242. Springer, Cham (2021). https://doi.org/10.1007/978-3-030-70629-6_20
38. Saber, Y.M., Abdel-Galil, H., El-Fatah Belal, M.A.: Arabic ontology extraction model from unstructured text. J. King Saud Univ. Comput. Inf. Sci. (2022)
39. Mikolov, T., Yih, W.-T., Zweig, G.: Linguistic regularities in continuous space word representations. In: Proceedings of the 2013 Conference of the North American Chapter of the Association for Computational Linguistics: Human Language Technologies, pp. 746–51 (2013)
40. Mikolov, T., Sutskever, I., Chen, K., Corrado, G.S., Dean, J.: Distributed representations of words and phrases and their compositionality. In: Advances in Neural Information Processing Systems, pp. 3111–3119 (2013)

Augmented Reality as a Learning Media to Improve Vocabulary Learning Among Preschoolers

Abdulrazak Yahya Saleh(✉) and Elaveni Kasirajan

FSKPM Faculty, University Malaysia Sarawak (UNIMAS), 94300 Kota Samarahan, Sarawak, Malaysia
ysahabdulrazak@unimas.my

Abstract. Technology is crucial to education since it has spurred numerous innovations and advancements in the educational system. At the moment, augmented reality (AR) is thought to be one of the technologies that could have a significant impact on the field of education. The visual and interactive aspects of augmented reality, which enable the projection of digital content into the user's field of view, have enormous promise as educational tools. For students, especially young ones, it might be helpful to design simulations that keep them attentive and interested in what they are studying. Therefore, the goal of this study is to enhance pre-schoolers' vocabulary learning through the use of augmented reality. To accomplish the goal of this study, an AR application called the Home AR is created. The application's focus is on teaching the children between the ages of 4 and 6; the names of everyday objects are in four distinct languages: English, Malay, Tamil, and Chinese. The furniture and appliances that are found in the living room, bathroom, bedroom, and kitchen are chosen to be displayed in this application. To display the images of home goods that can be scanned from the AR application, flashcards are created as markers. The performance of the students utilising the developed application and the conventional learning approach is compared in an experimental study. Additionally, a variety of evaluation techniques are used to examine how parents, teachers, and pre-schoolers feel about adopting augmented reality (AR) as a learning tool for vocabulary development. The findings gathered for this study demonstrated that pre-schoolers can efficiently acquire languages through AR. The results show that AR has a high potential for use as a revenue-generating medium in the future and can be used for learning.

Keywords: Augmented reality · AR flashcards · Household items · Learning media · Preschoolers · Vocabulary Learning

1 Introduction

Malaysia is a land of many races, the three largest of which are Malays, Chinese and Indians. The native Malays form the majority group of the nation, which corresponds to sixty percent of the population. Bahasa Melayu is the Malays' mother tongue, and it is also Malaysia's 'national language', which is often referred to as Bahasa Malaysia.

F. Saeed et al. (Eds.): ICACIn 2022, LNDECT 179, pp. 64–77, 2023.
https://doi.org/10.1007/978-3-031-36258-3_6

The Chinese are the second largest ethnic group, making up about twenty-five percent of the overall population; they speak various dialects of the Chinese language, primarily Mandarin and Hokkien. The Indians are the third largest ethnic group of the country, which comprise seven percent of the population. Like the Chinese, this group speaks a subvariety of a language that originated from India, but mostly the Tamil language [1].

The identity of an ethnic group in Malaysia is determined by the language spoken. People of Malaysia perceive a communal language as an essential indicator of a specific culture [2]. However, many people in the current generation not know that vocabulary is an essential tool to learn and speak a language fluently. Effective communication cannot take place without sufficient knowledge of vocabulary [3]. In order to master any language, it is necessary to acquire vast knowledge of vocabulary [4]. It also helps in understanding written and spoken text. Frequent exposure to vocabulary can build confidence in children to understand and interpret unknown terms of a text. Learning new words can help children know the different situations and contexts in which the words can be used. The concurrent and complicated process of extracting and building utterances using appropriate lexical combinations at the right time and place leads to language comprehension and production. It is important for children to learn any language as a tool that helps them to know the utterances that are useful to communicate efficiently, and realise what utterance is or is not suitable to be used in a certain context.

Preschool is a necessary phase for a young child as it plays a major role in vocabulary expansion [5]. Even though growing children's vocabulary is increasingly acknowledged as crucial for learning, it does not come naturally to every child [6, 7]. Whether it is intentional or not, children learn and acquire language through various contexts. For further language development, children learn through their daily encounters with people, things, or animals. They will process the sounds or words and store them in their memory, which eventually act as a basis for the development. Apart from that, a clear knowledge of children's cognitive processes is necessary to support vocabulary acquisition [8]. Thus, For children to improve their vocabulary development, a variety of learning materials and resources are supplied to them [9], which include flashcards– a popular tool used by instructors for vocabulary learning [10]. Flashcards can be categorised into few types: augmented reality (AR) flashcards, virtual flashcards, and traditional paper flashcards.

AR flashcards are more advanced than both traditional and virtual flashcards [11, 12]. AR flashcards give the experience of holding the card physically, similar to how the conventional flashcard method works. Children can also scan the flashcards with a mobile device to activate the virtual features. After scanning the flashcards, the appropriate 3D virtual visuals can be presented on the flashcards [13]. This allows the youngsters to engage with the virtual objects as if they were immersed in their world [14]. They can observe a specific perspective of the virtual object by controlling the mobile device [5]. In respect to this, many researches have been recommending applying AR tools in the preschool teaching process as in [15–19]. This paper shows how augmented reality application, which uses the flashcards as markers, can be used as a learning media to improve vocabulary learning among pre-schoolers. The objectives of the research are as below: design and develop an augmented reality application which can improve the vocabulary learning among pre-schoolers; evaluate the improvement of vocabulary level when using the augmented reality learning method as opposed to the traditional learning method; and finally analyse the perception of the pre-schoolers, preschool teachers, and the parents towards using the augmented reality as a learning media to improve vocabulary learning among pre-schoolers.

The organisation of this research paper is as follows: the details of materials and methodology are presented in Sect. 2; Sect. 3 explains the findings; while the conclusion is discussed in Sect. 4.

2 Materials and Methods

Methodology is a scientific and structured description of the techniques used in a field of research. It includes theoretical analyses of a variety of methods and concepts related to knowledge. In general, principles such as paradigm, analytical model, steps and quantitative or qualitative methods are included in methodology [20]. Methodology can be defined as a process because it does not give any solutions. Instead, methodology provides the theoretical basis for understanding which technique, collection of techniques or best practises may be adapted to a particular situation, such as estimating a specific outcome [21].

This section contains the methodological procedure used to develop the Home AR application. It provides a clear description of the application system development methodology, and the tools used to develop this application. It also contains a clear explanation of the research design, and the evaluation methods utilised to assess the performance of this application.

2.1 Research Methodology

For this research, an AR application called Home AR is developed to improve the vocabulary learning among the pre-schoolers. A specific mobile application development model is used to develop this application. For instance, each software development is distinct, and needs an effective system development life cycle (SDLC) approach to be followed, based on the internal and external factors. According to [22], there are numerous existing Software Development Life Cycle (SDLC) models that can be adapted to Mobile Application Development Life Cycle. These models are spiral process model, Iterative process model, Agile Methodologies for development of mobile applications, Mobile Application Development Lifecycle (MADLC) and Model-Driven Mobile Application Development. The appropriateness of the current process models that can be adapted to mobile application process models with respect to the development of mobile application, has been viewed in relation to such particular features.

After comparing the various process models adopted in mobile application development, a decision has been made to use the Mobile Application Development Lifecycle Model (MADLC) to develop the Home AR application.

There are many purposes for choosing MADLC to develop this application. The major reason is the activities and tasks mentioned in each stage are clearly described and elaborated by Vithani and Kumar [24]. MADLC consists of seven phases: identification, design, development, prototyping, testing, deployment, and maintenance as shown in Fig. 1 [23]. The very first step of the mobile application development life cycle, which

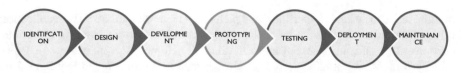

Fig. 1. Mobile Application Development Lifecycle Model (MADLC) [23]

deals with functional and non-functional elements is the identification phase. Ideas are gathered and categorised in this stage. The ideas can be suggested by the users as well as application developers [23]. On the other hand, identification also entails coming up with new ideas in order to solve a problem via a mobile application. Before moving on to the next level, the concepts must be thoroughly examined and the scope of applicability determined [25].

According to [24], the ideas generated by the developer in the previous phase will be transformed into an initial application design during this phase. The viability of designing the mobile application on all mobile platforms will be determined in this phase as well. The crucial aspect of the design process is to build a storyboard (see Fig. 2) that can clearly explain the user interface design flow of the application [25].

Moreover, the developer has also created a use case diagram to display the complete functional and technical view of the application. Figure 3. shows the complete use case diagram of the Home AR application.

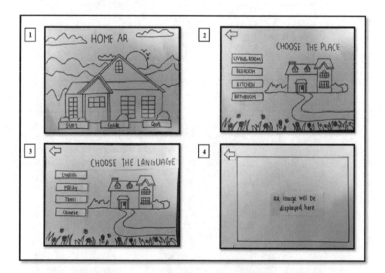

Fig. 2. The storyboard to view AR images and listen to the pronunciation

The initial interface design of the application will be combined with the programming language at this stage. Apart from that, the development process can be split into two different parts: programming for functional and user interface requirements. Functional programming describes the application's scope and function. On the other hand, user interface requirement programming describes the development process of multimedia components such as keys, hyperlinks, and pictures. The development of this application is done by using several software packages that contribute to the building of various functions. The software stacks used to develop the Home AR application are Unity 3D, Microsoft Visual Studio, Vuforia Engine, Audacity, SketchUp and Paint 3D. Further descriptions of the functions of these applications are provided. The programming language used to develop this application is C#, also known as C Sharp.

Furthermore, the 2D images and the 3D model are designed during this phase. Forty 2D and 3D images are required to be designed for this application which displays 10 images of things from four places: kitchen, living room, bedroom, and bathroom. Once the images are designed, the 2D images are uploaded to Vuforia SDK, and the 3D model is programmed in Unity 3D. Besides that, 40 flashcards which are the markers of the Home AR application are designed with the 2D images in this stage. Lastly, the pronunciation of the names of the things in 4 different languages, namely English, Malay, Tamil and Chinese are recorded using Audacity and programmed in Unity 3D. The Home AR application consists of six scenes. The first scene contains three buttons: start, guide and quit. The user must click on the start button to get into the application. In the next scene, there are four buttons which state living room, kitchen, bathroom and bedroom, and the users are required to choose one place. Once the users click on the place, the next scene will appear; the users must choose a language from four different languages shown: English, Malay, Tamil and Malay.

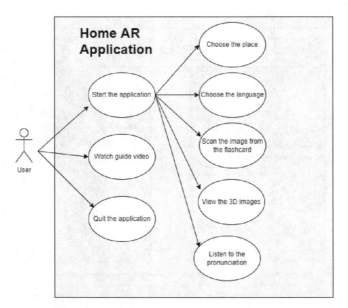

Fig. 3. Use case diagram of Home AR application

This scene is followed by the main scene of this application, which is the AR scene. Here the users are required to scan a flashcard that has the image of a marker and they will be able to view the image in 3D form and listen to the pronunciation of the name of the object.

On the other hand, new users can click the guide button on the home page and this scene will move to a scene with a complete guide video on how to use this application. Lastly, users can press the quit button on the home page and click yes in the following scene to quit the application, or click no to go back to the homepage. The user interface of the Home AR application is shown in Figs. 4–5 below.

Fig. 4. Homepage of the Home AR application

Fig. 5. The second and third scenes of the application

2.2 Evaluation

The evaluation of the Home AR application was carried out by using several methods such as user perception survey, system usability scale testing, and interview.

The user perception survey consists of 10 questions, as shown in Table 1; the Likert Scale is used and the participants can express the extent of agreement or disagreement with the statements in the survey. The questions of this survey are adapted from the research conducted by [26, 27]. In this research, the user perception survey would be carried out to gauge the perceptions of the pre-schoolers towards the Home AR Application. 10 pre-schoolers from the AR learning group in the experimental study were requested to answers the questions of the survey. Participants were required to rank the answer of every question from 1 to 5, based on their level of agreement for the given statements as shown below. Before the survey began, the researcher explained each and every question in detail to the pre-schoolers, so that they had a clear understanding of the questions, and hopefully as a result, the data collected would be valid.

Table 1. Survey about Home AR application for pre-schoolers

No	Questions	Strongly Disagree	Disagree	Neutral	Agree	Strongly Agree
1	I like this application	1	2	3	4	5
2	I feel that the application is easy to use	1	2	3	4	5
3	I have gotten used to the application quickly	1	2	3	4	5
4	I think I can learn better by using this application	1	2	3	4	5
5	I could see the images of the things from various positions	1	2	3	4	5
6	I have concentrated more on the words and pictures than on the tablet	1	2	3	4	5
7	I want to own this application	1	2	3	4	5
8	I would invite my friends to use this application	1	2	3	4	5
9	I would like to use this application to learn more vocabulary	1	2	3	4	5
10	I have had a good time while using this application	1	2	3	4	5

The interview protocol was used to ascertain the teachers' perception about using the Home AR application to improve vocabulary learning among pre-schoolers. It was designed with reference to the work of [5]. The interviews for this research were conducted face to face with two preschool teachers from Tadika Eceria, and a teacher from JJ International Preschool.

System usability scale (SUS) test was originally created by John Brooke in 1986. It can be used to evaluate a wide variety of products and services, including hardware, software, mobile devices, websites and applications [28]. It consists of a 10-item questionnaire with five response options for each question; respondents can select one response from Strongly agree to Strongly disagree [29].

3 Findings and Discussion

This section presents the results and discussions of the cognitive walkthrough, usability testing, performance of the pre-test and the post-test, user perception survey, interviews, and the system usability scale test. The results are evaluated based on the methods discussed in the section of materials and methods. Cognitive walkthrough was carried out at the prototyping phase with two main goals: the application must meet all the requirements; there are zero errors in the programming of the application. The results and feedbacks given by the experts during the cognitive walkthrough are stated in Tables 2–4.

Table 2. Results of task 1

No	Questions	Expert1	Expert2	Expert3
1	Will the user try to achieve the right action?	Yes, it has suitable interface and button for user	Yes, hopefully the user will try	Yes
2	Will the user notice that the correct action is available?	Yes, the correct action is available for user to click and explore the application	Yes	Yes
3	Will the user associate the correct action with the effect that the user is trying to achieve?	Yes	Yes	Yes
4	If the correct action is performed, will the user see that progress is being made toward solution of the task?	Yes	Yes	Yes

The results in Table 2. show that all the experts agreed with all the questions given by the researcher after conducting the walkthrough of the first task, which is viewing the AR image. This indicates that there are no errors in this task and the users will be able to understand, as well as carrying out the correct action to view the AR image while using the application.

Table 3. shows the results of the second task which is related to watching the guide video. The results of the second task also show that all the experts agreed with all the statements given, after carrying out the walkthrough of the second task. Hence, it is proven that the users will not face any difficulties watching the guide video when they use the Home AR application.

The third task of cognitive walkthrough is quitting the application. Results stated in Table 4. prove that all the questions were affirmed by the experts after performing the walkthrough of Task 3. The results indicate that the users will be able to carry out the correct action of quitting the application easily.

Table 3. Results of task 2

No	Questions	Expert1	Expert2	Expert3
1	Will the user try to achieve the right action?	Yes	Yes	Yes
2	Will the user notice that the correct action is available?	Yes	Yes	Yes
3	Will the user associate the correct action with the effect that the user is trying to achieve?	Yes	Yes	Yes
4	If the correct action is performed, will the user see that progress is being made toward solution of the task?	Yes	Yes	Yes

Table 4. Results of task 3

No	Questions	Expert1	Expert2	Expert3
1	Will the user try to achieve the right action?	Yes	Yes	Yes
2	Will the user notice that the correct action is available?	Yes	Yes	Yes
3	Will the user associate the correct action with the effect that the user is trying to achieve?	Yes	Yes	Yes
4	If the correct action is performed, will the user see that progress is being made toward solution of the task?	Yes	Yes	Yes

Table 5. Final feedbacks from the experts

Evaluator	Final Feedback
Expert1	Interesting project. My suggestion is to include voice buttons for kids especially for pages "choose the place and language". This is because some of the children 4-6 years maybe not yet proficient in reading
Expert2	It would be better if the buttons can pronounce the name of the places and the languages. This can make the kids to easily choose the correct action
Expert3	The menu (start, guide, quit) font size is too small and the location is at the bottom, which decreases the visibility and accessibility. Can enlarge the font size

The final feedbacks given by the experts are stated clearly in Table 5. For the final feedback, two experts gave the same comments: add the pronunciations to the buttons. For instance, if the user hovers over the button that states 'bedroom', the word 'bedroom' should be pronounced. This comment was taken into account and the pronunciation feature was added to all the buttons in the Home AR application. This added feature will helpful to the users, especially pre-schoolers who are not proficient in reading skills, and they can use this application efficiently and with ease. Apart from that, the size of the fonts on the home page was enlarged and the positions of the buttons were also brought upwards; this improvement was carried out based on the feedback of the third expert, after the cognitive walkthrough. System Usability Scale (SUS) test was carried out with 20 parents to gauge their perception towards the Home AR application. Apart from that, there is a specific method which can be employed to calculate the SUS form. For instance, for each odd-number question, the score is reduced by 1, and for each even-number question, 5 is subtracted from the score. Then these new scores are added and multiplied by 2.5. The raw scores and the final scores of the SUS test are presented in Table 6.

Table 6. Results of the SUS Test

Sample	Raw Score	Final Score	Sample	Raw Score	Final Score
1	36	90	11	36	90
2	34	85	12	31	77.5
3	36	90	13	36	90
4	32	80	14	32	80
5	37	92.5	15	37	92.5
6	37	92.5	16	39	97.5
7	31	77.5	17	36	90
8	35	87.5	18	33	82.5
9	35	87.5	19	34	85
10	35	87.5	20	35	87.5

An interview was conducted with 3 preschool teachers to find out their perception towards using the augmented reality as a learning media: how useful it is in improving vocabulary learning among pre-schoolers? The interview contains 5 questions, and the perceptions of the teachers are discussed in this section.

The interview started off with the first question which focuses on the teachers' prior knowledge about augmented reality. For the first question, 2 teachers said that they have not heard about AR; 1 teacher said that she has heard about AR but she is not sure how AR works as she has never used an AR application. This shows that all the interviewees have no prior knowledge about augmented reality and how it functions.

The second question is about the comparison of emotional engagement and the level of enjoyment when the pre-schoolers are using AR application. All the teachers gave similar answers to this question. The first respondent said that the children were definitely more emotionally engaged while using the AR learning technique. She felt that the AR application is a twenty-first century learning technique that can make children's learning more fun and interesting. Moreover, the second respondent gave this observation: there are many differences in the children's emotional engagement as well as different levels of enjoyment when using the AR application. The current generation of kids is more adapted to technology and using a new form of technology such as AR in learning improves the engagement of children; they are excited about the new educational device and enjoy the learning. The third respondent said that the children were happy and excited about using AR application while learning. Children nowadays are bored with the traditional learning methods, which use the normal flashcards. Hence, using AR as a learning media is something refreshing, and definitely makes a difference in the emotional engagement of the children.

The third question is about the advantages and the disadvantages of using Home AR application to learn vocabulary among pre-schoolers. The feedback from all the teachers for this question is quite similar. Here are some advantages cited by the teachers: the Home AR application is capable of grabbing the attention of the children and keep them focused while learning as it is a new and interesting learning method; the application enables the teachers to teach vocabulary to the children easily; it is helpful for the parents to teach the children at home. Moreover, the teachers believed that the children can learn more than one language from the application, and they can relate the knowledge gained to their real life. Using the Home AR application can also improve young children's creativity and thinking skills in addition to vocabulary acquisition. On the other hand, according to the preschool teachers, there are some disadvantages of using Home AR application in learning. For instance, operating this application for learning on the mobile phone is excellent, but the children might be tempted to open other applications in the mobile device such as games, and at the same time exposed to other unhealthy information. There will be a burdensome task for the teachers or parents to monitor the activity of the children when they are using the mobile phone. To avoid this problem, one of the teachers suggested installing this application on a tablet or an iPad that is dedicated for learning purposes only. Apart from that, using this application for learning will increase the children's screen time, which might eventually affect the children's eyesight. Some software issues may crop up when installing the Home AR application on mobile phones; for instance, the application might not be supported by all the mobile phones such as iPhones, as the application is not programmed for IOS devices. Lastly, the application cannot function continually for a very long period of time– the device battery may run out.

The fourth question of this interview focuses on the app's practical usage: would the teachers use the Home AR application to teach vocabulary to the students in the classroom? All the teachers interviewed gave an affirmative answer: yes, they would definitely use this application to teach vocabulary to the children in the classroom as it has many benefits and advantages – helpful, useful, easy to use, fun, and exciting. With the implementation of AR in the classroom, teachers can easily capture the attention of the students, and deliver the intended knowledge to the students efficiently and effectively.

The last question of the interview is about the overall perception of the Home AR application and the teachers' views for further improvement to the app. On the whole, the teachers opined that the concept of the Home AR application is excellent, as it is a very suitable and enjoyable tool for the pre-schoolers to learn vocabulary; the slow-paced pronunciation of the words are very clear, and easily understood by the children. Consequently, when the children listen to the clear pronunciation of the words, they can catch it and pronounce the words correctly. One of the teachers has this positive comment about this application: this is a good platform for the teachers to impart knowledge, as they cannot solely depend on books to educate the children. For further improvement to the application, the teachers recommended animation to be added to the 3D objects;

it will be an interesting feature that can whet the children's appetite for learning. Here are few other suggestions for improving the app: display the names of objects in 3D once the 3D image is triggered; add colourful pictures that focus more on the cartoon elements in the application.

Based on the answers of the interviewees, it can be concluded that the teachers are fairly confident and optimistic that augmented reality learning tools are practical and workable for improving the vocabulary learning among the pre-schoolers; both students as well as teachers will benefit greatly from the various features of the app.

4 Conclusion

The main goal of this article is to explore the potential of augmented reality as a learning media. Even though there are several other education-related AR applications in the market, it is necessary to continue the innovation and contribute to this relatively new field. Many possibilities in this field are waiting to be discovered, and it is believed that with increasing investigative works being done in AR that show promising results, the research in this field will gain traction and attract more interest from other developers. The successful completion of the application named Home AR has shown that is viable to develop an AR application that serves as a learning media for young children, in the present case, to learn vocabulary. The comments and feedback given by the pre-schoolers, teachers and parents have been very positive and encouraging. This shows that the potential for creating more sophisticated AR learning tools is tremendous, and the children today can look forward to learning in a fun and engaging way. For instance, the results of this study demonstrate that an augmented reality application is a highly effective instrument for pre-schoolers to learn vocabulary. This statement has been affirmed by the preschool teachers and pre-schoolers' parents. The content and presentation of the AR learning method easily draw the attention of the pre-schoolers, as the 3D images are striking and interesting to them. It improves the pre-schoolers' participation and keeps them engrossed in learning. While listening to the clearly pronounced words from the AR app, the children can grasp the phonetics of the words, and pronounce them easily and correctly. As a conclusion, this study has been conducted successfully and the developed application does fulfil the objectives of this research. Hardware and software constraints are the main limitations of this research. First, the Home AR app is not designed to work on IOS devices; therefore, it can only work on Android phones and tablets. The Home AR application's scope is fairly limited because it only concentrates on 10 items from each room of the house. This is due to software capacity constraint. The children will only learn a small number of words. It will be a step forward if future researchers or developers can expand the project scope of the Home AR application: include more objects like animals, automobiles, fruits, and veggies. Kids using the application will be able to learn a larger number of words. Additionally, this application may be configured to "speak" a variety of languages, including Iban, Bidayuh, Telegu, Malayalam, and many others. That will be fascinating and kids would love it!

Acknowledgements. This work was supported and funded by Universiti Malaysia Sarawak (UNIMAS), under the P.RAMLEE RESEARCH CHAIR (PRC) (F04/PRC/2174/2021). Moreover, we want to thank the Faculty of Cognitive Sciences and Human Development, Universiti Malaysia Sarawak (UNIMAS), for their support and funding of the publication.

References

1. Broadbent, J.T., Vavilova, Z.: Bilingual Identity: issues of self-identification of bilinguals in Malaysia and Tatarstan. 3L: Southeast Asian J. Engl. Lang. Stud. **21**(3), 141–150 (2015)
2. Kim, L.S., et al.: The English Language And Its Impact On Identities Of Multilingual Malaysian Undergraduates. GEMA Online J. Lang. Stud. **10**(1), 87–101 (2010)
3. Alqahtani, M.: The importance of vocabulary in language learning and how to be taught. Int. J. Teach. Educ. **3**(3), 21–34 (2015)
4. Viera, R.T.: Vocabulary knowledge in the production of written texts: a case study on EFL language learners. Revista Tecnológica-ESPOL **30**(3), 89–105 (2017)
5. Chen, R.W., Chan, K.K.: Using augmented reality flashcards to learn vocabulary in early childhood education. J. Educ. Comput. Res. **57**(7), 1812–1831 (2019)
6. Agustin, R.W., Ayu, M.: The impact of using instagram for increasing vocabulary and listening skill. J. Engl. Lang. Teach. Learn. **2**(1), 1–7 (2021)
7. Wong, K.M., Flynn, R.M., Neuman, S.B.: L2 vocabulary learning from educational media: the influence of screen-based scaffolds on the incidental–intentional continuum. TESOL J. **12**(4), e641 (2021)
8. Jazuli, A.J.M., Din, F.F.M., Yunus, M.M.: Using pictures in vocabulary teaching for low proficiency primary pupils via PI-VOC. Int. J. Acad. Res. Bus. Soc. Sci. **9**(1), 311–319 (2019)
9. Scott, W.A., Ytreberg, L.H.: Teaching English (1990)
10. Vanniarajan, S.: Language learning strategies: what every teacher should know by rebecca L. oxford. Issues Appl. Linguist. **1**(1) (1990)
11. Larchen Costuchen, A., Font Fernández, J.M., Stavroulakis, M.: AR-Supported Mind Palace for L2 Vocabulary Recall. Int. J. Emerg. Technol. Learn. **17**(13), 47–63 (2022)
12. Selvarani, A.G.: An interactive number learning augmented reality application for autistic preschool children. In: 2022 International Conference on Computer Communication and Informatics(ICCCI) IEEE (2022)
13. Carmigniani, J., Furht, B.: Augmented reality: an overview. Handbook of augmented reality p. 3–46 (2011)
14. Enyedy, N., Yoon, S.: Immersive environments: Learning in augmented+ virtual reality. In: International Handbook of Computer-Supported Collaborative Learning, pp. 389–405. Springer (2021)
15. Aydoğdu, F.: Augmented reality for preschool children: an experience with educational contents. Br. J. Edu. Technol. **53**(2), 326–348 (2022)
16. Aydogdu, F., Kelpšiene, M.: Uses of augmented reality in preschool education. Int. Technol. Educ. J. **5**(1), 11–20 (2021)
17. Chrisna, V., Satria, T.: Kotak Edu: an educational augmented reality game for early childhood. J. Phys. Conf. Ser. IOP Publishing **1844**, 12027 (2021)
18. Palamar, S.P., et al.: Formation of readiness of future teachers to use augmented reality in the educational process of preschool and primary education. CEUR Workshop Proceedings (2021)
19. Utami, F., et al.: Introduction to sea animals with augmented reality based flashcard for early childhood. Adv. Soc. Sci. Educ. Humanit. Res. **513**, 215–220 (2021)

20. Ishak, I.S., Alias, R.A.: Designing a strategic information system planning methodology For Malaysian institutes of higher learning (ISP-IPTA). Universiti Teknologi Malaysia Skudai, Malaysia (2005)
21. Igwenagu, C.: Fundamentals of research methodology and data collection. LAP Lambert Academic Publishing (2016)
22. Kaur, A., Kaur, K.: Suitability of existing software development life cycle (SDLC) in context of mobile application development life cycle (MADLC). Int. J. Comput. Appl. **116**(19), 1–6 (2015)
23. Raus, M., et al.: The development of FiTest for institution of higher learning using Mobile Application Development Lifecycle Model (MADLC): from identification phase to prototyping phase. Int. Acad. Res. J. Bus. Technol. **2**(2), 77–84 (2016)
24. Vithani, T., Kumar, A.: Modeling the mobile application development lifecycle. In: Proceedings of the International MultiConference of Engineers and Computer Scientists (2014)
25. Shanmugam, L., Yassin, S.F., Khalid, F.: Incorporating the elements of computational thinking into the Mobile Application Development Life Cycle (MADLC) model. Int. J. Eng. Adv. Technol. **8**, 815 (2019)
26. Tang, C.K.: Augmented reality apps for children to learn aphabet through storytelling (2012)
27. Calle-Bustos, A.-M., et al.: An augmented reality game to support therapeutic education for children with diabetes. PLoS ONE **12**(9), e0184645 (2017)
28. Jamaludin, M.H.: Augmented reality mobile app for children learning on colour. IRC (2015)
29. Che Hashim, N., et al.: User satisfaction for an augmented reality application to support productive vocabulary using speech recognition. Adv. Multimed. **2018**, 1–8 2018

Towards the Prediction of Students' Performance Using Educators' Domain Knowledge

Amina Ouatiq$^{(\boxtimes)}$ ⓘ, Sophia Faris, Khalifa Mansouri ⓘ, and Mohammed Qbadou ⓘ

M2S2I Laboratory, ENSET Mohammedia, Hassan II University, Mohammedia, Morocco
amina.ouatiq@gmail.com

Abstract. Predicting student performance, achievement, and learning outcomes is one of the primary applications of educational data mining. Identifying students at risk benefits all parties involved in the educational scenario. This leads to early intervention and more impact on the students' success. However, the role of teachers with domain expertise is minimized, despite being critical to the process, from understanding the domain and selecting features to interpreting results. In this paper, the possibility of benefiting from domain experts was investigated by allowing them to select the features used in prediction. Then, five data mining algorithms were used in two experiment to determine which one is the most accurate. KNN and ANN proved to be the most accurate models in the experiments and the teachers 'chosen feature proved to provide sufficient accuracies and have opportunity for development.

Keywords: Data Mining · Educational Data Mining (EDM) · Students' Performance Prediction · Teachers' perception

1 Introduction

In the information age; the 21st century, the technological evolution affected various areas of human life and society in general. Such as the medical domain, business, banking, insurance, transportation, and education, etc. [1]. This evolution resulted in large volumes of data. Therefore, to take benefits from this data and turn it into useful practical information [2], data mining was introduced to extract hidden information and help focus on important information for future decisions [3, 4].

Data mining tools and technique predict future trends and behaviors, help organizations make knowledge-driven decisions, and monitor users [5]. It has multiple applications for example: governments use it in election campaigns, social medias use it for marketing purposes, and in web content mining [2].

Analyzing patterns and predicting outcomes help the education domain to resolve educational issues. Like the high rates of dropout and failure. Which lead us to Educational Data Mining (EDM) as an interdisciplinary field of study that bridges between education and data mining [6]. EDM has several applications, Bakhshinategh et al. identified 13 categories of EDM tasks and applications, namely: Scientific inquiry, Evaluation,

© The Author(s), under exclusive license to Springer Nature Switzerland AG 2023
F. Saeed et al. (Eds.): ICACIn 2022, LNDECT 179, pp. 78–87, 2023.
https://doi.org/10.1007/978-3-031-36258-3_7

Adaptive and recommender systems, Planning and scheduling, course design and distribution, providing feedback and alerts, social network analysis, Profiling and grouping students, and finally Predicting student performance, achievement and behaviors [6].

One of the most useful applications is student exam performance prediction. It aims to provide an estimation of student achievements on exams [7]. It is beneficial for identifying students at risk of failing their final exam in order to provide them with additional assistance on time in order to prevent this from happening [8, 9]. Research has been conducted using different algorithms such as machine learning algorithms [10].

Providing educators, universities, and students with relevant information so that intervention could be planned. Teachers might offer to tutor, and adapt their courses, universities could recruit additional teaching staff, and improve curriculums. Students on their part should enhance their skills, ask for help and be more self-regulated.

Data mining converts data into usable information [11] and provides new insights to solve problems and improve how we address the issues. However, domain expertise should not be minimized in order to get useful insight. It plays a crucial role in the process, starting from understanding the domain and the data, choosing the features, and interpreting the results [12]

In this paper, the possibility of benefiting from teachers' expertise to predict the students' performance was investigated. We will use results from previous research, where teachers were asked about the indicators and features, they judge important to predict students' success or failure [13] to use as predicting features.

This paper will try to answer the following research question:

RQ: Does the factors chosen by the teachers predict students' performance?

To do this, we implement five ML algorithms namely: Decision Tree (DT), Naïve Bayes, Support Vector Machine (SVM), Artificial Neural Networks (ANN), and K-Nearest Neighbor (KNN) in two experiments. In the first experiment, the algorithms are applied on all the features of database. The second experiment apply the models only on features chosen by the domain experts (educators and teachers). The results shows that KNN and ANN are the top performant algorithms in both experiments, However, using all features outperformed the experiment where the only the chosen-features were applied.

The remainder of this paper is divided into four sections. The following section gives some related works. Section 3 describe the methodology. Finally, Sect. 4 displays the obtained results and discusses them, and Sect. 5 covers the conclusion and future works.

2 Related Work

EDM is the use of data mining techniques in the educational context to help us extend our understanding of the learning process, and solve problems of the domain, like the prediction of students' academic performance, in order identify students at risk and take preventive measures to avoid academic failure.

Based on the data of 486 students extracted from a Virtual Learning Environments (VLE) containing only the grades of assignments, quizzes, and exams. Evangelista proposed a hybrid machine learning framework using eight classification algorithms and three ensemble methods in order to predict students' performance and with the aim of comparing the prediction models, then choose the one with the highest predictive accuracy. They conclude that feature selection techniques combined with ensemble methods improved the prediction accuracy [14].

With the same goal, this study builds classification models with different algorithms like Naive Bayes, Decision Tree, Random Forest, Random Tree, REP Tree, J Rip, One R, Simple Logistic and Zero R. They found that classification algorithms like One R, REP Tree and Decision Tree (J48) have more than 76% accuracy. The results also shows that schools and study time affect the students' grades [15].

In another study, they were interested in predicting the final grades of the students and finding the factors that affect the students' results. They used a Deep Learning model and two open datasets containing 395 and 649 students, each with the same attributes. Therefore, they achieved an accuracy of 0.964 and the results showed that all the features of demographic or socio-economical type contribute significantly to the prediction [16].

Meanwhile, [17] and in order to help educators, perceive weak learners and reduce the failure rates. They proposed a prediction model based on both classification and clustering techniques, then tested it on a real data from various academic disciplines. The accuracy they obtained is 0.7547. And the results show a relation between learners' behavior and their academic performance.

Hasan and Aly discuss the case of predicting the students at risk in an undergraduate STEM core course. Despite the insufficient features early in the semester. They propose a novel hybrid Machine Learning based framework that executes multi-class classification. They used a pipeline to identify the most critical group at first then the less prone to risk [18].

The prevention of dropping out was the main goal of [19]. They focused on determining the key feature of dropout to identify students with a high probability of dropping out. Their results showed that simple algorithms achieve reliable levels of accuracy to identify predictors of dropout in order to decrease the dropout rate. Especially the decision tree model.

Despite several studies being made in predicting students' outcomes, there is still a lack of the use of teachers' expertise. This paper will apply the most used methods namely Bayesian Network (BN), Decision Tree (DT), Artificial Neural Networks (ANN), K-Nearest Neighbor (KNN), Support Vector Machine (SVM) [20], to predict students' performance and take teachers' knowledge and expertise into consideration by allowing them to choose the features and compare the results of the mentioned models on features chosen by the teachers, with the results of using them on all the features available in the dataset.

3   Methodology

The method presented in this study follows the process of knowledge discovery and data mining. The main steps are described below (Fig. 1):

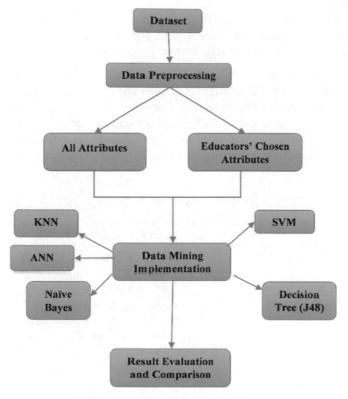

Fig. 1. Work flow of the methodology

Data Gathering. In this study, an open dataset has been obtained from Kaggle through a questionnaire conducted with 101 students [21]. It contains 30 attributes divided into three categories demographic (Age, Sex, Salary, High School Type, etc.), background (Mothers' and Fathers' Education, Mothers' and Fathers' jobs, Number of siblings, etc.), and educational data (Attendance to Lectures, Attendance to Seminar/Conference, taking notes, Effect of in-class Discussions, etc.). It has seven outputs as grades (Fail, DD, DC, CC, CB, BB, BA, AA). Therefore, we will consider:

- AA, BA, BB as top-performing,
- CC, CB as Average,
- Fail, DD, CD as under-performing.

Data Pre-processing. This stage focusses on transforming the data using pre-processing task to be appropriate for use by the EDM techniques. Such as eliminating outliers. Our data had no missing values, and almost no outliers (just 2 rows). All the data was numeric. It was scaled and changed from a numerical format that is variable in very different ranges to a nominal format. With the filter NumericToNominal in WEKA. Due to the problem of imbalanced data, re-balancing data has been applied to the data set. The solution to this problem is to balance the distribution of classes. The widely used oversampling algorithm SMOTE (Synthetic Minority Oversampling Technique) [22]. Which is a technique that uses the nearest neighbor rule to balance the data by introducing new synthetic elements to the minority class [23].

Feature Selection. In the first experiment, we use all the features for the prediction. Then in the second experiment, only the features chosen by the teachers will be used. The teachers were interested in the features that describe: The regularity and time spent on the system, The access to resources, Previous academic performance, Face-to-face performance, The number of help's requests, and Behavioral indicators [13]. Table 1 indicates the attribute that answers the teacher recommendations from the datasets.

Table 1. Features chosen by the teachers

Attributes chosen	Present in the dataset
Previous academic performance	• Cumulative grade point average in the last semester • Scholarship type
Face-to-face performance	• Attendance to classes • Preparation to midterm exams type • Preparation to midterm exams • Taking notes in classes • Listening in classes
Behavioral indicators (positive and negative)	• Weekly study hours • Reading frequency (scientific books/journals) • Attendance to the seminars/conferences related to the department • Discussion improves my interest and success in the course

Data-Mining. For the prediction of students' academic performance, two experiments and five data mining techniques were carried out Firstly, all the classification algorithms were executed using all 30 attributes. Formerly, only the 11 selected attributes were used. The models were built using 10-fold cross-validation, where 9 sets used for training the model, and the last one is used for testing [24].

To classify whether the student will be considered top-performing, average, or under-performing. We developed five prediction models by applying five classification techniques using the Waikato Environment for Knowledge Analysis (WEKA) tool [25], which is an open-source tool coded in Java, broadly used in EDM [26]. The algorithms are described below.

K-Nearest Neighbor (KNN) is an algorithm that search for x_0 K nearest neighbors and uses a majority vote to determine it class label. Euclidean distance is utilized to calculate the distances [27].

Artificial Neural Network (ANN) consists of a large number of processors linked by weighted connections. Each node output is determined by information that is locally available, internally or via the weighted connections [28].

Decision Tree: as the name indicates it has the flowchart of the algorithms looks like a tree with branches that represents an outcome of the test on attributes (node), and the class label is represented by each leaf node (or terminal node) [29].

Support Vector Machine (SVM) map items as points in a dimensional feature space, trained with learning algorithm that applies a statistical learning theory learning bias in order to find the best hyperplane [30].

Naive Bayes estimate the posterior probability P (y/x) of each class using Bayes theorem [31].

Evaluation. The performance of the models is determined by using the following concepts: (1) TP: the instances that are correctly predicted as positive. (2) FP: the instances that are incorrectly predicted as positive. (3) TN: the instances that are correctly predicted as negative. (4) FN: the instances that are incorrectly predicted as negative.

Accuracy is the fraction of correctly predicted results and is measured with (1):

$$Accuracy = \frac{(TP + TN)}{(TP + TN + FP + FN)} \tag{1}$$

Recall is the ratio between the correctly predicted positive results and all positive samples and is measured with (2):

$$Recall = \frac{TP}{(TP + FN)} \tag{2}$$

Precision is the percentage of correct positive samples and is measured with (3):

$$Precision = \frac{TP}{(TP + FP)} \tag{3}$$

F-Measure balances the recall and the precision, by highlighting the performance on common and rare categories. It is measured with (4):

$$F - Measure = \frac{2 * Recall * Precision}{(Recall + Precision)} \tag{4}$$

4 Results and Discussion

In the first experiment, five algorithms were applied using all attributes. In a second experiment, just the teacher-chosen features were considered. All the models namely Naïve Bayes, KNN, decision tree (J48), SVM, and ANN have been implemented by WEKA. Tables 2 and 3 show the accuracy and performance measures for each model in both experiments using all features in the first, and using only the teacher-chosen features in the second one.

As shown in Table 2 using all features gives accuracy up to 92.42% using the KNN model. And the lowest accuracy belonging to the decision tree (J48) model is 81.03%.

Table 3, exhibits that the SVM model has the lowest accuracy, which is equal to 69.54%. The most accurate model is the model built using the ANN technique, which has an accuracy of 75.86%. Although, the five algorithms gave satisfactory results.

KNN and ANN are the top accurate algorithms in both experiments.

Table 2. The outcome of the models on all features

Classifiers name	Accuracy	Precision	Recall	F-Score
Naïve Bayes	83.908	0.844	0.839	0.838
KNN	**92.241**	**0.925**	**0.922**	**0.923**
J48	81.034	0.812	0.810	0.811
SVM	86.493	0.866	0.865	0.865
ANN	88.218	0.883	0.882	0.883

Table 3. The outcome of the models on teachers' features

Classifiers name	Accuracy	Precision	Recall	F-Score
Naïve Bayes	70.115	0.699	0.701	0.698
KNN	74.712	0.744	0.747	0.744
J48	70.402	0.702	0.704	0.702
SVM	69.540	0.691	0.695	0.692
ANN	**75.862**	**0.757**	**0.759**	**0.757**

This preliminary study discovered that in order to predict student's performance, all features must be evaluated. However, the experiment with only the selected features gave sufficient accuracies up to 75.86%. It proved that it still has room for improvement. It could be due to dataset limitations and the absence of some crucial characteristics considered by teachers. While attempting to benefit from the teacher's knowledge and expertise in the education domain and detecting underperforming students to address their issues, a more comprehensive and extensive dataset including attributes from online and face-to-face setting, as suggested by the teacher, will be beneficial. As a result, the process must be discovered further using a vast and balanced dataset.

5   Conclusion

This preliminary study considers the original idea of using teachers' experience to predict student achievement. With the ultimate goal of identifying underperforming students and providing assistance to lower failure rates. Five machine learning algorithms were chosen and tested to see if the features selected by the teachers might help predict student outcomes. Namely the regularity and time spent on the system, the amount of access to resources, their previous academic performance, face-to-face performance, the number of requests for student help, behavioral indicators [13].

The results reveal that, despite the fact that the models' accuracy on all features beats those utilizing only the selected feature. It gives an opportunity for development in order to gain more experience with a larger dataset that contains all features mentioned by the educators. We can use the findings of this study to broaden the experience and experimenting with features from a blended educational environment to obtain a holistic view of the importance of features that teachers found important and consider when working with students on a daily basis.

For additional research, we aim to test with a larger and more inclusive dataset that includes more features and student characteristics, as well as ensemble methods such as bagging, boosting, and voting.

References

1. Kleissner, C.: Data mining for the enterprise. Proc. Hawaii Int. Conf. Syst. Sci. **7**, 295–304 (1998). https://doi.org/10.1109/HICSS.1998.649224
2. Ramzan, M., Ahmad, M.: Evolution of data mining: an overview. In: Proceedings of the 2014 Conference on IT in Business, Industry and Government: An International Conference by CSI on Big Data, CSIBIG 2014 (2014). https://doi.org/10.1109/CSIBIG.2014.7056947
3. Witten, I.H., Frank, E., Geller, J.: Data mining. ACM SIGMOD Rec. **31**, 76–77 (2002). https://doi.org/10.1145/507338.507355
4. Wirth, R., Hipp, J.: CRISP-DM: Towards a Standard Process Model for Data Mining
5. Daniel, T.L., Chantal, D.L.: Discovering Knowledge in Data: An Introduction to Data Mining. Google Livres
6. Bakhshinategh, B., Zaiane, O.R., ElAtia, S., Ipperciel, D.: Educational data mining applications and tasks: a survey of the last 10 years. Educ. Inf. Technol. **23**(1), 537–553 (2017). https://doi.org/10.1007/s10639-017-9616-z
7. Tomasevic, N., Gvozdenovic, N., Vranes, S.: An overview and comparison of supervised data mining techniques for student exam performance prediction. Comput. Educ. **143**, 103676 (2020). https://doi.org/10.1016/J.COMPEDU.2019.103676
8. Márquez-Vera, C., Cano, A., Romero, C., et al.: Early dropout prediction using data mining: a case study with high school students. Expert. Syst. **33**, 107–124 (2016). https://doi.org/10.1111/EXSY.12135
9. Mishra, T., Kumar, D., Gupta, S.: Mining students' data for prediction performance. In: International Conference on Advanced Computing and Communication Technologies, ACCT, pp. 255–262 (2014).https://doi.org/10.1109/ACCT.2014.105
10. Hussain, M., Zhu, W., Zhang, W., et al.: Using machine learning to predict student difficulties from learning session data. Artif. Intell. Rev. **52**(1), 381–407 (2018). https://doi.org/10.1007/S10462-018-9620-8
11. Devasia, T., Vinushree, T.P., Hegde, V.: Prediction of students performance using educational data mining. In: Proceedings of 2016 International Conference on Data Mining and Advanced Computing, SAPIENCE 2016, pp. 91–95 (2016). https://doi.org/10.1109/SAPIENCE.2016.7684167

12. Kopanas, I., Avouris, N.M., Daskalaki, S.: The role of domain knowledge in a large scale data mining project. In: Vlahavas, I.P., Spyropoulos, C.D. (eds.) Methods and Applications of Artificial Intelligence. LNCS (LNAI), vol. 2308, pp. 288–299. Springer, Heidelberg (2002). https://doi.org/10.1007/3-540-46014-4_26

13. Ouatiq, A., Riyami, B., Mansouri, K., Qbadou, M.: The Preferences and Expectation of Moroccan Teachers from Learning Analytics Dashboards in a Blended Learning Environment: Empirical Study, pp. 287–297 (2021). https://doi.org/10.1007/978-3-030-91738-8_27

14. Evangelista, E.D.: A hybrid machine learning framework for predicting students' performance in virtual learning environment. Int. J. Emerg. Technol. Learn. **16**, 255–272 (2021). https://doi.org/10.3991/IJET.V16I24.26151

15. Yass Salal, S.M.A., Mukesh, K.: Educational data mining: student performance prediction in academic 55. Int. J. Eng. Adv. Technol. **8**, 24–59 (2019)

16. Aslam, N., Khan, I.U., Alamri, L.H., Almuslim, R.S.: An improved early student's academic performance prediction using deep learning. Int. J. Emerg. Technol. Learn. **16**, 108–122 (2021). https://doi.org/10.3991/IJET.V16I12.20699

17. Francis, B.K., Babu, S.S.: Predicting academic performance of students using a hybrid data mining approach. J. Med. Syst. **43**(6), 1–15 (2019). https://doi.org/10.1007/s10916-019-1295-4

18. Hasan, M., Aly, M.: Get more from less: a hybrid machine learning framework for improving early predictions in STEM education. In: Proceedings – 6th Annual Conference on Computational Science and Computational Intelligence, CSCI 2019, pp. 826–831 (2019). https://doi.org/10.1109/CSCI49370.2019.00157

19. Perez, B., Castellanos, C., Correal, D.: Applying data mining techniques to predict student dropout: a case study. In: 2018 IEEE 1st Colombian Conference on Applications in Computational Intelligence, ColCACI 2018 – Proceedings (2018). https://doi.org/10.1109/COLCACI.2018.8484847

20. Qazdar, A., Er-Raha, B., Cherkaoui, C., Mammass, D.: A machine learning algorithm framework for predicting students performance: a case study of baccalaureate students in Morocco. Educ. Inf. Technol. **24**(6), 3577–3589 (2019). https://doi.org/10.1007/s10639-019-09946-8

21. Yılmaz, N., Sekeroglu, B.: Student performance classification using artificial intelligence techniques. In: Advances in Intelligent Systems and Computing 1095 AISC, pp. 596–603 (2020). https://doi.org/10.1007/978-3-030-35249-3_76

22. Viloria, A., et al.: Determination of dimensionality of the psychosocial risk assessment of internal, individual, double presence and external factors in work environments. In: Tan, Y., Shi, Y.H., Tang, Q.R. (eds.) Data Mining and Big Data, pp. 304–313. Springer International Publishing, Cham (2018). https://doi.org/10.1007/978-3-319-93803-5_29

23. Chawla, N.V., Bowyer, K.W., Hall, L.O., Kegelmeyer, W.P.: SMOTE: synthetic minority over-sampling technique. J. Artif. Intell. Res. **16**, 321–357(2002). https://doi.org/10.1613/JAIR.953

24. Refaeilzadeh, P., Tang, L., Liu, H.: Cross-validation. In: Liu, L., Özsu M. (eds.) Encyclopedia of Database Systems. Springer, New York, NY (2016). https://doi.org/10.1007/978-1-4899-7993-3_565-2

25. Hall, M., Frank, E., Holmes, G., et al.: The WEKA data mining software. ACM SIGKDD Explor. Newsl. **11**, 10–18 (2009). https://doi.org/10.1145/1656274.1656278

26. Mengash, H.A.: Using data mining techniques to predict student performance to support decision making in university admission systems. IEEE Access **8**, 55462–55470 (2020). https://doi.org/10.1109/ACCESS.2020.2981905

27. Song, Y., Huang, J., Zhou, D., Zha, H., Giles, C.L.: IKNN: informative K-nearest neighbor pattern classification. In: Kok, J.N., Koronacki, J., Lopez de Mantaras, R., Matwin, S., Mladenič, D., Skowron, A. (eds.) PKDD 2007. LNCS (LNAI), vol. 4702, pp. 248–264. Springer, Heidelberg (2007). https://doi.org/10.1007/978-3-540-74976-9_25

28. Dongare, A., Kharde, R.R., Kachare, A.D.: Introduction to artificial neural network. Int. J. Eng. Innov. Technol. **2**, 189–194 (2012)
29. Himani, S., Sunil, K.: A survey on decision tree algorithms of classification in data mining. Int. J. Sci. Res. **5**, 2094–2097 (2016)
30. Noble, W.S.: What is a support vector machine? Nat. Biotechnol. **24**, 1565–1567 (2006). https://doi.org/10.1038/nbt1206-1565
31. Webb, G.I.: Naïve Bayes. Encyclopedia of Machine Learning and Data Mining, pp. 895–896 (2017). https://doi.org/10.1007/978-1-4899-7687-1_581

Learners' Performance Evaluation Using Genetic Algorithms

Tariq Saeed Mian$^{(\boxtimes)}$ (iD)

Department of IS, College of Computer Science and Engineering, Taibah University,
Madinah Almunwarah, Saudi Arabia
`tmian@taibahu.edu.sa`

Abstract. In the education sector quality enhancement directly informs student performance. Educational institutions typically make decisions based on prior student performance. The literature demonstrates that Data Mining techniques perform a vital role in analyzing and predicting student performance. In this study, a questionnaire-based survey was conducted to collect data from Saudi secondary school students. The study allowed students to achieve target grades in higher education contexts at university level. Herein, the bioinspired feature selection technique Genetic Algorithms is used to develop a framework to predict student performance at university level. Student performance is categorized utilizing machine learning classifiers with pre-selected features. Student performance is measured in terms of accuracy, precision, recall, and the F1-measure. The Random Forest classifier outperformed the benchmark dataset delivering an accuracy 0.78% and F1_Score of 0.84%.

Keywords: Genetic Algorithms · Logistic Regression · Random Forest · Naïve Bayes (NB) · Accuracy · Precision

1 Introduction

A variety of different data analysis techniques have been employed in recent years to improve the quality of teaching and learning. Since the emergence of data mining and machine learning, its use in the education context (i.e., Educational Data Mining (EDM) has attained popularity among researchers [1]. EDM relies on the voluminous data acquired from several difference sources, including administrative records, exam results, student data, etc.

Predictive model approaches have made significant contributions emerging as valuable research trends in the field of educational performance improvement. Each year a large number of students complete their secondary education and enroll in universities. However, some students fail to complete their degrees or achieve their target grades, due in part to deficits in required skills and capabilities. Higher education institutions globally face an enormous challenge when seeking to predict and anticipate students' performance [2]. According to a USA report [3], 40% of graduates enrolled on a 4 year BSc program of study dropped out without completing the degree. Universities and higher

F. Saeed et al. (Eds.): ICACIn 2022, LNDECT 179, pp. 88–99, 2023.
https://doi.org/10.1007/978-3-031-36258-3_8

education institutions have developed and adopted multiple techniques to systemize students' performance, with the aim of reducing dropout rate to improve the educational institutions' quality ranking [4]. Early detection of student failure/underperformance can have many benefits, such as identifying students at risk, student behavior and habits, and course selection pathways. Data Mining is a beneficial approach to identify hidden patterns in data [5]. As mentioned above, universities worldwide are trying to maintain their rankings, which requires them to demonstrate improved student performance. The most challenging task for universities is to improve student performance through the early detection of potential failure [6].

Data mining and machine learning-based algorithms can be employed to analyze and predict student performance in higher education institutions [7]. Data mining techniques are very helpful in determining the hidden patterns within a dataset and can provide insights into our understanding of datasets. The timely classification and prediction of student performance will provide an early warning system to allow decision makers to develop strategies for early intervention to target students at risk and reduce dropout rates.

A thorough literature survey revealed that researchers have applied different ML techniques to forecast the learner performance. The authors of [8] used a dataset of 1407 records, employing an artificial neural network (ANN) to predict student performance. The cross-validation hold-out method was used to test the algorithms. In another study [9], the ANN model was employed to predict student performance, and 95% accuracy observed using the target test data. Meanwhile some other studies used machine learning models such as Naïve Bayes (NB) [10] and Decision tree [11] to predict student performance and reported an accuracy rate of 60%. Some researchers attempted to compare machine learning and data mining models with other models. In this study [12], several ML models designed to predict student performance were used, and it was observed that the Decision Tree outperformed other models. The authors of [13] assessed the performance of Logistic Regression, Decision Tree, NB, SVM, and KNN. In this study features selection was implemented to achieve higher accuracy. In this study [14], the author compared the performance of SVM, Random Forest, Logistic Regression, and XG Boost. In [15], the author presented differences in the performance of Artificial Neural Network, Logistic Regression XG-Boost, and Random Forest, and reported that XG-Boost outperformed the benchmark dataset in terms of accuracy.

When performing binary classification tasks supervised machine learning models are seen as the better choice to analyze data. In this study, we used a bio-inspired feature selection technique called Genetic Algorithms to select the optimal number of features from a larger set of features, and then passed the optimal features to traditional supervised machine learning models. The contributions made by the proposed study are given below:

- A review of the available literature revealed that the Genetic Algorithm (GA) as a feature selection has not been used in this domain, at least to the best of our knowledge, making this study novel.
- This study also attempted to compare the supervised machine learning models, with and without feature selection technique.

- The research further provides a framework based on machine learning techniques to automate the process of students' performance prediction.
- Student performance evaluation is a vital tool for perceiving student capabilities and distinguishing those areas that need to be improved.

2 Literature Review

Machine learning is increasingly widely used in all domains, and education is not an exception [16, 17]. In the field of education, machine learning algorithms are used for several different purposes. For example [18] used a look ahead fuzzy association rule mining algorithm to identify the relationship between questions, developing a final concept map [19] employing ML techniques to predict students' performance on their final examinations.

To justify the academic value of this work a thorough review of existing literature was performed. During this literature review research studies using different data mining techniques with statistical features selection techniques for student performance prediction were identified and reviewed. During the process of conducting the literature review, it emerged that predictions about dropout rate are frequently inaccurate and so could be further improved.

In higher education institutions academic advisors play a pivotal role in the success rates of the students. To develop a framework to facilitate student advisors to provide effective and timely advice to supervisees [20] used a C4.5 algorithm to develop decision trees and employed the accuracy measure, Kappa measure, and ROC area to evaluate the resultant decision trees. In [21] the author also used the Decision Tree to classify the given dataset to determine dropout with good classification results.

Early forecasting of learners' performance and final passing out grades is considered to be very helpful for an institution seeking to develop early intervention strategies. [22] generated two sets of classification models, based on the course grades and grade point average (GPA) of individual students, and employed popular machine learning algorithms. The study results demonstrated that course grades yield better accuracy if the prediction is done at the end of year 1 (before start of the third term), whereas GPA achieves better accuracy otherwise.

The authors in [23] used a hybrid regression model to improve the predictive accuracy of academic performance among learners, the presented model could also be used to predict the grades for different courses. Most importantly the study presented an optimized multi-label classifier to predict the factors that can influence individual learner's performance.

The NB model delivered good results in terms of its capacity to predict student dropout, as it contains an inductive mechanism that can determine student performance classification accuracy. The induct mechanism operates in two modes, i.e. training and testing. The proposed study of the NB model [24] predicts the student dropout ratio but cannot predict the student at risk of dropout. The authors of [25] presented a distance learning-based ML technique as a tool to predict student dropout. This study was innovative and helped to carve out a pathway for data mining in education. The algorithm was trained on demographic data and information from several project assignments rather than on class performance data as a way to predict student performance.

In [26], the authors presented a hybrid network of K-means clustering and an artificial neural network (ANN) to predict the success of higher education students where the medium of teaching was a foreign language. In this study, Neural Networking was designed to predict student performance first and this was then fitted to the clustering technique for the K-means algorithm. The proposed K-means clustering technique is a powerful tool to identify students' capabilities at a very early stage of their university life.

Some contemporary work has also used deep neural networks and demonstrated their effectiveness at predicting students' performance in higher education institutions. For example, in [27] an experiment was conducted evaluating students on five compulsory courses associated with two undergraduate programs, results of the study shows that the proposed model managed to achieve high accuracy in detecting the student at risk of failure. The author of [28] proposed an open-source recommendation system MyMediaLight to facilitate predictions of student performance. The author suggested a unique technique biased matrix function to predict performance when assessing student course selections. In this study [29], the author used Gray model and Taylor. By computing approximation multiple of times, the author improved the predictive power of the models. The outcome of the technique developed in this study could be employed, by education administrators and educators, "wishing to select ideal solutions to enhance student performance when students are struggling to adjust to the learning process".

In [30] the matrix factorization, collaborative filtering, and restricted Boltzmann machine were used to validate data collected from different academic institutions. The experimental study reported that restricted Boltzmann machines produced good results as opposed to matrix factorization and collaborative filtering.

It is evident from the above discussion that student performance predictions are of prime importance when seeking to demonstrate the effectiveness and success of a higher education institute. This type of performance prediction can help when developing strategies to target vulnerable students (i.e. students at-risk) [31]. Current literature shows that the researchers have made several attempts to forecast and evaluate the learners' performance using ML and DL approaches and using different sets of data retrieved from different sources, whereby each technique belongs to diverse areas like Machine Learning, Neural networks, and collaborative filtering.

3 Methodology

3.1 Data Set

This study is based on a dataset generating by using a modified version of a questionnaire developed by [32]. The study population comprises secondary school students in different regions of the Kingdom of Saudi Arabia. To collect data the study employed convenience sampling, which is a non-probability sampling technique. Convenience sampling is recommended when wishing to collect data from surveys speedily and efficiently [33]. Almost 400 questionnaires were distributed and 303 were returned; of these questionnaires 103 were rejected due to incomplete responses.

3.2   Feature Selection Technique – Genetic Algorithm

In this paper, feature selection technique Genetic Algorithm (GA) is used. GA is a heuristic technique implemented to solve problems involving search and optimization. The technique was based on Charles Darwin's theory of natural selection. The technique postulates that the fittest individual is selected for reproduction to produce offspring in the next generation. Usually, GA has five phases:

1. Initial Population
2. Fitness function
3. Selection
4. Crossover
5. Mutation.

The initial population starts with a set of individuals referred to as the population. Everyone gives a solution to a problem where everyone has a set of parameters characterized by Genes. A pair of genes are required to form a string called a Chromosome. The Fitness function provides information about how to fit an individual to compete with other individuals. It assigns a fitness score for everyone. Based on the fitness score, an individual is then selected for reproduction. In the selection phase, the fittest individuals are selected, and they pass their genes on to the next generation. Individuals with high fitness values have a greater chance of being selected for reproduction. The most important phase in GA is the Crossover. In this phase, each pair of parents are mated, and the crossover point is then selected randomly from the genes. As the new offspring are generated, some of them have a mutation phase with low random probability and some portions of the string can be flipped. The main theme of mutation is to maintain diversity between populations and prevent premature convergence [34].

The overall feature selection process in Genetic Algorithms is outlined in Fig. 1.

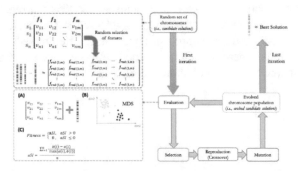

Fig. 1. Genetic Algorithm

It is apparent from the above discussion that a GA does not directly operate based on candidate solutions, which is the major difference between a GA and traditional techniques. Furthermore, a GA can produce several different optimal results (the probability of an optimal result in GA is therefore higher than in traditional algorithms) from different generations, whereas traditional algorithms can only generate a single result, which may not be optimal.

3.3 Logistic Regression

Logistic Regression is a supervised machine learning algorithm able to handle classification tasks only. Logistic Regression is a linear model composed of one more independent feature representing a relationship between a dependent variable and independent variables. The logistic regression uses a probability value of 0.5 to determine the output variable, which belongs to the appropriate category. We require knowledge of the distribution function to better understand the logistic regression target binary variable, which is modeled as follows:

$$\prod(x) = E(Y|x) \tag{1}$$

The above regression is bounded between the binary class instances 0 and 1. We can achieve the logistic distribution function represented by the author [35]:

$$\prod(x) = e\hat{x}(1 + e\hat{x}) \tag{2}$$

The logistic regression function ranges between 0 and 1, instead of the X value. The character of logistic regression is very valuable, so it can be used to represent the probabilities 1 and 0. Logistic regression takes the shape of an S, in which a value lower than x is close to 0 until a specified limit is met. The logistic regression function's value increases rapidly, around 1 for all the x's. Our proposed approach to multiple logistic regression is a linear sum calculated based on a constant value and unknown factors βi and determined by an exploratory features represented as $x_1, x_2, x_3 \ldots x_n$. The inverse of the logistic regression function can be expressed as:

$$\ln\left[\frac{\pi(x)}{1 - \pi(x)}\right] = \beta_0 + \beta_1 x_1 + \beta_2 x_2 + \cdots ..\beta_n x_n \tag{3}$$

The logistic regression use the sigmoid function as follows:

$$h_x = \frac{1}{1 + e^{-\theta Tx}} \tag{4}$$

To represent the binary variable outcome, the logistic regression employs the following equation.

$$p(y|x; 0) = \left(\frac{1}{1 + e^{-\theta Tx}}\right)^y \cdot \left(1 - \frac{1}{1 + e^{-\theta Tx}}\right)^{1-y} \tag{5}$$

3.4 Naive Bayes Model

The NB model is a classical conditional probability base Bayesian model [36]. The NB model does not apply an iterative technique like other classification models. Its approach is linear in time. The merit of using a NB model is that it works with even a small amount of data. The NB theory can be expressed as follows:

$$P(y|X) = \frac{P(x|y) * P(y)}{P(X)} \tag{6}$$

In the above equation, X is the explanatory variable and y is the target variable. The NB model assumes multiple variables are independent, provided by the classes of the dependent variable as shown below:

$$p(X|y) = p(X_1|y) * p(X_2|y) * \ldots p(X_n|y) \tag{7}$$

3.5 Random Forest Model

The Random Forest is an ensemble algorithm that can solve classification and regression tasks. The Random Forest is a combination of multiple root decision trees. It can have bagging and randomness feature used to create an uncorrelated forest of trees and generate more accurate results relative to individual trees. This model establishes outcomes based on decision tree predictions. The Random Forest model makes predictions by taking the mean of the output from various trees, and the average of the various tree. As the number of trees increases, the outcome accuracy of the Random Forest will increase (see Fig. 2).

Algorithm   *Random Forest for Regression or Classification.*

---

1. For $b = 1$ to B:

    (a) Draw a bootstrap sample $\mathbf{Z}^*$ of size N from the training data.

    (b) Grow a random-forest tree T_b to the bootstrapped data, by recursively repeating the following steps for each terminal node of the tree, until the minimum node size n_{min} is reached.

        i. Select m variables at random from the p variables.
        ii. Pick the best variable/split-point among the m.
        iii. Split the node into two daughter nodes.

2. Output the ensemble of trees $\{T_b\}_1^B$.

To make a prediction at a new point x:

Regression: $\hat{f}_{rf}^B(x) = \frac{1}{B} \sum_{b=1}^{B} T_b(x)$.

Classification: Let $\hat{C}_b(x)$ be the class prediction of the bth random-forest tree. Then $\hat{C}_{rf}^B(x) = $ *majority vote* $\{\hat{C}_b(x)\}_1^B$.

Fig. 2. Random Forest for Classification/Regression

4   Results and Discussion

Accuracy score, a confusion matrix, and classification report performance metrics are used to evaluate the effectiveness of the proposed model.

4.1   Accuracy Score

This measure gives the ratio of correct predictions from total number of predictions and is vital to obtain accurate results.

$$Accuracy_{score} = \frac{TP + TN}{TP + FP + FN + TN} \times 100 \tag{8}$$

4.2   Precision

Denotes the ratio of correct and true positive predictions out of the total true positive predictions. The precision metric can be expressed as follows:

$$Precision = \frac{TP}{TP + FP} \tag{9}$$

4.3   Recall

The ratio of correctly predicted positive predictions from all the observation in real class. Recall is also known as sensitivity.

$$Recall = \frac{TP}{TP + FN} \tag{10}$$

4.4   F1-Score

F1 Score is the weighted average for Precision and Recall. The F1-Score counts both false positives and false negatives to show the output value.

$$F1 - score = 2\frac{Recall * Precision}{Recall + Preicsion} \tag{11}$$

Optimal features are passed to proposed classifiers. We also use the three different machine learning classifiers in the proposed study. The Random Forest model shows the best result when compared to the Logistic Regression and NB model in terms F1_Score (Table 1).

Table 1. Results

Classifier	Accuracy Score	Precision	Recall	F1-Score
Logistic Regression	0.78	0.79	0.83	0.80
Naïve Bayes (NB)	0.75	0.75	0.81	0.72
Random Forest	0.78	0.90	0.79	0.84

We also compared the proposed classifier results for F1_Score output and showed the model's output through a bar graph. The bar graph clearly demonstrates that the Random Forest Model outperforms the benchmark dataset, excelling relative to the other two models (Fig. 3).

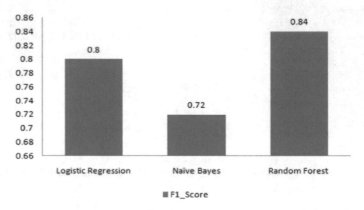

Fig. 3. F1-Score

We also show the results for the proposed models in terms of confusion matrix. The confusion matrix provides information about the model predictions for correct and incorrect instances.

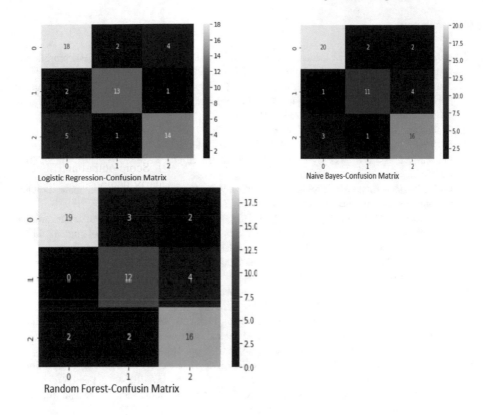

Logistic Regression-Confusion Matrix

Naïve Bayes-Confusion Matrix

Random Forest-Confusin Matrix

5 Conclusion and Future Work

We implemented a bio-inspired feature selection technique Genetic Algorithm with three supervised machine learning classifiers to predict students' performance in terms of grades ranging from A-C. Our proposed technique is suitable for teachers wishing to track the performance of students in their care, and for students wanting to keep an eye on their performance to further improve it. The proposed classifiers have some notable deficiencies, as they are not suited to handling a multiclass dataset, and so we are planning to change this strategy in a future study. However, Random Forest classifiers also deliver positive results in terms of an accuracy level of 0.78% and an F1-score of 0.84%. We are interested in conducting more research in this domain in the future to investigate the performance of students using different features selection, or a deep learning model. Thus, it is anticipated that educational institutes will then be able to access better options to improve their educational standards, and students will also be provided with accurate information regarding their studies and status with regard to their dropout risk.

References

1. Şahİn, M., Yurdugül, H.: Educational data mining and learning analytics: past, present and future. Bartın Univ. J. Faculty Educ. **9**(1), 121–131 (2020)
2. Manhães, L.M.B., da Cruz, S.M.S., Zimbrão, G.: Towards automatic prediction of student performance in STEM undergraduate degree programs. In: Proc. 30th Annu. ACM Symp. Appl. Comput. (SAC), pp. 247–253 (2015)
3. McFarland, J., et al.: The condition of education 2019 (NCES 2019–144) (2020)
4. Rastrollo-Guerrero, J.L., Gómez-Pulido, J.A., Durán-Domínguez, A.: Analyzing and predicting students' performance by means of machine learning: a review. Appl. Sci. **10**(3), 1042 (2020)
5. Ghorbani, R., Ghousi, R.: Predictive data mining approaches in medical diagnosis: a review of some diseases prediction. Int. J. Data Netw. Sci. **3**(2), 47–70 (2019)
6. Fernandes, E., Holanda, M., Victorino, M., Borges, V., Carvalho, R., Erven, G.V.: Educational data mining: Predictive analysis of academic performance of public school students in the capital of Brazil. J. Bus. Res. **94**, 335–343 (2019)
7. Baker, R.S., Yacef, K.: The state of educational data mining in 2009: a review and future visions. J. Educ. Data Mining **1**(1), 3–17 (2009)
8. Karamouzis, S.T., Vrettos, A.: An artificial neural network for predicting student graduation outcomes. In: Proc. World Congr. Eng. Comput. Sci., pp. 991–994 (2008)
9. de Albuquerque, R.M., Bezerra, A.A., de Souza, D.A., do Nascimento, L.B.P., de Mesquita Sá, J.J., do Nascimento, J.C.: Using neural networks to predict the future performance of students. In: Proc. Int. Symp. Comput. Educ. (SIIE), pp. 109–113 (2015)
10. Devasia, T., Vinushree, T.P., Hegde, V.: Prediction of students performance using educational data mining. In: Proc. Int. Conf. Data Mining Adv. Comput. (SAPIENCE), pp. 91–95 (2016)
11. Kovacic, Z.: Early prediction of student success: mining students' enrolment data. In: Proc. Informing Sci. IT Educ. Conf. (InSITE) (2010)
12. Acharya, A., Sinha, D.: Early prediction of students performance using machine learning techniques. Int. J. Comput. Appl. **107**(1), 37–43 (2014)
13. Marbouti, F., Diefes-Dux, H.A., Madhavan, K.: Models for early prediction of at-risk students in a course using standards-based grading. Comput. Educ. **103**, 1–15 (2016)
14. Hlosta, M., Zdrahal, Z., Zendulka, J.: Ouroboros: early identification of at-risk students without models based on legacy data. In: Proc. 7th Int. Learn. Anal. Knowl. Conf., pp. 6–15 (2017)
15. Kumar, V., Garg, M.L.: Comparison of machine learning models in student result prediction. In: Proc. Int. Conf. Adv. Comput. Netw. Inform, pp. 439–452 (2019)
16. Shafiq, D.A., Marjani, M., Ariyaluran Habeeb, R.A., Asirvatham, D.: Student retention using educational data mining and predictive analytics: a systematic literature review. IEEE Access (2022)
17. Ortiz-Lozano, J.M., Rua-Vieites, A., Bilbao-Calabuig, P., Casadesús-Fa, M.: University student retention: Best time and data to identify undergraduate students at risk of dropout. Innov. Educ. Teach. Int. **57**(1), 74–85 (2020)
18. Al-Sarem, M., Bellafkih, M., Ramdani, M.: Mining concepts' relationship based on numeric grades. Int. J. Comput. Sci. Issues **8**(4), 136 (2011)
19. Hashim, A.S., Akeel Awadh, W., Khalaf Hamoud, A.: Student performance prediction model based on supervised machine learning algorithms. In: IOP Conference Series: Materials Science and Engineering, vol. 928, no. 3, p. 032019. IOP Publishing (2020)
20. Al-Sarem, M.: Building a decision tree model for academic advising affairs based on the algorithm C 4-5. arXiv preprint arXiv:1511.04026 (2015)

21. Kelly, J.D.O., Menezes, A.G., de Carvalho, A.B., Montesco, C.A.: Supervised learning in the context of educational data mining to avoid university students dropout. In: 2019 IEEE 19th International Conference on Advanced Learning Technologies (ICALT), vol. 2161, pp. 207–208. IEEE (2019)
22. Tatar, A.E., Düştegör, D.: Prediction of academic performance at undergraduate graduation: course grades or grade point average? Appl. Sci. **10**(14), 4967 (2020)
23. Alshanqiti, A., Namoun, A.: Predicting student performance and its influential factors using hybrid regression and multi-label classification. IEEE Access **8**, 203827–203844 (2020)
24. Manhães, L.M.B., da Cruz, S.M.S., Zimbrão, G.: WAVE: an architecture for predicting dropout in undergraduate courses using EDM. In: Proceedings of the 29th Annual ACM Symposium on Applied Computing, pp. 243–247 (2014)
25. Moucary, C.E., Khair, M., Zakhem, W.: Improving student's performance using data clustering and neural networks in foreign-language based higher education. Res. Bull. Jordan ACM **2**(3), 27–34 (2011)
26. Yamashita, T.: Grit and Second Language Acquisition: Can Passion and Perseverance Predict Performance in Japanese Language Learning? University of Massachusetts, Amherst, MA, USA (2018)
27. Dunn, T.J., Kennedy, M.: Technology enhanced learning in higher education; motivations, engagement and academic achievement. Comput. Educ. **137**, 104–113 (2019)
28. Hug, N.: Surprise: a python library for recommender systems. J. Open Source Softw. **5**(52), 2174 (2020)
29. Gray, C.C., Perkins, D.: Utilizing early engagement and machine learning to predict student outcomes. Comput. Educ. **131**, 22–32 (2019)
30. Kotsiantis, S., et al.: Preventing student dropout in distance learning system using machine learning technique applied. Artif. Intell. **18**(5), 411–426 (2003)
31. Iqbal, Z., Qadir, J., Mian, A., Kamiran, F.: Machine learning based student grade prediction: a case study (2017)
32. Aljohani, S.: Predicting Student Performance in Academic Education, Using Machine Learning Techniques. MSc Thesis. College of Business Administration, Madinah Almunwarah (2017)
33. Etikan, I., Musa, S.A., Alkassim, R.S.: Comparison of convenience sampling and purposive sampling. Am. J. Theor. Appl. Stat. **5**(1), 1–4 (2017)
34. Gerges, F., Zouein, G., Azar, D.: Genetic algorithms with local optima handling to solve Sudoku puzzles. In: Proceedings of the 2018 International Conference on Computing and Artificial Intelligence, pp. 19–22 (2018)
35. Cox, D.R.: The Analysis of Binary Data. Nethuen London (1970)
36. Murphy, K.P.: Naive Bayes Classifiers, vol. 18. University of British Columbia (2006)

Tomato Plant Leaf Disease Identification and Classification Using Deep Learning

Hannia Tahir and Parnia Samimi[✉]

School of Computing and Digital Technology, Birmingham City University,
Birmingham B4 7XG, UK
Parnia.samimi@bcu.ac.uk

Abstract. Crops are an integral part of the agricultural industry in South Asian countries as they have fertile soil and advantageous weather conditions. However, the crops suffer from various plant diseases that influence the quality of the crops, which can be due to the lack of proper diagnosis and/or inefficient and untimely diagnoses of crop diseases. Diagnostic approaches of crop diseases are varied. Some can be diagnosed grossly, and some need especial laboratory efforts. Gross diagnosis approach is quite inaccurate and sometimes tricky. On the other hand, laboratory diagnosis is a time-consuming process and quite costly. In this project we developed an automated image processing method to enhance efficiency and accuracy in the gross assessment. This process accelerates the diagnosis process and can be considered as a preliminary method for diagnosis. We utilized deep learning to detect diseases in tomato crops using leaf images. The process involves building a convolutional neural network using a pre-trained VGG16 model that pre-processes the images according to its requirements and performs segmentation on images before training and testing the data. The model obtained an accuracy of 95%, taking only 30 min to run, train, test, and classify the 18,160 images to their respective 10 classes thus being very time-efficient and did not need a laptop with higher power processor, which makes the model accessible for other devices. Due to the low complexity of the model, it can be implemented on smaller devices without the need of a fast processor.

Keywords: VGG16 · Plant Disease Detection · Deep Learning · Tomato Plant

1 Introduction

Over the past century, the increase in the population has been enormous [1], thus a higher demand for food production is created on a global level. Certainly, it is essential to maintain the nutrition and quality of the food produced as well as to protect the ecosystem which is achieved by using procedures that lead to sustainable farming [2].

For human's, plants are the main source of energy as they contain various nutrients and are recognized as essential [3]. While trying to keep up with this demand, the problem of plant disease has had a major influence in the production of food and the development of human society [4]. Every year, plant diseases result in loss of farmer's income due

F. Saeed et al. (Eds.): ICACIn 2022, LNDECT 179, pp. 100–110, 2023.
https://doi.org/10.1007/978-3-031-36258-3_9

to loss in yield and the economy as well. According to Gunarathna, Rathnayaka, and Kandegama [5], the yearly production of fresh tomato crops on a global scale is 160 million tons, which is 3 and 6 times more than the production of potatoes and rice crops respectively. According to the statistics of the Census and Statistics Department of Sri Lanka, production of tomato crops between the period of 2000 till 2010 increased 71%. But farmers still have suffered from huge losses due to several tomato crop diseases. Due to insufficient research and facilities provided by the government, along with these complications, the agricultural sector's contribution towards the GDP (Gross Domestic Product) of Pakistan declined from 54% to 24% in the past 10 years and resulted in crop production less than its capacity [6].

Early identification and classification of plant diseases play a vital role in agriculture by preventing the loss of crops and improving their quality [7, 8]. Protecting plants from diseases or at least early diagnosis is a big challenge. Most of the farmers from developing countries rely on a traditional method of detecting the diseases which is through gross examination, which might be inaccurate. The development of Machine Learning and computer vision has made a huge impact on plant disease detection to develop automated systems using image recognition and classification. Many machine learning models have been applied in plant disease classification. However, nowadays deep learning is taking over in the computer vision field. Deep learning, a sub-field of machine learning and artificial intelligence, is a complex field that consists of neural networks, which are made up of several layers based on the complexity and severity of the problem. In the agricultural plant protection field, DL has become a great attraction for disease recognition, assessment of pest range etc. [9]. Convolutional Neural Network has proved to be very successful along with VGG16, AlexNet, etc. [5]. By using these models, an affordable model can be built to ease the problem faced by the farmers in developing countries like Pakistan, thus improving crop production and the economy [10].

Most current papers which utilize these techniques focus on the high accuracy of detection models, whereas only few discuss applicability in real life and how the time for training and testing can be a huge downside of these highly accurate and complex models in the real-world situation. Training these models efficiently requires expensive, high processing hardware such as Graphics Processing Card (GPU) and a huge Random Access Memory (RAM) [11]. If applied correctly, deep learning can be used to identify plant diseases simply by taking images from mobile phones and screening them through the model which can be integrated into a mobile application for easy access of the users. In this paper, deep learning is utilized to detect diseases in tomato crops using leaf images. The process involves building a convolutional neural network using a pre-trained VGG16 model that pre-processes the images according to its requirement and performs segmentation on images before training and testing the data. The rest of this paper is organized as follows. Section 2 consists of the related work done in the past years in plant disease detection. Section 3 shows the methodology used in the research which is further split between dataset, model, and experimental setup. Section 4 discusses the results obtained through the running of the model.

1.1 Related Works

Numerous studies have shown that image-based plant diagnosis methods are more precise compared with human visual detection. S. Ramesh et al. in 2018 [12] detected abnormalities in papaya leaves in the greenhouse and natural environment by using a Random Forest classifier to develop a model and trained it with 160 images of papaya leaves. The model resulted in 70% accuracy in predicting the abnormalities. In 2018, blast disease was detected by converting the images from the RGB to L*a*b and HSV color spaces which are then segmented using K-means clustering. Features like mean-value, standard deviation, and GLCM are extracted. ANN classifier is applied where 180 images are trained and 120 are used in the testing of the model. The classification has 86% of accuracy [13]. Ashqar et al. [14] used image-based plant disease detection on tomato plants using CNN with color and grayscale images. The accuracy of 99.84% and 95.54% was obtained by training and testing the model with 900 images. In 2019, rice plant diseases were detected such as leaf smut, bacterial leaf blight, and brown spot diseases using different algorithms including KNN, J48(Decision Tree), Naïve Bayes, and Logistics Regression. After training and testing, the accuracy of these algorithms was 91.66%, 97.9%, 50%, and 70.8% respectively [15]. Kumari et al. [16] classified leaf diseases using neural networks. RGB images were converted into grayscale and the grayscale correlation matrix was developed. Using K-mean clustering, the image was segmented, and the ANN classifier classified four diseases with an average of 92.5% accuracy. In 2019, classification and recognition of paddy leaf disease was performed using an optimized deep neural network. Images were obtained directly from the farm, then during pre-processing, the background was removed by converting RGB into HSV. Binary images were extracted to separate the diseased and non-diseased parts, depending upon hue and saturation. The classification was performed using the Optimized Deep Neural Network with Jaya Optimization Algorithm (DNN_JOA) which resulted in an accuracy of 98.9% [13].

Panigrahi et al. [17] used supervised machine learning techniques for maize plant disease detection. Techniques used were Naïve Bayes, Decision Tree, KNN, SVM, and RF. For image segmentation, the label edge detection method was used which calculates the gradient of image intensities at every pixel of the image. For feature analysis, grayscale pixel values were utilized and features such as shape, color, and texture were extracted from the images. RF had the highest accuracy of 79.23%. In 2020, a methodology was developed for plant diseases detection using integrated digital image processing techniques of machine learning Filters such as Average, Median, Linear, and Adaptive were applied for image pre-processing to remove any noise. Segmentation was performed by applying Otus's techniques and GLCM. Gabor Texture analysis was applied for feature extraction. Classification was performed using RF, SVM, KNN, and ANN and the accuracy obtained was 73.38%, 67.27%, 62.55%, and 64.76% respectively [18]. Sujatha et al. [3] compared the performance of machine learning and deep learning algorithms for the detection of citrus plant disease. Machine learning algorithms were SVM, RF, SGD, and deep learning algorithms were Inception-v3, VGG-16 and VGG-19. The accuracy of the experiment was 87%, 76.8%, 86.5%, 89%, 89.5%, and 87.4% respectively. Hassan et al. [19] developed a CNN model by using depth/separable convolution instead of standard, reducing the number of parameters and cost of computation. InceptionV3,

InceptionResNetV2, ModelNetV2, and EfficientNetB0 feature approaches were applied that resulted in the accuracy of 98.42%, 99.11%, 97.02%, and 99.56% respectively.

To sum up, based on the results of the research performed in the past, it was concluded that the deep learning models have the higher accuracy rates in plant disease detection and VGG16 produced outstanding results compared to other models. Therefore, in this study, the VGG16 model is utilized to classify the images. It provides a building block for a training model that predicts crop diseases. The main advantage of these trained models, especially when training for a specific application, is that new training layers can be added on top thus improving the performance which is called transfer learning.

2   Methodology

2.1   Dataset and Model

In the data set used in this paper, there are 18,160 images where 1591 are from healthy tomato plants whereas the rest are from diseased such as Bacterial Spot, Late Blight, Early Blight, Leaf Mould, Septoria Leaf Spot, Spider Mites (Two-spotted Spider Mite), Target Spot, Tomato Mosaic Virus, and Tomato Yellow Leaf Curl Virus. This unaugment dataset is obtained from [20]. Ten classes were recognized and classified in this research. Figure 1 shows samples of the dataset images.

Fig. 1. Sample Images from the Dataset

Images are converted to RGB channels as it is the requirement of the model that is developed. Since VGG16 pre-trained model is used, the image pre-processing is performed within the model itself and no separate pre-processing is required. For image segmentation, pre-trained Convnet is applied and ImageDataGenerator() is used for feature extraction. This can be used in cases where data augmentation is not applied. Classification is the last and the most important step in plant disease detection as this is where the disease is identified accurately so that it can be stopped at an early stage. VGG16 model is a CNN model which was first introduced by K. Simonyan et al. in their paper "Very Deep Convolutional Networks for Large-Scale Image Recognition". This model won the 2014 ImageNet competition, ILSVRC [21]. This VGG16 model has achieved an accuracy of 92.7% which is the top-5 test accuracy in ImageNet, where the dataset for ImageNet is over 14 million images with 1000 classes. The architecture of VGG16 is shown in Fig. 2.

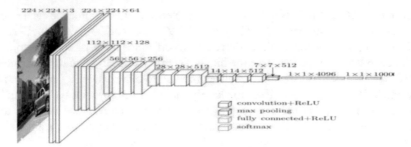

Fig. 2. VGG16 Model Architecture and Layers

In this Fig. 2, it is noticeable that there are many ConvNet layers. In the first con1 layer, the input is fixed at the size of 224 × 224, and the image is in RGB color space. Then this image goes over a pile of convolutional layers. In these layers, the filters are applied with a minor receptive field of 3 × 3. In another configuration, convolution filters are utilized at 1 × 1, and these are linear transformations. This linear transformation is of the input channel which is shadowed by non-linearity. In this pre-trained VGG16 model, the conv stride is set to 1-pixel, and the spatial padding of convolution layer input is set to 1 pixel for 3 × 3 convolution layers so that the spatial resolution is retained after convolution. Five max-pooling layers, which follow part of the Conv. Layers, do spatial pooling (not all the conv. Layers are followed by max-pooling). Over a 2 × 2 pixel window, max-pooling is conducted. Following a stack of convolutional layers (of varying depth in different architectures), three Fully-Connected (FC) layers are added: the first two have 4096 channels apiece, while the third performs 1000-way ILSVRC classification and hence has 1000 channels (one for each class). The soft-max layer is the final layer. In all networks, the completely connected levels are configured in the same way. The rectification (ReLU) non-linearity is present in all hidden layers. It should also be highlighted that, with the exception of one, none of the networks incorporate Local Response Normalization (LRN), which does not improve performance on the ILSVRC dataset but increases memory consumption and computation time. The input layer of the model, which is labeled as 7 × 7 × 512, to the last max-pooling layer is termed as the feature extraction part. The remaining model network is termed as classification part.

2.2 Experimental Setup

Jupiter and Anaconda were set up on the computer. All the important and necessary libraries, such as Keras, PIL, and TensorFlow, were downloaded and imported. Data was imported into Jupiter through Google Collab as it is much faster and more efficient. After importing data, each file is examined and the number of images in each file/ class is counted. Python Image Library (PIL) is used to read, load, and open the images. An example image is shown in Fig. 3. The data loaded was split into 3 parts, 70% for the training of the model, 15% for the testing of the model, and the rest 15% for the validation of the model. For this purpose, a split-folder was installed. The Fig. 3 shows the image loaded from the dataset without any pre-processing applied to it.

Fig. 3. Loaded Image from the Dataset without any Pre-processing

Images were converted to Keras generator format. The ImageDataGenerator.flow from directory() method returned a directory iterator that generated batches of normalised tensor image data from the directories specified, allowing the data to be fed to the Keras sequential model. The image data had already been pre-processed in the same way as the data sent through VGG16 had been. Figure 4 shows the images after applying pre-processing. The pre-processing was applied using ImageDataGenerator().

Fig. 4. Images after Pre-processing

The photos' dimensions were also set at 224 * 224. The image labels corresponding to the images were listed, and were one-hot encoded. In this model of VGG16, the last four layers were removed and replaced with new customized layers to avoid overfitting. Figure 5 shows the new added layers to the model.

```
flatten (Flatten)          (None, 25088)          0

dense (Dense)              (None, 1024)           25691136

dropout (Dropout)          (None, 1024)           0

dense_1 (Dense)            (None, 1024)           1049600

dropout_1 (Dropout)        (None, 1024)           0

dense_2 (Dense)            (None, 512)            524800

dropout_2 (Dropout)        (None, 512)            0

dense_3 (Dense)            (None, 10)             5130
==================================================================
Total params: 41,985,354
Trainable params: 27,270,666
Non-trainable params: 14,714,688
```

Fig. 5. New Model with Edited Layers

The model was compiled using Adam optimizer and SoftMax activation function was used as it was a multiclass classification. Figure 6 shows the details of accuracy, loss, time taken thus concluding the improvement in each epoch when the model runs. This information also helps to track any overfitting of the model.

The model was run over 30 epochs and 2 verbose. In terms of Keras, this method is stochastic gradient descent. The SoftMax activation function is used in the output layer

```
history = model.fit(x=train_batches,
                    steps_per_epoch=len(train_batches),
                    validation_data=valid_batches,
                    validation_steps=len(valid_batches),
                    epochs=30,
                    verbose=2
)
Epoch 1/30
1271/1271 - 95s - loss: 2.8977 - accuracy: 0.4926 - val_loss: 0.6678 - val_accuracy: 0.7926 - 95s/epoch - 75ms/step
Epoch 2/30
1271/1271 - 85s - loss: 0.9146 - accuracy: 0.7246 - val_loss: 0.5067 - val_accuracy: 0.8551 - 85s/epoch - 67ms/step
Epoch 3/30
1271/1271 - 89s - loss: 0.6356 - accuracy: 0.8096 - val_loss: 0.3517 - val_accuracy: 0.8974 - 89s/epoch - 70ms/step
Epoch 4/30
1271/1271 - 93s - loss: 0.4913 - accuracy: 0.8594 - val_loss: 0.2987 - val_accuracy: 0.9187 - 93s/epoch - 73ms/step
Epoch 5/30
1271/1271 - 87s - loss: 0.3894 - accuracy: 0.8883 - val_loss: 0.3050 - val_accuracy: 0.9132 - 87s/epoch - 68ms/step
Epoch 6/30
1271/1271 - 92s - loss: 0.3394 - accuracy: 0.9054 - val_loss: 0.2514 - val_accuracy: 0.9323 - 92s/epoch - 72ms/step
Epoch 7/30
1271/1271 - 93s - loss: 0.2605 - accuracy: 0.9233 - val_loss: 0.2587 - val_accuracy: 0.9298 - 93s/epoch - 74ms/step
Epoch 8/30
1271/1271 - 93s - loss: 0.2421 - accuracy: 0.9326 - val_loss: 0.2086 - val_accuracy: 0.9518 - 93s/epoch - 73ms/step
Epoch 9/30
1271/1271 - 93s - loss: 0.2021 - accuracy: 0.9428 - val_loss: 0.2275 - val_accuracy: 0.9397 - 93s/epoch - 73ms/step
Epoch 10/30
```

Fig. 6. Details of each run of the model

of the NN models. It is termed as a soft version of the argmax function. The SoftMax activation function predicts the multinomial probability distribution. In some uncommon examples, this function can be used for hidden layers in NN. In the output layer of the model, this function produces a single value as output for every node. The values of the output sum to 1.0 and signify probabilities. Figure 7 shows the formula for SoftMax Activation Function and its details.

$$\sigma(\vec{z})_i = \frac{e^{z_i}}{\sum_{j=1}^{K} e^{z_j}}$$

σ = softmax
$\vec{z}$ = input vector
e^{z_i} = standard exponential function for input vector
K = number of classes in the multi-class classifier
e^{z_j} = standard exponential function for output vector
e^{z_j} = standard exponential function for output vector

Fig. 7. SoftMax Function Formula [22]

After the model was trained and tested, losses and accuracy during the training process were plotted. Initially, when the VGG16 model was downloaded, it had 138, 357, 544 trainable parameters. Then the model was changed by adding some layers at the end, thus converting the VGG16 to a sequential model, meaning making it a linear stack of layers. Now, the trainable parameters were 27, 270, 666. The result obtained was recorded. Test images and labels were loaded to confirm that the test dataset was not shuffled (Fig. 8).

```
[[1. 0. 0. 0. 0. 0. 0. 0. 0. 0.]
 [1. 0. 0. 0. 0. 0. 0. 0. 0. 0.]
 [1. 0. 0. 0. 0. 0. 0. 0. 0. 0.]
 [1. 0. 0. 0. 0. 0. 0. 0. 0. 0.]
 [1. 0. 0. 0. 0. 0. 0. 0. 0. 0.]
 [1. 0. 0. 0. 0. 0. 0. 0. 0. 0.]
 [1. 0. 0. 0. 0. 0. 0. 0. 0. 0.]
 [1. 0. 0. 0. 0. 0. 0. 0. 0. 0.]
 [1. 0. 0. 0. 0. 0. 0. 0. 0. 0.]
 [1. 0. 0. 0. 0. 0. 0. 0. 0.1]]
```

Fig. 8. Images and Labels Plotted

3 Results and Discussion

By running the model with 30 epochs with training, testing, and the validating ratio of the dataset as 70%, 15%, and 15% respectively, the result is obtained and is shown in Table 1. The time taken to run this experiment was 30 min. Figure 9 shows the accuracy and the losses of the model over the 30 epochs.

Table 1. Accuracy and Losses during Training and Testing

	Loss	Accuracy
Training	0.074 or 7.4%	0.981 or 98.1%
Testing	0.302 or 30.2%	0.954 or 95.4%

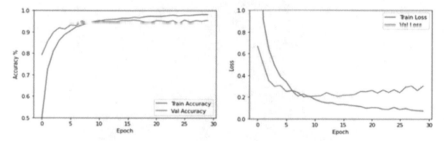

Fig. 9. Graphs plotted the for Loss and Accuracy

The evaluation of the model is performed to understand the results in visualization form for better understanding. A confusion matrix is calculated with actual values obtained in the experiment and the predicted values of the experimental result of the different class labels. To develop the confusion matrix, a prediction model is developed to predict the classification of the labels of the dataset. Prediction is made on the test dataset as Fig. 10. The confusion matrix can be plotted as normalized or without normalization. Moreover, the resulting matrix without normalization is shown in Fig. 10.

```
  predictions = model.predict(x=test_batches, steps=len(test_batches), verbose=0)


predicting on the test dataset


[ ]  np.round(predictions)

  array([[1., 0., 0., ..., 0., 0., 0.],
         [1., 0., 0., ..., 0., 0., 0.],
         [1., 0., 0., ..., 0., 0., 0.],
         ...,
         [0., 0., 0., ..., 0., 0., 1.],
         [0., 0., 0., ..., 0., 0., 1.],
         [0., 0., 0., ..., 0., 0., 1.]], dtype=float32)
```

```
Confusion matrix, without normalization
[[316   0   0   1   0   0   1   0   0   2]
 [  8 126   0   8   0   2   0   5   0   1]
 [  0   0 238   0   0   0   0   2   0   0]
 [  0   5   0 274   6   0   1   0   0   1]
 [  1   0   0   0 139   0   3   1   0   0]
 [  3   5   0   3   3 243   0   8   2   0]
 [  0   1   0   0   1   0 236  10   3   1]
 [  0   2   0   3   1   0   4 202   0   0]
 [  0   0   0   0   2   0   0   0  55   0]
 [  1   0   0   0   0   0   1   1   0 802]]
```

(a)   (b)

Fig. 10. (a) Plotting the Prediction Model, (b) Confusion Matrix

As shown in Table 1, the model achieved a training accuracy of 98.1% which is very high. While the testing accuracy is 95%, this indicates that the model did not overfit as the accuracy of the model is still high. The losses in the training have been very low but were high in the testing which can be reduced in future work. The graphs plotted indicate how the model accuracy increased over time in the training phase over the 30 epochs. Even though the training loss remained low, the validity loss still got high. The visualization of the matrix is shown in Fig. 11.

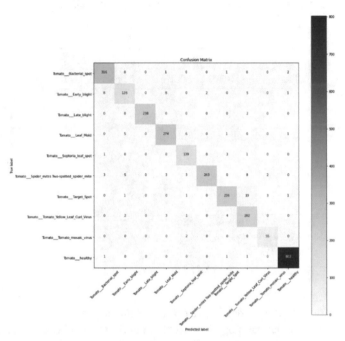

Fig. 11. Confusion Matrix with the Labelled Classes

From this image, the model misclassified a total of 103 images out of 2719 images from the training dataset. Most misclassified images were that of tomato early blight. The most correctly classified images were that of healthy tomatoes which also had the highest number of images in the dataset. The model indicates a very good accuracy rate, and the confusion matrix is dependable. It seems that the model is well prepared and can be integrated into a mobile application.

4　Conclusion

Deep learning methods provide an efficient approach in primary diagnosis of some plant diseases. In this study we assessed tomato leaves through deep learning method ?? to identify healthy plants from diseased ones. A classification model was produced using a pre-trained VGG16 model and by adding additional layers to it. The model successfully classified the images, belonging to 10 different classes of tomato plant leaf, with an accuracy of 95% without overfitting. It took 30 min to run the model, train, test, and classify the set of 18,160 images on a laptop with 16GB ram and 64-bit processor without GPU. Deep learning models require devices with higher processor and other hardware specifications and take longer on a laptop with a normal processor due to extensive computations. This makes the proposed model inexpensive and fast thus making it accessible to devices with fewer high specifications.

The two features of small-scale and accessibility of this prediction model makes it capable of being integrated into a mobile application. This method can benefit small scale agricultural economies as it can help farmers in early diagnosis of plant diseases without a need of professional knowledge. In future works, this model with high accuracy and the ability to run on systems with low hardware and specifics can be implemented on mobile so that farmers can access the trained model through their mobile phone application and identify different plant diseases. Another improvement that can be made on this model is that the time taken to classify the images can be reduced. Thus, it can be concluded that the model is a success that has room to be improved as well as to be implemented in further studies.

References

1. Kitzes, J., et al.: Shrink and share: humanity's present and future Ecological Footprint. Philos. Trans. R. Soc. B: Biol. Sci. **363**(1491), 467–475 (2007)
2. FAO: How to Feed the World in 2050. Food and Agriculture Organization of the United Nations, Rome, Italy (2009). https://www.fao.org/fileadmin/templates/wsfs/docs/expert_paper/How_to_Feed_the_World_in_2050.pdf. Accessed 10 Jan 2022
3. Sujatha, R., Chatterjee, J.M., Jhanjhi, N.Z., Brohi, S.N.: Performance of Deep Learning vs machine learning in plant leaf disease detection. Microprocess. Microsyst. **80**, 103615 (2021)
4. Palmgren, M., et al.: Are we ready for back-to-nature crop breeding? Trends Plant Sci. **20**(3), 155–164 (2015)
5. Gunarathna, M.M., Rathnayaka, R.M.K.T., Kandegama, W.M.W.: Identification of an efficient deep leaning architecture for tomato disease classification using leaf images. J. Food Agric. **13**(1), 33 (2020)
6. Malik, A.: Power Crisis in Pakistan: A Crisis of Governance? Pakistan Institute of Development Economics, Islamabad (2012)
7. Shrivastava, V., Pradhan, M., Minz, S., Thakur, M.: Rice plant disease classification using transfer learning of deep convolution neural network. ISPRS. In: International Archives of the Photogrammetry, Remote Sensing and Spatial Information Sciences, XLII-3/W6, pp. 631–635 (2019)
8. Phadikar, S., Sil, J., Das, A.K.: Rice diseases classification using feature selection and rule generation techniques. Comput. Electron. Agric. **90**, 76–85 (2013)
9. Li, L., Zhang, S., Wang, B.: Plant disease detection and classification by deep learning – a review. IEEE Access **9**, 56683–56698 (2021)
10. Rehman, A., et al.: Economic perspectives of major field crops of Pakistan: an empirical study. Pac. Sci. Rev. B: Hum. Soc. Sci. **1**, 145–158 (2016)
11. Krishnaswamy Rangarajan, A., Purushothaman, R.: Disease classification in eggplant using pre-trained VGG16 and MSVM. Sci. Rep. **10**, 2322 (2020)
12. Ramesh, S., et al.: Plant disease detection using machine learning. In: International Conference on Design Innovations for 3Cs Compute Communicate Control (ICDI3C), pp. 41–45 (2018)

13. Ramesh, S., Vydeki, D.: Rice Blast disease detection and classification using machine learning algorithm. In: 2nd International Conference on Micro-Electronics and Telecommunication Engineering (ICMETE), pp. 255–259 (2018)
14. Ashqar, B., Abu-Naser, S.: Image-based tomato leaves diseases detection using deep learning. Int. J. Acad. Eng. Res. **2**, 10–16 (2018)
15. Ahmed, K., Shahidi, T.R., Irfanul Alam, S.M., Momen, S.: Rice leaf disease detection using machine learning techniques. In: International Conference on Sustainable Technologies for Industry 4.0 (STI), pp. 1–5 (2019)
16. Kumari, C.U., Jeevan Prasad, S., Mounika, G.: Leaf disease detection: feature extraction with k-means clustering and classification with ANN. In: 3rd International Conference on Computing Methodologies and Communication (ICCMC), pp. 1095–1098 (2019)
17. Panigrahi, K.P., et al.: Maize leaf disease detection and classification using machine learning algorithms. In: Advances in Intelligent Systems and Computing, pp. 659–669 (2020)
18. Ganatra, N., Patel, A.: A multiclass plant leaf disease detection using image processing and machine learning techniques. Int. J. Emerg. Technol. **11**(2), 1082–1086 (2020)
19. Hassan, S.M., Maji, A.K., Jasiński, M., Leonowicz, Z., Jasińska, E.: Identification of plant-leaf diseases using CNN and transfer-learning approach. Electronics **10**, 1388 (2021)
20. Geetharaman, G., Arunpandian, J.: Identification of plant leaf diseases using a 9-layer deep convolutional neural network. Comput. Electric. Eng. **76**, 323–338 (2019)
21. Yang, H., Ni, J., Gao, J., Han, Z., Luan, T.: A novel method for peanut variety identification and classification by Improved VGG16. Sci. Rep. **11**, 15756 (2021)
22. Softmax Function. https://deepai.org/machine-learning-glossary-and-terms/softmax-layer. Accessed 05 Jan 2022

Using Deep Learning for the Detection of Ocular Diseases Caused by Diabetes

Asma Sbai[1]([✉]), Lamya Oukhouya[2], and Abdelali Touil[3]

[1] Laboratory of Bioscience and Health, FMPM, UCA, Marrakech, Morocco
asma.sbai@uca.ac.ma
[2] Laboratory LIMA, ENSA AGADIR, UIZ, Agadir, Morocco
l.oukhouya@uiz.ac.ma
[3] Laboratory of Engineering Science, FSA, UIZ, Agadir, Morocco

Abstract. Early diagnosis in ophthalmic field is the main key for many patients to avoid serious damage of the eye. In many cases ocular illnesses are caused by other health problems such as diabetes. In this article an investigation on the diagnosis, using deep learning, of ocular diseases caused by diabetes was conducted paying particular attention to cataract, glaucoma and diabetic retinopathy. The proposed approach to identify and classify these three diseases performed 96% accuracy on training, 89% on validation and 90.63% on testing. A deployment prototype of this model was also presented to build a suitable computer aided diagnosis tool.

Keywords: Deep Neural Network · Transfer Learning · Fundus images · ocular diseases · computer-aided diagnosis

1 Introduction

The healthcare industry has experienced an important revolution thanks to the implementation of computer vision [1]. It has been implemented in various medical procedures and contributed to significant advancements in medical image analysis which has become one of the most important fields of research. In addition, the demand for automated image processing has pushed forward the application of Artificial Intelligence (AI) for the interpretation of medical images and results, particularly in major advancements in Machine Learning (ML). Conventional practitioners of medical image analysis had to look for biomarkers in a patient's scans using their own prior knowledge and experience which impacted how the diagnosis was made and the outcome of the analysis. AI is the imitation of the diagnosis process made by the expert medics and aims to perform the same procedure utilizing an automated and objective approach. ML is a branch of AI and, as its name indicates, is the process of a computer learning from data and experience. The objective is to predict or decide depending on the available data loaded into the machine. Over the years, the increase of available data and both the computer's memory and processor evolution, contributed to encouraging more scientists and professionals to work with ML.

F. Saeed et al. (Eds.): ICACIn 2022, LNDECT 179, pp. 111–120, 2023.
https://doi.org/10.1007/978-3-031-36258-3_10

Deep Learning (DL) is a subcategory of ML known for its ability to learn on its own, unlike the ML algorithms [2]. DL has been applied in a variety of medical fields especially in radiology and extracting data from medical images that cannot be easily identified by medics.

Several medical image modalities are used to get data. Medical imaging is the overarching term used for methods and techniques employed to generate images of different parts of human body to support and lead diagnostic and prescribe treatments. It plays a noteworthy part in the detection of abnormality in different organs of the body. Medical imaging includes X-rays, Magnetic Resonance Imaging (MRI), Computed Tomography scans (CT), Ultrasounds and Nuclear medicine imaging and Positron-Emission Tomography (PET). To capture eye scans, fundus photography is the main modality used. It aims to show important parts of this organ to make sure the optic nerves, vitreous, retina, macula and blood vessels are healthy. By using a monocular camera on a 2D plan ophthalmologists can analyze many important biomarkers. Unlike other ophthalmic imaging modalities, fundus images can be obtained in a painless and non-expensive way which make this technique more suitable than OCT images and angiographs. Fundus images can be used to diagnose many eye diseases including glaucoma, Diabetic Retinopathy (DR), cataract, Retinopathy Of Prematurity (ROP) and Diabetic Macular Edema (DME).

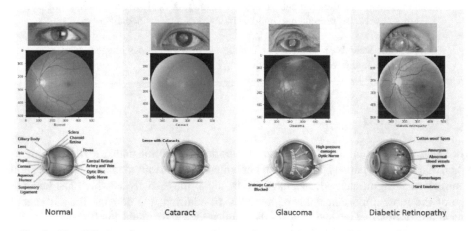

Fig. 1. The difference between a normal eye and eyes with abnormalities related to diabetes

Figure 1 represents a normal eye next to Glaucoma, Cataract and DR, which are among possible diabetes complications and have been the main concern of many studies in the field of machine learning and classification problems. The top of Fig. 1 shows images of the eyes and the corresponding fundus and state. DR occurs as a result of damage to blood vessels of the retina and can lead to blindness. It is one of the implications of diabetes resulting on morphological changes and anomalies in the fundus. These alterations cause microaneurysms, exudates, hemorrhages and the abnormal growth of the blood vessels [3]. The earlier signs of DR are the appearance of microaneurysms and hard exudates. Glaucoma is a chronic ophthalmic pathology that progressively deteriorates the nerve fibers and hence leads to progressive damage to the neuro-retinal rim and the optic nerve head (ONH). It involves a significant rise of intraocular pressure [4]. Furthermore, it appears to cluster vessels in the edge of the Optic Disc. Glaucoma is characterized by a significantly enlarged area of the optic disc, and optic cup can be observed in the fundus images. Once it starts to develop, the patient will progressively lose peripheral vision which is why early diagnosis is crucial to avoid any irreversible complications [5]. Cataract disease has a noticeable feature, glassy denseness, which is produced by the denaturation of protein, leading to blurring the fundus structures. Generally, it develops with aging and unfortunately are undetected until it blocks light [6]. Normal fundus is recognizable since the image does not show any flocculation or nodules [7]. It is also characterized by clear blood vessels.

This work investigates the detection of these three diseases from fundus images and classification of the eye into healthy, Glaucoma, Cataract and DR. We used a model for preparing images by applying the CLAHE filter and augmenting the fundus images, used transfer learning for feature extraction, then classification. The paper is organized as follows. Section 2 explores related work, Sect. 3 introduces the pipeline of the model including preprocessing data, feature extraction and classification. Section 4 presents the conducted experiments and the results obtained. Finally, Sect. 5 is the conclusion and the summary of this work.

2 Related Works

Recently, more studies have shed light on the application of Deep Learning to train algorithms to detect and classify abnormalities based on fundus photography and with those studies some public datasets have emerged. Table 1 gives a summary of some of the most popular datasets that exceed 1000 images used in the literature for experimentation.

DR is among the most studied ocular disorders in AI. Not only are models conceptualized to detect DR but also to classify it into proliferative DR, non-proliferative DR and diabetic macular ocdemic as well as its severity: mild, moderate and severe [17, 18]. In this study [19], 124 retinal photos were examined to identify four groups: normal plus three stages of DR groups. The classification was achieved using a three-layer feedforward neural network. Authors in [20] used Morphological Component Analysis (MCA) algorithm to distinguish normal from pathological retinal structures with an achievement of 92.01% sensitivity. The early detection of cataract is crucial to prevent complications, in [21] authors adopt a CNN algorithm to detect early stage of cataract. This technique performs a 70% exact integral agreement ratio. Another study [22] combined CNN to extract features then applied Support Vector Machine (SVM) to explore those features and finally a Softmax classifier. The results are up to 97% sensitivity and enable grading of the cataract, density and location. For the identification of Glaucoma, algorithms were designed to evaluate the optic disc thickness using OCT and evaluate its severity [23]. Researchers in [24] studied an algorithm of deep learning to detect glaucoma using fundus images which shows high sensitivity of 83.7% and specificity of 88.2% whereas the sensitivity performed by an ophthalmologist was between 61.3% and 81.6%. Recently, work was conducted on the multiclassification problem such as in [7], where authors claim that their model performed better than the original Xception network with an accuracy, precision, F1 value, and kappa score of the DSRA-CNN

Table 1. Some of the popular datasets in the literature that exceed 1000 images

Dataset	Number of images	Image size	Task
MESSIDOR [8]	1200	1440 × 960 2240 × 1488 2304 × 1536	OD/OC Detection Exudates Detection
MESSIDOR – 2 [9]	1748	1440 × 960 2240 × 1488 2304 × 1536	DR diagnosis
EYEPACS [10]	9963		DR diagnosis
DDR [11]	13673		Exudates Detection MA detection HE detection DR detection
REFUGE [12]	1200	2124 × 2056 1634 × 1634	Glaucoma Diagnosis
LAG [13]	11760	3456 × 5184	OD/OC Segmentation Glaucoma detection
ODIR [14]	10000	512 × 512	Classification into 8 diseases
AREDS [15]	206500		DR Diagnosis
ICHALLENGE-AMD [16]	1200		OD/OC segmentation DR Diagnosis

respectively 87.90%, 88.50%, 88.16%, and 86.17% experiments were made on the ODIR dataset [14] to classify fundus into 8 categories. In [25] authors shed light on the data unbalance in the ODIR dataset and suggest overcoming this problem by converting the multiclassification into binary classification and taking the same number of images for each classification using VGG-19.

3 Methodology

Many studies have been conducted on the detection of diseases of the eye and have the common goal of improving the accuracy of the classification of these illnesses. This paper aims to classify fundus images into four categories: Normal healthy eye, Cataract, Glaucoma and DR. The pipeline of the model we propose is represented in Fig. 2.

ODIR [14] is a dataset representing 5000 patients' left and right eyes from different types of cameras, categorized into 8 ocular diseases as represented in Fig. 3. For these experiments we picked only fundus images labeled with Normal, Cataract, Glaucoma and DR. The orange bars of Fig. 2 illustrate the four states we are investigating. The unbalance between the four categories is clear, with normal and DR class representing the biggest portions of the total images.

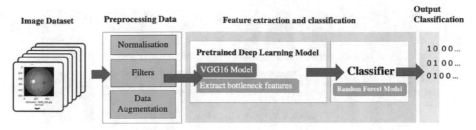

Fig. 2. Pipeline of the proposed model

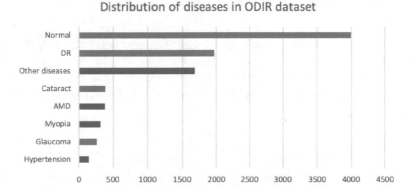

Fig. 3. Unbalanced disease representation in ODIR Dataset

Since we have an unbalanced dataset as the target, classes are not equally represented. We need to confront this situation that could generate bias toward highly represented class - the DR and Normal classes in this example - which will lead to over-predicting these classes. Sampling strategies resolve this issue by balancing the distribution in the dataset. We used the image augmentation technique to expand the size of our dataset using Keras ImageDataGenerator class by rotating 90°, flipping horizontally and vertically. We also implemented the Contrast-Limited Adaptive Histogram Equalization (CLAHE) on green channel to augment the contrast of images and thus achieving more visibility on some key features needed for the classification, namely the vascular information. All images were resized into 224 × 224. Figure 4 represents a sample of fundus images from the dataset before and after preprocessing and augmentation.

We used VGG16 as features extractor, (VGG stands for Visual Geometric Group referring to the group of scientists in Oxford who developed this architecture). This choice is based on the results of the study [26] where authors proved that this model performed the best among ResNet [27], InceptionV3 and MobileNet [28]. VGG16 is a CNN composed of 16 layers pretrained on ImageNet database for multi-classification purpose of objects. The dataset is divided into three datasets where 70% is used for training, 20% for validation and 10% for tests. To keep the same proportions of examples in each class we stratified a train-test split. Our approach involves using transfer learning and fine-tuning illustrated in Fig. 5. The activation function used is sigmoid combined

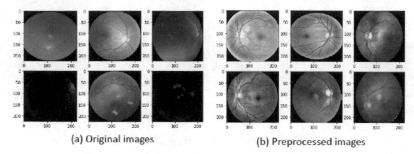

(a) Original images   (b) Preprocessed images

Fig. 4. Sample from the ODIR dataset (a) represent original images without augmentation (b) represent fundus images after preprocessing and augmentation

to binary cross-entropy loss function. The sigmoid function limits the prediction values between 0 and 1 and the loss function measure the distance between true and predicted value for each class. For each epoch we consider the averaged class-wise error and the weights are updated based on the loss. We use SGD optimizer and learning rate of 0.0001.

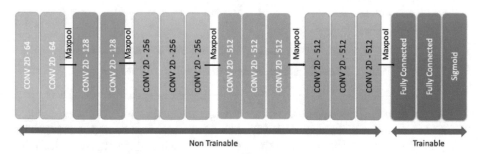

Fig. 5. Transfer learning and fine tuning of VGG16

As shown in the pipeline in Fig. 2, our approach consists of extracting the features and using them as inputs for Random Forest (RF) classifier. This is an ensemble learning method usually used for classification and regression [29].

4   Results and Discussion

To evaluate our proposed model, we used the following metrics: Accuracy, Recall, Precision and F1$_{score}$. Accuracy represent the ratio of correct prediction to the total of predictions. Recall is the ratio of true positives to the total positives in the ground truth. Precision is the proposition of positive findings that was actually correct. F1$_{score}$ combine precision and recall.

$$Accuracy = \frac{tp + tn}{tp + tn + fp + fn} \qquad (1)$$

$$Recall = \frac{tp}{tp + fn} \qquad (2)$$

$$Precision = \frac{tp}{tp + fp} \qquad (3)$$

$$F1_{score} = 2 \times \frac{Precision \times Recall}{Precision + Recall} \qquad (4)$$

After 15 epochs, we obtained the following results represented below in Table 2. On one hand, we have a 96% accuracy on training set and 89% on validation set. On the other hand, we attained on the testing set an overall accuracy of 90.63%

Table 2. Generated Classification report

	Accuracy	Precision	Recall	$F1_{score}$	Support
Cataract	0.906	0.83	0.71	0.77	12
Normal	0.984	0.94	1.0	0.97	16
Glaucoma	0.953	0.96	0.92	0.94	24
DR	0.968	0.83	1.0	0.91	12
Accuracy				0.9063	64

Table 3. Normalized Confusion Matrix

Predicted	Actual			
	Cataract	Normal	Glaucoma	DR
Cataract	83.33	0	16.67	0
Normal	6.25	93.75	0	0
Glaucoma	4.17	0	95.83	0
DR	16.67	0	0	83.33

We can deduce that the model performs efficiently. According to the confusion matrix in Table 3 inferred from test set, cataract and DR are the less predicted with 83.33% true Positives. Conversely, Glaucoma has 95.83% true positives. The quality and complexity of images certainly impact the results so one possible way to improve the recognition rate is to improve the quality of images and feed the model with a diverse but balanced dataset. Plus, one of the advantages of VGG16 is the number of its training parameters which is lower than many other usual models. This feature makes VGG16 a light model that is practically generalized for medical imaging use cases.

For the implementation of this model we suggest the use of TensorFlow serving which is a model deployment system that provides tools for maintenance and update of the model directly in the production environment. We propose to use TensorFlow

Serving with Docker and create a web interface using Flask web framework which is a lightweight Python framework to get predictions from the served TensorFlow model and enable the client side to use the model through API calls as illustrated in Fig. 6.

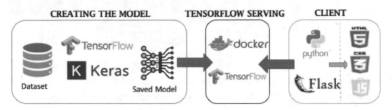

Fig. 6. Technologies used for the deployment of the model

5 Conclusion

In this paper we have proposed an approach combining two models for a multi-classification problem of three ocular diseases caused by diabetes, namely cataract, glaucoma and diabetic retinopathy. Our model performed an acceptable accuracy of 90.6% on the testing set. We concluded our work with a pipeline for the implementation of the proposed model because we are aware that developing programs for detecting and grading ocular diseases leads to more efficient screening and will empower practitioners to save more diabetic patient from blindness. For future work, we aim to improve the proposed approach to obtain better results and add stages of severity for each disease in the current classification.

References

1. Sarker, I.H.: Ai-based modeling: techniques, applications and research issues towards automation, intelligent and smart systems. SN Comput. Sci. **3**(2), 1–20 (2022)
2. Gill, M., et al.: Machine learning models outperform deep learning models, provide interpretation and facilitate feature selection for soybean trait prediction. BMC Plant Biol. **22**(1), 1–8 (2022)
3. Abbas, Q., Qureshi, I., Yan, J., Shaheed, K.: Machine learning methods for diagnosis of eye-related diseases: a systematic review study based on ophthalmic imaging modalities. Arch. Comput. Methods Eng. **29**(6), 3861–3918 (2022). https://doi.org/10.1007/s11831-022-097 20-z
4. Foster, P.J., Buhrmann, R., Quigley, H.A., Johnson, G.J.: The definition and classification of glaucoma in prevalence surveys. Br. J. Ophthalmol. **86**(2), 238–242 (2002)
5. Lazaridis, G.: Deep learning-based improvement for the outcomes of glaucoma clinical trials. (Doctoral dissertation, UCL (University College London)) (2022)
6. Silva, P.N.: Automatic detection of cataract in fundus images. (Doctoral dissertation, Universidade de Coimbra) (2019)
7. Yang, X.L., Yi, S.L.: Multi-classification of fundus diseases based on DSRA-CNN. Biomed. Signal Process. Control **77**, 103763 (2022)

8. Decencière, E., et al.: Feedback on a publicly distributed image database: the messidor database. Image Anal. Stereology **33**(3), 231–234 (2014)

9. Abràmoff, M.D., et al.: Automated analysis of retinal images for detection of referable diabetic retinopathy. JAMA Ophthalmol. **131**(3), 351 (2013). https://doi.org/10.1001/jamaophth almol.2013.1743

10. Diabetic Retinopathy Detection | Kaggle. Accessed 22 April 2022

11. Li, T., Gao, Y., Wang, K., Guo, S., Liu, H., Kang, H.: Diagnostic assessment of deep learning algorithms for diabetic retinopathy screening. Inf. Sci. **501**, 511–522 (2019)

12. Orlando, J.I., et al.: Refuge challenge: A unified framework for evaluating automated methods for glaucoma assessment from fundus photographs. Med. Image Anal. **59**, 101570 (2020)

13. Li, L., et al.: A large-scale database and a CNN model for attention-based glaucoma detection. IEEE Trans. Med. Imaging **39**(2), 413–424 (2019)

14. Peking university international competition on ocular disease intelligent recognition (ODIR-2019) (2019). https://odir2019.grand-challenge.org/. Accessed 29 March 2022

15. Age-Related Eye Disease Study Research Group: The age-related eye disease study (AREDS): design implications AREDS report no. 1. Control. Clin. Trials **20**(6), 573 (1999)

16. Fu, H., et al.. Adam: Automatic detection challenge on age-related macular degeneration (2020)

17. Ting, D.S.W., Cheung, C.Y., Lim, G., et al.: Development and validation of a deep learning system for diabetic retinopathy and related eye diseases using retinal images from multiethnic populations with diabetes. JAMA **318**(22), 2211–2223 (2017). https://doi.org/10.1001/jama. 2017.18152

18. Gulshan, V., Peng, L., Coram, M., et al.: Development and validation of a deep learning algorithm for detection of diabetic retinopathy in retinal fundus photographs. JAMA **316**(22), 2402–2410 (2016). https://doi.org/10.1001/jama.2016.17216

19. Yun, W.L., Acharya, U.R., Venkatesh, Y.V., Chee, C., Min, L.C., Ng, E.Y.K.: Identification of different stages of diabetic retinopathy using retinal optical images. Inf. Sci. **178**(1), 106–121 (2008)

20. Imani, E., Pourreza, H.R., Banaee, T.: Fully automated diabetic retinopathy screening using morphological component analysis. Comput. Med. Imaging Graph. **43**, 78–88 (2015)

21. Gao, X., Lin, S., Wong, T.Y.: Automatic feature learning to grade nuclear cataracts based on deep learning. IEEE Trans. Biomed. Eng. **62**(11), 2693–2701 (2015)

22. Liu, X., et al.: Localization and diagnosis framework for pediatric cataracts based on slit-lamp images using deep features of a convolutional neural network. PLoS ONE **12**(3), e0168606 (2017)

23. Omodaka, K., An, G., Tsuda, S., Shiga, Y., Takada, N., Kikawa, T.: Classification of optic disc shape in glaucoma using machine learning based on quantified ocular parameters. PLoS ONE **12**(12), e0190012 (2017)

24. Al-Aswad, L.A., et al.: Evaluation of a deep learning system for identifying glaucomatous optic neuropathy based on color fundus photographs. J. Glaucoma **28**(12), 1029–1034 (2019)

25. Khan, M.S., et al.: Deep learning for ocular disease recognition: an inner-class balance. Comput. Intell. Neurosci. **2022**, 1–12 (2022). https://doi.org/10.1155/2022/5007111

26. Gour, N., Khanna, P.: Multi-class multi-label ophthalmological disease detection using transfer learning based convolutional neural network. Biomed. Signal Process. Control **66**, 102329 (2021)

27. He, K., Zhang, X., Ren, S., Sun, J.: Deep residual learning for image recognition. In: Proceedings of the IEEE Conference on Computer Vision and Pattern Recognition, pp. 770–778 (2016)

28. Howard, A.G., et al.: Mobilenets: Efficient convolutional neural networks for mobile vision applications. arXiv preprint arXiv:1704.04861 (2017)
29. Breiman, L., Breiman, L., et al.: Random forests machine learning. J. Clin. Microbiol. **2**, 199–228 (2001)

Intelligent Chatbots for Electronic Commerce: A Customer Perspective

Norah Alrebdi[1]([✉]) [iD] and Mohammed Hadwan[1,2,3] [iD]

[1] Department of Information Technology, College of Computer, Qassim University,
Buraydah 51452, Saudi, Saudi Arabia
N.a.alrebdi@gmail.com, M.Hadwan@qu.edu.sa
[2] Intelligent Analytics Group (IAG), College of Computer, Qassim University, Buraydah,
Saudi Arabia
[3] Department of Computer Science, College of Applied Sciences, Taiz University, Taiz, Yemen

Abstract. Customer service plays a vital role in electronic commerce (EC) applications. However, the cost of providing customer service to meet customers' needs is high. The use of chatbots has helped to reduce this cost. A chatbot is an artificial intelligence application that imitates and processes human dialogue. Chatbot applications still have deficiencies in some aspects. Thus, we conducted this research to measure chatbot users' perspectives of EC websites and applications. We used a survey to elicit opinions about several aspects of EC chatbots: (i) their importance, (ii) performance, (iii) usage preference, and (iv) the area where they are most needed. Furthermore, we have introduced several positive and negative opinions as well as some suggestions from the customer's perspective. Our aim in this research is to assist stakeholders, including EC website and application owners and chatbot service developers, in improving the provided services and solving identified problems. The main output of this research is the analysis of users' expectations when dealing with chatbots, in addition to the strengths and weaknesses of chatbot performance, which can be used for chatbots improvement. The results showed that 69% of participants believed that fast response is the main advantage of chatbots. In contrast, 31% believed that the biggest problem is the lack of chatbots understanding of some customers' questions.

Keywords: Electronic commerce (EC) · Customer service · Artificial intelligence (AI) · Chatbot · Customer perspective · User experience

1 Introduction

Technology has contributed to the emergence of several modern fields that have attracted the owners and users of electronic commerce (EC) websites and applications. Statista (2020) reported that global electronic retail sales reached 3.53 trillion US dollars in 2019, while in 2022, electronic retail revenue can be expected to reach 6.54 trillion US dollars. This increase in electronic retail revenue is due to the advantages of EC, which traditional retail lacks. Several definitions of EC have been proposed. Some researchers define EC as an operation of buying, selling, or transporting any goods, services, or information

F. Saeed et al. (Eds.): ICACIn 2022, LNDECT 179, pp. 121–138, 2023.
https://doi.org/10.1007/978-3-031-36258-3_11

using the Internet [1]. Artificial intelligence (AI) applications, tools, and techniques are playing very important roles in the advancement of EC including big data analytics [2], Blockchain [3], visual search engine [4], social commerce [5], Sentiment analysis [6, 7] etc. Customer service is considered as one of the essential factors of EC's success where AI researchers pay attention to develop reliable intelligent chatbots. This is due to the huge efforts required for customer service departments and the large number of staff needed to avoid delays in responding to customer inquiries. Customer satisfaction is directly affected by the quality of customer service [8]. Given the importance of customer service, EC websites and applications have adopted chatbots to deal with customers efficiently instead of employing humans to provide customer service. The need for chatbot services is increasing under challenging circumstances, such as the COVID-19 pandemic [9]. The COVID-19 pandemic has caused huge increase in shopping online [10]. This increase makes it difficult for traditional customer service to respond to the vast number of inquiries in EC. A chatbot is a computer program designed to communicate intelligently with humans in ways similar to a human conversation [11]. Chatbots offer advantages such as availability anytime and anywhere [12]. Chatbots are also referred to as chat agents, smart agents, and virtual agents, among other terms [13]. Chatbots rely on AI, which refers to the system's ability to learn from data and interpret input to achieve specific goals [14]. Chatbots need a large body of vocabulary to provide a successful, intelligent conversation [15].

Many chatbot applications have been designed for customer service to improve the performance and solve faced problems. However, chatbots do not yet work perfectly as expected. Some researchers believe that chatbots face challenges due to the user's changing needs and desires [16]. Few studies have investigated the performance of EC chatbots from a customer perspective.

The main contribution of this study is to reduce the gap between the users (customers) and developers by provide an overview of the current state of the art of chatbots based on customers' views. Six research questions were formulated for this study, and a questionnaire was conducted to answer them. We sought to make this research as a guide for designers looking for improving the chatbot applications by clarifying the customers' perspectives. Additionally, this research will help website and application owners to make the right decision regarding the use of chatbots. Moreover, based on our review of the literature and the findings of this study, we offer some recommendations that will help in improving the development of chatbots applications.

This paper is organized as follows. Section 2 provides a literature review. Section 3 presents the research questions and describes the methodology followed. Section 4 reports the obtained results. Section 5 discusses the recommendations for chatbot developers and providers. Section 6 concludes this research.

2   Literature Review

This section reviews the available studies found in the literature based on two aspects to reduce the gap between the EC customers and chatbot developers. First, we review chatbot architectures. Second, we present several studies that illustrated the effect of chatbots on customer services according to questionnaires, interviews, or analytic studies that measure the extent of customer satisfaction for the performance of the chatbot in the current situation.

2.1 Chatbot Implementation

Researchers in [17] proposed a chatbot model that uses two types of answers. The first type is a pattern-based answer, which answers general questions using Artificial Intelligence Markup Language (AIML). The second type is a semantic-based answer, which uses latent semantic analysis to answer questions related to the provided services or questions that were not answered using AIML. The model has proven effective, with an accuracy of 0.97, although the responses to general questions need improvement. On the other hand, the authors in [18] proposed a pattern-based chatbot for Covenant University Shopping Mall. However, their pattern is constantly learning from the database's items. Thus, it performs better than the general pattern-based chatbots.

In [19], the authors proposed a system to answer customers' questions without human intervention. The system uses Telegram Server and two agents: a communication agent and an intelligent agent. The communication agent makes a Hypertext Transfer Protocol request periodically to Telegram Server to check if there is a question, which it sends to the intelligent agent. The intelligent agent uses Levenshtein distance[20]–[22] to determine the similarity between the received question and question-answer in corpora. Once the highest similarity is selected, the intelligent agent sends the communication agent's answer to the customer through Telegram Server. The authors tested the system in two respects: usability and performance. Despite typographical errors, the proposed system has proven its effectiveness.

The researchers in [23] provided an automated chatbot design to answer frequently asked questions in the Thai language. The chatbot uses deep learning to learn from big data. Moreover, it uses a recurrent neural network and long short-term memory to deal with sequence data. The chatbot was able to process 86.36% of questions with an accuracy of 93.2%.

In [24], the researchers provided a hybrid approach that uses a sequence-to-sequence structure. The approach includes a re-rank model and a model based on generation, combined with information retrieval. They implemented the new approach as a service on the Internet and named it AliMe Chat. AliMe Chat is integrated with an intelligent assistant in EC. The approach was tested on 878 questions and compared to a public chatbot. The proposed model achieved 37% better understanding than a public chatbot, but it was worse by 18%. Generally, the hybrid model was more effective than the information retrieval approach and generation approach individually.

The authors in [25] presented a chat system that they had implemented on their EC website to contact customers. When a customer submits a query, the chat system forwards the query to the AIML knowledge base to compare patterns using a matching algorithm. It then sends the appropriate answer to the customer through the website. Additionally, the authors tried to make the chatbot support the Bengali language by using a translation program that works as an intermediate between the AIML and customers. The proposed system showed a weakness in dealing with natural language properties.

In [26], the authors proposed an extension of EC websites called SuperAgent. Super-Agent is a customer service chatbot that uses data on EC websites to answer customers' questions. The agent depends on the data on the product page and automatically works on updating the data. When the user visits the page, SuperAgent appears automatically, allowing the user to submit queries. It then forwards the query to one of four different engines:

- The Fact Questions & Answers engine is designed to deal with product specifications.
- The FAQ search engine answers common questions in the EC field using several datasets.
- An engine uses customer reviews to extract information based on various approaches and methods.
- The Chit-Chat engine is designed to respond to greetings and questions unanswered by the previous three engines.
- Each engine responds to the meta engine to nominate the most appropriate answer depending on the tunable threshold.

The authors of [27] proposed a chatbot for the Android operating system called Cartbot, which includes an EC engine containing product catalogs. The proposed chatbot helps the user know about order placement and the order status after purchase, such as shipment information. Moreover, the chatbot can provide information about any item the user wants to know about. When the user submits a request, the chatbot compares it with some patterns to give a timely response. If the response is acceptable, the user can click the "Yes" button to add the product to a shopping cart.

In [28], the authors proposed an intelligent chatbot system for Indonesia's food sales. The chatbot works on the Telegram application. It uses an AIML-based EC assistant. Users' inputs are divided into three classes, general questions, stock checks, and calculations, which cover the ordering and payment process. The chatbot has been tested 300 times at different times using correct and formal questions. The system has proven its effectiveness by providing appropriate answers in an average time of 3.4 s.

Furthermore, the authors in [29] proposed an Indonesian chatbot called Bershca, which they developed for the hotel industry. It serves four functions: sales, facilities and services, reservations, and food. The Chabot's front end was designed in Google Flutter, and the back end was written in Python and AIML. The chatbot uses the Nazief–Adriani algorithm [30] to process the customers' queries. The design was tested using the technology acceptance model. The authors found that 85.7% of the users thought that the chatbot will enhance their job performance. Additionally, 84.33% of the users believe it is usable.

The researchers of [31] provided a Vietnamese EC chatbot framework for the retail sector to help customers contact sellers efficiently. The architecture includes two primary operations and uses deep neural networks instead of conventional structures to consider users' requests. The first operation is to classify user intent based on predefined categories. The second is to extract the context to clarify the problem and then give a solution. This operation considers three aspects of the context: product information, date and time information, and location. This operation applies in two layers: the first layer considers the context as one entity; the second layer considers each detail of the context as a separate entity. The authors observed that it was difficult to detect intent in some sentences of agreement and rejection compared with sentences of order information or addresses. They provided four new corpora on which to evaluate the performance of the proposed system. The results show that their system achieved an F-measure of 82.32% in detecting intent using convolutional neural networks. The results further proved the model's effectiveness in the extraction of context: F-measure of 90.91% for extracting product information and 79.33% for extracting address information. Moreover, it reached an F-measure of 85.98% for capturing dates and times in the first level. However, in the second level, the model achieved 93.08% in analyzing product descriptions, 90.58% in analyzing the address details, and 82.23% in the analysis of dates and times.

2.2 Chatbot Affect

The researchers in [32] conducted an online survey of 5,002 participants. The participants were from six countries: the US, UK, Australia, France, Japan, and Germany. One of the questions was whether the customer preferred to communicate immediately with a chatbot or to wait several minutes to speak to a human. In the US, UK, Australia, and France, most participants preferred to wait for a human, whereas, in Japan and Germany, most participants chose immediate service from a chatbot. The researchers then asked why the participants would prefer to wait for a human. Sixty percent of the participants believed that a human would understand their needs better, and 13% thought that humans would be empathetic. Moreover, 6% thought that a human would be faster than a chatbot, and 21% believed that a human would be more reliable.

The authors in [33] conducted an online questionnaire that included 146 participants from the US. It is important to note that this questionnaire was not about EC or customer service chatbots in particular, but rather about the reasons for chatbot usage in general. The questionnaire listed five categories of motivations to use a chatbot. Productivity was the main reason (68% of users). The term "productivity" was defined as the ease of use, speed of response, and convenience. The other reasons included entertainment, social reasons, and curiosity. A small proportion of the users' reasons did not fit any category, and so they were placed in a separate category called "Other." One reason that was placed in this category was that communicating with a chatbot is more comfortable than communicating with people, especially in solving personal problems. In contrast, a quantitative research method was applied in [34] to determine the suitability of customer services chatbots in business-to-business. The quantitative research illustrated that the chatbots perform sufficiently in the general inquiries and FAQs. However, despite this significant performance in duplicated inquiries, chatbots have a poor performance in delicate issues compared to human-to-human conversation.

Additionally, the authors in [35] interviewed 24 participants aged 18–76. The interviews consisted of three main topics: the interviewee's last interaction with a chatbot, their experience, and their thoughts about the future of chatbots. The results showed that more than half of the participants had previous experience with customer service chatbots. Most participants were aware of the limitations of chatbots. They mentioned that chatbots worked well with simple and concise questions. They also observed that the answers from the chatbot were understandable and straightforward. Twenty-three participants reported that their motivation for using a chatbot was adequate to support and accessibility. Regarding improvements, five participants suggested enabling chatbots to perform transactions rather than only providing information. Similarly, the authors in [36] conducted interviews with 24 participants, and they were asked several questions about customer service provided by a chatbot and a real person. Therefore, the participants' answers show that the ability to get efficient answers quickly is the most important feature of the chatbot. However, the participants illustrate that the chatbot sometimes does not understand the customers' questions.

In [37] the researcher focused on the relationship between chatbot personality and customer satisfaction. This research was based on studying how the personality of employees affects customer satisfaction. It sought to determine whether these personality factors are taken into account when designing chatbots. Reviewing the previous literature showed that extraversion, agreeableness, and conscientiousness are essential elements of personality that affect customer satisfaction. A survey was sent to companies that use chatbots. The results showed that the companies did not consider these factors important. Additionally, the companies ranked extraversion as the least important factor. On the other hand, a systematic review of multiple studies [38] illustrates that the efficient response of chatbots help business to achieve customers loyalty, mainly if the chatbots work like a human. Moreover, authors in [39] believed that the chatbot that considered the customer emotions positively affect customer satisfaction. Furthermore, an ecological brick chatbot contributed to increasing Lima company's transactions and sales processes [40].

In [41], interviews were conducted with 13 participants to explore the factors that influence users' trust in chatbots. The results showed the influencing factors could be divided into two types: factors related to the chatbot itself and factors associated with the context of services. In terms of the factors related to the chatbot itself, the most important was understanding and providing a proper response, followed by human-like responses. In terms of the factors related to the context of services, the brand and security were the most important.

According to a survey of 119 participants [42], four elements affect customer satisfaction with chatbot services in Indonesia. Each element contains several variables. The first element is usefulness which includes the ability and convenience of solving problems, saving time, and provide suitable answers. The second element is brand image, consisting of privacy, trust, and valuable information in understandable language. Personality is the third element and includes impression (funny, pleasant, etc.), character, friendliness, and effectiveness of the interaction and communication. The fourth element is the ease of use (including accessibility, understanding, and communication). In contrast, authors in [43] conduct a study on ten Romanian online stores. Generally, the study showed that the insufficient quality of the chatbot content has a negative effect on customer experience. On the other hand, authors in [44] apply a chatbot in digital marketing. However, using chatbots in marketing contributed to enhancing the capture of the probable customers of the provided services.

Previous studies have shown that chatbot services are an active and continuously developing area and use various technologies. Moreover, several influencing factors still require to be taken into consideration in the development of chatbots. Most of the reviewed surveys show that good performance is the most important factor in chatbot success. Thus, chatbot service providers need to improve chatbot performance to gain customer satisfaction. Furthermore, some results illustrate that human feeling is another important factor that is neglected in chatbots. Thus, it is possible to consider this need when providing chatbots as customer service by giving them names and training them to use some emotional phrases which provide convenience for the customers when they contact them. These improvement needs motivated us to investigate the viewpoints of EC customers on chatbot services to close the gap between customer opinions and chatbot service developers.

3   Research Questions and Methodology

For this study, a questionnaire was designed to elicit users' views of EC websites and applications. This section presents the research questions and the methodology that was used to conduct the questionnaire.

3.1   Research Question

This research aimed to answer several questions to help understand the current situation regarding customer service chatbots. Table 1 shows the aim of each research question.

Table 1. Aims of research questions.

Research Question	Aim
RQ1: Is the chatbot considered an important feature in electronic commerce websites?	To direct the website and application owners to use chatbot services in their EC channels
RQ2: Do the Chabot's answers match the user's expectations and questions?	To evaluate the current performance of chatbots in terms of responding to customer questions
RQ3: Do customers have a preference for using the chatbot over other ways to obtain answers?	To draw the attention of website and application owners to users' customer service preferences
RQ4: What topics do customers ask about most often?	To focus on training the chatbot more in the field
RQ5: What are the most significant benefits of a customer service chatbot?	To maintain and enhance the benefits of chatbot responses
RQ6: What are the biggest problems in customer service chatbots?	To resolve the issues of chatbot responses

3.2   Questionnaire Design

The questionnaire was designed to answer the research questions. We used the Google Forms service to structure the questionnaire. The questionnaire contained six questions: two open-ended questions, two close-ended questions, and two mixed questions. The two close-ended questions focused on customer preferences regarding chatbot services in EC applications and the service performance measurement. The two mixed questions were designed to measure the usage rate and determine the topic areas that are frequently asked about. The two open-ended questions were designed to determine the participants' views on the advantages and disadvantages of chatbots in EC applications. The answers to these questions were intended to provide an overview of how important chatbot services are in EC and how much the customer needs such services. As well as showing the general rate of chatbot responses, whether it is satisfactory or not. Finally, we sought to determine user opinions on the advantages and disadvantages of chatbots, which could be used to improve their development.

3.3 Questionnaire Sample and Distribution

The questionnaire was targeted at those involved in technology and EC. We wrote it in Arabic and published it on social media platforms, including WhatsApp, Twitter, Snapchat, and Telegram. The questionnaire was answered by 225 people.

3.4 Analysis of the Questionnaire Results

The close-ended questions did not need processing. The responses to those questions were recorded directly. In contrast, the open-ended and mixed questions required some processing, as shown in Fig. 1. After selecting a question, the responses were pre-processed. Given the participants' different terminology, we cleaned the data manually. The data cleaning involved deleting all unclear answers manually after reading all of the answers. Then we created multiple categories based on the repeated terms in the answers. To avoid overlooking any answer, we coded each answer manually using the appropriate category term. Finally, the Statistical Package for the Social Sciences program was used for descriptive analysis.

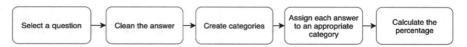

Fig. 1. Processing steps of open and mixed questions.

4 Results

This section presents the results of the questionnaire, which were used to answer the research question. Several outcomes are presented linked to the importance of chatbot services on EC websites and applications. Moreover, the questionnaire measured the user's need for this service and determined the information that customers frequently inquire about. Additionally, this study examined the performance of chatbot systems and identified positive and negative points. Each research question is discussed separately in the following subsections.

4.1   Addressing RQ1

The first research question, "Is the chatbot considered an important feature in EC web-sites?" aims to determine whether, from the user's point of view, a chatbot is an advantage in EC applications or an inessential addition. We started the questionnaire by asking the participants the following close-ended question: "Do you prefer the existence of a chat-bot service on the EC websites and applications you use?" Five options were presented: always, usually, sometimes, rarely, and never. These options were based on the Likert scale for measuring frequency [45]. The responses to this question are shown in Table 2. The most common response was always (34%), with sometimes the second-most com-mon (33%). The percentage of respondents who chose never was 13%. It is evident from the results, then, that most participants prefer to use a chatbot in an EC application.

Table 2. Frequency table for question 1.

Answer	Frequency	%
Always	77	34%
Usually	34	15%
Sometimes	74	33%
Rarely	11	5%
Never	29	13%
Total	225	100%

Table 3. Frequency table for question 2.

Answer	Frequency	%
Always	13	6%
Usually	61	27%
Sometimes	115	51%
Rarely	27	12%
Never	9	4%
Total	225	100%

4.2   Addressing RQ2

To answer the second research question, "Do the Chabot's answers match the user's expectations and questions?" this subsection presents an overview of the Chabot's per-formance based on the responses to the questionnaire. In the questionnaire, we asked the participants, "Do the answers provided by the chatbot match with your query?" This question also used a Likert scale of frequency [28] and had the same five options as question 1. Table 3 presents the results. More than half of the participants answered sometimes. This percentage does not reflect a positive view of the service. It reveals customer dissatisfaction with the answers provided by the chatbot.

4.3 Addressing RQ3

The third question in the questionnaire was "Would you prefer to learn the information by searching in the website's terms and conditions or asking the chatbot directly?" In this question, we focused on identifying participants' behavior and what they do if they need some information when using an EC website or application that has a chatbot service. In addition to the two options in the question, we added space below the question for the participants to provide other answers. The responses to this question were used to answer the third research question: "Do customers have a preference for using the chatbot over other ways to obtain answers?", Table 4 shows the results for this question. Sixty-six percent of participants reported that they preferred to ask the chatbot, 30% said they preferred to search by themselves on the EC website, and 9% added other answers, as presented in Fig. 2. One of the other answers was particularly interesting. The participant said that they would prefer to change sites if not all of the data they needed were evident. This answer highlights an important point. The information in an EC application must be shown in a clear way using simple language. Another participant said that they preferred to search on social media for any information they needed. Therefore, good services must be provided, which leads to customer satisfaction, and thus the customers may market well about these services in their accounts on social media platforms. Moreover, the EC websites and applications need to activate their accounts on social media such as Twitter, which can deal with customer complaints effectively.

Table 4. Frequency table for question 3.

Answer	Frequency	%
Using a chatbot	149	66%
Search in website	67	30%
Others	9	4%
Total	225	100%

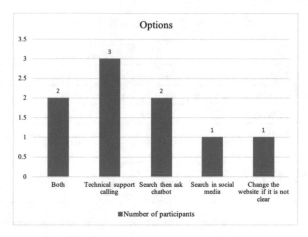

Fig. 2. Other answers to question 3.

4.4 Addressing RQ4

Chatbots usually handle common inquiries better by training them on similar questions. We asked the following question: "What information do you usually need to ask about on EC websites and applications?" The responses to this question were used to answer the fourth research question: "What topics do customers ask about most often?" which aims to help service providers enhance their chatbots in the domain most needed by customers. We included several options for this question, such as product information and specifications, delivery, exchange and returns, and price. We also added a section for the participants to state any other domain. Table 5 shows the answers to this question. The category most selected was "Information about Products and Specifications" (38%). This mean that, the product information usually provided by websites is inadequate. Hence, the chatbot providers need to consider this point and provide more information about the products. Additionally, it is necessary to develop chatbots that can answer the questions of this category. The second most common response was "Information about Returns and Exchanges" (30%). "Information about Payment Methods" was the least selected (1%), which can be explained by the fact that payment methods are usually standardized on many websites, and payment involves straightforward steps.

Table 5. Frequency table for question 4.

Answer	Frequency	%
Information about Products and Specifications	86	38%
Information about Returns and Exchanges	67	30%
Information about Delivery	44	19%
Information about Price	15	7%
All the Above	9	4%
Information about Payment Method	2	1%
Not Specified	2	1%
Total	225	100%

4.5 Addressing RQ5

The fifth research question was "What are the most significant benefits of a customer service chatbot?" which seeks to determine the features of chatbots that are most attractive to customers. To answer this question, we asked the participants this optional and open-ended question: "What are the advantages of the chatbot service?" We discarded some of the responses, however, due to lack of clarity, leaving 139 clear responses. The responses were then divided into six classes, as shown along with their percentages in Table 6.

Fast Response" was the most important feature (69%), which was expected. We believe that the most important reason for EC websites and applications to provide customer services chatbots is that customers need quick responses. The second most important advantage was "Saving Time and Effort" (11%). "Sufficient and Clear Answers" was a close third (10%). Several participants said that one of the essential benefits of chatbots is "Convenience," and they explained that they prefer dealing with chatbots more than dealing with human customer service representatives. This finding corresponds to that

Table 6. Frequency table for question 5.

Advantages	Frequency	%
Fast Response	96	69%
Providing Time and Effort	15	11%
Sufficient and Clear Answers	14	10%
24 h Availability	7	5%
Convenience	4	3%
Knowledgeable	3	2%
Total	139	100%

mentioned in [23]. This finding shows that chatbots can be an essential service for those who have difficulty interacting with others.

4.6 Addressing RQ6

To improve chatbots, the problems associated with them must be identified. For this reason, the sixth research question was "What are the biggest problems in customer service chatbots?" In the questionnaire, we asked the following question: "What are the disadvantages of chatbot services?" We obtained 110 clear answers and excluded the remaining ones for lack of clarity. Table 7 shows participants' opinions with percentages divided into 13 categories.

Misunderstanding customer queries was the highest negative factor in chatbots (31%), which leads to the third most mentioned negative factor, namely the chatbot providing inappropriate responses (22%). The second most reported negative factor was "Unable to answer some questions." This result shows that chatbots are not yet fully trained to cover all potential customer service questions. Moreover, chatbots may sometimes provide inaccurate answers, which is the fourth most mentioned factor (8%) by the participants in this questionnaire. Several participants made it clear that one disadvantage of chatbots is that they require specific formatting and cannot deal with the customer's natural or general language. This finding supports what was reported in [15]. Other participants pointed out that the lack of emotional interaction with chatbots is a major drawback.

Furthermore, 2% of the participants reported not understanding chatbots' responses. This misunderstanding is due to some chatbot models' use of official terminology that

Table 7. Frequency table for question 6.

Disadvantages	Frequency	Percentage
Misunderstanding a question	34	31%
Unable to answer a question	27	24%
Insufficient answers	25	22%
Inaccuracy of answers	9	8%
Requires the specific formulation of the question	3	3%
Lack of human communication and emotion	3	3%
Misunderstanding between the customer and the chatbot	2	2%
Lack of a credible impression	2	2%
Requires many details	1	1%
Unable to handle some problems	1	1%
Annoying appearance on the pages	1	1%
Failure to identify itself as a chatbot	1	1%
Cannot handle complicated questions	1	1%
Total	110	100%

not everyone may understand. Several users indicated that their lack of knowledge of the customer service provider, whether human or machine, was a negative factor. Others pointed out that having a chatbot conversation frequently appearing on the website page could negatively influence their experience. The disadvantages 9, 10, and 13 in Table 7 show chatbots' limited capabilities in specific and complex situations.

The previous subsections show that the responses to the questionnaire were sufficient to answer the research questions. Table 8 provides the answer to each question.

Table 8. Summary of the answers to the research questions.

Research Question	Answer
RQ1: Is the chatbot considered an important feature in electronic commerce websites	Yes (34% of 225 participants)
RQ2: Do the Chabot's answers match the user's expectations and questions?	Sometimes (51% of 225 participants)
RQ3: Do customers have a preference for using the chatbot over other ways to obtain answers?	Yes (66% of 225 participants)
RQ4: What topics do customers ask about most often?	Information about product and specifications (38% of 225 participants)
RQ5: What are the most significant benefits of a customer service chatbot?	Fast Response (69% of 139 participants)
RQ6: What are the biggest problems in customer service chatbots?	Misunderstanding a question (31% of 110 participants)

5   Recommendations and Discussion

This section gives several suggestions and recommendations reached by reviewing the previous studies and the participants' answers to the questionnaire. We believe these suggestions will help developers and chatbot providers to enhance the chatbots models in EC applications.

Although several studies have focused on developing chatbots' ability to handle natural language, the questionnaire results indicated that there is still a weakness in their understanding of some questions. Additionally, some participants indicated that they faced problems in formulating a question for a chatbot which may refer to the differences between the phrases used by the customers and the chatbot. Thus, training the chatbots using different real questions provided by customers will solve this problem.

To enhance the performance and usability, we suggest redoubling efforts to analyze natural language and train the chatbot to handle it, considering that the response time is not affected. Additionally, chatbot service providers should clarify Chabot's identity, as we believe that doing so will build trust between the customer and the service provider. Moreover, the website and application designers should consider how the chatbot conversation appears so that it does not cause the user inconvenience while shopping. Furthermore, to avoid a large number of inquiries from customers, which may be challenging to solve, it is vital to present all essential information clearly and explicitly on the website. On the other hand, the owners of EC applications and websites should consider operating special accounts on social media platforms to respond to inquiries and complaints and not be satisfied with their websites' chatbot service.

6 Conclusion

Chatbot services are increasingly used on EC sites and applications. Our contribution in this research help to bridge the gap between chatbot application providers and the customer use it in EC websites. We have identified the users' opinions of EC websites and applications via a questionnaire. The results have been used to answer the six research questions related to several aspects of chatbot that help to assess the Chabot's current status and what are the important requirements from a customer point of view. Moreover, the questionnaire was designed to elicit what the users thought were the most important advantages and disadvantages. The results showed that 34% of customers prefer to use chatbots when they need information rather than searching on EC websites and applications. Furthermore, the results revealed that the response speed is the most important characteristic of the chatbot service (mentioned by 69% of participants). The most common disadvantage reported was the Chabot's lack of understanding of the customer's inquiries (31%). We believe that, this research helps the chatbots' providers and developers to enhance their performance and provide a more convenient and reliable chatbot service to the EC users.

References

1. Turban, E., King, D., Lee, J.K., Liang, T.-P., Turban, D.C.: Overview of Electronic Commerce. In: Electronic Commerce. Springer Texts in Business and Economics, pp. 3–49. Springer, Cham (2015). https://doi.org/10.1007/978-3-319-10091-3_1
2. Alrumiah, S.S., Hadwan, M.: Implementing big data analytics in e-commerce: vendor and customer view. IEEE Access **9**, 37281–37286 (2021). https://doi.org/10.1109/ACCESS.2021.3063615
3. Mohammed, S., Fiaidhi, J., Ramos, C., Kim, T.-H., Fang, W.C., Abdelzaher, T.: Blockchain in eCommerce: a special issue of the ACM transactions on Internet of Things ACM Trans. Internet Technol. **21**(1), 11–55 (2021). https://doi.org/10.1145/3445788
4. Boriya, A., Malla, S.S., Manjunath, R., Velicheti, V., Eirinaki, M.: ViSeR: a Visual Search Engine for e-Retail. In: 2019 First International Conference on Transdisciplinary AI (TransAI). pp. 76–83 (2019). https://doi.org/10.1109/TransAI46475.2019.00021
5. Albelaihi, A., et al.: Social commerce in Saudi Arabia: a literature review. Int. J. Eng. Res. Technol. **12**(12), 3018–3026 (2019)
6. Hadwan, M., Al-Hagery, M.A., Al-Sarem, M., Saeed, F.: Arabic Sentiment Analysis of Users' Opinions of Governmental Mobile Applications. Comput. Mater. Continua (CMC) **72**(3), 4675–4688 (2022)

7. Hadwan, M., Al-Sarem, M., Saeed, F., Al-Hagery, M.A.: An improved sentiment classification approach for measuring user satisfaction toward governmental services' mobile apps using machine learning methods with feature engineering and SMOTE technique. Appl. Sci. (Switzerland) **12**(11), 1–25 (2022)

8. Cao, Y., Ajjan, H., Hong, P.: Post-purchase shipping and customer service experiences in online shopping and their impact on customer satisfaction: an empirical study with comparison. Asia Pac. J. Mark. Logist. **30**(2), 400–416 (2018). https://doi.org/10.1108/APJML-04-2017-0071

9. Spinelli, A., Pellino, G.: COVID-19 pandemic: perspectives on an unfolding crisis. Br. J. Surg. **107**(7), 785–787 (2020). https://doi.org/10.1002/bjs.11627

10. Elsayed, A.: The Effect of COVID-19 Spread on the e-commerce market: The case of the 5 largest e-commerce companies in the world.

11. Nithuna, S., Laseena, C. A.: Review on implementation techniques of chatbot. In: Proceedings of the 2020 IEEE International Conference on Communication and Signal Processing, ICCSP 2020, pp. 157–161 (2020). https://doi.org/10.1109/ICCSP48568.2020.9182168

12. Sanny, L., Susastra, A.C., Roberts, C., Yusramdaleni, R.: The analysis of customer satisfaction factors which influence chatbot acceptance in Indonesia. **10**(6), 1225–1232, (2020). https://doi.org/10.5267/j.msl.2019.11.036

13. Nguyen, T.: Potential effects of chatbot technology on customer support: a case study. Aalto University, Finland (2019)

14. Kaplan, A., Haenlein, M.: Siri, Siri, in my hand: Who's the fairest in the land? on the interpretations, illustrations, and implications of artificial intelligence. Bus. Horiz. **62**(1), 15–25 Elsevier Ltd (2019). https://doi.org/10.1016/j.bushor.2018.08.004

15. Ahmad, N.A., et al.: Review of Chatbots Design Techniques. Int. J. Comput. Appl. **181**(8), 975–8887 (2018)

16. Asbjørn, F., Brandtzaeg, P.B.: Users' experiences with chatbots: findings from a questionnaire study. Qual. User Exp. **5**, 3 (2020). https://doi.org/10.1007/s41233-020-00033-2

17. Thomas, N.T.: An e-business chatbot using AIML and LSA. In: 2016 International Conference on Advances in Computing, Communications and Informatics, ICACCI 2016. pp. 2740–2742 (2016). https://doi.org/10.1109/ICACCI.2016.7732476

18. Oguntosin, V., Olomo, A.: Development of an E-commerce chatbot for a University shopping mall. Appl. Comput. Intell. Soft Comput. **2021**(1), 1–14 (2021). https://doi.org/10.1155/2021/6630326

19. Bhawiyuga, A., Fauzi, M. A., Pramukantoro, E. S., Yahya, W.: Design of E-commerce chat robot for automatically answering customer question. In: Proceedings - 2017 International Conference on Sustainable Information Engineering and Technology, SIET vol. 2017, pp. 159–162 (2017). https://doi.org/10.1109/SIET.2017.8304128

20. Alhagree, S., Hadwan, M., Technology, I., Al-hagery, M.A., M. Al-sanabani, Alsurori, M.: Extended E-N-DIST Algorithm for Alias Detection.

21. Al-Hagree, S., Al-Sanabani, M., Hadwan, M., Al-Hagery, M.A.: An improved N-gram distance for names matching. In.: First International Conference of Intelligent Computing and Engineering (ICOICE). vol. 2019, pp. 1–7 (2019). https://doi.org/10.1109/ICOICE48418.2019.9035154

22. Hadwan, M.: Extended E-N-DIST algorithm for alias detection. IEEE Access **9**, 7952–7959 (2021)

23. Muangkammuen, P., Intiruk, N., Saikaew, K. R.: Automated Thai-FAQ chatbot using RNN-LSTM. (2018). https://doi.org/10.1109/ICSEC.2018.8712781

24. Qiu, M., et al.: "AliMe chat: A sequence to sequence and rerank based chatbot engine", in ACL 2017–55th Annual Meeting of the Association for Computational Linguistics. Proceedings of the Conference (Long Papers) **2**, 498–503 (2017). https://doi.org/10.18653/v1/P17-2079

25. Satu, M.S., Niamat, T.M., Akhund, U., Yousuf, M.A.: Online Shopping Management System with Customer Multi-Language Supported Query handling AIML Chatbot (2017). https://doi.org/10.13140/RG.2.2.10508.10885
26. Cui, L., Huang, S., Wei, F., Tan, C., Duan, C., Zhou, M.: SuperAgent: a customer service chatbot for E-commerce websites, In: Proceedings of the 55th Annual Meeting of the Association for Computational Linguistics-System Demonstrations. pp. 97–102 (2017). https://doi.org/10.18653/v1/P17-4017
27. Joshi, H.: Proposal of chat based automated system for online shopping. Am. J. Neural Netw. Appl. **3**(1), 1 (2017). https://doi.org/10.11648/j.ajnna.20170301.11
28. Nursetyo, A., Setiadi, D.R.I.M., Subhiyakto, E.R.: Smart chatbot system for E-commerce assitance based on AIML. In: 2018 International Seminar on Research of Information Technology and Intelligent Systems. ISRITI. vol. 2018, pp. 641–645 (2018). https://doi.org/10.1109/ISRITI.2018.8864349
29. Gunawan, D., Putri, F.P., Meidia, H.: Bershca: Bringing chatbot into hotel industry in Indonesia. Telkomnika (Telecommun. Comput. Electron. Control) **18**(2), 839–845 (2020). https://doi.org/10.12928/TELKOMNIKA.V18I2.14841
30. Tran, O.T., Luong, T.C.: Understanding what the users say in chatbots: a case study for the Vietnamese language. Eng. Appl. Artif. Intell. **87**, 103322 (2020)
31. Yudhana, A., Fadlil, A., Rosidin, M.: Indonesian words error detection system using Nazief Adriani Stemmer Algorithm. (IJACSA) Int. J. Adv. Comput. Sci. and Appl. **10**(12), 219–225 (2019)
32. LivePerson, How Consumers View Bots in Customer Care (2017)
33. Brandtzaeg, P.B., Følstad, A.: Why People Use Chatbots. In: Kompatsiaris, I., et al. (eds.) INSCI 2017. LNCS, vol. 10673, pp. 377–392. Springer, Cham (2017). https://doi.org/10.1007/978-3-319-70284-1_30
34. Behera, R.K., Bala, P.K., Ray, A.: Cognitive chatbot for personalised contextual customer service: behind the scene and beyond the hype. Inf. Syst. Front (2021). https://doi.org/10.1007/s10796-021-10168-y
35. Følstad, A., Skjuve, M.: Chatbots for customer service: user experience and motivation. (2019). https://doi.org/10.1145/3342775.3342784
36. van der Goot, M.J., Hafkamp, L., Dankfort, Z.: Customer Service Chatbots: A Qualitative Interview Study into the Communication Journey of Customers. In: Følstad, A., et al. (eds.) CONVERSATIONS 2020. LNCS, vol. 12604, pp. 190–204. Springer, Cham (2021). https://doi.org/10.1007/978-3-030-68288-0_13
37. de Hayco, H.: Chatbot Personality and Customer Satisfaction (2018)
38. Jenneboer, L., Herrando, C., Constantinides, E.: The impact of chatbots on customer loyalty: a systematic literature review. J. Theor. Appl. Electron. Commer. Res. **17**(1), 212–229 (2022). https://doi.org/10.3390/jtaer17010011
39. Song, S., Wang, C., Chen, H.: An Emotional Comfort Framework in the E - commerce Chatbot - AliMe. Naacl **2021**, 130–137 (2021)
40. Licapa-Rodriguez, R., Gomez-Ramos, J., Mauricio, D.: EcoBot: virtual assistant for e-commerce of ecological bricks based on Facebook Messenger In: Proceedings of the 2021 IEEE Engineering International Research Conference, EIRCON vol. 2021, pp. 1–4, (2021). https://doi.org/10.1109/EIRCON52903.2021.9613191
41. Følstad, A., Nordheim, C.B., Bjørkli, C.A.: What Makes Users Trust a Chatbot for Customer Service? An Exploratory Interview Study. In: Bodrunova, S.S. (ed.) INSCI 2018. LNCS, vol. 11193, pp. 194–208. Springer, Cham (2018). https://doi.org/10.1007/978-3-030-01437-7_16
42. Sanny, L., Susastra, A.C., Roberts, C., Yusramdaleni, R.: The analysis of customer satisfaction factors which influence chatbot acceptance in Indonesia. Manag. Sci. Lett. **10**(6), 1225–1232 (2020). https://doi.org/10.5267/j.msl.2019.11.036

43. Nichifor, E., Trifan, A., Nechifor, E.M.: Artificial intelligence in electronic commerce: casic chatbots and the consumer journey. Amfiteatru Econ. **23**(56), 87–101 (2021). https://doi.org/10.24818/EA/2021/56/87
44. Illescas-Manzano, M. D., López, N. V., González, N. A., Rodríguez, C. C.: Implementation of chatbot in online commerce, and open innovation. J. Open Innov. Technol. Mark. Complex. 7(2), 125 (2021). https://doi.org/10.3390/joitmc7020125
45. Sullivan, G.M., Artino, A.R.: Analyzing and interpreting data from Likert-Type scales. J. Grad. Med. Educ. **5**(4), 541–542 (2013). https://doi.org/10.4300/jgme-5-4-18

A Hybrid Sentiment Based SVM with Metaheuristic Algorithms for Cryptocurrency Forecasting

Nor Azizah Hitam[1,2]([✉]), Amelia Ritahani Ismail[1], Syed Zulkarnain Syed Idrus[3],
Muhammad Alfatih Muddathir Abdeurahim Eltayeb[4], Ruhaidah Samsudin[5],
Jari Jussila[6], and Eman H. Alkhammash[7]

[1] Department of Computer Science, KICT, International Islamic University Malaysia (IIUM),
Kuala Lumpur, Malaysia
nhitam@hct.ac.ae, amelia@iium.edu.my
[2] Computer Information Science – Higher Colleges of Technology, Abu Dhabi, UAE
[3] Faculty of Applied and Human Sciences, University Malaysia Perlis (UniMAP),
Perlis, Malaysia
[4] Department of Mechatronics, Kuliyyah of Engineering, International Islamic University
Malaysia (IIUM), Kuala Lumpur, Malaysia
[5] Faculty of Engineering, Universiti Teknologi Malaysia, Johor Bahru, Johor, Malaysia
[6] HAMK Design Factory, Häme University of Applied Sciences, Hämeenlinna, Finland
[7] Department of Computer Science, College of Computers and Information Technology,
Taif University, Taif, Saudi Arabia

Abstract. In 2016, cryptocurrency was reported to be active in terms of user adoption. Generally speaking, making the correct forecast is essential in any field, but it is more important in cryptocurrencies. Researchers have studied machine learning algorithms with cryptocurrencies, a concept that has been recently presented and has a great future as a financial instrument for investors. However, previous studies have ignored key variables like emotions and public opinion, which are vital in today's market. The next contribution to this project will be a hybrid sentiment-based support vector machine (SVM) with chosen optimization methods for bitcoin predictions. Additionally, we integrate a technical indicator, the Commodity Channel Index (CCI), which is utilised in conjunction with the machine learning approach to improve the results of time series forecasting. Particle swarm optimization (PSO) and moth-flame optimization (MFO) are effective at optimising functions. This work introduces a novel hybrid sentiment-based SVM optimised by particle swarm and moth-flame algorithms (SVMPSOMFO) to improve predicting accuracy. SVMPSOMFO optimises the model's parameter values by combining PSO and MFO, which increases search capacity and efficiency. The suggested algorithm's performance is compared to that of three optimization algorithms, SVMPSO, SVMMFO, and SVMALO. SVMPSOMFO outperforms other algorithms based on performance and comparative study.

Keywords: Cryptocurrency · machine learning · hybrid sentiment-based support vector machine optimized by particle swarm and moth-flame optimization algorithms (SVMPSOMFO) · Support Vector Machine (SVM)

F. Saeed et al. (Eds.): ICACIn 2022, LNDECT 179, pp. 139–147, 2023.
https://doi.org/10.1007/978-3-031-36258-3_12

1 Introduction

SVM is a well-known technique to machine learning for classification problems that has been successfully applied in a variety of areas, including agriculture, biometrics and chemicals [1, 2]. Classification models are built by training then on training data and then using the trained models to categorize the unknown sample. Two parameters in SVM, referred to as penalty parameters and kernel parameters, have a major impact on their performance. Numerous optimization techniques are used to tackle optimization challenges, including difficulties with parameter adjustment. To forecast daily stock prices, [3] advocated using Particle Swarm Optimization (PSO) for SVM parameter tuning, whereas [4] used the Cuckoo Search Optimizer (CS) for the same purpose. To improve boiler integration in thermal power plants, a combination of PSO and external optimization (EO) is employed [5]. Additionally, [6] resolved the well-known benchmarking functions using a hybrid PSO and Moth-Flame Optimization (MFO) method.

This paper is structured as follows: Sect. 2 detailed the introduction and Sect. 3: SVM, PSO and MFO algorithms literature and social media feed, Sect. 4 of this paper provides an explanation of the technique used in this work. Section 5 explains hybrid sentiment based SVM with PSO and MFO and Sect. 6 results and analysis followed by the Sect. 7 concludes with several research directions.

2 The Use of Machine Learning in Finance

Machine Learning (ML), a subset of Artificial Intelligence (AI) that use algorithms and neural network models to help computer systems in gradually improving their performance [3], gained prominence in finance beginning in the 1980s. Artificial Neural Networks begin to detect outlier claims/charges, which are eventually mirrored in human studies. Security Pacific National Bank established the Fraud Prevention Task Force in 1987 to prevent unauthorized use of Stewart and Watson debit cards (1985). In 2001, the age of robot manufacturing began, when robots powered by an IBM supercomputer defeated humans in a financial trade simulation war in which bots accounted for 7% more cash than humans [4]. Thus, machine learning contributed to the reduction of financial crimes and fraud by recognizing abnormal behavioral changes and abnormalities [5]. Market behavior prediction is one of the most difficult tasks each investor has to make the best option and maximize returns. This demonstrates the critical nature of knowledge, since one may acquire new information or abilities and apply them in any relevant sector [6], and this is also true for Machine Learning (ML), a critical element of current industry and research. Blockchain technology is one of the most researched areas of machine learning because it is used to generate a decentralized digital currency known as a cryptocurrency which is based on peer-to-peer transactions [7, 8].

3 Related Works

A supervised learning method known as a Support Vector Machine [9] is developed by [10] to overcome the problems of local minima and overfitting. SVM is known as the most flexible technique to construct the explicit and accurate boundaries [11] as well as sentiment analysis [9] SVM can handle a nonlinear classification problem and sparse data by mapping samples from low to high dimensional feature space with the use of predetermined kernel function [12]. The fundamental concept behind SVM is to use a nonlinear mapping function to transfer the training data into higher dimensional space, then use a predefined kernel function to conduct linear regression in higher dimensional space for data separation. The swarm refers to a population and particles to individuals. A PSO method will iteratively explore multidimensional search space with a swarm of particles to reach minimum or maximum global levels. Each particle flies according to its speed vector in the search space (see Fig. 1). Every iteration, the vector is tweaked to find the optimal location in terms of cognitive and social aspects, which is determined by particles within a certain neighborhood that function as attractors and attractor particles, respectively. PSO-SVM claimed to be simpler to adjust compared to other method named GA with minimum adjustment on its parameters. A study of reservoir annual inflow forecasting is very significant for reservoir management to ensure effectively used of water supply and resources. Result of the study shows that the optimum SVM-PSO model has outperformed the ANNs which then indicates that the optimized SVM-PSO model has better forecasting performance [15].

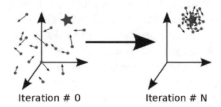

Iteration # 0   Iteration # N

Fig. 1. Iteration of particles in PSO

In [16], the authors chose to utilize a variety of technical analysis indicators, including correlations between multiple stock prices, to estimate the future stock price of a company's stock. The PSO algorithms are applied to choose the most advantageous input characteristics from all indicators accessible and the PSOSVM exceeded the individual SVM approach.

In other study, [17] have proposed a model for forecasting water usage that utilizes Least Square Support Vector Machines Optimized by a Hybrid Intelligent Algorithm (LS-SVM). Water consumption forecasting is accomplished by utilizing PSO algorithm characteristics such as parallelism and global convergence to avoid local optimums and by utilizing novel genetic algorithm concepts such as crossover and mutation operations to accelerate the search for the global optimum.

An optimizer known as Moth-Flame Optimization Algorithm (MFO) shows optimistic performance. It is said to be one of the meta-heuristic algorithms that could be applied in wide range of fields for specific optimization problems. [18, 19] had proposed a novel-inspired optimization navigation method of moths in nature called transverse orientation. Another research conducted by [20] had developed a new MFO algorithm that solved optimization problems, specifically for manufacturing industry. Their results had been proven that the MFO was able to effectively optimize and overcome manufacturing optimization problems.

[21] developed an MFO-based multi-layer perceptron to identify optimum MLP weights and biases. The MFO was used as a searching strategy in this research to discover the best solution with the lowest error rate. [22] used the MFO method to address the optimum power flow problem (OPF), which was identified as fuel cost reduction, active power loss minimization, and reactive power loss minimizing. Another study [23], created a novel MFO algorithm for solving optimization challenges in the manufacturing business.

Through instant messaging, Internet forums, and social networking, the Internet has enabled and expedited the development of new kinds of human connection [24]. Social media has become a necessity for the society, and it is growing in popularity. No one can predict how far we will be able to perceive the utilization of this type of technology, and to what extent the user will go in deceiving or preying on other users who are susceptible to being deceived or preyed on [24]. A key topic in this study is a cryptocurrency market forecast, which becomes one of the favorite field of research [25]. There is collective mood assessed [26, 27] in three states ('positive,' 'negative' and 'calm') which classify Twitter feelings. [27] provide proof that tweets published on the tweeter affect the investment choice and that the Dow Jones Industrial Average (DJIA) daily movements have a precision of 86.7%. This demonstrates that sentiment information has become vital information and provides a key insight into the popular perspective on a certain issue, as demonstrated by [28], where the sentiment does have a substantial influence on time series forecasting, as previously stated. This is discussed more [29] where the Web Crawler extracts the everyday feeling from the Internet and modifies the feeling of the investors on their following market (Saturday to Monday).

4 Methods

The specific design of this phase, including the processing processes and their sequence, is decided by exploratory data analysis, which is aimed at understanding data visualizations such as plots, scatterplots, and graphs. In the case of sentiment data, this will eliminate any unnecessary information, as well as NA values, data normalization, data transformation, and splitting. A common training and testing dataset are created and arrived for both market sectors and saved as a csv-file. Bitcoin dataset was extracted from the website https://coinmarketcap.com. The variables selected from historical data consist of the open, close, high, and low of the trading day. Figure 2 illustrates a data splitting of both training and testing dataset for bitcoin.

The sentiment data collection was derived from Twitter and includes tweets for the same time span as historical data mentioned above. However, not all tweets were included in the sample of this study. Only tweets which referenced to the 'Bitcoin' were chosen. The approach for identifying tweets that reference cryptocurrencies and foreign exchange (Forex) that are not generated from related works was also utilized during the Twitter data gathering [30–32]. This is done to narrow down the list of tweets and retrieve only the corresponding terms linked to the specified instruments. [25] in their paper searched for emoticons and keywords in their study on stock price prediction using sentiment analysis. To the searched terms, we use a similar technique and set of rules.

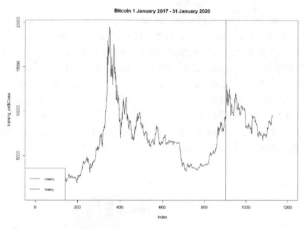

Fig. 2. Bitcoin Data splitting of Training and Testing dataset

As a benchmark model, a Support Vector Machine (SVM) is used in this study. The primary goal of employing this technique is to address modelling difficulties related to time series forecasting, therefore improving the stability and accuracy of the machine learning algorithm and producing superior performance outcomes. Furthermore, as previously mentioned in previous research [32, 33]. The main factors causing errors in machine learning models are over variance, bias, and noise, and the hybrid machine learning algorithm with metaheuristic algorithms can be used to handle over-fitting issues and thus improve the algorithm used by minimizing all highlighted factors. SVM pseudocode may be found in [34]. Furthermore, this study employs a hybrid SVM with metaheuristic algorithms, such as the Particle Swarm Optimization (PSO) and the Moth-Flame Optimization (MFO) algorithms. A research by [28] demonstrates that combining sentiment with SVM parameter adjustment can result in a more accurate outcome. We turn our attention to the hybridization of sentiment into the Support Vector Machine (SVM) and optimize the algorithm with the PSO and MFO for forecasting, because the proposed hybrid SVM with PSO and MFO provides greater forecast accuracy than previous research [35].

4.1 A Hybrid Sentiment Based SVM with PSO and MFO

In this study, the PSO and MFO algorithms are utilized in this work to optimize the SVM parameters C and γ by identifying the optimum values that minimize SVM root mean square (RMSE) during the training phase. All SVM parameters are recorded in the framework as a particle in the swarm, which represents a possible solution for the best SVM. Based on previous research in the field of financial analysis [41], two additional features were added to the basic feature set. We utilize a standard SVM with parameter optimization (C and γ) as well as an SVM method with optimization approaches to generate benchmark models against which the proposed model's performance can be compared. On the basis of the two methods and sentiment data, we propose SVMP-SOMFO (see Fig. 3). We employ these approaches to find a balance between PSO and

MFO exploration and exploitation. As noted in the literature, there are various optimization techniques for financial forecasting issues, including classical approaches and machine learning algorithms. In concept, optimising a location is typically treated as a non-linear mathematical optimization problem that prioritises different constraints and objective functions. To solve the models created in this research, Particle Swarm Optimization (PSO) and Moth Flame Optimization (MFO) are used. These two optimization algorithms, in addition to mathematical programming, are categorised as metaheuristic approaches since they are based on natural biological evolution or social interaction behaviour of organisms and have consistently produced acceptable results throughout time [42].

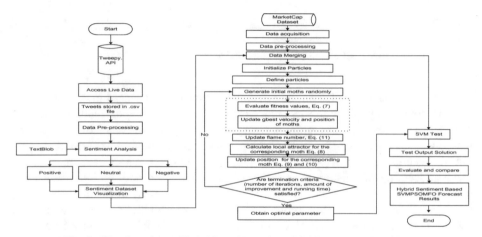

Fig. 3. The flowchart of hybrid sentiment based SVM with PSO and MFO

5 Performance Evaluation

A set of performance measurements based on regression model accuracy metrics [36] are used to evaluate the forecasting model's performance. Choosing and employing a range of forecasting methodologies is extremely beneficial to investors. It allows for comparisons of different forecasting models. The forecasting model's performance is assessed using the following statistical assessment metrics: This paper uses the RMSE as the metric to measure the performance of SVMPSOMFO. The better the forecast accuracy, the lower the values of the four measures, where x_i represents the actual values; x_j represents the forecasted values; and n indicates the sample count.

6 Results and Analysis

As shown in Table 1, the Hybrid Sentiment-based SVMPSOMFO approach beats the other three methods in terms of accuracy in bitcoin. The SVMPSO, SVMMFO, and SVMALO are the next three symbols. The results of the paired t-tests comparing SVMPSOMFO with the other three techniques are shown in Table 2. We can observe that the accuracy rate of SVMPSOMFO is higher than the accuracy rates of the other three techniques when the confidence level is set at 5%. The significance level for the t-test is set at 0.05. [43] point out that, although most researchers consider p-values to be a measure of statistical significance, they may also be seen as a continuous measure of the compatibility between data and the model that was used to generate it. As a result, it may be used to assess the model's fit to the data, with a value ranging from 0 to 1 representing full incompatibility and 1 representing perfect compatibility. Because the suggested model has a p-value that is higher than the level of significance (p-value > 0.05), the null hypothesis is unable to be rejected, suggesting that there is insufficient evidence to demonstrate that the prediction significantly differs from the actual data.

Table 1. Accuracy Rate

Model	RMSE
SVMPSOMFO	0.3856
SVMPSO	0.3871
SVMMFO	0.6265
SVMALO	0.6278

Table 2. Results of t-test

Models	t	p-value
SVMPSOMFO, SVMPSO	0.59046	0.555196
SVMPSOMFO, SVMMFO	0.050124	0.960047
SVMPSOMFO, SVMALO	0.054208	0.956795

7   Conclusion

In conclusion, the study aims at getting the accurate prediction of a digital currency to help the investors in decision making. This paper first proposes a hybrid sentiment based SVMPSOMFO for prediction the price of bitcoin. Since SVM is known for its capability in solving the parameter optimization and overfit concurrently. The results reveal the optimal hyperparameters and adding another two attributes, the error is minimized thus shows the comprehensiveness of the proposed model. Through hybridization with the optimization algorithms, the model is more predictive than other models. For comparison, SVMPSO, SVMMFO and SVMALO are applied and evaluated. Results shown that SVMPSOMFO has a very well classification accuracy. In addition, by adding a technical indicator, Composite Channel Index (CCI) into the data, this shows that the fluctuation of bitcoin is influenced by technical indicators and other factors too.

References

1. Hitam, N.A., Ismail, A.R., Saeed, F.: An Optimized Support Vector Machine (SVM) based on Particle Swarm Optimization (PSO) for cryptocurrency forecasting. Procedia Comput. Sci. **163**, 427–433 (2019). https://doi.org/10.1016/j.procs.2019.12.125
2. Tharwat, A., Gabel, T., Hassanien, A.E.: Parameter optimization of support vector machine using dragonfly algorithm. In: Hassanien, A.E., Shaalan, K., Gaber, T., Tolba, M.F. (eds.) AISI 2017. AISC, vol. 639, pp. 309–319. Springer, Cham (2018). https://doi.org/10.1007/978-3-319-64861-3_29
3. Foote, K.D.: A Brief History of Machine Learning – DATAVERSITY. Dataversity (2019)
4. BBC News|BUSINESS|Robots beat humans in trading battle (2001). http://news.bbc.co.uk/2/hi/business/1481339.stm
5. Schutzer, D.: CTO Corner: Artificial Intelligence Use in Financial Services – Financial Services Roundtable. Fsroundtable.Ccom, pp. 1–6 (2015)
6. Hassad, H., Hitam, N.A., Junainor, H.: The relationship of knowledge sharing behavior towards the organizational innovation capability (case study of ISP organization in Malaysia). Malaysian Technical University Networks, pp. 699–704 (2012). http://hdl.handle.net/123456789/29901
7. Madan, I., Saluja, S., Zhao, A.: Automated Bitcoin Trading via Machine Learning Algorithms, vol. 20, pp. 1–5 (2015)
8. Hitam, N.A., Ismail, A.R.: Comparative performance of machine learning algorithms for cryptocurrency forecasting. Indones. J. Electr. Eng. Comput. Sci. **11**(3), 1121–1128 (2018). https://doi.org/10.11591/ijeecs.v11.i3.pp1121-1128
9. Wang, W., Nie, X., Qiu, L.: Support vector machine with particle swarm optimization for reservoir annual inflow forecasting. In: Proceedings of the International Conference on Artificial Intelligence and Computer Intelligence, AICI 2010, vol. 1, pp. 184–188 (2010). https://doi.org/10.1109/AICI.2010.45
10. Toliyat Abolhassani, A., Yaghoobi, M.: Stock price forecasting using PSOSVM. In: 2010 3rd International Conference on Advanced Computer Theory and Engineering, pp. V3-352–V3-356 (2010). https://doi.org/10.1109/ICACTE.2010.5579738
11. Weilin, L., Kangning, C., Lina, L.: Prediction model of water consumption using least square support vector machines optimized by hybrid intelligent algorithm. In: Proceedings of the Second International Conference on Mechanic Automation and Control Engineering (MACE), pp. 3298–3301 (2011)
12. Shehab, M., Alshawabkah, H., Abualigah, L., AL-Madi, N.: Enhanced a hybrid moth-flame optimization algorithm using new selection schemes. Eng. Comput. **37**(4), 2931–2956 (2020). https://doi.org/10.1007/s00366-020-00971-7
13. Mirjalili, S., Lewis, A.: The whale optimization algorithm. Adv. Eng. Softw. **95**, 51–67 (2016). https://doi.org/10.1016/j.advengsoft.2016.01.008
14. Yildiz, B.S., Pholdee, N., Bureerat, S., Yildiz, A.R., Sait, S.M.: Robust design of a robot gripper mechanism using new hybrid grasshopper optimization algorithm. Expert Syst. **38**(3) (2021). https://doi.org/10.1111/exsy.12666
15. Syed Zulkarnain, S.I., Hitam, N.A.: Social media use or abuse : a review. J. Hum. Dev. Commun. (2014)
16. Kordonis, J., Symeonidis, S., Arampatzis, A.: Stock price forecasting via sentiment analysis on Twitter. In: ACM International Conference Proceeding Series (2016). https://doi.org/10.1145/3003733.3003787
17. Oliveira, N., Cortez, P., Areal, N.: On the predictability of stock market behavior using StockTwits sentiment and posting volume. In: Correia, L., Reis, L.P., Cascalho, J. (eds.) EPIA 2013. LNCS (LNAI), vol. 8154, pp. 355–365. Springer, Heidelberg (2013). https://doi.org/10.1007/978-3-642-40669-0_31

18. Bollen, J., Mao, H., Zeng, X.: Twitter mood predicts the stock market. J. Comput. Sci. **2**(1), 1–8 (2011). https://doi.org/10.1016/j.jocs.2010.12.007
19. Hitam, N.A., Ismail, A.R., Samsudin, R., Ameerbakhsh, O.: The influence of sentiments in digital currency prediction using hybrid sentiment-based Support Vector Machine with Whale Optimization Algorithm (SVMWOA). In: 2021 International Congress of Advanced Technology and Engineering. ICOTEN 2021, pp. 1–7 (2021).https://doi.org/10.1109/icoten 52080.2021.9493454
20. Ren, R., Wu, D.D., Wu, D.D.: Forecasting stock market movement direction using sentiment analysis and support vector machine. IEEE Syst. J. **13**(1), 760–770 (2019). https://doi.org/10.1109/JSYST.2018.2794462
21. Jain, A., Tripathi, S., Dhardwivedi, H., Saxena, P.: Forecasting price of cryptocurrencies using tweets sentiment analysis. In: 2018 11th International Conference on Contemporary Computing, IC3 2018 (2018). https://doi.org/10.1109/IC3.2018.8530659
22. Batra, R., Daudpota, S.M.: Integrating StockTwits with sentiment analysis for better prediction of stock price movement. In: 2018 International Conference on Computing Mathematics and Engineering Technologies Inventing Innovating Integrating Socioeconomic Development, iCoMET 2018 – Proceedings, pp. 1–5 (2018). https://doi.org/10.1109/ICOMET.2018.834 6382
23. El Rahman, S.A., Alotaibi, F.A., Alshehri, W.A.: Sentiment analysis of Twitter data. In: 2019 International Conference on Computer and Information Science, ICCIS 2019 (2019). https://doi.org/10.1109/ICCISci.2019.8716464
24. Wang, G.L.G., et al.: The performance of PSO-SVM in inflation forecasting. In: 2017 10th International Conference on Intelligent Computation Technology and Automation, vol. 1, no. 1, pp. 259–262 (2016).https://doi.org/10.1109/ICCSIT.2009.5234448
25. Dong, G., Fataliyev, K., Wang, L., Prediction, A.O.A.: One-Step and Multi-Step Ahead Stock Prediction Using Backpropagation Neural Networks, pp. 2–6 (2013)
26. Parmar, I., et al.: Stock market prediction using machine learning. In: ICSCCC 2018 – 1st International Conference on Secure Cyber Computing and Communications, pp. 574–576 (2018). https://doi.org/10.1109/ICSCCC.2018.8703332
27. Houssein, E.H., Hosney, M.E., Elhoseny, M., Oliva, D., Mohamed, W.M., Hassaballah, M.: Hybrid Harris hawks optimization with cuckoo search for drug design and discovery in chemoinformatics. Sci. Rep. **10**(1), 14439 (2020). https://doi.org/10.1038/s41598-020-715 02-z
28. Atsalakis, G.S., Atsalaki, I.G., Pasiouras, F., Zopounidis, C.: Bitcoin price forecasting with neuro-fuzzy techniques. Eur. J. Oper. Res. **276**(2), 770–780 (2019). https://doi.org/10.1016/j.ejor.2019.01.040
29. Wang, J.: An enhanced LGSA-SVM for S & P 500 index forecast, no. 2003, pp. 4094–4101 (2017)
30. Yang, Z., Shi, K., Wu, A., Qiu, M., Hu, Y.: A hybrid method based on particle swarm optimization and moth-flame optimization. In: Proceedings – 2019 11th International Conference on Intelligent Human-Machine Systems and Cybernetics, IHMSC 2019, vol. 2, pp. 207–210 (2019). https://doi.org/10.1109/IHMSC.2019.10144
31. Castoe, M.: Predicting Stock Market Price Direction with Uncertainty Using Quantile Regression Forest (2020)
32. Greenland, S., et al.: Statistical tests, P values, confidence intervals, and power: a guide to misinterpretations. Eur. J. Epidemiol. **31**(4), 337–350 (2016). https://doi.org/10.1007/s10654-016-0149-3

A Comparison Study of Machine Learning Algorithms for Credit Risk Prediction

Pica Salsabila Atmauswan[1], Shahrinaz Ismail[1], Nor Azizah Hitam[2], Akibu Mahmoud Abdullahi[1], and Mohammed Al-Sarem[3(✉)]

[1] School of Computing and Informatics, Albukhary International University, Jalan Tun Abdul Razak, Bandar Alor Setar, 05200 Alor Setar, Kedah, Malaysia
`pica.atmauswan@student.aiu.edu.my`, {`shahrinaz.ismail,`
`akibu.abdullahi`}`@aiu.edu.my`
[2] Faculty of Computer Information Science, Higher Colleges of Technology, Abu Dhabi, UAE
`nhitam@hct.ac.ae`
[3] College of Computer Science and Engineering, Taibah University, Medina, Saudi Arabia
`msarem@taibahu.edu.sa`

Abstract. The banking industry performs credit score analysis as an efficient credit risk assessment method to determine a customer's creditworthiness. In the banking industry, machine learning could be used for a variety of uses involving data analysis. A method of data analysis that is capable of self-regulation has been made possible by the development of modern techniques, such as classification approaches. The classification method is a form of supervised learning in which the computer acquires knowledge from the provided input data and then utilizes it to classify the dataset, which is used for training purposes. This study presents a comparative analysis of the various machine learning algorithms that are utilized to evaluate credit risk. The methods are used by utilizing the German Credit dataset that was collected from Kaggle, which consists of 1,000 instances and 11 attributes, all of which are used to determine if transactions are good or bad. The findings of data analysis using Logistic Regression, Linear Discriminant Analysis, Gaussian Naive Bayes, K-Nearest Neighbors Classifier, Decision Tree Classifier, Support Vector Machines, and Random Forest are compared and contrasted in this study. The findings demonstrated that the Random Forest algorithm forecasted credit risk effectively.

Keywords: Credit Risk · Banking · Machine Learning · Prediction · Features

1 Introduction

In 2007, the finance regulator applied the internal rating-based (IRB) technique for the first time to generate a new absolute credit risk assessment. As a result of Basel II regulations, the Basel Committee on Banking Supervision incorporated major credit risk components, which include Probability of Default (PD), Loss Given Default (LGD), and Exposure at Default. The 2007 subprime mortgage crisis and the 2008 global financial crisis heightened the significance of credit analysis for banks. Furthermore, banks and

F. Saeed et al. (Eds.): ICACIn 2022, LNDECT 179, pp. 148–159, 2023.
https://doi.org/10.1007/978-3-031-36258-3_13

financial organizations are enhancing their credit systems not just because of regulation but also because even little changes might result in enormous profits [1].

Credit risk is the loss incurred by lenders as a result of borrowers' failure to honor their credit obligations. Creditors rely on their judgments of credit approvals and interest rates in the management of their credit evaluations [2]. In the process of credit analysis, a credit scoring system is applied to increase the capacity for assessing the creditworthiness of consumers. Banks use credit ratings to determine applicants' creditworthiness and establish existing clients' interest rates and credit limits. The 5Cs of credit is the least five elements that lenders must consider when analyzing the creditworthiness of a potential borrower. The elements include Character, Capacity, Capital, Collateral, and Condition. Due to the system's capability to quickly gather relevant information, its application, and loan borrowing, the capacity to automate credit approval with correct analysis is the essential advantage of credit risk analysis, allowing the system to approve loans. Although an automatic credit evaluation cannot initiate credit, it can speed up the procedure if it must be refused or approved at some point [3].

Another concern among financial institutions is the scenarios that cause them to lose money, which is called a default [4]. This occurs when borrowers fail to repay their credit loans to the bank, resulting in a loss for the bank. Additionally, fraud is also a concern that many institutions encounter. Fraud is defined as an activity committed by a person that employs deceit to obtain an advantage, evades responsibilities, or inflicts financial or non-financial harm to others [5]. The fraudulent theft of money, assets, or other property belonging to or held by a financial institution is bank fraud. It is also deemed a fraud to collect funds from depositors by posing as a bank. As a result of this problem, credit risk becomes an increasingly important factor for banks to consider when deciding whether or not to offer credit. Credit analysts provide an individual assessment of each borrower's application using a credit score methodology, and the 5Cs are one of them.

Numerous banks do not have a single finance and risk management system that is automated, fully integrated, and comprehensive because it is challenging to develop a risk management system that is both comprehensive and scalable in order to predict customer risk scores [4]. However, there have been many different credit risk prediction systems made available. Review papers have attempted to classify them as statistical methods, intelligent systems, data mining techniques, and machine learning approaches, among other possible classifications. In order to differentiate between fraudulent behavior and legitimate activity, the machine learning method is widely recognized and considered to be the ideal instrument that can be used to comprehend the transaction patterns of banking customers. This is accomplished through the discovery of patterns in customer data using the machine learning method [4].

Machine learning (ML), which is common for commercial use, is a method for building analytical models that enable computers to "learn" from data and perform predictive analysis. It is a technology that was developed to expand the capacity of computers to "learn" and carry out predictive analytics. The importance of ML has increased in the era of big data, and it also has a number of applications. When deciding whether or not to extend credit to consumers, for instance, financial institutions must, on occasion, choose which risk variables to take into consideration. Even though the majority of the client qualities have little predictive power on creditworthiness or other potential customers, it is common practice to analyze a number of the client's characteristics. Machine learning is considered an appropriate tool that can be used sequentially to understand customer banking transaction patterns by identifying customer data patterns and differentiating between fraudulent activities and routine transactions. It could be accomplished through the use of a customer data pattern matching tool [6]. There are several commercial uses for machine learning, and it is seen as a significant factor in the development and transformation of the financial industry.

2 Related Works

Credit is described as the ability of a client to get goods or services prior to making payment, with the expectation that the amount would be made later. Credit risk is the probability that debtors may be unwilling or unable to fulfill their contractual obligations [7]. Relating to credit risk, the goal of credit scoring is to summarize a borrower's credit history by making use of a credit scoring model. Additionally, credit scoring is recognized as the essential instrument for analyzing credit risk in financial institutions [8]. Credit scoring models are decision support systems that accept a collection of predictor elements as input and output scores. Creditors use these systems to justify who is given credit and who is not given credit [9], often by combining five of the criteria or factors that credit applicants must meet, including the debt-to-income ratio, the quantity of money in hand, the purpose of credit, asset, and liability. The use of these variables allows for the determination and evaluation of a borrower's eligibility based on qualitative and quantitative measures, which provides a means to measure the borrower's financial strength, lower the likelihood of non-payment to a reasonable level, and estimate the probability of default from the borrower [10]. Credit scoring also aims to facilitate trust between various stakeholders, specifically lenders, and borrowers.

Implementing artificial intelligence (AI) in the banking industry allows reorganizing of business relationships based on objectivity and trust between banks and their stakeholders [11]. As a result, credit scoring was one of the very first applications of AI in the banking industry, mainly through the utilization of machine learning (ML). Risk management is one of the most important areas in which AI can improve banking operations. This can be accomplished by enhancing credit scoring, portfolio management, fraud detection, debt collection strategy optimization, rapid detection and interpretation of signals from weak borrowers, and the development of economic models. Based on a set of features, AI algorithms look at the available data to figure out how the model should work best [12]. It is critical that the independent variables in the model development process are properly selected since they define the features that impact credit score value. The importance of independent variables is generally collected through the application form [13].

Many machine learning algorithms have been developed for credit risk card prediction, for example, to evaluate the performance of decision trees, Naïve Bayes, and multi-layer perceptron (MLP) by predicting the credit risk [14]. The previous research used a German credit risk dataset to apply the algorithms, in which the algorithms were evaluated using the accuracy metric. Based on the result, the decision tree algorithm has lower accuracy with 69% compared to Naïve Bayes and MLP. Similarly, another study was conducted for credit risk prediction by comparing the performance of machine learning algorithms, which include eXtreem Gradient Boosting (XGBoost), Random Forest, MLP, Rpart, Rotation Forest, and Gradient Boosting Machine (GBM) [15]. The result shows that Random Forest gave the lowest accuracy of 64.65%, compared to the other algorithms.

In another study, a comparison was made on the performance of ML algorithms by predicting credit risk [16]. The algorithms include decision tree, K-Nearest Neighbor (K-NN), Random Forest, logistic regression, and Naïve Bayes. The result indicates that Random Forest and decision tree produced the highest accuracy with 96.53% and 94.68%, while Logistic Regression produced a lower accuracy with 81.05%. On the other hand, an artificial neural network (ANN) and Naïve Bayes algorithms were developed to compare their performance for credit risk prediction in 2020 [17]. ANN was found to have higher accuracy than Naïve Bayes with 81.85% and 81.32%, respectively.

Table 1. Comparison of Machine Learning Models Accuracy for Credit Risk Prediction.

Author	Dataset	Features	Model	Accuracy (%)
Qasem et al. (2020)	German credit risk	1,000 instances and 24 attributes	**Naïve Bayes**	**76.46**
			Decision Tree	**69.03**
			MLP	73.23
Aleksandrova (2021)	Lending club site	151 variables	XGBoost	64.68
			GBM	66.00
			Random Forest	**64.65**
			Rotation Forest	67.21
			Rpart	66.15
			MLP	65.49
Teles et al. (2020)	Financial institution	15 attributes and 1,890 records	**Naïve Bayes**	**81.32**
			Neural Network	81.85
Wang et al. (2020)	Loan information statistics of a commercial bank	N/A	KNN	81.51
			Decision Tree	**92.11**
			Random Forest	**94.57**
			Naive Bayes	**79.99**
			Logistic Regression	**81.05**

Table 1 shows the results from recent works on credit risk prediction using machine language (ML) algorithms. The ML models adopted in this research and the accuracy results from these previous works are emphasized in bold.

3 Methodology

3.1 Dataset

This research is based on a real-world credit card dataset obtained from the Kaggle website for the purpose of the comparison study. This collection of credit data was created by Professor Hofmann and titled "German Credit Dataset." Each element in the data list corresponds to the individuals who were receiving the bank credit. In order to protect the sensitive data of the financial institution's clients, the dataset does not contain any name. Depending on a set of established values, each individual will be placed into one of two credit risk categories, good or bad. This data collection from the German Credit Dataset contains 1,000 attributes, 700 of which are categorized as "creditworthy" and 300 as "non-creditworthy." This dataset has 11 features in total. Six attributes are "categorical," while the remaining five are "numerical." In subsequent studies employing this dataset, ten independent variables are set as input variables, and one dependent variable is set as an output variable.

3.2 Machine Learning Algorithms

Seven machine learning (ML) algorithms are adopted in this study, namely Logistic Regression, Linear Discriminant Analysis (LDA), Gaussian Naïve Bayes (NB), K-Nearest Neighbors Classifier (K-NN), Decision Tree Classifier, Support Vector Machine (SVM), and Random Forest (RF). The details of these algorithms are as follows.

Logistic Regression. Two prevalent statistical methods for assessing credit risk are linear discrimination analysis and logistic regression. As a basic parametric statistical model, linear discrimination analysis was one of the earliest credit scoring methods. However, it has been questioned due to the categorical character of the non-normally distributed credit data [2]. The covariance matrices of the classes with excellent and poor credit are asymmetrical. A more sophisticated statistical model emerges to address some shortcomings of the LDA method [18]. For credit scoring applications, Henley's logistic regression model is employed [1]. When prospective predictor variables are introduced, logistic regression may yield the likelihood of a discrete result.

Linear Discriminant Analysis. The discriminant analysis assumes different data classes using other Gaussian distributions [4]. Linear and quadratic discriminant algorithms are used for classification, while linear discriminant analysis (LDA) is the most used statistical method in credit scoring models. Some researchers criticized LDA due to its assumption of a linear relationship between dependent and independent variables, which is rarely true in practice, and its sensitivity to deviations from the multivariate normality assumption [19, 20]. This research takes up LDA to see whether it is still relevantly useful for credit risk prediction.

Gaussian Naïve Bayes. The Naive Bayes (NB) learning algorithm family is a series of supervised learning algorithms that implement Bayes' theorem, based on the "naïve" assumption that every pair of characteristics is independent of one another [21]. An NB classifier is a piece of software that determines the likelihood that a specific occurrence, or example, belongs to a particular category, in which the probability is expressed as a percentage. NB is adopted in this comparison study by looking at the characteristic independency possibility.

K-Nearest Neighbors Classifier. Parametric and non-parametric classification techniques exist. Parametric approaches estimate distribution parameters based on the assumption of a normally distributed population [22]. Non-parametric methods make no assumptions about distributions and are distribution-free [23]. The k-nearest neighbor (K-NN) classifier is non-parametric. K-NN classifiers look for the k training pattern in an unknown situation [24], as in a similar case to this research, which is the case of Tunisian banks. These k training instances are unknown cases of K-neighbors [25].

Decision Tree Classifier. Real-life trees have affected categorization and regression in machine learning. In decision analysis, a decision tree represents choices graphically. As the name implies, it makes choices in the form of tree-like, with branches and leaves. Decision trees, often termed classification trees, are easy-to-understand classifiers. Few data variances may be utilized to learn cause substantial model differences [26]. A decision tree approach organizes many observations into smaller groups depending on rules and desired attributes [27]. Previous work explored credit card scoring using a decision tree and multilayer perceptron neural network [24], proving that this algorithm is relevant for credit risk prediction. On a single data partition and neural network, the decision tree model and multilayer perceptron neural network have equivalent accuracy [28].

Support Vector Machine. Support Vector Machine (SVM) is a standard machine learning classification algorithm that has three advantages [2]. The first advantage is fewer assumptions regarding input variable distribution and continuity. Secondly, the ability to perform non-linear mapping, and finally, while addressing the maximizing issue, it learns the separation hyper-plane. In high-dimensional space, a linear discriminant function may replace non-linear separated features. This has brought the creation of hybrid SVM-based credit scoring models using neighborhood rough sets [29]. Using rough neighborhood sets and SVM-based hybrid classifiers could improve credit scoring.

Random Forest. A supervised ML technique known as Random Forest (RF) generates a forest and then randomly distributes its trees. The training of a forest, also known as a collection of Decision Trees, is accomplished via the bagging technique. RF generates a large number of decision trees and combines them in order to arrive at a consistent and reliable categorization [30]. The fact that the RF may be used for both regression analysis and classification research is the algorithm's primary strength.

3.3 Feature Selection

Certain factors are crucial to conceive the dataset employed in this research. It is essential to choose the independent variables, i.e., the variables that will determine the value, and which variable will change depending on the determining variables, i.e., the dependent variable, according to the research objectives. Before moving on to the data preprocessing step, this research determined the important features in the German Credit Dataset. It is noted that each variable and credit performance in the dataset must have a clear, rational, and explanatory relationship [31]. For example, each feature has a causal or explanatory relationship with credit performance, income, debt, and cost of living. "Prohibited Variables" that need to be adhered to, such as race, skin color, gender, marital status, and other discriminatory traits, are suggested to be removed from the credit risk assessment to produce an accurate prediction.

Feature selection is a technique to decrease the number of input variables in a model by using only relevant data and removing noise. Feature selection is a process that employs target variables to determine which variables can strengthen the model's performance. Therefore, independent and dependent variables must be separated during the feature selection process. This strategy is acknowledged for selecting appropriate features for ML models automatically based on the sort of issue that this research is attempting to solve. Significant characteristics are added or excluded without altering them during the process. This can assist in the reduction of data noise and input data size.

Table 2. Feature Selection for Credit Risk Prediction.

Feature	Description
Job	The job type is based on the capability of borrowers to pay the loan. One indicates unskilled jobs (wire pullers, shop assistants, etc.), 2 indicates skilled jobs (electricians, mechanics, etc.), and 3 shows highly skilled jobs (doctors, data scientists, etc.)
Housing	The house ownership that the borrowed currently resides, with the indicator of 0 for a free house, 1 for a rented house, and 2 for an owned house
Saving Account	Knowing how much money a borrower has on his/her account is necessary. It gives the lender reference on the capability of paying the loan. The dataset, German Credit Risk, uses the DM (Deutsche Mark) as its currency. Five categories are described as follows: (DM < 100), moderate (DM 100–500), quite rich (DM 500–1000), rich (DM > 1000), and NA (unknown/ no savings account)
Checking Account	This indicates the demand accounts or transactional accounts of customers. There are highly liquid accounts

(continued)

Table 2. (*continued*)

Feature	Description
Credit Amount	The amount of money that a borrower wants to borrow; the type of data used here is the whole number (integer)
Duration	The time span a borrower has to pay the loan; the unit used is months
Purpose	This is used to specify a borrower's reason behind the proposed loan. For instance, a borrower wants to purchase a car, send their children to school, run a business, or do something urgently
Risk	This is the dependent variable as an outcome of every record that has been processed. This determines the eligibility or worthiness of a borrower to get a credit loan. Only two selections are available, namely 'Good' and 'Bad'

Age and gender features are eliminated from the German Credit Risk dataset to avoid bias. Therefore, variables x (independent) are 'job,' 'housing,' 'saving account,' 'credit amount,' 'duration,' and 'purpose.' Meanwhile, the variable y (dependent) is 'risk,' which indicates the borrowers' creditworthiness. Table 2 shows the selected features from the dataset and their descriptions.

4   Results and Findings

This section covers the results and findings of the ML algorithms comparison using Python. The algorithms processed the dataset for credit risk prediction, and the details are as follows.

4.1   Data Pre-processing

Recently in the banking industry, the risk modeling process carried out by AI in the framework is now even clearer. Individual bias is perceived to give accurate results because it affects one's individual credit decision-making authority and rights, so this is rarely included in the assessment. The reason behind this is that credit managers often have emotional characteristics and biases in seeing the customer's situation [30]. However, it is not uncommon for some important individual judgments to be included for further consideration. The strong influence of characteristics and factors on decision-making and risk-taking has attracted the interest of economists and behavioral finance experts in the banking industry.

Through the dataset, several models are used in a number of different ways to see the approximate accuracy of the ML algorithm, and then one or two ways are chosen to solve it. The methods used produce different visualizations to show the average accuracy, variance, and other properties of the model's accuracy distribution. Identifying which features will be included in the final score model throughout the data pre-processing technique is essential. This research uses Numpy, Pandas, Seaborn, Matplotlib, and Scikit-learn (Sklearn). All these libraries will help machine learning's scientific computations.

Scikit-learn is implemented for data pre-processing since it supports most classification techniques. This research examines the null and non-null values in each row. The null value is only accessible for characteristics of savings and checking accounts. It was determined that the null value for savings accounts is 183, while the non-null value is 817. Therefore, the null value for the checking account is 394 and the value that is not null is 606. If a non-null value of 1,000, it means the remaining attributes are not null.

This research discovers that certain variables are categorical. There are six categorical features, which are gender, housing, savings account, checking account, purpose, and risk. If these feature variables are fed straight to the model during the data pre-processing, the model cannot comprehend them. All independent and dependent variables, or input and output characteristics, must be numeric for machines. This implies that if the data contains a category variable, during the data pre-processing, it must be converted to a numeric value before fitting the data to the model. The features of a customer's savings and checking accounts were selected as those that would characterize their creditworthiness. This research converts it depending on the specified value instead of having the machine perform the adjustment automatically. The value is encoded as "little" equal to 1; "moderate" equals 2; "quite rich" equals 3; "rich" equals 4, and the null value is considered as 0. The encoding continues with the value for the housing attributes: "free" equals 0, meaning that the customer does not own a house; "rent" equals 1, meaning that the customer rents a house; and "own" equals 2, meaning the customer owns a house. The other attributes, including gender, purpose, and risk, are encoded automatically.

4.2 Data Analysis

On the pre-processed dataset, different ML methods were evaluated following precise feature selection and suitable pre-processing. The primary aim was to identify the algorithm that provides the best accurate results in recognizing credit risk compared to its rivals. This was accomplished by determining the accuracy and standard deviation. The problem that has to be solved is known as a binary classification problem since the algorithms are designed to determine whether or not a person poses a credit risk.

Table 3. Cross Validation Results.

Algorithm	Accuracy	Standard Deviation (STDV)
Logistic Regression	69%	0.051
Linear Discriminant Analysis (LDA)	69%	0.054
Gaussian Naïve Bayes (NB)	69%	0.043
K-Nearest Neighbors Classifier (K-NN)	66%	0.045
Decision Tree Classifier	69%	0.038
Support Vector Machine	70%	0.067
Random Forest	**73%**	**0.033**

In this research, the three different random state integers are used to get different outputs during the data pre-processing. The first two integers, 0 and 1, give insignificant results with only a slight difference in the model accuracy and standard deviation. It is challenging to determine which model suits the best. Therefore, a random state value of 35 is used as a trial, which gives a result that satisfies the criteria required for this research. The results of ML algorithms' accuracy and standard deviation based on the assigned random state (i.e., 35) are shown in Table 3.

Table 3 summarizes the accuracy of each ML algorithm for credit risk prediction. The result shows that the best level of accuracy is 73%, using a Random Forest (RF) method. It also indicates that RF has the lowest standard deviation (STDV) with 0.033, which means that the individual data values are close to the center of the dataset, namely mean and median. STDV can also be used to measure the statistical confidence in the models. The comparison of ML algorithms' accuracy is visualized in Fig. 1.

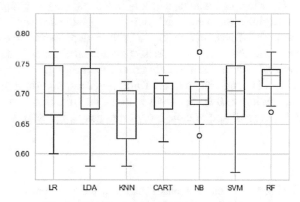

Fig. 1. Boxplot of the algorithm comparison.

As shown in Fig. 1, the line in the middle of the box plot of each result signifies the median values of the respective algorithm accuracy. Between those algorithm boxplots, RF has the highest median with a value of nearly 0.735. This is followed by the Support Vector Machine (SVM) in second place, with a median value of around 0.705, but it has quite a high STDV. Those values indicate that the RF algorithm has the highest median accuracy to use in credit risk assessment.

5   Discussion

After the completion of data pre-processing on the dataset relevant to German credit risk, the significance of hyperparameter tuning enables the implementation of various essential adjustments. Figure 1 demonstrates that the random state shifted in order to fulfill the predicted requirement. This was carried out identically to the prior discussion.

This research takes into consideration the precision of each model, as well as its standard deviation. The effectiveness of an algorithm may be evaluated by determining which features have the highest accuracy and which have the smallest STDV. The model with the highest accuracy is Random Forest, which has an accuracy rate of 73% and an STDV value of 0.033. On the other hand, the model with the lowest accuracy is the K-Nearest Neighbors Classifier (K-NN), which has an accuracy rate of 66% and an STDV value of 0.045.

Furthermore, a correlation between the dispersion and the distribution of the results is discovered in this research. As hypothesized, based on the boxplot in Fig. 1, the SVM approach produces a more extended boxplot than the rest. This is noticeable when comparing the boxplots side by side. According to the evidence shown here, the accuracy metrics for that approach are substantially more variant when compared to those for the other algorithms.

6 Conclusion

During the process of selecting features, it is concluded that the most important elements to use in the credit scoring system are 'job,' 'housing,' 'saving account,' 'credit amount,' 'duration,' and 'purpose.' These features serve as the independent variables in the analysis. On the other hand, the element of 'risk' becomes the dependent variable and is responsible for determining the creditworthiness of the borrowers. After reviewing the findings and results, it is concluded that the Random Forest algorithm is the most suitable for predicting credit risk. It is consistent with the fact that Random Forest is widely acknowledged to be a robust and reliable model that has a high degree of accuracy.

References

1. Hand, D.J., Henley, W.E.: Statistical classification methods in consumer credit scoring: a review. J. R. Stat. Soc. A. Stat. Soc. **160**(3), 523–541 (1997)
2. Nilsson, M., Shan, O.: Credit risk analysis with machine learning techniques in peer-to-peer lending market (2018)
3. Sakprasat, S., Sinclair, M.C.: Classification rule mining for automatic credit approval using genetic programming. In: IEEE Congress on Evolutionary Computation. IEEE (2007)
4. Turkson, R.E., Baagyere, E.Y., Wenya, G.E.: A machine learning approach for predicting bank credit worthiness. In: 3rd International Conference on Artificial Intelligence and Pattern Recognition (AIPR). IEEE (2016)
5. Ruin, J.E.: Internal auditing: Supporting risk management, fraud awareness management and Corporate Governance. Leeds Publications (M) (2009)
6. Sadok, H., Sakka, F., El Maknouzi, M.E.H.: Artificial intelligence and bank credit analysis: a review. Cogent Econ. Finan. **10**(1), 2023262 (2022)
7. Jorion, P.: The new benchmark for managing financial risk: value at risk (2001)
8. Fabozzi, F.J., Davis, H.A., Choudhry, M.: Introduction to Structured Finance, vol. 148. Wiley (2006)
9. Jentzsch, N.: Do we need a European Directive for Credit Reporting? CESifo DICE Report **5**(2), 48–54 (2007)
10. Eletter, S.F., Yaseen, S.G., Elrefae, G.A.: Neuro-based artificial intelligence model for loan decisions. Am. J. Econ. Bus. Administr. **2**(1), 27 (2010)
11. Leloup, L.: Blockchain: La révolution de la confiance. Editions Eyrolles (2017)

12. Chaisuwan, J., Chumuang, N.: Intelligent credit service risk predicting systems based on customer's behavior by using machine learning. In: 14th International Joint Symposium on Artificial Intelligence and Natural Language Processing (iSAI-NLP). IEEE (2019)
13. Islam, M., Lin, Z., Fei, L.: Application of artificial intelligence (artificial neural network) to assess credit risk: a predictive model for credit card scoring (2009)
14. Qasem, M.H., Nemer, L.: Extreme learning machine for credit risk analysis. J. Intell. Syst. **29**(1), 640–652 (2020)
15. Aleksandrova, Y.: Comparing performance of machine learning algorithms for default risk prediction in peer-to-peer lending. TEM J. **10**(1), 133–143 (2021)
16. Wang, Y., Zhang, Y., Lu, Y., Yu, X.: A comparative assessment of credit risk model based on machine learning – a case study of bank loan data. Procedia Comput. Sci. **174**, 141–149 (2020)
17. Teles, G., Rodrigues, J.J.P.C., Rabê, R.A., Kozlov, S.A.: Artificial neural network and Bayesian network models for credit risk prediction. J. Artific. Intell. Syst. **2**(1), 118–132 (2020)
18. West, D.: Neural network credit scoring models. Comput. Oper. Res. **27**(11–12), 1131–1152 (2000)
19. Karels, G.V., Prakash, A.J.: Multivariate normality and forecasting of business bankruptcy. J. Bus. Financ. Acc. **14**(4), 573–593 (1987)
20. Reichert, A.K., Cho, C.-C., Wagner, G.M.: An examination of the conceptual issues involved in developing credit-scoring models. J. Bus. Econ. Statist. **1**(2), 101–114 (1983)
21. Mitchell, D., Pavur, R.: Using modular neural networks for business decisions. Manage. Decision (2002)
22. Zhang, P.G., ed.: Neural Networks in Business Forecasting. IGI Global (2004)
23. Linoff, G.S., Berry, M.J.A.: Data Mining Techniques: For Marketing, Sales, and Customer Relationship Management. Wiley (2011)
24. Gupta, P.K.D., Krishna, P.R.: Database Management System Oracle SQL and PL/SQL. PHI Learning Pvt. Ltd. (2013)
25. Prince, N., et al.: Optimization of capacitive MEMS pressure sensor for RF telemetry. Int. J. Sci. Eng. Res. **2**(10), 1–4 (2011)
26. Tsymbal, A., Pechenizkiy, M., Cunningham, P.: Diversity in search strategies for ensemble feature selection. Inform. Fus. **6**(1), 83–98 (2005)
27. Bee, W.Y., Seng, H.O., Mohamed Husain, N.H.: Using data mining to improve assessment of credit worthiness via credit scoring models. Expert Syst. Appl. **38**(10), 13274–13283 (2011)
28. Davis, R.H., Edelman, D.B., Gammerman, A.J.: Machine-learning algorithms for credit-card applications. IMA J. Manag. Math. **4**(1), 43–51 (1992)
29. Ping, Y., Lu, Y.: Neighborhood rough set and SVM based hybrid credit scoring classifier. Expert Syst. Appl. **38**(9), 11300–11304 (2011)
30. Aithal, V., Jathanna, R.D.: Credit risk assessment using machine learning techniques. Int. J. Innov. Technol. Explor. Eng. **9**(1), 3482–3486 (2019)
31. Thomas, L.C., et al.: Readings in Credit Scoring: Foundations, Developments, and Aims. Oxford University Press on Demand (2004)

Forecasting Tourist Arrivals Using a Combination of Long Short-Term Memory and Fourier Series

Ani Shabri[1]([✉]), Ruhaidah Samsudin[2], Faisal Saeed[3], and Mohammed Al-Sarem[4]

[1] Department of Mathematical Sciences, Faculty of Science, Universiti Teknologi Malaysia, 81300 Skudai, Johor, Malaysia
ani@utm.my

[2] Faculty of Engineering, School Computing, Universiti Teknologi Malaysia, 81300 Skudai, Johor, Malaysia
ruhaidah@utm.my

[3] DAAI Research Group, Department of Computing and Data Science, School of Computing and Digital Technology, Birmingham City University, Birmingham B4 7XG, UK
faisal.saeed@bcu.ac.uk

[4] College of Computers and Engineering, Taibah University, Medina, Saudi Arabia
msarem@taibahu.edu.sa

Abstract. The sector that contributes most to the nation's economy nowadays is tourism. Policymakers, decision-makers, and organisations involved in the tourist sector can use tourism demand forecasting to gather important information for planning and making important decisions. However, it is difficult to produce an accurate forecast because tourism data is critical, especially when a periodic pattern, such as seasonality, trends, and non-linearity, is present in a dataset. In this research, we present a hybrid modelling approach for modelling tourist arrivals time series data that combines the long short-term memory (LSTM) with the Fourier series method. This method is proposed to capture the components of seasonality and trend in the data set. Various single models, such as ARIMA and LSTM, as well as a modified ARIMA model based on Fourier series, are evaluated to confirm the suggested model's accuracy. The efficiency of the proposed model is compared using monthly tourism arrivals data from Langkawi Island, which has a notable pattern and seasonality. The findings reveal that the proposed model is more reliable than the other models in forecasting tourist arrivals series.

Keywords: Fourier series · artificial neural network · long short-term memory · ARIMA · tourist arrival

1 Introduction

Tourism is a major source of economic activity, employment, and income generation. Accurate visitor arrivals forecasting is essential for stakeholders and researchers to make operational decisions such as providing appropriate operational funds, investing, money planning, and predicting future risks.

© The Author(s), under exclusive license to Springer Nature Switzerland AG 2023
F. Saeed et al. (Eds.): ICACIn 2022, LNDECT 179, pp. 160–170, 2023.
https://doi.org/10.1007/978-3-031-36258-3_14

For tourist forecasting, various statistical time series prediction approaches have been proposed in the past, including autoregressive integrated-moving average (ARIMA) [1–9], exponential smoothing [1–5], Nave [1–3] and moving average (MA) [4, 5]. Because of its high predicted precision and flexibility in representing a wide range of time series, the ARIMA model is one of the most widely used statistical models. However, due to its linearity, the ARIMA model is insufficient for modelling complex real-world time series with significant seasonality [10].

Artificial intelligence (AI) methods including support vector machines [1, 11, 12], fuzzy time series [14] and artificial neural networks (ANN) [1, 3, 5, 6, 8, 13] have been more popular in tourist arrival forecasting during the recent year. These AI-based models are capable of explaining non-linear data. Finding the best network model and training algorithms for ANN apps, on the other hand, remains a challenge [15]. In addition, more parameters must be created in order to develop the ANN model, which can lead to over-fitting and consequently bad predicting abilities [16].

Lately, it was discovered that Recurrent Neural Network (RNN) designs are better suited to coping with the intricacy of time series than ANN [17]. However, because to the issue of vanishing gradients in RNN training, new RNN variants, such as long short-term memory (LSTM), have been proposed. Because of this property, the LSTM model has been used to solve forecasting problems [18–20].

Despite the obvious advantages of ARIMA, ANN, and LSTM models, it is not always adequate to use any single model to identify time series data with complexities and seasonal fluctuations. Several previous research have recommended combining several prediction models to improve the performance of an single statistical model or artificial intelligence model. The combined model's main goal is to overcome the inadequacies of single models and establish a predictive synergy. Several studies have also shown that combining prediction models is an excellent approach in time series analysis, such as combining the SVR-ANN [2], ARIMA-ANN [1, 9, 21], ARIMA-regression [22], ARIMA-deep learning model [23], ANN-Grey–Markov [24], ARIMA-GARCH [21], Grey-Fourier series [25, 26] and ARIMA-Fourier series [27, 28].

As a result, the primary goal of this research is to assess the efficacy of a combined prediction model based on the LSTM and Fourier series in terms of improving prediction accuracy. A sample of monthly tourism arrivals on Malaysia's Langkawi Island was chosen for a data analysis study to assess the efficacy of the proposed approach.

2   Forecasting Models

This section describes forecasting models such as ARIMA, LSTM, and proposed combining models.

2.1   Arima Model

The ARIMA model was among the most commonly used time series models [29]. The structures of ARIMA(p, d, q)(P, D, Q)s for non-seasonal and seasonal compounds are as follows:

$$\phi_p(B)\Phi_P\left(B^s\right)(1 - B)^d\left(1 - B^s\right)^D x_t = \theta_q(B)\Theta_Q\left(B^s\right)a_t \tag{1}$$

where d and D denote the degree of non-seasonal and seasonal differentiation, respectively, s is the season's length and a_t is the errors. $\varphi_p(B)$ and $\theta_q(B)$ are the order p and q parameters of the autoregressive and moving average, respectively.; $\Phi_P(B^s)$ and $\Theta_Q(B^s)$ are the order P and Q parameters of the seasonal autoregressive and moving average, respectively; The ARIMA was developed in four stages, according to reference [29]: identification, estimating parameter, diagnostic procedures, and forecasting.

The ARIMA models in this study were discovered with R software, which was built on the Forecast package [30]. The seasonal unit root test and the Osborn-Chui-Smith-Birchenhall test [31] are used to determine the number of d and D. The Akaike Information Criterion (AIC) is used to determine the order values of p, q, P, and Q. The best ARIMA model for modelling is the one with the lowest AIC value.

2.2 LSTM Model

Hochreiter and Schmidhuber introduced the LSTM model [32] and has been used to achieve well-known results for many sequential data problems. LSTMs learn long-term dependencies using a mechanism known as gates. These gates can decide which data in a sequence should be kept and which should be disposed. An LSTM's three gates are input, forget, and output. The construction of an LSTM cell is shown in Fig. 1 [33]. The cell state is the horizontal line between C_{t-1} and C_t. The heart of the LSTM model, where data is added and deleted from memory via pointwise addition and multiplication. The LSTM block's input and forget gates, as well as the output "tanh" activation function, are used to conduct these tasks. The following are the computations inside the LSTM neurons:

$$\text{Forget gate: } f_t = \sigma(W_f.\left[h_{t-1}, x_t\right] + b_f) \tag{2}$$

$$\text{Input gate: } i_t = \sigma(W_i.\left[h_{t-1}, x_t\right] + b_i \tag{3}$$

$$\text{Process input: } \tilde{C}_t = tanh(W_C.\left[h_{t-1}, x_t\right] + b_C \tag{4}$$

$$\text{Cell update: } C_t = f_t \times C_{t-1} + i_t \times \tilde{C}_t \tag{5}$$

$$\text{Output gate: } o_t = \sigma(W_o.\left[h_{t-1}, x_t\right] + b_o \tag{6}$$

$$\text{Output:} h_t = o_t tanh(C_t) \tag{7}$$

where b_f, b_o, b_i and W_f, W_o, W_i are the bias and weight matrices of the forget, output and input gates, respectively. b_C and W_C are the bias and weights of the cell state. h_t calculates the final outputs σ is sigmoid function, the previous cell state's output is h_{t-1}, the current cell state's input is x_t and o_t determines whether part of the cell state should be exported.

The forget gates are formed by combining the present time step input with the prior time step's hidden state, as shown in Eq. 2. Equation 3 is an input gate that directs cell state adjustments to all neural net cells. Equation 4 and 5 upgrade the cell states by transferring fresh information and the previous cell state from Eq. 2 and 3 to Eq. 5. Equation 6 and 7 are being used to generate the current time step's hidden state by combining the hidden and updated cell states from the previous time step from Eq. 5.

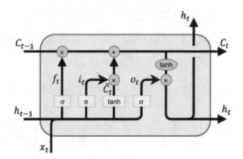

Fig. 1. Structure of LSTM model

The loss function was mean squared error, and the "Adam" optimization technique [33] has been used to identify the best weight values for the networks. The data was normalised to a 0 to 1 scale. The function used to normalise the dataset in this investigation is described by Eq. (8):

$$y_t = \frac{x_t}{x_{max}} \tag{8}$$

where x_t is the actual value, y_t is the normalized time series, and x_{max} is the maximum actual value.

2.3 The Fourier LSTM Model

Fourier series have been applied to improve forecast model precision by modifying forecast model residuals. The methodology for developing a modified forecasting method known as the Fourier LSTM (FLSTM) model is as follows:

Step 1: Fit the LSTM model to a data set y_t.
Step 2: Create the residual series, $-e_t = y_t - \hat{y}_t$ based on the LSTM predicted values $\hat{y}_t$.
Step 3: Fit the Fourier series to a residual e_t as shown below.

$$e_t = \frac{a_0}{2} + \sum\nolimits_{t=1}^{M} a_t cos\left(\frac{2tk\pi}{n-1}\right) + b_t sin\left(\frac{2tk\pi}{n-1}\right) \tag{9}$$

where $M = \frac{n-2}{2} - 1, e = AB,$

$A = (0.5P_1 \ldots P_k \ldots P_F),$

$$B = (a_0, a_1, b_1, a_2, b_2, \ldots, a_M, b_M)^T$$

$$A_k = \begin{pmatrix} cos\left(\frac{2\times 2\pi k}{n-1}\right) & sin\left(\frac{2\times 2\pi k}{n-1}\right) \\ cos\left(\frac{3\times 2\pi k}{n-1}\right) & sin\left(\frac{3\times 2\pi k}{n-1}\right) \\ \vdots & \vdots \\ cos\left(\frac{n\times 2\pi k}{n-1}\right) & sin\left(\frac{n\times 2\pi k}{n-1}\right) \end{pmatrix} \tag{10}$$

The ordinary least squares (OLS) method is used to estimate the parameters B.

$$B = (A^T A)^{-1} A^T e^T \tag{11}$$

The estimated residual by the Fourier series can be calculated using the following equation.

$$\hat{e}_t = \frac{a_0}{2} + \sum_{t=1}^{M} a_t cos\left(\frac{2tk\pi}{n-1}\right) + b_t sin\left(\frac{2tk\pi}{n-1}\right) \tag{12}$$

Step 4: Finally, the predicted values of the FLSTM model can be calculated by adding the predicted values of the LSTM model to the estimated residual from the Fourier series.

$$\tilde{y}_t = \hat{y}_t + \hat{e}_t \tag{13}$$

3   Examining The Performance

The proposed model's efficiency was measured using the mean absolute percentage error (MAPE) and root mean square error (RMSE).

$$MAPE = \frac{1}{n} \sum_{i=1}^{n} \left| \frac{y_i - \hat{y}_i}{y_i} \right|$$

$$RMSE = \sqrt{\frac{1}{n} \sum_{i=1}^{n} (y_i - \hat{y}_i)^2}$$

where y_i, $\hat{y}_i$ and n denote the actual, predicted, and overall amount of datasets, respectively. The best forecasting model is the one with the lsmallest MAPE and RMSE values.

3.1   Analysis Data

This paper analysed tourist numbers arrivals to Langkawi Island in Malaysia, as reported on the Langkawi Development Authority (LADA) Malaysia website. This study focuses on examining the monthly tourism arrivals from January 2002 to December 2016. The training data covers the period from January 2002 to December 2015, whereas the testing data covers the period from January 2016 to December 2016. Figure 3 shows the monthly tourism arrivals to Langkawi Island (January 2002-December 2016). The monthly tourism arrivals curve, as seen in Fig. 3, has a trend, is non-stationary, and has a seasonal pattern (Fig. 2).

ARIMA, LSTM, and FARIMA models were used in this study to evaluate the efficacy of the proposed model.

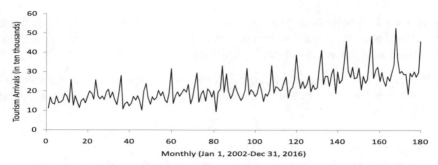

Fig. 2. Monthly visitor arrivals on Malaysia's Langkawi Island from 2002 to 2016.

3.2 The ARIMA Model Fit to the Data

The ARIMA model was identified and optimised using R Forecast package [37]. By default, this package specifies the maximum order for d, P and Q to 2, D to 1 and p and q to 5. The ARIMA model parameters were estimated using the maximum likelihood estimate (MLE) and the Ljung-Box (LB) test was used to confirm that the model selected was suitable for the data. The AIC is used to choose values for p, P, q, and Q, while the Kwiatkowski-Phillips-Schmidt-Shin (KPSS) is applied to choose values for d and D.

The AIC and p-value of the Ljung Box test for numerous ARIMA models are summarised in Table 1. The five models considered were found to be acceptable by the Ljung-Box test, with ARIMA(1,0,2)(0,1,1)12 having the smallest AIC values. The model can be expressed as follows:

$$x_t = 941.745 + x_{t-12} + 0.771x_{t-1} - 0.771x_{t-13} + (1 - 0.810B + 0.287B^2)(1 - 0.330B^{12})a_t$$

The ARIMA model's AR term is as follows:

$$x_t = f(x_{t-1}, x_{t-12}, x_{t-13})$$

Table 1. AIC and p-values of Ljung-Box test for ARIMA models

No	ARIMA	p-value of LB Test			
	(p,d,q)(P,D,Q)12	lag = 12	lag = 24	lag = 36	AIC
1	$(2,0,2)(0,1,1)_{12}$	0.977	0.996	0.219	3406.7
1	$(2,0,3)(0,1,1)_{12}$	0.999	0.999	0.281	3405.6
2	$(1,0,2)(0,1,2)12$	0.997	0.996	0.226	3406.4
4	$(1,0,2)(0,1,1)_{12}$	0.997	0.996	0.226	3404.3
5	$(1,0,3)(0,1,1)_{12}$	0.997	0.996	0.207	3406.4

Bold values indicate the smallest AIC.

Input Time Lag Selection: The selection of significant input variables is one of the most critical processes in the LSTM model generation process. The input variables were determined using three ways in this investigation.

i. Packard et al. [34] suggested in the selection of the input variables, the number of nodes in the input layer equals the number of lag variables. The y_t output is given as

$$y_t = f(y_{t-1}, y_{t-2}, \ldots, y_{t-p})$$

where p is a the number of period for seasonal.

ii. The lagged variables of the ARIMA model are the most significant variables to use as input variables for AI models [35]. The AR term from the ARIMA model is defined in Table 2 as

$$x_t = f(x_{t-1}, x_{t-12}, x_{t-13})$$

iii. Prastyo et al. [36] suggested PACF in determination input variables of AI models.

Figure 4 and 5 show the decomposition plot and PACF graph. Through observing Fig. 4, it is obvious that the data has a seasonal pattern with seasonal period is 12. Figure 5 shows that the PACF significance at lags 1, 2, 3, 6, and 11 to 15.

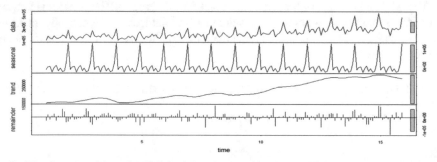

Fig. 3. The decomposition plot (Original data (top) and its three additive components (bottom)).

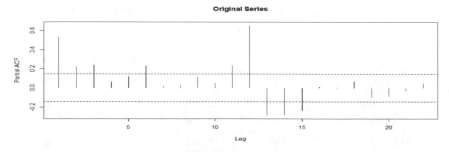

Fig. 4. PACF of original data

As a result of the decomposition plot, PACF and AR terms, the LSTM model input may be expressed as shown in Table 2.

Table 2. Input variables for LSTM model

Model	Method	Input
LSTM1	Period 12	$x_t = f(x_{t-1}, x_{t-2}, \ldots, x_{t-12})$
LSTM2	AR	$x_t = f(x_{t-1}, x_{t-12}, x_{t-13})$
LSTM3	PACF	$x_t = f(x_{t-1}, x_{t-2}, x_{t-3}, x_{t-6}, x_{t-11},$
		$x_{t-12}, x_{t-13}, x_{t-14}, x_{t-15})$

3.3 LSTM Model Fit to the Data

During the training phase of this study, we examined all possible combinations to find the best parameters for LSTM1, LSTM2, and LSTM3. The number of epochs is 300, the batch size is 4, and the Adam optimization algorithm [37] with a default-learning rate of 0.01 is used throughout this study. The mean square error is used as the loss function.

The testing dataset gave RMSEs of 44367.33, 61411.02, and 55492.73 for the LSTM1, LSTM2, and LSTM3 models, respectively. LSTM1, LSTM2, and LSTM3 have MAPEs of 11.462, 19.407, and 16.549, respectively. The LSTM2 model has the lowest RMSE and MAPE values and show performs better in forecasting.

3.4 The LSTM and Fourier Series Fit to the Data

The residual series from the best ARIMA and LSTM models were then combined with Fourier series to form the Fourier-ARIMA (FARIMA) and Fourier-LSTM (FLSTM) models, respectively. Table 3 demonstrates the RMSE and MAPE performance of the ARIMA and LSTM models, as well as their Fourier series. In Table 3, the ARIMA, LSTM, FARIMA, and FLSTM models had RMSEs of 49319.52, 44367.35, 34113.20, and 29148.60, respectively. ARIMA, LSTM, FARIMA, and FLSTM models had MAPEs of 14.64%, 11.46%, 10.92%, and 9.28%, respectively.

The forecasting accuracy of the single ARIMA and LSTM models improved when the Fourier residual estimation was used, as shown in Table 3. The FLSTM model had the best overall testing phase performance (RMSE = 29148.60 and MAPE = 9.28%). This shows that the FLSTM model has fewer errors than single ARIMA and LSTM models, as well as a combination model, the FARIMA model. According to the RMSE and MAPE criteria, the FLSTM model has a strong forecasting ability. According to the findings above, the FLSTM model received the highest ranking of all the models considered.

The real data versus the ARIMA, FARIMA, LSTM, and FLSTM models' forecasted values are presented in Fig. 5. In comparison to other approaches, the FLSTM model performed excellently and closely followed the true data. The findings suggest that the FLSTM models can accurately anticipate the amount of tourists arrivals. The empirical findings indicate that the FLSTM model may be successful in improving model forecasting accuracy in a series of seasonal monthly tourist arrival to Langkawi Island.

Table 3. Model performance for monthly tourism arrivals data

Model	RMSE	MAPE (%)
ARIMA	49319.52	14.64
LSTM	44367.35	11.46
FARIMA	34113.20	10.92
FLSTM	29148.60	9.28

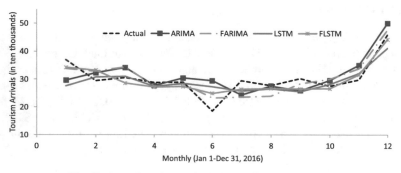

Fig. 5. Actual tourism arrivals versus forecasted values

4 Conclusion

In this study, we set out to forecast the number of visitors to the Langkawi region using two traditional methods. To solve the shortcomings of each method, we then combined the two methods in hybrid models utilising Fourier series. The results demonstrated that the LSTM performed better than the ARIMA technique, which produced substantial output values and accurately represented the pattern of the actual data. The hybrid model that combined LSTM and Fourier series was consistently better accurate when compared to other combination technique and the traditional techniques. These findings show that the ARIMA or LSTM model alone cannot be used to forecast visitor arrivals. However, as each technique captures a certain characteristic of the data, researchers should combine different techniques for results with high significance in order to overcome each technique's limitations. More study might examine additional hybrid models, such as those that employ alternative weighting techniques when combining more than two models in an ensemble learning technique. Other research might compare various ensemble learning methodologies when employing mixed-frequency data, which has emerged as the new standard in forecasting tourist demand.

Acknowledgment. The authors are grateful to Universiti Teknologi Malaysia (UTM) and the Ministry of Higher Education Malaysia (MOHE) for their support of this project through the Fundamental Research Grant Scheme (R.J130000.7854.5F271).

References

1. Chen, K.Y.: Combining linear and nonlinear model in forecasting tourism demand. Expert Syst. Appl. **38**(8), 10368–10376 (2011)
2. Cang, S., Yu, H.: A combination selection algorithm on forecasting. Eur. J. Oper. Res. **234**(1), 127–139 (2014)
3. Xu, X., Law, R., Chen, W., Tang, L.: Forecasting tourism demand by extracting fuzzy Takagi Sugeno rules from trained SVMs CAAI. Trans. Intell. Technol. **1**(1), 30–42 (2016)
4. Önder, I., Gunter, U.: Forecasting tourism demand with Google Trends for a major European city destination. Tour. Anal. **21**(2–3), 203–220 (2016)
5. Hassani, H., Silva, E.S., Antonakakis, N., Filis, G., Gupta, R.: Forecasting accuracy evaluation of tourist arrivals. Ann. Tour. Res. **63**, 112–127 (2017)
6. Díaz, M.A., Nadal, J.R.: Forecasting British tourist arrivals in the Balearic Islands using meteorological variables. Tourism Econ. **16**(1), 153–168 (2010)
7. Park, S., Lee, J., Song, W.: Short-term forecasting of Japanese tourist inflow to South Korea using Google trends data. J. Travel Tour. Mark. **34**(3), 357–368 (2017)
8. Li, S., Chen, T., Wang, L., Ming, C.: Effective tourist volume forecasting supported by PCA and improved BPNN using Baidu index. Tour. Manage. **68**, 116–126 (2018)
9. Nor, M.E., Nurul, A.I., Rusiman, M.S.: A hybrid approach on tourism demand forecasting. J. Phys. Conf. Ser. **995**(1), 1–11 (2018)
10. Neto, P.S.G., Cavalcanti, G.D.C., Madeiro, F.: Nonlinear combination method of forecasters applied to PM time series. Pattern Recogn. Lett. **95**(1), 65–72 (2017)
11. Hong, W.C., Dong, Y., Chen, L.Y., Wei, S.Y.: SVR with hybrid chaotic genetic algorithms for tourism demand forecasting. Appl. Soft Comput. **11**(2), 1881–1890 (2011)
12. Chen, R., Liang, C.Y., Hong, W.C., Gu, D.X.: Forecasting holiday daily tourist flow based on seasonal support vector regression with adaptive genetic algorithm. Appl. Soft Comput. **26**, 435–444 (2015)
13. Law, R., Au, N.: A neural network model to forecast Japanese demand for travel to Hong Kong. Tour. Manage. **20**(1), 89–97 (1999)
14. Hadavandi, E., Ghanbari, A., Shahanaghi, K., Abbasian-Naghneh, S.: Tourist arrival forecasting by evolutionary fuzzy systems. Tour. Manage. **32**(5), 1196–1203 (2011)
15. Maier, H.R., Dandy, G.C.: Application of artificial neural networks to forecasting of surface water quality variables: issues, applications and challenges. In: Govindaraju, R.S., Ramachandra Rao, A. (eds.) Artificial Neural Networks in Hydrology, pp. 287–309. Springer Netherlands, Dordrecht (2000). https://doi.org/10.1007/978-94-015-9341-0_15
16. Sun, W., Zhang, X.: Application of self-organizing combination forecasting method in power load forecast. In: Proceedings of the 2007 International Conference on Wavelet Analysis and Pattern Recognition, Beijing, China (2007)
17. Barbounis, T.G., Teocharis, J.B., Alexiadis, M.C., Dokopoulos, P.S.: Long-term wind speed and power forecasting using local recurrent neural network models. IEEE Trans. Energy Convers. **21**(1), 273–284 (2006)
18. Brownlee, J.: Time Series Prediction with LSTM Recurrent Neural Networks in Python with Keras (2016). https://machinelearningmastery.com/time-series-prediction-lstm-recurrent-neural-networks-python-keras/
19. Gers, F.A., Schmidhuber, J., Cummins, F.: Learning to Forget: Continual Prediction with LSTM. Neural Comput. **12**(10), 2451–2471 (2000)
20. Hochreiter, S., Schmidhuber, J.: Long short-term memory. Neural Comput. **9**(8), 1735–1780 (1997)
21. Coshall, J.T.: Combining volatility and smoothing forecasts of UK demand for international tourism. Tour. Manage. **30**(4), 495–511 (2009)

22. Chu, F.L.: "Forecasting tourism: a combined approach. Tour. Manage. **19**(6), 515–520 (1998)
23. Chen, Y., He, K., Tso, G.K.F.: Forecasting crude oil prices: a deep learning based model. Procedia Comput. Sci. **122**, 300–307 (2017)
24. Hu, Y.C., Jiang, P., Lee, P.C.: Forecasting tourism demand by incorporating neural networks into Grey–Markov models. J. Oper. Res. Soc. (2018). https://doi.org/10.1080/01605682.2017.1418150
25. Wang, Z.X., Pei, L.L.: Forecasting the international trade of Chinese high-tech products using an Fourier Nash nonlinear grey Bernoulli model. In: 2014 Seventh International Joint Conference on Computational Sciences and Optimization (2014)
26. Chia-Nan, W., Van-Thanh, P.: An improved nonlinear grey bernoulli model combined with fourier series. Mathem. Probl. Eng. ID 740272, pp. 1–7 (2015)
27. Huang, T.F., Chen, P.J., Nguyen, T.L.: Forecasting with fourier residual modified ARIMA Model- an empirical case of inbound tourism demand in New Zealand. Recent Res. Appl. Econ. Manage. **2**, 61–65 (2013)
28. Nguyen, T.L., Chen, P.J., Shu, M.H., Hsu, B.M., Lai, Y.C.: Forecasting with fourier residual modified arima model- the case of air cargo in Taiwan. In: Proceeding of 2013 International Conference on Technology Innovation and Industrial Management, Phuket Thailand, 2013
29. Box, G.E.P., Jenkins, G.M.: Time Series Analysis: Forecasting and Control. Holden-Day, San Francisco (1976)
30. Hydman, R.J., Khandakar, Y.: Automatic time series forecasting: the forecast package for R. J. Stat. Softw. **27**(3), 1–22 (2008)
31. Osborn, D.R., Smith, A.P.L., Birchenhall, C.R.: Seasonality and the order of integration for consumption. Oxford Bull. Econ. Statist. **50**(4), 361–377 (1988)
32. Patterson, J.: Deep Learning: A Practitioner's Approach, OReilly Media (2017)
33. Wegayehu, E.B., Muluneh, F.B.: Short-term daily univariate streamflow forecasting using deep learning models. Adv. Meteorol. **1860460**, 1–21 (2022)
34. Packard, N.H., Crutchfield, J.P., Farmer, J.D., Shaw, R.S.: Geometry from a time series. Phys. Rew. Lett. **45**(9), 712–716 (1980). https://doi.org/10.1103/physrevlett.45.712
35. BuHamra, S., Smaoui, N., Gabr, M.: The Box-Jenkins analysis and neural network: prediction and time series modeling. Appl. Math. Model. **27**, 805–815 (2003)
36. Prastyo, D.D., Nabila, F.S., Suhartono, M.H., Lee, N.S., Fam, S.F.: VAR and GSTAR-based feature selection in support vector regression for multivariate spatio-temporal forecasting. In: Yap, B.W., Mohamed, A.H., Berry, M.W. (eds.) Soft Computing in Data Science: 4th International Conference, SCDS 2018, Bangkok, Thailand, August 15-16, 2018, Proceedings, pp. 46–57. Springer Singapore, Singapore (2019). https://doi.org/10.1007/978-981-13-3441-2_4
37. Zhang, R., Song, H., Chen, Q., Wang, Y., Wang, S., Li, Y.: Comparison of ARIMA and LSTM for prediction of hemorrhagic fever at different time scales in China. PLoS ONE **17**(1), 1–14 (2022)

Age and Gender Classification from Retinal Fundus Using Deep Learning

Tareq Obaid[1]([⊠]) [iD], Samy S. Abu-Naser[1] [iD], Mohanad S. S. Abumandil[2] [iD],
Ahmed Y. Mahmoud[1] [iD], and Ahmed Ali Atieh Ali[3] [iD]

[1] Faculty of Engineering and IT, Alazhar University, Gaza, Palestine
{tareq.obaid,abunaser,ahmed}@alazhar.edu.ps
[2] Faculty of Hospitality, Tourism and Wellness, Universiti Malaysia Kelantan, Kota Bharu,
Malaysia
moha-nad.ssa@umk.edu.my
[3] School of Technology and Logistics Management, Universiti Utara Malaysia UUM,
Kedah 06010 Sintok, Malaysia

Abstract. Since the rise of social media, age and gender classification have become increasingly important in a growing number of applications. Existing approaches, however, still have low accuracies when applied to real-world retinal fundus images. In the current study, we show that using Deep Learning (DL) to train representations can result in a considerable improvement in the age and gender classification performance. We present a DL model based on the pretrained model named Xception. 26,000 retinal fundus images from the Kaggle library were collected for training the model suggested. The data was preprocessed before being divided into three parts (training, validation, and testing). The DL Xception model was assessed using the test data once it had been trained and cross-validated. The test results indicate that the ROC measure is 1.0, precision is 98.62%, recall is 98.62%, and f1-score is 98.61%, whilst accuracy is 98.62%. There is a prevalent non-awareness among clinicians regarding the changes in retinal variable variances among age and gender, emphasizing on the necessity of model explain ability of the age and gender classification of the images of retinal fundus. DL may assist clinicians to uncover new visions and illness biomarkers using the proposed method.

Keywords: Gender · Age · Deep Learning · Classification · Retinal Fundus

1 Introduction

DL is an Artificial Intelligence (AI) category which allows the autonomous learning and improvement of systems without depending on explicit programming. Machine learning entails the development of computer models which can autonomously access and utilize data. The learning starts by annotations or data like direct commands or experience that enable the identification of data patterns towards making better future decisions by the presented samples. The main objective is to teach a computer to absorb and perform like people, and eventually enable them to improve their learning on their own, by feeding those facts and information from observations and interactions in the real world [1, 2].

© The Author(s), under exclusive license to Springer Nature Switzerland AG 2023
F. Saeed et al. (Eds.): ICACIn 2022, LNDECT 179, pp. 171–180, 2023.
https://doi.org/10.1007/978-3-031-36258-3_15

Deep learning is an AI-based machine learning which reproduces the way humans acquire knowledge. Deep learning exists in data science, also encompassing statistics and predictive modeling. Data scientists benefit tremendously from deep learning when they are tasked to gather, analyze, and interpret large volumes of data as it expedites and streamlines the whole procedure [3, 4].

The optic disc, retina, macula, posterior pole, and fovea are all part of the eye fundus that are located opposite the lens of the eye. Ophthalmoscopy and/or fundus photography can be used to study the fundus. The fundus color varies between and within species. Human fundus (from blond people) was red in a study of primates only whereas the rest are yellow, blue, red, orange and green. The regularity and size of the macular area edge, the shape and size of the optic disc, retina seeming texturing, and pigmentation of retina were all noticed as noteworthy distinctions among the "higher" primate species [5, 6].

Fundus photos of the right eye, image on the left, and left eye the image on the right, is viewed from the forward-facing (nose of the person lie between both images if they were looking at the viewer). There are no signs of illness or pathology in any of the Fundus. Each image has in the middle the macula, and optic disc located towards the nose, because they are staring into the camera. Both optic discs show pigmentation across the lateral border, which is not considered abnormal [7]. The left photo reveals brighter patches around greater vessels, which has been thought to be a common occurrence in people that are younger (Fig. 1).

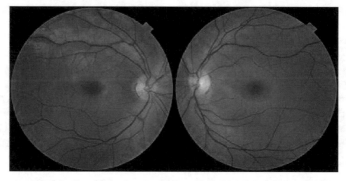

Fig. 1. Photographs of fundus for both eyes (right eye is the left image) and (left eye is the right image)

DL have recently demonstrated accuracy that is physician like level in a variety of health tasks that involves images, including radiology [7], dermatology [8], pathology [9], and ophthalmology [10]. Furthermore, DL has been found to perform well in activities that are difficult for clinicians to accomplish, such as properly predicting gender from retinal pictures [11]. Ophthalmologists are not specifically trained for this activity because it is rarely therapeutically relevant. However, ophthalmologists' low performance shows that age-gender variations in fundus images are neither visible nor significant. Even though the macula, retinal blood vessels and optic disc have been suggested as possible locations for gender-related structural variations in fundus images by saliency maps [7] and follow-up investigations [12, 13], definitive proof is still lacking.

As a result, DL's strong gender prediction ability has piqued the interest of the community of medical imaging, as DL hopes to uncover biomarkers which are difficult to be detected by humans[18–20].

On a vast set of images of retinal fundus gathered from Kaggle library, we trained the customized pre-trained Xception DL model [21–24].

2   Related Work

2.1   Gender Classification

Previously, researchers have utilized either basic CNN architectures or simple logistic regression aside from expert-delineated characteristics to detect gender from fundus images. For instance, [7] used the UK Biobank dataset to train Inception-v3 networks for predicting from fundus images risk factors cardiovascular and discovered that the CNNs can estimate the gender of a patient with (AUC = 0.97). [8] employed similar network. The authors of both research used saliency maps that is post hoc to investigate the variables that drive the decisions of network. [7] discovered that the vessels, optic disc,, and other non-specific areas in the images that were commonly underlined in a sample that consists of 100 attention maps. Yet, this appears to be the issue of practically all dependent features, making testing hypotheses for gender differences extremely difficult. Similarly, [9] manually examined occlusion maps sample and concluded that CNNs can detect gender using geometrical characteristics of the vessels of the blood near the disc of optic. [13] showed that the model of CNNs can identify gender from retinal fundus images and from OCT scans too, with the pit region of foveal appearing to be the most informative based on gradient-oriented saliency maps. [8] employed specialist image variables in a basic LR model to take a new approach. Despite their model's poor performance AUC (0.78), and it was discovered that several metrics based-color-intensity and the angle among particular arteries of retinal were meaningful predictors, nonetheless the utmost of the influenced sizes were minor.

2.2   Age Classification

Previous age classification studies have relied on facial images rather than retinal fundus images. Twenty years ago, researchers began researching age classification and estimation using facial photos. Early reviews focused on facial analysis techniques that involved determining correlations between a face's geometric elements [14]. Face characteristics such as nose, eyes, mouth, and chin, among others, are first determined, followed by the calculation of feature sizes and distances, as well as the ratio between them, for use in facial image categorization. These approaches are unsuccessful in age categorization into precise age groups since they can classify general age categories like children, adults, and older men. The work in [15], on the other hand, proposed AGES, a new paradigm for automatic age estimation. The idea was to create an aging model based on a set of individual face images organized by time. Projection onto the subspace was used to derive the aging pattern of the hidden facial picture. [16] employed a new paradigm in which facial contours were defined as points on a Grassmann manifold. This method used landmark-based shape investigation and produced excellent results in face verification and age estimation [17]. Deep learning algorithms have recently undergone a significant advancement, making it feasible to represent and extract the properties of any image or dataset, regardless of size.

3   Used Methods

3.1   Preprocessing of Dataset

Kaggle platform [18] provides access to a massive and multimodal archive that are related to datasets. 26,000 retinal fundus images were collected as a result of this. All pictures were reduced to 300 pixels by 300 pixels in size (see Fig. 2).

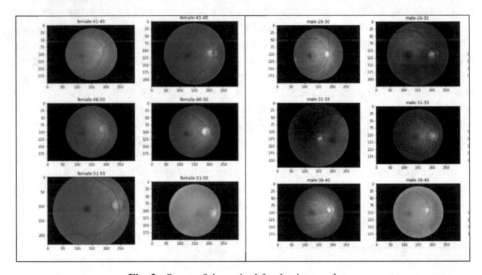

Fig. 2. Some of the retinal fundus image classes

The dataset was split into three parts: train, valid, and test, with a ratio of 60:20:20 percent of the photos in each. Age was divided into groups. Except for the first group, which contains ten years, each group has five years in order (0–10). Subsequently, as shown in Table 1, the groups 16–20 in female and 0–10 in female were eliminated since they were empty (no photos). As a result, our dataset has 26 classes.

3.2   Architecture of the Network

Keras implementation of the Xception network was used [2] as suggested by Francois Chollet. The Xception model was a development of the Inception model that depends on Separable Convolutions depth-wise to take over the normal Inception model. The Xception model consists of a 71-layer deep CNN. Pre-training was conducted on the network to enable it to classify 1000 images of different objects like mice, an eraser,

Table 1. Datasets distribution

Gender	Age group in years	Training size	Validation size	Testing size
Female	0–10	600	200	200
	11–15	600	200	200
	21–25	600	200	200
	26–30	600	200	200
	31–35	600	200	200
	36–40	600	200	200
	41–45	600	200	200
	46–50	600	200	200
	51–55	600	200	200
	56–60	600	200	200
	61–65	600	200	200
	66–70	600	200	200
	71- and-more	600	200	200
Male	11–15	600	200	200
	16–20	600	200	200
	21–25	600	200	200
	26–30	600	200	200
	31–35	600	200	200
	36–40	600	200	200
	41–45	600	200	200
	46–50	600	200	200
	51–55	600	200	200
	56–60	600	200	200
	61–65	600	200	200
	66–70	600	200	200
	71- and-more	600	200	200
Total	26 classes	15,600	5,200	5,200

and various types of animal. Consequently, the network was trained to learn numerous representations of various images.

At first, the images go through the entrance course, followed by the middle course eight times, and lastly via the exit course. Traditionally, Xception has outclassed VGG-16, ResNet, and Inception V3 (it is noteworthy that batch normalization follows all the Convolution and Separable Convolution layers) (see Fig. 3) [3].

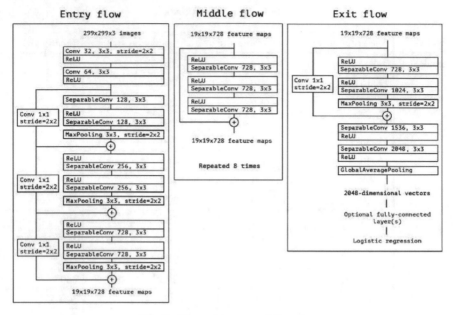

Fig. 3. The architecture of Xception

The developers of the Xception networks pre-trained them using ImageNet [36]. We updated the top layer which represents the classifier which classifies 1000 objects with a new classifier that classifies 26-objects with a softmax output layer for our current Age-Gender classification problem. We used the retinal fundus images to train only the output layer for 20 epochs. After that, all layers were fine-tuned for 100 cycles. Adam optimization algorithm was used, that is stochastic gradient descent extension widely adopted at present for DL applications in natural language processing and computer vision of which was designed explicitly in the training of deep neural networks (DNN). The control of adaptable methods of learning rates is leveraged by these algorithms in finding each parameter's individual learning rates. It has the benefits of Adagrad that performs fine in situations by gradients of the sparse but face difficulties in nonconvex neural network optimization and RMSprop that addresses some Adagrad's issues and performs well in online environments [4].

4   Result and Discussion

The proposed model was trained and validated for 120 epochs. Figures 4 and 5 demonstrate the results for train accuracy (100%), train loss (0.0001), valid accuracy (98.83%), and Valid Loss (0.0730) of the proposed model for the final 20 epochs.

The roc curve of each class of (Age-Gender) in the model obtained 100% for all 26 classes. The following is the model's classification report whereby Accuracy (98.62%), Recall (98.62%), Precession (98.62%), and F1-Score (98.61%) (See Table 2).

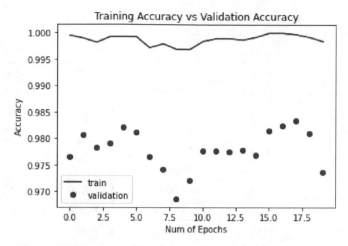

Fig. 4. Training and validating accuracies for the proposed model

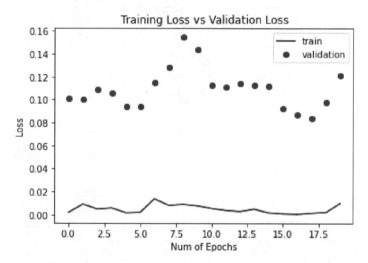

Fig. 5. Training and validating losses for the proposed model

The accuracy of our suggested model was good in terms of F1-score, Recall, and Precision; however, the dataset we used in this investigation was not the same as that used in earlier studies.

Table 2. Report classification of the proposed model

Class	Precision	Recall	F1-score	Support
female-0–10	100.00%	100.00%	100.00%	200
female-11–15	100.00%	100.00%	100.00%	200
female-21–25	100.00%	100.00%	100.00%	200
female-26–30	100.00%	100.00%	100.00%	200
female-31–35	100.00%	100.00%	100.00%	200
female-36–40	99.50%	100.00%	99.75%	200
female-41–45	99.50%	100.00%	99.75%	200
female-46–50	99.50%	99.50%	99.50%	200
female-51–55	97.40%	93.50%	95.41%	200
female-56–60	97.04%	98.50%	97.77%	200
female-61–65	97.40%	93.50%	95.41%	200
female-66–70	99.49%	98.50%	98.99%	200
female-71-more	97.09%	100.00%	98.52%	200
male-11–15	100.00%	100.00%	100.00%	200
male-16–20	100.00%	100.00%	100.00%	200
male-21–25	100.00%	100.00%	100.00%	200
male-26–30	100.00%	100.00%	100.00%	200
male-31–35	100.00%	99.50%	99.75%	200
male-36–40	99.50%	100.00%	99.75%	200
male-41–45	99.50%	99.00%	99.25%	200
male-46–50	95.65%	99.00%	97.30%	200
male-51–55	98.49%	98.00%	98.25%	200
male-56–60	93.81%	91.00%	92.39%	200
male-61–65	95.00%	95.00%	95.00%	200
male-66–70	96.15%	100.00%	98.04%	200
male-71-more	99.00%	99.00%	99.00%	200
Accuracy	98.62%			5200
Macro avg	98.62%	98.62%	98.61%	5200
Weighted avg	98.62%	98.62%	98.61%	5200

5   Conclusion

This study is the first of its kind for predicting Age-Gender by using retinal fundus photos collected from Kaggle library. In prior studies, the model performed better than others. Our proposed model demonstrated that DL learning can effectively distinguish

Age and Gender using retinal fundus photos. ROC of the proposed model in all classes is 1.0, overall precision (98.62%), recall (98.62%), f1-score (98.61%) and accuracy (98.62%). There is a prevalent unawareness among clinicians regarding the differences in retinal feature variations between Age and Gender, underlining the significance of model explain ability for Age-Gender classification using retinal fundus photos. The proposed deep learning approach could help clinicians find new visions and disease biomarkers.

References

1. Abunasser, B.S., AL-Hiealy, M.R.J., Zaqout, I.S., Abu-Naser, S.S.: Breast cancer detection and classification using deep learning Xception algorithm. Int. J. Adv. Comput. Sci. Appl. (IJACSA) **13**(7), 78–83 (2022)
2. Abunasser, B.S., AL-Hiealy, M.R.J., Barhoom, A.M., Almasri, A.R., Abu-Naser, S.S. Prediction of instructor performance using machine and deep learning techniques. Int. J. Adv. Comput. Sci. Appl. (IJACSA) **13**(7), 223–228 (2022)
3. Elsharif, A.A.E.F., et al.: Retina diseases diagnosis using deep learning. Int. J. Acad. Eng. Res. (IJAER) **6**(2), 1–8 (2022)
4. Arqawi, S., Atieh, K.A.F.T., Shobaki, M.J.A.L., Abu-Naser, S.S., Abu Abdulla, A.A.M.: Integration of the dimensions of computerized health information systems and their role in improving administrative performance in Al-Shifa medical complex. J. Theor. Appl. Inf. Technol. **98**(6), 1087–1119 (2020)
5. Waghmare, V.K., Kolekar, M.H.: Brain tumor classification using deep learning. In: Chakraborty, C., Banerjee, A., Kolekar, M.H., Garg, L., Chakraborty, B. (eds.) Internet of Things for Healthcare Technologies. SBD, vol. 73, pp. 155–175. Springer, Singapore (2021). https://doi.org/10.1007/978-981-15-4112-4_8
6. Obaid, T. et al.: Factors contributing to an effective E- government adoption in palestine. In: Saeed, F., Mohammed, F., Ghaleb, F. (eds.) Advances on Intelligent Informatics and Computing. IRICT 2021. Lecture Notes on Data Engineering and Communications Technologies, vol. 127. Springer, Cham (2022). https://doi.org/10.1007/978-3-030-98741-1_55
7. MacGillivray, T.J., et al.: Retinal imaging as a source of biomarkers for diagnosis, characterization and prognosis of chronic illness or long-term conditions. Br. J. Radiol. **87**, 20130832 (2014)
8. Abu Naser, S.S.: Evaluating the effectiveness of the CPP-Tutor, an intelligent tutoring system for students learning to program in C++. J. Appl. Sci. Res. **5**(1), 109–114 (2009)
9. Albatish, I.M., Abu-Naser, S.S.: Modeling and controlling smart traffic light system using a rule based system. In: Proceedings - 2019 International Conference on Promising Electronic Technologies, ICPET 2019, pp. 55–60, 8925318 (2019)
10. Wendland, J.P.: The relationship of retinal and renal arteriolosclerosis in living patients with essential hypertension. Am. J. Ophthalmol. **35**, 1748–1752 (1952)
11. Elzamly, A., Messabia, N., Doheir, M., Al-Aqqad, M., Alazzam, M. Assessment risks for managing software planning processes in information technology systems. Int. J. Adv. Sci. Technol. **28**(1), 327–338 (2019)
12. Naser, S.S.A.: JEE-Tutor: an intelligent tutoring system for java expressions evaluation. Inf. Technol. J. **7**(3), 528–532 (2008)
13. Wong, T.Y., et al.: The prevalence and risk factors of retinal microvascular abnormalities in older persons: the Cardiovascular Health Study. Ophthalmology **110**, 658–666 (2003)
14. Naser, S.S.A.: Developing visualization tool for teaching AI searching algorithms. Inf. Technol. J. **7**(2), 350–355 (2008)

15. Elzamly, A., Hussin, B., Naser, S.A., Selamat, A., Rashed, A.: A new conceptual framework modelling for cloud computing risk management in banking organizations. Int. J. Grid Distrib. Comput. **9**(9), 137–154 (2016)
16. Normando, E.M., et al.: The retina as an early biomarker of neurodegeneration in a rotenone-induced model of Parkinson's disease: evidence for a neuroprotective effect of rosiglitazone in the eye and brain. Acta Neuropathol. Commun. **4**, 86–94 (2016)
17. Naser, S.S.A.: Developing an intelligent tutoring system for students learning to program in C++. Inf. Technol. J. **7**(7), 1051–1060 (2008)
18. Abu Ghosh, M.M., Atallah, R.R., Abu Naser, S.S.: Secure mobile cloud computing for sensitive data: Teacher services for palestinian higher education institutions. Int. J. Grid Distrib. Comput. **9**(2), 17–22 (2016)
19. Cheung, C.Y., et al.: Retinal vascular fractal dimension and its relationship with cardiovascular and ocular risk factors. Am. J. Ophthalmol. **154**, 663-674.e1 (2012)
20. Naser, S.S.A.: Intelligent tutoring system for teaching database to sophomore students in Gaza and its effect on their performance. Inf. Technol. J. **5**(5), 916–922 (2006)
21. Ting, D.S.W., et al.: Development and validation of a deep learning system for diabetic retinopathy and related eye diseases using retinal images from multiethnic populations with diabetes. JAMA **318**, 2211–2223 (2017)
22. Mady, S.A., Arqawi, S.M., Al Shobaki, M.J., Abu-Naser, S.S.: Lean manufacturing dimensions and its relationship in promoting the improvement of production processes in industrial companies. Int. J. Emerg. Technol. **11**(3), 881–896 (2020)
23. www.kaggle.com. Accessed 15 Mar 2022
24. Abu-Naser, S.S., El-Hissi, H., Abu-Rass, M., El-khozondar, N.: An expert system for endocrine diagnosis and treatments using JESS. J. Artif. Intell. **3**(4), 239–251 (2010)

Heart Disease Prediction Using a Group of Machine and Deep Learning Algorithms

Samy S. Abu-Naser[1](✉) (iD), Tareq Obaid[1] (iD), Mohanad S. S. Abumandil[2] (iD), and Ahmed Y. Mahmoud[1,1] (iD)

[1] Faculty of Engineering and IT, Alazhar University, Gaza, Palestine
{abunaser,tareq.obaid,ahmed}@alazhar.edu.ps
[2] Faculty of Hospitality, Tourism and Wellness, Universiti Malaysia Kelantan, Kota Bharu, Malaysia
moha-nad.ssa@umk.edu.my

Abstract. Heart disease is on the rise at an alarming speed, and thus it is critical to forecast heart disease early. Diagnosing heart disorders is not an easy process that requires accuracy and proficiency. The goal of this research is to figure out which patients are more prone to develop heart disease depending on a variety of medical characteristics. Using the medical attributes of the person, we created a prediction model for heart disease to determine if a person is possibly to be detected with heart disease or not. We employed a variety of machine learning (ML) algorithms and one Deep Learning (DL) method for predicting and classifying whether patients have heart disease or not. To govern how the proposed DL model may be utilized for the improvement of the accuracy of heart disease prediction in any patient, a very useful strategy was applied. The proposed model's strength was highly satisfactory, as it was capable of predicting evidence of having a heart condition in a specific patient utilizing DL and the ML Model (Random-Forest-Classifier) that had high accuracies when compared to other employed classifiers. The proposed DL methodology for predicting heart disease is going to improve medical care while lowering costs. This research provides important information that can help us anticipate who will develop heart disease conditions. The data was obtained from the Kaggle repository, and the model was developed in Python language.

Keywords: Heart Disease · Machine Learning · Deep Learning · Prediction

1 Introduction

ML is an Artificial Intelligence (AI) subfield that focuses on developing systems that learn from the data they are given and boost their performance. Artificial intelligence is a catch-all phrase for systems or gadgets that mimic human intelligence. Artificial intelligence and machine learning are typically deliberated together, and the terms are occasionally used interchangeably, although they are not the same thing. It is worth noting that, while all machine learning techniques are AI, not all AI is machine learning [1, 2]. A machine learning system uses a range of strategies to train computers to perform jobs

F. Saeed et al. (Eds.): ICACIn 2022, LNDECT 179, pp. 181–196, 2023.
https://doi.org/10.1007/978-3-031-36258-3_16

for which there is no totally adequate answer. When there are many possible answers, one approach is to classify some of the correct answers as correct. This information can then be utilized as training data to assist the computer in improving its algorithm(s) for finding proper responses [3, 4].

The machine learning method known as DL (hierarchical learning or deep structured learning) entails the learning representations of the data as contrasting to task-based procedures. According to [5], learning can come in three forms: supervised, semi-supervised, or not supervised. In each deep learning layer, all received data are transformed to become more abstract and complex. In the first layer, pixels are abstracted and edges are encoded. In the second layer, the edge arrangements are constructed and encoded. In the third layer, the nose and eyes are encoded. And finally, in the fourth layer, the image may be detected to form a face. In the deep learning process, learning occurs in placing traits to the correct levels [6].

There is a high prevalence of Cardiovascular Diseases (CVD) at present, covering multiple conditions affecting a person's heart. The Centers for Disease Control and Prevention (CDC) asserts that diseases of the heart are among the principal reasons of passing away in the US specifically among white people, African Americans, American Indians, and Alaskan Natives. Over 47 percent of Americans are affected by the three main risk features of heart disease namely high cholesterol, smoking and high blood pressure. Table 1 shows further vital factors namely diabetes, sedentary lifestyle, obesity and extreme alcohol intake. It is critical to identify and end the risk factors for the disease of the heart. Today, a patient's likelihood of developing heart disease can be predicted using deep and machine learning [7]. The patient's medical history and traits are collected from the Kaggle depository. Patients with heart disease can be predicted using this dataset. There are 18 medical attributes of a patient contained in the dataset [13].

2 Related Work

This current study is driven by a broad range of research on Cardiovascular Heart Disease diagnosis via Machine Learning. Among the strategies that have been used to successfully predict cardiovascular heart disease are Logistic Regression, KNN, Random Forest Classifier, and others. Each strategy has been proven to be powerful in achieving the set targets.

Muktevi Srivenkatesh conducted a study to compare the accuracy of utilizing rules to each findings of SVM, RF, NB classifier, and LR on a dataset collected in an area to offer a precise cardiovascular disease predicting model. The machine learning algorithms utilized in that study had an accuracy range of 58.71 percent to 77.06 percent in predicting cardiovascular disease in patients. When compared to other ML procedures, LR has the top accuracy (77.06%) [7]. [8] provided different heart disease variables and a model developed using supervised learning methods including NB, DT, KNN, and RF. The required data was derived from the UCI repository of heart disease database in Cleveland, covering 303 examples and 76 features. Out of the 76 criteria, only 14 were tested regardless of their significance in demonstrating the effectiveness of various algorithms. Based on the results, K-nearest neighbor showed the highest score for accuracy i.e. 90.79 percent. [9] compared the findings of the UCI ML Disease of the Heart dataset utilizing various machine learning methods and deep learning. A total of 14 key features in the dataset were utilized for analysis. The number of possible outcomes were validated using the accuracy and confusion matrix. Insignificant features were tackled using the Random Forest (RF). To achieve better results, the data was subsequently standardized, ultimately leading to a 94.2 percent accuracy via the usage of the deep learning method. [10] utilized nine machine learning classifiers in the study including ET, AB, LR, MNB, XGB, LDA, SVM, RF, and CART. They also performed some preprocessing, dataset standardization, and hyper parameter tuning on the regular dataset of heart disease to ensure accuracy. They also used the conventional K-fold cross-validation technique for training and validating the ML models. Lastly, the experimental findings showed that hyper parameter tweaking enhanced prediction classifier accuracy, and that data normalization and hyper parameter tuning of ML classifiers produced remarkable outcomes. [11] advocated using attributes to predict cardiovascular disease using machine learning techniques. They employed BMI as one of the key features for prediction. In terms of predicting cardiovascular disease, BMI was crucial. The study's major focus was on the impact of BMI on cardiovascular disease prognosis. Different characteristics, as well as regression and classification algorithms, have been proposed in the model. They came to the conclusion that BMI was a significant predictor of cardiovascular disease.

To increase the accuracy of machine learning systems, [12] used evolutionary methods like GA and PSO to select the features. For feature selection, PSO and GA were integrated with NB, J48 and SVM. Following the selection of significant features, the efficiency for the selection of the feature process was assessed using ML techniques on both the reduced dataset and the full dataset. The RF, LR, DT, NB and SVM algorithms, as well as five other machine learning methodologies, were employed for predicting disease of the heart and so test the efficiency of the selection of feature methods. The GA was shown to be the most successful method for feature selection, as it improves prediction accuracy.

Following a careful review of the past studies, it appears that the dataset utilized was UCI ML of the disease of heart dataset that differs from Kaggle repository dataset [13]. For training, evaluating, and testing these algorithms, we will employ familiar machine learning methods as well as a novel deep learning algorithm.

3 Dataset

Individuals from an Organized Dataset were chosen based on their history of cardiac issues and other medical concerns. Heart disease refers to a variety of illnesses that affect the heart. Cardiovascular disorders are the principal cause of death in middle-aged persons, according to WHO. The medical history of 319795 patients of various ages were obtained as part of the collection. All the required data are provided in this dataset namely the medical features of the patients including their BMI, Age, Smoking, Stroke, Alcohol Drinking, Mental Health, Diff Walking and Physical Health so on which facilitate the detection of patients diagnosed with any heart disease (see Table 1). The dataset has 18 medical features from 304 person that help us identify if the patient is at danger of having heart disease or not, as well as categorize patients who having risk and those who are not having risk. Training, Validating, and Testing were the three portions of the dataset. There are 319795 rows and 18 columns in this dataset, with each record corresponding to one person [13]. Table 1 lists all of the features.

Table 1. Features of the dataset

Feature	Type of Feature
HeartDisease	The output [Y/N]
BMI	Input [Y/N]
Smoking	Input [Y/N]
AlcoholDrinking	Input [Y/N]
Stroke	Input [Y/N]
PhysicalHealth	Input Numeric [0 to 30] days
MentalHealth	Input Numeric [0 to 30] days
DiffWalking	Input [Y/N]
Sex	Input [Female/Male]
AgeCategory	Input Category
Race	Input Category
Diabetic	Input [Y/N]
PhysicalActivity	Input [Y/N]
GenHealth	Input Category
SleepTime	Input Numeric
Asthma	Input [Y/N]
KidneyDisease	Input [Y/N]
SkinCancer	Input [Y/N]

4   Methodology

This study analyzed numerous machine learning algorithms including a variety of ML methods and one DL model which facilitate the accurate diagnosis of Heart Diseases. This research looked at journals, published articles, and a cardiovascular disease dataset. The approach is a set of actions that transform a given dataset into understandable data for users. The steps in the suggested methodology (as shown in Fig. 1) are as follows: The data is collected in the first stage, significant values are extracted in the second step, and the data is explored in the third step i.e. pre-processing. Based on the employed procedure, data preprocessing tackles missing values, data cleaning, and standardization [8]. The pre-processed data was then classified using each of the classifiers. Finally, we tested the suggested model using several performance indicators to assess its correctness and performance. Using various classifiers, an operational system for Heart Disease forecasting has been established in this model. For prediction, this model considers 18 medical factors including smoking tendency, age, alcohol consumption BMI, physical health, Diff Walking, stroke history, mental health and so on [9].

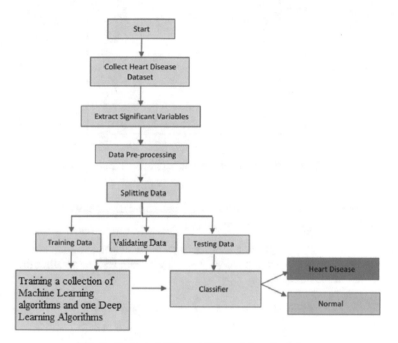

Fig. 1. Proposed ML and DL model methodology

4.1 Data Cleaning

There are approximately 319795 entries in the dataset, and consists of 18 variables. There are no null-values, and there are 14 numerical features and four categorical features. We can convert the binary properties with only two exclusive values, but firstly there is a need to ensure that no abnormal values are present. Certain attributes may have more than 2 exclusive values; we utilized One Hot Encoder later in the preprocessing phase.

4.2 Exploratory Analysis

Categorical Features. Figure 2 presents case distributions with heart disease (Yes or No) based on Sex. The value 1 denotes men and value 0 denotes women at the x-axis. Based on the Figure, it can be found that:

- Most heart disease cases inflict men.
- Most non-cases of heart disease involve women

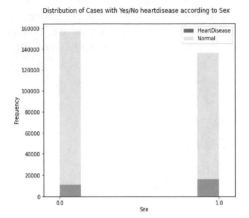

Fig. 2. Distribution of target class (heart disease) according to Sex

Figure 3 presents case distributions with Yes/No heart disease based on Smoking status. Based on the figure, it can be found that:

- Smokers form the largest group suffering from heart disease.
- Other cases of heart disease not caused by smoking are related to other factors.

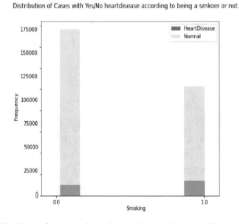

Fig. 3. Distribution of target class (heart disease) according to Smoking status

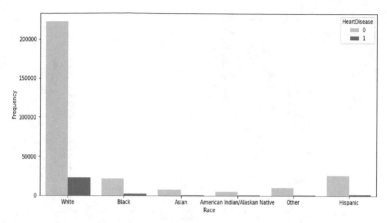

Fig. 4. Distribution of target class (heart disease) according to Race

Figure 4 presents case distributions with heart disease (Yes or No) based on the feature Race whereby 1 denotes having heart disease and value 0 denotes not having. Based on the figure, it can be found that:

- People who are white are more vulnerable to disease of the heart.

Figure 5 presents case distributions with Yes/No heart disease based on Age Category. Based on the figure, it can be found that:

- People with 80 years or more are more susceptible to diseases of the heart.

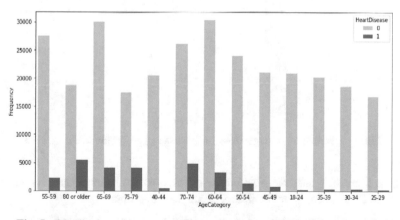

Fig. 5. Distribution of target class (heart disease) according to Age Category

Figure 6 presents case distributions with Yes/No heart disease based on Kidney Disease feature. Based on the figure, it can be found that:

- People not having Kidney Disease are more susceptible to diseases of the heart.

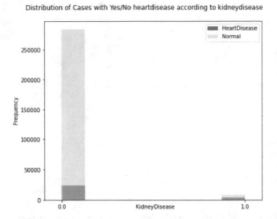

Fig. 6. Distribution of target class (heart disease) according to Kidney Disease

Figure 7 presents case distributions with (Yes or No) heart disease based on feature Skin Cancer. Based on the figure, it can be found that:

- People not having Skin Cancer are more susceptible to diseases of the heart.

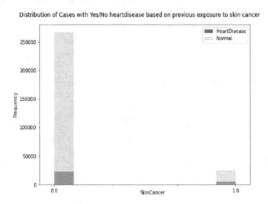

Fig. 7. Distribution of target class (heart disease) according to Skin Cancer

Figure 8 presents case distributions with (Yes or No) heart disease based on past exposure to Stroke. Based on the figure, it can be found that:

- People who experience Stroke are less susceptible to getting Heart Diseases.

Distribution of Cases with Yes/No hartdisease based on previous exposure to Stroke

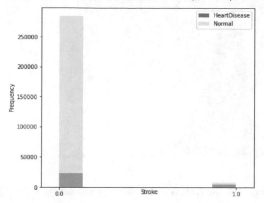

Fig. 8. Distribution of target class (heart disease) according to Stroke

Figure 9 presents case distributions with (Yes or No) heart disease based on past exposure to Diabetic. Based on the figure, it can be found that:

- People who experienced Diabetes are less susceptible to getting Heart Diseases.

Distribution of Cases with Yes/No hartdisease based on previous exposure to Diabetic

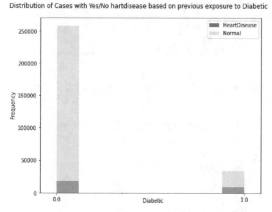

Fig. 9. Distribution of target class (heart disease) according to past exposure to Diabetic

Visualization of Numerical Features. Individuals weighing 40 kg and higher have a greater likelihood of developing heart diseases, as shown in Fig. 10. Based on Sleep Time measurements. Figure 11 depicts the case distributions with Yes/No heart disease. Figure 12 depicts the case distributions with (Yes or No) cardiac disease according to their current physical health. Figure 13 depicts the case distributions with Yes/No cardiac disease for the previous 30 days according to mental health status.

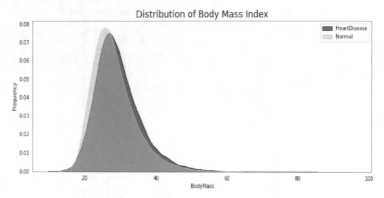

Fig. 10. Distribution of target class (heart disease) according to Body Mass Index

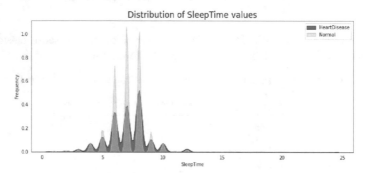

Fig. 11. Distribution of target class (heart disease) according to SleepTime values

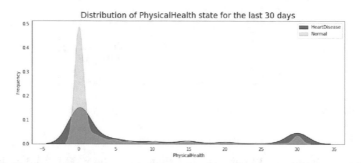

Fig. 12. Distribution of target class (heart disease) according to Physical Health state for the past 30 days

4.3 Data Preprocessing

The Standardization Scale method was used to normalize these features: Mental Health, BMI, Physical Health, and Sleep Time. The One Hot Encoding technique was used for these features: Age Category, Race, and General Health. The ratio of the class (Heart

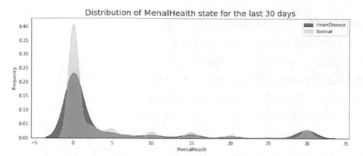

Fig. 13. Distribution of target class (heart disease) according to Mental Health state for the past 30 days

Disease) is 91 percent and (health) to 9% (Yes represent having Heart Disease). Hence, the dataset is imbalanced.

Oversampling and under sampling are two typical strategies for balancing a dataset. Increasing the lower attribute so that it would be equal to the higher attributes can be done by utilizing the under sampling approach, which downsizes the higher variable to the same size as the lower variable data. This study utilized both oversampling and under sampling, and the outcomes for both were compared.

4.4 Splitting Dataset

The dataset was divided after being grouped into 3 datasets namely: Training, Validating and Testing with a ratio of 60:20:20 percent respectively.

4.5 Building of the Proposed DL Model

The structure of the proposed deep learning model consists of one input layer, one output layer and 5 hidden layers. The input layer consist of 38 features, the output layer has one output, and the five hidden layers (256, 128, 64, 32, 16) nodes as can be seen in Figs. 14 and 15. Adam optimizer, sigmoid activation function, and learning rate (0.0001) were used. We trained the proposed model using the training data for 200 epochs and we used cross valuation. Once the training was done, we tested the proposed deep learning model using the testing data.

4.6 Experiment I

We used under sampling in the first experiment to balance the dataset. A combination of ML methods and a customized DL method to train the dataset. The accuracy, precision, recall, f1-score, and training duration were used to record the outcomes [14–24]. Table 2 shows the results of each algorithm.

```
from tensorflow.keras.optimizers import Adam
inputs = tf.keras.Input(shape=(X.shape[1],))
x = tf.keras.layers.Dense(32, activation='relu')(inputs)
x = tf.keras.layers.Dense(64, activation='relu')(x)
x = tf.keras.layers.Dense(128, activation='relu')(x)
x = tf.keras.layers.Dense(256, activation='relu')(x)
x = tf.keras.layers.Dense(256, activation='relu')(x)
outputs = tf.keras.layers.Dense(1, activation= 'sigmoid')(x)
model = tf.keras.Model(inputs, outputs)
model.compile(Adam(lr=0.0001), loss='binary_crossentropy', metrics=['accuracy'])
model.summary()
```

Fig. 14. Python code of the structure of the proposed Deep Learning model

```
Model: "model"

 Layer (type)                Output Shape              Param #
=================================================================
 input_1 (InputLayer)        [(None, 38)]              0

 dense (Dense)               (None, 256)               9984

 dense_1 (Dense)             (None, 128)               32896

 dense_2 (Dense)             (None, 64)                8256

 dense_3 (Dense)             (None, 32)                2080

 dense_4 (Dense)             (None, 16)                528

 dense_5 (Dense)             (None, 1)                 17

=================================================================
Total params: 53,761
Trainable params: 53,761
Non-trainable params: 0
```

Fig. 15. Structure of the proposed DL model

4.7 Experiment Two

In experiment two, we used over sampling to balance the dataset. We used the same set of ML techniques and the same DL model that we built from to train the model on the dataset. The following measures were used to record the results: accuracies, precisions, recalls, f1-scores, and training time in seconds. Table 3 shows the results of every model.

5 Discussions

Although most researchers use diverse processes such as KNN, DT, RFC, LR and SVC to detect people identified with heart disease, these results show that with a dissimilar dataset than the one used in this current study, the highest accuracy other researchers achieved was 88.5 percent. The algorithms we utilized are more accurate, more cost-effective, and faster than the algorithms used by earlier studies. Furthermore, Deep Learning achieved the highest performance with under sampling: F1-scores (79.61%), accuracies (78.77%), recalls (82.60%), precisions (76.83%) and training and validation time (81 s). Table 2 shows the remaining outcomes of the other algorithms.

Table 2. Outcome of all methods using under sampling procedure

Models name	Accuracies	Precisions	Recalls	F1_scores	Time in aeconds
Gaussian-Mixture	0.4981	0.0000	0.0000	0.0000	0.49
Perceptron	0.7028	0.7095	0.6904	0.6998	0.17
Nearest-Centroid	0.7288	0.7478	0.6935	0.7196	0.12
Multinomial-NB	0.7457	0.7528	0.7344	0.7435	0.10
Logistic-Regression-CV	0.7632	0.7542	0.7832	0.7685	7.99
Linear-SVC	0.7627	0.7532	0.7839	0.7683	1.48
Linear-Discriminant-Analysis	0.7628	0.7533	0.7841	0.7684	0.34
Label-Propagation	0.7117	0.7185	0.6997	0.7090	58.03
Calibrated-Classifier-CV	0.7626	0.7545	0.7810	0.7675	8.267
Extra-Tree-Classifier	0.6810	0.6822	0.6820	0.6821	0.15
SGD-Classifier	0.7529	0.7393	0.7841	0.7611	0.17
Quadratic-Discriminant-Analysis	0.7316	0.6890	0.8476	0.7601	0.58
Gaussian-NB	0.7494	0.7054	0.8595	0.7749	0.14
Random-Forest-Classifier	0.7442	0.7447	0.7442	0.7441	5.93
Complement-NB	0.7539	0.7646	0.7363	0.7502	0.09
MLP-Classifier	0.7611	0.7359	0.8171	0.7744	57.11
Bernoulli-NB	0.7528	0.7669	0.7290	0.7475	0.15
Bagging-Classifier	0.7211	0.7318	0.7013	0.7162	1.79
LGBM-Classifier	0.7669	0.7500	0.8032	0.7757	0.68
Ada-Boost-Classifier	0.7663	0.7638	0.7736	0.7687	1.78
KNeighbors-Classifier	0.7366	0.7361	0.7406	0.7383	7.83
Logistic-Regression	0.7681	0.7615	0.7830	0.7721	0.82
Gradient-Boosting-Classifier	0.7659	0.7532	0.7934	0.7728	4.61
Proposed DL model	**0.7876**	**0.7682**	**0.8260**	**0.7961**	81.00

Table 3. Outcome of all methods using over sampling procedure

Models name	Accuracies	Precisions	Recalls	F1_scores	Time in seconds
Gaussian-Mixture	0.6910	0.6389	0.8780	0.7396	11.73
Perceptron	0.6568	0.6280	0.7689	0.6913	2.11
Nearest-Centroid	0.7169	0.7328	0.6826	0.7068	0.78
Multinomial-NB	0.7454	0.7425	0.7512	0.7468	0.71
Logistic-Regression-CV	0.7619	0.7439	0.7988	0.7704	77.16
Linear-SVC	0.7621	0.7411	0.8050	0.7720	31.91
Linear-Discriminant-Analysis	0.7619	0.7399	0.8070	0.7722	3.31
Label-Propagation	0.7590	0.7370	0.8054	0.7697	1.31
Calibrated-Classifier-CV	0.8520	0.8462	0.8608	0.8534	2.89
Extra-Tree-Classifier	0.7606	0.7452	0.7919	0.7678	155.57
SGD-Classifier	0.7327	0.7067	0.7955	0.7485	2.03
Quadratic-Discriminant-Analysis	0.7273	0.6740	0.8802	0.7634	0.91
Gaussian-NB	0.9021	0.9022	0.9021	0.9021	85.51
Random-Forest-Classifier	0.7427	0.7410	0.7462	0.7436	0.71
Complement-NB	0.8025	0.7735	0.8553	0.8123	461.56
MLP-Classifier	0.7415	0.7493	0.7257	0.7373	1.02
Bernoulli-NB	0.8963	0.9043	0.8864	0.8952	34.48
Bagging-Classifier	0.8805	0.8883	0.8704	0.8793	10.20
LGBM-Classifier	0.8048	0.7873	0.8351	0.8105	22.25
Ada-Boost-Classifier	0.8324	0.7987	0.8886	0.8413	954.00
KNeighbors-Classifier	0.7601	0.7426	0.7960	0.7684	6.00
Logistic-Regression	0.8284	0.8104	0.8572	0.8332	72.00
Gradient-Boosting-Classifier	0.8706	0.8669	0.8756	0.8712	4.70
Proposed DL model	**0.9235**	**0.9083**	**0.9420**	**0.9249**	600.00

The findings of the same algorithms utilizing over sampling, on the other hand, produced by the same technique (Deep Learning) but with higher accuracy. Over sampling provided the best results in terms of F1-scores (92.49 percent), accuracies (92.35 percent), recalls (94.20 percent), precisions (90.84 percent), and training and validation time (600 s). Random Forest Classifier had the second highest accuracy: F1-scores (90.21 percent), accuracy (90.21 percent), precisions (90.22 percent), recalls (90.21 percent) and training and validation time (85.51 s). Table 3 shows the remaining outcomes of the other methods.

In conclusion, our accuracy measurements outperformed earlier investigations. Our research also found that DL and RFC were the most effective in predicting patients with heart disease. This demonstrates that DL and RFC are more effective in detecting diseases of the heart.

6 Conclusion

A cardiac disease diagnosis model was created utilizing a combination of ML classification modeling approaches and a DL model. This study predicts persons who will develop cardiac disease by taking patient medical history that takes you to a deadly heart disease from a dataset that includes information such as smoking, stroke, alcohol use, physical, mental health, BMI and so on. This model for Heart Disease Detection aids a patient according to the medical info from a previous heart disease diagnosis. Deep Learning, Random Forest Classifier, and a variety of additional classifiers were used to create the presented model. Our models have a peak accuracy of 92.23%. The more training data we have (as a result of dataset balancing) means the model has a better chance of correctly predicting whether or not a given person has heart disease. We can anticipate the patient faster and better using these tools, and the cost can be significantly decreased. We can work on a variety of medical databases since ML models are better and can anticipate outcomes better than humans, which benefits both patients and doctors. As a result, this study assists us in predicting patients diagnosed with heart problems by cleaning the dataset and applying deep learning. Our deep learning model has an accuracy of 92.23%, which is higher than the prior models' accuracy of 85%. Furthermore, it has been determined that DL has the top accuracy among all algorithms that we have utilized.

References

1. Abunasser, B.S., AL-Hiealy, M.R.J., Zaqout, I.S., Abu-Naser, S.S.: Breast cancer detection and classification using deep learning Xception algorithm. Int. J. Adv. Comput. Sci. Appl. (IJACSA) **13**(7), 78–83 (2022)
2. Abunasser, B.S., AL-Hiealy, M.R.J., Barhoom, A.M., Almasri, A.R., Abu-Naser, S.S.: Prediction of instructor performance using machine and deep learning techniques. Int. J. Adv. Comput. Sci. Appl. (IJACSA) **13**(7), 223–228 (2022)
3. Waghmare, V.K., Kolekar, M.H.: Brain tumor classification using deep learning. In: Chakraborty, C., Banerjee, A., Kolekar, M.H., Garg, L., Chakraborty, B. (eds.) Internet of Things for Healthcare Technologies. SBD, vol. 73, pp. 155–175. Springer, Singapore (2021). https://doi.org/10.1007/978-981-15-4112-4_8
4. Obaid, T., et al.: Factors contributing to an effective e-government adoption in Palestine. In: Saeed, F., Mohammed, F., Ghaleb, F. (eds) Advances on Intelligent Informatics and Computing. IRICT 2021. Lecture Notes on Data Engineering and Communications Technologies, vol. 127. Springer, Cham (2022). https://doi.org/10.1007/978-3-030-98741-1_55
5. Arqawi, S., Atieh, K.A.F.T., Shobaki, M.J.A.L., Abu-Naser, S.S., Abu Abdulla, A.A.M.: Integration of the dimensions of computerized health information systems and their role in improving administrative performance in Al-Shifa medical complex. J. Theor. Appl. Inf. Technol. **98**(6), 1087–1119 (2020)

6. Muktevi, S.: Prediction of cardiovascular disease using machine learning algorithms. Int. J. Eng. Adv. Technol. (IJEAT) **9**(3), 1–10 (2020)
7. Shah, D., Patel, S., Bharti, S.K.: Heart disease prediction using machine learning techniques. SN Comput. Sci. **1**(6), 1–6 (2020). https://doi.org/10.1007/s42979-020-00365-y
8. Bharti, R., et al.: Prediction of heart disease using a combination of machine learning and deep learning. Comput. Intell. Neurosci. (2021). https://doi.org/10.1155/2021/8387680
9. Saboor, A., et al.: A method for improving prediction of human heart disease using machine learning algorithms. Mob. Inf. Syst. (2022). https://doi.org/10.1155/2022/1410169
10. Mady, S.A., Arqawi, S.M., Al Shobaki, M.J., Abu-Naser, S.S.: Lean manufacturing dimensions and its relationship in promoting the improvement of production processes in industrial companies. Int. J. Emerg. Technol. **11**(3), pp. 881–896 (2020)
11. Abu Naser, S.S.: Evaluating the effectiveness of the CPP-Tutor, an intelligent tutoring system for students learning to program in C++. J. Appl. Sci. Res. **5**(1), 109–114 (2009)
12. Albatish, I.M., Abu-Naser, S.S.: Modeling and controlling smart traffic light system using a rule based system. In: Proceedings - 2019 International Conference on Promising Electronic Technologies, ICPET 2019, pp. 55–60, 8925318 (2019)
13. Elzamly, A., Messabia, N., Doheir, M., Al-Aqqad, M., Alazzam, M.: Assessment risks for managing software planning processes in information technology systems. Int. J. Adv. Sci. Technol. **28**(1), 327–338 (2019)
14. Naser, S.S.A.: JEE-Tutor: an intelligent tutoring system for java expressions evaluation. Inf. Technol. J. **7**(3), 528–532 (2008)
15. Nikam, A., Bhandari, S., Mhaske A., Mantri, S.: Cardiovascular disease prediction using machine learning models. In: 2020 IEEE Pune Section International Conference (PuneCon), pp. 22–27 (2020). https://doi.org/10.1109/PuneCon50868.2020.9362367
16. Naser, S.S.A.: Developing visualization tool for teaching AI searching algorithms. Inf. Technol. J. **7**(2), 350–355 (2008)
17. Aleem, A., Prateek, G., Kumar, N.: Improving heart disease prediction using feature selection through genetic algorithm. Advanced (2022)
18. Elzamly, A., Hussin, B., Naser, S.A., Selamat, A., Rashed, A.: A new conceptual framework modelling for cloud computing risk management in banking organizations. Int. J. Grid Distrib. Comput. **9**(9), 137–154 (2016)
19. Naser, S.S.A.: Developing an intelligent tutoring system for students learning to program in C++. Inf. Technol. J. **7**(7), 1051–1060 (2008)
20. Network Technologies and Intelligent Computing, ANTIC 2021. Communications in Computer and Information Science, vol. 1534. Springer, Cham (2021). https://doi.org/10.1007/978-3-030-96040-7_57
21. www.kaggle.com. Accessed 15 Mar 2022
22. Abu Ghosh, M.M., Atallah, R.R., Abu Naser, S.S.: Secure mobile cloud computing for sensitive data: teacher services for Palestinian higher education institutions. Int. J. Grid Distrib. Comput. **9**(2), 17–22 (2016)
23. Naser, S.S.A.: Intelligent tutoring system for teaching database to sophomore students in Gaza and its effect on their performance. Inf. Technol. J. **5**(5), 916–922 (2006)
24. Abu-Naser, S.S., El-Hissi, H., Abu-Rass, M., El-khozondar, N.: An expert system for endocrine diagnosis and treatments using JESS. J. Artif. Intell. **3**(4), 239–251 (2010)

Tweet Recommendation System Based on TA Algorithm and Natural Language Processing

Kaoutar El Handri[1,2(✉)], Mohamed Rachdi[1], and Karim El Bouchti[2]

[1] LIMIE Laboratory, Higher Institute of Engineering and Business, 393 Rte d'El Jadida, Casablanca, Morocco
kaoutar.elhandri@isga.ma
[2] IPSS Laboratory, Faculty of Sciences, University Mohammed V, Rabat, Morocco

Abstract. Information retrieval of different types uses many techniques to rank query processing. End users are most interested in the relevant answer in various application domains. Top-k processing connects to many areas of database research, including query optimization, indexing methods, and query languages. Therefore, various emerging applications warrant efficient support for top-k queries. In this work, we faced these challenges by applying the Artificial Intelligence methods combined with the (TA) Threshold Algorithm in our model. Using the spark library, PCA, and WordTovec as Natural Language Processing tools to collect tweets. The result shows our approach's performance while enhancing the runtime and achieving similar results to state of the art based on a comparison with the well-known technology in big data and streaming processing.

Keywords: Top-k query · Recommendation Systems · TA algorithm · Word embeddings · Natural Language Processing NLP

1 Introduction

In recent years, social networks have developed into a viral phenomenon that provides many real-time messages to users. Tweets, for example, are categorized chronologically, and users browse the timelines of the people they follow to find what interests them. However, the problem of information overload has over-whelmed users and decision-makers alike, especially those with large followings and thousands of tweets coming in each day [1]. Consequently, selecting only relevant information, recommendation algorithms are the best solution. A popular paradigm for tackling this issue is using top-k queries, i.e., ranking the results and returning the k results with the highest scores. The answer to a top-k query [3, 11, 14] is an ordered set of tuples, where the ordering is based on how closely the tuples match the given query. Thus, the answer to a top-k query does not include all tuples that "match" the query, but only the best k such tuples. [8].

In this paper, we aim to apply the discussed algorithms in another field of recherche than that presented before. Hence, we propose a combined approach to tweet selection that uses one of the most widely used algorithms in the field of RSs: the TA and Wordtovec

F. Saeed et al. (Eds.): ICACIn 2022, LNDECT 179, pp. 197–203, 2023.
https://doi.org/10.1007/978-3-031-36258-3_17

algorithms. The primary purpose of this combination is to select the best tweet based on the personal behavior on the social network of the tweeter that presents the profile of the systems users [13]. Selection has been made based on the Big data technology emerging in the field of recommendation, namely Spark Framework [15], to take advantage of its enriching libraries that manage machine learning and deep learning in an intelligent and relevant way [4] to select these tweets. We present the main related works in the first section of this paper. In the second section, we offer the proposed approach, and in the third section, the result and interpretation. Finally, we conclude this work by highlighting the main future perspectives.

2 Proposed Approach

Threshold Algorithm (TA). Among the top-k algorithms, the Threshold Algorithm, or TA, is the best-known example because of its simplicity and memory requirements [2]. Algorithm 1 shows how it works in general, then to understand the approach of this algorithm, take the two steps in the following example (Fig. 1).

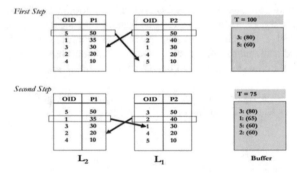

Fig. 1. AN example of Topk query processing [12]

In the first step, the algorithm retrieves the principal object from each list and probes the value of its other scoring predicate in the other list, revealing the exact scores for the top items. The item seen is buffered in the order of their partitions. A threshold value, T, for the unseen object scores is calculated by applying F to the last seen scores in the two lists, resulting in $50 + 50 = 100$. Since both seen objects have scores below T, no results can be reported. In the second step, T drops to 75, and object three can be safely notified since its score is above T. The algorithm continues until either k objects are registered or the sources are exhausted [8, 9].

The process begins by storing in a file called Spark store. The configuration of spark streaming has been taken into account to receive the data via and keep them in RDDs (resilient distributed data frames with a time interval of 10 s). These RDDs are sent to hdfs. 2nd file is to be executed in parallel with the tweepy streamer. Once access to tweets is received, the system stops the execution of the two files (1st then 2nd). After the algorithm reads the tweets stored in the folder, the system opts for preprocessing to clean them up. Before this step is finalized, the application of the Word2vec algorithm starts by considering 200-dimensional vectors. Then, the system uses the PCA principal component analysis method for dimensionality reduction, which is 3 in our case. The following step is recording the result in a table SQLContext of 5 attributes, namely the id, the cleaned text, dimension 1, dimension2, and dimension 3. After the algorithm reads, the tweets stored in the folder, the system opts for preprocessing to clean them up. Before this step is finalized, the application of the Word2vec algorithm starts by considering 200-dimensional vectors. Then, the system uses the PCA principal component analysis method for dimensionality reduction, which is 3 in our case (Fig. 2).

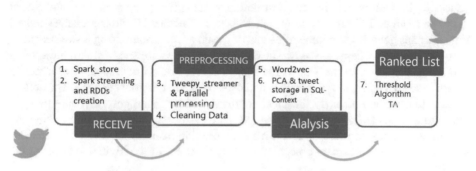

Fig. 2. Tweet Processing using the presented approach

Algorithm 1 TA

1: **Do** parallel *SA* on all *m* lists object *x* seen under SA in a list
2: fetch its scores from other lists by *RA*
3: compute overall score
4: **if** |*Buffer*| < k **then**
5: add *x* to *Buffer*
6: **else**if score(x) ≤ *k* th score in the *Buffer*
7:   replace bottom of buffer with (*x*, *score(x)*)
8:   toss
9:   Stop when threshold ≤ *k* th score in the *Buffer*
10: Threshold := *t* (worst score seen on $L_1,...,L_m$)
11: **end if**
12:
13: **Output** the Top_k objects & scores (in *Buffer*)

Finally, the system selects the data from the table, creating three data frames. Namely, with id and a sorted by descending order of the dimension and converting them into data frames thanks to the library pandas of python, then pass them to the algorithm TA see the (Algorithm 1) the object of the discussed combination, to exploit this algorithm for a relevant recommendation. This rec ommendation which is the result of a selected tweet by saving it in a file CSV to facilitate its exploitation has to end to evaluate the performance of the found results of this selection.

Word2Vec Word2Vec is an Estimator that takes sequences of words representing documents and forms a Word2Vec Model. The model associates each word with a unique vector of fixed size. The Word2Vec Model transforms each document into a vector using the average of all words in the document; this vector can then be used as features for prediction, document similarity calculations, etc.

PCA It is a statistical procedure that uses an orthogonal transformation to convert a set of observations of possibly correlated variables into values of linearly uncorrelated variables called principal components. A class PCA trains a model to project vectors into a low-dimensional space using PCA.

Tweepy Tweepy facilitates using the Twitter streaming API by handling au- thentication, login, session creation and destruction, reading incoming messages, and partial message routing. Twitter streaming API is used to download Twitter messages in real time. It helps get a high volume of tweets or create a live stream using a site stream or a user stream. In Tweepy, a weepy. The Stream instance establishes a streaming session and routes messages to the Stream Listener instance. The on data method of a stream listener receives all message functions and calls based on the message type. The default Stream Listener can classify the most common Twitter messages and route them to appropriately named methods, but these methods are just stubs. Therefore, using the streaming API has three steps: Create a class inheriting from Stream Listener, then Use this class to create a Stream object, and Finlay Connect to the Twitter API using the stream.

Spark. Spark libraries for machine learning used:

- RegexTokenizer: Tokenization is taking a text (such as a sentence) and breaking it into individual terms (usually words).
- StopWordsRemover Stopwords are words that should be excluded from the entry, usually because the words appear frequently and do not have meaning. StopWordsRemover takes as input a sequence of strings (for example, the output of an output of a tokenizer) and removes all stop words from the input sequences. Input sequences [5, 7, 10].

3 Results and Discussions

This work was performed on a Core i7 machine with 12GB of RAM using Cloudera quickstart VM version 5.13.0 and java version: java-1.8.0-OpenJDK. Hadoop Hdf installation, Hadoop version: hadoop2.6, Spark version: spark-2.4.5, PyS- park: Python 3.6.7 for Spark 2.4.5. We also use Pycharm and connect with PySpark and add different packages like Tweepy, pandas, and other functional modules. Create a class inheriting from Stream Listener, then Use this class to create a Stream object and Finlay Connect to the Twitter API using the stream. In the first step, we created our Twitter application allowing us to collect the tweets, and we got the four keys (tokens) to access it. (tokens) allowing its access. In the next step, we used the Tweepy library, which manages the authentication and the connection to the streaming API of Twitter streaming API. We create the TweetsListener class, which inherits from the StreamListener module in Tweepy, and we replace the method on data () method to be able to send tweets (considering only the id of the tweet and the text of the tweet without emojis) via the socket. We searched for the words **covid19, covid-19,covid 8.** Which were chosen because it was trendy on Twitter during the experimentation, and we can get tweets quickly. It is triggered only when it receives the call from the Spark Streaming module and sends the tweets to the Spark engine via the TCP socket.

Following, we localized Spark on our local machine. Then we import the necessary packages from Pyspark. We run SparkContext () and receive data via socketTextStream. SparkContext, which is the entry point for all Spark functionality. When we run a Spark application, a driver program starts, which has the main function, and SparkContext is initialized here. A SparkContext denotes the connection to a Spark cluster and can be used to create RDDs, accumulators, and broadcast variables on that cluster. SparkContext uses Py4J to launch a JVM and creates a JavaSparkContext (source). It is important to note that only one SparkContext can run in each session (Table 1).

Table 1. Results of Top 3 tweets after using TA algorithm.

Number of K	id of tweet	Tweet score based on TA algorithm
k = 1	1311610705624223744	**1.9117217709878922**
k = 2	1311582708389249029	**1.9076745654546885**
k = 3	1311637033945792520	**1.7953153436910463**

After that, we launch the StreamingContext() with 10-s batch intervals. Consequently, the input streams will be split into batches every 10 s during the streaming and sent to Hadoop hdfs. In the existing folder in called TWEETS, we have subfolders, each representing a batch of 10 tweets. And inside each subfolder, we have the famous SUCCESS file and a part-00000 file of hdfs containing the tweets. After collecting and storing thousands of tweets, we prepared these tweets to apply the threshold algorithm. For this, we proceeded to the cleaning, stemming, and applications of word2vec and PCA. To apply the threshold algorithm TA [9], we created three data frames (converted into pandas data frames), each having as columns the tweet id and a PCA dimension. These data frames are sorted in descending order of the PCA dimension as required by the Threshold algorithm model and passed as parameters to the algorithm. We applied the algorithm to obtain the top-3 tweets. The runtime (in seconds) is **44.56 s.** Finally, to evaluate the proposed approach using the TA algorithm, we decided to compare it to a SQL query (spark SQL), respecting the exact condition of the experimentation. To make this possible, we created temporal tables equivalent to the data frames we applied the algorithm. And we got the following top-3 tweets presented in Fig. 3 and a runtime of **240 s.** It is noticed that the ids of the top-2 tweets in the 2 cases are the same.

Consequently, to see the difference between the remaining tweet of each of the results in the 2 cases. We noticed that they have the same content. Therefore, we have identical effects. But the TA algorithm is much faster (**44s**) than the SQL query (**240s**). Finally, we accessed the top-3 tweets on Twitter for more observations. It is also noticed that the 3rd top tweet in TA, which is different from the 3rd top tweet of SQL, is not a retweet

(a) First tweet   (b) Second tweet   (c) Third tweet

Fig. 3. Top 3 tweet based on TA algorithm using the proposed approach

but an update of the other in number. Since the cleaning level, the numbers are removed, and both have been treated similarly.

4 Conclusion and Future Works

The advantages of the proposed approach come mainly from the tweepy processing tools, which allow several selections and filtering of tweets before their storage. On the other hand, it comes from the spark streaming, SQL, and MLlib library while using hdfs storage. Although it still has some challenges to solve because it cannot limit the number of tweets collected for the algorithm TA. Also, the use of Random accesses to compute a tweet's global score. In addi- tion, a monotonic function is simply the sum of dimensions (vectors). Thus, the approach can be improved in future work by testing other query optimization techniques, such as non-random access to the score and sorted and controlled access. And it is improving the score functions, which can give better results, using Topk-ws [6] rather than TA to improve the execution time and the results' quality. In particular, the switch from tweepy to TA requires significant resources for experimentation, but the results are still promising compared to other work.

References

1. Ben-Lhachemi, N., et al.: Using tweets embeddings for hashtag recommendation in twitter. Proc. Comput. Sci. **127**, 7–15 (2018)
2. EL Handri, K., Idrissi, A.: Efficient topkws algorithm on synthetics and real datasets. Under Rev. Int. J. Artif. Intell. IJ-1I, submitted 14 mai 2020
3. El handri, K., Idrissi, A.: Comparative study of topk based on fagins algorithm using correla- tion metrics in cloud computing QoS. Int. J. Int. Technol. Secured Trans. (IJITST) (in press) (2018)
4. El Handri, K., El Ahrach, S., Touil, M.: A new synesthesia-based approach for AI in healthcare. (IJRE) Int. J. Res. Ethics **5**(1) (2022). (ISSN 2665-7481)
5. El Handri, K., Idrissi, A.: E´tude comparative de topk bas´ee sur l'algorithme de fagin en utilisant des m´etriques de corr´elation dans la qualit´e de service de cloud computing. In: EGC, pp. 359–360 (2019)

6. El Handri, K., Idrissi, A.: syst'eme collaboratif d'aide a' la d'ecision a' base des recomman-dations multi crit'eres, 03 September 2020, MA Patent 50776
7. El Handri, K., Idrissi, A.: Parallelization of top_k algorithm through a new hybrid recommendation system for big data in spark cloud computing framework. IEEE Syst. J. **15**(4), 4876–4886 (2021). https://doi.org/10.1109/JSYST.2020.3019368
8. Fagin, R., Lotem, A., Naor, M.: Optimal aggregation algorithms for middleware. In: Proceedings of the Twentieth ACM SIGMOD-SIGACT-SIGART Symposium on Principles of Database Systems, PODS 2001, ACM, New York, NY, USA, pp. 102–113 (2001). https://doi.org/10.1145/375551.375567
9. Fagin, R., Lotem, A., Naor, M.: Optimal aggregation algorithms for middleware. J. Comput. Syst. Sci. **66**(4), 614–656 (2003)
10. Handri, K.E., Idrissi, A.: Comparative study of top_k based on Fagin's algorithm using correlation metrics in cloud computing QoS. Int. J. Int. Technol. Secur. Trans. **10**(1–2), 143–170 (2020)
11. Idrissi, A., Elhandri, K., Rehioui, H., Abourezq, M.: Top-k and skyline for cloud services research and selection system. In: Proceedings of the International Conference on Big Data and Advanced Wireless Technologies, pp. 1–10, p. 40. ACM (2016)
12. Ilyas, I.F., Beskales, G., Soliman, M.A.: A survey of top_k query processing techniques in relational database systems. ACM Comput. Surv. (CSUR) **40**(4), 1–58 (2008)
13. Pitsilis, G.K., Ramampiaro, H., Langseth, H.: Detecting offensive language in tweets using deep learning. arXiv preprint arXiv:1801.04433 (2018)
14. Sedhai, S., Sun, A.: Hashtag recommendation for hyperlinked tweets. In: Proceedings of the 37th International ACM SIGIR Conference on Research and Development in Information Retrieval, pp. 831–834 (2014)
15. Zaki, N.D., Hashim, N.Y., Mohialden, Y.M., Mohammed, M.A., Sutikno, T., Ali, A.H.: A real-time big data sentiment analysis for iraqi tweets using spark streaming. Bulletin of Electrical Engineering and Informatics **9**(4), 1411–1419 (2020)

Deep Learning Based for Cryptocurrency Assistive System

Muhammad Zakhwan Mohamed Rafik[1(✉)], Noraisyah Mohamed Shah[1],
Nor Azizah Hitam[2], Faisal Saeed[3(✉)], and Shadi Basurra[3]

[1] Universiti Malaya, 50603 Kuala Lumpur, Malaysia
zakhwanedu@gmail.com, noraisyah@um.edu.my
[2] Computer Information Science - Higher Colleges of Technology, Abu Dhabi, UAE
azizah.hitam@aiu.edu.my
[3] DAAI Research Group, Department of Computing and Data Science, School of Computing
and Digital Technology, Birmingham City University, Birmingham B4 7XG, UK
{faisal.saeed,shadi.basurra}@bcu.ac.uk

Abstract. Cryptocurrency is branded as a digital currency, an alternative
exchange currency system with significant ramifications for the economies of rising nations and the global economy. In recent years, cryptocurrency has infiltrated
almost all financial operations; hence, cryptocurrency trading is frequently recognized as one of the most popular and promising means of profitable investment.
Lately, with the exponential growth of cryptocurrency investments, many Alternative Coins (Altcoins) resurfaced to mimic the fiat currency. There are several
methods to forecast cryptocurrency prices that have been widely used in forecasting fiat and stock prices. Artificial Intelligence (AI),Machine Learning(ML) and
Deep Learning(DL) provide a different perspective on how investors can estimate
crypto price trend and movement. In this paper, as cryptocurrency price is time-
dependent, Recurrent Neural Network (RNN) is presented due to RNN's nature,
which is well suited for Time Series Analysis (TSA). The topology of the proposed RNN model consists of three stages which are model groundwork, model
development, and testing and optimization. The RNN architecture is extended
to three different models specifically Long Short-Term Memory (LSTM), Gated
Recurrent Unit (GRU), and Bi-Directional Long Short-Term Memory (LSTM).
There are a few hyperparameters that affect the accuracy of the deep learning
model in predicting cryptocurrency prices. Hyperparameter tuning set the basis
for optimizing the model to improve the accuracy of cryptocurrency prediction.
Next, the models were tested with data from different coins listed in the cryptocurrency market. Then, the model was experimented with different input features to
figure out how accurate and robust these models in predicting the cryptocurrency
price. GRU has the best accuracy in forecasting the cryptocurrency prices based
on the values of Root Mean Square Error (RMSE), Mean Absolute Percentage
Error (MAPE), and Executional Time, scoring 2.2201, 0.8076, and 200s using the
intraday trading strategy as input features.

Keywords: Deep Learning · Cryptocurrency Prediction · Time Series Analysis
(TSA) · Alternative Coins (Altcoins)

© The Author(s), under exclusive license to Springer Nature Switzerland AG 2023
F. Saeed et al. (Eds.): ICACIn 2022, LNDECT 179, pp. 204–217, 2023.
https://doi.org/10.1007/978-3-031-36258-3_18

1   Introduction

Cryptocurrencies have been dubbed a digital currency, an alternative exchange currency system with substantial implications for emerging nations and the global economy [1]. The excitement around cryptocurrency is undeniable, particularly in recent years, as it has permeated virtually all financial activities. As a result, cryptocurrency trading is often regarded as one of the most popular and promising forms of successful investing.

Nonetheless, compared to the traditional fiat market, this ever-expanding financial industry is characterized by high volatility and price swings over time. Chowdhury et al. Nowadays, bitcoin forecasting is widely regarded as one of the most challenging time-series prediction issues due to the vast number of unknown variables and the extreme volatility of cryptocurrency values, which results in complex temporal dependencies [2].

The basic approach to forecasting cryptocurrency prices is to look for patterns, or what one can refer to as price fluctuations, in the market. Cryptocurrency analysis is extrapolating previous data to forecast future cryptocurrency prices. With the fast advancement of technology, particularly in Artificial Intelligence (AI), experts' educated estimates are made by machines.

Many firms implement machine learning and deep learning methods to analyze and forecast data. Nowadays, all financial analysts, crypto market analysts, and scientist are eager to find the most accurate ways to forecast cryptocurrency price movement. Due to its peculiarities and volatile nature, bitcoin price data is more difficult to anticipate than financial time series data. Support Vector Machines (SVM) and Artificial Neural Networks (ANN), for example, are commonly employed, which is evident in [3] to forecast crypto prices and movements. Every algorithm has a different method for learning patterns and then predicting them [4]. [5] conducted prediction on Bitcoin Price, focusing on 3 input variables, Close Price, Gold Price and Tweets (sentiment). The latter found that GRU outperformed CNN with an RMSE of 179.23, however LSTM was the best model with 151.67. [6] studied the precision with which the direction of the Bitcoin price in United States Dollar (USD) can be predicted. Besides feature selection, they also used Bayesian optimization to select LSTM parameters. The Bitcoin dataset ranged from the 19th of August 2013 to 19th of July 2016. The latter used multiple optimization methods to improve the performance of deep learning methods. The primary problem of their work is overfitting. Elsewhere GRUs offer additional benefits due to having a more straightforward structure [7], predicting the future price using Open, High, Low, Close and Volume Price of historical data which results in GRU having a quite low RMSE at 0.2113. To evaluate the possibility of outperforming the market, this paper pays particular attention to deep learning topics for cryptocurrency price predictions.

So, the main objective of this paper is to examine cryptocurrency prediction algorithms using artificial intelligence and propose a suitable model for prediction, acquire relevant input features affecting cryptocurrency prices to achieve an accurate result when predicting, develop and optimise the deep learning-based algorithm for cryptocurrency close price prediction, analyse the effect of different trading days and various input features combination on deep learning models prediction and evaluate the performance of the proposed cryptocurrency prediction model.

2   Methods

2.1   Overview

In this paper, the techniques and methods used in identifying specific parameters or processes is described. Proper selection of certain parameters and specific processes is essential in any research because every chosen method must have a valid justificaion and referencing. So, typically, developing a neural network model for cryptocurrency forecasting involves many processes and methods, this can be seen in the flow shown in Fig. 1 below. This research requires many steps, activities and processes before delivering the result. Figure 1 shows the phases involved and deliverables.

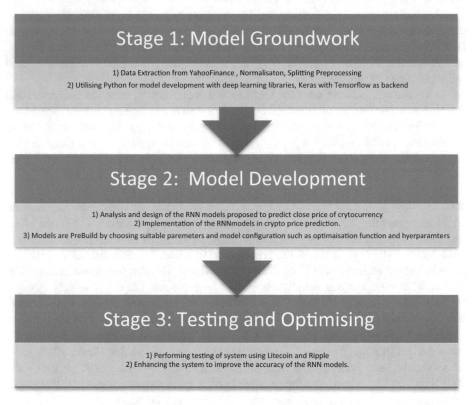

Fig. 1. Overview of the research Methodology

The process for deploying the RNN model in predicting cryptocurrency price involves three stages. In Stage 1, to train the model, dataset was collected from YahooFinance containing historical price information from a rank 10 cryptocurrency: Litecoin. Dataset is gathered based on daily prices staring from 29th April 2013–27th February 2021. Data normalization is performed to increase the model's efficiency and accuracy.

In Stage 2, RNN-LSTM will be deployed as the predictive model. The model will be split into three parts which are training, validating, and testing. Models are then tuned to achieve optimum prediction by tuning the hyperparameters. In Stage 3, models are then tested, comparing the actual and predicted price. Enhancing is done by feeding different input features combination to the proposed models.

2.2 Model Groundwork

Data Extraction. There are a lot of steps taken during Stage 1. Firstly, data is extracted from YahooFinance. YahooFinance is a cryptocurrency exchange platform where we could obtain the data of crypto price freely and easily. However, it is limited to certain periods for certain data of crypto price. The data extracted from YahooFinance is required to be sorted and normalised so that it could be fitted to the RNN model that is used as well providing valid output result. The dataset was downloaded with. csv format which have some features like Date Open, High, Low, Close, Volume (OHLCV) and Marketcap. The Dataset of 2862 rows of which the row is based on the number of days, totaling up to 16,902 data points to be trained. Input features for the model set up is Close Price as the targeted predicted output is Close Price.

Table 1. Features Description of Litecoin LTC and Ripple XRP

Features	Description
Date	Date of observation
Open	Opening price on the given day
High	Highest price on the given day
Low	Lowest price on the given day
Close	Closing price on the given day
Volume	Volume of transactions on the given day

The Correlation Analysis is a method of analyzing the linear relationship between two variables. The two variables can be independent or correlated, and the strength of the relationship between two variables is called correlation. The correlation analysis uses the Pearson correlation coefficient. The Pearson correlation coefficient is a measure of the linear correlation between two variables. The Pearson correlation coefficient has a value between $+1$ and -1 due to the Kosi-Schwartz inequality, where $+1$ is a perfect positive linear correlation, 0 being no correlation, and -1 is a perfect negative linear relationship.

Data Normalization. The goal of normalization is to change quantitative values in the data set to a common scale, provided zero changes in the original range of values. The main idea behind normalization/standardization is always the same. Normalization/standardization is always based on the same principle. Variables measured on varying scales may not contribute equally to the model fitting & model learnt function, which may result in a bias. Before fitting a model to data from [0, 1] normalisation techniques such as Min max scaling are typically employed to address this possible issue.

Data Splitting. According to [8], data splitting are divided into train 70%, remaining 30% for both validation and test. Dataset is divided into a training set: observations between 29 April 2013–21 October 2018, which is 1988 trading days, a validation from 22 October 2018–23 December 2019 having 415 days and testing from 24 December 2019–27 February 2021 also consist of 415 trading days. Similar data splitting is done for different altcoin, which is ripple (XRP), whereby, dataset is divided into a training set: observations between 5 August 2013–20 November 2018 which is 1934 Trading days, a validation from 7 November 2018–9 January 2020 (400 trading days) testing from 10 January 2020–27 February 2021 having 400 trading days.

2.3 Model Development

A RNN model is developed and modified by referring online sources. The most important function in the RNN is as below: the best combination of parameters. Hyperparameters are tuned to achieve the optimum predictive model.

Number of Epochs. In terms of artificial neural networks, an epoch refers to one cycle through the full training dataset. Usually, training a neural network takes more than a few epochs. In other words, if a neural network is fed with the training data for more than one epoch in different patterns, a better generalization is hoped when given a new "unseen" input (test data). For this work, the number of epochs are determined based on the commonly used values from existing research coupled with a trial and error values in determining the optimal value.

Adaptive Optimization Algorithm. Optimization algorithms are used to update weights and biases of a model to reduce error. Optimization algorithms can be divided into two main categories, which are constant learning rate algorithm and adaptive learning algorithm. The common first order optimization functions are Stochastic Gradient Descent (SGD), RMSProp and Adam.

Stochastic Gradient Descent (SGD) provides the foundation for several other learning algorithms, such as Adam and RMSProp; however, these algorithms have an adaptable learning rate. Indeed, the learning rate is a crucial hyperparameter in neural networks, as the loss function can be responsive or insensitive in certain directions of the parameter space.

For instance, gradients might get stalled at local minima or flat areas. The objective of the RMSProp method, a modified version of the AdaGrad algorithm [9], is to improve performance with non-convex functions. RMSProp modifies the gradient, g, by dividing the learning rate, η, by an exponentially declining average of squared gradients, θ which represents the error gradient.

$$\theta = \theta - \frac{\eta}{\sqrt{\mathbb{E}[g^2] + \epsilon}} g \tag{1}$$

Adam developed by [10] is another adaptive algorithm and is nowadays one of the most used optimization algorithms. It is a combination of RMSProp and momentum SGD algorithms. β_1, β_2 represents initial decay rate,

$$
\begin{aligned}
m &= \beta_1 m + (1 - \beta_1)g \\
v &= \beta_2 v + (1 - \beta_2)g^2
\end{aligned}
\tag{2}
$$

where m and v are estimates of the first moment (mean) and the second moment vectors of the gradient, g. These estimations are biased, so the authors compute a bias-correction at time step t

$$
\begin{aligned}
\hat{m} &= \frac{m}{1 - \beta_1^t} \\
\hat{v} &= \frac{v}{1 - \beta_2^t}
\end{aligned}
\tag{3}
$$

Hence, the update rule for given iteration:

$$
\theta = \theta - \frac{\eta}{\sqrt{\hat{v} + e}} \hat{m}
\tag{4}
$$

Dropout Rate. Dropout is a strategy that is designed to handle 2 major concerns over-fitting, and bigger number of neurons. It prevents overfitting and enables the efficient combination of an exponentially large number of distinct neural network topologies [11]. The word "dropout" refers to units in a neural network that are no longer active (both hidden and apparent). By dropping a unit from the network, we mean temporarily disconnecting it from all of its incoming and outgoing connections. The units to be dropped are chosen at random. In the simplest instance, each unit is preserved with a fixed probability p independent of other units, where p can be determined using a validation set or set to 0.5, which appears to be near to optimum for a broad variety of networks and tasks.

Batch Size. The batch size restricts the amount of samples displayed to the network prior to a weight change. When making predictions with the fitted model, the same constraint applies. In particular, the batch size employed while fitting. The model that determines how many forecasts must be made simultaneously. this becomes problematic when fewer forecasts are made than the batch size. Ideal results with a big batch size can be achieved however, predictions made for a single observation at a time while solving a problem involving a time series or sequence, may result in longer execution time [12]. Utilises a batch size of 32, whereas [13] uses a batch size of 128.

2.4 Testing and Enhancing

Initially, the data gathered from YahooFinance in forecasting Litecoin (LTC). The LSTM model is tested with trial and error to obtain the best model which results in the highest accuracy or least average root mean square error in terms of share price. Then, the LSTM model is applied to other data from other coin which is Ripple (XRP).

Table 2 depicts different combination of input features to further examine the performance of forecasting for all models. Finally, the accuracy of LSTM GRU and Bi-LSTM model is compared and analyzed (Table 1).

Table 2. Set of input features for testing and enhancing of predictive models

Input fetaures combination	Number of features	Targeted output
Close Price	1	Close Price
Open Price, High Price, Low Price and Close Price (OHLC)	4	
Open Price, High Price, Low Price,Close Price and MarketCap (OHLCM)	5	
Open Price, High Price, Low Price,Close Price Volume and MarketCap (OHLCVM)	6	

3 Results and Discussion

3.1 Overview

In this paper, the preliminary results of the proposed model and the data extraction method for the training validation and testing of the model were being discussed. As mentioned, the proposed model is RNN, and the parameters chosen for the model as mentioned previously were being applied and analyzed. A comparative analysis of RNN models is examined. There are 4 hyperparameters to be manipulated in the LSTM, GRU and Bi-LSTM model. Validation Loss and Training Loss are performed to overcome overfitting issue.

The first parameter is number of epochs, followed by Adaptive Optimization Algorithm, Batch Size and Dropout Rate. Then, each model is tested with different combination of input features to further observe the robustness of each model. The score indicators are RMSE, MAPE and Executional Time.

3.2 Evaluation of LSTM, GRU and Bi-LSTM Model

Determine Optimum Number of Epochs in LSTM, GRU and Bi-lSTM Model. In this experiment, the parameter of the model whereby the number of epochs is analyzed. The hyperparameter is tested to the training set of LTC data. The number of epochs used are 20,40,60,80, 100.

Table 3. Effect of number of epochs on LSTM, GRU and Bi-LSTM RMSE

Models	Number of epoch	RMSE
LSTM	20	5.1693
	40	3.226
	60	0.1993
	80	**0.00736**
	100	0.6743
GRU	20	5.7259
	40	3.7321
	60	2.0293
	80	0.9216
	100	**0.0693**
Bi-LSTM	20	3.2946
	40	2.0293
	60	0.5294
	80	**0.4331**
	100	0.7723

Based on Tables 3 and 4 above, it could be concluded that the optimum number of epochs are 80,100 and 80 with RMSE of 0.00736, 0.0693 and 0.7723 for LSTM, GRU and Bi-LSTM respectively (Tables 5, 6 and 7).

Determine Optimum Optimization Algorithm in LSTM, GRU and Bi-lSTM Model In this experiment, the parameter of the model whereby the adaptive optimizer is analyzed. The hyperparameter is tested to the training set of LTC data. The optimizer used are Adam and RMSprop.

Table 4. Effect of optimizers on LSTM,GRU and Bi-LSTM RMSE

Models	Optimiser	RMSE
LSTM	**Adam**	**2.0982**
	RMSProp	2.6613
GRU	**Adam**	**1.6474**
	RMSProp	3.3573
Bi-LSTM	**Adam**	**0.8732**
	RMSProp	3.2668

Based on table above, it could be concluded that the optimum optimiser are Adam algorithm with RMSE of 2.0982, 1.6474 and 0.8732 for LSTM, GRU and Bi-LSTM respectively.

Determine Optimum Batch Size in LSTM, GRU and Bi-lSTM Model. In this experiment, the parameter of the model whereby the batch size is analysed. The hyperparameter is tested to the training set of LTC data. The batch size used are 32,64 and 128.

Table 5. Effect of batch sizes on LSTM,GRU and Bi-LSTM RMSE

Models	Batch size	RMSE
LSTM	32	2.3060
	64	2.9178
	128	**0.1993**
GRU	**32**	**0.9216**
	64	1.8813
	128	2.6249
Bi-LSTM	32	1.8765
	64	2.6425
	128	**0.8834**

Based on table above, it could be concluded that the optimum batch sizes are 128,32 and 128 with RMSE of 0.1993, 0.9216 and 0.8834 for LSTM, GRU and Bi-LSTM respectively.

Determine Optimum Dropout Rate in LSTM, GRU and Bi-lSTM Model. In this experiment, the parameter of the model whereby the batch size is analysed. The hyperparameter is tested to the training set of LTC data. The dropout rate used are 0.1,0.2,0.4, 0.5 and 0.7

Based on table above, it could be concluded that the optimum dropout rates are 0.1, 0.1 and 0.2 with RMSE of 1.8503, 1.6014 and 0.7782 for LSTM, GRU and Bi-LSTM respectively.

Evaluate Performance of LSTM, GRU and Bi-lSTM Model with Different Input Features on LTC and XRP. The combinations are selected based on the correlation weight of features towards the predicted output. Open Price, High Price, Low Price, Close Price (OHLC) are selected as a prediction benchmark which is similarly used by [14] which the latter used similar models to analyse their performances when predicting the close price of cryptocurrency.

In this experiment, the models are tested with different combination input features using the LTC and XRP dataset. The experiment will be carried out with four different input features combinations according to the cases namely, Close Price for a univariate model, Open Price, High Price, Low Price, Close Price (OHLC), Open Price, High

Table 6. Effect of dropout rates on LSTM, GRU and Bi-LSTM RMSE

Models	Dropout rate	RMSE
LSTM	**0.1**	**1.8503**
	0.2	1.8666
	0.4	2.9457
	0.5	2.9968
	0.7	1.9969
GRU	**0.1**	**1.6014**
	0.2	1.9527
	0.4	3.3642
	0.5	3.0203
	0.7	1.6015
Bi-LSTM	0.1	1.4321
	0.2	**0.7782**
	0.4	4.0991
	0.5	0.9963
	0.7	0.8765

Price, Low Price, Close Price and Market Cap (OHLCM)and Open Price, High Price, Low Price, Close Price, Volume and Market Cap (OHLCVM). The overall performance of all predictive models are tabulated a follows;

From Fig. 2, GRU outperforms LSTM and Bi-LSTM when predicting the price of LTC. Based on previous work done by [14] also found that GRU outperforms LSTM and BILSTM, with 2.201 RMSE value in this paper when OHLC price is treated as input features which justifies the development and optimization of this model. The RMSE value obtained from this paper deviates by 1.671 in RMSE. The latter also obtained a RMSE value of 3.069 for LSTM (deviation of 16.44% from this paper) and 4.307 for Bi-LSTM (deviation of 10.82% from this paper) which suggests a closer value to the RMSE obtained in this paper and [14] findings for LSTM and Bi-LSTM models. The MAPE scored by [14] and this paper is not far off; LSTM of 0.874 and 0.8893 respectively, GRU of 0.2216 and 0.4888 respectively and Bi-LSTM of 2.332 and 0.8864 respectively. The above figure also shows that when the input features increase, GRU has an inconsistent result as the RMSE value fluctuates but still managed to outperform both LSTM and Bi-LSTM when all 6 features are fed. On the other hand, Bi-LSTM and LSTM shows a positive impact when the features increase.

Although, the RMSE value spiked when OHLC is tested out, they both shows gradual reduction in terms of RMSE value when OHLCM and OHLCVM are experimented. This shows that Bi-LSTM and LSTM are more robust and accurate as more input data are being fed to the model.

Table 7. Summary of Performance Evaluation for RMSE, MAPE and Execution Time of all models for LTC and XRP closing prices.

Reference	Input features	Model	Result					
			RMSE		MAPE (%)		Time(s)	
			LTC	XRP	LTC	XRP	LTC	XRP
Hameyal et. al 2021	OHLC Price	LSTM	3.069	-	0.8474	-	NA	-
		GRU	**0.825**	-	**0.2116**	-	NA	-
		Bi-LSTM	4.307	-	2.332	-	NA	-
This paper	Close Price	LSTM	2.5642	0.1260	0.8893	0.8893	480	480
		GRU	**2.4960**	**0.0237**	0.4888	0.4888	200	200
		Bi-LSTM	4.7732	0.1307	0.8664	0.8664	1200	1200
This paper	OHLC Price	LSTM	3.8869	0.0390	1.3596	1.5791	640	640
		GRU	**2.2201**	**0.0089**	0.8076	0.6620	200	200
		Bi-LSTM	4.9831	0.0513	0.9352	0.8237	4320	4320
This paper	OHLC Price and MarketCap	LSTM	3.2258	**0.0125**	0.7282	0.6020	800	800
		GRU	**3.0567**	0.0192	0.6357	1.4997	600	600
		Bi-LSTM	3.9137	0.0367	1.441	1.4537	4640	4640
This paper	OHLC Price Volume and MarketCap	LSTM	2.2237	0.0073	0.7782	3.4628	800	800
		GRU	**0.9589**	0.0338	0.6659	1.3054	600	600
		Bi-LSTM	2.5738	**0.0046**	1.2090	0.8723	4640	4640

Figure 3 depicts the RMSE value when XRP are used as dataset which underwent similar experiment from [15] successfully verified that Bi-LSTM network is the most effective model when predicting the close price of XRP. The latter however used a different input feature which involves OHLC Price and Volume.

Bi-LSTM outperforms LSTM and GRU when all 6 features OHLCVM are treated as input, in contrast to when LTC is utilised. Bi-LSTM scored an astonishing 0.0046 in RMSE value scoring better than [15] findings; 0.979 in RMSE. In this paper, LSTM is not far off with only 0.0073 of RMSE while GRU performed the worst scoring a mere 0.0338 as for the RMSE.

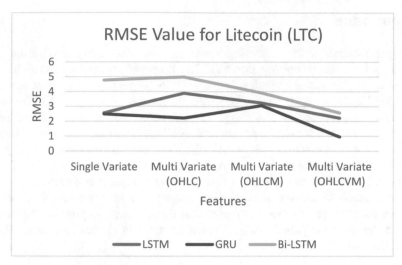

Fig. 2. Line Chart Representation on the effect of different input features on LSTM, GRU and Bi-LSTM models for LTC

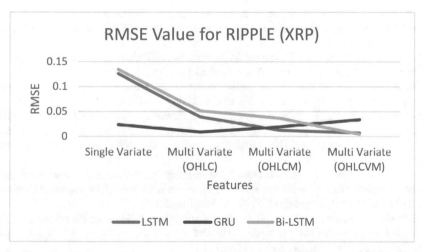

Fig. 3. Line Chart Representation on the effect of different input features on LSTM, GRU and Bi-LSTM models for LTC

Again, for single input features GRU performs the best with the lowest RMSE score of 0.0237, which is logical since GRU having a faster computational time and accomplished better result due to having only update and reset gate. As a matter of fact, a simpler model like GRU, caters for a smaller dataset size while high complexity model namely LSTM and Bi-LSTM are superior when dataset size broadens.

4 Conclusion

This paper discusses on the forecasting cryptocurrency prices using deep learning models as a tool for cryptocurrency investors. The proposed forecasting model has been made based on the studied reviews which were RNN ecxtensions. Performance scores – RMSE,MAPE and computational time - were calculated for LTC and XRP to test the accuracy of the proposed models. Based on these outcomes, GRU were exceptional in terms of performance for LTC for every different input features. This model is considered the best model however, LSTM and Bi-LSTM models showed superiority when the number of input features fed increased, indicating the memory capacity of the bi-directional architecture in predicting a time large time series data. For the extension of this work, sentiment analysis should be considered as a factor on how they influence the cryptocurrency price as well as performing dimensionality reduction technique to further experiment the performance of higher complexity models such as LSTM and Bi-LSTM.

References

1. Nasir, M.A., Huynh, T.L.D., Nguyen, S.P., Duong, D.: Forecasting cryptocurrency returns and volume using search engines. Finan. Innov. **5**(1), 1–13 (2019). https://doi.org/10.1186/s40854-018-0119-8
2. Livieris, I.E., Pintelas, E., Stavroyiannis, S., Pintelas, P.: Ensemble deep learning models for forecasting cryptocurrency time-series. Algorithms **13**(5), 121 (2020)
3. Hitam, N.A., Ismail, A.R.: Comparative performance of machine learning algorithms for cryptocurrency forecasting. Ind. J. Electr. Eng. Comput. Sci **11**(3), 1121–1128 (2018)
4. Selvamuthu, D., Raj, R., Sahu, K., Aggarwal, A.: Tradeoff between performance and energy-efficiency in DRX mechanism in LTE-A networks. In: 2019 IEEE International Conference on Advanced Networks and Telecommunications Systems (ANTS), pp. 1–5. IEEE, December 2019
5. Aggarwal, A., Gupta, I., Garg, N., Goel, A.: Deep learning approach to determine the impact of socio economic factors on bitcoin price prediction. In: 2019 Twelfth International Conference on Contemporary Computing (IC3), pp. 1–5. IEEE (2019)
6. McNally, S., Roche, J., Caton, S.: Predicting the price of bitcoin using machine. In: 26th Euromicro International Conference on Parallel, Distributed and Network-based Processing (PDP), Cambridge, UK, pp. 339–343 (2018)
7. Tanwar, S., Patel, N.P., Patel, S.N., Patel, J.R., Sharma, G., Davidson, I.E.: Deep learning-based cryptocurrency price prediction scheme with inter-dependent relations. IEEE Access **9**, 138633–138646 (2021)
8. Dutta, A., Kumar, S., Basu, M.: A gated recurrent unit approach to bitcoin price prediction. J. Risk Finan. Manag. **13**(2), 23 (2020)
9. Hinton, G., Srivastava, N., Swersky, K.: Neural networks for machine learning lecture 6a overview of mini-batch gradient descent. Cited on 14(8), 2 (2012)
10. Kingma, D.P., Ba, J.: Adam: A method for stochastic optimization. arXiv preprint arXiv:1412.6980 (2014)
11. Srivastava, N., Hinton, G., Krizhevsky, A., Sutskever, I., Salakhutdinov, R.: Dropout: a simple way to prevent neural networks from overfitting. J. Mach. Learn. Res. **15**(1), 1929–1958 (2014)

12. Ranawat, K., Giani, S.: Artificial intelligence prediction of stock prices using social media. arXiv preprint arXiv:2101.08986 (2021)
13. Koo, E., Kim, G.: Prediction of Bitcoin price based on manipulating distribution strategy. Appl. Soft Comput. **110**, 107738 (2021)
14. Hamayel, M.J., Owda, A.Y.: A novel cryptocurrency price prediction model using GRU, LSTM and bi-LSTM machine learning algorithms. AI, **2**(4), 477–496 (2021)
15. Birim, S.O.: An analysis for cryptocurrency price prediction using LSTM, GRU, and the bi-directional implications (2022)

Networking and IoT

Profile of Fitness and Diet App Users in Online Wellness Social Communities in Malaysia: Cross-Sectional Pilot Study

Rasha Najib Aljabali[✉] and Norasnita Ahmad

Universiti Technologi Malaysia, Johor, Malaysia
rashagabaly@gmail.com, norasnita@utm.my

Abstract. This study explores users' characteristics, motivations, and usage of fitness and diet apps in online wellness social communities in Malaysia. 169 online wellness social communities' users aged 18+, answered a web-based survey for this pilot study. Results showed a high rate of usage (128/169, 77%) of fitness and diet apps among online wellness social communities' users. Multivariate logistic regression revealed that gender, education level, occupation, income, and weekly exercise influenced the usage of fitness and diet apps. Paired t-test showed that the frequency usage of fitness apps is significantly higher than ($P < 0.01$) diet apps among participants. Exercise tracking ($P < 0.001$) was the primary motivation behind using fitness apps. Tracking daily diet activities ($P < 0.01$) was the significant motivation for using diet apps. The main contribution of this pilot study was that online wellness social communities were an ideal environment to promote fitness and diet apps.

Keywords: mHealth App · Fitness · Diet · Mobile Health · Physical Activity · Online Wellness Communities

1 Introduction

Unhealthy diet and physical inactivity have a significant association with increased risk factors of not only numerous non-communicable diseases such as cardiovascular disease, diabetes, and obesity but also mortality. Globally, an estimated 5.3 and 3.2 million deaths were attributable to physical inactivity in 2008 and 2010, respectively [1]. Similarly, in 2010, researchers estimated 4.9, 4.0, and 2.5 million deaths related to diets low in fruits, high in sodium, and low in nuts and seeds, respectively. The consequences of unhealthy diet and physical inactivity are global public health concerns across all age groups; Malaysia is no exception [2]. Meanwhile, internet behavior, seeking health information online, and social media use have significantly changed to impact everyday life. Mobile technology is therefore becoming an integral part of the identity of each individual, affecting their lives, decisions, and health behavior on a daily basis [3]. The Malaysian Communications and Multimedia Commission estimated that 88.8% of the Malaysian population in 2021 were internet users [4] and 93% of them were online for

F. Saeed et al. (Eds.): ICACIn 2022, LNDECT 179, pp. 221–232, 2023.
https://doi.org/10.1007/978-3-031-36258-3_19

social purposes. Thus, Social Networking Sites (SNSs) have grown rapidly in the past decade and have become an informal source of health education and promotion [5]. The importance of social presence in health contexts, where sociability promotes trust and desire for interaction about personal and often anxiety-producing information, is heightened [6]. Thus, individuals with fitness and diet goals use online wellness social communities' platforms such as Facebook, Instagram, YouTube, and Telegram to share fitness and diet achievements, find innovative fitness and diet products, seek coaching and professional support, and gain peer support. Additionally, social presence online can lead to changes in eating habits and increased physical activity on the offline side or lead to acquire supported tools such as fitness and diet apps [6]. According to prior research, there is evidence that the use of online wellness communities for health information and mobile health apps for managing healthy behaviors such as diet and exercise is simultaneous [7].

Undoubtedly, fitness and diet apps are cost-saving tools and have shown promising results in promoting physical activity and a healthy lifestyle [8, 9]. It was recently reported that there are currently 325,000 health and fitness apps available on major app stores [9]. Fitness apps include aerobic exercise (e.g., walking, jogging, or running), muscle building, and body line shaping [10–14]. Other fitness apps can introduce diet and weight loss functions such as providing nutritional values and vegetarian choices [15, 16]. Mhealth apps from the diet perspective (e.g., MyFitnessPal) are connected to meal planning, meal recipes, physical activity, water intake, calorie checker, weight tracking, and weight goal setting [17]. Therefore, the Ministry of Health of Malaysia (MOH) has set a goal to promote healthy behaviors using innovative and evidence-based tools such as Web-based health promotion, including fitness and diet apps, in government and private health sectors. Efforts made by healthcare practitioners to develop, endorse, and promote such tools might be wasted if dissimilarities in individuals' characteristics such as sociodemographics, health, physical activity and eating behaviors are ignored. Studies found that individuals differed widely in utilizing fitness and diet apps. 9.23% of individuals may intend to use fitness apps and 7% may intend to use diet apps but are not yet current users whereas 15% of expected users have decided to drop fitness apps and 14% have decided to drop diet apps [14]. Similarly, the mobile app developers may invest in designing fitness and diet apps that might not be used because of the lack of consideration of local statistics related to the targeted audience. Developers may depend on the predominant United States (US) or United Kingdom (UK) statistics regarding the characteristics of diet and fitness apps' users [18–20]. However, these results cannot be generalizable to developing countries' context due to the distinction in culture, technology usage rate, social support in achieving health outcomes through technology, and mobile health consumer behavior.

Hence, there is insufficient evidence about the actual fitness and diet app users' characteristics in Malaysia and the motivations behind their usage. In addition, few studies have examined characteristics of fitness and diet app users [21]. Fitness and diet apps need to be better understood in order to increase acceptance and adoption, as well as to encourage engagement for continued use. Therefore, this study is a cross-sectional survey for Malaysian users to further explore the diet and/or fitness apps users and non-users in online wellness social communities. Thus, this study examines the impact of sociodemographic, health-related information, international recommendations for physical activity, and vegetable and fruit consumption of the diet and fitness app users and non-users. In addition, this study compares the usage of diet apps and fitness apps and investigates the motivations behind using diet and fitness apps.

2 Methods

2.1 Participants

Data was collected from the 12th of January 2022 to the 8th of March 2022. Participants eligible for this study were Malaysians aged 18 or older. This study used the purposive sampling technique. Therefore, respondents were recruited from fitness and diet online wellness social communities on platforms such as Facebook, Telegram, and Instagram. A web-based questionnaire was distributed using a Google Doc URL link in public and private fitness and diet Facebook groups in Malaysia. The fitness groups involved provided exercise, body reshaping, yoga, running, cycling, Zumba, and hiking support. Diet groups included intermediate fasting, ketogenic diet, tracking calorie intake, vegetarian diet, etc. In addition, the survey was shared by fitness and diet Instagram influencers to their followers. Participants were informed about the aim of the study and the confidentiality of the raw dataset. The dataset was saved in a secured password-protected electronic format. Moreover, data was stored anonymously. The online survey settings were configured to not share the participants' IP addresses or email addresses. The survey was headed with written, electronic consent, and participants voluntarily took the survey. If participants changed their mind about taking the survey, they could quit the survey without saving their initial responses.

2.2 Measures

This study used an online survey to identify the fitness and diet app users and non-users in online wellness social communities and compare frequency of diet and fitness app usage. In addition, this survey assessed the motivations behind the usage of diet and fitness apps. Therefore, the research team drafted the questionnaire in English in accordance with prior literature. The authors recruited a professional translator to translate the survey into Malay. The second author reviewed the survey as bilingual in English and Malay. The questionnaire was pre-tested and distributed to random respondents to assess the accuracy, readability, understanding, and time taken to complete the questionnaire.

The survey contained 26 questions which comprised eight divisions: sociodemographic characteristics (6 items), health-related characteristics (4 items), international recommendations of physical activity and daily vegetable and fruit portions (3 items), smartphone/tablet ownership (1 item), downloading fitness and diet apps (1 item), current number of fitness and diet apps (1 item), usage frequency of fitness and diet apps (2 items), and motivations for using fitness and diet apps (8 items). The researchers used height (m) and weight (kg) to calculate the Body Mass Index (BMI) ($kg/m2$) of respondents. The questions' types are various: categorical (e.g., age, education level, work, income, race, BMI, and chronic disease), continuance (e.g., frequency of weekly exercise in hours was calculated by multiplying the number of exercise days by the exercise duration per day), and dichotomous (e.g., gender, smoker/non-smoker, smartphone/tablet ownership and downloading fitness and diet apps). Subsequently, the downloading fitness and diet apps question is the first dependent variable. While participants who had installed fitness and diet apps were followed up to complete the survey, participants who did not download any fitness and diet apps were eliminated. The fitness and diet apps characteristics question included the number of fitness and diet apps downloaded. A five-point Likert scale from 1 (Never) to 5 (many times per day) was used to measure the frequency of usage of diet apps; it is the second dependent variable that measured the actual usage of diet apps. Similarly, the frequency usage of fitness apps is the third dependent variable that measures the actual usage of fitness apps [17, 22]. The motivations behind using the fitness and diet apps were measured using eight questions representing the motivations extracted from prior literature [12, 23].

2.3 Statistical Analysis

Data was extracted and reviewed in the form of a Microsoft Excel file. Descriptive statistics were used to explain the proportions of the dependent variable, downloading fitness and/or diet apps. In addition, descriptive statistics were used to describe the frequency and percentage of each category in sociodemographic characteristics, health-related factors, and international recommendations of physical activity and daily vegetable and fruit portions based on the first dependent variable (see Table 1). Logistic regressions were performed to explain relationships between independent variables (categorical or continuance) and categorical dependent variable (i.e., users who used a fitness and diet app). Multivariate logistic regression analysis was used to explore the association between characteristics and fitness and diet apps usage (see Table 1). For fitness and diet apps of current users, a paired t-test was undertaken to compare the actual usage of diet apps and fitness apps (see Table 2). Multiple regression analysis was conducted to assess users' motivations to use both fitness apps and diet apps (see Table 3). The statistically significant levels used in the analysis were (P < 0.05) with 95% CI Confidence Intervals. All statistical analysis was performed using R statistical package version 4.1.2.

3   Results

3.1   Participants' Characteristics

Table 1 describes the descriptive statistics of the characteristics of sociodemographic, health-related, and international recommendations (frequency of weekly exercise, daily vegetable and fruit portions) based on fitness and/or diet apps users and non-users. As this study is a pilot study, the sample size is convenient for understanding the population characteristics for the primary data collection. A total number of 171 participants took the survey. However, two participants clicked to disagree with the electronic consent; therefore, the number of participants reduced to 169, thus, the response rate of this study is 98.8%. Results showed that an estimated (128/169) 77% of Malaysians who have smartphones and tablets have downloaded fitness and/or diet apps. In comparison, an estimated (41/169) 23% did not download fitness and/or diet apps. Based on the sample distribution described in Table 1, most participants' ages belonged in two age group compositions; 25–34 years (38.46%) and 35–44 years (33.14%). Overall, 65.68% of the participants were female, and 81% were from the Bumiputera ethnic group. The majority of the respondents were employed/self-employed (67.49%), had achieved a bachelor's degree (48.52%), postgraduate degree (40.24%), and had monthly income more than MYR4000 (40.83%). Overall, a large proportion of participants were free from chronic disease (86%), were non-smokers (94.67%), and of normal weight based on BMI (kg/m^2) $> = 18.5$ and $< = 24.9$ (48.52%). Regarding the international recommendations of physical activity, participants took an average of 2.67 h of exercise weekly. Moreover, participants had low consumption of the daily vegetable and fruit portions, as the majority (30%) consumed one portion (80 g) of vegetables and fruit daily.

3.2 Indicators Associated with the Fitness and Diet App Usage

The multivariate logistic regression analysis results have been described in Table 1. Logistic regression was conducted to explore the fitness and diet apps usage indicators for 169 participants. The independent variables involved in the nominal logistic analysis were sociodemographic, health-related, and international recommendations (weekly physical activity and daily vegetables and fruit portions). The results revealed that fitness and diet app usage is influenced by gender, education level, occupation, monthly income, and weekly exercise. As shown in Table 1, the odds ratio (OR) with 95% Confidence Interval (95% CI) of the categories for each variable were reported. Subsequently, the male individuals compared to females were 0.31 times less likely (OR: 0.31, 95% CI: 0.10 – 0.87, P: 0.03*) to download and use fitness and diet apps. Individuals with basic education levels such as secondary school compared to individuals with postgraduate (master's, PhD, MD, etc.) were 0.06 times less likely (OR: 0.06, 95% CI: 0.00 – 0.69, P: 0.027*) to download and use fitness and diet apps. Regarding occupation, student respondents (OR: 6.8, 95% CI:1.11 – 47.60, P: 0.044*) were more likely to use fitness and diet apps compared to employed respondents. Respondents who received monthly income less than MYR1000 were 0.13 times less likely (OR: 0.13, 95% CI: 0.02 – 0.75, P: 0.025*) to use fitness and diet apps compared to respondents with monthly income more than MYR4000. Results disclosed that individuals who followed the international recommendations of the weekly exercise frequency have 1.47 times higher (OR: 1.54, 95% CI 1.16 – 1.93, P: 0.003**) usage of fitness and diet apps. However, results found that age, smoking, chronic disease, BMI, race, consumption of daily vegetable and fruit portions have an insignificant relationship with usage of fitness and diet apps.

Table 1. Descriptive Statistics for the participants and Multivariate Logistic Regression Analysis of the study (N = 169)

Characteristics	Participants (n = 169)	Users (n = 128,77%)	Non-Users (n = 41,23%)	Odds Ratio (OR)	(95% CI)	P value
Statistics						
Sociodemographic characteristics **Age n (%)**						
18–24	11 (6.52%)	8 (5%)	3 (1.52%)	0.52	0.06 – 4.79	0.547
25–34 (ref)	65 (38.46%)	51 (30.46%)	14 (8%)	-	-	-
35–44	56 (33.14%)	42 (25.14%)	14 (8%)	0.76	0.24 – 2.40	0.648
45–54	28 (16.57%)	21 (12.57%)	7 (4%)	1.81	0.42 – 8.40	0.436
54+	9 (5.33%)	6 (3.7%)	3(1.5%)	0.25	0.02 – 2.87	0.246
Gender n (%)						
Female (ref)	111 (65.68%)	86 (51%)	25 (15%)	-	-	-
Male	58 (34.32%)	42 (25.5%)	16 (9%)	0.31	0.10 – 0.87	0.03*
Race n (%)						
Malay(ref)	134 (81%)	104(61%)	34 (20%)	-	-	-
Indian	10 (6%)	7 (4.5%)	3 (1.5)	0.66	0.11 – 4.59	0.652
Chinese	17 (10%)	14 (9%)	3 (1.5)	0.8	0.18 – 4.29	0.777
Others	4 (2%)	3 (1.5%)	1 (0.5)	2.59	0.20 – 75.54	0.498
Education n (%)						
High school	5 (2.96%)	2 (1.5%)	3(1.5)	0.06	0.00 – 0.69	0.027*
College/vocational school/apprenticeship	14 (8.28%)	11 (6.5%)	3 (1.5)	1.37	0.25 – 9.08	0.726
Bachelor's degree	82 (48.52%)	63 (37.5%)	19 (11.5%)	1.15	0.41 – 3.27	0.793
Graduate degree (master's, PhD, MD, etc.) (ref)	68 (40.24%)	52 (31%)	16(9%)	-	-	-
Occupation n (%)						
Employed/Self-Employed (ref)	114 (67.49%)	86 (51%)	28 (17%)	-	-	-

(*continued*)

Table 1. (*continued*)

Characteristics	Participants (n = 169)	Users (n = 128,77%)	Non-Users (n = 41,23%)	Odds Ratio (OR)	(95% CI)	P value
Unemployed	15(9%)	10(6%)	5(3%)	1.7	0.23 – 13.29	0.606
Student	40 (23.69%)	32 (18.60%)	8 (5%)	6.8	1.11 – 47.60	0.044*
Income n (%)						
Below MYR 1000	36 (21.3%)	27 (16%)	9 (5%)	0.26	0.04 – 1.62	0.146
MYR 1001- MYR 2000	22 (13.02%)	15 (9%)	7 (4%)	0.13	0.02 – 0.75	0.025*
MYR 2001- MYR 3000	21 (12.43%)	15 (9%)	6 (4%)	0.35	0.07 – 1.61	0.17
MYR 3001- MYR 4000	21 (12.43%)	16 (10%)	5 (3%)	0.26	0.04 – 1.62	0.146
More than MYR 4000 (ref)	69 (40.83%)	55 (32.8%)	14 (8%)	-	-	-
Health-Related Characteristics Chronic Disease n (%)						
No chronic disease (ref)	145 (64%)	109(64%)	36 (22%)	-	-	-
With chronic disease	24 (14%)	19 (11%)	5 (3%)	0.99	0.27 – 4.23	0.993
Smoker n (%)						
Non-Smoker(ref)	160 (94.67%)	121 (72%)	39 (23%)	-	-	-
Smoker	9 (5.33%)	7 (4%)	2 (1%)	1.04	0.16 – 9.64	0.966
Body Mass Index (BMI) kg/m2 n (%)						
Underweight < 18.5	7 (4%)	5 (3%)	2 (1%)	1.38	0.15 – 17.61	0.785
Normal weight > = 18.5, < = 24.9 (ref)	82 (48.52%)	62 (36.52%)	20 (12%)	-	-	-
Overweight > = 25, < = 29.9	47 (27.81%)	36 (20.81%)	11 (7%)	1.64	0.59 – 4.78	0.348
Obese > = 30	33 (19.52%)	25 (14.5%)	8 (5%)	2.48	0.72 – 9.59	0.165
Physical Activity, fruit and vegetable consumption international recommendations						
Weekly Exercise in (Hours) mean (SD)	2.67(2.45)	3.00(2.51)	1.64(1.92)	1.47	1.16 – 1.93	0.003**
Number of portions of vegetables and fruit n (%) consumed daily						

(*continued*)

Table 1. (*continued*)

Characteristics	Participants (n = 169)	Users (n = 128,77%)	Non-Users (n = 41,23%)	Odds Ratio (OR)	(95% CI)	P value
None	9 (4.14%)	8 (4%)	2 (1%)	1.49	0.23 – 14.18	0.694
Half portion per day	34 (21%)	21 (13%)	13 (8%)	0.61	0.20 – 1.83	0.376
One portion per day (ref)	50 (30%)	37 (22%)	13 (8%)	-	-	-
Two portions per day	38 (23%)	33 (19%)	7 (4%)	2.09	0.62 – 7.61	0.244
Three portions or more per day	34 (20%)	29 (17%)	6 (3%)	2.16	0.57 – 8.96	0.266

* $P < 0.05$; ** $P < 0.01$; 95% CI: 95% Confidence Interval

3.3   Frequency Usage Comparison Between Fitness Apps and Diet Apps

Eight respondents (6.8%) answered 1, (None) on the current fitness and/or diet apps in their smartphone or tablet; so, the number of current diet and fitness apps participants declined to 120. Table 2 shows the paired t-test results for comparing the means of the two dependent variables, frequency usage of diet apps and frequency usage of fitness apps for 120 participants. Both variables were measured using a five-point Likert scale from 1 (Never) to 5 (many times per day). The paired t-test values revealed a statistically significant difference ($P < 0.01$) between the two means. As a result, respondents used fitness apps more frequently than diet apps.

Table 2. Paired T-test for Means of the frequency usage of Diet and Fitness Apps (N = 120)

	Frequency usage of diet apps	Frequency usage of fitness apps
Mean (SD)	2.35 (1.46)	2.76 (1.25)
Paired test result	t = -2.841, df = 119, p-value = 0.005292**	

** $P < 0.01$

3.4   Motivations Behind Fitness and Diet Usage

Multiple Regression analysis results are shown in Table 3. The analysis has been conducted to investigate relationships between various motivations to use diet and/or fitness apps and the frequency of usage. The results revealed that respondents use diet apps to track food intake, daily calorie intake and receive healthy nutrition recommendations with a statistically significant level ($P < 0.01$). However, respondents use fitness apps due to the tracking information of their physical activity such as the number of steps walked and calories burned, pace, and GPS information with a statistically significant level ($P < 0.001$). However, the results revealed that the remaining motivations did not have a statistically significant relationship with the usage of diet and fitness apps.

Table 3. Multiple Regression Analysis for motivations behind Fitness and Diet Apps Usage (N = 120)

Independent variables	Frequency usage of diet apps				Frequency usage of fitness apps			
	Estimate	Std.Error	t value	P value	Estimate	Std.Error	t value	P value
(Intercept)	−0.99	0.77	−1.29	0.20	1.38	0.72	1.91	0.06
Number of Apps	0.12	0.10	1.21	0.23	0.16	0.09	1.67	0.10
Weight management goals	0.19	0.16	1.19	0.24	−0.17	0.15	−1.11	0.27
Tracking daily fitness activities	0.07	0.13	0.55	0.58	0.49	0.12	3.93	***
Tracking daily diet activities	0.41	0.12	3.31	**	0.12	0.12	1.00	0.32
Social influence	0.13	0.10	1.28	0.20	0.10	0.09	1.13	0.26
Desire for an improved appearance	0.18	0.16	1.14	0.26	−0.19	0.15	−1.24	0.22
Desire for improved well-being and health	−0.07	0.18	−0.37	0.71	−0.27	0.17	−1.61	0.11
Stress relief and mental health management goals	−0.11	0.12	−0.87	0.38	0.17	0.11	1.44	0.15
Doctor /Physician recommendation	0.08	0.14	0.56	0.58	0.01	0.13	0.10	0.92
R^2	0.34				0.20			
Adjusted R^2	0.30				0.14			

** $P < 0.01$, *** $P < 0.001$

4 Discussion

Using a cross-sectional pilot study, 169 online wellness social communities' users answered the questionnaire. The research team explored the sociodemographic characteristics, health-related characteristics, international recommendations for physical activity, vegetable and fruit consumption that influence fitness and diet app usage in Malaysia. Moreover, this study compared frequency of usage of both fitness and diet app. Furthermore, this study revealed the relationships between motivations and current fitness and diet apps usage. The proportion of the users of fitness and diet apps in online social communities users (n = 129/169, 77%) is higher than the users in the = general population in prior research 60.4% [19], 58.23% [18], and 50.23 [20]. The findings of this study unveiled the characteristics of fitness and diet apps' users. Surprisingly, age did not have a statistical relationship with fitness and diet apps usage. This result is inconsistent with prior research; resistance to using health apps among individuals in the United States decreases with younger age [19, 20, 24]. The difference in results might be because of the sample composition, older online social communities' users usually have health and fitness interests, high level of smartphone and apps self-efficacy [7]. Results found that more females than males used and downloaded the fitness and diet apps. Findings revealed a consistent result with previous research [25] that higher education levels significantly impact the increased usage of physical activity and diet tracking apps. A higher number of student respondents than employed individuals used fitness and diet apps. In addition, the monthly income variable affects the usage of innovative tools to achieve fitness and diet goals [25].

Consequently, respondents with higher monthly income have higher usage of fitness and diet apps. However, ethnicity has no significant impact on fitness and diet usage. Results indicated that health-related characteristics such as chronic disease, BMI, and smoking status did not influence the usage of fitness and diet apps for online social communities. Regarding regular exercise supported by the international recommendations, results found that an increase in weekly exercise hours significantly influenced the usage of fitness and diet apps. Individuals who have regular physical activity, e.g., running, walking, swimming, and aerobics, were higher in the utilization of health and fitness apps compared to those with inactive exercising behaviors. In previous research, eating behaviors were considered powerful indicators of using fitness and diet apps. This study indicated that Malaysians did not consume the recommended number of portions of vegetables and fruit (5 portions (80 g) per day). Consequently, consuming the recommended number of vegetable and fruit portions did not influence the fitness and diet apps usage. Findings revealed that Malaysians frequently use the fitness apps such as tracking steps and GPS-based apps more than diet apps. The reason for such distinction between the usage of diet apps and fitness apps is that individuals may use fitness apps several times a week, but nutrition app usage is typically on a daily basis. Another reason is that new smartphone features track physical activity automatically using smartphone sensors [12] or wearables without interference from users, whereas food needs to be logged manually. To add to the fitness apps pros, they support exercise in suitable locations without the necessity of going to the gym [11]. Moreover, several studies have highlighted that excessive data entry is time-consuming and can hinder the usage of Mhealth apps [18]. Therefore, fewer users want to monitor their dietary behaviors compared with physical activity behaviors. As a result, several nutrition apps have embedded features to reduce the efforts of food journaling, such as a barcode scanner, digital weight and grouping food into lists. Findings revealed that individuals use the fitness apps because they wanted to track and monitor their physical activity, calories burned, pace etc. The tracking information is considered a motivational aspect; therefore, users can share their achievements in their online social communities [26]. Similarly, calorie tracking, calorie checking, and water tracking is the motivation for using diet apps.

5 Limitation

Building on the persuading results from this pilot study, future studies may use more rigorous research designs to address potential impact of fitness and diet apps usage. As a pilot study, the sample size was relatively small and came from a few wellness online communities in Malaysia that shade a light on the comprehension of the users' characteristics. This study would be more generalizable if the sample size and diversity were increased.

6 Conclusion

This research contributes to understanding the characteristics of fitness and diet app usage among online social communities' users in Malaysia. Therefore, social media may be an ideal environment to promote diet and fitness apps for app developers, healthcare practitioners, and e-sport providers. This pilot study gives inspiration to the researchers about the role of social media in M-health app usage. This study is not without limitations. Results may not be generalized because this study was designed as a pilot study. The small sample size could be suitable for understanding the characteristics of diet and fitness users; however, a larger sample could give more insightful and generalized findings.

References

1. Lim, S.S., et al.: A comparative risk assessment of burden of disease and injury attributable to 67 risk factors and risk factor clusters in 21 regions, 1990–2010: a systematic analysis for the Global Burden of Disease Study 2010. The Lancet **380**(9859), 2224–2260 (2012)
2. Mohammadi, S., et al.: Determinants of diet and physical activity in Malaysian adolescents: a systematic review. Int. J. Environ. Res. Publ. Health **16**(4) (2019)
3. Balapour, A., et al.: Mobile technology identity and self-efficacy: implications for the adoption of clinically supported mobile health apps. Int. J. Inf. Manage. **49**, 58–68 (2019)
4. Malaysian Communications and Multimedia Commission. Internet Users Survey 2020 (2020) 01 Mar 2022
5. Marks, R.J., De Foe, A., Collett, J.: The pursuit of wellness: social media, body image and eating disorders. Child. Youth Serv. Rev. **119** (2020)
6. Lazard, A.J., et al.: Cues for increasing social presence for mobile health app adoption. J. Health Commun. **25**(2), 136–149 (2020)
7. Elavsky, S., Smahel, D., Machackova, H.: Who are mobile app users from healthy lifestyle websites? analysis of patterns of app use and user characteristics. Trans. Behav. Med. **7**(4), 891–901 (2017). https://doi.org/10.1007/s13142-017-0525-x
8. Dallinga, J.M., et al.: App use, physical activity and healthy lifestyle: a cross sectional study. BMC Publ. Health **15**, 833 (2015)
9. Yin, Q., et al.: Understanding the effects of self-peer-platform incentives on users' physical activity in mobile fitness apps: the role of gender. Inf. Technol. People (2021). ahead-of-print (ahead-of-print)
10. Zhang, X.X., Xu, X.G.: Continuous use of fitness apps and shaping factors among college students: a mixed-method investigation. Int. J. Nurs. Sci. **7**, 580–587 (2020)
11. Li, A., Yang, X., Guo, F.: Which is more important in fitness apps, continuance, satisfaction, or attitude loyalty? Int. J. Technol. Human Interact. **17**(1), 105–122 (2021)

12. McGloin, R., Embacher, K., Atkin, D.: Health and exercise-related predictors of distance-tracking app usage. Health Behav. Policy Rev. **4**(4), 306–317 (2017)
13. Cheng, L.K., Huang, H.-L., Lai, C.-C.: Continuance intention in running apps: the moderating effect of relationship norms. Int. J. Sports Market. Sponsor. (2021). ahead-of-print (ahead-of-print)
14. König, L.M., et al.: Describing the process of adopting nutrition and fitness apps: Behavior stage model approach. JMIR mHealth and uHealth **6**(3) (2018)
15. Vinnikova, A., et al.: The use of smartphone fitness applications: the role of self-efficacy and self-regulation. Int. J. Environ. Res. Publ. Health **17**(20) (2020)
16. Huang, C.-Y., Yang, M.-C.: Empirical investigation of factors influencing consumer intention to use an artificial intelligence-powered mobile application for weight loss and health management. Telemed. e-Health (2020)
17. Akdur, G., Aydin, M.N., Akdur, G.: Adoption of mobile health apps in dietetic practice: case study of Diyetkolik. JMIR mHealth and uHealth **8**(10) (2020)
18. Krebs, P., Duncan, D.T.: Health app use among US mobile phone owners: a national survey. JMIR Mhealth Uhealth **3**(4), 107–119 (2015)
19. Bhuyan, S.S., et al.: Use of mobile health applications for health-seeking behavior among US adults. J. Med. Syst. **40**(6), 153 (2016)
20. Carroll, J.K., et al.: Who uses mobile phone health apps and does use matter? a secondary data analytics approach. J. Med. Int. Res. **19**(4) (2017)
21. Vasiloglou, M.F., et al.: Perspectives and preferences of adult smartphone users regarding nutrition and diet apps: web-based survey study. JMIR Mhealth Uhealth **9**(7), e27885 (2021)
22. Ndayizigamiye, P., Kante, M., Shingwenyana, S.: An adoption model of mHealth applications that promote physical activity. Cogent Psychol. **7**(1) (2020)
23. Molina, M.D., Myrick, J.G.: The 'how' and 'why' of fitness app use: investigating user motivations to gain insights into the nexus of technology and fitness. Sport Soc. (2020)
24. Bender, M.S., et al.: Digital technology ownership, usage, and factors predicting downloading health apps among Caucasian, Filipino, Korean, and Latino Americans: the digital link to health survey. JMIR Mhealth Uhealth **2**(4), e43 (2014)
25. Cho, J., Kim, S.: Personal and social predictors of use and non-use of fitness/diet app: application of random forest algorithm. Telematics Inform. (2020)
26. Li, J., et al.: Users' intention to continue using social fitness-tracking apps: expectation confirmation theory and social comparison theory perspective. Inform. Health Soc. Care **44**(3), 298–312 (2019)

Software Defined Networking Concept (SDN) Based on Artificial Intelligence: Taxonomy of Methods and Future Directions

Rihane Abderrahman[1]($\boxtimes$), Faysal Bensalah[2], Yassine Maleh[3], Baddi Youssef[4], and Youness Bouzekri[1]

[1] SEL, National School of Applied Sciences, Ibn Tofail University, Kenitra, Morocco
abdrrahmane.rihane@ucd.ac.ma
[2] LERSEM Lab, ENCG School, Chouaib Doukkali University, El Jadida, Morocco
[3] SMIEEE, National School of Applied Sciences, Khouribga, Morocco
[4] STIC Lab, Chouaib Doukkali University, El Jadida, Morocco

Abstract. SDN is essential to keep up with the rate at which technology evolves to securely link users to applications, as well as the quick changes in traffic patterns. New tools must be created and deployed on a continuous basis to automatically detect and analyze abnormal network behavior prior to it affecting end users. To minimize and prevent network problems before they affect users, networking professionals must quickly grasp and implement these techniques. Machine learning (ML), present a promising technologies enhancement for networking, which can significantly reduce the time required to deploy changes, more efficiently handle networking issues, and assist in constantly and automatically adapting to new scenarios. AI and machine learning are currently being used in production for a wide variety of purposes, including security. They are rapidly infiltrating IT operations and fundamentally altering how humans engage with technology, allowing for a more proactive and automated approach. Today's upcoming SDN systems with machine learning algorithms respond far more swiftly to changes in your environment than human involvement can. In this paper we present a unique survey covers a wide range of ML applications. The reader will gain a complete understanding of the various learning paradigms and ML approaches applied to SDN. This survey also identifies future potential for ML and SDN applications.

Keywords: Machine learning · Network security · Software-Defined Networking · Survey · meta-learning

1 Introduction

The architecture and administration of network systems provide substantial hurdles for many enterprises. As conventional, network-based physical hardware hardly meet the requirements of modern enterprises, visible Infrastructure-as-a-Service (Iaas) issue are increasingly used. Like well-known infrastructures, these cloud solutions that give end users access to virtualized IT resources are distinguished by a high degree of flexibility and an exceptional level of cost control: unlike a permanent hardware structure, the needed resources can be added at any moment.

F. Saeed et al. (Eds.): ICACIn 2022, LNDECT 179, pp. 233–241, 2023.
https://doi.org/10.1007/978-3-031-36258-3_20

Most of the time, software tools on both the client and provider sides are used to provision and scale virtual resources instead of requiring human access to specific physical network elements. The network concept under discussion stands for software-defined networking (SDN).

The remaining sections are grouped as follows. In part 2, we define SDN topologies and ML in SDN technologies, as well as provide context and terminology. In Sect. 3, we show various prior art efforts that have utilized machine learning to SDN-based network solutions. Finally, we finish the article in Sect. 4.

2   Background and Motivations

2.1   SDN

Software-based network management is made possible by a network architecture that is described by SDN. This indicates that the control plane, which is typically implemented in hardware components or control logic, has been separated from the hardware; the firmware is referred to in this sense. The separation of the infrastructure from its configuration is the essence of the SDN idea.

A network design that allows for entirely software-defined network administration is referred to as software-defined networking. This means that the hardware's intelligence, which is nothing more than its own operating software, has been abstracted away from the control layer or control logic that is standardized in hardware components (firmware). SDN, put another way, refers to the idea of separating infrastructure from configuration [1–3].

On the other hand, the data plane continues to be a single network device (e.g. all routers, switches and firewalls integrated in the network). However, because SDNs are solely used for packet transport, relatively little computational power is needed. The benefit of this is that it is far less expensive than other network approaches and that the firmware does not need to be changed.

The duties of the abstract control plane are more intricate; for example, it is in charge of ensuring that data traffic in the SDN architecture is accurate and correct so that all relevant analysis can be carried out. However, it is far more programmable and flexible than previous architectures in terms of network management because it is decoupled from the hardware and implemented in centrally managed software.

A special communication interface between the data and controller layers is required which makes the SDN software running on the control layer can dispatch appropriate packet flow commands to the integrated network components. This communication protocol, managed by the Open Networking Foundation (ONF), is the first standardized interface between the control and data layers. In numerous SDN networks, it also replaces individual interfaces of the network equipment, making it less dependent on hardware manufacturers.

Managers can immediately gain network insight using the control layer and the associated SDN software after communication has been established. They can also manage network devices using centralized software control. In contrast to networks where each component has its own control logic, this greatly improves the management of data flows, which makes virtualization and resource scaling simpler. Additionally, rather than being distributed among all routers, routing and topology data is gathered in one place.

2.2 Machine Learning in SDN

Human error is a leading cause of unanticipated network disruption. While automation reduces human error, it does not remove all errors if people continue to make final judgments. Time spent identifying and resolving network issues adds significantly to a network's total cost of ownership.

When AI and machine learning are integrated into an SDN system, the network gains vast data processing capabilities as well as a more complete awareness of network and application performance. Essentially, the network grows more intelligent. Automation, machine learning, and artificial intelligence work in concert to create a self-healing, self-driving network one that can monitor, analyze, correct, and modify itself with minimal human intervention.

Nowadays, machine learning techniques are becoming more prevalent. These techniques are believed to be superior to standard algorithms, particularly for processing and analyzing massive amounts of data. Researchers are focusing their attention on the application of these strategies in the field of networking.

Additionally, in the realm of SDN, machine learning has been applied to a variety of applications, including traffic engineering [4, 5], resource management [6, 7], intrusion detection systems [8, 9], and other security-related applications [10, 11]. For example, Mijumbi et al. [12] use it to alter virtual networks and manage resources in virtualized networks via the control plane, while Akyildiz et al. [13] establish the state of the art for traffic engineering in SDN/OpenFlow networks.

As a result, the role of machine learning in SDN has lately increased due to its numerous applications. SDN's architectural logic is more compatible with machine learning algorithms than with classical techniques. Numerous research findings, in example, combine machine learning approaches with SDN to optimize routing. Additionally, machine learning is viewed as a critical technology development for 6G and beyond [14].

3 Machine Learning on SDN Networks-Based Solutions

The ML approach is based on induction and synthesis rather than deduction. Different machine learning methods have varying classification standards. Regression, and structured learning models are all types of machine learning models. Based on their parameters, machine learning models are classed as linear or nonlinear. A linear model can be assigned a role. Additionally, it supports nonlinear models. Both ML and DL are included in the nonlinear model (deep learning).

Both Supervised/unsupervised learning, and RL are the three categories of machine learning approaches. From a tagged training sample, these methods generate a mathematical model. Unsupervised learning algorithms performs by learning from unlabeled data. With variation sof supervised learning, algorithms like semi-supervised learning also generate models using unlabeled input. In contrast to prior machine learning techniques, reinforcement learning considers how agents should behave in a particular environment in order to maximize a cumulative reward. Due to the breadth of this subject, it has been researched in a variety of other fields, including statistics and genetic algorithms.

Machine learning techniques are frequently employed to solve categorization and prediction problems [15]. In this section, we will discuss various classical machine learning approaches that have been used to SDN networks area based on the classification of machine learning algorithms.

3.1   Used ML Algorithms

Support Vector Machine (SVM): SVMs are a subset of machine learning algorithms that are used to address classification, regression, and anomaly detection issues. They are renowned for their robust theoretical assurances, high level of flexibility, and usability—even for people with no prior experience with data mining.

Naive Bayes (NB): The Bayesian classifier is a machine learning algorithm based on the Bayes theorem. It is a subset of the Bayesian network, and it is a model that is based on probability. All features in the NB network are conditionally independent. As a result, adjustments to one feature have no effect on another. The Naive Bayes technique is well suited for classifying large datasets. Conditional independence is used in the classifier algorithm. Conditional independence requires that the value of an attribute is unrelated to the values of other characteristics in a class.

K-nearest Neighbor (KNN): The k-nearest neighbor (k-NN) algorithm is a non-parametric approach that can be used for regression as well as classification. It classifies unknown objects using the feature space's nearest k training samples. The parameter k is critical for the kNN's success in discovering the nearest neighbors. The closest neighbors are determined using a variety of distance functions, including Euclidian, Hamming, Manhattan, and Minkowski. kNN classification does not require a learning model in general. As a result, it is also classified as a lazy learning algorithm, as it does not require the storage of the entire training dataset. Consistency of training data is critical.

Although it is a basic classifier that may be used in a wide variety of recognition applications, it is considered too sluggish for real-time prediction, especially when the data set is large.

Decision Tree C4.5 comes from the ID3 algorithm, which is a straightforward decision tree algorithm. It splits trees based on the information gain ratio. It takes data as input and outputs a decision tree. This algorithm generates trees with a single node. Classification rules are expressed using decision trees. When the split is less than a certain threshold value, the splitting of trees is halted. It employs error-based pruning and is an excellent algorithm for dealing with numeric attributes.

K-means: One of the most widely used algorithms for clustering. By classifying "similar" data into groups (or clusters), a data set with descriptive features can be analyzed.

The "distance" between descriptors may be used to measure the connection between two collections of data. Consequently, two highly comparable data are two data with extremely similar descriptors. In accordance with this description, data division entails locating K "prototype data" that may be merged with additional data.

These prototype data are called centroids; in practice the algorithm associates each data with its closest centroid, in order to create clusters. On the other hand, the averages of the descriptors of the data in a cluster define the position of their centroid in the descriptor space: this is the origin of the name of this algorithm (K-means).

After initializing its centroids by taking random data from the dataset, K-means alternates several times between these two steps to optimize the centroids and their clusters:

- Group each object around the nearest centroid.
- Replace each centroid according to the average of the descriptors in its group.

After a few iterations, the algorithm finds a stable division of the dataset: we say that the algorithm has converged.

3.2 Application of Machine Learning in SDN

DDOS Attack Prediction. In Ref [16], authors proposed Predis is a privacy-preserving cross-domain attack detection model for SDNs that combines disruptive cryptography and data encryption to detect anomalies using the computationally simple and efficient kNN for Protection technique. Additionally, They increased kNN's efficiency. Through theoretical research and extensive simulations, Predis is capable of detecting attacks efficiently and accurately while protecting each domain's critical information.

On the basis of the Floodlight controller in SDN, authors of [17] designed a DDoS detection system. The Attack Detection Trigger Module uses PACKET IN messages to shorten response time to attacks and reduce controller workload.

A flow-table feature-based DDoS attack detection method is produced by the flow table entry collecting module by fusing OpenFlow protocol with DDoS attack features. The appropriate flow characteristics are extracted using the FS method after statistical analysis of the OpenFlow switch flow table entries, and the overlap feature is eliminated. To build a detection model for identifying DDoS attacks, the attack detection module uses a classifier to train inner samples.

FADM [18] is a fast and lightweight framework for detecting and mitigating DDoS flooding assaults in an SDN context. To begin, in order to improve the accuracy of data gathering, both CT-based and sFlow-based methods are employed, depending on the network environment. Then, the entropy-based technique are used to quantify network feature changes and the SVM classifier to determine if the present network state is normal or not. To ensure that the network can continue to operate normally in the event of a DDoS attack, authors offer an efficient attack mitigation system based on white-listing and dynamic updating of forwarding rules. By incorporating a mitigation agent into the network, the attack traffic is moved in a timely manner while continuing to forward benign traffic normally. The testing findings demonstrate that FADM is capable of properly detecting and effectively mitigating several DDoS flooding attacks, allowing the network to quickly recover. Additionally, the results demonstrate that FADM has a low overhead. In [19], authors discussed the security of SDN controllers. It demonstrates how DDoS attacks can be identified using machine learning algorithms to classify incoming requests. IDS has performed better when implemented in SDN utilizing machine learning methods. The proposed technique is implemented utilizing Mininet and the Ryu controller. The results demonstrate the solution's efficiency by experimenting with various topologies. The SVM method is a superior choice for implementing IDS in SDN since it has the highest accuracy and recall and a reasonable precision value among the three algorithms studied.

Link Failures Prediction. In the presence of mobility in an SDN-based WMN, a new method for traffic control was provided [20]. Author uses the centralized SDN control features for the circumvent difficulties associated with mobility management in WMN. In particular, a two-level SVM-based link failure prediction approach and an alternate routing system based on centralized traffic engineering in the control layer were developed. The entire system was validated and suggested on the ns-3 simulator, where it outperformed standard mobile WMN systems.

In work of [21], authors provided an introduction to the benefits of incorporating failure prediction into the architecture of 5G systems. In detail, they examined how an optical link failure prediction module may be included into the recovery process for optical cloud services. Then they measured the benefits in terms of resource efficiency and availability of cloud services. With the aforementioned goal in mind, authors suggested the PPR + SR method, which proactively relocates cloud services that traverse projected failure connections. As a result, fewer cloud services are affected by failures, resulting in fewer cloud service relocations and (as a result) improved cloud service availability performance.

This article of [22] proposes a novel way for improving the software-defined network's fault tolerance mechanism. To depict the theoretical element, authors built a new network model based on set and graph theory. Additionally, they constructed a new framework based on the generated model to depict the practical aspects of this work. The experimental results illustrate how the proposed strategy improved the performance of reactive failure recovery by splitting the network topology into a specified number of non-overlapping cliques. The proposed strategy has not been investigated previously, implying that this research is the first to propose partitioning as a technique for improving the restoration mode of the SDN's fault tolerance.

Anomalies Detection. The authors of [23] describe a novel network architecture that combines the advantages of deep learning and SDN technology with a DNN-based system for categorizing network traffic. The system may allocate unique network resources to unique applications, greatly enhancing QoS. The system includes VNF to help with traffic identification to lessen the workload on the SDN controller. As a result, the network's operations are decoupled, enabling independent operation of each module. The high load issue that arises when SDN controllers apply Deep Learning for traffic classification is resolved by the introduction of VNF.

In reference [24], authors investigated data mining techniques for identifying network faults by modeling network faults using three popular data mining algorithms: K-Means, Fuzzy C Means, and Expectation Maximization. The datasets were taken from an actual network that had various types of traffic, including regular traffic, link failure, server crash, broadcast storm, and babbling node. Four datasets were used in the testing since they were produced from two distinct scenarios, each of which had a different amount of traffic stress on the server and router. The outcomes show that the three algorithms perform slightly differently. Faster than FCM, K-Means and EM perform better in terms of accuracy and recall, however at the sacrifice of processing time. A cost-benefit analysis has been done to compare performance and time requirements. As a result, the FCM performs more accurately than K-Means and EM.

End hosts in an SDN environment [16] provide an application-aware framework that is a common communication channel for SDN controllers.

As a result, SDN controllers can increase the end-view hosts of the connected network and expand the application of policy to the edge of the network. Controllers may also learn more about the needs unique to a given application. The suggested example expands the functionality of the OpenDaylight SDN controller by enabling communication with end-hosts that utilize the NEAT networking stack. They show a scenario where the controller exchanges policies and path data to govern significant amounts of low-latency traffic.

Flow Classification. SDN is crucial for creating networks that can successfully and appropriately adapt to shifting traffic patterns, network policies, and security constraints. Implementing security policies that take into consideration the many operating scenarios and apps that run on the network can be a challenging undertaking, even using high-level extraction languages based on responsive programming.

By applying ML approaches for traffic flow categorization and building high-level SDN rules using the produced flow classes, the authors of [25] provide a way that reduces complexity. With pre-trained models for diverse types of traffic, both supervised and unsupervised learning techniques are applied. A flow grouping algorithm then determines which flows are commonly observed concurrently after the flows have been identified. a C4.5 model combination for supervised learning with per-flow variables including inter-packet arrival time, packet size, packet count, and flow tuple. In the unsupervised circumstance, they also apply the k-means method to the same set of features.

After obtaining traffic flow data resulting from machine learning, their integration into an SDN controller is studied, and an overview of the necessary hardware and software architecture is presented. The potential for leveraging this data to identify network anomalies, botnets, and traffic redirection to a network honeypot is also looked at from a security standpoint.

The authors of ref [26] showed how elephant flow might be detected using cost-sensitive SDN. The strategy is based on cost-aware decision trees. To increase detection rates and decrease misclassification rates related to elephant flows, cost-sensitive decision trees' metrics are given. The usage of the Internet and data center tracking is used to explain the concepts and procedures.

A conducted examinations using a wide variety of data sets and a wide variety of circumstances. The experiment's outcomes also show that the proposed solutions to the elephant flow detection issue work well.

4 Conclusion

ML is recognized as one of the most promising AI approaches for autonomous systems integration and administration due to its robust capacity to extract knowledge from data and the theoretical development of machine learning frameworks. SDN is increasing power and vitality, especially in light of the current expansion of networks and big data in support of future advanced customer needs.

Machine learning applications of SDN concept networks are investigated in this study, from two different perspectives: that of machine learning algorithms, as well as the standpoint of SDN network applications. We investigated the most extensively used machine learning algorithms, as well as their applicability in both supervised and unsupervised learning environments. Additionally, we discussed the concept of meta-learning and how it may be used to improve the performance of SDNs.

References

1. Sebbar, A., Boulmalf, M., El Kettani, M-D-E.-C., Baddi, Y.: Detection MITM attack in multi-SDN controller. In: 2018 IEEE5th International Congress on Information Science and Technology (CiSt), pp. 583–587 (2018)
2. Abbou, A.N., Baddi, Y., Hasbi, A.: Software defined net-works in internet of things integration security: challenges and solutions. In: 2018 6th International Conference on Wireless Networks and Mobile Communications (WINCOM), pp. 1–6 (2018)
3. Zemrane, H., Baddi, Y., Hasbi, A.: Improve IoT ehealth ecosystem with SDN. In: Proceedings of the 4th International Conference on Smart City Applications, SCA 2019. Association for Computing Machinery, New York, NY, USA (2019)
4. Amaral, P., Dinis, J., Pinto, P., Bernardo, L., Tavares, J., Mamede, H.S.: Machine learning in software defined networks: data collection and traffic classification. In: 2016 IEEE 24th International Conference on Network Protocols (ICNP), pp. 1–5. IEEE (2016)
5. Phan, T.V., Islam, S.T., Nguyen, T.G., Bauschert, T.: Q-data: enhanced traffic flow monitoring in software-defined networks applying q-learning. In: 2019 15th International Conference on Network and Service Management (CNSM), pp. 1–9. IEEE (2019)
6. Kim, S.I., Kim, H.S.: Dynamic service function chaining by resource usage learning in SDN/NFV environment. In: 2019 International Conference on Information Networking (ICOIN), pp. 485–488. IEEE (2019)
7. Xu, J., Wang, J., Qi, Q., Sun, H., He, B.: Iara: an intelligent application-aware VNF for network resource allocation with deep learning. In: 2018 15th Annual IEEE International Conference on Sensing, Communication, and Networking (SECON), pp. 1–3. IEEE (2018)
8. Schueller, Q., Basu, K., Younas, M., Patel, M., Ball, F.: A hierarchical intrusion detection system using support vector machine for SDN network in cloud data center. In: 2018 28th International Telecommunication Networks and Applications Conference (ITNAC), pp. 1–6. IEEE (2018)
9. Somwang, P., Lilakiatsakun, W.: Computer network security based on support vector machine approach. In: 2011 11th International Conference on Control, Automation and Systems, pp. 155–160. IEEE (2011)
10. Kaur, G., Gupta, P.: Hybrid approach for detecting ddos attacks in software defined networks. In: 2019 Twelfth International Conference on Contemporary Computing (IC3), pp. 1–6. IEEE (2019)

11. Prakash, A., Priyadarshini, R.: An intelligent software defined network controller for pre-venting distributed denial of service attack. In: 2018 Second International Conference on Inventive Communication and Computational Technologies (ICICCT), pp. 585–589. IEEE (2018)
12. Mijumbi, R., Serrat, J., Rubio-Loyola, J., Bouten, N., De Turck, F., Latre, S.: Dynamic resource management in SDN-based virtualized networks. In 10th international conference on network and service management (CNSM) and workshop, pp. 412–417. IEEE (2014)
13. Akyildiz, I.F., Lee, A., Wang, P., Luo, M., Wu, C.: A roadmap for traffic engineering in sdn-openflow networks. Comput. Networks **71**, 1–30 (2014)
14. Mourad, A., Yang, R., Lehne, P.H., de la Oliva, A.: Towards 6g: evolution of key performance indicators and technology trends. In: 2020 2nd 6G wireless summit (6G SUMMIT), pp. 1–5. IEEE (2020)
15. Nanda, S., Zafari, F., DeCusatis, C., Wedaa, E., Yang, B.: Predicting network attack patterns in SDN using machine learning approach. In: 2016 IEEE Conference on Network Function Virtualization and Software Defined Networks (NFV-SDN), pp. 167–172. IEEE (2016)
16. Zhu, L., Tang, X., Shen, M., Xiaojiang, D., Guizani, M.: Privacy-preserving DDOS attack detection using cross-domain traffic in software defined networks. IEEE J. Sel. Areas Commun. **36**(3), 628–643 (2018)
17. Yao, Y., Guo, L., Liu, Y., Zheng, J., Zong, Y.: An efficient SDN-based DDOS attack detection and rapid response platform in vehicular networks. IEEE Access **6**, 44570–44579 (2018)
18. Hu, D., Hong, P., Chen, Y.: Fadm: DDOS flooding attack detection and mitigation system in software-defined networking. In: GLOBECOM 2017–2017 IEEE Global Communications Conference, pp. 1–7. IEEE (2017)
19. Barki, L., Shidling, A., Meti, N., Narayan, D.G., Mulla, M.M.: Detection of distributed denial of service attacks in software defined networks. In: 2016 International Conference on Advances in Computing, Communications and Informatics (ICACCI), pp. 2576–2581. IEEE (2016)
20. Bao, K., Matyjas, J.D., Hu, F., Kumar, S.: Intelligent software-defined mesh networks with link-failure adaptive traffic balancing. IEEE Trans. Cogn. Commun. Network. **4**(2), 266–276 (2018)
21. Natalino, C., Coelho, F., Lacerda, G., Braga, A., Wosinska, L., Monti, P.: A proactive restoration strategy for optical cloud networks based on failure predictions. In: 2018 20th International Conference on Transparent Optical Networks (ICTON), pp. 1–5. IEEE (2018)
22. Fan, R., Chang, Y.: Machine learning for black-box fuzzing of network protocols. In: Qing, S., Mitchell, C., Chen, L., Liu, D. (eds.) Information and Communications Security. ICICS 2017. LNCS, vol. 10631. Springer, Cham (2018). https://doi.org/10.1007/978-3-319-89500-0_53
23. Xu, J., Wang, J., Qi, Q., Sun, H., He, B.: Deep neural networks for application awareness in SDN-based network. In: 2018 IEEE 28th International Workshop on Machine Learning for Signal Processing (MLSP), pp. 1–6. IEEE (2018)
24. Qader, K., Adda, M., Al-Kasassbeh, M.: Comparative analysis of clustering techniques in network traffic faults classification. Int. J. Innov. Res. Comput. Commun. Eng. **5**(4), 6551–6563 (2017)
25. Comaneci, D., Dobre, C.: Securing networks using SDN and machine learning. In: 2018 IEEE International Conference on Computational Science and Engineering (CSE), pp. 194–200. IEEE (2018)
26. Xiao, P., Qu, W., Qi, H., Xu, Y., Li, Z.: An efficient elephant flow detection with cost-sensitive in SDN. In: 2015 1st International Conference on Industrial Networks and Intelligent Systems (INISCom), pp. 24–28. IEEE (2015)

A Review of WBAN Intelligent System Connections for Remote Control of Patients with COVID-19

Suhad Ibraheem Kadhem[1], Intisar A. M. Al Sayed[2], Thuria Saad Znad[3],
Jamal Fadhil Tawfeq[4], Ahmed Dheyaa Radhi[5], Hassan Muwafaq Gheni[6],
and Israa Al-Barazanchi[7(✉)]

[1] Computer Science Department, Baghdad College of Economic Sciences University, Baghdad, Iraq
[2] Ashur University College-Medical Instrumentation Techniques Engineering, Baghdad, Iraq
[3] Computer Science, Electronic Computer Center, AlNahrain University, Jadriya, Baghdad, Iraq
[4] Department of Medical Instrumentation Technical Engineering, Medical Technical College, Al-Farahidi University, Baghdad, Iraq
[5] College of Pharmacy, University of Al-Ameed, PO Box 198, Karbala, Iraq
[6] Computer Techniques Engineering Department, Al-Mustaqbal University College, Hillah, Iraq
[7] Computer Engineering Techniques Department, Baghdad College of Economic Sciences University, Baghdad, Iraq
Israa44444@gmail.com

Abstract. Novel coronavirus infection (COVID-19) has severely threatened public health. Frequent hospital visits are required for COVID-19 therapy and monitoring, which raises the load on hospitals and patients. Current advancements in wearable sensors and communication protocols are contributing to improving the healthcare system and will soon transform healthcare delivery. Future patient monitoring and health information delivery systems will be profoundly impacted by current and upcoming communication breakthroughs and microelectronics and embedded systems technology improvements. Bandwidth constraints, power consumption, and skin or tissue protection are significant obstacles. This study offers a comprehensive analysis of wireless body area networks, details their use in the COVID-19 treatment process, and proposes a paradigm for integrating body area networks into telemedicine systems. This study also addresses current developments, the WBAN-telemedicine system, and the scope of future research.

Keywords: COVID-19 · Blockchain · Wireless body area networks · Telemedicine · Body sensor networks

1 Introduction

The December 2019 epidemic of a new coronavirus illness (COVID-19) in Wuhan, China, has become a significant public health emergency. Five hundred nineteen million individuals have been infected as of May 1, 2022, and 6.26 million have died as a

F. Saeed et al. (Eds.): ICACIn 2022, LNDECT 179, pp. 242–254, 2023.
https://doi.org/10.1007/978-3-031-36258-3_21

consequence [1]. The government has committed many human and material resources to combat the pandemic since disease prevention and control have become an essential priority. According to the fundamental idea of infectious disease control, the most successful strategies include removing the source of infection, severing the transmission channel, and safeguarding the vulnerable population. Consequently, using wireless body area networks (WBAN) in clinical practice is crucial, particularly in COVID-19 pandemic prevention and control activities. The Latest Royal Pneumonia Not solely is WBAN machine administration integral for patients, but additionally for the ordinary public. Individuals may also create their views using wearable sensors, Independently monitoring, observing, and reporting their respiratory rate, coronary heart rate, physique temperature, and physiological data each day. Changes in acute symptoms may also be understood promptly, even in isolation. In addition, the software of WBAN in the scientific frontline can also motivate the improvement of innovative, cutting-edge healthcare, such as faraway screening, clever diagnostics, and far-flung fundamental care, considered full-size advances in the prevention and manipulation of ordinary epidemics. Remote screening reduces the chance of viral exposure by eliminating the need to screen large numbers of individuals in emergency departments. Intelligent diagnostics may assist in resolving the issue of sluggish and erroneous manual interpretation of image scan data. As a supplementary tool, smart diagnosis may assist frontline physicians in determining if a patient is infected with COVID-19, isolating the patient immediately, and initiating treatment. Teleintensive care is suitable for men and women troubled with the New Coronavirus. Even if these healed sufferers are launched from the hospital, medical doctors may also use 5G technology, Wi-Fi, and different third-party cell gadgets to display modifications in their critical signs and symptoms and make critical hints and coaching as the phase of the evolution of telemedicine. Moreover, in the event of a scarcity of medical personnel or facilities, telemedicine may assist mitigate this issue without the risk of human contact and cross-contamination. Not solely does WBAN play a significant position in medical treatment, but additionally in the administration of public society. WBAN can assist the whole human population in resisting the newly identified coronavirus.

1.1 Outbreak of COVID-19: Analysis

In late 2019, the first instance of a new coronavirus was reported in Wuhan, China. Since January 2020, it has been rapidly expanding across the world's population, and its expansion is accelerating exponentially. On May 1, 2022, roughly 519 million individuals were infected, and 6.26 million died from COVID-19. COVID-19 is classified by the World Health Organization as an epidemic [2]. On March 27, 2020, the United States was the first to surpass 100,000 cases, quadrupling over the following five days and reaching 1 million cases with a 5.8% death rate [3]. The globe is facing an exceedingly grave health crisis: a respiratory coronavirus (COVID-19) epidemic. COVID-19 may have had a higher effect than severe acute respiratory syndrome (SARS) in 2003. Numerous nations have reacted to this new problem by extending standard infection control and public health procedures to manage the COVID-19 pandemic. It varies from draconian isolation measures (such as in China and India) to thorough contact tracking involving thousands of contact tracers (South Korea and Singapore) [4]. Large-scale isolation (quarantine) of possibly infected persons is required to combat unexpected outbreaks of coronavirus illness [5]. This method is ineffective for preventing CoVID-19 outbreaks when the embargo affects the economy of several nations. New digital technologies may be able to aid the CoVID-19 system. An effective strategy should be required immediately. We evaluated four possible uses of digital technologies for enhancing coVID-19 response strategies: blockchain, artificial intelligence, public-key cryptosystems, and the Internet of Medical Things. Now, we pose a few questions about the CoVID-19 outbreak:

1. A new post-crown illness. Researchers have discovered that the worst-case situation for a virus-infected individual is post-neocrown. Those who recover from the virus are more prone to develop cardiovascular illness, hypertension, pulmonary disease, and diabetes, among other conditions.
2. Inadequate medical resources. Secondly, because of the COVID-19 pandemic, medical equipment, hospitals, and medical workers have been in short supply.
3. Harm to the world economy. However, to overcome this obstacle, they blocked whole nations, resulting in their financial loss. Therefore, the devastation of the world economy is the third concern related to COVID-19.
4. Confidentiality of patient information. Each stakeholder, including patients and hospitals, is concerned with the confidentiality of patient information, such as identification, health data, suggested providers, and treatment type.

2 A New Approach to COVID-19 Prevention and Control: Wearable Devices

In the 1960s, MIT laboratories pioneered wearable science, combining applied sciences such as multimedia, sensors, and Wi-Fi conversation into garments. As COVID-19 spreads internationally, wearable units based on AIoT structures are utilized to analyze bodily indicators associated with Neoconiosis, such as aspiratory price and physique temperature. Muhammad et al. [6] The positioning system resides in the intelligent helmet. The system reacts automatically when a temperature over normal is detected. The positioning system module quickly identifies the geographic location and sends a GSM-based alert to a selected smartphone. Thus, medical workers can get body temperature information promptly. As a mobile communication system of the second generation, the Global System for Mobile Communications (GSM) falls behind the emerging 5G era. In China, 5G technology is evolving gradually. At the most recent China International Import Expo, Lightning Technology debuted a 5G-based intelligent helmet. According to the data, the intelligent helmet's innovative low-power design reduces system power usage by 85 per cent. Simultaneously, the machine's standby duration exceeds 72 h, while the recognition function may operate for six hours. It is anticipated that following the implementation of the intelligent helmet, 2 out of every 10,000 individuals would be able to recognize and respond to potential dangers. In places where 5G technology has not yet been created, how might the performance of intelligent helmets be enhanced? The combination of the GSM module and a specialized mobile application is offered. Every element is proven through the cell application, which receives the records thru a GSM module. Only in this manner can the person be thoroughly monitored. A group headed by using Fyntanidou [7] has created a wrist-worn wearable system that combines state-of-the-art digital sign processing algorithms to constantly extract coronary heart rate, oxygen saturation, and physique temperature estimations among different data. The wrist-worn wearable may hastily switch processed statistics to software or a third-party cell port through a cell verbal exchange protocol like an intelligent helmet. Primarily, they improve an app for emergency sufferers connected to the patient's wrist to put on a wearable machine that detects adjustments in the patient's fundamental symptoms in a well-timed way and may also be utilized to help doctors make scientific preferences about the patient. On the other hand, the device's flaws, such as its memory of the usage of volatile organic compounds and its battery's operating capacity, restrict its growth. Figure 1 depicts the innovative edge system developed in [8] that depends on wearable sensors to identify individuals at risk of infection. Utilizing a wearable module with infrared and pulse sensors, the device measures physique temperature and coronary cardiac cost in actual time. In the past, non-portable modules had been frequently mounted in airports and retail malls. The module shows data on the patient's respiratory and blood pressure. The two modules inform the excellent authorities when a suspected occasion is observed in a public space. However, there is no assurance that the non-wearable module will be deployed in a public area where the suspect has been.

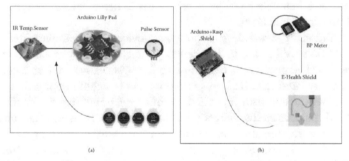

Fig. 1. Wearable and non-wearable device modules: (a) wearable devices; (b) non-wearable devices.

Additionally, the module location offers specific safety issues. The lookup described in [9] used a soft, embedded sensors gadget related to the physique or implanted in a sternotomy to seize statistics such as respiration and coronary heart rate, which can also be used to pick out physiological changes. Given the received laboratory data, this article's effects may also be too optimistic. It is predicted that more extraordinary researchers will be capable of screening this particular form of COVID-19. Embedded sensors grant a novel way to diagnose neo-coronary pneumonia primarily based on non-stop monitoring of modifications in coronary heart rate, breathing, coughing, and physique temperature, all of which indicate coronavirus infection. These applied sciences and information networks rely notably upon Bluetooth, Wi-Fi, and 5G technology. However, relying solely on these applied sciences for non-stop surveillance can result in accuracy and record loss worries.In the remedy of neocoronary pneumonia, wearable applied sciences are automatically used to look at the fitness reputation of perhaps contaminated sufferers and to perceive physiological modifications at some stage in isolation. Wearable gadgets employ GP data to track location, allowing doctors to readily monitor a patient's health. For instance, the helmet described in Smart [6] is equipped with a thermal imaging technology that automatically and without human involvement identifies coronavirus from thermal pictures. This might save time and minimize the number of human encounters, accelerating the transmission of coronaviruses.

Wearable technology, like clever watches, may accumulate self-reported symptom monitoring information to differentiate between advantageous and poor situations amongst contaminated persons. Using essential gadgets like headphones, cell phones, and headwear devices for detecting and monitoring COVID-19 symptoms may aid in diagnosing respiratory and heart rate abnormalities. Wearables on the wrist may be used expressly for emergencies, such as identifying and prioritizing urgent patients. The lookup specified in [10] improved a wearable machine that can predict the severity of the novel coronavirus in patients. The device employs an algorithm to assume the transmission of the virus from one stage to the next, analyzes the patient's fitness information, and warns the clinician when the contamination is expected to develop to the subsequent stage. This method may keep a patient's health from deteriorating. The gadget is precious for self-isolated patients who are not notified of their status in a timely way, even if they lack medical expertise. In addition, the Wearable prototype in 3D format via Bassam et al. [11] carries a wearable physique sensor, a net utility programming interface layer, and an acellular front-end layer.

The integration of wearable transportable devices into computerized healthcare devices is additionally included. One benefit of the healthcare machine is its ability to realize new cardiac pneumonia early. Figure 2 shows the system's architecture. Temperature, coronary heart rate, oxygen saturation, and coughing approach are measured employing the wearable sensor layer. This technology, not unlike the intelligent aspect machine of [8], publicizes the patient's GPS function to the healthcare unit in actual time, permitting extra immediate and environment-friendly monitoring of sufferers and suspected patients. It applies to emergency scientific treatment. When an affected person breaches self-isolation boundaries, the system sends an immediate alarm and notification signal. Future wearable gadgets based totally on AToT technological know-how will supply an array of modern coronavirus detection, sensing, and monitoring applied sciences that have substantial promise and are attainable for the healthcare system. Table 1 enumerates the preceding wearable gadgets.

3   WBAN-Based Remote Screening

Here, we highlight developing solutions lately garnered attention for addressing COVID-19-related challenges. The Internet of Medical Devices IoMT attaches many implanted or wearable sensors (with limited processing power and data storage) to a patient to constantly detect and transmit data to a personal server (smartphone) [12]. The personal server gathers data from each sensor and assists WBAN in Internet-based data transmission to the medical server. After receiving the data, the medical server evaluates and saves it before recommending the appropriate doctor based on the patient's severity. The medical server is a core database; thus, if the system is hacked, it will cause more severe harm. To tackle this issue, we use blockchain (a public ledger of transactions) to

Table 1. Overview of the advantages and applications of wearable devices

Authors	Advantages	Applications
Mohammed et al. [6]	Coronavirus can be detected automatically from thermal images	Observe the selecting procedure
Fyntanidou's team [7]	Vital signs may be continually checked	Prioritize emergency patients
Ashraf et al. [8]	It is possible to warn individuals in advance when they have a fever	Remote monitoring of patients
Lonini et al. [9]	For optimal body fit, the sensor may be positioned on the sternum	Screening for high-risk individuals
Bassam et al. [11]	COVID-19 patients' health and recovery may be followed	Predict the situation and issue alerts
Dhadge and Tilekar [10]	The gadget may broadcast warnings to prevent the infection from spreading	Predicting disease progression

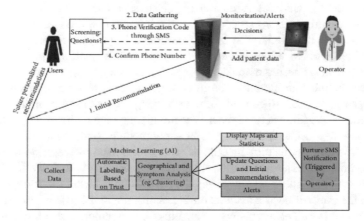

Fig. 2. Platform architecture

create a decentralized records database and distribute data across all participants. In the context of a new epidemic of crown pneumonia, screening affected people straight away is a vital method for outbreak prevention and management. Indeed, guide screening is wasteful. However, far-flung screening applied sciences have the practicable to amplify the screening's velocity and effectiveness. Consequently, IoT-based far-off screening options are required. Schinkothe's crew [13] created a complimentary WBAN caregiver cockpit (C19CC). The person talks with fitness care vendors and then categorizes them based on their description. The severity of the candidate is indicated by employing colour codes. For instance, a red code indicates a COVID-19 patient. Therefore applicants with a red code are instantly disqualified and noted with their personal information and recent activity range. Not only may C19CC be utilized for remote screening and a variety of other applications, such as remote monitoring and hospital wards. C19CC performs a non-contact pre-screening of patients to rapidly identify those in urgent need of therapy. COVID-19 is tested or diagnosed utilizing a lung ultrasound imaging classifier in addition to typical remote screening techniques. Tan and Liu [14] suggested a rudimentary face recognition-based technique. The fundamental goal of this method is to appoint thermal imaging to remotely display suspected patients, observed employing face consciousness applied sciences to become aware of their shut contacts and keep apart from them in time to stop the unfold of the viruses. Facial cognizance can be utilized for a far-flung preliminary examination. However, it is inefficient in contrast to different processes [15]. The platform obtains ultrasound photos of the lungs through a mobile scanner, as advised by Hou et al. [16].

Before categorizing them into a subspace network, the platform analyses the data—this sequence of processes lets in screening persons with newly identified coronary pneumonia. Such classifiers may additionally be used extensively in nursing homes, reducing the chance of contamination amongst aged sufferers who sometimes go to the hospital, supplied sufficient funding is available. Since COVID-19 patients typically exhibit fever symptoms, face recognition technology will also be used to identify patients over the Internet or mobile devices and notify hospital platforms to isolate and further diagnose the patient. The research in [17] presents a technique for assessing neck pictures and experiments with a coupled network-based one-time learning framework and recognizing pneumonia. The lookup described in [18] developed a like-minded web-based platform with low-end server architecture. The platform provides apps for screening, a facts-gathering module, a computing device getting-to-know module, and a consumer notification module. The filter software used to be developed for smartphone apps and websites to go well with the filtering necessities of countless customers simultaneously. Additionally, the records accumulating module helps a massive variety of concurrent users. The computer gets to know module solutions to infection-related inquiries. The user notification module delivers customized messages to alert users of findings on demand. The notification module alerts persons with COVID-19 signs of illness progression. There is still an urgent need to prevent the platform from being abused in the event of hostile assaults on the platform and the loss of user data. Additionally, wearable gadget technology is capable of doing remote screening. For instance, Lonini et al. [9] created a gentle wearable machine comprised of safer, softer, reusable wearable sensors that gather facts by sensing minor vibrations brought about by pulse and breathing. To verify the physiological modifications added on by using neocoronary pneumonia, cardiac and respiratory parameters have been investigated. The techniques and examples of faraway screening described in this learn are summarized in Table 2.

4 Tele-Intensive Care for Patients with COVID-19

COVID-19 is primarily harmful because it is extensively disseminated and challenging to cure. Therefore, the-intensive care is a smart way to limit needless interaction between medical workers and patients. Consequently, how can we attain comprehensive tele-intensive care? Some researchers have pioneered IoT-based artificial intelligence recognition systems. Face detection techniques are used to identify and recognize

Table 2. Mention the method and results of the remote screening

Authors	Methods	Results	Features
Schinkothe et al. [13]	The cockpit (C19CC)	Identifying telemedicine options, which can quickly improve patient care	Improving care and safety for people with COVID-19
Tan and Liu [14]	Facial recognition system using thermal imaging	Identification and tracking of patients	Helping to control the spread of COVID-19
Hou et al. [16]	Portable scanner + subspace multilayer networks	Test data accuracy rate of 96% or more	Improving energy efficiency
Mulchandani et al. [17]	The one-shot learning framework based on the Siamese network	Diagnosing coronavirus from tonsillitis effectively for mass-screening	Used to detect susceptible patients at a very early stage
Chilipirea et al. [18]	Scalable COVID-19 screening platform	Capacity to gather and analyze data from more than 200,000 individuals per minute	The utilization of a dispersed, lightweight design is a key aspect
Lonini et al. [9]	Use of wearable soft devices	Rapid Identification of COVID-19	Reduce the risk of pneumonia among medical personnel

human faces automatically. When a COVID-19 patient is admitted, the beginning and personal information are immediately examined to see whether the information is kept in a database, and the patient will be remotely watched continually. This technique bridges the gap between standard surveillance systems, facial recognition, and real-time remote monitoring. Using deep learning (CNN) methods, [19] offers a detection and verification system for detecting faces. The system detects faces using a DPSSD face detector and localizes them with an integrated CNN. Moorthy et al. [20] later suggested using facial detection algorithms for remote surveillance and tracking. Therefore, remote monitoring of patients via face recognition systems is still challenging owing to the vast variances in facial expression, position, skin tone, and location across faces. There are several facial similarities between twins and other individuals. In addition, What significantly diminished the accuracy and precision of the face recognition technology utilized to monitor critically sick patients during COVID-19 due to the masking?

Regarding the capacity of face monitoring systems to cover all active patient locations, developing nations and backward regions continue to encounter problems. Wearable and inconspicuous sensors are included in a WBAN-based patient tale critical care healthcare platform in the research described in [21]. The platform's primary function is to collect and analyze affected person statistics to furnish speedy cure interventions.

Experiments using the PAR structure have proven the platform's consistency and adaptability for far-flung integral care. Moorthy et al. It is fundamental to consider whether or not algorithms or face consciousness techniques on their own would possibly lead to misdiagnosis or omissions in far-flung monitoring. Figure 3 depicts similar situations of sensors and applications.

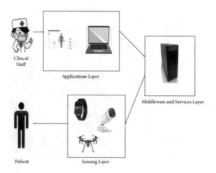

Fig. 3. Sensor and application examples

Tele-intensive care for COVID-19 sufferers results from advances in telemedicine science [22, 23]. In reality, this approach no longer helps clear up the shortage of essential care personnel and decrease morbidity and mortality. However, it additionally reduces charges and expenditures. According to telemedicine, far-off sufferers receiving crucial care can hire wearable sensor units to reveal imperative signs and symptoms and do "preliminary categorization" based on the first statistics collected. Tuyle et al. [24] encouraged deploying a wearable sensor community device for far-flung emergency therapy of patients based on the symptoms discovered. Figure 4 shows the system's architecture. The LabVIEW software program is created mainly for far-off monitoring applications.

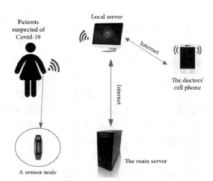

Fig. 4. A remote critical care design using a network of wearable sensors

The sensor node detects the patient's physique temperature and transfers the records thru Wi-Fi or Bluetooth to a close-by server for analysis. When a strange physique temperature is detected, the physician is alerted. In this system, the medical doctor is the primary server, and all sufferers who get far-flung fundamental care serve as sub-servers. The doctor facet of this gadget is the principal server, while the affected person facet receiving far-off emergency therapy is the secondary server. Changes in a patient's imperative signs and symptoms are communicated to the medical doctor by replacing statistics between the sub-servers and the central server. Each prognosis and monitoring of the physician's cell device is accomplished on the physician's cell device casting off the want for oblique transmission and enjoying a vast section in the telemedicine monitoring of suspected cases. Government policies and constraints hinder the increase of quintessential faraway care for patients, regardless of their colossal potential. According to the lookup in [25], the federal authorities of the United States have adopted the required criminal and regulatory modifications to prevent the virus from spreading during an outbreak, ensuing to an extent in the use of telemedicine to grant affected person treatment. Before the modern cardiac pneumonia epidemic emerged, telemedicine development was once insignificant. A literature survey [26] uniquely presented telemetry structures. This technological know-how allows the switch of information besides human intervention via a network. The article [27] also provided a blood series package for pulse samples as a far-flung screening approach. With IoT technology, pneumoconiosis victims may get far off severe bodily remedy through clever cell devices. Rajasekar [28, 29] created an IoT and RFID-based device for pneumoconiosis case tracking. The model's shape is proven in Fig. 5. RFID tags or cell units are employed to discover feasible contacts.

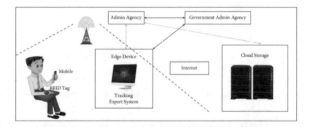

Fig. 5. Structure of the RF recognition model

Due to the worldwide expansion of Neoconiosis, it is crucial to validate the precision of the face recognition systems for tale critical concerns before employing them to offer tale critical care. Additionally, face recognition technology will be more effective at apprehending offenders and locating missing individuals, lowering the policing burden. Moreover, using WBAN-based sensors or telemedicine technological principles for tale critical care is entirely viable. Even when the epidemic has ended, these technologies will play an increasingly vital role in healthcare delivery. They will continue to depend on remote operations, industrial automation, manufacturing, and artificial intelligence for various applications [30, 31].

5 Conclusion

Even though New Crown Pneumonia has struck devastation in more than a few international locations and areas, the complete information is that the twenty-first century is a time of scientific advancements. Researchers strive to enhance New Coronary Pneumonia screening, diagnosis, and monitoring. Innovative options primarily based on WBAN structures have proven to be a valuable resource for functions ranging from wearable technological know-how to faraway screening and wise diagnostics to far-flung integral care for sufferers of New Coronary Pneumonia. Location-based solutions enable wireless body area network devices to assess individual health status and provide immediate data input to hospital departments. Remote screening eliminates interaction between health care personnel and screeners and enhances screening efficiency, resulting in early patient discovery and isolation. Telecritical care, enabled by the advancement of telemedicine, may address the issue of timely treatment and efficiently prevent medical resource waste. With the ongoing development of WBAN, the technology may assist public social management and businesses in resuming work and production after an outbreak. Countries have sought measures to restore to levels before the outbreak. WBAN has also shown its viability. To fight epidemics in the future, new WBAN-based clever gadgets will be created; for instance, cloud science will be merged with COVID-19 screening, diagnosis, and surveillance, and built-in algorithms will be utilized to forecast sickness recurrence - diseases requiring on-the-spot attention. To fight world illnesses, it is a greater likelihood that a unified infrastructure consisting of WBAN and cloud computing will be established.

References

1. Novel Coronavirus(2019-nCoV) Situation Report – 38. 2020 Feb p. ttps://www.who.int/docs/default-source/coronaviruse/situation-reports/20200227-sitrep-38-covid-19.pdf?sfvrsn=9f98940c_2
2. Singh, V., Mishra, V.: Coronavirus disease 2019 (COVID-19): current situation and therapeutic options. Coronaviruses. **2**(4), 481–491 (2021)
3. COVID-19 CORONAVIRUS PANDEMIC. Worldometer; 2020 May p. https://www.worldometers.info/coronavirus/
4. Lau, H., Khosrawipour, V., Kocbach, P., Mikolajczyk, A., Schubert, J., Bania, J., et al.: The positive impact of lockdown in Wuhan on containing the COVID-19 outbreak in China. J. Travel Med. (2020)
5. Parmet, W.E., Sinha, M.S.: Covid-19—the law and limits of quarantine. N. Engl. J. Med. **382**(15), e28 (2020)
6. Mohammed, M., Syamsudin, H., Al-Zubaidi, S., AKS, R.R., Yusuf, E.: Novel COVID-19 detection and diagnosis system using IOT based smart helmet. Int. J. Psychosoc. Rehabil. **24**(7), 2296–303 (2020)
7. Fyntanidou, B., Zouka, M., Apostolopoulou, A., Bamidis, P.D., Billis, A., Mitsopoulos, K., et al.: IoT-based smart triage of Covid-19 suspicious cases in the Emergency Department. In: 2020 IEEE Globecom Workshops (GC Wkshps, pp. 1–6. IEEE (2020)
8. Ashraf, M.U., Hannan, A., Cheema, S.M., Ali, Z., Alofi, A., et al.: international conference on electrical, communication, and computer engineering (ICECCE). IEEE **2020**, 1–6 (2020)

9. Lonini, L., Shawen, N., Botonis, O., Fanton, M., Jayaraman, C., Mummidisetty, C.K., et al.: Rapid screening of physiological changes associated with COVID-19 using soft-wearables and structured activities: a pilot study. IEEE J. Transl. Eng. Health Med. **9**, 1–11 (2021)
10. Dhadge A, Tilekar G. Severity monitoring device for COVID-19 positive patients. In: 2020 3rd International Conference on Control and Robots (ICCR), p. 25–29. IEEE (2020)
11. Al Bassam, N., Hussain, S.A., Al Qaraghuli, A., Khan, J., Sumesh, E., Lavanya, V.: IoT based wearable device to monitor the signs of quarantined remote patients of COVID-19. Inform. Med. Unlocked **24**, 100588 (2021)
12. Sun, W., Cai, Z., Li, Y., Liu, F., Fang, S., Wang, G.: Security and privacy in the medical internet of things: a review. Secur. Commun. Networks. **2018** (2018)
13. Schinköthe, T., Gabri, M.R., Mitterer, M., Gouveia, P., Heinemann, V., Harbeck, N., et al.: A web-and app-based connected care solution for COVID-19 in-and outpatient care: qualitative study and application development. JMIR Public Health Surveill. **6**(2), e19033 (2020)
14. Tan, W., Liu, J.: Application of face recognition in tracing COVID-19 fever patients and close contacts. In: 2020 19th IEEE International Conference on Machine Learning and Applications (ICMLA), pp. 1112–1116. IEEE (2020)
15. Zhang, Y., Wang, Q., Yuan, S.H.: Introduction to the special issue on computer modelling of transmission, spread, control and diagnosis of COVID-19. Comput. Model. Eng. Sci. **127**(2), 385–387 (2021)
16. Hou, D., Hou, R., Hou, J. :Interpretable Saab subspace network for COVID-19 lung ultra-sound screening. In: 2020 11th IEEE Annual Ubiquitous Computing, Electronics & Mobile Communication Conference (UEMCON), pp. 0393–0398. IEEE (2020)
17. Mulchandani, H., Dalal, P., Ramwala, O.A., Parikh, P., Dalal, U., Paunwala, M., et al.: Tonsillitis based early diagnosis of COVID-19 for mass-screening using one-shot learning framework. In: 2020 IEEE 17th India Council International Conference (INDICON), p. 1–6. IEEE (2020)
18. Chilipirea, C., Morogan, L., Toma, Ş.A.: A scalable COVID-19 screening platform. In: 2020 IEEE Globecom Workshops (GC Wkshps), p. 1–6. IEEE (2020)
19. Ranjan, R., Bansal, A., Zheng, J., Xu, H., Gleason, J., Lu, B., et al.: A fast and accurate system for face detection, identification, and verification. IEEE Trans. Biomet. Behav. Identity Sci. **1**(2), 82–96 (2019)
20. Moorthy, R., Upadhya, V., Holla, V.V., Shetty, S.S., Tantry, V.: CNN based smart surveillance system: a smart IoT application post covid-19 era. In: 2020 Fourth International Conference on I-SMAC (IoT in Social, Mobile, Analytics and Cloud) (I-SMAC), pp. 72–77. IEEE (2020)
21. de Morais Barroca Filho, I., Aquino, G., Malaquias, R.S., Girão, G., Melo, S.R.M.: An IoT-based healthcare platform for patients in ICU beds during the COVID-19 outbreak. IEEE Access **9**, 27262–27277 (2021)
22. Chand, R.D., Kumar, A., Kumar, A., Tiwari, P., Rajnish, R., Mishra, S.K.: Advanced communication technologies for collaborative learning in telemedicine and tele-care. In: 2019 9th International Conference on Cloud Computing, Data Science & Engineering (Confluence), pp. 601–605. IEEE (2019)
23. Beckhauser, E., Petrolini, V.A., von Wangenheim, A., Savaris, A., Krechel, D.: Software quality evaluation of a low-cost multimodal telemedicine and telehealth station. In: 2018 IEEE 31st International Symposium on Computer-Based Medical Systems (CBMS), pp. 444–445. IEEE (2018)
24. Touil, M., Bahatti, L., El Magri, A.: Telemedicine application to reduce the spread of Covid-19. In: 2020 IEEE 2nd International Conference on Electronics, Control, Optimization and Computer Science (ICECOCS), pp. 1–4. IEEE (2020)
25. Wu, Z., Chitkushev, L., Zhang, G.: A review of telemedicine in time of COVID-19. In: 2020 IEEE International Conference on Bioinformatics and Biomedicine (BIBM), pp. 3005–3007. IEEE (2020)

26. Wongsaroj, W., Shoji, N., Takahashi, H., Kikura, H.: Telemetry system for experimental study of ultrasonic measurement on COVID-19 situation. In: 2020 IEEE International IOT, Electronics and Mechatronics Conference (IEMTRONICS), pp. 1–5. IEEE (2020)

27. Hidayat, A., Wardhany, V.A., Nugroho, A.S., Hakim, S., Jhoswanda, M., Syamsiana, I.N., et al.: Designing IoT-based independent pulse oximetry kit as an early detection tool for Covid-19 symptoms. In: 2020 3rd International Conference on Computer and Informatics Engineering (IC2IE), pp. 443–448. IEEE (2020)

28. Rajasekar, S.J.S.: An enhanced IoT based tracing and tracking model for COVID-19 cases. SN Comput. Sci. **2**(1), 1–4 (2021)

29. Al_Barazanchi, I., Hashim, W., Alkahtani, A.A., Abdulshaheed, H.R.: Survey: the impact of the Corona pandemic on people, health care systems, economic: Positive and negative outcomes. J. Appl. Soc. Inform. Sci. **1**(1) (2022)

30. Niu, Y., Kadhem, S.I., Al Sayed, I.A.M., Jaaz, Z.A., Gheni, H.M., Al Barazanchi, I.: Energy-saving analysis of wireless body area network based on structural analysis. In: 2022 International Congress on Human-Computer Interaction, Optimization and Robotic Applications (HORA) [Internet], pp. 1–6. IEEE (2022). https://ieeexplore.ieee.org/document/9799972/

31. Al-Barazanchi, I., Hashim, W., Ahmed Alkahtani, A., Rasheed Abdulshaheed, H., Muwafaq Gheni, H., Murthy, A., et al.: Remote Monitoring of COVID-19 patients using multisensor body area network innovative system. In: Al-Sarem, M. (ed.) Comput Intell Neurosci [Internet]. 2022 Sep 15, pp. 1–14 (2022). https://www.hindawi.com/journals/cin/2022/9879259/

A Recent Review of Video Anomaly Detection for Smart Cities

Marouane Ghoulami[1](✉), Mounia Miyara[1], Najat Messaoudi[1], Zouhair Chiba[1], Hamza Toulni[2], and Mohcine Boudhane[3]

[1] Faculty of Science Ain Chock, University Hassan II, Casablanca, Morocco
marouane.ghoulami2-etu@etu.univh2c.ma
[2] Computer Science and Automation Department, EIGSICA, Casablanca, Morocco
[3] Faculty of Socio-Technical Systems Engineering, Vidzeme University of Applied Sciences, Valmiera, Latvia

Abstract. According to Maslow's hierarchy of needs, safety is a crucial component that must be satisfied before individuals can attend to higher up needs. Therefore, video surveillance systems have become more popular and heavily used in urban areas, expressing people's need for the safety of their lives and goods. However, video surveillance cameras alone cannot serve efficiently this purpose. Thusly, the urge for an automated video surveillance system has emerged to timely detect anomalies and raise early alarms to ensure public safety and security. In this paper, we will be focusing mainly on safety and security in smart cities, as we will be reviewing more precisely recent works related to video anomaly detection in public spaces, we will also provide the reader with a brief dataset benchmarking along with a comparative analysis of state-of-the-art literature.

Keywords: Big Data · IoT · Smart City · Anomaly · Safety · Security

1 Introduction

As of 2021, more than 57% of the world's population was living in urban areas due to globalization and industrialization that have played a major role in the evolution of population growth in urban areas [1]. The United Nations has predicted that 66% of the world's population will be living in cities as of 2050 [1]. Moreover, this is more likely going to rise the tensity of current problems related to health, public safety, and resources usage, as well as increasing the complexity and management challenges for concerned authorities to properly approach related issues.

In order to solve problems caused by urbanization and related to crime, congestion, diseases, etc., it became a necessity for cities to get smarter by adapting technology in all fields in favor of citizens and deploying innovative and cutting-edge solutions [2]. Smart City is relatively a new concept that started gaining a wider attention to address cities common issues. By adopting a design of new strategies that can ensure more livable, competitive, and self-reliant cities to support a better Quality of Life (QoL).

Big Data and IoT (Internet of Things) are considered pivotal technologies and even cornerstones for building and evolving of smart cities. According to [3], implementing

F. Saeed et al. (Eds.): ICACIn 2022, LNDECT 179, pp. 255–264, 2023.
https://doi.org/10.1007/978-3-031-36258-3_22

Big Data applications supports smart city components to reach the required level of sustainability. However, studies done in [4, 5] showed that the field of safety and security receives less focus in development strategies comparing to other fields. On the other hand and according to Maslow's hierarchy of needs, safety and security are crucial components of life quality in every city [6]. Hence, we can say that every smart city must be a safe city as well. In addition, the increasing usage of CCTV (Closed-Circuit Television) cameras is another reason that proves why safety and security topic should be given more focus in future works. Moreover, video surveillance market statistics has predicted 11% growth between 2019 and 2021 [7]. This is showing how people feel about safety and security in their environments.

In recent years, many research has emerged related to the usage of AI (Artificial Intelligence) to process images streamed by already installed cameras across cities. But developing such systems that can detect and predict anomalies on time has always been a challenging task. Urban anomalies are typically unusual events occurring in urban areas, such as fighting, crowd gathering, theft, violence, loitering, etc. Which poses immense threats to public safety and security if not efficiently handled. To prevent serious incidents from happening or to reduce their severity, it is of a great value to detect or even predict near-future anomalies accurately. In this paper, we will go through state-of-the-art literature related to anomalies detection in public spaces, we will introduce definitions, classifications, used algorithms and performance evaluation metrics, as well as a datasets brief-benchmarking in Sect. 2. Section 3 will provide comparative analysis and limitations of the cited methods and algorithms. Section 4 will conclude and present some perspectives.

2 Literature Survey

In this section, various definitions and anomalies classifications were given for better understanding of their complex nature, we will also discuss traditional as well as deep learning-based methods and their limitations, along with a brief-benchmarking of publicly available datasets and some of performance evaluation metrics.

2.1 Definitions and Anomalies Classifications

In literature, multiple synonyms were found describing anomalies, such as outliers, deviants, abnormalities, and unusualness. According to [8, 9], an anomalous activity is defined as "an activity executed at an unusual location or at an unusual time". Most of the definitions that were given agrees that an anomaly is an event that deviates significantly from the rest of the observations whether in context, appearance, and motion.

Most early studies as well as current works focus on anomalies found among people. Some authors have suggested that this type of anomalies has been classified in three categories. Sparsely crowded, moderately crowded and densely crowded [10]. Others have studied other types, such as traffic anomalies, unexpected crowds and individual anomalies, such as fighting, loitering, running, etc., along with environmental anomalies, such as floods, earthquakes and fire. [11]. Although, traffic anomalies such as accidents and congestion could be placed in the crowd category if we considered humans, cars and animals are just moving objects. There have been also numerous studies to investigate road-surface anomalies, in order to ensure the safety of motorists, such as potholes, manholes, and traces of repairs to damaged road surfaces [12].

2.2 Smart Video Surveillance System Approaches

A series of studies on SVSS (Smart Video Surveillance Systems) has indicated that these systems operate mostly in two different modes, i.e., off-line mode and online or real-time mode. According to [13], authors have defined two types of approaches depending on the application mode. An Accuracy-Oriented Approach (AO) and Processing-Time Oriented Approach (PO). As the studies suggests, to achieve a higher detection accuracy, the processing time will be significantly higher too. In other words, this approach is mostly used in offline surveillance applications. On the other hand, processing-time oriented approach gives more importance to speed on the account of accuracy, which makes this approach more suitable for real-time surveillance applications, as it achieves a competitive level of accuracy with minimum processing-time.

2.3 Problem formulation for Video Anomaly Detection

Previous research showed that problem formulation is a crucial step in VAD (Video Anomaly Detection), it involves narrowing down a broader application area into a specific and targeted application, as well as available resources, along with fixed performance evaluation metrics. According to [14], problem formulation for video anomaly detection can be carried out as follows:

1. Assumption: Anomalous observations occur rarely and they're significantly different in appearance, in motion or both comparing to other observations.
2. Inputs: The available anomalies datasets covered different scenarios and challenges, including illumination, occlusion, resolution, and different camera views. Due to the imbalance factor between normal and abnormal data, data split must be defined based on data availability and the nature of the problem in hand.
3. Objectives: The critical task is twofold. Therefore, video anomaly is composed of VAD, as well as VAL (Video Anomaly Localization).
4. Methods: Based on inputs, objectives and application mode, the selection of the appropriate methods contributes substantially to the results obtained.
5. Field trials: Test obtained results based on application modes. For online mode, testing is done with live feed images. For offline mode, testing is done with stored video samples.

2.4 Traditional Feature-Based Methods Used for VAD

Several authors, and before the emergence of deep learning, relied mainly on feature extraction. As this has been previously reported in literature, classic VAD methods followed mostly three phases:

1. Feature extraction.

2. Learning from extracted features to distinguish the distribution of normal and abnormal observations.
3. And finally, identification of outliers from isolated clusters as anomalous data points.

Different techniques have been employed extensively in feature extraction phase, such as low-level trajectories, image coordinates and regular patterns. However, these approaches faced many difficulties related to environmental changes [15]. As a result of these challenges, Low-Level Spatial-Temporal features-based techniques have been introduced taking advantage of Spatial-Temporal features to learn from normal observations patterns and capture patterns distribution. They included Histogram of Oriented Gradients, Histogram of Oriented Flow and Markov Random Field, as well as Mixture of Probabilistic Principal Component Analysis and Exponential Distribution of the Optical Flow, along with Gaussian Mixture Model-based techniques, Dictionary Learning and Sparse Coding [15]. Other studies of face recognition in [16], have used Histogram Equalization and Contrast Limited Adaptive Equalization, as well as a Combination of Spatial and Bilateral Correlation Filters. Although results appear not sufficient when it comes to recognition at night, and in rainy weather.

Other research in literature were conducted on video/image processing, such as Video-to-Still (V2S), Still-to-Video (S2V), and Video-to-Video (V2V). Authors in [17] have described used methods in this topic as follows, such as Point-to-Set Correlation Learning, Simultaneous Inter/Intra-Video Distance Learning and Face Pair Matching, along with Ensemble String Matching Scheme that have been used as Pattern Matching algorithms. However, these techniques have been previously assessed only to a limited extent because of the challenges that are related to the reidentification between videos due to large variations in datasets, unavailability of any prior information to be matched with a predefined pattern and the inability to capture clear facial images due to low illumination.

2.5 Deep Learning-Based Methods Used for VAD

In the current era, deep learning-based models have achieved great success in numerous domains of nonlinear high dimensional data, such as activity recognition and video summarization, among many others [15]. In this section, we will introduce state of the art deep learning-based methods used for VAD.

According to the authors in [18], Trajectory-based methods focus on detecting objects and tracking them across all video frames in order to generate objects-trajectories. They are more suitable for sparsely crowded environment. However, they are affected by crowd density and not yet suitable for moderately nor densely crowded environment. As well as they lack contextual information when deciding abnormalities.

According to [19], Representation Learning is the process of extracting useful features based on prior knowledge of a particular problem. These models help eliminate the curse of dimensionality, process frames faster and use less memory, as well as they are generalized, along with their easy implementation. But one of their major drawbacks is that we must define the problem first.

Predictive models on the other hand are used based on the assumption that video frames are spatial-temporal signals with a particular order. Therefore, the main objective of these models is to predict the current frame based on the previous ones. They are widely used for employing both spatial and temporal features. However, training requires high computation power and a significantly large memory [20].

The purpose of Deep Generative models is to learn joint probability P (X, Y), where Y is the label of the observation X. Subsequently, they calculate conditional posterior probability P(X/y) where y is the predicted label. These models give actual distribution of each class and generate similar data to input images. Recently, Deep Generative models have been widely used for VAD for their ability to address data scarcity as well as data imbalance problems in video anomaly detection. However, training demands high computation power due to the use of large datasets [20].

In summary, even deep learning-based methods are incomplete when dealing with video anomaly detection. Therefore, researchers have developed Deep Hybrid models, combining between multiple models to get efficient results. The output of Deep Neural Network models became the input for ML (Machine Learning) methods such SVM (Support Vector Machines). Although this hybrid type is more scalable and computationally efficient, as well as effective when dealing with the curse of dimensionality, it is suboptimal due to the zero-influence on the representation learning [21].

2.6 Performance Evaluation Metrics

Several authors have well acknowledged that video anomaly detection is measured based on two performance evaluation metrics. Such as evaluation criteria and performance metrics including quantitative and qualitative analysis. In this section, we will discuss each one of them in brief details.

VAD is generally evaluated using three evaluation criteria such as Frame Level, Pixel Level and Dual-Pixel Level. As well as other evaluation criteria related to IOU (Intersection Over Union), such as Object Level, Track Level and Region Level.

According to [22], frame level is when an anomaly is detected in at least one pixel, the frame containing the same pixel is considered an anomaly as well. As to Pixel level is when 40% or more of the evaluated pixels matched the pixel level ground truth, the frame containing those pixels is considered an anomaly as well. Subsequently, this prediction is then compared to pixel level ground truth to determine TP (True Positive) and FP (False Positive) frames. As for Dual-Pixel level's definition, a frame is considered an anomaly if two conditions are satisfied. Firstly, if the frame satisfied pixel level evaluation criterion. Secondly, if $\beta\%$ or more of pixels, where β is a threshold, are detected as anomalies in a frame and are common to the anomaly's ground truth.

Quantitative analysis is a technique that uses mathematical and statistical modelling, measurement, and research to understand model's behavior. It includes various metrics, such as Error Matrix, Receiving Operating Characteristics Curve and Precision-Recall Curve, as well as Equal Error Rate, Detection Rate and Reconstruction Rate, along with Anomaly Score, Regularity Score and Peak Signal-to-Noise Ratio [23].

As for Qualitative Analysis, it uses subjective judgment to analyze model's results value or prospects based on a non-quantifiable information, such as Visualization of Frame Regularity and T-Distributed Stochastic Neighbor Embedding, as well as Explainable Deep Learning [23]. These latter have been introduced as they can generate visual explanations on what is really going on during predictions. They are a way of making deep neural networks more comprehensible and approachable for visualizing the decisions that are being made in each hidden layer [24].

3 Datasets Benchmarking

Different datasets have been used in various branches of VAD fields. In this section, we will discuss some of the publicly available and frequently used ones and their weaknesses, including real life as well as staged data.

3.1 Real-Life Scenarios

For this category, the most used datasets for VAD problem are classified to at least 8 classes as shown in Table 1.

Table 1. Real life captured datasets.

Name	Description	Limitations
i-Lids [25]	Scenarios of abandoned objects	Slight camera shake
		Dirty training set
LV [26], ShanghiTech [27]	Diverse anomalous scenes	Changing illumination
UCF-Crime [15], Violent flows [28]	Violent and non-violent activities	Shadows and reflections are detected as moving objects
Behave [29]	Various behaviors and interactions of people	Cluttered and textured background
Caltech-USA [24], KITTI [30], KITTI-360 [31]	Videos recorded from a vehicle driving through regular traffic	Variant duration and resolution Unavailability of ground truth for some datasets
QMUL Junction [32], MIT Traffic [33]	Traffic anomalies	Insufficient targeted application
WWW Crowd [34], JHU Crowd [35, 36], Web [37], PETS 2009 [38]	diverse scenes of crowd gathering	
Mall [39], CAVIAR [40], CHUCK Avenue [41], UCSD Pedestrian [42]	Videos captured from public spaces, such as malls, commercial centers, and campus	

At first glance, Table 1 shows that most of the publicly available datasets are about crowd and traffic. We can also see less datasets describing other types of anomalies, such as abandoned objects and single entities (one or two-persons interaction). This can only be explained by the more the density of objects (people, cars, etc.) the more anomalous events occur, and vice versa. As a result, large datasets could be biased towards some types of anomalies on behalf of others.

As for the challenges, numerous datasets lack ground truth as they are recorded using public CCTV and it demands a great number of workforces to perform annotations.

Real life datasets are captured using different cameras with different setups which may result variety of duration, resolution, and different environmental changes such as lighting, reflections, shadows, etc.

3.2 Staged Scenarios

For this category, the most frequently used datasets for video anomaly detection are classified to at least 3 classes as shown in Table 2.

Table 2. Staged anomalies footage datasets.

Name	Description	Limitations
UR FALL [43], Lab [44]	Videos of staged fall events	Different camera setups
Weizmann [45], UMN [37, 46]	Covers anomalous patterns in images and videos	Staged scenes Ideal environments Unavailability of ground truth for some datasets
UT-Interaction [47], BOSS [48]	Contains videos of continuous human interactions	Insufficient targeted application Variant duration and resolution

Staged and simulated datasets cover only a limited number of realistic anomalous behaviors, resulting insufficient targeted applications. These scenes are recorded using predefined scripts and mostly ideal environments which may not be enough due to the variety of anomalous observations in real life scenarios.

4   Conclusions and Perspectives

In summary, this paper reviewed video anomalies various definitions and classifications in related works. As well as different SVSS approaches, along with detailed problem formulation. We have presented also traditional feature-based and deep learning-based methods as well as their limitations. We have provided performance evaluation metrics and finally we have categorized publicly available datasets, including real life and staged scenarios, gathered from various related works.

Overall, the emergence of deep learning in VAD is promising but it still suffers from identifiable limitations that are mainly related to anomalies nature, followed approach to solve the problem as well as to datasets unbalance, variety, and availability. VAD is also challenged by incompleteness of methodology, meaning that there isn't a single method that can do a complete job efficiently and correctly.

In future works we plan on approaching this topic from a different angel, we plan on using video description followed by Natural Language Processing model to analyze and filter detected activities and label them whether normal or anomalous using the gathered spatial-temporal features such as places, time periods, motion, appearances, context, people description and counting, etc. for further processing using NLP and ML techniques to achieve an end-to-end pipeline. This approach already exists but to our knowledge it is not yet applied for detecting nor predicting video-based anomalies specifically. The main idea is to use only video description ignoring the ambiguous nature of anomalies to be able to exploit different available datasets at once for the purpose of building a generalized, scalable, and efficient models that can treat various scenarios at once.

References

1. Division, U.N.P.: Urban population (% of total population. https://bit.ly/39IhMSK. Accessed 28 Sept 2022
2. Hollands, R.G.: Will the Real Smart City Please Stand Up? Tylor & Francis Group (2008)
3. Al Nuaimi, E., Al Neyadi, H., Mohamed, N., Al-Jaroodi, J.: Applications of big data to smart cities. J. Internet Serv. Appl. 6(1), 1–15 (2015). https://doi.org/10.1186/s13174-015-0041-5
4. Ristvej, J., Lacinák, M., Ondrejka, R.: On smart city and safe city concepts. Mobile Networks Appl. 25(3), 836–845 (2020). https://doi.org/10.1007/s11036-020-01524-4
5. Lacinák, M., Ristvej, J.: Smart city, safety and security. In: International Scientific Conference on Sustainable, Modern and Safe Transport (2017)
6. Trivedi, A.J., Mehta, A.: Maslow's hierarchy of needs - theory of human motivation. Int. J. Res. All Subjects Multi Lang. 7 (2019)
7. Krishna, A., Pendkar, N., Kasar, S., Mahind, U., Desai, S.: Advanced video surveillance system. In: 2021 3rd International Conference on Signal Processing and Communication (ICPSC), pp. 558–561 (2021). https://doi.org/10.1109/ICSPC51351.2021.9451694
8. Varadarajan, J., Odobez, J.-M.: Topic models for scene analysis and abnormality detection. In: IEEE 12th International Conference on Computer Vision Workshops, Kyoto (2009)
9. Popoola, O.P., Wang, K.: Video-based abnormal human behavior recognition—a review. IEEE Trans. Syst. Man Cybern. 42, 11 (2012)
10. Herbert, J.: To count a crowd. Columbia J. Rev. 6, 37 (1967)
11. Zhang, M., Li, T., Yu, Y., Li, Y., Hui, P., Zheng, Y.: Urban anomaly analytics: description, detection and prediction. IEEE Trans. Big Data 28, 04 (2020)
12. Lee, T., Chun, C., Ryu, S.-K.: Detection of road-surface anomalies using a smartphone camera and accelerometer. Sensors. 21 (2021)
13. Leyva, R., Sanchez, V., Li, C.-T.: Video anomaly detection with compact feature sets for online performance. IEEE Trans. Image Process. 26, 3463–3478 (2017)
14. Nayak, R., Pati, U.C., Das, S.K.: A comprehensive review on deep learning-based methods for video anomaly detection. Image Vis. Comput. 106 (2021)
15. Ullah, W., Ullah, A., Hussain, T., Khan, Z.A., Baik, S.W.: An efficient anomaly recognition framework using an attention residual LSTM in surveillance videos. Sensors 21, 2811 (2021)
16. Azis, F.M.A., Nasrun, M., Setianingsih, C., Murti, M.A.: Face recognition in night day using method eigenface. In: International Conference on Signals and Systems (ICSigSys.), Bali, Indonesia (2018)
17. Huang, Z., et al.: A benchmark and comparative study of video-based face recognition on COX face database. IEEE Trans. Image Process. 24, 5967–5981 (2015)

18. Morris, B.T., Trivedi, M.M.: A survey of vision-based trajectory learning and analysis for surveillance. IEEE Trans. Circuits Syst. Video Technol. **18**, 1114–1127 (2008)
19. Bengio, Y., Courville, A., Vincent, P.: Representation learning: a review and new perspectives. IEEE Trans. Pattern Anal. Mach. Intell. **35**, 1798–1828 (2013)
20. Parakkal, B.R.K.M.T.: An overview of deep learning based methods for unsupervised and semi-supervised anomaly detection in videos. J. Imag. **4**, 36 (2018)
21. Pang, G., Shen, C., Cao, L., Hengel, A.V.D.: Deep learning for anomaly detection: a review. ACM Comput. Surv. **54**, 1–38 (2022)
22. Sabokrou, M., Fayyaz, M., Fathy, M., Klette, R.: Deep-cascade: cascading 3D deep neural networks for fast anomaly detection and localization in crowded scenes. IEEE Trans. Image Process. **26**, 1992–2004 (2017)
23. Wiktorski, T., Demchenko, Y., Belloum, A., Shirazi, A.: Quantitative and qualitative analysis of current data science programs from perspective of data science competence groups and framework. In: 2016 IEEE International Conference on Cloud Computing Technology and Science (CloudCom), pp. 633–638 (2016). https://doi.org/10.1109/CloudCom.2016.0109
24. Chi, C., Zhang, S., Xing, J., Lei, Z., Li, S.Z., Zou, X.: Relational Learning for Joint Head and Human Detection. arXiv:1909.10674 [cs]. (2019)
25. Buch, N., Orwell, J., Velastin, S.A.: Urban road user detection and classification using 3D wire frame models. IET Comput. Vis. **4**, 105–116 (2010). https://doi.org/10.1049/iet-cvi.2008.0089
26. Leyva, R., Sanchez, V., Li, C.-T.: The LV dataset: a realistic surveillance video dataset for abnormal event detection. In: 2017 5th International Workshop on Biometrics and Forensics (IWBF), pp. 1–6 (2017). https://doi.org/10.1109/IWBF.2017.7935096
27. Wu, J., Li, Z., Qu, W., Zhou, Y.: One shot crowd counting with deep scale adaptive neural network. Electronics **8**, 701 (2019). https://doi.org/10.3390/electronics8060701
28. Hassner, T., Itcher, Y., Kliper-Gross, O.: Violent flows: real-time detection of violent crowd behavior. In: 2012 IEEE Computer Society Conference on Computer Vision and Pattern Recognition Workshops, pp. 1–6 (2012). https://doi.org/10.1109/CVPRW.2012.6239348
29. Blunsden, S., Fisher, R.B.: The BEHAVE video dataset: ground truthed video for multi-person (2009)
30. Geiger, A., Lenz, P., Stiller, C., Urtasun, R.: Vision meets robotics: the KITTI dataset, vol. 32 (2013)
31. Liao, Y., Xie, J., Geiger, A.: KITTI-360: A Novel Dataset and Benchmarks for Urban Scene Understanding in 2D and 3D. arXiv:2109.13410 [cs]. (2021)
32. Russell, D.M., Gong, S.: Exploiting periodicity in recurrent scenes. In: Proceedings of the British Machine Vision Conference 2008, pp. 71.1–71.10. British Machine Vision Association, Leeds (2008). https://doi.org/10.5244/C.22.71
33. Wang, X., Ma, X., Grimson, W.E.L.: Unsupervised activity perception in crowded and complicated scenes using hierarchical Bayesian models. IEEE Trans. Pattern Anal. Mach. Intell. **31**, 539–555 (2009). https://doi.org/10.1109/TPAMI.2008.87
34. Shao, J., Kang, K., Loy, C.C., Wang, X.: Deeply learned attributes for crowded scene understanding. In: 2015 IEEE Conference on Computer Vision and Pattern Recognition (CVPR), pp. 4657–4666 (2015). https://doi.org/10.1109/CVPR.2015.7299097
35. Sindagi, V., Yasarla, R., Patel, V.: Pushing the frontiers of unconstrained crowd counting: new dataset and benchmark method. In: 2019 IEEE/CVF International Conference on Computer Vision (ICCV), pp. 1221–1231 (2019). https://doi.org/10.1109/ICCV.2019.00131
36. Sindagi, V.A., Yasarla, R., Patel, V.M.: JHU-CROWD++: Large-Scale Crowd Counting Dataset and A Benchmark Method. arXiv:2004.03597 [cs]. (2020)
37. Mehran, R., Oyama, A., Shah, M.: Abnormal crowd behavior detection using social force model. In: 2009 IEEE Conference on Computer Vision and Pattern Recognition, pp. 935–942 (2009). https://doi.org/10.1109/CVPR.2009.5206641

38. Ferryman, J., Shahrokni, A.: PETS2009: dataset and challenge. In: 2009 Twelfth IEEE International Workshop on Performance Evaluation of Tracking and Surveillance, pp. 1–6 (2009). https://doi.org/10.1109/PETS-WINTER.2009.5399556

39. Loy, C.C., Chen, K., Gong, S., Xiang, T.: Crowd counting and profiling: methodology and evaluation. In: Ali, S., Nishino, K.,Manocha, D., Shah, M. (eds.) Modeling, Simulation and Visual Analysis of Crowds. TISVC, vol. 11, pp. 347–382. Springer, New York (2013). https://doi.org/10.1007/978-1-4614-8483-7_14

40. CAVIAR: Context Aware Vision using Image-based Active Recognition. https://homepages.inf.ed.ac.uk/rbf/CAVIAR/. Accessed 28 Sept 2022

41. Avenue Dataset. http://www.cse.cuhk.edu.hk/leojia/projects/detectabnormal/dataset.html. Accessed 28 Sept 2022

42. UCSD Anomaly Detection Dataset. http://www.svcl.ucsd.edu/projects/anomaly/dataset.html. Accessed 28 Sept 2022

43. UR Fall Detection Dataset. http://fenix.univ.rzeszow.pl/~mkepski/ds/uf.html. Accessed 31 Oct 2021

44. Multiple cameras fall dataset. http://www.iro.umontreal.ca/~labimage/Dataset/. Accessed 28 Sept 2022

45. Detecting Irregularities in images and in Video. https://www.wisdom.weizmann.ac.il/~vision/Irregularities.html. Accessed 28 Sept 2022

46. Monitoring Human Activity – Home. http://mha.cs.umn.edu/. Accessed 28 Sept 2022

47. SDHA 2010 High-level Human Interaction Recognition Challenge. https://cvrc.ece.utexas.edu/SDHA2010/Human_Interaction.html. Accessed 28 Sept 2022

48. Demiröz, B.E., Ari, İ., Eroğlu, O., Salah, A.A., Akarun, L.: Feature-based tracking on a multi-omnidirectional camera dataset. In: 2012 5th International Symposium on Communications, Control and Signal Processing, pp. 1–5 (2012). https://doi.org/10.1109/ISCCSP.2012.6217867

Motivations, Development Challenges and Project Desertion in Public Blockchain: A Pilot Study

Alawiyah Abd Wahab⬥, Shehu M. Sarkintudu(✉) ⬥, and Huda Hj Ibrahim⬥

School of Computing, Universiti Utara Malaysia, 06010 Sintok, Kedah, Malaysia
{alawiyah,huda753}@uum.edu.my, stjabo@gmail.com

Abstract. Blockchain technology has attracted tremendous interest from the software development community. For instance, GitHub currently hosts many active blockchain projects. As a result, information systems research has investigated these projects and their developer community. Although the number of these projects is growing rapidly, the motivations and development challenges of developers remain a mystery. Also, there is a perception that the scarcity of development tools, which is obvious for new technology, can influence project turnover. Therefore, the primary objective of this study was to evaluate the reliability and validity of an instrument developed from the identified motivations and development challenge factors that might lead to project desertion derived from open-source software (OSS) literature and social cognitive theory. This pilot study examined a small sample of data in order to assess the study's construct validity and dependability. The survey method was used to collect the data. The results of 66 actual surveys from blockchain developers were compiled using the stratified random sample method and the data were analyzed using IBM SPSS version 28.0.0.0 for Windows. According to the results, the adapted instruments used in the pilot study are reliable and valid.

Keywords: Blockchain Project · Open Source · GitHub · Developers · Turnover

1 Introduction

Blockchain technology exhibits core properties of digital information system (IS) artifacts that were initially developed for cryptocurrency [1]. However, the technology has potential uses, to go beyond cryptocurrencies, such as supply chain management [2], identity management [3], smart contracts [2], halal product assurance [4], and the Internet of things (IoT) [5]. Blockchain projects are often based on open-source software (OSS), which refers to software developed by diverse communities. The prominent examples of blockchain projects that are OSS are Bitcoin and Ethereum [6]. For example, Bitcoin is a public decentralized database project whose source codes are often free to modify, similar to the traditional OSS concepts found in the literature.

The interest in blockchain projects has been increasing [7] since the introduction of Bitcoin in 2008 by anonymous individuals and groups of developers [8]. Following

F. Saeed et al. (Eds.): ICACIn 2022, LNDECT 179, pp. 265–277, 2023.
https://doi.org/10.1007/978-3-031-36258-3_23

that, the global open-source software (OSS) community's efforts are what ultimately drive its evolution [9–11]. Blockchain relies upon the voluntary collaborative actions of thousands of developers [10, 11], and the idea of OSS as software that has diverse individual developers is discussed in [12]. Therefore, developers constitute essential critical actors in the successful evolution of blockchain projects [8, 13].

Currently, there are more than 3,000 blockchain projects currently being hosted on the GitHub development platform [14]. Nevertheless, unlike traditional OSS, blockchain developers need to be cautious about the security and reliability of the software by designing efficient and reliable protocols for immutable distributed databases [6]. Many scholars have considered how crucial it is to understand developer turnover concerns, specifically project desertion [15, 16], given that blockchain has the ability to disrupt a wide range of industries, notably in the financial sectors [17]. Despite the disrupting nature of blockchain projects, the failure rate still silences the success rate [18]. Some researchers suggest that the success of blockchain projects consists of community innovative activities [8, 15], which are constrained due to the lack of necessary tools specifically for blockchain development projects [19]. Developers working on blockchain projects face unique challenges that have contributed to a high rate of developer attrition. These difficulties include a lack of specialized tools because the tools currently in use are for traditional OSS projects, a lack of cryptographic expertise, higher costs associated with defects, a hostile and decentralized environment, technological complexity, and difficulty upgrading the software after it has been released due to Blockchain's immutable nature [8, 10, 18]. As a result, even though the community of software developers has also shown great interest in the technological revolution, fundamental adjustments are necessary for the design, development, and implementation of blockchain.

Prior studies on developer contribution and turnover mainly focused on what motivates developers to join a project and their performance [20]. These studies were carried out within traditional OSS projects, such as Mozilla, Firefox, GNU/Linux, and LibreOffice [21, 22], which are mature and have sufficient developers to achieve better performance [23]. In addition, several empirical studies [8, 15] have focused on blockchain projects, with a few studies on developer turnover issues [24–26]. However, studies on the relationship between development challenges (such as contributed code decoupling, code testing task, and blockchain system integration) and project desertion in a blockchain project are still lacking. According to prior studies, the majority of blockchain projects are, by definition OSS, even though it is known how development issues affect OSS developer turnover [11, 27, 28]. However, blockchain projects differ from traditional OSS development in that they are more innovative and place a higher priority on security and dependability than traditional OSS projects [8, 15]. As a result, these differences become the main sources of difficulties for developers. These factors can cause turnover in the development community, which results in experienced developers abandoning or leaving the project. Project abandonment results in the loss of the developers, who have knowledge of the project specifics and project ideology [24, 25].

In blockchain projects, developers are operating traditionally, despite that many auxiliary tools that can aid their development activities are either yet to be available [13] or are immature and unreliable [8]. The lack of supporting tools and documentation are the sources of challenges for many developers [8]. Development tasks such as testing frameworks and debugging are difficult and complex activities that lead to a high project desertion rate in blockchain projects [8, 29]. In addition, the conclusions from prior OSS studies [21, 28] cannot be immediately applied to the blockchain setting for two reasons. First, project innovative updates, known as blockchain forks, differ from traditional OSS forks in that their execution requires consensus among the development community, which is difficult to achieve due to developers' decisions [11]. Second, developer disengagement in a blockchain project is widely known as costly and can directly affect project sustainability and investment [10, 30]. Thus, to verify the instruments' dependability, content validity, and internal consistency while studying in the context of blockchain projects, it is necessary to conduct a pilot test. The current study sought to foresee potential issues and made adjustments before beginning the actual data collection.

According to [31], validity measures the magnitude to which a particular instrument is measuring what it is supposed to measure. In contrast, reliability measures the extent to which a tool is consistent, free from error, and stable across scale items. Similarly, this research aimed to investigate the extent to which the integration of various forms of developer motivations and development challenges resulted in project desertion, as well as the moderating effect of decision rights delegated to developers. This paper presents the findings of a pilot test on the factors that influence development issues and project desertion in the application of blockchain technology.

2   Literature Review

Project desertion is a developer's decision to stop working on open-source projects [8], which is primarily caused by a lack of responsiveness, availability, and motivation on the part of developers. Scholars have identified project desertion as one important issue of developer turnover [32, 33]. It is well recognized that project desertion by OSS developers is expensive and a severe problem [10, 21], thereby negatively impacting a project's viability. Due to the interdependencies between projects, an OSS project's developer turnover problems can affect the overall environment [24, 26]. This persistent issue harms OSS tasks. Preceding literature has identified many issues related to OSS projects in the context of staff turnover, including complete request abandonment [32], developer disengagement [26], reasons developers take breaks [24], and senior developers abandoning a software project [34, 35]. For example, 41% of OSS projects failed due to the evolution of the developer team [32]. Since these problems highlight the causes and predicative elements of established project desertion, they are regarded as significant challenges in traditional OSS projects. In blockchain projects, numerous studies have recognized the main issues facing traditional OSS initiatives, and these issues have significantly contributed to projects' premature demise [12]. For example, [36] examined 350 pull requests to explore why developers stopped contributing to projects. The authors discover that contributions are not integrated because of issues with the distributed development process, not because of simple technical considerations. While other studies focus on other factors, this study examined the relationships between project desertion, development problems, and motivation.

Developer project turnover in OSS is significantly influenced by motivation [37, 38]. Similar to this, other researchers argue that intrinsic motivation, such as motivation to learn, is linked to the intent to leave. A person's motivation refers to the individual need to feel capable and in control [8], which includes entertainment, hobbies, self-interest, and a sense of competence [39]. There are additional intrinsic motivations, such as the potential for job advancement, benefits from software use [40], and external rewards, which direct or indirect incentives are monetary remuneration [41]. For instance, experts believe that creating a system that can provide advantages to those willing to take part is associated with the duration of their attachments to particular projects [8]. Prior studies also describe that ideology is also a form of motivation as it includes the norms, beliefs, and values shared especially among the developers of an OSS project, and is crucial for OSS success or failure [42, 43].

In addition, project expertise and skills are vital for new project start-ups [43–45]. In the context of OSS, skill is conceptualized as a contributor's ability to execute project tasks according to requirements, and knowledge level is expected to increase with time spent on the project. In the context of blockchain, developers require network and networking security knowledge [13]. Knowledge specific to a particular domain is the quality of an individual possessing knowledge relevant to work [46], with higher levels of knowledge associated with the levels of turnover behavior in a domain [47]. As blockchain projects are complex and require extensive developer knowledge of cryptography, which is hard to master, lack of knowledge is associated with the turnover intention [48].

Expertise is crucial to formulate project resources, plans, and operational goals and objectives. In the context of blockchain, developers like to create a payment system that regulates itself [49–51]. Studies find that future blockchain developers are likely to have an idea about the characteristics of the existing community who are involved in the industry with a wide variety of ethical considerations associated with a project's failure and success. As bugs are expensive, users demand a highly reliable blockchain project. Therefore, blockchain developers face several difficulties due to these scenarios and the unique features of blockchain [52], as well as are under intense pressure from investors in this area [13].

A principal element of innovative applications of blockchain, such as cryptocurrency and smart contracts, is the lack of central authority; where the decision right is delegated to individual developers due to a lack of formal power in the project [53]. Decision right delegation is the degree to which developers have the authority to make specific contributions to design decisions [50]. They encompass decisions about project features, functionality, design, and implementation. This conceptualization builds on decision rights as the allocation of decision-making authority [54–56]. Decision rights, therefore, signify the locus of control on contributions [14, 27, 57], which reside entirely with developers rather than project owners (i.e., wholly delegated). Despite developer autonomy leads to disagreement within the development community [10], proper decision delegation promotes the innovative practice, enabling the blockchain development landscape to change rapidly with projects fiercely competing with traditional technologies to emerge as market leaders [58].

In traditional OSS, switching is costly once developers have committed to projects. Decision rights delegation helps to reduce hold-up problems that are particularly prevalent in traditional OSS projects (centralized projects) by allowing "project owners" (e.g., Mozilla Apps) to hold up developers, for instance, by implementing changes that make their previous investments obsolete [9]. On the other hand, decentralized projects, such as blockchain, drastically reduce the potential for hold up since developers can always decide to desert or leave the project [9]. Decision rights delegation encompasses activities such as adding special features specific to a blockchain to project specification and laying expectations by establishing project deliverables [9]. As a result, in general, with proper decision delegation in a decentralized project, each developer can refuse to contribute or make an available change, which protects them from hold-up. However, incentives to adapt in a coordinated manner are limited [27].

In sum, studies reveal that the existence of decision right delegation in blockchain projects significantly correlate with developer turnover [24–26, 34]. Previous empirical research investigations in information systems suggest that decision right delegation is the leading cause of various motivations, development issues, and project desertion. In this study, decision right delegation contributes as a moderator between motivations, development challenges, and project desertion.

3   Methodology

This pilot study aimed to assess the opinion of blockchain project developers on the reliability and validity of a set of questionnaires. As a pilot study's sample size was relatively small [59], the current pilot study distributed 87 questionnaires to blockchain developers via email. An email reminder was sent a week following the initial request to check if they had responded and, if not, to ask them to respond. A total of 87 blockchain developers were contacted via the GitHub platform, but 21 emails were automatically returned due to incorrect addresses or security protection.

In order to increase the validity and reliability of the developed instrument, this study adopted a multi-stage techniques as suggested by [60]. The research process consists of three main phases: 1) instrument development, 2) content validity, and 3) pilot study.

3.1   Instrument Development

The social cognitive theory (SCT) [61] in the IS literature and research goal are the key points for the identification of the project desertion factors in the current study. The objective of prior work was to establish the full potential of SCT, which had still not been revealed due to a lack of consideration of the complete SCT model. The birth of the Internet has prompted the management and adoption of web-based technologies, resulting in the SCT being employed from a different perspective, with the maintenance of internet-based applications such as blockchain being modelled as an evolving process. Such processes are erudite by an individual in which their behavior, cognitive and emotional activities, and technological environment are interrelated, mutually impacting one another. Based on the SCT, a systematic literature review (SLR) was conducted to identify factors that may influence blockchain project desertion. In addition to the SLR, an analysis of the relevant literature was performed to develop the instrument.

3.2 Expert Review

A panel of academic experts and a small sample of blockchain practitioners were asked to give their comments and input on the appropriateness of the adopted items. Academic experts include associate professors and assistant professors in information systems, software engineering, and computer science areas from University Teknikal Malaysia in Malaysia, the University of Phoenix in the US, Federal University Dutse and Bayero University Kano in Nigeria, the University of Nottingham in the UK, and several blockchain practitioners on the GitHub OSS development platform.

To aid understanding of the instrument, factors and their items that should be considered when evaluating the identified constructs were added. Before the pilot study was conducted, certain factors and their measurement items were reworded to ensure that all the constructs adhere to the experts' recommendations. The constructs, shown in Table 2, were listed in the questionnaire.

4 Results

4.1 Instrument Development Result

Nearly 210 papers (from IEEE, 53; Springer, 35; Elsevier, 29; AIS, 13; Emerald, 38; Taylor & Francis, 42) reporting research on developing the various constructs of the theory supporting its authority were identified. The papers, which have integrated a review of blockchain and OSS literature, were carefully selected based on the significant pieces of evidence for identifying the factors in the review. However, the decision was to identify and extract factors that influenced project desertion from prior studies in blockchain or OSS projects using the following criteria:

- To ensure a high-quality study, only publications published in the Web of Science indexes journals or Scopus between 2010 to 2020 were considered.
- The selected papers encompassed studies related to blockchain or OSS projects written in English, as well as empirical, conceptual frameworks, and systematic reviews.

Out of 210 papers, 84 were considered irrelevant due to a lack of abstracts. Furthermore, abstracts' assessment of the remaining 126 papers was made which resulted in the removal of 45, leaving 81 papers. The full text was thoroughly examined against the inclusion criteria, leading to the rejection of 47 more papers. The quality of the remaining 34 was then appraised, based on publications, resulting in the exclusion of another 13 papers. Only the remaining 21 papers that met the inclusion criteria were chosen for the identification of project desertion factors. The study scrutinized and mined 12 factors and one project desertion construct. These include the intention to learn, financial gain intention, blockchain project leadership, technical contribution norm, contributed code decoupling, code testing task, system integration, network management knowledge, expertise heterogeneity, developer involvement, decision right delegation, blockchain archetype, and project desertion.

The adopted definitions and their conformity with the employed measurement were evaluated to ensure that the project desertion factors were similar to those investigated by previous researchers [45, 50]. The 12 chosen factors highlighted their relationships with the project desertion of the respective organizations, which was similar to the context of the current study. The complete instrument is made up of three sections. Section 1 contains items related to the 12 factors and one desertion constructs. Section 2 includes items related to project characteristics. Meanwhile, Sect. 3 comprises of items related to respondents' demographic.

4.2 Expert Review Results

The analysis of the data collected from experts found all constructs were important. A detailed discussion about the instrument development and expert review phases are awaiting to be published in the Interdisciplinary Journal of Information, Knowledge, and Management (IJIKM).

4.3 Pilot Study Results

Respondents Demographic. The descriptive analysis in Table 1 shows that 66.7% of respondents opted to contribute to highly decentralized blockchain projects, whereas 33.3% contributed to low decentralized projects. Furthermore, the data collected reveals that 42.4% of the blockchain projects were 1 to 2 years old, 36.4% were 2 to 3 years old, and 6.1% were 3 to 4 years old. Meanwhile, 30.3% of the blockchain projects had reached development maturity status, 21.2% were at the planning stage, 13.6% were at the alpha and beta stages and the remaining 12.1% and 9.1% were at pre-alpha and production stages, respectively. In addition, the male respondents made up 72.7% of the sample, while the female made up 27.3%. Meanwhile, 51.5% had a 20–21 age bracket, 45.5% were between the ages of 30 to 39, and only 3% were between the ages of 40 to 49.

Table 1. Respondent's demographics

Blockchain archetype	Frequency	Percentage
Low decentralised	22	33.3
Highly decentralised	44	66.7
Blockchain project age		
<1 year	7	10.6
1–2 years	28	42.4
2–3 years	24	36.4
3–4 years	4	6.1
4–5 years	2	3.0

(*continued*)

Table 1. (*continued*)

Blockchain archetype	Frequency	Percentage
>5 years	1	1.5
Development status		
Planning	14	21.2
Pre-alpha	8	12.1
Alpha	9	13.6
Beta	9	13.6
Production/stable maturity	6	9.1
Maturity	20	30.3
Gender		
Male	48	72.7
Female	18	27.3
Age		
20–29	34	51.5
30–39	30	45.5
40–49	2	3.0

Reliability Test. The reliability test shows that all constructs had high-reliability values which ranged from 0.699 to 0.886 (see Table 2). Cronbach's alpha value of 0.60 is considered average reliability, while a coefficient of 0.70 (or higher value) shows the higher reliability of instruments [31]. After pilot testing, the results of the reliability test show that Cronbach's alpha value for the thirteen constructs were all above 0.60. Thus, it can be concluded that all the constructs of motivations and development challenges are reliable, and therefore, there is no need to remove any item from the instruments.

Table 2. Reliability Statistics

Constructs	Cronbach's Alpha	No. of items
Intention to learn	0.780	3
Financial gain intention	0.699	4
Blockchain project leadership	0.805	8
Expertise heterogeneity	0.762	3
Network management knowledge	0.886	6
Technical contribution norms	0.706	6
System integration	0.827	4
Code testing task	0.868	5

(*continued*)

Table 2. (*continued*)

Constructs	Cronbach's Alpha	No. of items
Contributed code decoupling	0.767	5
Developer involvement	0.799	5
Decision right delegation	0.778	5
Project desertion	0.832	5

5 Conclusion, Limitation, and Future Research

In order to prepare for the main research investigation, the goal of this study was to conduct a pilot study to assess and represent the validity and reliability of the factors for the actual data collection stage. The results of the pilot study indicate that all of the constructs under consideration have a Cronbach's alpha value larger than 0.60. Given the absolute threshold of 0.60, it may be said that all constructs related to motivation and developmental challenges are trustworthy. There is no need to alter, separate, or rewrite any particular item. Hence, the questionnaire is considered valid and reliable to measure motivations and development challenges such as intention to learn, system integration, code testing task, contributed code decoupling, and decision right delegation. To fully test the instrument, a larger study needs to be carried out against further scenarios. This needs to be addressed in the next stages of research. Although the constructs of the instrument are derived from open-source software, blockchain, and information systems literature, they have been thoroughly tested and accepted.

Acknowledgment. This research was supported by the Ministry of Higher Education (MoHE) of Malaysia through Fundamental Research Grant Scheme (FRGS/1/2020/ICT03/UUM/02/4).

References

1. Schweizer, A., Schlatt, V., Urbach, N., Fridgen, G. :Unchaining social businesses - blockchain as the basic technology of a crowdlending platform. In: 38th International Conference on Information Systems, pp. 1–21 (2017)
2. Korpela, K., Hallikas, J., Dahlberg, T.: Digital supply chain transformation toward blockchain integration. In: Proceedings of the 50th Hawaii International Conference on System Sciences | 2017 Digital, pp. 4182–4191 (2017)
3. Biais, B., Bisière, C., Bouvard, M., Casamatta, C.: The blockchain folk theorem. Rev. Financ. Stud. **32**(5), 1662–1715 (2019). https://doi.org/10.1093/rfs/hhy095
4. Katuk, N.: The application of blockchain for halal product assurance: A systematic review of the current developments and future directions. Int. J. Adv. Trends Comput. Sci. Eng. **8**(5), 1893–1902 (2019)
5. Firoozjaei, M.D., Lu, R., Ghorbani, A.A.: An evaluation framework for privacy-preserving solutions applicable for blockchain-based internet-of-things platforms. Secur. Priv. **3**(6), 1–28 (2020)

6. Lindman, J.: What open source software research can teach us about public blockchain (s)?—lessons for practitioners and future research. Front. Hum. Dyn. **3**(October), 1–7 (2021)
7. Katuk, N., Ku-Mahamud, K.R., Zakaria, N.H., Jabbar, A.M.: A scientometric analysis of the emerging topics. J. Inf. Commun. Technol. **19**(4), 583–622 (2020)
8. Bosu, A., Iqbal, A., Shahriyar, R., Chakraborty, P.: Understanding the motivations, challenges and needs of blockchain software developers: a survey. Empir. Softw. Eng. **24**(4), 2636–2673 (2019)
9. Arruñada, B., Garicano, L.: Blockchain: the birth of decentralized governance. Pompeu Fabra University, Economics and Business Working Paper Series Apr 10, p. 1608 (2018)
10. Islam, N., Mäntymäki, M., Turunen, M.: Understanding the role of actor heterogeneity in blockchain splits: an actor-network perspective of bitcoin forks. In: Proceedings of the 52nd Hawaii International Conference on System Sciences, vol. 6, pp. 4595–4604 (2019)
11. Kiffer, L., Levin, D., Mislove, A.: Stick a fork in it: analyzing the Ethereum network partition Lucianna. In: Proceedings of the 16th ACM Workshop on Hot Topics in Networks - HotNets-XVI, March, pp. 94–100 (2017)
12. Reboucas, M., Santos, R.O., Pinto, G., Castor, F.: How does contributors' involvement influence the build status of an open-source software project? In: IEEE International Working Conference on Mining Software Repositories, pp. 475–478 (2017)
13. Chakraborty, P., Shahriyar, R., Iqbal, A., Bosu, A.: Understanding the software development practices of blockchain projects. In: Proceedings of the 12th ACM/IEEE International Symposium on Empirical Software Engineering and Measurement, pp. 1–10 (2018)
14. Das, A., Uddin, G., Ruhe, G.: An empirical study of blockchain repositories in gitHub. In: ACM International Conference Proceeding Series. Association for Computing Machinery, pp. 211–20 (2022)
15. Reijers, W., O'Brolcháin, F., Ledger, P.H.: Governance in blockchain technologies and social contract theories: open review. LedgerjournalOrg **5980**(1), 134–151 (2016)
16. Ferreira, F., Silva, L.L., Valente, M.T.: Turnover in open-source projects: the case of core developers. In: Proceedings of the 34th Brazilian Symposium on Software Engineering, pp. 447–56 (2020)
17. Riasanow,, T., Setzke D.S.: The generic blockchain ecosystem and its strategic implications. In: Proceedings of the 24th Americas Conference on Information Systems [Internet], pp. 1–10 (2018). www.crunchbase.com
18. Rastogi, A., Nagappan, N.: Forking and the sustainability of the developer community participation - an empirical investigation on outcomes and reasons. In: 2016 IEEE 23rd International Conference on Software Analysis, Evolution, and Reengineering, pp. 102–11 (2016)
19. Porru, S., Pinna, A., Marchesi, M., Tonelli, R.: Blockchain-oriented software engineering: Challenges and new directions. In: Proceedings - 2017 IEEE/ACM 39th International Conference on Software Engineering Companion, ICSE-C 2017, pp. 169–71 (2017)
20. Rastogi, A., Nagappan, N., Gousios, G., van der Hoek, A.: Relationship between geographical location and evaluation of developer contributions in github. In: Proceedings of the 12th ACM/IEEE International Symposium on Empirical Software Engineering and Measurement, pp. 1–8 (2018)
21. Lee, A., Carver, J.C., Bosu,, A.: Understanding the impressions, motivations, and barriers of one time code contributors to FLOSS projects: a survey. In: IEEE/ACM 39th International Conference on Software Engineering (ICSE), pp. 187–97 (2017)
22. Machado, F.S., Raghu, T.S., Sainam, P., Sinha, R.: Software piracy in the presence of open source alternatives. J. Assoc. Inf. Syst. **18**(1), 1–2 (2017)
23. Seker, A., Diri, B., Arslan, H., Amasyalı, M.F.: Open source software development challenges: a systematic literature review on gitHub. Int. J. Open Source Softw. Process **11**(4), 1–26 (2020)

24. Iaffaldano, G., Steinmacher, I., Calefato, F., Gerosa, M., Lanubile, F.: Why do developers take breaks from contributing to OSS projects? A preliminary analysis. In: Proceedings - 2019 IEEE/ACM 2nd International Workshop on Software Health, pp. 9–16 (2019)
25. Fagerholm, F., Guinea, A.S., Münch, J., Borenstein, J.: The role of mentoring and project characteristics for onboarding in open source software projects. In: International Symposium on Empirical Software Engineering and Measurement, pp. 1–10 (2017)
26. Miller, C., Widder, D.G., Kästner, C., Vasilescu, B.: Why do people give up FLOSSing? A study of contributor disengagement in open source. In: IFIP Advances in Information and Communication Technology, pp. 116–29 (2020)
27. Walton, R.: What do the consequences of environmental, social and governance failures tell us about the motivations for corporate social responsibility? Int. J. Financ. Stud. **10**(1), 1–19 (2022)
28. Xiao, X., Lindberg, A., Hansen, S., Lyytinen, K.: Computing requirements for open source software: a distributed cognitive approach. J. Assoc. Inf. Syst. **19**(12), 1217–1252 (2018)
29. Kosba, A., Miller, A., Shi, E., Wen, Z., Papamanthou, C.H.: The blockchain model of cryptography and privacy-preserving smart contracts. In: Proceedings - 2016 IEEE Symposium on Security and Privacy, pp. 839–58 (2016)
30. Nyman, L., Lindman, J.. Perspectives on code forking and sustainability in open source software. Technol. Innov. Manag. Rev. **3**(1), 7–12 (2018)
31. Sekaran, U.: Research methods for business a skill building approach. In: Edition F (ed.) The Encyclopedia of Research Methods in Criminology and Criminal Justice, volume II: Parts 5–8, pp 537–545. Wiley (2003)
32. Cheng, C., Li, B., Li, Z.Y., Zhao, Y.Q., Liao, F.L.: Developer role evolution in open source software ecosystem: an explanatory study on GNOME. J. Comput. Sci. Technol. **32**(2), 396–414 (2017)
33. Izquierdo-Cortazar, D., Robles, G., Ortega, F., Gonzalez-Barahona, J.M.: Using software archaeology to measure knowledge loss in software projects due to developer turnover. In: Proceedings of the 42nd Annual Hawaii International Conference on System Sciences, IIICSS, pp. 1–10 (2009)
34. Dias, L.F., Santos, J., Steinmacher, I., Pinto, G.: Who drives company-owned OSS projects : Employees or volunteers ? In: Workshop on Software Visualization, Evolution and Maintenance, VEM, pp. 1–8 (2017)
35. Daniel, S.: Loop learning: linking free/libre open source software (FLOSS) developer motivation, contribution, and turnover intentions. ACM SIGMIS Database DATABASE Adv. Inf. Syst. **51**(4), 68–92 (2020)
36. Li, Z., Yu, Y., Wang, T., Yin, G., Li, S., Wang, H.: Are you still working on this an empirical study on pull request abandonment. IEEE Trans. Softw. Eng. **48**(6), 1–17 (2021)
37. Yang, N., Ferreira, I., Serebrenik, A., Adams, B.: Why do projects join the apache software foundation? In: Proceedings - International Conference on Software Engineering. Association for Computing Machinery, pp. 161–71 (2022)
38. Kaur, R., Kaur, C.K., Saini, M.: Understanding community participation and engagement in open source software projects: a systematic mapping study. J. King Saud Univ. – Comput. Inf. Sci. **34**(7), 4607–4625 (2020)
39. Daniel, S.L., Maruping, L.M., Cataldo, M., Herbsleb, J.: The impact of ideology misfit on open source software communities and companies. MIS Q. **42**(4), 1069–1096 (2018)
40. Avelino, G., Constantinou, E., Valente, M.T., Serebrenik, A.: On the abandonment and survival of open source projects: an empirical investigation. In: ACM/IEEE International Symposium on Empirical Software Engineering and Measurement (ESEM), pp. 1–12 (2019)
41. Xu, B., Jones, D.R., Shao, B.: Volunteers' involvement in online community based software development. Inf. Manag. **46**(3), 151–158 (2009)

42. Choi, N.C.: Loyalty, ideology, and identification: an empirical study of the attitudes and behaviors of passive users of open source software. J. Assoc. Inf. Syst. **16**(8), 674–706 (2015)
43. Kazan, E., Tan, C.W., Lim, E.T.K.: Value creation in cryptocurrency networks: towards a taxonomy of digital business models for bitcoin companies. In: PACIS 2015 Proceedings, pp. 1–15 (2015)
44. Lin, B., Robles, G., Serebrenik, A.: Developer turnover in global, industrial open source projects: insights from applying survival analysis. In: Proceedings - 2017 IEEE 12th International Conference on Global Software Engineering, ICGSE, pp. 66–75 (2017)
45. Huang, L.C., Shiau, W.L.: Factors affecting creativity in information system development: insights from a decomposition and PLS-MGA. Ind. Manag. Data Syst. **117**(3), 496–520 (2017)
46. Walch, A.: In code(rs) we trust: Software developers as fiduciaries in public blockchains. In: Hacker, P., Lianos, I., Dimitropoulous, G., Eich, S. (eds.) Regulating Blockhain Tecno-Social and Legal Challenges, pp. 58–81 (2019)
47. Calefato, F., Gerosa, M.A., Iaffaldano, G., Lanubile, F., Steinmacher, I.: Will you come back to contribute? Investigating the inactivity of OSS core developers in gitHub. Empir. Softw. Eng. **27**(3), 1–41 (2022)
48. Hayes, A.: What factors give cryptocurrencies their value: an empirical analysis. SSRN Electron J. **20**(12), 1–7 (2018)
49. Hacker, P.: Corporate governance for complex cryptocurrencies? A framework for stability and decision making in blockchain-based organizations. In: Hacker, P., Lianos, I., Dimitropoulous, G., Eich, S. (eds.) Regulating Blockhain Tecno-Social and Legal Challenges. Oxford University Press, pp. 140–66 (2019)
50. Tiwana, A.: Platform desertion by app developers. J. Manag. Inf. Syst. **32**(4), 40–77 (2015)
51. Howell, B.E., Potgieter, P.H., Sadowski, B.M.: Governance of blockchain and distributed ledger technology projects. SSRN Electron J. 1–24 (2019)
52. Valiev, M., Vasilescu, B., Herbsleb, J.: Ecosystem-level determinants of sustained activity in open-source projects: a case study of the PyPI ecosystem. In: ESEC/FSE 2018 - Proceedings of the 2018 26th ACM Joint Meeting on European Software Engineering Conference and Symposium on the Foundations of Software Engineering, pp. 644–55 (2018)
53. Manolache, M.A., Manolache, S., Tapus, N.: Decision making using the blockchain proof of authority consensus. In: The 8th International Conference on Information Technology and Quantitative Management Elsevier B.V, pp. 580–588. (2021)
54. Singh, P.K., Singh, R., Nandi, S.K., Ghafoor, K.Z., Rawat, D.B., Nandi, S.: An efficient blockchain-based approach for cooperative decision making in swarm robotics. Internet Technol. Lett. **3**(1), 1–7 (2020)
55. De Filippi, P., Loveluck, B.: The invisible politics of Bitcoin: governance crisis of a decentralized infrastructure. Internet Policy Rev. **5**(4), 1–32 (2016)
56. Pelt, R.V., Jansen, S., Baars, D., Overbeek, S.: Defining blockchain governance: a framework for analysis and comparison. Inf. Syst. Manag. **38**(1), 21–41 (2020)
57. Deshpande, A., Start, K., Lepetit, L., Gunashekar, S.: Distributed ledger technologies/blockchain: challenges, opportunities and the prospects for standards [Internet]. British Standards Institute (2017). https://www.bsigroup.com/LocalFiles/zh-tw/InfoSec-newsletter/No201706/download/BSI_Blockchain_DLT_Web.pdf
58. Khatoonabadi, S., Costa, D.E., Abdalkareem, R., Shihab, E.: On wasted contributions: understanding the dynamics of contributor-abandoned pull requests. ACM Trans. Softw. Eng. Methodol. (arXiv:2110.15447), 1–38 (2022)
59. Hill, R.: What sample size is "enough" in internet survey research. Interperson. Comput. Technol. Electron. J. 21st Century **6**(3–4), 1–12 (1998)
60. Titah, R., Barki, H.: Nonlinearities between attitude and subjective norms in information technology acceptance: a negative synergy? MIS Q **33**(4), 827–844 (2009)

61. Carillo, Kévin. D.: Social cognitive theory in IS research – literature review, criticism, and research Agenda. In: Prasad, S.K., Vin, H.M., Sahni, S., Jaiswal, M.P., Thipakorn, B. (eds.) Information Systems, Technology and Management, pp. 20–31. Springer Berlin Heidelberg, Berlin, Heidelberg (2010). https://doi.org/10.1007/978-3-642-12035-0_4

Performance Analysis of Scheduling Algorithms for Virtual Machines and Tasks in Cloud Computing

Hind Mikram[1], Said El Kafhali[1(✉)] ⓘ, and Youssef Saadi[2]

[1] Faculty of Sciences and Techniques, Computer, Networks, Modeling, and Mobility Laboratory (IR2M), Hassan First University of Settat, B.P. 539, 26000 Settat, Morocco
{h.mikram,said.elkafhali}@uhp.ac.ma
[2] Data Science for Sustainable Earth Laboratory, Faculty of Sciences and Techniques, Sultan Moulay Slimane University, Beni Mellal, Morocco
y.saadi@usms.ma

Abstract. Scheduling virtual machines (VMs) and tasks in the appropriate location is a fundamental challenge integral to the consolidation process in cloud data centers. Therefore, many optimization methods have been developed as optimal solutions to assign VMs to host or tasks to VM. The most popular algorithms are SJF (Short Job First), FCFS (First Come First Served), MinMin, and MaxMin. This work introduces a comparative study by evaluating those algorithms with PSO (Particle Swarm Optimization) and ACO (Ant Colony Optimization) using the CloudSim toolkit. The performance metrics used are degree of imbalance and makespan. The simulation results showed that ACO outperforms the evaluated scheduling algorithms in terms of the degree of imbalance and VM makespan.

Keywords: Server Consolidation · Cloud Computing · Resource Efficiency · Virtual Machines Placement · Energy Consumption

1 Introduction

Cloud computing is the modern economic backbone that allows entrepreneurs and researchers in various fields to lease and host computing resources such as storage, software, servers and networking over the Internet. It has emerged as an attractive platform for users from all over the world based on the pay-as-you-go model [1]. In recent decades, cloud providers applied server consolidation techniques in data centers to reduce costs while satisfying customers' demands [2]. These techniques use a virtualization environment to execute several applications and operating systems in the same physical machine [3]. The main objective of the server consolidation process is resource management, and to facilitate this, scheduling algorithms are introduced. First, scheduling is distributing users' tasks into available shared resources and balancing the load in the cloud. Second, adapting the algorithm objective to the dynamic environment and users' needs. In addition, the goal of load balancing and task scheduling is to reduce the time taken to complete tasks. On the other hand, an incorrect task assignment strategy can imbalance the load between virtual machines, leading to a situation where the virtual machines are loaded while others are idle [4], thus increasing task completion, waiting times, and the overall makespan of the system. As a result, efficient load balancing among VMs becomes the primary element for improving the performance of resources and maintaining stability in the cloud environment [5]. Therefore, it is critical to create an algorithm that can increase system performance by balancing workload between virtual machines and reducing makespan [6].

Several basic algorithms that solve these problems include FCFS, SJF, MaxMin, and MinMax. Moreover, metaheuristic methods treat multi-objective problems and give a near-optimal solution in a reasonable time, which leads to the popularity of these techniques in scheduling approaches. We choose PSO and ACO as metaheuristics algorithms. The PSO algorithm, a type of iterative optimization algorithm, offers several advantages (such as convergent, resilient, etc.) in obtaining great solutions for dynamic objects and is appropriate for the study of population behavior. An extremely popular distributed technique for solving NP-hard combinatorial optimization problems is the ACS (Ant Colony System). Its basic model is based on how real ants choose the shortest route to food by assessing the density of pheromones left along the path [7].

Task scheduling is the main challenge of cloud computing data centers and considered an NP-hard combinatorial optimization problem [8]. The tasks submitted to the cloud environment must be rapidly completed with the available resources. The existing comparative studies use different metrics to discuss and compare the classic algorithms. This work presents and discusses the abovementioned scheduling algorithms in a time-shared and space-shared manner using CloudSim toolkit, VM makespan and degree of imbalance as parameters for the evaluation. This article focuses on task scheduling policies on a virtual machine in the cloud environment.

The remaining sections are organized as follows. Section 2 presents some related work. Section 3 presents the scheduling algorithms used in this paper in the cloud computing environment. Section 4 provides the implementation and analysis of the respective policies. Finally, Sect. 5 concludes the article with a summary of our future work.

2   Related Work

Several comparative works have been presented for scheduling problems related to task distribution across available resources or balancing the cloud load. For example, the papers [9] and [10] compare RR, SJF, and FCFS using time-shared policies and space-shared policies in the CloudSim tool. Those mechanisms are used to manage tasks into VM and VMs into host according to each algorithm priority. Each paper chooses different metrics to evaluate these basic algorithms. The authors of [10] analyze the deadline of cloudlets according to the net gain and the task penalty, whereas [9] and [11] discuss the turnaround time and the waiting time to compare the execution time of cloudlets. The comparison of RR, SJF, and FCFS in [9] showed that RR is better than SJF and FCFS at both cloudlet-scheduling policies. On the other hand, the authors in [11] compare RR, Priority scheduling, FCFS, SJF, SRTF (Shortest Remaining Time First),

LJF (Longest Job First), and HRRN (Highest Response Ratio Next). The results showed that the process completion and overall waiting times are the shortest for SRTF. The LJF strategy performed poorly compared to the other studied strategies. A comparison of the average waiting time of each of these algorithms is discussed in [12] to determine which algorithm is best for some processes without addressing priority between tasks. The result showed that FCFS is effective for short periods. If the process enters the CPU concurrently, the SJF is preferable. The Round Robin method is the best one for adjusting the required average waiting time. Workflow datasets are used in [13] to compare scheduling methods based on makespan, including Round Robin, First Come First Served, MaxMin, and MinMin. The performance of the proposed method in [14] was evaluated by the makespan, response time, and degree of imbalance using CloudSim as a simulator. This evaluation showed that by effectively dividing requests among VMs, the suggested strategy reduces communication overheads and increases the degree of load balancing. Moreover, the completion time of cloudlet is a significant factor in identifying attacks in the cloud. The authors in [15] consider the number and length of cloudlets to analyze the effect of time-shared and space-shared allocation strategies in the normal and attack nodes. The results demonstrated that, when the number of cloudlets varies, the execution of the cloudlets under normal and attack situations takes longer under the space-shared allocation policy compared to the time-shared allocation approach. The same result was obtained when the cloudlet length varies.

Some existing algorithms used PSO to solve the scheduling problems such as [16] that combined PSO with levy flight algorithm to minimize the utilization rate of physical machines (PMs), however, the authors do not consider the multi-objective problems, as well as dynamic migration. In [17], the authors proposed an improved PSO that adjusts the number of active servers using the minimization fitness function to reduce the number of overloaded hosts. Its effectiveness is measured in terms of host shutdowns, energy consumed, ESV, and VM migrations but only addresses continuous problems. The algorithm presented in [18] increased the utilization of resources and minimized the execution time of the task. Its performance is evaluated by applying standard datasets to achieve scalable algorithms that can handle both small and large workloads. This study introduced a novel threshold SD value (SD ≤ 5) for load balancing throughout the scheduling sequence.

In [19], the authors modeled workload consolidation as an instance of multidimensional bin packing (MDBP) issue. The ACO algorithm is used to design a new nature-inspired workload consolidation. The results obtained showed that ACO achieved superior energy gains and better server utilization. Therefore, this method does not account for the workload characteristics, which leads to the degradation of service performances. In [20], the authors used ACO to balance the total load of the system while aiming to decrease the time for a certain set of tasks. The proposed method outperforms ACO and FCFS algorithms and can handle the variation in the number of tasks number but the researchers do not consider the task types, which means unrealistic of the environment. The authors in [21] presented a multi-objective algorithm built on a modified version of ACO. They used energy consumption, makespan, degree of imbalance, and turnaround time as performance parameters. However, the developed algorithm performs inadequately when it relates to load balancing with a high number of tasks.

3 Scheduling Algorithms in Cloud Computing

Resource scheduling is a consolidation process-building component that determines how resources are shared over time and how they are made available to customers in the most efficient way possible. It offers the highest quality of services to customers while also making the most effective use of resources to maximize returns. Changing user needs are unexpected in terms of time, making it challenging to meet resource requirements in a dynamic environment [22]. Users' tasks to VMs and VMs to the PMs are the two types of scheduling in cloud computing. The first scheduling ensures service quality by respecting response time, execution cost, and makespan, whereas the secondary scheduling is responsible for migrations, resource usage, load balancing, and power consumption [23]. For that, many algorithms are proposed; in this section, we will highlight four popular scheduling algorithms and two metaheuristic methods, namely PSO and ACO.

3.1 FCFS (First Come First Served)

FCFS (or FIFO First In First Out) checks the CPU availability first and then executes the process that arrives first. Moreover, the execution of the process started cannot be interrupted until the method is finished, so this operation is named non-preemptive. The problem with this algorithm is the long waiting time because the arrival process in the queue must wait until all the existing process finish their execution [11].

3.2 Short Job First (SJF)

The process in SJF executes if it has the least burst time. This algorithm is non-preemptive and the preemptive version is called Shortest Remaining Time First (SRTF). The comparison of burst time is between existing processes. However, when several processes have the same burst time, FCFS will be used. The drawback of this algorithm is that a long burst time process may wait in the queue for a long time or may never be executed by the system [12].

3.3 MinMin and MaxMin

The MinMin algorithm assigns the task with the shortest execution time to the machine with the shortest execution time for that task [24]. The MaxMin algorithm assigns the task with the longest execution time to the resource with the shortest execution time for that task.

3.4 Particle Swarm Optimization (PSO)

The standard PSO considers a population of p particles, where each particle is an optimal solution in continuous space. Each particle has its parameters, such as position, velocity, best position, and global position. Its objective is to adjust the direction of the swarm (population of particles) based on the particle's best previous information and the neighbor's best previous information [25]. The following formulae calculate the new population velocity:

$$V_{i+1}^t = \omega.V_i^t + c_1r_1.\left(pBest_i^t - X_i^t\right) + c_2r_{21}.\left(gBest_i^t - X_i^t\right) \tag{1}$$

where ω is the inertia factor that influences the algorithm's local and global abilities; V_i^t and X_i^t are the ith velocity vectors and i^{th} position vector's at i iterations; c_1 and c_2 are weights influencing cognitive and social factors, respectively; and r_1 and r_2 are random numbers. $pBest_i^t$ and $gBest_i^t$ are the local best position obtained by t particle and the global best position, respectively.

The path planning of this method can be divided into various phases, as illustrated in the flow diagram in Fig. 1.

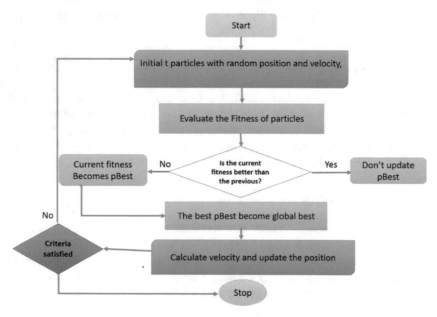

Fig. 1. Flow chart of the POS algorithm.

3.5   Ant Colony Optimization (ACO)

ACO is a metaheuristic method based on the natural feeding behavior of real ants, which inspired this system development. As ants' vision and hearing abilities are limited, they have developed a type of indirect communication by using a pheromone (chemical substance). Each ant distributes this chemical on the path that it follows, and the pheromone disappears over time. Other ants can detect the quantity of this chemical and choose pathways based on the amount of pheromone placed on the path around them. Eventually, the entire ant colony converges on the shortest path to the food source. This short path was determined when the ant returned faster to its initial location. As a result, this route will have a higher pheromone concentration, making it more attractive to ants following behind [18]. Figure 2 presents the flow chart of this method.

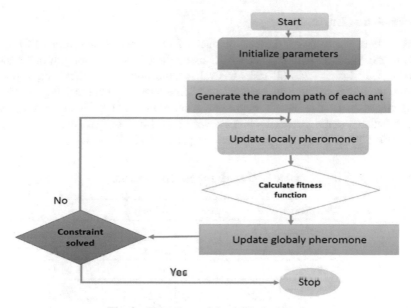

Fig. 2. Flow chart of the ACO algorithm.

4 Simulations Results and Analysis

4.1 The Performance Metrics

The objectives of scheduling in the cloud are balancing the load across all VMs as well as reducing the completion time of tasks [20]. In this article, the makespan time, the degree of imbalance, and the processing speed are considered mathematical parameters for evaluation.

Degree of Imbalance (DI) determines how much load is distributed between virtual machines, with a lower degree of imbalance indicating that the load is successfully balanced. DI is expressed mathematically in Eq. 2.

$$DI = \frac{T_{max} - T_{min}}{T_{avg}} \tag{2}$$

where T_{max}, T_{min} and T_{avg} are the maximum, minimum, and average execution times of all VMs, respectively. As a result, the scheduling problem aims to decrease the degree of imbalance. The use of DI throughout the allocation process would help eliminate VM workload imbalances [20].

Makespan defined as the total amount of time necessary to complete all of the tasks. The resources receive efficient and effective task planning as a result of the shorter makespan. Hence, the makespan is expressed mathematically in Eq. 3.

$$Makespan = max_{1 \leq i \leq m}\{completionTime_i\} \tag{3}$$

where m is the number of VMs.

4.2 Simulation Environment

In this section, we examine the performance of the classical algorithms FCFS, SJF, MinMin, and MaxMin with PSO and ACO using the CloudSim tool. We first consider 300 tasks, 500 tasks, and 800 tasks to evaluate the abilities of these algorithms. In addition, we compare the Space-shared and Time-shared scheduling algorithms. We consider makespan and imbalance degree as factors of evaluation. CloudSim architecture has four fundamental components that are very useful in setting up a cloud computing system. The CloudSim system parameters are presented in Table 1.

Table 1. CloudSim system parameters.

Parameters	Values
System Architecture	X86
Operating System	Linux
Time Zone	10
Processing Cost	3
Memory Cost	0.05
Storage Cost	0.001
Bandwidth Cost	0.0

We tested the efficacy of scheduling algorithms with the help of some Cloudsim components. The classes in question are the Datacenter, DatacenterBroker, VM, and Cloudlet. The Datacenter is responsible for offering to cloud users hardware-level services. VMs are created and destroyed by DatacenterBrokers based on Cloudlet requirements. DatacenterBrokers also defend the user from VM control. Cloudlet is currently running as a virtual machine job. Table 2 shows the cloudlet parameters.

Table 2. Cloudlet parameters.

Parameters	Values
Cloud Cloudlets	300 or 500 or 800
Length	2000
Pes	1

The virtual machine executes the tasks with the cloudlet scheduler's policies. Tables 3 and 4 show the settings for hosts and virtual machines, respectively.

4.3 Results and Discussion

The configuration and scenario described above will be discussed in terms of both VM scheduling using imbalance degree, and VMs makespan as performance metrics.

Table 3. Host parameters.

Parameters	Values
Number of PMs	3
Number of Brokers	2
Processing Speed (MIPS)	500000
Pes	5
Memory (MB)	500000
Storage	1000000
Bandwidth	1000000

Table 4. VM parameters.

Parameters	Values
Number of VMs	3
Processing Speed (MIPS)	2000
Pes	5
Memory (MB)	512
Storage	10000
Bandwidth	1024

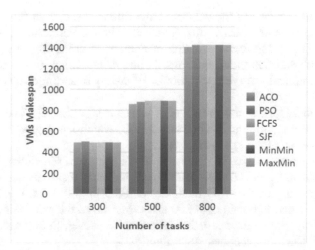

Fig. 3. VM makespan versus the number of tasks with time-shared of tasks and space shared of VMs

Figure 3 presents the VMs makespan for scheduling algorithms in 300, 500, and 800 tasks. The graphs indicate almost the same results between all algorithms at each number of tasks, but the metaheuristics outperform other algorithms. In other words, ACO and PSO have the lowest value in 500 and 800 tasks. However, PSO has the highest value in 300 tasks. On the other hand, the VMs Makespan increases when the number of tasks is augmented.

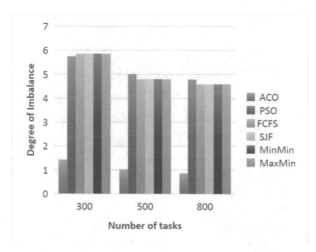

Fig. 4. Degree of Imbalance versus the number of tasks with time-shared of task and space shared of VMs

The imbalance degree of the compared algorithms is shown in Fig. 4. The results show that ACO is more effective than other algorithms. Furthermore, the degree of imbalance decreases during the variation of number of tasks. However, PSO's value increases in 500 and 800 tasks. For Figs. 3, and 4, we follow those metrics for the compared algorithms by varying the number of tasks for tasks time-shared and space shared of VMs.

The VMs makespan is shown in Fig. 5, where it is increasing. The ACO method outperforms other algorithms in contrast to PSO, which has a greater value, which means it is the worst algorithm in this case.

Figure 6 shows the degree of imbalance of the compared scheduling algorithms in the tests of 300, 500, and 800 tasks. The Figure indicates that the value of this metric decreased, especially in the ACO method. For PSO in 500 tasks, the DI value was augmented compared to obtained in 800 tasks. Moreover, the PSO values in 500 and 800 tasks are superior to the heuristic algorithms. However, in 300 tasks, PSO has a small value compared to other heuristics algorithms.

Figures 4 and 6 show how the VM time-shared scheduling strategy outperforms the VM space-shared scheduling approach for the degree of imbalance. The DI shows more difference in both VM scheduling policies, but the difference between VMs Makespan is smaller. The same trend for ACO shows a significant difference compared to PSO and heuristic algorithms in terms of imbalance degree.

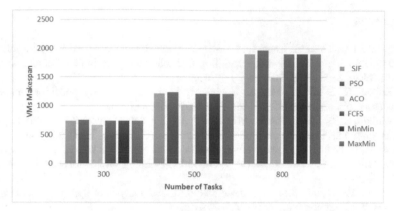

Fig. 5. VMs makespan versus the number of tasks with time-shared of tasks and time-shared of VMs

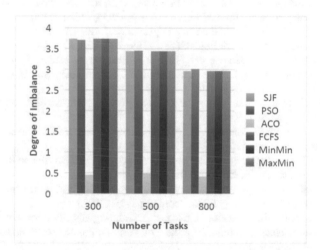

Fig. 6. Degree of Imbalance versus the number of tasks with time shared of task and time shared of VMs

5   Conclusion and Future Work

In this work, we compared four basic algorithms and two metaheuristic algorithms to solve scheduling problems. Based on the obtained results, the classic algorithms offered optimal solutions in a short time and of a smaller size. The ACO performed far better than the PSO and other algorithms in terms of the degree of imbalance and makespan. Furthermore, the VMs makespan and imbalance degree obtained by the metaheuristic methods are lower than those of other algorithms, especially ACO. The low value means that the load is more equally distributed, indicating that ACO outperforms others in terms of VMs makespan and Degree of Imbalance. In the future, a scheduling issue might be addressed with a variety of objectives and a combination of algorithms.

References

1. Mapetu, J.P.B., Chen, Z., Kong, L.: Low-time complexity and low-cost binary particle swarm optimization algorithm for task scheduling and load balancing in cloud computing. Appl. Intell. **49**(9), 3308–3330 (2019)
2. El Kafhali, S., El Mir, I., Salah, K., Hanini, M.: Dynamic scalability model for containerized cloud services. Arab. J. Sci. Eng. **45**(12), 10693–10708 (2020)
3. El Kafhali, S., El Mir, I., Hanini, M.: Security threats, defense mechanisms, challenges, and future directions in cloud computing. Arch. Comput. Meth. Eng. **29**(1), 223–246 (2022)
4. Saadi, Y., El Kafhali, S.: Energy-efficient strategy for virtual machine consolidation in cloud environment. Soft. Comput. **24**(19), 14845–14859 (2020). https://doi.org/10.1007/s00500-020-04839-2
5. Salah, K., El Kafhali, S.: Performance modeling and analysis of hypoexponential network servers. Telecommun. Syst. **65**(4), 717–728 (2017)
6. Adhikari, M., Amgoth, T.: Heuristic-based load-balancing algorithm for IaaS cloud. Future Gener. Comput. Syst. **81**, 156–165 (2018)
7. Surekha, P., Sumathi, S.: PSO and ACO based approach for solving combinatorial Fuzzy Job Shop Scheduling Problem. Int. J. Comp. Tech. Appl **2**(1), 112–120 (2011)
8. El Kafhali, S., Salah, K.: Modeling and analysis of performance and energy consumption in cloud data centers. Arab. J. Sci. Eng. **43**(12), 7789–7802 (2018)
9. Pratap, R., Zaidi, T.: Comparative study of task scheduling algorithms through cloudsim. In: 2018 7th International Conference on Reliability, Infocom Technologies and Optimization (ICRITO), pp. 397–400. IEEE (2018)
10. Sidhu, H.S.: Comparative analysis of scheduling algorithms of Cloudsim in cloud computing. Int. J. Comput. Appl. **975**, 8887 (2014)
11. Pirani, M., Ranpariya, D., Vaishnav, M.: A comparative review of CPU scheduling algorithms. Int. J. Sci. Res. Eng. Trends **7**(4), 2446–2452 (2021)
12. Siahaan, A.P.U.: Comparison analysis of CPU scheduling: FCFS, SJF and Round Robin. Int. J. Eng. Develop. Res. **4**(3), 124-132 (2016)
13. Hamid, L., Jadoon, A., Asghar, H.: Comparative analysis of task level heuristic scheduling algorithms in cloud computing. J. Supercomput. 1–19 (2022)
14. Ebadifard, F., Babamir, S.M.: Autonomic task scheduling algorithm for dynamic workloads through a load balancing technique for the cloud-computing environment. Cluster Comput. **24**(2), 1075–1101 (2021)
15. Mehmi, S., Verma, H.K., Sangal, A.L.: Comparative analysis of cloudlet completion time in time and space shared allocation policies during attack on smart grid cloud. Procedia Comput. Sci. **94**, 435–440 (2016)
16. Fatima, A., et al.: Virtual machine placement via bin packing in cloud data centers. Electronics **7**(12), 389 (2018)
17. Ibrahim, A., Noshy, M., Ali, H. A., Badawy, M.: PAPSO: a power-aware VM placement technique based on particle swarm optimization. IEEE Access **8**, 81747–81764 (2020)
18. Sahana, S.K.: Ba-PSO: A Balanced PSO to solve multi-objective grid scheduling problem. Appl. Intell. **52**(4), 4015–4027 (2022)
19. Feller, E., Rilling, L., Morin, C.: Energy-aware ant colony based workload placement in clouds. In: 2011 IEEE/ACM 12th International Conference on Grid Computing, pp. 26–33. IEEE (2011)
20. Li, K., Xu, G., Zhao, G., Dong, Y., Wang, D.: Cloud task scheduling based on load balancing ant colony optimization. In 2011 Sixth Annual ChinaGrid Conferenc, pp. 3–9. IEEE (2011)
21. Elsedimy, E., Algarni, F.: MOTS-ACO: an improved ant colony optimiser for multi-objective task scheduling optimisation problem in cloud data centres. IET Networks **11**(2), 43–57 (2022)

22. Ouammou, A., Tahar, A.B., Hanini, M., El Kafhali, S.: Modeling and analysis of quality of service and energy consumption in cloud environment. Int. J. Comput. Inform. Syst. Indust. Manage. Appl. **10**, 098–106 (2018)

23. Kumar, M., Suman: Hybrid cuckoo search algorithm for scheduling in cloud computing. CMC-Comput. Mater. Continua **71**(1), 1641–1660 (2022)

24. Er-Raji, N., Benabbou, F., Eddaoui, A.: Task scheduling algorithms in the cloud computing environment: survey and solutions. Int. J. Adv. Res. Comput. Sci. Softw. Eng. **6**(1), 604–608 (2016)

25. Poli, R., Kennedy, J., Blackwell, T.: Particle swarm optimization. Swarm Intell **1**(1), 33–57 (2007)

A Survey of Mobile Ad-Hoc Networks Based on Fuzzy Logic

Ibrahim Ahmed Alameri[1,2,3]($\boxtimes$) (iD), Jitka Komarkova[2] (iD), and Tawfik Al-Hadhrami[3] (iD)

[1] Computer Science Department, Nottingham Trent University, Nottingham NG1 4FQ, UK
[2] Faculty of Economics and Administration, University of Pardubice, Studentska 95, 532 10 Pardubice, Czech Republic
st61833@upce.cz
[3] Jabir Ibn Hayyan Medical University, Najaf, Iraq

Abstract. Wirelessly connected mobile devices with self-configuration need to have reliable communication. The deployment of routing techniques in dynamic Ad-Hoc networks has enhanced traffic management. In mobile Ad-Hoc networks (MANETs), each node freely moves from one place to another and acts as a router. However, mobile networks are also considered Ad-Hoc wireless networks. Wireless technologies are merged in MANETs to improve communication among the nodes. MANETs can be easily installed in civil and military applications. This paper presents a survey study on MANETs using fuzzy logic-based techniques based on reviewed literature. Also, some applications are discussed, which include forestry, search and rescue operations, and digital libraries. In addition, different types of routing protocols are incorporated, like proactive, reactive, hybrid, swarm intelligence, and fuzzy logic-based routing. However, a brief overview of fuzzy systems and different real-time applications that use fuzzy technology as a controller are given preference. Besides, investigating fuzzy logic in MANETs, especially in routing protocols, has attracted researchers to this field of study.

Keywords: AODV · DSDV · VANETs · Fuzzy Logic

1 Introduction

Mobile Ad-Hoc networks (MANETs) are incredibly adaptive regarding their nodes' ability to join and depart networks. No central core has proven its ability to monitor and control the movement of network nodes. In addition, the nodes in such networks are self-organizing [1]. The structure is not fixed and has a relatively low range of motion. Every intermediate node also has a hop-to-hop connection due to the frequent changes in MANETs' physical structure, which directly influences routing challenges. However, Solmaz et al. designed a fuzzy- logic-based reliable routing protocol that identifies the optimal path to overcome queuing delays in the network [2]. The most appealing research topic is mobile Ad-Hoc networks, which can be easily implemented in laptops, mobile phones, and sensors that interact via wireless communication links. In complex networks, nodes transmit more data packets and use energy efficiently. While maintaining the battery charged and replacing the battery in the environment, power management is a significant concern in extending the network's life.

F. Saeed et al. (Eds.): ICACIn 2022, LNDECT 179, pp. 290–299, 2023.
https://doi.org/10.1007/978-3-031-36258-3_25

Furthermore, infrastructure-less behavior allows MANETs to operate with real- time applications such as rescue missions, battlefields, and natural disasters. To choose the optimum way, an energy-aware fuzzy controllable route is proposed to select the optimum way. The primary goal is to estimate latency and enhance residual energy by evaluating route stability [3].

To improve the capabilities of MANETs, different routing protocols have been developed. Communication protocols severely influence wired connectivity, which magnifies overhead concerns caused by networks. Compared with Ad-Hoc On- Demand Distance Vector Routing (AODV) and Destination-Sequenced Distance Vector Routing (DSDV), the Q-learning algorithm can quickly enhance several network metrics such as bandwidth, connection quality, power, and optimal distance. In addition, Fuzzy Logic Q-Learning Based Asymmetric Link Aware and Geographic Opportunistic (FQ-AGO) has a much higher throughput and a reduced end-to-end delay [4]. However, the transmission control protocol is utilized to provide a reliable channel to solve the problem of end-to-end connectivity. From source to destination, secure communication channels must form hand- shake patterns. The Ad-Hoc transport protocol was designed exclusively for MANETs to ensure smooth data transmission. This paper suggests an enhanced transport protocol that uses a fuzzy logic controller to increase the MAC layer's data rate. In simulation results, ITP has performed well.

The decentralized nature of an Ad-Hoc network makes it ideal for a wide range of applications. Without proper infrastructure, mobile nodes communicate with one another. Due to mobility, there has been a high channel error rate, delay, and packet loss. MANETs have proposed a unique concept called "vehicle Ad-Hoc networks." Apart from that, proactive and reactive routing protocols can improve balance problems, node localization, and power-control-based techniques to classify power-aware optimization in the network [5].

On the other hand, the MANET relies on locally linked wireless nodes that transfer data directly to a ground base station. In Ad-Hoc networks, there is no concept of fixed nodes. In mobile networks, routing transports data from one location to another. A protocol is a method of exchanging data between routers. When implementing routing protocols, the network layer is essential for gathering information about the status of the inter-network.

The main characteristics of MANETs include dynamic network physical structure, multi-hop, limited battery power, easy deployment, and are infrastructure-less. The main problem of scalability is observed [6]. Two metrics—battery power and trustworthy nodes—can help in reliable path-finding. Because it works on a demand basis, AODV is one of the most popular routing protocols for reducing broadcast routes. This method determines the shortest path by sending and broadcasting RREQ messages from source to destination [7]. Venkanna et al. proposed a novel idea called "cooperative routing." Since malicious nodes influence route selection in mobile Ad-Hoc networks, trustworthy packets, energy, and availability time define the network level. Therefore, fuzzy logic-based trust AODV (FLBT-AODV) is designed for this purpose. So, the suggested method quickly addresses the challenges faced by MANETs. Ambient intelligence joins MANETs to provide a new concept anytime and anywhere to facilitate connections. However, for intelligent decision-making, a distributed evolving fuzzy modelling framework installs an unsupervised online one-pass fuzzy clustering mechanism to learn friend and stranger nodes. Also, traffic patterns identify malicious nodes in the network as well [8].

In mobile Ad-Hoc networks, reinforcement learning solves energy-aware routing challenges. Using this method, each node tries to learn from its surroundings to choose the next hop for easier routing. The core idea is that of reward and punishment. In addition, the feedback signals are implemented. Adaptive fuzzy logic is employed in the RREQ method to choose the best route. Dynamic fuzzy energy state-based AODV (DFES-AODV) and FSARSA-AODV demo comparable simulation results in mobile Ad-Hoc networks achieving high energy-aware routing [9].

The latest addition to AODV routing is the secure Ad-Hoc on-demand distance vector protocol. Cryptographic signatures are employed for authentication using chain hash to secure the hop-count field in the network. As a result, every node uses an essential management technique to get public keys directly employed in mobile Ad-Hoc networks. Top hash, signature, and hash are among the Secure Ad hoc On-demand Distance Vector (SAODV) protocol fields. This cryptographic method will be capable of meeting all security needs [10].

This paper aims to provide a conceptual and theoretical framework based on the MANET application, routing protocols, and MANET issues. In addition, this work seeks to summarize the most recent routing protocols enhanced based on the fuzzy approach. This paper shows the fuzzy concept, which is essential for improving network performance through routing protocols like the AODV. Finally, the literature study shows that the AODV-modified by the fuzzy approach presents better results than the other MANET routing protocols.

The rest of the article is structured as follows: In Sects. 2 and 3, explains an overview of MANET and the routing protocols of MANETs. An overview of fuzzy logic and its essential role in routing protocols are explained in Sect. 4. The conclusion and future work are concluded in the Conclusion Sect. 5.

2   MANET Overview

Due to the rapid rise of smart devices that employ wireless communication technology as a backbone, Ad-Hoc networks are becoming increasingly popular. Ad-Hoc networking develops without infrastructure, and the current framework is relatively expensive. Smart devices may maintain connections and add or delete sensor nodes/workstations at any moment in mobile Ad-Hoc networks. Furthermore, MANETs have various applications that are either dynamic or static networks with low-power backups. The following are some of the most remarkable MANET applications:

2.1   Forestry

Mobile Ad-Hoc networks may be easily installed to acquire environmental information to safeguard the forest. The fundamental networking of MANETs in the forest, which includes sensors linked to the access point, is visualised. Backbone node is connected directly to the base station. Fires in the forest environment can harm animal life and damage nature. Therefore, fire may quickly be detected to balance the protection of the network system in the forest. The most sophisticated technology for detecting intruder attacks or undesirable incidents happening in the surrounding area is the mobile Ad-Hoc network.

2.2 Search and Rescue Operations

In mobile Ad-Hoc networks, an Android-based application is implemented. It allows for emergencies such as disasters and search and rescue operations to be quickly attended to. A new software application has been developed that only operates in a 50-m radius. The prototype has text messaging controls that help users transmit information from one mobile device to another.

2.3 Digital Libraries

In the standard system, the library's supervisor often stands at the back or end of the building to verify each reader's card. On the other hand, some book readers use web-based services to reserve their books. Although they have made an online reservation, they must still wait at the admission gate to have their cards checked. Following this, they must request that the librarian take the book reservation online. Mobile Ad-Hoc networks have changed the system's dynamics, making it possible for the network to respond intelligently to the devices in the hands of readers. When a student or reader enters the library with a mobile device linked to a wireless access point, the mobile device instantly begins communication with the system. The reader does not need to wait in line because they have booked the book online. Wireless RF tags have also been developed to be programmed to update the serial number via a wireless interface [11].

3 Routing Protocols of MANET

Sensor nodes are connected using IoT-based wireless technologies in MANET. There are three different forms of wireless-MANETs: mesh, star, and device-to-device networking. Routing protocols are necessary for effective communication between various nodes. In addition, throughput, end-to-end delay, packet delivery, and overhead are used to check the performance of a network. According to this study, network simulator-2 (NS2) is considered the most commonly used tool for simulation. Besides, proactive, reactive, hybrid, swarm intelligence-based routing, and fuzzy-logic-based routing techniques are discussed below.

3.1 Proactive Routing

Although proactive routing approaches are primarily concerned with updating the routing table, this strategy is proactive [12]. Every sensor node must be able to maintain a routing table, and nodes must update routing information automatically when the network topology changes. The most widely used routing methods are Fisheye state routing (FSR), optimal link-state routing, and source tree adaptive routing. Proactive routing is distinct from reactive routing and has many advantages and limitations.

3.2 Reactive Routing

Nodes establish communication on a demand basis in this mechanism. This method will easily overcome flooding in mobile Ad-Hoc networks, which will help decrease overhead concerns. However, the primary disadvantage of this approach is that the nodes are not always updated on time, which causes significant difficulty throughout the network's lifetime. Some reactive routing protocols include dynamic source routing, AODV, and Ad-Hoc on-demand multipath distance vector (AOMDV).

3.3 Hybrid Routing

The combination of proactive and reactive features forms hybrid protocols. Anindya Kumar Biswas et al. formulated a secure hybrid routing technique that uses proactive and reactive approaches to establish communication routes. However, the physical structure of MANETs and minimum spanning trees is being proactively developed. In comparison, digital certificates are used to secure communication channels. The security association method between source and target is set up to send information confidentially over the network. Diffie-Hellman or Shamir's threshold scheme can be utilised in the hybrid approach for network safety purposes. Also, the zone routing protocol is considered a hybrid routing approach [12]. Furthermore, the routing protocols' classification is present in Fig. 1.

4   Overview of Fuzzy Logic

Fuzzy logic is the sub-part of artificial intelligence that has gained much attention in recent decades. Japanese scientists and engineers initially utilized fuzzy logic, which boosted the electric industry. However, the rule-based inference is quite simple and low-cost. Also, fuzzy systems give solutions to non-linear problems. The concept of fuzzy systems was formulated for the first time by L.A.Zadeh. Neuro and fuzzy systems are used in conventional soft computing for industrial applications [13]. In Ad-Hoc networks, the quality-of-service metrics needs to be enhanced. Where QoS is used to evaluate the network's lifetime, multiple parameters like end-to-end delay, packet loss, and overhead quantitatively improve the network resources. Therefore, the Fuzzyvan-QoS model is designed and applied to multiple protocols, while in network design, two factors, including priority and weights, have positively improved the core systems' QoS [14]. Wireless mobile networks face many problems, like frequent disruption and a lack of routes from source to target. In delay-tolerant networks (DTN), connectivity is a severe problem in transferring information from one node to another. Reinforcement learning is one of the optimal solutions to balance routing production and cost. Multi-path routing needs reliable routing in the dynamic structure of sensor nodes. For this purpose, a fuzzy-logic-based double Q-learning approach is implemented to quickly learn from the surrounding environment. In this case, the main focus is node activity, like interval and movement speed. Due to this mechanism, a buffer overflow can be easily managed in delay tolerant networks. While nodes are dynamic, selecting a reliable path in an Ad-Hoc network is difficult. The fuzzy logic-based routing method improves network usage, and autonomous agents assist in determining the best path. Traditional AODV routing is compared with the proposed algorithm. To find possible solutions for decision-making, a fuzzy system is used to measure the importance of different attributes.

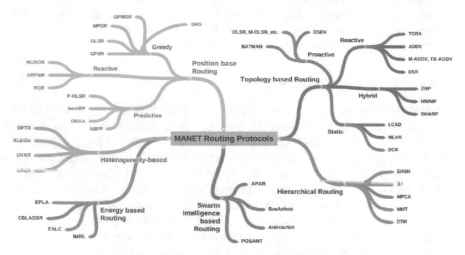

Fig. 1. MANET protocols - Protocol Hierarchy.

Nowadays, as the world is advancing in technology, the study of fuzzy logic has revolutionized several fields, including control systems, UAVs, electric machines, and communication transmission channels. The use of fuzzy systems at the low level of machines strongly emphasizes product engineering. A Fuzzy system is a type of reasoning based on Boolean logic and may be used in various applications such as medicine, business models, automobiles, and natural language processing. To accomplish cognitive abilities, fuzzy systems are also employed as controllers in artificial intelligence-based approaches. Fuzzy logic can help engineers, scientists, and practitioners clear up doubts and uncertainties so they can make better decisions.

4.1 Fuzzy Logic-Based Routing

Due to the dynamic topology of mobile nodes with respect to the fixed base station, multipath routing is complicated in mobile ad hoc networks. Fuzzy controllers are utilized in multipath routing algorithms to reduce the reconstruction of routes in MANETs. The fuzzy logic controllers consist of feedback, which works as adaptive control, using fuzzy logic to calculate each data packet index for scheduling. Expiry time, data rate, and queue length are used to be fuzzified, while the novel protocol FMRM is compared with AOMDV and AODVM.

In contrast, the novel protocol fuzzy controllers based multipath routing algorithm (FMRM) is compared with AOMDV and Ad-Hoc on-demand distance vector multipath (AODVM) [15]. An on-demand multicast routing protocol can establish high-quality interconnected nodes within the network. For this purpose, fuzzy logic is incorporated to control the flooding of data packets, which creates congestion. Overhead, end-to-end delay, packet delivery, and power consumption are all easily controlled by the original algorithms ODMRP and AAMRP [16]. Quality of service is considered in the diverse environment of mobile networks. For a high level of throughput and reduced end-to-end delay, a new routing protocol is introduced: the integration of ant colony and fuzzy system optimization. The main focus is to improve the quality of experience in MANETs. However, to improve the energy level of mobile networks, a multicast tree-based fuzzy marginal energy disbursed is proposed for multi-path routing.

4.2   Fuzzy Logic-Based Routing

A wireless network of linked mobile devices operates without a fixed base station. Because of the changing topological structure, the sensor nodes directly influence each other. Optimized link-state routing is a proactive strategy for delivering broadcast data. Multi-point relays are used to distribute information, controlling the message data before passing it on to neighbors. As the efficiency of OLSR is influenced by the change in the hello packet, throughput increases automatically. For this purpose, fuzzy logic is incorporated, which improves the transmission of the hello packet in the communication channel because the protocol is named FLBHIT-PEOLSR [17]. The mechanism computes hello interval time based on mobile network inputs such as size and mobility pattern. The symbol channel is connected with low, medium, or high mobility. Therefore, the proposed technique has shown better delay, throughput, and load balancing.

The 5th generation (5G) cellular communication is increasingly becoming popular. The deployment of 5G in MANETs will change the dynamics and reduce end-to-end delay, as the ultra-dense network is becoming the leading technology to provide high data rates. Due to the dynamic pattern in ad hoc networks, the quality of experience needs to be enhanced. To design the channel state information feedback method, fuzzy logic-based activation strategy selection is used to activate the threshold. This approach adapts different traffic load management techniques to improve communication channels in cellular networks.

Fuzzy logic is utilized in MANETs as controllers to evaluate the stability of routes. Fuzzy systems have input and output variables applied to inference for fuzzification. The proposed route consists of three metrics: movements of nodes, bandwidth, and energy. However, trapezoidal membership functions are used. Ad-Hoc on-demand fuzzy routing is being developed to improve packet delivery, throughput, and average end-to-end delay [2]. At the same time, AODV routing is table-driven and used to establish multi-path from source to destination. A fuzzy modified multipath approach using AODV to control data packets. The path that will not be utilized for communication will be discarded.

While AODV routing is table-driven and used to establish multipath from source to destination, a fuzzy modified multi-path approach using AODV controls data packets. The path that will not be utilized during communication will be discarded. For searching, path data packets are broadcast on an RREQ to find the nearby node. Fuzzy logic is used in AODV to activate the lifetime of each node. Also, fuzzy linguistic rules are applied, as the membership function for hope count and lifetime is used [18]. Heuristic-based routing is used in ad hoc networks to utilize data packets in the network. DSR, DSDV, TORA, and AODV can be deployed in the network for scalability and the removal of location errors. Fuzzy-based on-demand routing is used to increase the battery lifetime and channel stability [19]. By exchanging information in a mobile network, the approach continuously learns from the physical structure. In a mobile Ad-Hoc network, almost every sensor node tries to transmit data packets to the entire network. Because of this, the main problem found is overhead, which can be found by using the MPR idea. This method controls hello packets, topology, and multiple interfaces to give better performance in metrics like delay, load and throughput [20]. To look at how the best path is chosen, DSR and AODV, which are common routing algorithms for mobile ad hoc networks, are used. MANETs have many limitations due to special features. Fuzzy stochastic routing upgrades multiple parameters in terms of route discovery. The routing protocols based on fuzzy logic lead to better results in parameters such as throughput, network delay, route discovery, packet delivery ratio, and routing overhead. Different simulation tools were used in those studies, including OPNET, NS2 and OMNET + +. Furthermore, the mobile vehicular routing problem is solved by using fuzzy logic. This is the immediate issue in hospitals transmitting medical materials.

Several limitations still exist with the new versions of AODV, such as broadcasting issues, insufficient path discovery, data packet duplication, longer routes, and no control of information distribution. Also, the network delay and overhead in the related work were noticed due to the discovery routing operation. This paper's extension work will investigate develop AODV-based Fuzzy logic.

5 Conclusion

The presented paper briefly studies fuzzy logic applications in mobile Ad-Hoc networks for route selection. Fuzzy logic is a soft computing-based technique to determine the degree of truthfulness of an event. It is based on fuzzy set theory. The current study discovered that fuzzy logic had enhanced routing performance in mobile Ad-Hoc networks. Several fuzzy-based routing protocols have been designed to overcome the shortcomings of legacy proactive, reactive, and hybrid routing protocols. However, the study also reveals that fuzzy-based route selection approaches are prone to many limitations. In this investigation, several papers compared the proposed model with legacy routing techniques. Most researchers did not implement their algorithms with state-of-the-art routing protocols. This can generate weak algorithms that can affect the network's performance. Many researchers did not pay considerable attention to the mobility model. With the advent of faster and more reliable communication interventions, mobility scenarios are also changing, becoming an obstacle to efficient routing. Newer, prompt communication technologies can create more extensive MANET. In the future, this study will look into how to improve the AODV routing protocol by using a fuzzy approach. This could lead to a change in the way the protocol routes packets to deal with unstable link quality.

Acknowledgment. We appreciate the time and effort put in by the reviewers and editor of this conference, as well as the valuable comments and suggestions they provided us with as we prepared this study for submission.

This work is supported by SGS University of Pardubice project No. SGS_2022_008.

References

1. Alameri, I.A., Komarkova, J.: Network routing issues in global geographic information system. In: SHS Web of Conferences (2021)
2. Ghasemnezhad, S., Ghaffari, A.: Fuzzy logic based reliable and real-time routing protocol for mobile ad hoc networks. Wireless Pers. Commun. **98**(1), 593–611 (2017)
3. Raza, N., Umar Aftab, M., Qasim Akbar, M., Ashraf, O., Irfan, M.: Mobile ad-hoc networks applications and its challenges. Commun. Netw. **08**(03), 131–136 (2016)
4. Helen, D., Arivazhagan, D.: A stable routing algorithm for mobile ad hoc network using fuzzy logic system. Int. J. Adv. Intell. Paradig. **14**(3–4), 248–259 (2019)
5. Alshehri, A., Badawy, A.-H.A., Huang, H.: FQ-AGO: fuzzy logic Q-learning based asymmetric link aware and geographic opportunistic routing scheme for MANETs. Electronics **9**(4), 576 (2020)
6. Kanellopoulos, D., Cuomo, F.: Recent developments on mobile ad-hoc networks and vehicular ad-hoc networks. Electron. **10**(4), 1–3 (2021)
7. Al-Absi, M.A., Al-Absi, A.A., Sain, M., Lee, H.: Moving ad hoc networks—a comparative study. Sustain. **13**(11) (2021)
8. Mutalik, P., Nagaraj, S., Vedavyas, J., Biradar, R.V., Patil, V.G.C.: A comparative study on AODV, DSR and DSDV routing protocols for Intelligent Transportation System (ITS) in metro cities for road traffic safety using VANET route traffic analysis (VRTA). In: 2016 IEEE Int Conf Adv Electron Commun Comput Technol ICAECCT 2016, pp. 383–6 (2017)
9. Chettibi, S., Chikhi, S.: Dynamic fuzzy logic and reinforcement learning for adaptive energy efficient routing in mobile ad-hoc networks. Appl. Soft. Comput. **38**, 321–328 (2016)
10. Moudni, H., Er-rouidi, M., Mouncif, H., El Hadadi, B.: Secure routing protocols for mobile ad hoc networks. In: 2016 International Conference on Information Technology for Organizations Development (IT4OD), pp. 1–7 (2016)
11. Hsu, K.-K., Tsai, D.-R.: Mobile ad hoc network applications in the library. In: 2010 Sixth International Conference on Intelligent Information Hiding and Multimedia Signal Processing, pp. 700–3 (2010)
12. Alameri, I.A.: A novel approach to comparative analysis of legacy and nature inspired ant colony optimization based routing protocol in MANET. J. Southwest Jiaotong Univ. **54**(4) (2019)
13. Biswas, A.K., Dasgupta, M.: A secure hybrid routing protocol for mobile ad-hoc networks (MANETs). In: 2020 11th Int Conf Comput Commun Netw Technol ICCCNT 2020 (2020)
14. Liu, C., Wu, J., Kohli, P., Furukawa, Y.: Raster-to-vector: revisiting floorplan transformation. In: Proceedings of the IEEE International Conference on Computer Vision, pp. 2195–203 (2017)
15. Mchergui, A., Moulahi, T., Nasri, S.: QoS evaluation model based on intelligent fuzzy system for vehicular ad hoc networks. Computing **102**(12), 2501–2520 (2020). https://doi.org/10.1007/s00607-020-00820-x
16. Pi, S., Sun, B.: Fuzzy controllers based multipath routing algorithm in MANET. Phys Procedia [Internet]. **24**, 1178–85 (2012). https://doi.org/10.1016/j.phpro.2012.02.176
17. Mani Kandan, J., Sabari, A.: Fuzzy hierarchical ant colony optimization routing for weighted cluster in MANET. Clust. Comput. **22**(4), 9637–9649 (2017)
18. Vikkurty, S., Shetty, S.P.: Design and implementation of fuzzy logic based OLSR to enhance the performance in mobile ad hoc networks. In: International Conference on Intelligent Computing and Communication Technologies, pp. 424–33 (2019)

19. Su, B.-L., Wang, M.-S., Huang, Y.-M.: Fuzzy logic weighted multi-criteria of dynamic route lifetime for reliable multicast routing in ad hoc networks. Expert Syst. Appl. **35**(1–2), 476–484 (2008)
20. Tabatabaei, S., Teshnehlab, M., Mirabedini, S.J.: Fuzzy-based routing protocol to increase throughput in mobile ad hoc networks. Wireless Pers. Commun. **84**(4), 2307–2325 (2015)

The Influence of Node Speed on MANET Routing Protocol Performance

Ibrahim Alameri[1,2](✉) ⓘ, Jitka Komarkova[2] ⓘ, Tawfik Al-Hadhrami[1] ⓘ,
and Raghad I. Hussein[3] ⓘ

[1] Computer Science: School of Science and Technology Nottingham, Trent University,
Nottingham, UK
st61833@upce.cz
[2] Faculty of Economics and Administration, University of Pardubice, Pardubice II,
Czech Republic
[3] University of Kufa, Najaf, Iraq

Abstract. The Mobile Ad-Hoc Network, also known as MANET, is one of the most popular networks that have arisen in the area of technology in recent years. MANET has several routing protocols to transmit data between the source and destination nodes. A MANET protocol must be appropriately configured to guarantee effective data transfer. There must be a suitable routing protocol in place to accomplish this goal. As a result, the correct routing protocol must be selected. MANET relies heavily on adequately using parameter values in its routing protocols. MANET is a network of battery-powered nodes that communicate with each other. Routing protocols are essential in MANET because MANET nodes rely on them to transmit data. While all the routing protocols perform the same function in the network, their performance varies significantly. A network simulator (NS-2) is used in this study to evaluate the performance of four routing protocols at various node speeds. The impact of node speed on the performance of MANETs is essential. This study examined AODV, DSDV, DSR, and ZRP metrics to compare the most popular routing protocols. Comparative experiments confirmed their validity for throughput, packet loss, end-to-end delay, and energy consumption. More than 40 simulation scenarios were run to ensure a network performance result, and four performance parameters were analyzed and compared across four different network speeds to provide a thorough investigation. The finding has confirmed that routing protocols' performance is influenced by node speed variety.

Keywords: MANET · Routing Protocols · Routing Metrics

1 Introduction

Integrated networks are now possible due to the advancements in wireless communication that have revolutionized data networking and telecommunications. Mobile communication systems, such as wireless LANs, mobile radios, and cellular technologies, enable the undertaking of completely dispersed mobile computers and communications, regardless of time and place. Wireless networks can be categorized into mobile Ad-Hoc

F. Saeed et al. (Eds.): ICACIn 2022, LNDECT 179, pp. 300–309, 2023.
https://doi.org/10.1007/978-3-031-36258-3_26

networks (MANET) and cellular wireless networks. Cellular Wireless Networks (CWNs) are networks that rely on infrastructure to function. In this scenario, communication between nodes takes place through a base station [1].

In contrast, multi-hop radio transmission is used in Ad-Hoc wireless network systems, which can operate without infrastructure or infrastructure-less [2]. Unlike cellular networks in MANET, there is no central coordinator, in this case, making routing more challenging. However, Ad-Hoc networks have to deal with various challenges, such as transmission with a limited range, mobility, restricted energy, and routing topology that can change dynamically. In addition, MANET is a self-forming and self-organizing computer network that can relocate and re-configure itself to adapt to topology changes. Unlike an infrastructure network, MANET does not depend on a pre-existing system of fixed links, such as standard carrier cables or telephone transmission lines. It operates within existing systems linking one wire-less node with another, such as cellular networks or wireless fidelity (WiFi). It is also called an Ad-Hoc wireless network or an instant network. A MANET comprises several mobile routers (also known simply as "nodes") [3]. Each node performs both routing and forwarding functions for its connected links. Since nodes may have multiple links, the router may receive the same packet from different next-hop neighbors [4]. The set of neighbors over which a node routes packets is called its neighbor set or "adjacent list". As network size increases, the cost of maintaining a state for every connection becomes impractical. Therefore, MANETs use either active or passive approaches to routing, in both cases, only selected nodes store state information about the active topology at any given time. The dynamic topology of MANET is a big challenge. The selection of routing protocols is essential in MANET configuration to address dynamic topology variations and achieve a quality, reliable connection. Accordingly, an effective routing protocol is essential to improve MANET connectivity. Selecting the proper routing protocol acts as an essential key to network efficiency.

The paper primarily aims to investigate and analyze the effects of various routing protocols (proactive, reactive, and hybrid) on network performance by testing multiple parameters. The current study examines routing protocols' scalability, mobility of routing protocols, and network parameters. The paper primarily aims to investigate and analyze the effects of various routing protocols (proactive, reactive, and hybrid) on network performance by testing multiple parameters. The current study examines routing protocols' scalability, mobility of routing protocols, and network parameters. A primary concern of the current investigation is the various node speeds evaluated in the experimental results using network simulator 2 (NS-2) to evaluate the optimal routing protocol. Eventually, the optimal protocol is to operate more nodes at a changeable speed. The AODV and DSR are reactive routing protocols, while the DSDV is proactive. In contrast, ZRP acts as a hybrid protocol.

The remaining sections of this paper are organized as follows: Sect. 2 will discuss the Routing and Protocols Overview. Section 3 discusses the materials and techniques approach to assessment. Results and discussion were concluded in Sect. 4 and were followed by the conclusion of Sect. 5.

2 Routing Protocols Classification

Routing is when nodes determine the optimal route/path to forward the packets toward the destination. If a device or node receives a packet, and the target destination is not the destination for which the packet was sent, it must route it. All intermediate nodes in Ad-Hoc networks need to make routing decisions for each packet by using routing table lookup. Routing protocols populate the routing table. Routing protocols play a significant role in MANET, particularly when mobile nodes and network dynamics change at run-time. The literature on routing interventions is grouped into different classes based on how they operate, build, and maintain the routing table. The taxonomy of routing protocols is presented in Fig. 1. Figure 1, shows an overview of a few protocols sharing standard functionality [5]. However, the network's routing protocols could have different types [6].

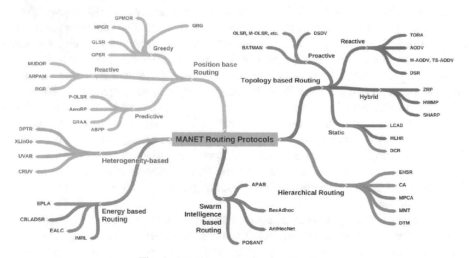

Fig. 1. Routing protocols classification.

- Position-based Routing Protocols.
- Energy-based Routing Protocols.
- Heterogeneity-based Routing Protocols.
- Hierarchical Routing Protocols.
- Swarm Intelligence-based Routing Protocols.
- Routing Protocols based-Topology.

Routing protocols based on topology techniques use network topology data to build a routing table. This class of routing approach is most widely utilised in Ad-Hoc networks. Topology-based protocols are divided into the following groups: proactive, reactive, hybrid routing, and static. In MANET, a static technique to build the routing table is not often used and recommended due to its static nature.

2.1 Proactive Routing Protocols

A complete routing table is built by all proactive routing protocols in advance at the start of operation. This table aims to establish and maintain a path to every destination node alive in the network/topology. All devices/nodes that speak any proactive routing protocol start to exchange network information initially, also known as the setup phase. In the setup, nodes exchange routing messages and calculate the best next hop for every destination. Nodes exchange complete routing tables periodically and at each change in the topology because it consumes heavy network bandwidth and computational resources of the nodes. However, all nodes in the MANET network carry an entire "map". Therefore, the best route to any target is readily available, reducing the discovery time for each packet. These protocols should not be used in large networks or where the topology changes more often, including the Destination Sequenced Distance Vector (DSDV) routing protocol.

2.2 On-demand or Reactive Routing Protocols

If the on-demand protocols are used, the routes are checked only when required, meaning the paths are only checked once. When any nodes need to connect, the discovery process of the When a path is discovered, or when no path is discovered, In this context, these characteristics make it a reactive article. Several MANET routing protocols implement reactive techniques, including Dynamic Source Routing Protocol (DSR) and Ad-Hoc On-Demand Distance Vector (AODV) routing protocols.

2.3 Hybrid Routing Protocols

Network delays and overhead in the network discovery routing operations can be reduced using hybrid routing, which merges similar reactive routing protocols with proactive routing protocols. Higher reliability and scalability provide the contributions of this type of protocol. The drawback of these protocols' new routes is that connectivity issues are presented within a network's latency. The Zone Routing Protocol (ZRP) is one of the significant protocol types.

3   Materials and Techniques

The current section is concerned with the methodology used for this study. The presented paper uses a simulation approach and parameters configured to execute the work successfully. Therefore, the study demonstrates the importance of using NS-2 to carry out the simulation. A significant amount of functionality is provided for modelling various protocols over wired and wireless networks by Network NS-2. All types of network components, protocols, traffic, and routing can be simulated on this platform. In order to accommodate various network features, protocols, traffic, and routing types, it uses a perfect modular framework for both types of networks, whether wired or wireless simulations. The current section is concerned with the methodology used for this study. The presented paper uses a simulation approach and parameters configured to execute the work successfully. In order to accommodate various network features, protocols, traffic, and routing types, it uses a perfect modular framework for both types of networks, whether wired or wireless simulations. NS-2 employs the Object-Oriented Tool Command Language (OTCL) to develop the simulation topology, while the central core is written in C++. The presented paper uses the latest version of NS-2, which is 2.35. It generates two types of trace files, which are (i) simulation trace and (ii) nam trace. The simulation trace file is further used for data analysis. In contrast, the Nam trace can be fed into the network animator (Nam) utility to view how the simulation is carried out. Figure 2, shows how the simulation is carried out using NS-2.

In addition, graphical representations and analyses of the trace file were generated using the MATLAB programming language and formulas. The proposed paper creates four different testbeds, or scenarios, for our study. In all three scenarios, the basic simulation parameters are the same. However, the node's speed is the main difference in all three testbeds.

The current paper conducts four scenarios. In scenario one, the simulation configuration of topology is mentioned in Table 1. The number of nodes in this scenario was configured at 300, and the node speed was 20 m per second. In contrast, the node speed was 50 in Scenario Two. While scenarios three and four used 100 and 150 node speeds for the same number of nodes, as shown in Table 1.

Table 1. Simulation setup.

Network Parameters	Value
Simulator	NS2
Simulation Time	300 s
Area	1000m * 1000m
Number of Nodes	300 nodes
Max-Speed	20, 50, 100, 150 m/s
Routing Protocols	AODV, DSDV, DSR, ZRP

The proposed paper performed ten iterations of simulation for each simulation scenario. This repeated exercise aims to reduce statistical anomalies/discrepancies in the outcome. Therefore, 40 total rounds/iterations of the simulation were carried out during this study. The time duration of the simulation contributes to an essential role in studying the behaviour of any phenomenon.

4   Results and Discussion of the Simulation

The effectiveness and performance of each routing protocol were evaluated using four different metrics in this study.

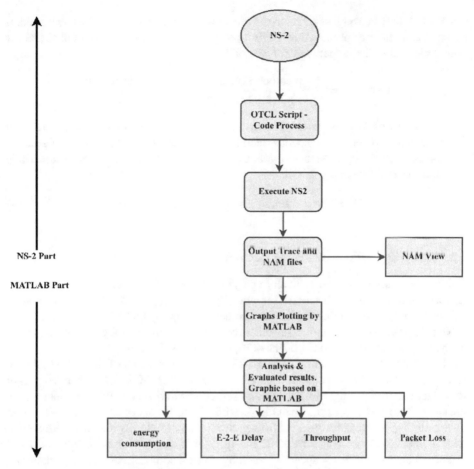

Fig. 2. NS-2 Structure & Code operations.

4.1 Network Parameters

Throughput: The efficiency of a protocol is measured by its throughput. Higher performance rates indicate optimal results, while low throughput indicates restricted activity in the network. The following equation calculates throughput.

$$\text{Throughput} = \frac{\text{total of the packet sent}}{\text{total data sending time}} \tag{1}$$

End to End Delay (E-2-E Delay): The current experiment investigating the average E-2-E delay takes data packets to travel from one end to the other. An E-2-E delay refers to the time it takes for a packet to reach its final destination. The following equation calculates the E-2-E delay.

$$\text{Delay} = \frac{\text{time packet received} - \text{time packet sent}}{\text{total package received}} \tag{2}$$

Packet loss: Packet loss refers to the rate at which data packets are lost over time. It counts the number of packets that are dropped per second. Another essential signal for overloaded networks is "network congestion".

$$PacketLoss = \frac{\sum \text{ Number of packets received at destination}}{\sum \text{ Number of packets send by node}} \tag{3}$$

Energy Consumption (EC): The residual or energy consumption correlates with the efficiency of a routing protocol's operations. For each node that has completed the simulation, there is an amount of residual energy left over. The unit of measurement is the joule. The initial energy is 75 J.

$$EC = \text{Nodes Initial Energy - Consumed Energy} \tag{4}$$

4.2 Analyzing and Discussing the Outcomes

The current subsection will critically analyse the result. These are significant results that will be described and discussed sequentially. Figure 3, provides the results obtained from the analysis of routing protocols. The network throughput of the AODV routing protocol gives a better result than the other routing protocols. Although it started to drop as the node speed increased from 20 to 150, the result decreased.

In contrast, the ZRP routing protocol was the worst in terms of throughput. At the same time, the DSDV has given acceptable results after AODV. In terms of packet loss fraction, AODV outperforms all other protocols at all speeds while maintaining a packet delivery fraction, as shown in Fig. 4. While the result of packet loss for DSDV was disappointing, At the same time, the DSR performance was closely matched with ZRP's, with a slight outperformance for ZRP. In Fig. 5, it is observed that the ZRP protocol is vulnerable to high node speeds. Generally, the outcome of ZRP fails to maintain performance. AODV performs better than DSR and DSDV, making it adept at running at high speeds. The observed superiority of the AODV routing protocol is due to an obvious consequence of the working nature of the algorithm.

While Fig. 6, shows outcomes similar to previous results, protocol ZRP was more efficient in terms of energy consumed than other routing protocols, AODV, DSDV, and DSR, throughout the simulation. The result of AODV was close to the DSR results. The DSDV routing protocols show slightly higher performance than AODV and DSR, where it maintains the energy till the simulation ends. The results conclude that the routing protocols consume much energy when the node's speed increases.

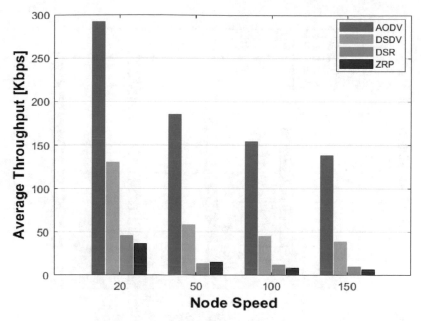

Fig. 3. Throughput plot by varying nodes speed.

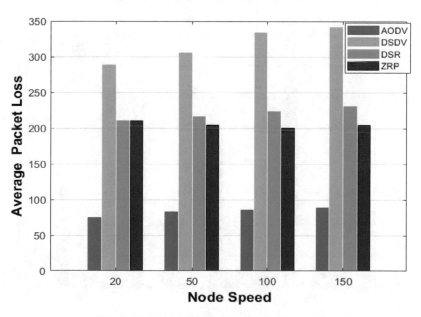

Fig. 4. Packet loss plot by varying nodes speed.

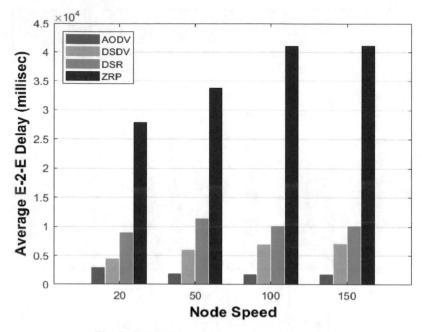

Fig. 5. E-2-E Delay plot by varying nodes speed.

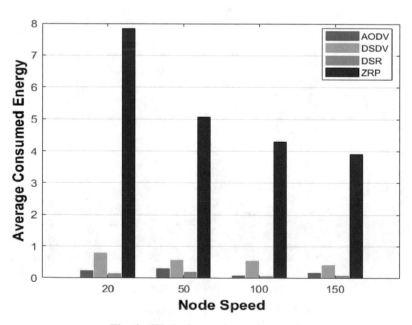

Fig. 6. CE plot by varying nodes speed.

5 Conclusion

Different evaluation matrices have been used to evaluate the routing protocol's performance, including E–2–E delay, packet loss, energy consumption, and throughput. The current work was done with the NS-2 simulation and the analysis of graphics made with MATLAB. The work was done to evaluate routing protocols under different node speeds and show the impact of node speed. Different speeds of nodes are used in this study to show how node speed affects how quickly the topology changes. Various routing protocols have been used to obtain the simulation results, such as AODV, DSR, DSDV, and ZRP. From this empirical study, it can be concluded that the AODV routing protocol provides the best metric performance. On the other hand, the ZRP performs better than AODV in energy consumption. Finally, this study confirms that the performance of routing protocols decreases when the node speed increases. Hence, node speed was considered when changing from 20, 50, and 100 to 150.

This study will, in the future, investigate the use of a fuzzy method to determine how the AODV routing protocol might be improved. This may result in a modification to how the protocol routes packets to accommodate for variable link quality.

Acknowledgment. We appreciate the reviewers' and editors' time, effort, and helpful comments and recommendations as we prepared this study for submission.

This work is supported by SGS University of Pardubice project No. SGS_2022_008.

References

1. Oe, K., Koyama, A., Barolli, L.: Proposal and performance evaluation of a multicast routing protocol for wireless mesh networks based on network load. Mob Inf Syst. 2015 (2015)
2. Li, J., Wang, M., Zhu, P., Wang, D., You, X.: Highly reliable fuzzy-logic-assisted AODV routing algorithm for mobile ad hoc networks. Sensors (Basel). **21**(17) (2021)
3. Huang, J., Fan, X., Xiang, X., Wan, M., Zhuo, Z., Yang, Y.: A clustering routing protocol for mobile ad hoc networks. Math Probl Eng. 2016 (2016)
4. Wood, J.D.G.: The effects of the distribution of mortgage credit on the wage share: Varieties of residential capitalism compared. Comp. Eur. Polit. **15**(6), 819–847 (2017)
5. Alameri, I., Komarkova, J., Ramadhan, M.K.: Conceptual analysis of single and multiple path routing in MANET network. Int Conf Inf Digit Technol 2021, IDT 2021. 235–44 (2021)
6. Alotaibi, E., Mukherjee, B.: A survey on routing algorithms for wireless Ad-Hoc and mesh networks. Comput Networks [Internet] **56**(2), 940–65 (2012). https://doi.org/10.1016/j.com net.2011.10.011

Validation of Measuring Item in Dealing Negative E-WOM for Online Reputation

Rizka Dhini Kurnia[1]($\boxtimes$), Halina Mohamed Dahlan[2], and Samsuryadi[1]

[1] Fakultas Ilmu Komputer, Universitas Sriwijaya, Palembang, Indonesia
samsuryadi@unsri.ac.id
[2] Information System Department, Azman Hashim International Business School (AHIBS), Universiti Teknologi Malaysia (UTM), Skudai, Johor, Malaysia
halina@utm.my

Abstract. Online reputation is a set of procedures that are incorporated into organizations and have evolved into a critical component of the online world. Negative e-WOM, on the other hand, poses a threat to the organization. This occurs when unfavorable messages spread over multiple online platforms have a detrimental impact on other people's views and behaviors, as well as a negative reputation for the organization. In order to overcome negative e-WOM, the positive feedback is significantly needed to deal and maintain the online reputation of the organization. This research aims to validate and confirm the items to evaluate the extracted constructs towards dealing the threats of online reputation because of the worst effects by the negative e-WOM. The method to find the items that is from the previous model, then confirmed through content validity and get some valid instruments. The instrument can be used by other researchers in the same context area to assess online reputation in the future.

Keywords: Online Reputation · e-WOM · Negative e-WOM

1 Introduction

The electronic word-of-mouth (e-WOM) refers to the online conversations between users which able to influence other users behaviour [1]. e-WOM involved any positive or negative statement that made by users about business issues through online networks which influenced the changing attitudes [2]. This situation of e-WOM brought the challenges to the business more critical when the e-WOM involved with greater scalability, speed of diffusion, persistency, accessibility, measurability, and quantifiability [2].

There are two types of e-WOM which are positive and negative e-WOM. The positive e-WOM refers to the product or service which able to increase the likelihood of its purchase. Meanwhile, the negative e-WOM decreases the likelihood of the purchase behaviours [2]. The positive e-WOM brought the profits to the business and company but the negative e-WOM is significantly on the serious threats towards the business. The fast spreads of information diffusion through online networks through all online platforms site give the negative e-WOM reach a lot of people easily and influenced them towards the negative opinions and damage the business online reputation.

Online reputation is really important as a lot of business used online network. Online platforms also allow customers to give personal ratings to each of the online items and this make the online reputation able to be rated by people in easy way. If people are not satisfied with the reputation of the business either on their products or services, or even the company's itself; people tends to spread the negative e-WOM and give the profit lost towards the businesses.

2 Background of Study

The negative e-WOM brought out several problems towards online reputation. As mention in literature review, online reputation is significantly important in dealing with customers through online network as nowadays many businesses used online platforms as their main business platforms in reaching more customers. This situation brought the problems where the negative messages been spread faster and wider through many platforms, such as blogs, search engines, virtual communities, and consumer review systems, among others [3].

In addition, testimonials are frequently published on social media channels [1]. The negative testimonials is a major reason why negative e-WOM been spreads among individuals all over the worldthrough social media [4]. The negative words can swiftly spread and harm a company's reputation. When negative e-WOM goes viral on social media sites, the problem becomes worst. Furthermore, consumers are primarily focused on existing customer reviews and how customers' participation in e-WOM would effect un-favorable information both directly and indirectly [5].

The negative e-WOM has an impact on the customer-brand identification, which relies on electronic e-WOM to attain high resilience to negative information [5]. Negative e-WOM will make it difficult for customers to engage with the company. As a result, the organization's internet reputation will suffer.

Previous studies also explicitly show that customers tend to post the negative e-WOM because of anger, regret, or disappointment through their experience towards the services they used [2]. Organization should not ignore the customer feelings and dissatisfaction that can lead to the negative e-WOM.

The previous researchers also developed several negative e-WOM model design theory, which significantly related and needed for this study. Even though, only a few models been developed in dealing negative e-WOM for maintaining online reputation of an organization either brick-and-mortar or online organization. Due to that, this research is conducted to deal with negative e-WOM in maintaining online reputation of business organization.

3 Methodology

The information searching method which used in this study is through a systematic literature review (SLR) technique. SLR was performed to satisfy this study's objectives. In a comprehensive research procedure, the SLR was used to locate and summarize all the related sources.

SLR and meta-analyses are critical techniques for properly and reliably summarizing data. This study adopted a four-stages approach of SLR included understanding problems, identifying construct, identifying the measurement survey items, survey instrument development, and survey instrument confirmation through content validity.

The SLR review protocol was described and depicted. Then, the inclusion and exclusion criteria were explained. After that, the search strategy was discussed. The study selection process was defined. Thereafter, the quality assessment was clarified. The data extraction and synthesis were expounded. The methodology of the present study encompasses the steps included identified construct, measurement items development, and content validity.

4 Identified Constructs

This section presents the defining process of each related constructs for this study. As refers to the SLR process for this study, ten constructs are selected in dealing the negative e-WOM. Table 1 is presenting the identified constructs along with the sources reference.

5 Measurement Items Development

Several models of study have refined the first construct, which is the Apology Statement. If the corporation apologizes right away, reconciliation will be simple. An apology statement should be delivered personally rather than in bulk to achieve consumer reconciliation. Customers will be loyal if businesses do not hesitate to apologize to them. In order to achieve customer reconciliation, apologies in the customer's native language are easier to accept and understand. Social media apologies may help to improve a company's reputation. The refining procedure for Apology Statement is shown in Table 1.

The next step is to explain the problem. In order to accomplish customer reconciliation, customers must be properly informed about any problems that happen. The customer is more tolerant of the apology if the problem is explained in detail. Customers assume that the business organization is trustworthy after hearing about the situation. Furthermore, customers are less hesitant to interact with the same business organization after hearing about the current condition. After the situation has been resolved, the company will make every effort to prevent repeating the mistake. The refining procedure for Problem Explanation is shown in Table 2.

Customer dissatisfaction can be lessened by sending an official message with a positive message delivered via empathy. Offering good solutions to clients might assist businesses in achieving customer reconciliation. Customers will return if the company understands their problems. Customers gain access to efficient solutions that can assist the company boost its image. The refining procedure for the Positive statement is shown in Table 3.

Offering a monetary reward can improve the likelihood of a customer reconciliation. The most critical aspect of customer reconciliation is compensation. Customers will be more satisfied if compensation is comparable to the loss value. Customers who are compensated are more inclined to provide good testimonials for a company. If customers are

Table 1. Constructs Identifying

Construct	Definition	Source
Apology statement	Refers to the reconciliation which easily achieved apology statement been completed in short time and focus personally instead in bulk. Apologies in the customer's native language are easier to accept and understand in order to gain customer reconciliation	[6, 7]
Problem explanation	Refers to detailed explanation of the problem that occurs. Explanation of the actual situation makes customers believe that the company is worthy of trust	[8]
Positive statement	Refers to offering good solutions to customers and help the company enhance their reputation	[9]
Compensation	Refers to offering compensation as the most important thing to get customer reconciliation. Compensation equal in losses value will make customers more satisfied	[9]
Appreciation	Refers to rewarding customers and make them feel appreciated. Customers will return and will promote to others if they are appreciated	[10]
Customer Service	Refers to handling customers complaints while gain their trust. Customer service which responsive leads to positive testimonials about the organization	[11]
Communication	Refers to sharing valuable information with customers and makes the company earn positive feedback. The company's reputation is enhanced by sharing valuable information with its customers	[11]
Respond	Refers to the action of providing feedback to customers. Responding rapidly to consumer complaints will encourage them to return	[12]
Reconciliation	Refers to customers who are reconciled with the company are more likely to give positive testimonials. Reconciliation will make customers to promote good things about the company. Customer reconciliation can enhance a company's reputation	[13]

(continued)

Table 1. (*continued*)

Construct	Definition	Source
Maintaining Online Reputation	Refers to maintaining an online reputation, by controlling the negative news through online platform. Customers will not respond to negative company news if the positive company's online reputation is maintained	[14]

Table 2. Refining the Apology Statement Item

Section	No	Item Code	Refined Items
Apology statement	1	AS1	Reconciliation will be easily achieved if the company apologizes immediately
	2	AS2	To achieve customer reconciliation, an apology statement should be provided personally rather than in bulk
	3	AS3	Customers will return when companies do not hesitate to apologize
	4	AS4	Apologies in the customer's native language are easier to accept and understand in order to gain customer reconciliation. Apologies delivered on social media might enhance a company's reputation
	5	AS5	To achieve customer reconciliation, an apology statement should be provided personally rather than in bulk

compensated, they are more inclined to return. When it comes to customer reconciliation, customers want material value recompense. The Compensation refining process is shown in Table 4.

Companies can achieve customer reconciliation by rewarding customers. If customers feel valued, they will give good feedback. Customers will return if the company expresses gratitude or appreciation. If customers feel appreciated, they will recommend to others. The refinement process for Appreciation is shown in Table 5.

Customers received notice of concerns are more likely to believe that their issues have been received. Customers appreciate responsive customer service, and it makes customer reconciliation a breeze for businesses. In order to achieve customer reconciliation, the employees assigned to explain the difficulties which occurs. Responsive customer service leads to positive testimonials about the company. The refining process for Customer Service is shown in Table 6.

Responsive customer service leads to great testimonials about the company. Customer service that is responsive makes it easier for the organization to complete customer reconciliation. Customers believe in the company which easy to communicate. Companies that engage in active client interaction are more likely to accomplish customer

Table 3. Refining the Problem Explanation

Section	No	Item Code	Refined Items
Problem explanation	6	**PE1**	Customers must be fully informed about any problems that arise in order to achieve customer reconciliation
	7	PE2	Detailed explanation of the problem that occurs makes the customer more accepting of the apology
	8	PE3	Explanation of the actual situation makes customers believe that the company is worthy of trust
	9	PE4	Explanation of the actual situation makes customers not afraid to transact with the same company When the issue has been fixed, the company will make every effort to avoid making the same problem again
	10	PE5	Customers must be fully informed about any problems that arise in order to achieve customer reconciliation

Table 4. Refining the Positive statement

Section	No	Item Code	Refined Items
Positive statement	11	**PS1**	Customer dissatisfaction can be reduced by delivering empathy by official message
	12	PS2	Offering good solutions to customers can help companies get reconciliation with customers
	13	PS3	I believe that my friend can influence me to violate security of classified information
	14	PS4	Customers will return if the company empathize with them
	15	PS5	Customers benefit from good solutions, which can help the company enhance their reputation

reconciliation. When a corporation shares useful information with its clients, it receives favourable feedback. Customer reconciliation is made easy by the speed with which each firm communication channel responds. Sharing useful information with customers improves the company's reputation. The Communication refining procedure is shown in Table 7.

Customers expect responses to customer complaints in less than 24 h. Customers will give good feedback if the company responds to their problems quickly. Responding quickly to customer complaints will entice them to come back. Customers will recommend your business to others if you reply quickly responding. If companies reply to every customer complaint, they may be able to easily achieve customer reconciliation. The refining procedure for Respond is shown in Table 8.

Table 5. Refining the Compensation

Section	No	Item Code	Refined Items
Compensation	16	CP1	Offering compensation can increase the chances of customer reconciliation
	17	CP2	Compensation is the most important thing to get customer reconciliation
	18	CP3	Compensation equal in losses value will make customers more satisfied
	19	CP4	Customers are more likely to give positive testimonials for a company if they are compensated
	20	CP5	Customers are more likely to return if they are compensated
	21	CP6	Customers prefer material value compensation in terms of achieving customer reconciliation

Table 6. Refining the Appreciation

Section	No	Item Code	Refined Items
Appreciation	**22**	**AP1**	Rewarding customers enables companies to achieve customer reconciliation
	23	AP2	Customers will give positive feedback if they feel appreciated
	24	AP3	Customers will return if company show appreciation
	25	AP4	Customers will promote to others if they are appreciated

Table 7. Refining the Customer Service

Section	No	Item Code	Refined Items
Customer Service	**26**	**CS1**	Notice to customer complaints makes customers more trusting that their complaints are received
	28	CS2	Responsive customer service is very helpful for customers and makes it simple for companies to accomplish customer reconciliation
	29	CS3	The staff assigned to explain the problems must be someone with the capacity for the task in order to get customer reconciliation
	30	CS4	Customer service that is responsive leads to positive testimonials about the organization
	31	CS5	Responsive customer service makes company easier to accomplish customer reconciliation

Table 8. Refining the Communication

Section	No	Item Code	Refined Items
Communication	32	CO1	The ease of communicating with the company makes customers believe in the company
	33	CO2	Companies will achieve customer reconciliation more easily if they are active in interacting with customers
	34	CO3	Sharing valuable information with its customers makes the company earn positive feedback
	35	CO4	Speed of reaction each company communication channel makes it easier to get customer reconciliation
	36	CO5	The company's reputation is enhanced by sharing valuable information with its customers

Table 9. Refining the Respond

Section	No	Item Code	Refined Items
Respond	37	RP1	Attending to customer complaints less than 24 h is what customers expect
	38	RP2	Customers will give positive testimonials if company respond immediately to their complaints
	39	RP3	Responding rapidly to consumer complaints will encourage them to return
	40	RP4	Customers will refer company to others if you respond fast to their issues
	41	RP5	Companies may easily reach customer reconciliation if they actually respond to every customer complaint

Customers who have reached an agreement with the company are more inclined to provide good feedback. Customers will promote the company's beneficial qualities if it is reconciled. A company's reputation can be improved via customer reconciliation. Customers will recommend your business to others if you fast reply to their concerns. Customer reconciliation can help you keep your online reputation intact. The refining procedure for Reconciliation is shown in Table 9.

Customers can be retained if the company's online reputation is keep going. If the company's online reputation is maintained, customers will not respond to negative corporate news. If customer reconciliation can be accomplished, online reputation can be preserved. The refining method for Maintaining Online Reputation is shown in Table 10.

The constructions that have been chosen as primary items contained in this study are apologetic statement, problem explanation, positive statement, compensation, appreciation, customer service, communication, respond, reconciliation, and sustaining online

Table 10. Refining the Reconciliation

Section	No	Item Code	Refined Items
Reconciliation	42	RC1	Customers who are reconciled with the company are more likely to give positive testimonials
	43	RC2	Reconciliation will make customers to promote good things about the company
	44	RC3	Customer reconciliation can enhance a company's reputation
	45	RC4	Customers will refer company to others if you respond fast to their issues
	46	RC5	Customer reconciliation can maintain online reputation

Table 11. Refining the Maintaining Online Reputation

Section	No	Item Code	Refined Items
Maintaining Online Reputation	47	MOR1	By maintaining an online reputation, the company can retain customers
	48	MOR2	Customers will not respond to negative company news if the company's online reputation is maintained
	49	MOR3	Online reputation can be maintained if customer reconciliation can be achieved
	50	MOR4	I believe that my awareness of IT risk on classified information level has positive impacts on the protection of classified information in my university

reputation, as previously said. The major items in the questionnaire survey instrument would be these constructs.

The items were tested for both face and content validity in order to validate the questionnaire. Face validity was utilized to analyze the appropriateness of the questionnaire's items, as well as the questionnaire's language and appearance. Before constructing the items, the past relevant literature and validated instruments were assessed, and then experts were recruited to examine the items as part of the content validity process.

6 Content Validity

This section is to confirm the measurement items through content validity. The extent to which a test measures a representative sample of the subject matter. Content validity pinpoints the degree of item acceptance in the measurement based on its reflective operational definition of constructs. There are different ways of carryout content validity.

However, in this research, Content Validity Index (CVI) was used to validate this research instrument. CVI approach is flexible and it requires a minimum of 3 experts as panel members of interrater. They further ague that CVI practically applicable in terms of time and less cost.

Therefore, the items were validated in term of its relevancy in content. 4-point ranking scale were used for each item to measure relevancy in content. In addition, there was a provision for expert to comment in each of the item. The CVI labels are as follows: "Not Relevant = 1; Relevant but not important = 2; Relevant but need review = 3; Highly Relevant = 4". The proportion agreement is 3 and 4. From their answers, the CVI were computed as presenting in Table 11.

Table 12. List of Content Validity Experts

Serial	Status of Experts	No of Expert invited	Numbers of experts that respondent
1	Professors	4	3
2	Doctors	6	2
Total		10	**5**

Therefore, a total of 5 ratters were valid and was used for the ranking which is above the minimum number. As results, the S-CVI as calculates the content validity of the overall scale is 1. This means the survey items have excellent content validity. It is also supported by the value of AS-CVI. The AS-CVI value for this study is 0.986 which clearly stated that mention on the value should be than 0.9 to get having excellent content validity (Table 12).

7 Conclusion

This study developed valid measurement items for the constructs extracted from the SLR process in order to investigate the factor to solve the negative e-WOM towards the organization. The benefits of the survey instrument are match with expanding this study toward developing the research model in future. The survey instrument items also give benefits to other researchers who are study in the same area of e-WOM. Next step is to prepare for pilot study and arranging the main survey data collection process in order to get the relevant data for testing each of the main factors. The validated items will be spread among 589 respondents that involved in this study.

Acknowledgment. We would like to thank the visiting lecturer, Assoc. Prof. Dr. Ab Razak Che Hussin from Universiti Teknologi Malaysia (UTM) to Universitas Sriwijaya, Palembang, Indonesia (UNSRI) and keep assisting us in understanding the concept of this study. It is involved several meetings in investigating the right flow and context of this research study. This study is one of a great cooperative work between us and UTM.

References

1. Kim, S.J., Wang, R.J.H., Maslowska, E., Malthouse, E.C.: 'understanding a fury in your words': the effects of posting and viewing electronic negative word-of-mouth on purchase behaviors. Comput. Human Behav. **54**, 511–521 (2016)
2. Nam, K., Baker, J., Ahmad, N., Goo, J.: Determinants of writing positive and negative electronic word-of-mouth: empirical evidence for two types of expectation confirmation. Decis. Support Syst. **129**(2019), 113168 (2020)
3. Chang, H.H., Wu, L.H.: An examination of negative e-WOM adoption: brand commitment as a moderator. Decis. Support Syst. **59**(1), 206–218 (2014)
4. Balaji, M.S., Khong, K.W., Chong, A.Y.L.: Determinants of negative word-of-mouth communication using social networking sites. Inf. Manag. **53**(4), 528–540 (2016)
5. Augusto, M., Godinho, P., Torres, P.: Building customers resilience to negative information in the airline industry. J. Retail. Consum. Serv. **50**(May), 235–248 (2019)
6. Xuehe, Z., Cui, Y., Peng, X.: How to use apology and compensation to repair competence-versus integrity based trust violations in e-commerce. Electronic Research App. **32**, 3748 (2018)
7. Bakar, R.M., Hidayati, N., Giffani, I.R.: Apology and compensation: impact on customer forgiveness and negative word-of-mouth (WOM). Jurnal Manajemen dan Kewirausahaan (2019)
8. Sengupta, S., Ray, D., Trendel, O., Van Vaerenbergh, Y.: The effects of apologies for service failures in the global online retail. International Journal of Electronic Commerce (2018)
9. Bozic, B., Kuppelwieser, V.G.: Customer trust recovery: an alternative explanation. J. Retailing Consumer Services **49**, 208218 (2019)
10. Gilsa, S., Horton, K.E.: How can ethical brands respond to service failures? understanding how moral identity motivates compensation preferences through self-consistency and social approval. Journal of Business Research (2018)
11. Jean, L., Walker, H.: The critical role of customer forgiveness in successful service recovery. J. Business Res. **95**, 376391 (2019)
12. Sparks, B.A., Fung So, K., Bradley, G.L.: Responding to negative online reviews: The effects of hotel responses on customer inferences of trust and concern. Tourism Management J. **53**, 7485 (2016)
13. Honora, A., Chih, W.-H., Wang, K.-Y.: Managing social media recovery: The important role of service recovery transparency in retaining customers. Journal of Retailing and Consumer Services **64**, 102814 (2022)
14. Perez-Cornejo, C., Quevedo-Puente, E., Delgado-García, J.B.: How to manage corporate reputation? the effect of enterprise risk management systems and audit committees on corporate reputation. European Management J. **37**(4), 505515 (2019)

Leveraging Deep Learning for MmWave Channel Impulse Response Prediction

Mohd. Sharique[1], Mohammad Samar Ansari[2(✉)], and Chirag Gangal[1]

[1] Aligarh Muslim University, Aligarh, India
[2] University of Chester, Chester, UK
m.ansari@chester.ac.uk

Abstract. In communication systems research, wireless channel estimation is a challenging problem due to the dual requirements of real-time implementation and high estimation accuracy. This work presents a Long-Short Term Memory (LSTM) based deep learning (DL) model for the prediction of mmWave channel response for real-time and real-world non-stationary channel scenarios. The DL model trains on the predefined history of channel impulse response (CIR) data along with two other features *viz.* root-mean-square delay spread values and transmitter-receiver update distance, which are also varying in time with the CIR. The objective is to generate an estimate of CIRs using prediction through the DL model. For training the model, a sample dataset is generated through the open-source channel simulation software NYUSIM which produces samples of CIRs using measurement-based channel models based on various multipath channel parameters. From the DL model test results, it is observed that the proposed approach provides a viable lightweight solution for root-mean-square delay spread values channel prediction.

Keywords: Channel Estimation · Deep Learning · Long Short-Term Memory (LSTM) · mmWave Communication · Wireless Communication

1 Introduction

To address the demand of increasing high speed data rate, the 5G wireless systems use the millimeter wave (mmWAve) frequency spectrum. The mmWave system will provide at least $10 \times$ higher bandwidth than the current 4G systems to support higher data rate [15, 16]. The demand for wireless application in high mobility application is another challenge for researchers. The physical layer performance of the wireless network is limited by the prediction accuracy of wireless channel response which is commonly non-stationary and rapidly varying in high mobility applications [5]. A common approach to obtain the channel response at the receiver is to perform the estimation operation from the received observation with sending the pilot signal or blind estimation approach. The accuracy of the channel estimation plays an important role in the performance of the wireless communication systems. In addition, the implementation complexity of the estimation process is one of the limiting factors in system design for high-speed applications. Also, the number of pilot signals required for estimation increases the overhead

© The Author(s), under exclusive license to Springer Nature Switzerland AG 2023
F. Saeed et al. (Eds.): ICACIn 2022, LNDECT 179, pp. 321–330, 2023.
https://doi.org/10.1007/978-3-031-36258-3_28

in the system [8]. In next generation networks, the implementation complexity of the channel prediction grows exponentially as the number of antenna elements increases. In case of pilot-aided estimation process, the pilot signal requirement is an overhead, and pilot signal design to avoid pilot contamination is another pressing issue. Therefore, there have been active research efforts to devise approaches which reduce/eliminate the requirement of a pilot signal. The blind estimation process does not require the pilot signal, however the performance is inferior to pilot-aided method.

There have been numerous recent efforts to leverage the impressive performance offered by state-of-art deep learning models to modern day applications such as computer vision [23], cybersecurity [2], and finance [6]. The fact that deep learning models are able to extract complex underlying inter-relationships between large number of features (given enough data, and a suitably trained model) has led to the application of deep learning models to several other important use-cases in the real world – including, but not limited to, communication systems [12], agriculture [21], sports [14], and drug discovery [3]. Machine learning and deep learning based channel response prediction methods are examples of such approaches [13]. Efficient deep learning algorithms to solve the channel estimation and symbol detection problems in next generation wireless systems with low complexity implementation are very desirable [1, 9, 18].

A common assumption in the detection of the symbol from received observation is that the perfect channel state information is available at the receiver. The channel state information at the receiver is obtained through channel estimation method based on either pilot-aided approach or blind estimation method. It can be easily inferred that the detection accuracy also depends on the channel estimation accuracy. To make an accurate channel prediction, the problem is formulated as an optimization problem aiming to minimize the mean square error (MSE) value between the actual and predicted. In order to obtain a low complexity solution providing the minimum square error in the $\hat{h}$ values, many conventional techniques as well as well deep learning based approaches have been described in the literature. However, the complexity of the channel estimation technique grows exponentially with increase in the antenna count in MIMO systems. A wide range of DL-based approaches are developed to improve the estimation accuracy of massive MIMO structures. DNN-based approaches, CNN-based approaches, and DL blended with traditional techniques are the three types of schemes [11, 22]. The channel prediction for massive MIMO systems on real world data using ML is presented in [18]. The challenges and solutions in the channel estimation for mmWave communication is presented in detail [4] (and other related references). The deep learning based channel estimation and tracking is proposed in [10].

This paper attempts to harness the performance of a well-known recurrent DL model (*viz.* the Long Short-Term Memory (LSTM)) for solving the channel prediction task. LSTM was selected for this task over conventional Recurrent Neural Networks since the latter suffer from the vanishing gradient problem, which an LSTM is capable of mitigating, and also due to the fact that LSTM networks are able to extract long-term dependencies in the data by virtue of their ability to learn from longer sequences (which a typical RNN does not manage to do).

This paper is organised as follows. Section 2 contains a discussion on the channel modelling problem and the generation of the dataset used in this work. Section 3 contains details of the proposed deep learning model for the CE problem. This is followed by Sect. 4 which presents the results of the proposed approach. This section also presents a discussion on the different training scenarios for the deep learning model. Lastly, Sect. 5 contains some concluding remarks.

2 mmWave Channel Model and Dataset Generation

There are several wireless channel simulators which have been developed and used by many researchers, such as SIRCIM (Simulation of Indoor radio Channel Impulse response Models) developed by Rappaport and Seidel for the indoor channels [17], and SMRCIM (Simulation of Mobile Radio Channel Impulse Response Models) which is an open-source RF propagation simulator developed for simulating the outdoor channels [20].

The mmWave channel dataset to be used in this work, is generated using the open source software NYUSIM[1], and is also available with a platform-independent graphical user interface [7]. NYUSIM can be used to generate realistic channel responses for the practical physical 5G/6G scenario which incorporate the temporal as well as spatial variations. The NYUSIM simulator is based upon the statistical spatial channel model for the mmWave wireless communication. This simulator can be used for a wide range of carrier frequencies (500 MHz to 100 GHz) and a range of radio frequency bandwidths (0 to 800 MHz). There are different operating scenarios available in the NYUSIM software such as UMi (Urban microcell), UMa (Urban macrocell) and RMa (Rural macrocell), and it also has the capability to incorporate multiple-input multiple-output (MIMO) antenna arrays at the transmitter side and receiver side [7]. One of the most important aspects of the NYUSIM is that it was built using wideband propagation channel measurements in the real world at numerous millimeter-wave (mmWave) frequencies spanning from 28 to 73 GHz in varied outdoor situations. As a result, NYUSIM provides an accurate supply of real-time and real-space channel parameters [19].

The parameters used for data generation are listed in Table 1, and the scenario considered for the wireless channel variation is presented in Table 2.

3 Proposed Model

In a typical RNN, the gradient may reduce to a very small (negligible) value during the back propagation process; this may lead to no parameters being significantly updated. This inherent issue is referred to as the *vanishing gradient problem.*

To rectify this problem of vanishing gradients in RNNs, Long Short-Term Memory was proposed, LSTM also enables the model to learn longer sequences (in comparison to a what a typical RNN is capable of learning).

---

[1] NYUSIM Version 3.1 available at: https://wireless.engineering.nyu.edu/nyusim/

Table 1. The parameters considered for the dataset generation

Parameter	Value(s)	Parameter	Value(s)
NumBsElements	1	Tx Power (dBm)	10
NumUTElements	1	Barometric Pressure (mBar)	1013.25
Num Rx Locations	1	Temperature (C)	20
Carrier Frequency (GHz)	52	Polarisation	"Co-Pol"
Bandwidth (MHz)	800	Foliage Loss	"No"
BS Height (m)	10	Humidity (%)	50
UT Height (m)	1.5	Scenario	"Urban Macro"
Distance Range (m)	100	O2I Penetration Loss	"No"

Table 2. The scenario considered for the dataset generation

Parameter	Value(s)	Parameter	Value(s)
Environment	LOS	Correlation Distance	10 m
Tx-Rx Separation, Min	50 m	UT Track Type	Hexagon
Tx-Rx Separation, Max	200 m	Side Length	10 m
Track Distance	100 m	UT Velocity	1 m/s
Update Distance	0.1 m	Moving Direction	45
Orientation	Clockwise	Human Blockage	"No"

The proposed LSTM-based deep learning model to output predictions of the channel response ($\hat{h}$) is presented in Fig. 1. The model employs 4 layers: three LSTM layers and one Dense layer with different number of units in each layer. The number of units in each layer of the proposed model is mentioned in Fig. 1 (in red). Details about the number of units in each layer as well as other pertinent parameters (such as Optimizer and the Loss Parameter) are also included in Table 3.

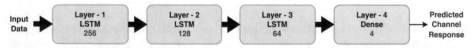

Fig. 1. Architecture of the proposed deep learning model for the prediction of $\hat{h}$

The approach works by training the model using a sequence input comprising of a 'history' vector of past h values (*real* and *imaginary* components of h being considered as individual features), along with other relevant feature such as transmitter-receiver distance and the delay spread. The output of the model is the 'future' value of h i.e., $\hat{h}$). Once the model is sufficiently trained (which is decided by the loss metric, mean squared

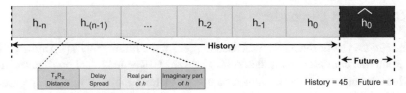

Fig. 2. Depiction of the input (past values of h) and the output (predicted future values of h) *i.e.*, $\hat{h}$

error, attaining a certain minimum threshold) the model can be used to output predicted values of $\hat{h}$) given the past h values and other required data (transmitter-receiver distance and the delay spread). An illustration of the arrangement of the training data sequence is shown in Fig. 2. For the work embodied in this paper, the 'history' and 'future' windows are kept at 45 and 1 respectively. This implies that the trained DL model can predict the future value of $\hat{h}$) (along with the future value of delay spread, and transmitter receiver distance) given a sequence of the past 45 values of h. Other pertinent details of the training process are included in Table 3 from where it can be noted that the model is trained for 100 epochs with a batch size of 2. A train-test split of 0.8 was used to partition the data (sequentially, not randomly) into the training and testing sets. The Adam optimizer was used for the model, and the performance metric was chosen to be the Mean Square Error (MSE). The model took 314.39 s to train to 100 epochs.

Figure 3 depicts the plots of the variation (progression) of the Accuracy and the Loss parameters with progressing epochs during the training process. It can be seen that the performance of the model improves significantly during the first few epochs in the training, but the relative performance improvement slows down with further training epochs. In fact, should a shorter training duration be desired in some application scenario, the training could be stopped at around 50 epochs without sacrificing the model performance to any significant extent.

Table 3. Details of parameters and number of units for the different layers in the proposed model of Fig. 1

Parameter	Value	Layer	Units
Loss parameter	Mean Squared Error	LSTM-1	256
Batch size	2	LSTM-2	128
Epochs	100	LSTM-3	64
Train-Test split	0.8	Dense	4
Optimizer	Adam	Trainable parameters	514,052
History	45	Non-trainable parameters	0

4   Results and Discussion

The trained LSTM-based model was used to predict the future values $(\hat{h})$ of the CIR, and the output was compared with the (known) actual CIR values. It should be mentioned that since h (and therefore, $\hat{h}$) values comprise of both real and imaginary parts, two different comparisons are needed to compute the overall accuracy of the predictions – one for the real part, and another one for the imaginary part of the CIR. Figure 3(left, in blue) and Fig. 3(right, in red) depict the training time accuracy and loss respectively. It can be seen that the loss gets to a very low value and saturates at around.

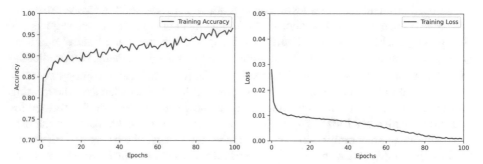

Fig. 3. Accuracy and loss plots for the deep learning model

100 epochs. Figure 4 depict the plots of the actual values and predictions for the real and imaginary parts respectively, of h (in blue color) and $\hat{h}$ (in orange color). It can be seen that the model performance remains quite close to the actual values, with the overall calculated value of test data MSE equal to 0.019, and test data Mean Absolute Error (MAE) being 0.097.

Another interesting observation to note from Fig. 4 is that for the indices in the predictions where the model output matches the actual values in the real component, the output in the imaginary parts also is very close to the actual imaginary component. Similarly, for indices where the real part of $\hat{h}$ exhibits large error, the imaginary part is also distant from the ground truth of imaginary part of h. This observation needs

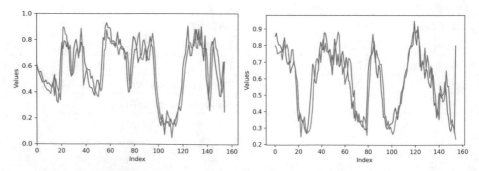

Fig. 4. Plots of ground truths and predictions for the real and imaginary parts of h

more thorough investigation, and perhaps *separate* training of two models to predict the real and imaginary parts separately would result in an overall more accurate ensemble model – however, that is beyond the scope of this paper and is an avenue for future research work.

The value of '45' chosen for the History parameter in the previous section was somewhat arbitrary. Indeed, the model can be offered a longer (or shorter) past sequence to learn from. To explore the performance of the proposed deep learning model over History values other than the value chosen in the previous section (History = 45), the

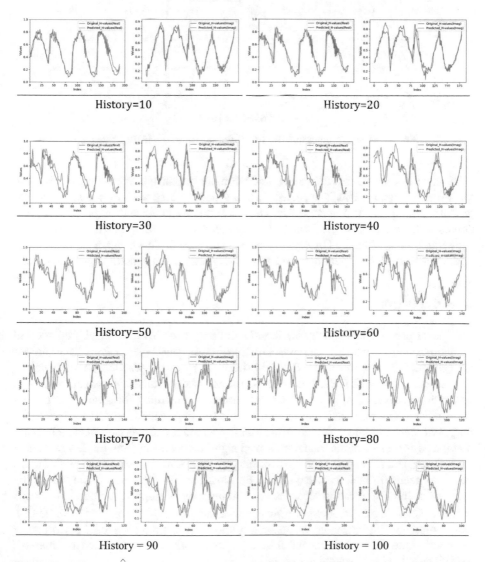

Fig. 5. Results of the $\hat{h}$ predictions obtained from the model for History values ranging from 10 through 100. The real and imaginary values of $\hat{h}$ are plotted separately.

same model was further trained and tested for different scenarios ranging from History = 10 through History = 100, and the prediction results are presented in Fig. 5.

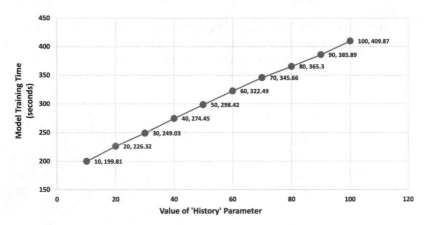

Fig. 6. Model training time for the different values of History parameter

It can be observed that the model prediction performance improves with increasing the value of History. This is expected since a longer history window allows the model to learn from a longer past sequence. However, this improvement in the performance comes with a trade-off in the form of linearly increasing training time for the model, as shown in Fig. 6, when the History parameter is increased. This is again expected since with increasing History parameter, the model is being trained to learn longer sequences.

5 Conclusion

This paper presents a deep learning based solution for mmWave channel prediction in realistic scenario scenario for the conventional wireless transceiver systems. The data sets are generated from open source NUYSIM simulator by taking practical prameter values at carrier frequency 52 GHz. The detail about parameters values and scenario for data generation is given. The proposed LSTM-based deep learning for channel prediction is described. The mmWave dataset is used for training, validation and testing of the proposed deep learning model. From the training accuracy and loss plots, it was observed that the performance of the model improved significantly during the first few epochs in the training, and the overall model performance improvement slowed down with further training epochs. In fact, in case a shorter training duration (for faster deployments) be required in some application scenario, the training may be stopped at around 50 epochs without sacrificing the model performance to a large extent. The predicted value of $\hat{h}$ was compared with the actual values, and it was seen that the DL model performance remained quite close to the actual h values. This work can be extended for massive MIMO based 5G and beyond networks.

References

1. Alrabeiah, M., Alkhateeb, A.: Deep learning for TDD and FDD massive MIMO: Mapping channels in space and frequency. In: 2019 53rd Asilomar Conference on Signals, Systems, and Computers, pp. 1465–1470 (2019)
2. Ansari, M.S., Bartos, V., Lee, B.: GRU-based deep learning approach for network intrusion alert prediction. Fut. Gener. Comput. Syst. **128**, 235–247 (2022)
3. Gupta, R., Srivastava, D., Sahu, M., Tiwari, S., Ambasta, R.K., Kumar, P.: Artificial intelligence to deep learning: machine intelligence approach for drug discovery. Mol. Divers. **25**(3), 1315–1360 (2021)
4. Hassan, K., Masarra, M., Zwingelstein, M., Dayoub, I.: Channel estimation techniques for millimeter-wave communication systems: achievements and challenges. IEEE Open J. Commun. Soc. **1**, 1336–1363 (2020). https://doi.org/10.1109/OJCOMS.2020.3015394
5. Heath, R.W., González-Prelcic, N., Rangan, S., Roh, W., Sayeed, A.M.: An overview of signal processing techniques for millimeter wave MIMO systems. IEEE J. Sel. Top. Sig. Process. **10**(3), 436–453 (2016). https://doi.org/10.1109/JSTSP.2016.2523924
6. Jiang, Y.: Application and comparison of multiple machine learning models in finance. Sci. Program. **2022** (2022)
7. Ju, S., Kanhere, O., Xing, Y., Rappaport, T.S.: A millimeter-wave channel simulator NYUSIM with spatial consistency and human blockage. In: 2019 IEEE Global Communications Conference (GLOBECOM), pp. 1–6. IEEE (2019)
8. Li, X., Alkhateeb, A.: Deep learning for direct hybrid precoding in millimeter wave massive MIMO systems. In: 2019 53rd Asilomar Conference on Signals, Systems, and Computers, pp. 800–805 (2019)
9. Mattu, S.R., Chockalingam, A.: Learning-based channel estimation and phase noise compensation in doubly-selective channels. IEEE Commun. Lett. **26**(5), 1052–1056 (2022)
10. Moon, S., Kim, H., Hwang, I.: Deep learning-based channel estimation and tracking for millimeter-wave vehicular communications. J. Commun. Netw. **22**(3), 177–184 (2020). https://doi.org/10.1109/JCN.2020.000012
11. Naeem, M., De Pietro, G., Coronato, A.: Application of reinforcement learning and deep learning in multiple-input and multiple-output (MIMO) systems. Sensors **22**(1) (2022). https://doi.org/10.3390/s22010309
12. Ozpoyraz, B., Dogukan, A.T., Gevez, Y., Altun, U., Basar, E.: Deep learning-aided 6G wireless networks: a comprehensive survey of revolutionary PHY architectures. arXiv preprint arXiv: 2201.03866 (2022)
13. O'Shea, T., Hoydis, J.: An introduction to deep learning for the physical layer. IEEE Trans. Cogn. Commun. Netw. **3**(4), 563–575 (2017)
14. Rangasamy, K., As'ari, M.A., Rahmad, N.A., Ghazali, N.F., Ismail, S.: Deep learning in sport video analysis: a review. Telkomnika **18**(4), 1926–1933 (2020)
15. Rappaport, T.S., MacCartney, G.R., Samimi, M.K., Sun, S.: Wideband millimeter wave propagation measurements and channel models for future wireless communication system design. IEEE Trans. Commun. **63**(9), 3029–3056 (2015). https://doi.org/10.1109/TCOMM.2015.2434384
16. Rappaport, T.S., Xing, Y., MacCartney, G.R., Molisch, A.F., Mellios, E., Zhang, J.: Overview of millimeter wave communications for fifth generation (5G) wireless networks—with a focus on propagation models. IEEE Trans. Antennas Propag. **65**(12), 6213–6230 (2017). https://doi.org/10.1109/TAP.2017.2734243
17. Rappaport, T., Seidel, S., Takamizawa, K.: Statistical channel impulse response models for factory and open plan building radio communicate system design. IEEE Trans. Commun. **39**(5), 794–807 (1991)

18. Shehzad, M.K., Rose, L., Wesemann, S., Assaad, M.: Ml-based massive mimo channel prediction: does it work on real-world data? IEEE Wirel. Commun. Lett. **11**(4), 811–815 (2022)
19. Sun, S., Ju, S., Rappaport, T.S.: NYUSIM user manual version 1.6.1. New York University and NYU WIRELESS (2018)
20. Sun, S., MacCartney, G.R., Rappaport, T.S.: A novel millimeter-wave channel simulator and applications for 5G wireless communications. In: 2017 IEEE International Conference on Communications (ICC), pp. 1–7. IEEE (2017)
21. Wang, D., Cao, W., Zhang, F., Li, Z., Xu, S., Wu, X.: A review of deep learning in multiscale agricultural sensing. Remote Sens. **14**(3), 559 (2022)
22. Wang, Z., Pu, F., Yang, X., Chen, N., Shuai, Y., Yang, R.: Online LSTM-based channel estimation for HF MIMO SC-FDE system. IEEE Access **8**, 131005–131020 (2020)
23. Zaidi, S.S.A., Ansari, M.S., Aslam, A., Kanwal, N., Asghar, M., Lee, B.: A survey of modern deep learning based object detection models. Digit. Sig. Process., 103514 (2022)

Energy-Efficient Clustering Protocol Using Particle Swarm Algorithm for Wireless Sensor Networks

Ahmed A. Jasim[1,4]([✉]), Noor Riyadh Issa[2,5], Ghufran Saady Abd Al-Muhsen[3], Mohd Yamani Idna Idris[2], Saaidal Razalli Bin Azzuhri[2], and Ali M. Muslim[1]

[1] Department of Computer Science, Dijlah University College, Baghdad, Iraq
ahmed.abdulhadi@duc.edu.iq, ahmed.abdhadi@coeng.uobaghdad.edu.iq
[2] Department of Computer System and Technology, Faculty of Computer Science and Information Technology, The University of Malaya, 50603 Kuala Lumpur, Malaysia
[3] Department of Computer Science, Madenat Al Elem University College, Baghdad, Iraq
[4] College of Engineering, University of Baghdad, Baghdad, Iraq
[5] Ministry of Education, Directorate of Education Karkh 1, Albutoula high school for girls, Abu Graib, Iraq

Abstract. The energy of sensor nodes in the network consumes more energy due to the sending and receiving of data among them before being forwarded to the server or the sink. Therefore, the remains energy of nodes is one of the main concerns in Wireless Sensor Networks (WSNs) that should be addressed and improved. Many approaches and methods are proposed to decrease the consumption of node energy in the network, one solution has been proposed to solve this problem such as the Particle Swarm Optimization Based Energy Efficient Clustering Protocol (PSOEEC). However, this method only concentrates on utilizing initial energy and residual energy among nodes to cluster head nodes, while for the nodes close to the base station not much attention is given to enhancing it to reduce the consumption of energy. Therefore, this paper proposes an Energy-Efficient Particle Swarm Algorithm (EEPSA) to address the hotspot problem and extend the life of the network. Our method produces uneven clusters dependent on the radius of competition. The responsibility of this function is to generate unequal nodes in the clustering protocols and is capable to determine the sizing of cluster nodes from the sink or base station. Besides, the delay time with the fitness value of sensor nodes calculates based on high residual energy, average residual-energy, and the minimum and maximum distance from nodes to the base station to choose the optimal cluster head in the network. Finally, we have concluded with potential future works out of this effort.

Keywords: Wireless Sensor Networks · Clustering Protocol · Particle Swarm Algorithm · Energy Consumption

1 Introduction

The WSN is composed of numerous fixed or mobile sensor nodes that form a wireless network using self-organizing and multi-hop processes. Its aim is to discover, process, and broadcast monitoring information for an object within a network coverage area that

F. Saeed et al. (Eds.): ICACIn 2022, LNDECT 179, pp. 331–342, 2023.
https://doi.org/10.1007/978-3-031-36258-3_29

consists of many nodes and sub-nodes. These nodes obtain the information from their nearby environment and transmit it to it is final destination server or the base station. Many applications have been used in WSNs, such as eHealth, forest fire tracking, military purpose, catastrophe management, and security [1–3]. The communication link between nodes helps to transfer data to the sink or server. [4, 5]. The WSN can sense lots of environmental conditions, such as pollution, noise, mechanical pressure strength, temperature and humidity, vehicle movement, airflow, and speed with low-cost implementation. However, the sensor nodes have a power supply unit which considers as it is the main limitation that might affect the performance of the sensor nodes, which may lead to reduce the overall network lifetime. Therefore, based on these issues, the managing of energy consumption is one of the major issues during work.

Figure 1 is shown the architecture of WSNs. In the beginning idea, the sensor nodes collect the data from neighbors nodes and send it to the base station in a traditional way. As a consequence, this scenario may lead to poor communication due to the long distances and it will consume more energy. Therefore, collisions may occur through data transmission. Thus, this scenario will also lead to data retransmission and higher power consumption. To solve this issue and extend the lifetime of the network, a rethink of routing and aggregation is needed [6]. In WSNs, energy issues are now considered a major problem, so, the best and most popular technique to solve this issue is the clustering technique. [7, 8]. In the mechanism of clustering in the network, the clusters have divided the sensing of nodes into numerous cluster members. The cluster is elected by the sensor nodes as called a Cluster Head (CH) and the nodes in this cluster head are called Cluster Members (CMs) [9, 10].

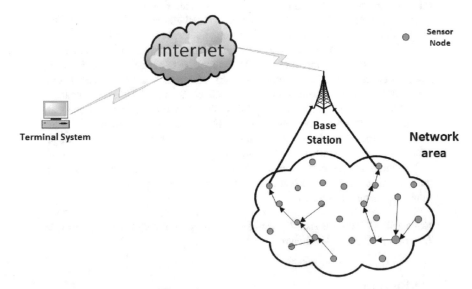

Fig. 1. The Network Architecture

The procedure of cluster members is to collect the data from the surrounding environment and transmit it to the main cluster head (CH). When the clustering receives data from the member nodes, the cluster head aggregates the data, sends a packet, and forwards the data to the BS. The advantages of aggregation methods are to avoid redundancy, traffic, and delay. The advantages of the clustering approach are scalability, balanced energy consumption, and improved network lifetime [11–13]. Nevertheless, the clustering has some limitations, such as sometimes the data when transmitted to the sink rises the energy consumption due to the distance from the CMs or the CHs farthest from the BS, as a result, it will drain more energy for transition to the BS. Therefore, to preserve the energy consumption of this case preferred utilizes multi-hop communication mechanisms.

However, the CHs closer to the base station will take more forwarding tasks for transmission, aggregation, and receiving in the multihop communication, which results in a high overhead of CHs, and the energy of CHs will consume and might run out sooner. This situation of loss of communication and a breakdown of the clusters, so the breakdown is called a hot spot problem.

The occurrence of a hot spot issue when the distance of the cluster head is closer to the BS, so the energy of nodes will increase the consumption of it by the aggregation of data and transmission process. Therefore, optimal approaches are significant to select the optimal CHs and increase the lifetime of the network. [14–16]. Due to the importance of the optimization problems for WSNs, so, the PSO technique is the base one to solve them, and also is a part of swarm intelligence. [17, 18].

In this paper, we propose an energy Efficient Using Particle Swarm Algorithm (EEPSA) to avoid the hot spot problem and grow the network lifetime. Our method produces uneven clusters dependent on the radius of competition. The responsibility of this function is to generate unequal nodes in the clustering protocols and is capable to determine the sizing of cluster nodes from the sink or base station. In addition, we are following the (PSO-EEC) method for the selection of the optimal cluster heads (CHs) rather than utilizing only the fitness function for selecting CHs. The delay time for selecting the optimal CHs in the networks has improved by utilizing the function to improve it. In this paper, we define the contributions as follows:

1. We propose an energy Efficient Using Particle Swarm Algorithm (EEPSA) for wireless sensor networks.
2. We improve the selection the optimal of cluster heads (CHs) by utilizing the fitness function and delay time to improve and increase the network lifetime.
3. Calculate the computational radius to solve the hot spot problem.

The rest of this paper is structured as follows. Section 2 presents some particle swarm algorithms for WSNs to be described in section related work. Sections 3 and 4 describe the methodology of the proposed method. Section 5 summarizes this study and outlines future work.

2 Related Work

In the clustering technique, several researchers have proposed methods and techniques to decrease energy draining that aim to extend the network lifetime. The authors proposed a Low-Energy Adaptive Clustering Hierarchy (LEACH) protocol. This protocol aimed to preserve energy draining and to choose the optimal CHs. However, the hot spot problem was not addressed and the selection of CHs was random, which lead to more energy draining by the sensor node [19]. Many methods and techniques are proposed to improve the LEACH protocol, such as [18] unequal clustering and routing algorithms (nCRO-UCRA). In this method, the selection of CHs utilizes the fitness function based on the distance and energy of nodes in the network. However, the residual energy on sensor nodes does not much attention, entails the energy consumption will increment during the receiving and transmission tasks. In addition, the LEACH-C algorithm was proposed based on a hybrid firefly with the PSO technique to find the optimal positioning of CHs [20].

Moreover, the author in [21] proposed a linear programming method based on the mentioned above, the PSO technique is the best technique to solve the clustering issue. On the other hand, the Harmony Search with PSO algorithm was proposed to prolong the network lifetime, for the selection of optimal clustering it utilizes the dynamic nature of PSO with harmony matrix [22]. The authors have proposed a sailfish optimization (SFO) technique for WSNs to optimize the network lifetime [23]. The proposed technique uses the multi objectives to select optimal cluster heads by using the best optimal path for data transmission. However, this technique focuses on minimizing energy but the distance between nodes and CHs was not addressed.

On the other hand, Gupta, V., & Pandey [24] proposed Enhanced Energy Recognition Distributed Unequal Clustering Protocol (EADUC) to enhance CH selection based on information and node numbering and address hotspot issues. The bat optimization algorithm was defined by [25] to optimize the unequal clustering before selecting the primary cluster heads in the network. This algorithm reduces the energy consumption of nodes, however, this method addresses the network lifetime but the distance of the node from the base station was not considered, this leads to draining the energy of nodes faster and will die sooner.

3 Proposed EEPSA Method

To illustrate the hot spot problem and increase the network lifetime, we proposed Energy Efficient Using Particle Swarm Algorithm (EEPSA). The main objective of our method is to improve the selection of CHs to decrease the energy draining by the nodes for WSNs. In the next sub-section, we will describe more detail regards the energy tasks of nodes before beginning to describe our protocol.

3.1 Energy Model

The energy consumption of sensor nodes based on clustering protocols can be analyzed in the energy model to evaluate our proposed EEPSA method for WSNs. In the energy model of this paper, we followed and use the same as the energy model of the LEACH protocol defined by [1, 19]. The consumption of energy occurs when the sensor nodes are communicated between there and will start with sensing, sending, and receiving data. Therefore, in this case, the sensor nodes during the processing are more energy will be consuming. So, in this study, the energy for transmitting and receiving we are considered for communication as the energy consumption. The formula of calculates the energy consumption when transmitting l-bit data can be described as follows:

$$E_{TX} = \begin{cases} k \times E_{elec} + k \times efs \times d^2 & when\, d \leq d_0 \\ k \times E_{elec} + k \times emp \times d^4 & when\, d \geq d_0 \end{cases} \tag{1}$$

where E_{TX} denotes the energy consumption during the transmission tasks, k is the range of data transfer, E_{elec} denotes the energy consumed when sending or receiving 1-bit data via the circuit of it, efs s free space mode, emp is a multi-path mode, d is the distance of nodes, and d_0 is the threshold. The threshold distance value d_0 can be calculated as:

$$d_0 = \sqrt{\frac{efs}{emp}} \tag{2}$$

K-bit data transfer consumes energy, so we calculated it as follows:

$$E_{RY}(j) = k\, F_{elec} \tag{3}$$

where E_{RX} denotes the energy consumption during the receiving, and E_{elec} denotes the energy consumed when sending or receiving 1-bit data via its circuit of it.

3.2 Clustering Phase

In this section, we describe overall our proposed Energy Efficient Using Particle Swarm Algorithm (EEPSA) for WSNs. The main objective of this study is to reduce energy consumption in order to increase the network lifetime. The CHs closer to the base station will take more forwarding tasks for transmission, aggregation, and receiving in the multihop communication, which results will in a high overhead of CHs, and the energy of CHs will consuming and might run out sooner. This situation of loss of communication and a breakdown of the clusters, so the breakdown is called a hot spot problem. Therefore, we propose the EEPSA method to resolve the hot spot problem in the networks. Our method produces uneven clusters dependent on the radius of competition. The responsibility of this function is to generate unequal nodes in the clustering protocols and is capable to determine the sizing of cluster nodes from the sink or base station. In addition, we are following the (PSO-EEC) method for the selection of the optimal cluster heads (CHs) but rather than utilizing only the fitness function for selecting CHs, we also employ the delay time function to improve the selection CHs in the networks and increase the network lifetime.

The proposed EEPSA method is consist of two phases: the first one is the clustering setup phase. This phase describes the fitness value and the delay time when calculating to choose the optimal CHs and assigning the cluster members. While the second phase is the data transmission phase, the transmitting data among sensor nodes occur in this phase.

3.3 Clustering Setup Phase

This sub-section introduces the selection of the optimal node to become the main CHs based on delay time and fitness function. Before discussing the selection of CHs, we first generate unequal clustering nodes in the network dependent on calculating the competition radius to determine the node size of the cluster with a base station node and the delay time to choose the optimal CHs.

3.4 Clustering Competition Radius

To determine the sizing of the sensor nodes from the sink and to generate unequal clustering, should be to calculate the competition radius. Our EEPSA method calculates the competition radius based on the lowest and highest interspace from nodes from the sink or the base station, the distance of all the nodes to the sink as well, and the remaining energy of sensor nodes in the network. In this paper, we followed a similar function to calculate the competition radius as defined by[1]. The competition radius R_c can be calculated as follows:

$$R_c(i) = \left[1 - a\left(\frac{D_{max} - d_{i,BS}}{d_{max} - d_{min}}\right) - b\left(1 - \frac{E_{rem}(i,r)}{E_{max}}\right)\right]RL_{max} \tag{4}$$

where $R_c(i)$ represents the competition radius of node (i), D_{max} represent the furthest interspace from nodes to the base station, $d_{i,BS}$ represents the interspace from node (i) to the base station, $E_{rem}(i,r)$ represents the remaining-energy of node (i) at round r, E_{max} represents the highest degree of node energy, RL_{max} represents the maximum competition radius for suitable CH, and (a,b) is the element among [0,1].

On the other hand, once calculating the radius of each node, we calculate the delay time of sensor nodes in the network to announce being a CH. The delay time of sensor nodes can be calculated as follows:

$$D_t(i) = \left(1 - \frac{E_{rem(i)}}{E_{avg(i)}}\right) * W_t + R_v \tag{5}$$

where $D_t(i)$ is the delay time of node (i), $E_{rem(i)}$ is the remaining-energy of node (i), W_t is the time of CH, and R_v denote the random value.

In this study, the fitness function helps us to select the optimal CHs with delay time. The procedure of this function considers the initial energy of nodes, the distance among nodes, and the node degree for the generation of the fitness value of the node. The goal of this function is to preserve the energy draining and increase the network lifetime. For a more detailed fitness function, the researcher can find it in [3].

$$F(N) = a1 * \sum_{i=1}^{N}\frac{E_{ini}}{E_{ini} - E_i} + a2 * \frac{1}{aj}\sum_{j=1}^{aj}D_{max} + a3 * \frac{1}{\sum_{I=1}^{N}NodeD(i)} \tag{6}$$

where $F(N)$ represents the fitness function, $a1$, $a2$, and $a3$ re the weight coefficient for the fitness function, E_{ini} represents the initial energy of node i, E_i represent the current energy of node i, and $NodeD(i)$ represents the node degree of node i. If the node with high residual energy and less distance to the base station and the delay time is close to zero, the node will become CH and send a broadcast message to all cluster members including node ID and information of nodes in the cluster. The CH creates a Time

Division Multiple Access (TDMA) schedule for the cluster members to transfer their data to CH to avoid data collision.

Algorithm 1 is described in our proposed EEPSA method for the selection CHs. The input-output parameters (N, ID, RLmax, the starting energy of the node, R_v, W_t) of the algorithm is specified in steps 1–2. Steps 3–7 calculate the $R_c(i)$, the average energy of nodes $E_{avg(i)}$, delay time $D_t(i)$, and fitness value $F(N)$ to check the weather of the cluster head of nodes. If the nodes with close to zero delay time and low distance with high residual energy, the node will become a cluster head (CH) as shown in steps 8–13. The cluster heads of the list and cluster members are specified in steps 14–19.

Algorithm 1. Clustering Setup Phase

1. Input (N, ID, RLmax, the initial energy of node, R_v, W_t)

2. output (optimal cluster head (CH)

3. **For** Node (i) & Node (j)   **do**

4. Calculate the competition radius according to Equation (4)

5. Calculate $E_{avg(i)}$

6. Each node calculates $D_t(i)$ according to Equation (5)

7. Calculate the Fitness value of the node (i) according to Equation (6)

8.   **If** Node i- $D_t(i)$ = close to 0 && distance of Node i ≤ Node j && the E of Node i ≥ Node j **then**

9.   Node i == CH

10.   CH count= CH+1

11.   CH broadcast to all cluster member nodes

      Else

12. Node j == CH && Node i = normal node

13. **end if**

14. Compute the lowest distance from CMs to CHs to choose CH-list

15.   **If** the E of Node (i) ≥ E_avg(i) && neighbor of node (i) ≤ neighbor of node j **then**

16.   CH list = CH (i)

17.   **end if**

18. **end for**

19. **end**

3.5 Data Transmission Phase

Generally, the sending and receiving data process occurs inside the sensor nodes, the sensor nodes collect data by a data aggregation algorithm and send it to the main clusters of nodes. Then the main clusters will forward data to the base station or sink. However, this procedure leads to increased overheads of nodes and energy consumption of nodes during the process. Due to the drain of energy among nodes for sending and transmitting data to the sink, so, we should consider this issue to prolong the network lifetime for

WSNs. Therefore, this section describes the data transmission from nodes to cluster heads (CHs), and the CHs forward to the base station node.

In this paper, we focus on the unused energy of nodes and the length of distance from the sensor nodes to the server. Therefore, our work calculates the average energy consumption, unused energy of nodes, and the minimum, and maximum length of distance from the sensor nodes to the base station. Accordingly, we utilize the residual energy of nodes and the distance from cluster members to the base station. The description method can be written as follows:

If the remind energy is less than the threshold value and the length of distance is greater than the threshold value, the node will collect the data from the environment and send information to the sink node, whereas, if the CHs closest to the base station will send the information and forward it directly to the BS with a single hop. This leads to preserving the energy when data is transmitted between nodes to the sink. This leads to prolonging the network lifespan and reduces the energy consumed by the nodes in the network.

4   Simulation Results

This section describes the performance results of our proposed EEPSA Method for WSNs. The main objective of our method is to raise the network lifespan and preserve the energy consumption of nodes in the network. Due to calculating the fitness value and delay time for selection CHs, the simulation results of the proposed EEPSA method show are high as compared with other methods such as nCRO-UCRA, and PSO techniques. In addition, we handled the hot spot problem and the distance length from cluster members to the BS node. We use Matlab 2022 for the simulation method and compare it with other methods and techniques.

Figure 2 plots the life nodes in the network to investigate the network lifespan, including the First Node Dies (FND), Half Node Dies (HND), and the Last Node Dies (LND). We established (100) sensor nodes in a $200\,m^2 \times 200\,m^2$ to check the situation of the sensor nodes when sending and receiving data from the environment. We can see the performance results of our method in Fig. 2 compared with existing methods because our method EEPSA chooses the optimal cluster heads depending on fitness function and delay time of nodes, also the unequal clustering technique is developed by calculating the competition radius in the clustering protocol. In addition, the competition radius is achieved based on the unused energy of nodes, distance length from nodes to the base station, and the nearest and farthest distance among nodes to cluster head nodes. Moreover, the transmission round of our method helps to prolong the network lifespan because if the node is farthest from the base station this means the remaining energy is less than the energy threshold and if the distance length is greater than the distance threshold, will be sent to the neighbor's nodes, and the neighbors will send it to the base station.

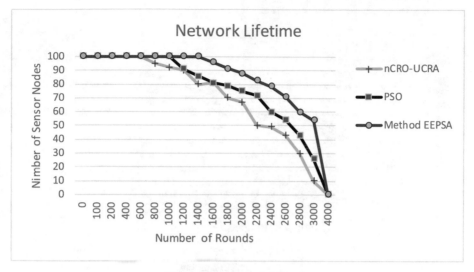

Fig. 2. The Network Lifespan

On the other hand, Fig. 3 plots the energy consumption in joules of sensor nodes. The results show our EEPSA is less than the other methods namely nCRO-UCRA, and PSO. The reason is that our method calculates the minimum and maximum distance from nodes to the base station. In addition, the fitness function helps to reduce energy consumption because if the fitness value of nodes is high and the delay time is not close to zero, so this leads to increases consuming energy during sending and receiving process in the network. Therefore, we calculated the fitness value with the delay time to avoid the long distance and to choose the optimal main cluster heads.

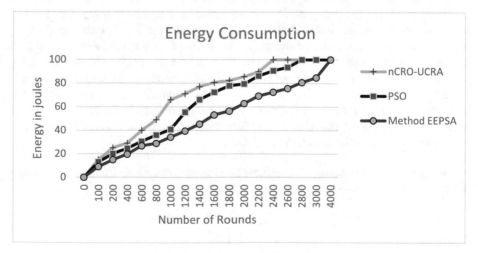

Fig. 3. The Energy Consumption

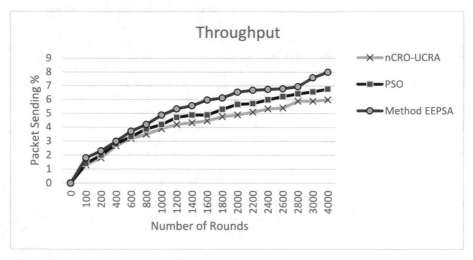

Fig. 4. The Throughput

Apart from the energy consumption, we evaluated also the throughput of data transmission to the base station as shown in Fig. 4. From this evaluation, the EEPSA method results show the throughput is greater than other methods compared in this paper. The reason is due to the selection of optimal cluster heads based on fitness function and delay time, and the unequal clustering proposed to avoid the hot spots problem.

5 Conclusion

In this paper, we have proposed an Energy Efficient Using Particle Swarm Algorithm (EEPSA) to handle the hot spot problem and raise the network lifetime. Our method produces uneven clusters based on the radius of competition. The responsibility of the competition radius is to generate an unequal clustering for each node and is capable to determine the sizing of cluster nodes from the sink or base station. Besides, the delay time with the fitness value of sensor nodes calculates based on high residual energy, average residual energy, and the minimum and maximum distance from nodes to the base station to choose the optimal cluster head in the network. In future work, we will improve our method by proposing a new algorithm to preserve the energy consumption of the sensor nodes in the networks for WSNs.

Acknowledgment. This Research is Supported by a Research Grant from the Ministry of Higher Education of Malaysia Grant (FP103–2020).

References

1. Jasim, A.A., et al.: Energy-efficient wireless sensor network with an unequal clustering protocol based on a balanced energy method (EEUCB). Sensors **21**(3), 784 (2021)

2. Jasim, A.A., et al.: Secure and energy-efficient data aggregation method based on an access control model. IEEE Access **7**, 164327–164343 (2019)
3. Rawat, P., Chauhan, S.: Particle swarm optimization-based energy efficient clustering protocol in wireless sensor network. Neural Comput. Appl. **33**(21), 14147–14165 (2021)
4. Zhu, F., Wei, J.: An energy-efficient unequal clustering routing protocol for wireless sensor networks. Int. J. Distrib. Sens. Netw. **15**(9), 1550147719879384 (2019)
5. Chauhan, V., Soni, S.: Energy aware unequal clustering algorithm with multi-hop routing via low degree relay nodes for wireless sensor networks. J. Ambient. Intell. Humaniz. Comput. **12**(2), 2469–2482 (2021)
6. Farsi, M., et al.: Deployment techniques in wireless sensor networks, coverage and connectivity: a survey. IEEE Access **7**, 28940–28954 (2019)
7. Fanian, F., Rafsanjani, M.K.: Cluster-based routing protocols in wireless sensor networks: a survey based on methodology. J. Netw. Comput. Appl. **142**, 111–142 (2019)
8. Arjunan, S., Pothula, S.: A survey on unequal clustering protocols in wireless sensor networks. J. King Saud University-Computer and Inf. Sci. **31**(3), 304–317 (2019)
9. Shagari, N.M., et al.: Heterogeneous energy and traffic aware sleep-awake cluster-based routing protocol for wireless sensor network. IEEE Access **8**, 12232–12252 (2020)
10. Bozorgi, S.M., Bidgoli, A.M.: HEEC: a hybrid unequal energy efficient clustering for wireless sensor networks. Wireless Netw. **25**(8), 4751–4772 (2019)
11. Afsar, M.M., Tayarani-N, M.-H.: Clustering in sensor networks: a literature survey. J. Netw. Comput. Appl. **46**, 198–226 (2014)
12. Amodu, O.A., Raja Mahmood, R.A.: Impact of the energy-based and location-based LEACH secondary cluster aggregation on WSN lifetime. Wireless Networks **24**(5), 1379–1402 (2018)
13. Priyadarshi, R., Rawat, P., Nath, V., Acharya, B., Shylashree, N.: Three level heterogeneous clustering protocol for wireless sensor network. Microsyst. Technol. **26**(12), 3855–3864 (2020)
14. Edla, D.R., Kongara, M.C., Cheruku, R.: SCE-PSO based clustering approach for load balancing of gateways in wireless sensor networks. Wireless Netw. **25**(3), 1067–1081 (2018)
15. Beloglazov, A., Abawajy, J., Buyya, R.: Energy-aware resource allocation heuristics for efficient management of data centers for cloud computing. Futur. Gener. Comput. Syst. **28**(5), 755–768 (2012)
16. Anand, V., Pandey, S.: New approach of GA–PSO-based clustering and routing in wireless sensor networks. Int. J. Commun Syst **33**(16), e4571 (2020)
17. Datta, A., Nandakumar, S.: A survey on bio inspired meta heuristic based clustering protocols for wireless sensor networks. In: IOP Conference Series: Materials Science and Engineering. IOP Publishing (2017)
18. Srinivasa Rao, P.C., Banka, H.: Novel chemical reaction optimization based unequal clustering and routing algorithms for wireless sensor networks. Wireless Netw. **23**(3), 759–778 (2016)
19. Heinzelman, W.R., Chandrakasan, A., Balakrishnan, H.: Energy-efficient communication protocol for wireless microsensor networks. In: Proceedings of the 33rd annual Hawaii international conference on system sciences. IEEE (2000)
20. Pitchaimanickam, B., Murugaboopathi, G.: A hybrid firefly algorithm with particle swarm optimization for energy efficient optimal cluster head selection in wireless sensor networks. Neural Comput. Appl. **32**(12), 7709–7723 (2019)
21. Azharuddin, M., Jana, P.K.: PSO-based approach for energy-efficient and energy-balanced routing and clustering in wireless sensor networks. Soft. Comput. **21**(22), 6825–6839 (2016)
22. Shankar, T., Shanmugavel, S., Rajesh, A.: Hybrid HSA and PSO algorithm for energy efficient cluster head selection in wireless sensor networks. Swarm Evol. Comput. **30**, 1–10 (2016)
23. Mehta, D., Saxena, S.: MCH-EOR: Multi-objective cluster head based energy-aware optimized routing algorithm in wireless sensor networks. Sustainable Computing: Informatics and Syst. **28**, 100406 (2020)

24. Gupta, V., Pandey, R.: An improved energy aware distributed unequal clustering protocol for heterogeneous wireless sensor networks. Eng. Science and Technol., an Int. J. **19**(2), 1050–1058 (2016)
25. Sahoo, B.M., Amgoth, T.: An improved bat algorithm for unequal clustering in heterogeneous wireless sensor networks. SN Computer Sci. **2**(4), 1–10 (2021)

Data Science

A Single Channel EEG-Based Algorithm for Neonatal Sleep-Wake Classification

Awais Abbas[1], Saadullah Farooq Abbasi[2,4], Muhammad Zulfiqar Ali[3],
Saleem Shahid[1(✉)], and Wei Chen[4]

[1] Department of Electrical and Computer Engineering, Air University, Islamabad, Pakistan
saleem.shahid@mail.au.edu.pk
[2] Department of Biomedical Engineering, Riphah International University, Islamabad, Pakistan
[3] James Watt School of Engineering, University of Glasgow, Glasgow, UK
[4] Department of Electronic Engineering, Fudan University, Shanghai, China

Abstract. Sleep is categorized as an arrangement of modifications occurring in our body inside our brain, muscles, working its way through our eyes (occipital lobe), respiratory along with cardiac activity. It makes the human body fresh and ready for the next day. In neonates, it is essential for brain and physical development. Polysomnography is the gold standard for determining and classification of sleep stages. However, it is expensive and requires human intervention. Therefore, over the past two decades, researchers proposed multiple algorithms for automatic neonatal sleep stage classification. All the previous studies used multichannel EEG recordings for classification. Not every intensive care unit contains a multichannel EEG extraction device. For this reason, a single channel automatic neonatal sleep-wake classification algorithm, using a support vector machine, has been proposed in this paper. 3525 30-s training and testing were used to train and test the network. The proposed algorithm can reach sleep-wake classification accuracy of 77.5% with mean kappa 0.55 using single channel EEG. The results were extracted using five-fold cross-validation and the mean has been reported in this paper. Experimental results and statistical analysis show that single channel EEG can be used for neonatal sleep classification with notable accuracy.

Keywords: Support vector machine · Neonatal · Sleep · Electroencephalography

1 Introduction

Neonates are born during a crucial stage in the development of the brain and nervous system. Because babies spend a significant amount of time sleeping, their brain develops and works in an innate manner. This has been discovered to be critical for brain maturation and neuronal survivability at a young age [1, 2]. Furthermore, sleep patterns reveal any potential dangers of brain growth that could lead to cognitive, psychomotor, and behavioral issues [2, 3]. As a result, using daily monitoring techniques to measure neonatal sleep patterns offers the necessary support to doctors in neonatal intensive care units (NICU) for optimum nourishment and neonatal care phases [2]. As per the American Academy of Sleep Medicine (AASM) [4] around 10% of infants in the United States

F. Saeed et al. (Eds.): ICACIn 2022, LNDECT 179, pp. 345–352, 2023.
https://doi.org/10.1007/978-3-031-36258-3_30

(US) require critical care in the NICU. Furthermore, by the age of five years, newborns with unstructured sleep have impaired mental and emotional development, as well as an increased chance of sudden infant death syndrome (SIDS) and sleep apnea [5, 6].

Polysomnography (PSG) is known to be a gold standard for sleep classification. It comprises of many biophysiological signals and is used by clinicians to determine sleep stages [7]. Due to its simplicity, single-channel EEG has become popular in sleep monitoring in recent years. In particular, PSG segments have been divided into one of three phases: active sleep (AS), quiet sleep (QS), or awake. All the existing algorithms classified neonatal sleep using multichannel EEG which is not available in every NICU. This may be considered a limitation in the existing algorithms.

In the proposed study, a single-channel EEG (F4-C4) has been used for neonatal sleep-wake classification. Twelve prominent features were extracted from 19 EEG recordings. These features are categorized into two types: Frequency-domain and Time-domain. Then, a support vector machine (SVM) has been deployed for neonatal sleep-wake classification. The proposed algorithm achieved an accuracy of 77.5% for sleep-wake classification. Which, to date, is the highest reported accuracy for neonatal sleep-wake classification using single channel EEG. The novelty of the proposed study is given as:

1. This is the first time a single-channel EEG has been deployed for neonatal sleep.
2. In existing algorithms, AS and awake stage were amalgamated into a low voltage irregular (LVI) state. In the proposed study, sleep and wake have been classified separately.

2   Related Work

To categorize EEG data into relevant sleep stages, several researchers have used traditional machine learning algorithms. Feature extraction and sleep stage classification are usually the first two steps in these approaches. There are certain maturational changes that can be examined only during QS. For this reason, Turnbull et al. [8] discovered the Trace Alternant (TA) EEG-pattern in 2001. Although the suggested technique is effective at classifying TA, it is difficult to identify the entire QS using this approach. Certain brain modifications, which indicate effects on the brain, can only be delivered in QS [9–11]. Anneleen Dereymaeker et al. [12] suggested an automatic QS detection technique based on cluster based adapted sleep staging (CLASS). The key advantage of CLASS is that it can classify QS in preterm infants quickly. Ninah Koolen et al. proposed employing radial basis function support vector machines (RBF-SVM) to identify QS. The system was trained and tested using a sum of 57 components. For QS identification, an RBF-SVM-based algorithm can achieve an accuracy of up to 85% [13].

Authors categorized newborn EEG into four sleeping phases: anterior dysrhythmia (AS I), low voltage irregular (LVI) or AS II, high voltage slow (HVS), and Trace Alternant (TA)/Trace Discontinue (TA/TD) (TD). Pillay et al. [14] proposed a generative modeling strategy based on HMMs and GMMs to classify various sleep stages. Massive 112 extracted features have been used to test the proposed framework. For four-stage categorization, this approach had a cohen's kappa of 0.62. With a kappa of 0.64 [15], a convolutional neural network (CNN)-based algorithm surpassed all other algorithms in

2020. Primary Data collected across 113 recordings were fed into a convolutional neural network.

The research stated above either classified QS and AS or classified neonatal sleep into four stages i.e. low voltage irregular signals (LVI), AS II, high voltage slow (HVS) and Trace Alterant (TA)/Trace Discontinua. These LVI signals contain both sleep and awake states. This amalgamation corrupts 50–60% of the overall dataset. For this reason, Fraiwan et al. proposed a scheme for neonatal sleep-wake classification using deep autoencoders [16]. The proposed scheme achieved good results for overall sleep stage classification however, the accuracy for awake classification was limited to merely 17%. In 2020 [17], Saadullah et al. proposed an algorithm using a multilayer perceptron (MLP) neural network. The propounded algorithm extracted 12 prominent features from multichannel bipolar EEG. After feature extraction, MLP was applied for training and testing the neural network. The proposed scheme achieved an accuracy of 82.53% for neonatal sleep-wake classification.

All the existing algorithms used multichannel EEG for neonatal sleep. These multi-channel electrodes can affect the quality of neonatal sleep. Also, not every NICU contains a multichannel EEG recording device. For this reason, a single-channel EEG recording is used for sleep-wake classification in this study. The proposed study possesses good results i.e. 77.5% for neonatal sleep-wake classification.

3 Dataset and Preprocessing

NicoletOne IoT device has been used to record a maximum of 19 bipolar EEG records. EEG readings have been taken in Fudan Children's Hospital in Shanghai, China's NICU. The Fudan Children's Hospital Research Ethics Committee gave their consent (Approval No. (2017) 89) [17–21]. Each EEG record included at least two sleep cycles. The Nicole-tOne device was used to capture 9 bipolar EEG networks. The NicoletOne IoT device's electrodes were placed using standard 10–20 system [22] (Fig. 1).

The EEG recordings were analyzed at 500 Hz, which would be the initial sample frequency. These EEG signals become polluted by noise and artifacts during tracking. Prior to actual training and testing, these artifacts should be deleted. In the proposed study, preprocessing is mainly divided into three steps:

- Baseline noise, powerline noise and motion distortions were eliminated from neonatal EEG using finite impulse response filter (FIR) having cutoff frequencies 0.3–35 Hz.
- After cleaning the EEG recordings, the EEG data was divided into 4560 30-s segments and their associated labels.
- Finally, the "artifacts" epochs were carefully removed. For network training and testing, 3525 segments were set aside. The size 108 input vector was created by extracting 8 time as well as 4-frequency domain characteristics from 9-EEG channels. The handcrafted features incorporated in the proposed technique are depicted in Fig. 1.

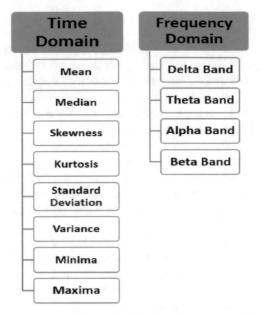

Fig. 1. Extracted handcrafted features from neonatal EEG

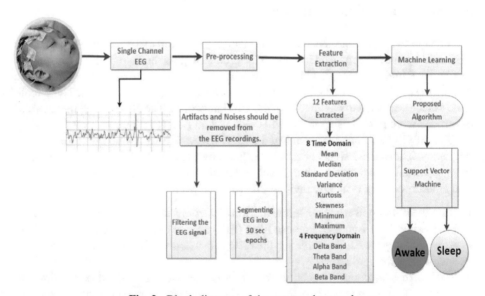

Fig. 2. Block diagram of the proposed network

4 Methodology

The block diagram is illustrated in Fig. 2. For classification, the proposed study used a support vector machine (SVM). SVM algorithm, developed by Vapnik with colleagues later in 1970's, is among the commonly utilized reference implementation learning algorithms in such a diversity of machine learning techniques. SVM is an efficient classifier-building technique. Its objective is to establish an outcome boundary between different categories that allows labels to be predicted through different feature vectors. The classifier, or set point, is oriented so that it would be much farther away from the nearest data points of the categories. Support vectors are indeed the spots that are nearest. Given a labeled training sample, perform the following:

$$(x_1, y_1), ..., (x_n, y_n), x_i \in R_d \text{ and } y_i \in (-1, +1) \tag{1}$$

While x_i seems to be a feature vector description and y_i is training compound's categorical variable (negative or positive). Appropriate hyperplane therefore can be formulated as:

$$wx^T + b = 0 \tag{2}$$

where w denotes mass, x denotes input feature vector, while b denotes bias. The standard inequalities must be incorporated into the w and b for all parts of the training set:

$$wx_i^T + b \geq +1 \text{ if } y_i = 1 \tag{3}$$

$$wx_i^T + b \leq -1 \text{ if } y_i = -1 \tag{4}$$

Finding w and b such that the hyperplane divides the input and maximizes its margin would be the goal of programming a SVM algorithm.

5 Results

All the results have been processed in MATLAB 2022 intel CORE i5. The proposed algorithm achieved an accuracy of $77.5 \pm 1.5\%$ for neonatal sleep stage classification. The confusion matrix of the proposed study is given in Fig. 3.

From the confusion matrix, the performance matrices i.e. accuracy, kappa, sensitivity, specificity are calculated. These matrices are shown in Table 1. Also, it is important to note that the proposed dataset is imbalanced. For this reason, the kappa coefficient should be considered as the main parameter for performance evaluation.

The proposed scheme has been compared with other machine learning algorithms like Logistic Regression, K-nearest neighbors, and Ensemble (Fig. 4). From Fig. 4, it is evident that the proposed algorithm outperformed other machine learning algorithms for neonatal sleep-wake classification.

Similarly, the proposed study has been compared with deep learning algorithms. The comparison is shown in Table 2. There are some algorithms which presented the results for neonatal sleep-wake classification using single channel EEG. This study compared

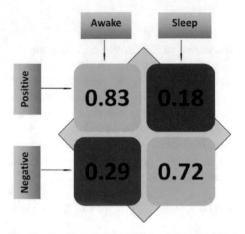

Fig. 3. Confusion matrix of the proposed SVM

Table 1. Performance parameters of the proposed SVM

	Accuracy	Kappa	Sensitivity	Specificity	F1 Score	PPV	NPV
SVM	77.5	0.55	73.5	80.13	77.93	82.18	71.29

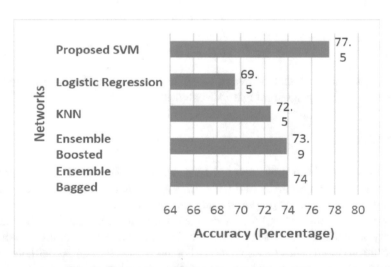

Fig. 4. Comparison with machine learning algorithms

the results of the proposed SVM with these algorithms and the results are shown in Table 2 (Saadullah et al., Frawan et al.).

For a real-time application, it is very important for a good network to test the network in minimum time. The proposed network is fast and efficient in this regard. For training, the proposed scheme requires up to 2 min and 13 s. For testing 880 30-s segments,

Table 2. Comparison with deep learning networks

	Accuracy	Kappa	Publication Year
Auto encoders [16]	17	-	2017
MLP	71	0.52	2020
CNN	69	-	-
RNN	65	-	-
Proposed SVM	**77.5**	**0.5445**	**2022**

the proposed network requires only 31.24 s. It is evident that the presented algorithm requires less time for classification and didn't use any post-processing step therefore, it can be used for real-time clinical applications.

6 Conclusion

In this paper, an automatic sleep-wake classification algorithm has been proposed using single channel EEG. Twelve prominent EEG features are extracted from time and frequency domain. Then, a support vector machine has been deployed for classification. The proposed algorithm achieved an accuracy of 77.5% for sleep wake classification. Statistical results show that single channel EEG can be used for neonatal sleep classification. To conclude, we can say that the proposed algorithm can be used as a real-time application for sleep-wake cycling in a neonatal intensive care unit.

References

1. Ansari, A., et al.: Quiet sleep detection in preterm infants using deep convolutional neural networks. J. Neural Eng. **15**(6) (2018)
2. Dereymaeker, A., et al.: Review of sleep-EEG in preterm and term neonates. Early Hum Dev **113**, 87–103 (2017)
3. Fraiwan, L., Lweesy, K., Khasawneh, N., Fraiwan, M., Wenz, H., Dickhaus, H.: Time-frequency analysis for automated sleep stage identification in fullterm and preterm neonates. J Med Sys **35**(4), 693–702 (2011)
4. American Academy of Sleep Medicine (AASM) Homepage. https://aasm.org/. Accessed 2018
5. Weisman, O., Magori-Cohen, R., Louzoun, Y., Eidelman, A.I., Feldman, R.: Sleep-wake transitions in premature neonates predict early development. Pediatrics **128**, 706–714 (2011)
6. Levy, J., et al.: Impact of hands-on care on infant sleep in the neonatal intensive care unit. PediatrPulmonol **52**(1), 84–90 (2017)
7. Memar, P., Faradji, F.: A novel multi-class EEG-based sleep stage classification system. IEEE Trans. Neural Syst. Rehabil. Eng. **26**(1), 84–95 (2018)
8. Turnbull, J.P., Loparo, K.A., Johnson, M.W., Scher, M.S.: Automate detection of tracé alternant during sleep in healthy full-term neonates using discrete wavelet transform. Clinical Neurophysiology **112**(10), 1893–1900 (2001)
9. Holmes, G.L., Lombroso, C.T.: Prognostic value of background patterns in the neonatal EEG. Journal of Clinical Neurophysiology **10**(3), 323–323 (1993)

10. Watanabe, K.: Neurophysiological aspects of neonatal seizures. Brain and Development **36**(5), 363–371 (2014)
11. Kidokoro, H., Inder, T., Okumura, A., Watanabe, K.: What does cyclicity on amplitude-integrated EEG mean. Journal of Perinatology **32**(8), 565–569 (2012)
12. Dereymaeker, A., Pillay, K., Vervisch, J., Huffel, S.V., Naulaers, G., et al.: An automated quiet sleep detection approach in preterm infants as a gateway to assess brain maturation. Int. J. Neural Syst. **27**(06), 1750023 (2017)
13. Koolen, N., Oberdorfer, L., Rona, Z., Giordano, V., Werther, T., et al.: Automated classification of neonatal sleep states using EEG. Clin. Neurophysiol. **128**(6), 1100–1108 (2017)
14. Pillay, K., Dereymaeker, A., Jansen, K., Naulaers, G., Huffel, S.V., et al.: Automated EEG sleep staging in the term-age baby using a generative modelling approach. Journal of Neural Engineering **15**(3) (2018)
15. Ansari, A.H., De Wel, O., Pillay, K., Dereymaeker, A., Jansen, K., et al.: A convolutional neural network outperforming state-of-the-art sleep staging algorithms for both preterm and term infants. J. Neural Eng. **17**(1), 016028 (2020)
16. Fraiwan, L., Lweesy, K.: Neonatal sleep state identification using deep learning autoencoders. In: IEEE 13th International Colloquium on Signal Processing & its Applications, pp. 228–231 (2017)
17. Abbasi, S.F., Jamil, H., Chen, W.: EEG-based neonatal sleep stage classification using ensemble learning. Comput. Mater. Contin. **70**, 4619-4633 (2022)
18. Abbasi, S.F., Ahmad, J., Tahir, A., Awais, M., Chen, C., et al.: EEG-based neonatal sleep-wake classification using multilayer perceptron neural network. IEEE Access **8**, 183025–183034 (2020)
19. Abbasi, S.F., Awais, M., Zhao, X., Chen, W.: Automatic denoising and artifact removal from neonatal EEG. The Third International Conference on Biological Information and Biomedical Engineering, Hangzhou, China, pp. 1–5 (2019)
20. Awais, M., Long, X., Yin, B., Chen, C., Akbarzadeh, S., et al.: Can pre-trained convolutional neural networks be directly used as a feature extractor for video-based neonatal sleep and wake classification. BMC. Res. Notes **13**(1), 1–6 (2020)
21. Awais, M., Chen, C., Long, X., Yin, B., Nawaz, A., et al.: Novel framework: face feature selection algorithm for neonatal facial and related attributes recognition. IEEE Access **8**, 59100–59113 (2020)
22. Homan, R.W., Herman, J., Purdy, P.: Cerebral location of international 10–20 system electrode placement. Electroencephalography and Clinical Neurophysiology **66**(4), 376–382 (1987)

An Intelligent Machine Learning-Based System for Predicting Heart Disease Using Mixed Feature Creation Technique

Abdelrahman Elsharif Karrar[1][(✉)] [iD] and Rawia Elarabi[2]

[1] College of Computer Science and Engineering, Taibah University, Medina, Saudi Arabia
akarrar@taibahu.edu.sa
[2] Computer Science Department, Jazan University, Jazan, Saudi Arabia
relarabi@jazanu.edu.sa

Abstract. The quality of healthcare outcomes is influenced by the reliability and accuracy of computational analysis techniques applied to clinical data. Advanced classification systems improve the precision and speed of medical diagnosis, providing critical decision insights for doctors and resource optimization. Classification techniques based on machine learning have been applied as effective and reliable non-surgical techniques for the efficacious diagnosis and treatment of heart disease patients. This paper demonstrates the application of the Mixed Feature Creation (MFC) approach to classify a heart-disease dataset from UCI Cleveland using techniques such as the Recursive Feature Elimination with Random Forest (RFE-RF) feature selection and Least Absolute Shrinkage and Selection Operator (LASSO). Parameters from each technique are optimized through grid-search and cross-validation methods. Further, classifier performance models are used to determine the classification techniques' F1-scores, precision, sensitivity, specificity, and accuracy based on independent measures such as RMSE and execution time.. The findings suggest that ML-driven classification algorism can be used to develop reliable predictive models for the accurate diagnosis of heart diseases.

Keywords: Machine Learning · Classification Algorithms · Mixed Feature Creation · Heart Disease Prediction

1 Introduction

Data mining is a computational technique used to extract meaningful insights from unstructured data by applying various association forecasting methods, classification algorithms, and clustering models [1]. Supervised learning algorithms based on Machine Learning (ML) are used to classify historical data by assigning validation labels for training and testing sets. The selection of supervised learning algorithms over other ML approaches is based on simplifying feature selection models using prior knowledge of dataset attributes such as class labels [2], which significantly improve classification accuracy [3].

© The Author(s), under exclusive license to Springer Nature Switzerland AG 2023
F. Saeed et al. (Eds.): ICACIn 2022, LNDECT 179, pp. 353–367, 2023.
https://doi.org/10.1007/978-3-031-36258-3_31

Modern healthcare systems collect abundant amounts of patient data, which is computationally analyzed to provide insights for diagnosis and treatment plans. However, a critical gap still exists in the deployment of data-driven decision support systems, which utilize artificial intelligence (AI) and machine learning techniques to improve the efficiency of diagnosis and treatment processes. Research findings suggest that the data mining techniques such as Ensembles, logistic regression analysis, decision trees, support vector machines (SVM), and artificial neural networks deliver superior efficiency and accuracy when applied to health data [4–7].

Heart disease is among the globally leading chronic health conditions in terms of incidence and mortality. Studies report that heart disease contributes significantly to overburdening global healthcare systems due to diagnostic and treatment complexities, which subject patients to prolonged hospital stays, especially in third-world countries [8]. Factors such as low integration of diagnostic technology, underfunding, and inadequate skilled personnel are the key factors contributing to the care complexities for heart disease patients. Further, the European Cardiology Society (ESC) reports that the global annual incidence of heart disease is 3.6 million, and 50% of the patients die within the first two years of diagnosis [9].

The invasive diagnosis and treatment methods for heart disease have been positively correlated to high error risks, exposing patients to life-threatening incidents of misdiagnosis and delays during the analysis [10]. While technology deployment in such procedures is research-intensive, research evidence suggests that the data-driven prediction models based on ML and AI techniques could lower the mortality rates significantly when applied to heart disease management programs [4, 7, 11–13]. However, the efficiency and reliability of ML and AI models are based on the accuracy of testing and training datasets used for the classification [13, 14].

The present study suggests the adoption of approaches such as LASSO, RFE-RF, and Relief selection to improve the accuracy of prediction models. The computational data classification techniques have been applied as the basis of developing intelligent clinical decision support systems for heart disease. The techniques were applied on the UCI Heart-Disease/Cleveland Dataset [15] for feature extraction and target prediction through the following supervised learning models;

- Gradient Boosting
- Logistic Regression
- AdaBoost Classification
- Random Forest Classification
- Decision-Tree Classification
- Stochastic Gradient Descent
- Support Vector Machines
- K-Nearest Neighbor

Estimation parameters are optimized through grid search and cross-validation methods during algorithm implementation, while classification evaluation metrics were used to estimate the F1-scores, precision, sensitivity, specificity, and accuracy of our model, considering factors such as RMSE and the execution time. The present study computationally classifies the heath care dataset through the following steps;

Evaluation of classifier results to extract full features, including F1-score, precision, sensitivity, specificity, and accuracy of the two classification models alongside RMSE and execution time.

The combination of best features, RFE-RF, and LASSO techniques for selecting correlated features with strongly correlated features significantly influence the predicted outcomes and the resolution of underfitting and overfitting issues in machine learning [16].

Evaluate the suitable classifier and algorithms for the development of intelligent diagnosis and treatment systems for heart disease through precision classification of clinical data.

The rest of the paper is structured in Sects. 2, 3, 4, and 5, which provide a literature review, methodology, results and discussion, and conclusion, respectively.

2 Related Work

The integration of advanced computational analytics capabilities in healthcare systems is responsible for the development of hybrid models that combine ML and AI capabilities, a promise of superior accuracy and efficacy in disease diagnosis and treatment [17]. This section contains information about some existing systems on the UCI Heart-Disease/Cleveland Dataset.

According to Mohan et al. [12] Experimental trials on the application of algorithms built on advanced classifiers such as Naïve Bayes, Random Forest, and Multilayer Perceptron have established about 85.48% accuracy in the detection of heart disease.

A study by Haq et al. [9] applied a hybrid intelligent system built on machine learning to predict heart disease from a patient dataset by applying 10-fold cross-validation. Applying feature-selection algorithms such as LASSO and Minimal-Maximal-Relevance (mRMR) extracted heart disease-related attributes with 94% accuracy when using KNN algorithms, 82.2% using Bayesian algorithms, 88.9% using logistics network model, and 77.9% using decision trees [9].

The findings from a study by Ali et al. [13] suggest that applying the hybrid model to feature selection of a patient dataset attained 87.41% when machine learning techniques such as KNN, Neural Networks, SVM, and LR were used. A hybrid intelligent classification system that combines SVM and Mean Fisher Score Feature Selection (MFSFS) machine learning algorithms yielded 88.68% accuracy and 72.92% sensitivity when used to obtain a disease-related feature subset from the Cleveland dataset [11].

Hybrid intelligent ML systems have advanced capability to overcome potential performance bottlenecks resulting from multilayer categorization. A study [18] reports that the integration of the Probabilistic Principal Component Analysis (PPCA) model on hybrid intelligent systems enhances their detection capacity for the most critical heart disease predictors. The Probabilistic principal component analysis provides a reliable statistical framework for the classification of pre-defined disease attributes when applied to ML-based classification.

An investigation ML-based classification technique on the Cleveland dataset using the Factor Analysis of Mixed Data [FAMD] to extract relevant features and predict disease by Gupta et al. [19] achieved 93.44% precision, 96.96% specificity, and 89.28%

sensitivity. Further integration of LASSO and feature selection algorithms to the hybrid intelligent system with multiple boosting and bagging techniques delivered 99.05% classification accuracy when tested on RFB with relief features. Evidence from the extensive review of findings from studies that applied ML-based classification algorithms on the Cleveland Heart Disease dataset cannot be statistically generalized since it contains limited cases and features. The dataset also contains mixed categorical and numerical variables implying the need for Mixed Feature Creation (MFC) techniques to improve classification efficiency and accuracy for the reduction in mortality and morbidity caused by heart disease.

Table 1 shows the machine learning approaches followed in the systems presented by these studies through the statement of accuracy, recall, and specificity for each;

Table 1. Accuracy, Recall, and Specificity for some existing UCI Heart-Disease/Cleveland Dataset systems.

Source	Approach	Accuracy	Recall	Specificity
[18]	PPCA	82.18	75	90.57
[20]	PSO with SVM	84.36	-	-
[9]	Relief + LR	89	77	98
[9]	mRMR + NB	84	77	90
[9]	LASSO + SVM	88	75	96
[11]	RBF kernel-based SVM	81.19	72.92	88.68
[12]	HRFLM	88.7	92.8	82.6
[13]	L1 Linear SVM + L2 Linear RBFSVM	92.22	82.92	100
[21]	Hybrid Approach + Majority vote with NB, BN, RF, and MP	85.48	-	-
[19]	FAMD + RF	93.44	89.28	96.96

3 Methodology

3.1 Data Description and Dataset Specification

Implementation of the proposed model requires a detailed description of the proposed dataset to understand its unique qualities and characteristics. The data used in this study was obtained from the ML repository at UCI Cleveland Cardiology department. It contains 14 attributes such that the "target" attribute is the output, labeled into the 'normal' and 'heart patient' categories, while 13 are diagnostic inputs. The dataset analysis revealed zero missing values and one duplicated row. The data features and characteristics are shown in Table 2.

Statistical properties such as correlation coefficients among the variables are illustrated in a scaled heatmap in which blue indicates a positive correlation while red indicates a negative correlation between the features, as shown in Fig. 1.

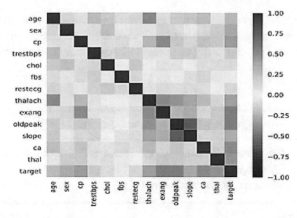

Fig. 1. Pearson Correlation Coefficient Heat Map of Cleveland CHD Features.

Table 2. Characteristics and features of the Cleveland dataset

	count	mean	std	min	25%	50%	75%	max
age	302	54.42	9.05	29	48	55.5	61	77
sex	302	0.68	0.47	0	0	1	1	1
cp	302	0.96	1.03	0	0	1	2	3
trestbps	302	131.6	17.56	94	120	130	140	200
chol	302	246.5	51.75	126	211	241	274.8	564
Fbs	302	0.15	0.36	0	0	0	0	1
restecg	302	0.53	0.53	0	0	1	1	2
thalach	302	149.6	22.9	71	133	153	166	202
exang	302	0.33	0.47	0	0	0	1	1
oldpeak	302	1.04	1.16	0	0	0.8	1.6	6.2
slope	302	1.4	0.62	0	1	1	2	2
ca	302	0.72	1.01	0	0	0	1	4
thal	302	2.31	0.61	0	2	2	3	3
target	302	0.54	0.5	0	0	1	1	1

3.2   Feature Modelling

Mixed Feature Creation (MFC). The Cleveland dataset contains numeric and categorical features, which are statistically quantitative and qualitative, respectively. The MFC technique adopts a factorial approach in the extraction of numerical attributes through factor analysis and concatenation of the categorical attributes [22]. This approach scales numerical features down to unit variances and transforms categorical features into disjunctive pseudo-codes expressed as Algorithm 1, from which 61 features are extracted.

Algorithm 1. Mixed Feature Creation (MFC)	
Input	Cleveland CHD dataset (D), the features F divided into two sets: C_i categorical features where $1 \leq i \leq$ n, and N_j numerical features where $1 \leq j \leq$ m

Process	begin
	    $C_i \leftarrow$ categorical feature and $i \leftarrow$ 1,2,...,n
	    $N_j \leftarrow$ numerical feature and $p \leftarrow$ 1,2,...,m
	    $P_j \leftarrow$ Factor analysis of N_j
	    for j =1 to m do
	        Q_j= new numerical feature N_j/P_j
	    end for
	    for R in the list of categorical features $[C_i]$ do
	        for J in the list of new numerical features $[Q_j]$do
	        MFC= $[C_i]$+ $[Q_j]$
	        end for
	    end for
	    return mixed feature creation (MFC)
	end

Output	Return a new set of features where i $1 \leq i \leq$ n of dataset D by concatenating the categorical and the new numerical features, which is an MFC dataset. Then apply this data to the classifiers after selecting the best features by different selection techniques to obtain high-accuracy results.

Data Cleaning and Pre-processing. The next step is pre-processing the data to identify missing values of duplicates. Label encoding and standardization methods were further applied on the dataset to control unbalancing effects of illness classes by z-score normalization and the conversion to standard deviation 1Σ () and mean $0(\mu)$, respectively [23].

Methodological Framework of the Proposed Solution. The proposed intelligent ML-based system for the diagnosis of heart disease aims at optimizing the predictive accuracy and precision, which is crucial for early detection. The sequence of tasks involved in the model development is illustrated in Fig. 2. The present study utilized RFE-RF and LASSO techniques for the extraction of specific data features, which are labeled as 'training' and 'testing' categories. The proposed system applied all traditional classifiers and ensemble models compared and validated using different metrics.

Evaluation metrics. The performance efficiency of the proposed model is evaluated based on performance indicators such as ROC and AUC characterized by features such

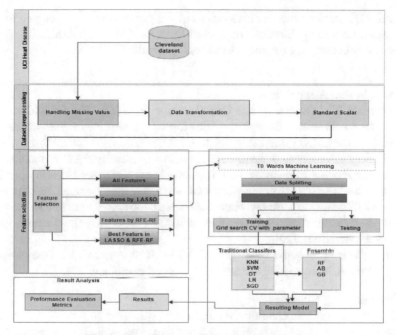

Fig. 2. Flow Diagram of The Proposed Model.

as True Positives (TP), True Negatives (TP), False Positives (FP), and False Negatives (FN), which are modeled as a confusion matrix illustrated using Eq. (1), Eq. (2), Eq. (3), Eq. (4) and Eq. (5);

$$\text{ClassificationAccuracy}(AC) = \frac{TP + TN}{+TNTP + FP + FN}x100\% \qquad (1)$$

$$\text{Precision} = \frac{TP}{TP + FP}x100\% \qquad (2)$$

$$\text{Sensitivity} = \frac{TP}{TP + FN}x100\% \qquad (3)$$

$$\text{F1} - \text{Score} = \frac{2(PrecisionxRecall)}{(Precision + Recall)}x100\% \qquad (4)$$

$$\text{Specificity} = \frac{TN}{TN + FP}x100\% \qquad (5)$$

TP = Correct identification of heart disease.
TN = Correct identification of opposite classes, i.e., true identification of patients having no heart disease.
FP = Incorrect identification of heart disease patients.
FN = Incorrect identification of opposite classes, i.e., identifying heart disease patients as normal patients.

ROC and AUC performance metrics represent the curve area of the degree to which classifiers are accurately classified within class instances [24]. High AUC values improve the classification accuracy for normal and heart patients.

Feature Selection Approaches

Least Absolute Shrinkage and Selection Operator (LASSO). The LASSO technique refines feature coefficients through modification of the absolute values whereby zero coefficients are removed from feature subsets. LASSO delivers high performance given low coefficient values through multiple assessments to identify frequent features [25].

Recursive Feature Elimination with Random Forest Features Selection (RFE-RF). This technique is based on the implementation of machine learning models used to compute coefficient values for the worst and best-performing features from the dataset. The computation is repeated several times, with the best-performing features grouped in subsequent iterations. [26].

Best features (Combined BLR and RFE-RF). In this technique, best-performing features from LASSO and RFE-RF are extracted and combined [27].

Machine Learning Classifiers

Logistic Regression. This classification algorithm was used to predict the value of a variable y in binaries [0, 1], whereby 1 is a positive class while 0 is the negative class. The binary classes are classified based on the hypothesis $h(\theta) = g(\theta^T X)$, such that when $h(\theta) < 0.5$, a person is healthy, and if $h(\theta) > 0.5$, they suffer heart disease [28].

Decision Trees. The decision tree approach utilizes nodes labeled leaf and decision nodes presented in a tree shape such that branch nodes are test outcomes and interior nodes represent the testing property when applied to regression and classification problems [29].

Random Forest. Random classifiers combine multiple classification algorithms, which utilize ensemble learning to predict outcomes by extracting features through bootstrapping aggregation [30]. The approach is considered highly effective and reliable when testing datasets with high computation complexities by assigning upper and lower limits to dataset values. Feature classification based on the RF model is illustrated in Eq. (6);

$$j = \frac{1}{B} \sum\nolimits_{b=1}^{B} f_b(x^i) \tag{6}$$

Support Vector Machine. The SVM technique is applied to classification problems that require binary classifications using a hyperplane, which separates instances $w^T x + b = 0$ such that d and w represent dimensional coefficient vectors, which have a normality property relative to the hyperplane [31, 32]. Langrangian multipliers are used to classify data points in the form W and b, whose linear discriminant function is as shown in Eq. (7);

$$g(x) = sgn(\sum\nolimits_{i=1}^{n} \alpha_i y_i x_i^T x + b) \tag{7}$$

K-Nearest Neighbors. This technique is used to categorize data features in supervised learning through a nonparametric comparison of features from two datasets [23, 33].

KNN solves classification, recognition, and regression problems, which compare closest classes by learning from training sets. The dimensional space between the data points A and B are estimated on a two-dimensional space. The Euclidean distance between previous data $B(x_2, y_2)$ and new data $A(x_1 y_1)$ is defined by the formula shown in (8);

$$\sqrt{(x_2 - x_1)^2 + (y_2 - y_1)^2} \qquad (8)$$

AdaBoost Algorithm. This classification algorithm integrates multiple weak classifiers to create a powerful classification model based on the decision stump [34]. Dataset entries are assigned weights $W(X_i) = 1/N$ given, N represents training sample frequency, and x_i represents ith training set. The classification accuracy is calculated by Eq. (9);

$$Error = (Correct - N)/N \qquad (9)$$

When the classification is boosted using multiple trainers, the final classifier is represented using Eq. (10);

$$H_k(p) = +/ - (\sum_{k=1}^{k} a_k h_k(p)) \qquad (10)$$

4   Results and Discussion

The feature selection techniques (LASSO, RFE-RF, and BLR) effectively separated the 25 best features from the dataset and computed their respective correlation coefficients, as illustrated in Fig. 3, Fig. 4, and Fig. 5. The heat map was scaled to depict feature correlation on a scale of −1.00 (red color)– 1.00 (blue color).

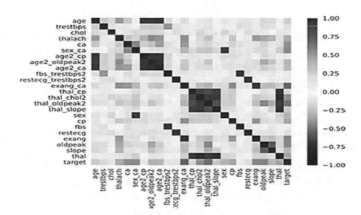

Fig. 3. A Heat Map of Correlation Coefficients Using LASSO Feature Selection Technique.

The classification techniques achieved varying levels of accuracy when applied to the UCI dataset. However, AdaBoost and SVM delivered the highest (90.16%) feature

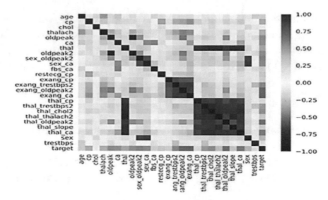

Fig. 4. A Heat Map of Correlation Coefficients Using RFE-RF Feature Selection Technique.

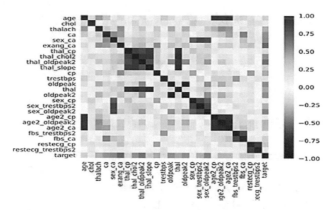

Fig. 5. A Heat Map of Correlation Coefficients Using BLR Feature Selection Technique.

detection accuracy, while KNN delivered 88.52%, and LR attained 86.89%. RF technique achieved the best classification accuracy (96.72%) when applied to the data features selected by LASSO, while KNN, AdaBoost, SVM, and LR attained relative accuracy of 88.52%, 90.16%, 91.80%, and 93.44%, respectively. The application of classification algorithms on features selected using LASSO showed significant improvements in the detection accuracy from 81.97% to 96%, as illustrated in Fig. 6;

When compared on the basis of precision, the classifiers produced varying rates within a range of 84%–92%. When applied to features detected using LASSO, the RF technique delivered 100% precision while LR delivered 94% precision. Combined RFE-RF classification attained 94% precision, while KNN delivered 86%. Further, AdaBoost delivered 94% precision when applied to BLR feature classification. The relative precision of the classification techniques is illustrated in Fig. 7;

Further comparisons of the classification techniques based on the sensitivity scores revealed that KNN attained a 92% score while RF attained 82%. KNN and LR algorithms had a considerably high score (95% +) sensitivity performance when applied to the

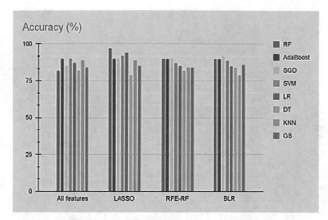

Fig. 6. The Relative Accuracy of ML-based Classification Methods.

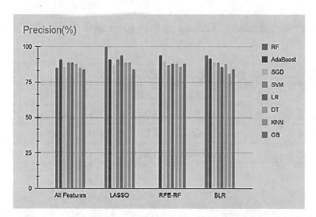

Fig. 7. Relative Precision of The Classification Techniques.

features identified using the RFE-RF technique. However, KNN attained a low sensitivity score when applied to BLR features, as shown in Fig. 6.; (Fig. 8)

A similar level of consistency in the classification accuracy was observed when the algorithms were compared on the basis of specificity, ROC-AUC score, and Runtime metrics. Most classification algorithms' approximate score range was within 80%94%. For instance, results from the analysis of runtime showed that RT had the shortest runtime while RT had the longest score of 56.8%, as shown in Table 3;

Figure 9. Shows the comparison between the proposed system and some of the current systems on the UCI Heart-Disease/Cleveland dataset, reviewed in Table 1, in terms of accuracy, recall, and specificity.

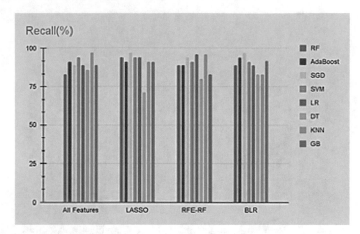

Fig. 8. Sensitivity scores of different classification methods.

Table 3. Comparative Classifier Accuracy and Performance Using Different Classification Techniques

Classifier	RF	AdaBoost	SGD	SVM	LR	DT	KNN	GB
All	54.5	2.27	0.725	0.688	0.747	0.748	0.713	54.1
LASSO	48.8	1.93	0.569	0.941	2.05	0.639	0.639	31.2
RFE-RF	52.6	2.03	0.658	3.31	0.670	0.663	0.681	34.7
BLR	56.8	2.33	0.674	0.981	0.727	0.705	0.717	38.3

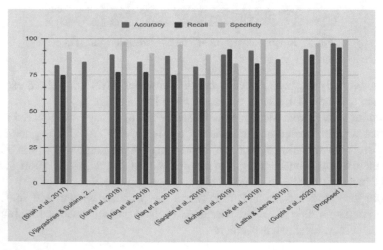

Fig. 9. Comparison of the proposed system's accuracy, recall, and specificity with some existing systems on the UCI Heart-Disease/Cleveland dataset

5 Conclusion

The objective of this paper was to develop a smart machine learning system with superior accuracy in the diagnosis of heart disease. The proposed system modeled advanced computational techniques in the classification of datasets to identify and separate heart disease patients from healthy individuals. Mixed feature classification properties were applied to the UCI Heart Disease/Cleveland dataset using BLR, RFE-RF, and LASSO techniques. The study findings suggest that the LASSO technique delivers high performance with most ML algorithms, while RF is best suited for high-impact properties, and RF has 96.72% classification accuracy.

Future research on the application of multiple feature selection algorithms on larger datasets is proposed to improve the predictive accuracy of ML-based classification algorithms for the diagnosis of heart disease.

References

1. Karrar, E.; A Novel Approach for Semi Supervised Clustering Algorithm. International Journal of Advanced Trends in Computer Science and Engineering **6**(1), 1–7 (2017). http://www.warse.org/IJATCSE/static/pdf/file/ijatcse01612017.pdf
2. Al-Sarem, M., Saeed, F., Boulila, W., Emara, A.H., Al-Mohaimeed, M., Errais, M.: Feature selection and classification using catboost method for improving the performance of predicting parkinson's disease. In: Saeed, F., Al-Hadhrami, T., Mohammed, F., Mohammed, E. (eds.) Advances on Smart and Soft Computing. AISC, vol. 1188, pp. 189–199. Springer, Singapore (2021). https://doi.org/10.1007/978-981-15-6048-4_17
3. AlAfandy, K.A., Omara, H., Lazaar, M., al Achhab, M.: Machine Learning, pp. 83–113 (2022). https://doi.org/10.4018/978-1-7998-9831-3.ch005
4. Kausar, N., Palaniappan, S., Samir, B.B., Abdullah, A., Dey, N.: Systematic analysis of applied data mining based optimization algorithms in clinical attribute extraction and classification for diagnosis of cardiac patients. In: Applications of intelligent optimization in biology and medicine, Springer, pp. 217–231 (2016). https://doi.org/10.1007/978-3-319-21212-8_9
5. Aljanabi, M., Qutqut, M.H., Hijjawi, M., Al-Janabi, M.I.: Machine learning classification techniques for heart disease prediction: a review. Review Article in Int. J. Eng. Technol. **7**(4), 5373–5379 (2018). https://doi.org/10.14419/ijet.v7i4.28646
6. Khan, Y., Qamar, U., Yousaf, N., Khan, A.: Machine learning techniques for heart disease datasets: a survey. In: ACM International Conference Proceeding Series, vol. Part F148150, pp. 27–35 (2019). https://doi.org/10.1145/3318299.3318343
7. Mythili, T., Mukherji, D., Padalia, N., Naidu, A.: A heart disease prediction model using SVM-decision trees-logistic regression (SDL). International Journal of Computer Applications (0975), **68**(16), 11–15 (2013)
8. Ghwanmeh, S., Mohammad, A., Al-Ibrahim, A.: Innovative artificial neural networks-based decision support system for heart diseases diagnosis. J. Intell. Learn. Syst. Appl. **05**(03), 176–183 (2013). https://doi.org/10.4236/jilsa.2013.53019
9. Haq, A.U., Li, J.P., Memon, M.H., Nazir, S., Sun, R.: A hybrid intelligent system framework for the prediction of heart disease using machine learning algorithms. Hindawi Mobile Information Systems accuracy, vol. 2018 (2018)
10. Singh, D., Samagh, J.S.: A comprehensive review of heart disease prediction using machine learning. Journal of Critical Rev. **7**(12), 281–285 (2020)

11. Saqlain, S.M., et al.: Fisher score and Matthews correlation coefficient-based feature subset selection for heart disease diagnosis using support vector machines. Knowl. Inf. Syst. **58**(1), 139–167 (2018)
12. Mohan, S., Thirumalai, C., Srivastava, G.: Effective heart disease prediction using hybrid machine learning techniques. IEEE Access **7**, 81542–81554 (2019)
13. Ali, L., et al.: An optimized stacked support vector machines based expert system for the effective prediction of heart failure. IEEE Access **7**, 54007–54014 (2019)
14. Amin, M.S., Chiam, Y.K., Varathan, K.D.: Identification of significant features and data mining techniques in predicting heart disease. Telematics Inform. **36**, 82–93 (2019)
15. Janosi, A., Steinbrunn, W., Pfisterer, M., De-trano, R.: UCI Heart-Disease/Cleveland Dataset | ML Dataset. Kaggle (2021). https://doi.org/10.34740/KAGGLE/DSV/2641256
16. Mawarni, M., Utaminingrum, F., Mahmudy, W.F.: The effect of feature selection on gray level co-occurrence matrix (GLCM) for the four breast cancer classifications. J. Biomimetics, Biomaterials and Biomedical Eng. **55**, 168–179 (2022). https://doi.org/10.4028/p-09g3n8
17. Abdeldjouad, F.Z., Brahami, M., Matta, N.: A hybrid approach for heart disease diagnosis and prediction using machine learning techniques. In: International conference on smart homes and health telematics, pp. 299–306 (2020)
18. Shah, S.M.S., Batool, S., Khan, I., Ashraf, M.U., Abbas, S.H., Hussain, S.A.: Feature extraction through parallel probabilistic principal component analysis for heart disease diagnosis. Physica A **482**, 796–807 (2017)
19. Gupta, A., Member, S., Kumar, R.: MIFH : A Machine Intelligence Framework for Heart Disease Diagnosis. no. Ml, pp. 14659–14674 (2020)
20. Vijayashree, J., Sultana, H.P.: A machine learning framework for feature selection in heart disease classification using improved particle swarm optimization with support vector machine classifier. Program. Comput. Softw. **44**(6), 388–397 (2018)
21. Latha, B.C., Jeeva, S.C.: Informatics in Medicine Unlocked Improving the accuracy of prediction of heart disease risk based on ensemble classification techniques. Inform Med Unlocked, **16**, no. July, p. 100203 (2019). https://doi.org/10.1016/j.imu.2019.100203
22. Umair, M., et al.: Main path analysis to filter unbiased literature. Intelligent Automation and Soft Comput. **32**(2) (2022). https://doi.org/10.32604/iasc.2022.018952
23. Karrar, A.E.: The effect of using data pre-processing by imputations in handling missing values. Indonesian Journal of Electrical Engineering and Informatics (IJEEI), **10**(2), (2022). https://doi.org/10.52549/ijeei.v10i2.3730
24. Giudici, P., Raffinetti, E.: A Generalised ROC Curve (2021). https://ssrn.com/abstract=388 3422
25. Zhou, Wieser, A.: Jaccard analysis and LASSO-based feature selection for location fingerprinting with limited computational complexity. In: LBS 2018: 14th International Conference on Location Based Services, pp. 71–87 (2018)
26. Demir, S., Şahin, E.K.: Assessment of feature selection for liquefaction prediction based on recursive feature elimination. European Journal of Science and Technology (2021). https://doi.org/10.31590/ejosat.998033
27. Wei, P., et al.: Extraction of Kenyan grassland information using PROBA-V based on RFE-RF algorithm. Remote Sens (Basel), **13**(23) (2021). https://doi.org/10.3390/rs13234762
28. Lu, Q., Zhou, S., Tao, F., Luo, J., Wang, Z.: Enhancing gene expression programming based on space partition and jump for symbolic regression. Inf Sci (N Y) **547**, 553–567 (2021). https://doi.org/10.1016/j.ins.2020.08.061
29. Abri Aghdam, K., et al.: A novel decision tree approach to predict the probability of conversion to multiple sclerosis in Iranian patients with optic neuritis. Mult Scler Relat Disord, **47** (2021). https://doi.org/10.1016/j.msard.2020.102658
30. Genuer, R., Poggi, J.-M.: Random Forests with R. In: USE R, Springer, Cham, pp. 33–55 (2020). https://doi.org/10.1007/978-3-030-56485-8_3

31. Karrar, A.E.: A proposed model for improving the performance of knowledge bases in real-world applications by extracting semantic information. International Journal of Advanced Computer Science and Applications **13**(2) (2022). https://doi.org/10.14569/IJACSA.2022.0130214
32. Elarabi, R., Alqahtani, F., Balobaid, A., Zain, H., Babiker, N.: COVID-19 analysis and predictions evaluation for KSA using machine learning. In: 2021 International Conference on Software Engineering & Computer Systems and 4th International Conference on Computational Science and Information Management (ICSECS-ICOCSIM), pp. 261–266 (2021). https://doi.org/10.1109/ICSECS52883.2021.00054
33. Rezaei, N., Jabbari, P.: K-nearest neighbors in R. In: Immunoinformatics of Cancers, Rezaei, N., Jabbari, P. (eds.) Academic Press, pp. 181–190 (2022). https://doi.org/10.1016/B978-0-12-822400-7.00006-3
34. Karrar, A.E.: Investigate the ensemble model by intelligence analysis to improve the accuracy of the classification data in the diagnostic and treatment interventions for prostate cancer. International Journal of Advanced Computer Science and Applications, **13**(1) (2022). https://doi.org/10.14569/IJACSA.2022.0130122

Relu Dropout Deep Belief Network for Ontology Semantic Relation Discovery

Fatima N. AL-Aswadi[1,2] ⓘ, Huah Yong Chan[1(✉)] ⓘ, and Keng Hoon Gan[1] ⓘ

[1] School of Computer Sciences, Universiti Sains Malaysia, 11800 Gelugor,
Pulau Pinang, Malaysia
`fnsa15_com016@student.usm.my`, {`hychan,khgan`}`@usm.my`
[2] Faculty of Computer Science and Engineering, Hodeidah University, P.O. Box 3114,
Hodeidah, Yemen
`fatima_aswadi@hoduniv.net.ye`

Abstract. This paper presents the learning strategy and the environment for Ontology Learning (OL) relations discovery task for the scientific publications domain by adjusting Deep Belief Network (DBN). The adjusted DBN is called Relu Dropout DBN (Re-DDBN). This paper elaborates on the adjusted Re-DDBN configuration, its structure, hyper-parameters, and functions. In addition, the adjusted Re-DDBN was compared with traditional DBN and other comparative models (e.g., Support vector machine (SVM) and Naïve Bayes (NB)) for ontology semantic relations discovery. The outcomes revealed that the adjusted Re-DDBN displayed the best performance when compared to the other models. The SemEval-2018 task 7 dataset was applied in this study.

Keywords: Activation Function · Deep belief Network · Deep Learning · Dropout Strategy · Ontology Learning · Semantic Relation

1 Introduction

Ontology learning (OL) refers to an automatic procedure that extracts and represents knowledge in a machine-understandable format. Ontology is a fundamental scheme that organises knowledge as a collection of concepts and their relations within a domain. Ontologies have been beneficial and efficient in a variety of applications, including Question-Answering (QA), Semantic Searching, Decision-Support, and Automated Fraud Detection [1–4]. Relations discovery is the backbone task for an OL system.

Relations discovery techniques denote the combination of Natural Language Processing (NLP) and Machine Learning (ML), such as lexico-syntactic patterns, semantic templates, association rules, syntactic structure analysis, dependency structure analysis, clustering, classification, and logical inference. These relations discovery techniques are classified into four classes. They are (i) linguistic-based approaches that prioritise the use of syntactic analysis and patterns, (ii) statistical-based approaches that are based on associations and hierarchical analysis and rules, (iii) logic-based approaches that explore logic theories to infer or derive new relations, and (iv) hybrid approaches that blend

F. Saeed et al. (Eds.): ICACIn 2022, LNDECT 179, pp. 368–378, 2023.
https://doi.org/10.1007/978-3-031-36258-3_32

conventional techniques with new approaches (e.g., artificial neural network (ANN)). The reviews detailed in [5–7] provide in-depth descriptions and explanations of the OL techniques and approaches regarding each task.

The existing OL systems have many limitations [6, 8, 9], such as reliance on a large amount of predefined patterns to extract relations. Although manually predefined patterns have reasonable precision, they result in a very low recall [5, 10–12]. This signifies the pressing need to develop efficient automatic methods that improve relations discovery task for OL so that relations can be extracted automatically with minimal reliance on predefined patterns. Dual stages are required to develop this automatic method for relations discovery task. The first one is adjusting and configuring the appropriate learning environment and strategy that can automatically learn or detect relations. Next, the second stage is for developing relations extraction and classification technique based on the established learning strategy from the first stage. As such, this study reports the stabilisation analysis and experiments for the first stage of this automatic relation discovery method. DBN promises to provide better solution for this problem type. However, with imbalanced data and low characteristic dimensions on it, the traditional DBN would not perform well. Thus, the DBN was adjusted for relations discovery by incorporating the dropout strategy, as well as by changing the DBN activation function from the traditional sigmoid activation function to Relu activation function with using the stochastic gradient descent (SGD) optimisation algorithm. This approach called Re-DDBN is elaborated in Sect. 3. Moreover, selecting the right hyper-parameters and building an appropriate network are also main challenges.

The remaining of this paper is organised as follows: Sect. 2 presents prior work related to relations discovery, while Sect. 3 elaborates on the adjusted DBN. Section 4 details the experiments and evaluations. Section 5 concludes this study.

2 Related Work

Many studies have looked into the OL relation discovery task. Sombatsrisomboon et al. [13] provided a simple approach that only used *"NP is a/an NP"* pattern to identify taxonomic relationships between pairs of terms. However, this approach frequently failed to obtain hypernyms for common nouns [13]. Specia, Motta [14] proposed a hybrid of traditional technologies to derive the semantic relations between pairs of named entities using data mining and linguistic technologies.

Sánchez, Moreno [15] introduced unsupervised approaches to detect non-taxonomic relations by using domain-relevant verb phrases, as well as statistical and linguistics analyses. El-Kilany et al. [16] introduced an unsupervised clustering-based relations discovery technique to create a raw dataset of relations, which was utilised to generate and identify legible relations from the news dataset. This proposed technique performed better in the recall measure, but performed worse in the precision measure when compared to traditional methods.

Many studies have used hybrid techniques for relations extraction. Minard et al. [17] used SVM to identify eight relations of clinical reports, while [18] applied TF-IDF matrix and SVM for features extraction and relations classification that classified explicit music content into topic, mood, and genre relations. Sureshkumar, Zayaraz [19] used NB to extract attributes or taxonomic relations. Zhang et al. [20] developed the Simultaneously Entity and Relationship Extraction (SERE) model to extract attributes based on Conditional Random Fields (CRF). Etzioni et al. [21] built the TextRunner tool based on CRF to extract attributes between two named entities in each sentence.

Zhong et al. [22] and Chen et al. [23] developed deep learning (DL) models to extract the attributes of named entity. The DL is a type of ANN that has multiple hidden layers. In the traditional DBN, a study [23] presented an information extraction model to determine and pair one of five sorts of relations between each two Chinese named entities. Meanwhile, an unsupervised Entity Attribute Extraction model based on traditional DBN (EAEDB) was developed to extract Chinese named entity attributes of persons, organisations, and locations [22].

3 Adjusted Deep Belief Network

A significant move towards deep analysis was noted in the last few years, whereby deeper techniques portrayed better handling. In light of DL and ML models, the use of DBN for relations discovery of OL has certainly yielded promising results. However, the main challenge of using DBN for relations discovery task is to build an efficient network that selects the right hyper-parameters. This section presents the comprehensive details of Re-DDBN applied for the relations discovery task.

3.1 Deep Belief Network

The DBN is one of the milestone models in DL - a probabilistic generative feed-forward network composed of an organised stack of Restricted Boltzmann Machine (RBM). The learning algorithm of DBN has dual stages: pre-training and fine-tuning. The pre-training process is performed through building blocks for each layer and training it from input to top layers. On the contrary, the fine-tuning process is from top to bottom layers. Figure 1(a) illustrates the DBN structure and the learning process with three-RBM stack, whereas Fig. 1(b) presents the RBM structure. The energy function is presented in Eq. (2), while the joint configuration (v,h) of visible and hidden units is expressed in Eq. (1).

$$E(v, h) = -\sum_{i=1}^{n} a_i v_i - \sum_{j=1}^{m} b_j h_j - \sum_{i=1}^{n} \sum_{j=1}^{m} v_i w_{ij} h_j \qquad (1)$$

$$p(v, h) = \frac{1}{Z} e^{-E(v,h)} \qquad (2)$$

where v_i refers to the visible unit i of visible layer vector, h_j denotes the hidden unit j of the hidden layer vector, while w_{ij} signifies the weight connecting between visible unit v_i and hidden unit h_j. Next, a_i is the bias of the visible unit v_i and b_j refers to the bias of the hidden unit h_j. Meanwhile, n and m are the number of visible units and hidden units, respectively. Z refers to the partition function.

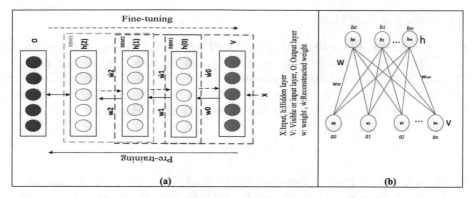

Fig. 1. The Basic Structure and Learning Process of DBN

Partition function Z, which sums over all possible pairs of visible and hidden vectors, is expressed in Eq. (3). The probability that the network assigns to a visible vector v is given in Eq. (4).

$$Z = \sum_{v,h} e^{-E(v,h)} \tag{3}$$

$$p(v) = \frac{1}{Z} \sum_{h} e^{-E(v,h)} \tag{4}$$

The updating of the weight parameters of RBM is calculated by using Eq. (5). The activation probability of the j hidden layer unit is given in Eq. (6), while the activation probability of the i visible layer unit is presented in Eq. (7).

$$\Delta w_{ij} = \epsilon \left(\langle v_i h_j \rangle_{\text{data}} - \langle v_i h_j \rangle_{\text{recon}} \right) \tag{5}$$

$$p\left(h_j = 1 \mid v\right) = \sigma \left(b_j + \sum_i v_i w_{ij} \right) \tag{6}$$

$$p(v_i = 1 \mid h) = \sigma \left(a_i + \sum_j h_j w_{ij} \right) \tag{7}$$

where ϵ refers to the learning rate, $\sigma(x)$ is the activation function (commonly used logistic sigmoid function in Eq. (9)), Δ refers to the change in the referred parameter, and the angle brackets are the expected changes of parameters v and h using contrastive divergence (CD). The training process of the CD algorithm needs only K steps (usually $K = 1$) Gibbs sampling to obtain good results.

The loss function for the DBN actual output y and target output $\bar{y}$ is estimated by using Eq. (8).

$$E = -y \times \log(\bar{y}) \tag{8}$$

3.2 Activation Functions

Sigmoid Activation Function. Traditional DBN uses the sigmoid activation function for hidden layers (see Eq. (9)). Sigmoid function is a nonlinear function that ranges at [0,1]. A general problem of the sigmoid function is that it saturates - small values snap to 0 and large values snap to 1 [24]. Once saturated, it becomes difficult for the learning algorithm to adapt the weights to enhance the model performance [25].

$$f(x_i) = \frac{1}{1 + e^{x_i}} \tag{9}$$

Rectified Linear Activation Function. Since 2011, more studies have concentrated on the use of Rectified Linear Activation (Relu) function for DL. It was commonly applied with Convolutional Neural Network (CNN) in recent studies. Only a handful of studies have assessed DBN, in comparison to studies that examined the traditional DBN with sigmoid functions. For instance, a study used DBN for power transformer fault diagnosis of gas analysis for insulating oil [25]. In light of relations discovery, we did not come across study which has applied the Relu-DBN despite its efficiency in addressing vanishing problem. Relu function given by Eq. (10) is nonlinear because the output is 0 when $x_i < 0$. It avoids and rectifies the vanishing gradient problem, while computationally more cost-effective than Sigmoid.

$$f(x_i) = \max(0, x_i) = \begin{Bmatrix} x_i, x_i > 0 \\ 0, x_i \leq 0 \end{Bmatrix} \tag{10}$$

3.3 Stochastic Gradient Descent Optimisation

Stochastic gradient descent (SGD) is an iterative method that minimises an objective function for optimisation. It is used for fine-tuning the DBN. The local gradient for an output layer δ_i is equal to the product of the corresponding error er and the derivative of the activation function $\overline{f}(x_i)$, while the local gradient for hidden layer δ_j denotes the changes required in weights. The derivative of the Relu activation function, the local gradient for the output layer, and the local gradient for the hidden layer are expressed in Eqs. (11), (12), and (13), respectively.

$$\overline{f}(x_i) = \begin{Bmatrix} 1, x_i > 0 \\ 0, x_i \leq 0 \end{Bmatrix} \tag{11}$$

$$\delta_i = er \times \overline{f}(x_i) \tag{12}$$

$$\delta_j = \overline{f}(x_j) \times \sum_{i}^{n} \delta_i \times w_{ij} \tag{13}$$

3.4 Dropout Strategy for DBN

Deep network is primarily used to analyse large amounts of data samples or with high characteristic dimensions data. For samples with low characteristic dimensions, an overfitting problem may occur in the network. A dropout strategy was developed by [26] to overcome this overfitting issue. It is applied to DBN as follows: for the training network, the weights of some hidden layer units are randomly inactivated with a certain probability p (see Eq. (14)). This equation is a modification of Eq. (6) using the dropout strategy. The equation presents how the output of the jth unit h_j of the hidden layer $h^{(l)}$ is calculated.

$$h_j^{(l)} = X_j \sigma \left(b_j + \sum_{i=1}^{m} w_{ij} v_i \right) = \begin{cases} \sigma \left(b_j + \sum_{i=1}^{m} w_{ij} v_i \right), & \text{if } X_j = 1 \\ 0 & , \text{other else } X_j = 0 \end{cases} \quad (14)$$

where $P(X_j = 0) = p, p$ is the dropout percentage of hidden units that are randomly inactivated, X_j indicates if the jth unit is active, l is the number of hidden layer, and j is the certain unit.

Inactive (discarded) units can be temporarily treated as not being part of the network structure. Its weight is kept, but not updated. In the next iteration, these inactive units could be re-used for training. For low characteristic dimensions samples or small sample training data, executing many iterations could lead to interdependencies among units. This dropout strategy hinders the interdependence between units.

3.5 Adjusting Re-DDBN

Numerous parameters affect the performance of Rc-DDBN. The DL models do not have a fixed structure; the number of hidden layers and their nodes are varied. The DL model has many varying versions (respective variations) based on the experiments and the assigned tasks. Hundreds of scenarios and experiments have been conducted to determine the Re-DDBN hyper-parameters and dropout percentage. Table 1 tabulates the Re-DDBN hyper-parameters.

Table 1. Re-DDBN Hyper-parameters

Hyper-parameter	Value/Type	Hyper-parameter	Value/Type
Hidden Layers	3	Batch size	Small dataset: 100–180 Large dataset: 180–270
Neurons/nodes	800,600,500	epochs	50–80
Hidden Layers Activation	Relu	Dropout	0.4
Output Layers Activation	Softmax	Optimisation algorithm	SGD
learning rate	0.2,0.3		

The sampling (reconstruction) for units in the hidden layer was computed by using the modified Eq. (14), instead of the original Eq. (6). Next, the activation function used for sampling visible and hidden layers was the Relu Eq. (10), instead of using the sigmoid function in Eq. (9). The pre-training phase of Re-DDBN refers to unsupervised training, while training for fine-tuning phase denotes supervised training.

4   Experiments and Evaluation

4.1   Dataset and Performance Evaluation Measurements

"SemEval-2018 task 7"[1] is a collection of two corpora, ACLRelAcS and ACL RD-TEC 2.0, within the computational linguistics domain. They are based on the ACL Anthology Reference Corpus (a digital archive of published papers in both journals and conferences for NLP and computational linguistics domain). This collection of dataset, which had six relation classes, was applied for relation training and classification; but considered imbalanced as it had huge variances in the sample size for each relation (e.g., one relation had 51 samples, while other relations had more than 400 samples). It appeared to be a noisy dataset that part of it was annotated manually and the rest was annotated automatically for varied definitions of concepts.

Accuracy (A) is the main performance metric used to assess ML and DL models. It is used to determine which model is best at identifying relationships and patterns (see Eq. (15)). The k-fold cross-validation is a procedure used to examine ML and DL models by dividing the dataset into k smaller sets called fold. Then each time respectively, one of these folds was used as the test dataset, while the rest was used as training dataset; this process was repeated k times for all folds. The k-fold cross-validation ensures that the models test the entire dataset [27]. The result of this k-fold cross-validation was calculated using the macro function. The macro A is expressed in Eq. (16).

$$A = \frac{TP + TN}{TP + TN + FP + FN} \tag{15}$$

$$Macro\ A = \frac{\sum_i^n A_i}{n} \tag{16}$$

where n refers to experiments number, TP is true-positive, TN is true-negative, FP is false-positive, and FN is false-negative. True-positive denotes the number of correct relations identified, whereas true-negative means the number of correct cases identified as no relation. False-positive indicates the number of incorrect relations identified, while false-negative is the number of incorrect cases left unidentified.

4.2   Experiments

To pre-process the dataset, all non-ASCII symbols and numbers were deleted. Next, the tokenizing step was performed to build the sentences list from all documents, while parsing the sentences was performed to get the POS and the phrases of sentences for concept extraction. Then, a binary bag of word (BOW) was built to represent the data. The sentence elements were binary represented (0 or 1 for each element). The size

---

[1] https://lipn.univ-paris13.fr/~gabor/semeval2018task7/.

of the BOW is the number of corpus elements (e.g., concepts, verbs and etc.). The SemEval-2018 task 7 was divided into two sizes; the small fraction had 800 samples and the large part contained 1800 samples. Figure 2 illustrates the different percentages of dropout applied for Re-DDBN and the average of the corresponding accuracy obtained in small and large sampling. The outcomes revealed that 40% dropout (=0.4) had the best results for both small and large sampling. To compare the developed Re-DDBN with other models, a comparative analysis was conducted 10 times with 3-fold cross-validation (= 30) for each model to obtain the results. The results were calculated using the macro function of k-fold cross-validation. Table 2 presents the comparison between the developed Re-DDBN and the other comparative models. These comparative models include the traditional DBN used in [22, 28], the Re-DBN used in [25], and the Sig-DDBN applied in [29]. The Re-DBN is a variation of DBN that used Relu activation function, whereas the Sig-DDBN is a variation of DBN that used dropout strategy and sigmoid activation function. Hundreds of scenarios and experiments were conducted to select the hyper-parameters for these variations of DBN. Many experiments were performed to determine the learning rate (0.05 to 0.5), the number of epochs (10 to 100), the size of mini-batch (10 to 100), and the dropout percentage if required (0.2 to 0.6). The developed Re-DDBN was also compared with the linear SVM using the same parameters applied in [18] and multinomial NB with the same parameters applied in [19].

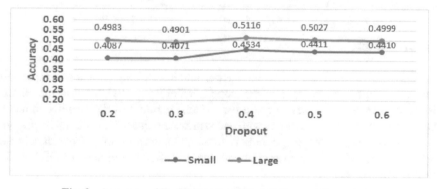

Fig. 2. Accuracy of Re-DDBN for Different Dropout Percentages

4.3   Results and Discussion

Table 2 shows that the developed Re-DDBN obtained the best accuracy result among the comparative models in small and large samples. The second highest accuracy was obtained by SVM for large samples and by NB for small samples. The sigmoid function showed weaker accuracy result than other models due to the vanishing problem, which let the small values snap to 0 and the large values snap to 1. As for the imbalanced dataset and low characteristic dimensions of samples, the traditional DBN with sigmoid function did not work well when compared with other models.

Table 2. Comparison of the Developed Re-DDBN, Traditional DBN, Re-DBN, Sig-DDBN, SVM, and NB for Small and Large Samples

DL/ML	Model	Hidden Layers Activation	With/without Dropout	Dataset size	Accuracy
DL	Re-DBN	Relu	without	Small	0.4310 ± 0. 0701
	Re-DDBN	**Relu**	**with**	**Small**	**0.4620 ± 0.0805**
	Traditional DBN	Sigmoid	without	Small	0.4093 ± 0.0690
	Sig-DDBN	Sigmoid	with	Small	0.4227 ± 0.0498
	Re-DBN	Relu	without	Large	0.4575 ± 0.1632
	Re-DDBN	**Relu**	**with**	**Large**	**0.5244 ± 0.1407**
	Traditional DBN	Sigmoid	without	Large	0.4070 ± 0.1305
	Sig-DDBN	Sigmoid	with	Large	0.4106 ± 0.1026
ML	SVM	Linear	--	Small	0.4428 ± 0.1379
	SVM	Linear	--	Large	0.5058 ± 0.0839
	NB	Multinomial	--	Small	0.4563 ± 0.0690
	NB	Multinomial	--	Large	0.4868 ± 0.0839

5   Conclusion

A Re-DDBN model that uses a dropout strategy and Relu activation function is proposed in this study. Upon comparison, the developed model Re-DDBN outperformed the traditional model DBN and other comparative models. In future work, we will conduct the comparison of using Relu functions with and without batch normalization with applying the dropout strategy or applying without it. In addition, the upcoming paper shall present the development of relations extraction and classification techniques for OL based on the Re-DDBN model.

Acknowledgments. This research was partially supported by Universiti Sains Malaysia (USM) under Grant Account No. 203.PKOMP.6777003.

References

1. Franco, W., et al.: Ontology-based question answering systems over knowledge bases: a survey. In: 22nd International Conference on Enterprise Information Systems (ICEIS), pp. 532–539. SCITEPRESS Digital Library (2020)
2. AL-Aswadi, F.N., Chan, H.Y., Gan, K.H.: Extracting semantic concepts and relations from scientific publications by using deep learning. In: Saeed, F., Mohammed, F., Al-Nahari, A. (eds.) IRICT 2020. LNDECT, vol. 72, pp. 374–383. Springer, Cham (2021). https://doi.org/10.1007/978-3-030-70713-2_35

3. Ahmed, I.A., Al-Aswadi, F.N., Noaman, K.M.G., Alma'aitah, W.Z.: Arabic Knowledge Graph Construction: A close look in the present and into the future. J. King Saud Univ.-Comput. Inf. Sci. **34**, 6505-6523 (2022)
4. Tiwari, S., Al-Aswadi, F.N., Gaurav, D.: Recent trends in knowledge graphs: theory and practice. Soft. Comput. **25**(13), 8337–8355 (2021). https://doi.org/10.1007/s00500-021-057 56-8
5. Wong, W., Liu, W., Bennamoun, M.: Ontology learning from text: a look back and into the future. ACM Comput. Surv. **44**(4), 1–36 (2012). https://doi.org/10.1145/2333112.2333115
6. Al-Aswadi, F.N., Chan, H.Y., Gan, K.H.: Automatic ontology construction from text: a review from shallow to deep learning trend. Artif. Intell. Rev. **53**(6), 3901–3928 (2019). https://doi.org/10.1007/s10462-019-09782-9
7. Zhao, Z., Han, S.-K., So, I.-M.: Architecture of knowledge graph construction techniques. Int. J. Pure Appl. Math. **118**(19), 1869–1883 (2018)
8. Alma'aitah, W.Z., Talib, A.Z., Osman, M.A.: Opportunities and challenges in enhancing access to metadata of cultural heritage collections: a survey. Artif. Intell. Rev. **53**(5), 3621–3646 (2020). https://doi.org/10.1007/s10462-019-09773-w
9. Saber, Y.M., Abdel-Galil, H. Belal, M.-F.: Arabic ontology extraction model from unstructured text. J. King Saud Univ.-Comput. Inf. Sci. **34**(8), 6066–6076 (2022). https://doi.org/10.1016/j.jksuci.2022.02.007
10. Sathiya, B., Geetha, T.V.: Automatic ontology learning from multiple knowledge sources of text. Int. J. Intell. Inf. Technol. **14**(2), 1–21 (2018)
11. Arefyev, N., et al.: Neuralgranny at semeval-2019 task 2: A combined approach for better modeling of semantic relationships in semantic frame induction. In: Proceedings of the 13th International Workshop on Semantic Evaluation, pp. 31–38 (2019)
12. Gillani Andleeb, S.: From text mining to knowledge mining: An integrated framework of concept extraction and categorization for domain ontology. Department of Information Systems, p. 146. Budapesti Corvinus Egyetem, Budapest (2015)
13. Sombatsrisomboon, R., Matsuo, Y., Ishizuka, M.: Acquisition of hypernyms and hyponyms from the WWW. In: Proceedings of the 2nd International Workshop on Active Mining (2003)
14. Specia, L., Motta, E.: A hybrid approach for relation extraction aimed at the semantic web. In: Larsen, H.L., Pasi, G., Ortiz-Arroyo, D., Andreasen, T., Christiansen, H. (eds.) Flexible Query Answering Systems, pp. 564–576. Springer Berlin Heidelberg, Berlin, Heidelberg (2006). https://doi.org/10.1007/11766254_48
15. Sánchez, D., Moreno, A.: Learning non-taxonomic relationships from web documents for domain ontology construction. Data Knowl. Eng. **64**(3), 600–623 (2008)
16. El-Kilany, A., Tazi, N.E., Ezzat, E.: Building Relation Extraction Templates via Unsupervised Learning. In: Proceedings of the 21st International Database Engineering & Applications Symposium, pp. 228–234. United Kingdom ACM, Bristol (2017)
17. Minard, A.-L., Ligozat, A.-L., Grau, B.: Multi-class SVM for relation extraction from clinical reports. In: Recent Advances in Natural Language Processing, Hissar, Bulgaria (2011)
18. Bergelid, L.: Classification of explicit music content using lyrics and music metadata. Trita-eecs-ex. Stockholm, Sweden: kth Royal Institute of Technology (2018)
19. Sureshkumar, G., Zayaraz, G.: Automatic relation extraction using naïve Bayes classifier for concept relational ontology development. Int. J. Comput. Aided Eng. Technol. **7**(4), 421–435 (2015)
20. Zhang, J., Liu, J., Wang, X.: Simultaneous entities and relationship extraction from unstructured text. Int. J. Database Theory Appl. **9**(6), 151–160 (2016)
21. Etzioni, O., Banko, M., Soderland, S., Weld, D.S.: Open information extraction from the web. Commun. ACM. **51**(12), 68–74 (2008)

22. Zhong, B., Liu, J., Du, Y., Liaozheng, Y., Pu, J.: Extracting attributes of named entity from unstructured text with deep belief network. Int. J. Database Theory Appl. **9**(5), 187–196 (2016)
23. Chen, Y., Li, W., Liu, Y., Zheng, D., Zhao, T.: Exploring deep belief network for chinese relation extraction. In: Proceedings of the Joint Conference on Chinese Language Processing (CLP'10), pp. 28–29 (2010)
24. Feng, J., Shengnan, Lu.: Performance analysis of various activation functions in artificial neural networks. J. Phys.: Conf. Ser. **1237**(2), 022030 (2019). https://doi.org/10.1088/1742-6596/1237/2/022030. IOP Publishing
25. Dai, J., Song, H., Sheng, G., Jiang, X.: Dissolved gas analysis of insulating oil for power transformer fault diagnosis with deep belief network. IEEE Trans. Dielectr. Electr. Insul. **24**(5), 2828–2835 (2017)
26. Srivastava, N., Hinton, G., Krizhevsky, A., Sutskever, I., Salakhutdinov, R.: Dropout: a simple way to prevent neural networks from overfitting. J. Mach. Learn. Res.earch **15**(1), 1929–1958 (2014)
27. Nematzadeh, Z., Ibrahim, R., Selamat, A.: Comparative studies on breast cancer classifications with k-fold cross validations using machine learning techniques. In: 2015 10th Asian Control Conference (ASCC), pp. 1–6 (2015)
28. Wang, H.: Semantic Deep Learning. University of Oregon (2015)
29. Huang, J., Guan, Y.: Dropout deep belief network based Chinese ancient ceramic non-destructive identification. Sensors **21**(4), 1318 (2021)

Financial Sentiment Analysis on Twitter During Covid-19 Pandemic in the UK

Oluwamayowa Ashimi$^{(\boxtimes)}$, Amna Dridi, and Edlira Vakaj

School of Computing and Digital Technology,
Birmingham City University, Birmingham B4 7XG, UK
oluwamayowa.ashimi@mail.bcu.ac.uk

Abstract. The surge in Covid-19 cases seen in 2020 has caused the UK government to enact regulations to stop the virus's spread. Along with other aspects like altered customer confidence and activity, the financial effects of these actions must be taken into account. This later can be studied from the user generated content posted on social networks such as Twitter. In this paper, we provide a supervised technique to analyze tweets exhibiting bullish and bearish sentiments, by predicting a sentiment class positive, negative, or neutral. Both machine learning & deep learning techniques are implemented to predict our financial sentiment class. Our research highlights how word embeddings, most importantly word2vec may be effectively used to conduct sentiment analysis in the financial sector providing favorable solutions. In addition, comprehensive research has been elicited between our technique and a lexicon-based approach. The outcomes of the study indicate how well Word2Vec model with deep learning techniques outperforms the others with an accuracy of 87%.

Keywords: Financial sentiment analysis · Covid-19 · Deep learning

1 Introduction

The term "Sentiment analysis" defines the contextual mining of different types of texts and helps to identify and extract the primary as well as subjective information that is present in the source material [1]. As per the observation of [2], sentiment analysis helps business organizations as well as a country to understand the economic effect with the help of online conversations by monitoring appropriately. The research paper discusses the "financial sentiment analysis on Twitter text through the pandemic of Covid-19 in the UK". Covid-19 became a global pandemic that not only affected the personal lives of humans but also affected their professional lives and financial conditions.

The pandemic affected the economy of the UK on a large scale. When the pandemic was at its single peak, the economy declined by approximately 13.5%. On the other hand, the economic value from medical burden increased during the second peak of the pandemic. Therefore, it is observed that the economic value of the UK was affected approximately by 29.2% during the pandemic of Covid-19 [3]. However, the global pandemic has affected Italy, France, Europe, Canada, and many other countries' economies on a large scale.

F. Saeed et al. (Eds.): ICACIn 2022, LNDECT 179, pp. 379–389, 2023.
https://doi.org/10.1007/978-3-031-36258-3_33

The main reason why we are carrying out this research work is because during the lockdown period, people have to stay inside for their safety. As a result, they felt bored as well as worried about the situation that had been risen due to the Covid19 pandemic [4]. As there is no other way to aware people of the influence of Covid-19 on their welfare along with the financial condition of humans, different types of social media platforms were chosen to aware people. In addition to that, people shared their concerns and daily activities during this lockdown period. According to the discussion of [5], Twitter played an important role to share people's thoughts and awareness with the help of tweets and re-tweets. This has provided the basis for our problem definition which goes thus?

How was customer confidence, conducts and financial activity affected amid the UK's Covid-19 pandemic?

How well does word2vec model perform when compared to other models when performing sentiment analysis with financial datasets?

2 Related Work

Due to the enormous growth of social media over the past few decades, it has begun to contribute immensely in gauging public sentiment. The technique for performing text analysis adopts "Natural Language Processing (NLP)" and the algorithms "Machine Learning (ML)" in allocating weighted sentiment points to different types of factors such as topics, themes, categories, and entities that are present in a phrase or a sentence. As proposed by [6], it helps data analysts to gauge public opinion and conduct different types of market research, monitor the collect data, and understand the opinion of the collected data. [6] analyses SA utilising the Lexicon-based method to examine sentiments in six distinct nations from March 15 to April 15, 2020, collected from Twitter. Further studies highlight deep learning techniques provides better result when compared to Lexical based approach for analysis.

[7] "use bag-of words or n-grams as features and apply generative or discriminative classifier models for predicting sentiments. Internet message postings were used by [8] to classify financial text as buy, hold or sell. [9] pointed out that investors are trading stocks publicly due to inattention. Records of prior earnings and large revenues have a big impact on trading.

The impacts of investment advisor on market gains were assessed by [10] which is a very exciting scientific paper to review their ideas more often. Preliminary studies on the financial effect of Covid-19 have been two-sided. This study evaluates the market's reactions in prior comparable pandemic events and aim to examine the implications of the outbreak on different industries. If comparable pandemic outbreaks recur, economic knowledge learnt during COVID 19 would be useful in shaping future risk-decisions and financial sector probes [11]. [12] made a significant addition to Twitter sentiment classification. They use polarity estimates across web pages as loud labels for training, and testing.

The process of getting the test data wasn't highlighted. They suggest fusing syntactic components of twitter posts like retweets, hashtag, punctuation points & characteristics such as past polarity of phrases & keyword's POS. According to the discussion of [5], the Twitter dataset requires to be divided into two parts. The first part includes both the positive and negative classes on the other hand; the second part consists of three classes known as neutral, positive, and negative. Reactions of people can be analyzed with the help of the tweets so that it becomes easy to predict whether people are happy with the decision of the Government or not. The tweet posts include both comforting and panicky words that are similar to the positive and negative sentiment regarding financial crisis. For example, fear can be considered as the predominant sentiment, whereas sadness can be related to the outbreak of the disease. In addition to that, anger can be compared with the prevalent sentiment and related to quarantine. Text-based methodologies were employed by [13] in concluding that COVID-19 wrecked the stock prices because of policy reaction to the emergence of the pandemic. The writers examined historical resources from 1900 to reach their results. In its examination of current news reports, the article highlights how important fluctuations in the financial-market to pandemic related events in prior times were not addressed as much like the Covid-19 outbreak. Markets were affected by the magnitude of the epidemic, quicker information dissemination, the interconnectivity of the global market, state policies, limits on travel, social distancing, resulting into halt in economic progress [14]. A prior study [15] found a connection involving social networking sentiment and stock price volatility, then presented forward a technique to gauge how investors would respond to corona virus-related news and developments using textblob. This study is limited to stock markets trends with Numerical data and not used to detect financial sentiments of users to earnings.

[16] conducted an imperative study in China before and after the lockdown was implemented using paired sample t-test by SPSS to measure the emotional sentiment of Weibo post users. Results are expected to be unbalanced as Weibo primary young people make up the majority of users, highlighting only their sentiment. [17] study used a sentiment analysis of Covid-19 twitter posts to demonstrate how social media is being impacted by popularity. To emphasize how tweets failed to inform users during the COVID-19 outbreak, ensemble methods were used. A limitation of this study is that the datasets utilized for this experiment consisted of tweets with at least 1000 retweets, which gives an imbalanced and objective assessment of each user's post.

As per the observation of [18], the geographic database is beneficial for providing information of different locations completely and accurately with the help of the names of the city, states, streets, or even postal codes. After that, the list of locations is utilized as a "Gazetteer list" in a pipeline of GATE. After utilizing the list of the geographic database, each word of the tweets is matched with the list. The name of the cities, states, or countries is attributed as a feature of the tweet. This paper conducted a sentiment analysis on covid 19 news on major outlets from four countries but limited it's study by not considering financial news which is a primary source of worry as earnings were limited after lockdown was imposed. [19] also analyzed the sentiments of Indians using R and word cloud, after the authorities instituted a lockdown. Not many posts were still deemed to be from that nation. The study emphasizes how Indians responded well to the government's method for enforcing the closure.

This paper proposes the use of SVM, Random Forest and deep learning techniques such as LSTM and FFNN classifiers to train and test our datasets with TF-IDF & word2vec to extract our features, proving to be very efficient in conducting sentiment analysis on financial data delivering outstanding results when compared to lexical based approaches.

3   Preprocessing

Data Collection. This section explains how we collect the financial data from twitter during Covid 19 outbreak. Getting data from twitter involve the following process:

1. Navigate to the twitter developer page and sign up for a new application, then enter your keys and token. We collect the secret key and API key for authentication, and mine our dataset
2. We import our various modules and use the search query with our financial keywords like #UKfinance or #ukdebtfreecommunity.
3. After getting the tweet from the API, we initialize the data frame, populate it and export it as a CSV file. Below is a table showing a detail of tweets download.

Table 1. Statistics of Financial Tweets

Datasets	Tweets	Positive	Negative	Neutral
Twitter	12525	6560	1493	4472

Data Pre-processing. Is essential to clean our data in order to obtain the accurate data. Noisy characters will be eliminated from our text in this step so that it can be further analyzed. A typical tweet includes a variety of words, emojis, user-mentions, hash-tags, and so on. The pre-processing step's major purpose is to standardize the text into a usable format for determining the user's thoughts.

The steps followed for converting text into meaningful data for classification are:

Cleansing Data. This is the process of noise reduction from our datasets by removal of unimportant words such as hashtag (#).

Tokenization. This divides huge textual material into smaller parts called tokens, the division of sentences into groups of words, and paragraphs into groups of phrases. We are using nltk word tokenizer for this study.

Stop-Word Removal. Another essential pre-processing procedures to be carried out on our datasets is the elimination of stop words, since it is used to weed out irrelevant characters.

Sentiment Detection. In order to detect sentiment, textual processing is used to extract feedback from reviews. Phrases from the text are separated, and the subjectivity of each statement is examined.

Data Labelling. In this research, the frames of our datasets are being classed into three various labels tagged as positive, negative, & neutral highlighting polarity and sentiment of our tweet data. Labeling was conducted manually through the corpus. Languages expressed in tweets are sometimes informal, hence it was necessary to manually label our datasets.

Confusion Matrix. It analyzes experimental results with reported true values in a table format. The confusion matrix is being used to assess how well my classification algorithm performed.

3.1 Financial Data Description

Bullish/Positive values are marked on tweets reflecting a positive pattern in mood, bearish/negative values are labeled on twitter posts that suggest a negative pattern and neutral when sentiment value is 0. For this paper, 12525 tweet data have been pre-processed; 6560 labelled positive, 1493 annotated negative, and 4472 was neutral.

3.2 Sentiment Score Granularity

In this paper, we adopted using SVM regression in calculating inferential sentiment score. It is crucial to identify the positive and negative linked phrases. Table 2 below highlights our word-score correlation measure detecting our positive and negative linked terms. Our words could be unigram, bigram, or three-gram. These statistic shows how the occurrence of words significantly influences the average deviation of a text's score from its mean. Our top seven terms that are most strongly and weakly connected across all tweets are displayed in our table.

Table 2. The top 5 positive and negative correlation words

Positive	Score	Negative	Score
Long	0.0252	Short	0.0420
Stock	0.0019	Sell	0.0303
Up	0.0018	Down	0.0242
Buy	0.0171	$Spy	0.0191
Call	0.0168	Downgrade	0.0132

4 Methodology

4.1 Feature Extraction

Word Embedding. Along with LSTM is our proposed approach for this paper because, when using word embeddings, in a preset vector space, each term is denoted as real value vectors [20, 21]. Every word is mapped to a single vector, while its weights gets processed in a neural network-like fashion, allowing it to measure sentiment with more precision. Though there are many techniques in word embedding, but for the purpose of this dissertation Word2Vec technique is employed. Word2Vec is a method to effectively creates word embeddings, utilizing a two-layer neural network [22]. Word2Vec operates by turning texts into a numerical form that deep neural networks can understand [23]. In this paper, word embedding will primarily be used to process our datasets with the LSTM classifier.

Term Frequency Inverse Document Frequency (TF-IDF). It can be defined as the calculation of how relevant a word in a series or corpus is to a text. The meaning increases proportionally to the number of times in the text a word appears but is compensated by the word frequency in the corpus [24]. It is defined by Eq. 1 as follows:

$$tf - idf(t, d) = tf(t, d) \times idf(t) \tag{1}$$

where: tf refers to the *Normalised Term Frequency* and it is defined by $tf(t,d) =$ count of t in d/number of words in d, where t refers to *term* and d refers to *document*, and idf refers to *Inverse Document Frequency*, it tests how relevant the word is, and it is defined by $idf(t) = \log(N/df(t))$, where N is the total number of documents and $df(t)$ is the document frequency of a term t. The approach suggested in this paper aims to accept comments on twitter (Tweets) as input then estimate the emotion score for every tweet that is cited in such text instance. Sentiment levels must be ranging from - 1 (highly negative) till 1 (extremely positive), with 0 being neutral. In this paper, we've chosen to offer the whole sentiment a real-valued score in order to provide exact and accurate sentiment evaluations in the financial language.

4.2   Classification Techniques

Long Short-Term Memory (LSTM). This is a recurrent neural network well suited in learning from sequential data. In this project, We used LSTM to classify a dataset of tweets. Unlike traditional RNNs, LSTM networks have a special kind of unit, called memory cells, which remembers information over a lengthy period [23, 25]. This makes LSTM networks very effective at learning from sequences of data, such as text. The tweets were labeled with either 0 (negative) or 1 (positive). Datasets were divided into training and test sets. Tweets were vectorized, using a technique called word embedding. This technique converts words into vectors, which are then input into the LSTM model. The LSTM model was able to learn the sequence of words in the tweets and classify them with high accuracy.

We chose to use an LSTM network for this work because of the sequential nature of the data. We wanted our model to be able to learn the dependencies between words in a tweet, as well as the overall context of the tweet (Fig. 1).

Feed Forward Neural Network (FFNN). The most basic type of neural networks is feedforward NN. The wave's front propagation is activated by a classifying algorithm. There aren't any response loops in feedforward NN, which prevents the model's results from being sent back [26]. In the FFNN, the information only moves forward from the source to the destination after passing through the hidden layers. As a result, no data

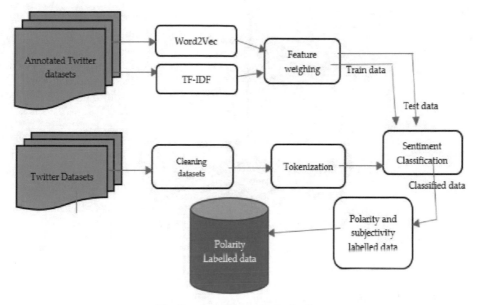

Fig. 1. sentiment analysis pipeline

will be sent through. Consequently, each text in the phrase needs its own input node in order to be modified. In this paper, we use FFNN with our word2vec model and achieved 87.3% accuracy.

Random Forest. Uses ensemble classification techniques. Using ensemble approaches, different iterations of the same algorithm are merged to create a more accurate prediction model [27]. The use of the random forest technique supports regression and classification tasks. We used Random Forest to classify our twitter dataset. We first created a TF-IDF vectorizer and bag of words model, which is a type of word embedding. This model converts words into vectors, which can be used as input to the Random Forest algorithm. We then trained the Random Forest algorithm on the vectors, and it was able to learn the relationship between the words and the labels. Finally, we tested the algorithm on a new set of tweets, and it was able to accurately predict the labels.

Support Vector Machine (SVM). The SVM is a supervised ML technique notably employed to carry out classification and regression tasks. This technique is trained on a dataset of labeled examples, where each example is a vector in feature space (Fig. 2).

In this project, we used it for classification, specifically to label our datasets from twitter as positive or negative. To train SVM, we first converted each tweet into a vector of features. We did this by using a TF-IDF vectorizer. We then used these feature vectors, along with the labels, to test and train SVM.

The SVM algorithm locates the decision boundary even when the data is not linearly detachable as its algorithm performs data mapping onto high dimensional features where it becomes linearly separable, using a kernel function [28]. After training, we tested SVM on a held-out set of tweets. It achieved good results.

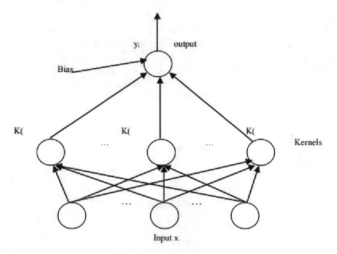

Fig. 2. svm structure

5 Evaluation and Results

Using datasets in Table 1, we assessed our suggested three sentiment analysis approaches. We conducted an experimental study on the dataset using the original content of the text as stated in Sect. 4. For each, we've explored the various feature representations mentioned in Sect. 4. We have compared the supervised approaches using LSTM, FFNN, Random Forest, and SVM as our classifiers, with Word2Vec and TF-IDF as our NLP techniques. Implementation was done, and below are the results table (Figs. 3 and 4 and Table 3).

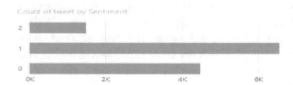

Fig. 3. Data visualization before preprocessing

The obtained findings show that deep learning techniques with Word2Vec model which has an accuracy score of 87.3% and 87% respectively is outperforming the Random Forest and SVM with TF-IDF model with accuracy score of 78% and 69% respectively, since the sentiment precision is more precise when handling lists of strings which represent sentiments in tweet posts. This is a significant advance from text-level classification to sentence-level classification (TF-IDF). LSTM is a powerful tool for training and testing datasets. It can be used in improving the performance of our model when dealing with large data as it possesses the capability of remembering and forgetting data in a seamless conduct.

From the results above, LSTM with an accuracy score of 87%, better precision and f1-score outperforms SVM with an accuracy score of 69%, lower precision and frequency

tweet	Sentiment	CleanTweet
$DG #Dollar General #Options #maxpain Chart, O...	1	dg, general, chaopen, interest, chaupdate
"Education is not only a ladder of opportunity...	0	", education, is, not, only, a, ladder, of, op...
$GCLT news coming soon #fintech #finance https...	0	gclt, news, coming, soon
That amazing feeling when you finally close a ...	1	that, amazing, feeling, when, you, finally, cl...
From OHM to SHIB: 5 Most Impressive Altcoins o...	1	from, ohm, to, shib, :, most, impressive, altc...

Fig. 4. Data visualization after preprocessing

Table 3. Result Evaluation and Accuracy Score

Word2Vec	precision	f1-score	Accuracy
FFNN	0.85	0.86	**87.3%**
LSTM	0.86	0.86	**87%**
TF-IDF	precision	f1-score	Accuracy
SVM	0.77	0.65	**69%**
Random forest	0.78	0.77	**78%**

score when training datasets with a large number of features. This is because LSTM is less likely to overfit the training data, more robust to noise, can learn from data with missing values, able to learn long-term dependencies and can handle long sequences of data, while SVM struggles with long sequences which is essential for datasets with a lot of information.

Using the TF-IDF model, SVM produced a precision of 0.77, frequency score of 0.65 and an accuracy of 69% while Random Forest gave a precision of 0.78, frequency score of 0.77 with an accuracy of 78% after training and testing.

As a result of our findings, FFNN had an accuracy of 87.4% but took 116 ms/step and 232 ms/epoch while LSTM with 87% accuracy took 35 ms/step and 2 s/epoch. We chose LSTM as the best model as when all parameters are compared. Furthermore, the results show that using relatively basic text-extraction algorithm to compute important regions of our posts, possibly made significant gains above message-based sentiment categorization. This highlights and concludes that Word2Vec model performs better than TF-IDF. Datasets and codes are available online at GitHub.[1]

---

[1] https://github.com/lordmayor231/UPDATED-Financial-sentiment-analysis-during-Covid19-in-the-UK.git.

6 Conclusions and Future Work

This study accurately tackled our challenge on financial sentiment analysis by proposing a methodology that generates the real-value sentiment score for every financial tweet referenced throughout the text of Twitter in our dataset, utilizing sentiment Analysis. This work will help the public in framing the ideas in a significant manner which provide an in-depth knowledge about analysis of financial data during Covid 19 in the UK. The application of the aforementioned models has played a major role in order to understand the impact of an investor's role in bringing change over the user interface of Twitter. Using the technique of thematic analysis helps in comparing the authentic information from various types of verified journal articles and articles from websites to develop a proper report.

We intend to develop a method for improving sentiment identification accuracy in the future by using not just text data but also visual, videography, and sound classification. We also intend to solve the challenges of sarcasm, in which sarcastic phrases use good terms to indicate a negative judgement about a target. For instance, "Nice fragrance, it's essential that you marinade in it." Despite the fact that the statement comprises solely good words, it conveys a negative message.

References

1. Sudhir, P., Suresh, V.D.: Comparative study of various approaches, applications and classifiers for sentiment analysis. Glob. Transitions Proc. **2**(2), 205–211 (2021)
2. Costola, M., Iacopini, M., Santagiustina, C.R.M.A.: Google search volumes and the financial markets during the COVID-19 outbreak. Finan. Res. Lett. **42**, 101884 (2021). https://doi.org/10.1016/j.frl.2020.101884
3. Keogh-Brown, M.R., Jensen, H.T., Edmunds, W., Smith, R.D.: The impact of covid-19, associated behaviours and policies on the UK economy: a computable general equilibrium model. SSM – Popul. Health **12**, 100651 (2020)
4. Presti, G., Mchugh, L., Gloster, A., Karekla, M., Hayes, S.: The dynamics of fear at the time of covid-19: a contextual behavioral science perspective. Clin. Neuropsychiatry **17**, 65–71 (2020)
5. Basiri, M.E., Nemati, S., Abdar, M., Asadi, S., Acharrya, U.R.: A novel fusion-based deep learning model for sentiment analysis of covid-19 tweets. Knowl.-Based Syst. **228**, 107242 (2021). https://doi.org/10.1016/j.knosys.2021.107242
6. Hota, H.S., Sharma, D.K., Verma, N.: Lexicon-based sentiment analysis using twitter data: a case of covid-19 outbreak in India and abroad. In: Data Science for COVID-19, pp. 275–295. Elsevier (2021)
7. Huang, A.H., Zang, A.Y., Zheng, R.: Evidence on the information content of text in analyst reports. Accounting Review **89**(6), 2151–2180 (2014)
8. Antweiler, W., Frank, M.Z.: Is all that talk just noise? The information content of internet stock message boards. J. Finance **59**(3), 1259–1294 (2004). https://doi.org/10.1111/j.1540-6261.2004.00662.x
9. Barber, B.M., Odean, T., Zheng, L.: Out of sight, out of mind: The effects of expenses on mutual fund flows. J. Bus. **78**(6), 2095–2120 (2005)
10. Jiang, F., Lee, J., Martin, X., Zhou, G.: Manager sentiment and stock returns. J. Financ. Econ. **132**(1), 126–149 (2019)

11. Kwon, K.T., Ko, J.-H., Shin, H., Sung, M., Kim, J.Y.: Drive-through screening center for COVID-19: a safe and efficient screening system against massive community outbreak. J. Korean Med. Sci. **35**(11), e123 (2020). https://doi.org/10.3346/jkms.2020.35.e123

12. Barbosa, L., Feng, J.: Robust sentiment detection on twitter from biased and noisy data. In: Coling 2010: Posters, pp. 36–44 (2010)

13. Baker, S.R., Bloom, N., Davis, S.J., Kost, K., Sammon, M., Viratyosin, T.: The unprecedented stock market reaction to covid-19. Rev. Asset Pricing Stud. **10**(4), 742–758 (2020). https://doi.org/10.1093/rapstu/raaa008

14. Eachempati, P., Srivastava, P.R., Panigrahi, P.K.: Sentiment analysis of covid-19 pandemic on the stock market. Am. Bus. Rev. **24**(1), 141–165 (2021). https://doi.org/10.37625/abr.24.1.141-165

15. Biswas, S., Sarkar, I., Das, P., Bose, R., Roy, S.: Examining the effects of pandemics on stock market trends through sentiment analysis. Xi'an Dianzi Keji Daxue Xuebao/Journal of Xidian University **14**, 1–14 (2020)

16. Li, S., Wang, Y., Xue, J., Zhao, N., Zhu, T.: The impact of COVID-19 epidemic declaration on psychological consequences: a study on active weibo users. Int. J. Environ. Res. Public Health **17**(6), 2032 (2020). https://doi.org/10.3390/ijerph17062032

17. Chakraborty, K., et al.: Sentiment analysis of covid-19 tweets by deep learning classifiers—a study to show how popularity is affecting accuracy in social media. Appl. Soft Comput. **97**, 106754 (2020)

18. Ghasiya, P., Okamura, K.: Investigating covid-19 news across four nations: a topic modeling and sentiment analysis approach. IEEE Access **9**, 36645–36656 (2021)

19. Barkur, G., Vibha, Kamath, G.B.: Sentiment analysis of nationwide lockdown due to covid 19 outbreak: evidence from India. Asian J. Psychiatry **51**, 102089 (2020). https://doi.org/10.1016/j.ajp.2020.102089

20. Mikolov, T., Chen, K., Corrado, G., Dean, J.: Efficient estimation of word representations in vector space. In: Proceedings of Workshop at ICLR, 2013 (2013)

21. Sithole, V.: Fine-tuning semantic information for optimized classification of the internet of things patterns using neural word embeddings. J. Adv. Comput. Netw. **8**(1), 26–30 (2020)

22. Mikolov, T., Chen, K., Corrado, G., Dean, J.:. Efficient estimation of word representations in vector space. arXiv preprint arXiv:1301.3781 (2013)

23. Muhammad, P.F., Kusumaningrum, R., Wibowo, A.: Sentiment analysis using Word2vec and long short-term memory (LSTM) for Indonesian hotel reviews. Procedia Comput. Sci. **179**, 728–735 (2021). https://doi.org/10.1016/j.procs.2021.01.061

24. Jing, L.-P., Huang, H.-K., Shi, H.-B.: Improved feature selection approach tfidf in text mining. In: Proceedings of the International Conference on Machine Learning and Cybernetics, vol. 2, pp. 944–946 (2002)

25. Morin, F., Bengio, Y.: Hierarchical probabilistic neural network language model. In: Robert, G.C., Ghahramani, Z. (eds.) Proceedings of the Tenth International Workshop on Artificial Intelligence and Statistics, volume R5 of Proceedings of Machine Learning Research, pp. 246–252. PMLR, 06–08 Jan 2005. Reissued by PMLR on 30 March 2021

26. Etaiwi, W., Suleiman, D., Awajan, A.: Deep learning based techniques for sentiment analysis: a survey. Informatica **45**(7) (2021)

27. Breiman, L.: Random forests. Mach. Learn. **45**, 5–32 (2001)

28. Agrawal, A.K., Chakraborty, G.: On the use of acquisition function-based Bayesian optimization method to efficiently tune SVM hyperparameters for structural damage detection. Struct. Control. Health Monit. **28**(4), e2693 (2021)

Automatic Classification for ADHD Disorder Using Deep Learning Techniques

Nouf Alharbi[✉], Reham Al-Johani, Maram Al-Ahmadi, Nuha Al-Refaai, Atheer Al-Sharif, and Yara Al-Aqeel

College of Computer Science and Engineering, Taibah University, Al Madinah Al Munawwara, Saudi Arabia
{nmoharbi,reham9090,marram369,Nuha-saleh,tu3755963, tu3950311}@taibahu.edu.sa

Abstract. The effective classification of attention deficit hyperactivity disorder (ADHD) using imaging and functional brain biomarkers would have a significant impact on public health. However, this classification remains a challenge because the underlying mechanism behind ADHD is not completely understood and ADHD's diagnosis is mostly based on some questionnaires and behaviour analysis. Recently, with the advancement of computing resources and deep learning methods, several studies have been conducted to classify ADHD. In these studies, resting-state functional magnetic resonance imaging has emerged as a promising common neuroimaging modality for the diagnosis of brain disorders. We narrow down our research to focus only on ADHD classification through functional connectivity analysis. In this paper, we propose an approach that produces an automatic classification between ADHD and typically developing control individuals based on the functional connectivity of brain regions using deep learning techniques. Our proposed approach is able to achieve a classification accuracy of 87.5%.

Keywords: rs-fMRI · Deep learning · ADHD · Functional connectivity · Convolutional neural networks · Biomarkers

1 Introduction

Attention deficit hyperactivity disorder (ADHD) is considered one of the most common behavioural disorders that happen in childhood, according to the Centers for Disease Control and Prevention [1]. ADHD is a neurodevelopmental disorder that begins in childhood as a result of a deficiency or defect in some neurotransmitters in the frontal cortex (frontal lobe) that allow cells to carry out their functions and link parts of the brain to one another, leading to hyperactivity, impulsivity, and distraction. ADHD continues to maturity and its symptoms vary between individuals. Doctors and scientists have been unable to determine the cause of this disorder. This inability makes ADHD difficult to diagnose. ADHD can be misdiagnosed since many of its symptoms overlap with those of other disorders, such as depression, bipolar disorder, and anxiety [2]. Misdiagnosing

© The Author(s), under exclusive license to Springer Nature Switzerland AG 2023
F. Saeed et al. (Eds.): ICACIn 2022, LNDECT 179, pp. 390–398, 2023.
https://doi.org/10.1007/978-3-031-36258-3_34

children with ADHD can have grave consequences on their lives and may mask other serious psychological problems, leaving the root of the problem unaddressed. Misdiagnosing children may also lead to their exposure to wrong medications, which can further complicate children's ability to function. ADHD's prevalence across lifetime is estimated at 8% in the Kingdom of Saudi Arabia, according to the Saudi National Mental Health Survey for the year 2019 [3]. In the healthcare sector, computer-aided diagnosis and detection is an area that is progressively growing; and when it comes to performance, unlike medical staff, computers will not be affected by overwork, distraction, or emotions. Computers can also analyse large amounts of data in a quick and accurate manner. This is one reason why computer diagnosis is important and also why more knowledge about it is needed. With the strengthening of automation, the advancement of technology in the current era, and in accordance with Vision 2030's goal of establishing a social health system and vibrant life for people with ADHD, it has become critical to raise community awareness and knowledge about this disorder as well as find appropriate ways to deal with it and identify individuals who suffer from it. Therefore, we used a deep learning technique, inspired by the structure of the human brain, to build a model that can detect ADHD in individuals who have it and distinguish them from those who do not have it. Our model is based on the functional connectivity of brain regions and uses the resting-state functional magnetic resonance imaging (rs-fMRI) dataset with some clinical information regarding subjects [11]. rs-fMRI is a neuroimaging technique for determining activity in healthy and diseased human brains by measuring changes in blood oxygenation level-dependence (BOLD) [4]. BOLD is considered the most suitable method for determining the functional activity of brain regions [5] because it is non-invasive and displays a high level of spatial resolution.

2 Problem Definition

There is no doubt that ADHD is a real problem that will not disappear. The issue is that misdiagnosing ADHD seems set to continue. The impact of misdiagnosing ADHD can have serious consequences for individuals in all aspects of their lives. Furthermore, misdiagnosis wastes a considerable amount of money and may pose a potential risk for high future disease and mental illness rates. Physicians mostly use the cookbook approach to diagnostics [6]: when they have a list of symptoms for a particular diagnosis, they just check the list of what they suspect and ask patients whether they have symptoms or not. The problem with using this approach only is that no matter what diagnoses they start with, if physicians check enough and get a diagnosis, the diagnosis ends there. Physicians may also use Conners assessment, which is a questionnaire that asks questions about things such as behaviour, work, or academic and social life; or the Vanderbilt ADHD Diagnostic Rating Scale (VADRS) assessment, which is a psychological assessment tool for ADHD symptoms. Both assessments are used to rate how frequently specific behaviours happen. For example, in bipolar disorder, the wording of symptoms can be misleading, as we have mentioned before. Bipolar mood changes do not happen over a period of minutes or hours to truly match the criteria for either a hypomanic or depressive episode – they have to last for several days to weeks and not be caused by anything in the environment. While an individual with ADHD may experience a mood swing due to an event in their life, the mood change will be short. Therefore, if physicians do not evaluate ADHD carefully and if they perform an assessment fast, they will probably miss it altogether.

3 Contributions

- Created an efficient and accurate diagnostic tool and increased the possibility of diagnosing many individuals in a short period of time.
- Proved the effectiveness of functional connectivity in diagnosis.
- Increased awareness of ADHD and the suffering of its patients.

4 Dataset

We used the ADHD-200 publicly accessible dataset that is available on the Nilearn library – a useful python module for visualizing neuroimaging data. ADHD-200 is composed of rs-fMRI and anatomical data collected from several research centres. Nilearn has data for only 40 subjects [7].

5 Methodology

Through our research and by evaluating the advantages and disadvantages of each diagnostic method, we have found that functional connectivity is a new measurement of the brain. Instead of looking at how neurons in the brain are structurally connected, we measure how different parts of the brain are functionally connected (a statistical relationship) because resting-state data contain valuable and essential information that is usually diagnostically relevant and is easy to obtain [8]. Our work proposes an approach for the detection and diagnosis of ADHD. It consists of three stages: image preprocessing, functional connection to a network, and neural networking. The first stage applies cleaning techniques. The second stage involves identifying the functional connectivity between brain regions. The goal of the third stage is to extract characteristics or features from an individual brain region's preprocessed time-series signals and to train a model.

6 Flowchart

In Fig. 1, we see a flowchart that we have used to present and illustrate the individual stages of an ADHD detection process framework in sequential order. Our flowchart begins with taking rs-fMRI data from a dataset as input and ends with a classification or no classification of ADHD as output. The first process is preprocessing, which involves a series of steps to clean and standardize data prior to statistical analysis. The second process determines the similarity between brain regain functions, and the last process is for classification.

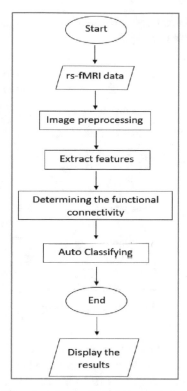

Fig. 1. Flowchart for an ADHD diagnosis framework.

6.1 Preprocessing Stage

The dataset from the Nilearn library had already been preprocessed and was ready for use. The software used to prepare the preprocessing pipeline are docker and fMRIPrep. The primary steps in rs-fMRI preprocessing are slice timing correction, motion correction or realignment, and co-registration of structural and functional elements.

6.2 Functional Connectivity (FC) Network Stage

The brain network is composed of multiple interconnected brain regions. When overseeing and performing the different functions of the body, the brain regions are in a continuous state of efficient coordination with each other, causing a complicated brain connectivity pattern. Recent research shows that brain connectivity is a promising diagnostic source of the illnesses that are associated with unusual functional organization of the brain [9]. The FC stage measures similarities or dependencies between brain regions based on BOLD signals by taking a sequence of images acquired over time and studying how things change. This is considered a noninvasive technique for studying brain activity because there are no known negative effects of taking frequent fMRI scans. The human brain is considered a great and sophisticated network that controls and monitors the body's functions. The input to this stage is preprocessed re-fMRI images, and the output is the functional connectivity matrix of a brain's regions. The steps for generating the FC network are illustrated in Fig. 2.

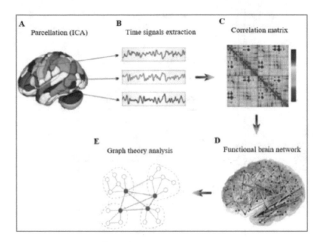

Fig. 2. FC analysis steps

Other methods, such as summation-based synergetic [10] and hybrid 2-dimensional convolutional neural network–long short-term memory [11] have been used to diagnose ADHD. However, FC, in contrast to these methods, reveals more than just a diagnosis – its shows the interaction of regions with each other; and this allows scientists to make conclusions about functional interactions and their causes.

After loading the entire dataset, the FC coefficients were extracted and analysed (the values representing how strong functional connections between brain regions are) [12]; but first, we needed to conduct an independent component analysis (ICA) on the raw rs-fMRI data. The rs-fMRI images are inherently four-dimensional since they provide a sequence of whole-brain volumes over time, and our approach does not directly rely on neuroimaging scans because we are only interested in resting-state networks and vectorising FC measures. To remove the irrelevant data, we used the ICA method which separates a signal with many sources into BOLD subcomponents (to determine regions of interest). Also, we chose a 20-component decomposition based on the criteria provided in the Nilearn documentation for the selected dataset. We extracted FC correlation coefficients between each region of the brain. Also, we averaged each correlation matrix across the control and ADHD subjects.

One of the connectivity matrices that we extracted from subject 1 are shown above in Fig. 3. There are 20 rows and 20 columns for each of the 20 ROIs. Each square's colour represents the value of the FC correlation coefficient.

To better visualize the connections and differences, we used a functional connectome – a map of FC represented as edges and nodes of all the connections for 20 regions. This is shown in Fig. 4.

There were some differences that we observed in Fig. 4 between the control and ADHD groups. First, the ADHD connections do not appear as dense as the control

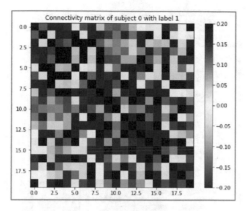

Fig. 3. FC matrix

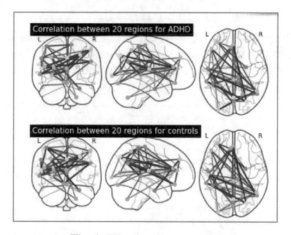

Fig. 4. Functional connectome

connections. This might be related to the idea of decreased FC associated with ADHD [13]. Second, there are few connections in the superior parietal cortex in the ADHD group which we think are implicated in inattention [14]. Although these are small differences, they suggest that finding these functional biomarkers is key because the currently available methods of diagnosis are entirely dependent on symptoms, which makes it impossible to differentiate between healthy controls and people with ADHD.

6.3 The Neural Network Stage

After obtaining correlation matrices and coefficients that provide a vectorised measure of FC, we can use them as input data for our neural network to extract features and train a model. Before building our model, we split the data into a training set (80%) and a testing set (20%). The model was built using a sequential approach, which has 3 dense layers, as illustrated in Fig. 5.

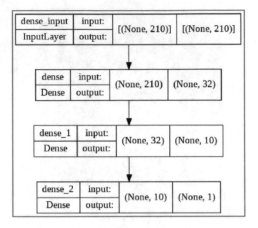

Fig. 5. NN architecture

To measure the performance of the neural network, we made an AUC curve as seen in Fig. 6 which shows a prediction of 90%. This means that our model performs well, but there is a slight overfitting due to the size of the dataset.

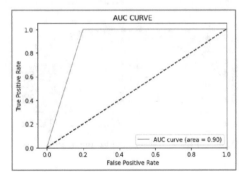

Fig. 6. AUC curve

7   Results and Discussion

We performed and presented the performance comparison of our proposed model using several metrics after testing it. We were able to accomplish the majority of the desired goals. Table 1 shows the performance metrics that we used to test and validate the model.

Also, we have discussed the performance comparison of our proposed model after testing it. We were able to achieve the majority of the desired goals. We have made comparisons between a neural network (NN) and long short-term memory (LSTM). The dataset was trained using these two models. The training results for each model associated with loss function are presented in Table 2 below.

Table 1. Classification results

Our model	Accuracy	Precision	Recall	F1-score
	87.5%	1	0.80	0.89

Table 2. Comparison of results between models

	Accuracy	Loss function
NN	87.5%	0.0124
LSTM	75%	0.6055

Based on previous results, we found that the NN model is more efficient than the LSTM model. LSTM requires many parameters which makes the training process slow, and NN is far faster than LSTM. Therefore, we concluded that speed, early predictive power, and robustness of NNs should pave the way for their use in the process prediction of many disorders. A previous study used an end-to-end model to diagnose ADHD based on FC, and the model achieved an accuracy of 73.1% [15]; we outperformed this model using a pipeline model with an NN architecture that achieved an accuracy of 87.5%. We noticed, as shown in Fig. 3, that there are few connections in the superior parietal cortex which we thought to be implicated in attention.

8 Conclusion

Combating ADHD misdiagnosis should be made a global health priority. Several long-term consequences would be avoided if ADHD misdiagnoses were not so common worldwide. In this paper, we have proposed and established the different stages of our model. The proposed model takes preprocessed time-series signals of rs-fMRI as input for healthy subjects and subjects with ADHD and learns to predict the classification label across the FC matrix. As far as we know, this is one of the first projects in Saudi Arabia that addresses the issue of misdiagnosing ADHD. We have demonstrated the effectiveness of FC in diagnosis and have been able to classify ADHD with an accuracy rate of 87.5% for testing and 100% for training.

References

1. National Center on Birth Defects and Developmental Disabilities. Attention-Deficit/Hyperactivity Disorder (ADHD). https://www.cdc.gov/ncbddd/adhd/index.html (2021)
2. Milberger, S., Biederman, J., Faraone, S.V., Murphy, J., Tsuang, M.T.: Attention deficit hyperactivity disorder and comorbid disorder: issues of overlapping symptoms. Am. J. Psychiatry **152**, 1793–1799 (1995)
3. Murfet, D.: Saudi national mental health survey. http://www.healthandstress.org.sa/Results/Saudi%20National%20Mental%20Health%20Survey%20-%20Technical%20Report.pdf

4. Li, K., Guo, L., Nie, J., Li, G., Liu, T.: Review of methods for functional brain connectivity detection using fMRI. Comput. Med. Imaging Graph. **33**(2), 131–139 (2009)
5. Huettel, S.A., Song, A.W., McCarthy, G.: Functional magnetic resonance imaging. Chapter Four and Five. Sinauer Associates, Inc. Publishers Sunderland, Massachusetts, USA, p. 75 (2004)
6. Campesato, O.: Artificial Intelligence, Machine Learning, and Deep Learning. Stylus Publishing LLC (2020)
7. Zhang, T., et al.: Separated channel attention convolutional neural network (SC-CNN-attention) to identify ADHD in multi-site rs-fMRI dataset. Entropy **22**(8), 893 (2020)
8. Andrews-Hanna, J.R., et al.: Disruption of large-scale brain systems in advanced aging. Neuron **56**(5), 924–935 (2007). https://doi.org/10.1016/j.neuron.2007.10.038
9. Varoquaux, G., Baronnet, F., Kleinschmidt, A., Fillard, P., Thirion, B.: Detection of brain functional-connectivity difference in post-stroke patients using group-level covariance modeling. In: Jiang, T., Navab, N., Pluim, J.P.W., Viergever, M.A. (eds.) Medical Image Computing and Computer-Assisted Intervention – MICCAI 2010: 13th International Conference, Beijing, China, September 20-24, 2010, Proceedings, Part I, pp. 200–208. Springer Berlin Heidelberg, Berlin, Heidelberg (2010). https://doi.org/10.1007/978-3-642-15705-9_25
10. Peng, J., Debnath, M., Biswas, A.K.: Efficacy of novel summation-based synergetic artificial neural network in ADHD diagnosis. Mach. Learn. Appl. **6**, 100120 (2021)
11. Khullar, V., Salgotra, K., Singh, H.P., Sharma, D.P.: Deep learning-based binary classification of ADHD using resting state MR images. Augmented Hum. Res. **6**(1), 1–9 (2021). https://doi.org/10.1007/s41133-020-00042-y
12. Chen, H., Song, Y., Li, X.: Use of deep learning to detect personalized spatial-frequency abnormalities in EEGs of children with ADHD. J. Neural Eng. **16**(6), 066046 (2019)

High Accuracy Feature Selection Using Metaheuristic Algorithm for Classification of Student Academic Performance Prediction

Al Farissi[1]([✉]) [iD], Halina Mohamed Dahlan[2], Zuraini Ali Shah[3], and Samsuryadi[1]

[1] Fakultas Ilmu Komputer, Universitas Sriwijaya, Palembang, Indonesia
{alfarissi,samsuryadi}@unsri.ac.id
[2] Information Systems Department, Azman Hashim International Business School,
Universiti Teknologi Malaysia, Skudai, Johor, Malaysia
halina@utm.my
[3] Faculty of Computing, Universiti Teknologi Malaysia, Skudai, Johor, Malaysia
zuraini@utm.my

Abstract. This paper was aimed at producing optimal features and improve classification accuracy to predict student academic performance using feature selection based on metaheuristic algorithms and ensemble classifiers. The method improves the feature selection technique in the data pre-processing process because it can find solutions in the total search space and use global search capabilities, significantly improving the ability to find high-quality solutions in a reasonable time and such that it can obtain the optimal classifier features to improve prediction performance. The proposed method's performance was evaluated by using standard procedures: Accuracy, Precision, Recall, and F-Measure. In addition, ten fold cross-validation was used to evaluate the proposed model with three public data sets with multiclass and binary class labels. The experimental work found that the superiority of the proposed method over the traditional feature selection technique was confirmed by a 10% increase in the accuracy on three benchmark public data sets. The promising part of the method is the improvement shown in the GAFSRF accuracy of 94.9%.

Keywords: Student Academic Performance Prediction · Genetic Algorithm · Feature Selection · Ensemble Classifier

1 Introduction

Educational Data Mining (EDM) is a methodology that aims to extract beneficial information and behavior patterns from enormous educational databases to assist students and lecturers achieve adequate performance [1]. EDM's function is to develop, research, and apply a data mining model by extracting large amounts of data to attain knowledge [2]. However, technological development in the education system is increasing due to more influential data in the education database. Therefore, prediction and analysis of student academic achievement have an essential role in student academic development. Student

F. Saeed et al. (Eds.): ICACIn 2022, LNDECT 179, pp. 399–409, 2023.
https://doi.org/10.1007/978-3-031-36258-3_35

academic performance (SAP) prediction is the most widely popular EDM application [3]. Student academic performance prediction aims to estimate the unknown value of a variable that describes the student's performance. Consequently, it could ensure strategic programs to enlighten or guide the students for a better performance that may lead them to a better future. Various EDM methods have been used in predicting student academic performance, such as classification, clustering, grouping, regression, and estimation of density to predict student activity and link student interactivity.

Classification is the popular EDM method for predicting student academic performance. Many studies have been conducted to predict student academic performance using classification techniques. However, there are still problems in research regarding student academic performance prediction related to accuracy. However, the dataset's irrelevant features in student academic performance prediction affect the learning model's performance with unsatisfactory accuracy. It has been empirically proven that applying feature selection techniques to select relevant features has improved accuracy. Moreover, an ensemble classification algorithm can achieve high accuracy [4].

Many researchers use features that are more focused on quantitative data because they have real value. In general, there are two approaches to feature selection techniques, namely: the Wrapper Based approach and the Filter Based approach. However, these two approaches are in the whole process of finding a solution by searching locally. Therefore, it makes this challenging to approximate the optimal solution. Nevertheless, the metaheuristic optimization method as a feature selection technique in the data pre-processing process may find a solution in the whole search space and use a global search ability, significantly increasing the ability to find high-quality solutions within a reasonable time [5]. Furthermore, it was shown in a study conducted by [6] that the accuracy results obtained were better than the non-evolutionary feature selection technique and able to enhance the performance of data mining algorithms and proven to improve accuracy.

Additionally, selecting a machine learning algorithm that is appropriate for a certain domain will influence the outcomes of predictive performance. Each learning algorithm has some bias. It performs well in some domains, but sometimes it performs poorly in other domains. Several studies indicate that the application of base classification algorithms in student academic performance prediction is insufficient to guarantee it works appropriately compared to the ensemble method [7]. In addition to improving the performance of the student performance prediction approach for better performance compared to base classification, the ensemble method is used to give accurate feature evaluations that may affect student performance prediction.

Therefore, this study combines metaheuristic methods for feature selection and ensemble classifiers to overcome the prediction accuracy problem. This study proposes a metaheuristic-based feature selection and an ensemble classifier to improve student academic performance prediction accuracy.

The rest of this paper is organized into several parts: Sect. 2 presents related work in the education data mining algorithms field. Then, experimental designs are presented in Sect. 3, and a discussion of experimental evaluations and the results are detailed in Sect. 4. In the final part of this paper, some conclusions are drawn.

2   Related Work

Research in educational data mining, especially on student academic performance prediction, has been widely carried out, especially in increasing predictive accuracy. Rohani et al. [8] proposed a classification algorithm based on a metaheuristic algorithm to predict students' academic performance. In their study, they combined the GA and SA algorithms as classifiers. G-SA was implemented along with the Information-Gain feature selection technique. The method improved by 24.93% compared to other metaheuristic methods.

Ghorbani et al. [9], in their study used various machine learning classifiers: Random Forest, K-Nearest Neighbor, Neural Networks, XG-Boost, Support Vector Machine (RadialBasisFunction), Decision Tree, Logistic Regression, and Naïve Bayes with SMOTE to address the dataset imbalance problem. Experiments show that SVM-SMOTE achieves efficient resampling for unbalanced datasets, which can improve accuracy performance results.

Research [10] applies rule extraction to the DIMLP and DTs algorithm ensembles. Their experimental results showed an increase of four times by GB, three times by the DIMLP ensemble, and once by RF for the highest mean predictive accuracy values.

Ragab et al. [11] recommended an ensemble classifier to attain a superior performance model. Based on the consecutive experiments, the accuracy value increased from 90.4% to 91.4%, recall from 90.4% to 91.4%, and precision from 90.5% to 91.5%. They stated that the proposed bagging method with two different classifiers could improve the model's performance to predict student academic performance.

Al Duhayyim et al. [12] introduced the IEAFSS-NFC model to predict student academic performance. They stated that the proposed model has efficiency in learning time. The proposed model has two stages, where feature selection is carried out using the Chaotic Whale Optimization Algorithm (CWOA) to produce a subset of highly related features to improve classification accuracy. Furthermore, the Neuro-Fuzzy Classification (NFC) technique was used to classify the datasets. Based on the results of the benchmark dataset from the UCI repository, the proposed model got a maximum accuracy of 92.74%.

Researchers in [13] proposed hybrid feature selection using Chi-square and SFS with an SVM classification algorithm. They claimed that the combined method was excellent in terms of accuracy, precision, recall, and F-Measure. Their study stated that they succeeded in identifying a suitable feature selection algorithm for optimal feature identification with hybrid feature selection. The method could determine the optimal features while yielding adequate accuracy, precision, recall, and f-measure than the existing hybrid feature selection. Furthermore, the proposed hybrid feature selection was robust by ten fold cross-validation on benchmark datasets with a different number of features and the number of different instances.

Based on previous relevant studies, the authors hypothesize that applying feature selection and using a classification ensemble can improve the model's performance in predicting students' academic performance; feature selection can select relevant features and eliminate useless and redundant features. Moreover, previous studies have shown that ensemble classification has advantages over base classifiers. Therefore, it is divined to improve the classification of student academic performance prediction. The ensemble method can contribute to a substantial improvement in the prediction performance of learning algorithms. In addition, in EDM studies, the researchers used the UCI-Student-Performance and xAPI-student-academic performance datasets to predict current student academic performance precisely. Table 1 summarizes the research predicting student academic performance for the last three years.

Moreover, although many researchers use feature selection techniques, there is still limited discussion regarding metaheuristics in feature selection in predicting students' academic performance. In this study, we applied a genetic algorithm feature selection approach with ensemble classification techniques to increase the value of predictive accuracy in predicting student academic performance.

Table 1. Summary Related Work

Ref.	Year	Feature Selection	Classification Algorithm	Dataset	Model Validation	Accuracy (%)
[8]	2020	Info-Gain	G-SA	UCI [14]	10 fold-CV	DS3 (94.84%), DS2 (93.52%)
[9]	2020	–	RF, K-NN, ANN, XG-boost, SVM, DT,LR, NB	UCI [14]	10 fold-CV	DS2 (81.27%)
[10]	2021	–	Discretized Interpretable Multi-Layer Perceptron	UCI [14]	10 fold-CV	DS2 (92.2%)
[11]	2021	–	Bagging-DT+MLP & Bagging-DT	xAPI-Edu-Data & UCI [14, 15]	10 fold-CV	DS1 (80.83%), DS2 (91.4%)
[12]	2022	(CWOA)	Neuro-Fuzzy Classification (NFC)	UCI [14]	10 fold-CV	DS3 (92.74%)
[13]	2022	Chi-square + SFS	SVM	xAPI-Edu-Data & UCI [14, 15]	10 fold-CV	DS1 (75%), DS2 & 3 (92.91% & 90.91%)

3 Methodology

This section discusses the research methodology, dataset, experimental configuration, parameters for each of the following: feature selection techniques and classification algorithms, and performance measurement.

3.1 Research Framework

This study proposes a predictive model with a classification technique to predict students' academic performance. The research has been conducted on a comparison of feature selection techniques for predicting student academic performance. According to [1, 7], feature selection techniques to predict student academic performance give excellent results. On the other hand, studies conducted [16] stated that the ensemble technique could improve model performance in predicting student performance.

Generally, process methodology combines feature selection techniques and classification algorithms. The research methodology carried out is shown in Fig. 1. The first step in this study was to choose a benchmark for a publicly sourced dataset. The UCI Student Performance Dataset [14] and xAPI-Student-academic-performance [15] datasets on student academic performance were employed because many researchers have used them. The process's initial step is by selecting potential basic classification algorithms (DT, NB, k-NN, and ANN) and (Bagging, Boosting, Voting, and Random Forest), then combining them with feature selection techniques (PCA, Information gain, Gain ratio, PSO and GA). The next step is training the model with all FS and classifier combinations, then validating with ten fold cross-validation to evaluate performance using performance metrics: accuracy, precision, recall, and F-measure.

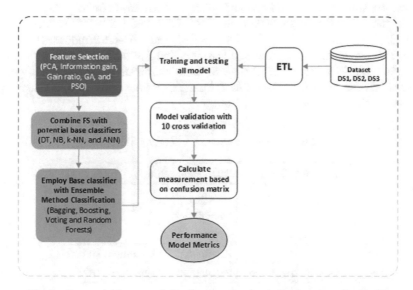

Fig. 1. Research Framework for Student Academic Performance Prediction

3.2 Dataset

In this study, applying a combination of feature selection techniques with classification algorithms to predict students' academic performance, all models were tested for classification prediction with datasets consisting multiclass and binary class targets. Therefore, this study used three datasets collected from two public dataset sources. Dataset 1 (DS1) is an xAPI-Student-academic-performance dataset from The University of Jordan. The dataset is collected from a learning management system (LMS) called Kalboard 360 and consisted of 480 features and 16 features. The attributes were categorized into demographic, academic, and learning behavior. This dataset was multiclass-targeted by grouping targets into three classes: 'low', 'medium', 'high' ('L', 'M', 'H'), based on the rating range: "L" (0–59), "M" (60–89), and "H" (90–100). Regarding the distribution of the target class, 142 are for class H, 211 are for class M, and 127 are data for class L.

Furthermore, mathematical dataset (DS2) & Portuguese dataset (DS3) were collected from the Student Performance Dataset of secondary education of two Portuguese schools from the University of California Irvine Machine Learning Repository. Two datasets were used here. The first was the mathematical dataset with 396 samples and 33 attributes; the other was the Portuguese dataset with 650 records and 33 features. These two datasets had the same target binary class ('pass', 'fail') attribute class target G3.

3.3 Experimental Setup

The first step in the experiment was to combine each model with feature selection and base classifiers, namely DT, NB, k-NN, and ANN. Furthermore, ensemble classifiers: Bagging, Boosting, Voting, and Random Forest, were employed for the next phase. Every phase is conducted for the three datasets: DS1, DS2, and DS3. This experiment used a 1.8 GHz fifth generation i7 CPU, 12 GB RAM, and a 64-bit Windows 10 operating system. This experiment compared six classifiers and five feature selection techniques: information gain, PCA, gain ratio, Particle Swarm Optimization, and Genetic Algorithm. Table 2 shows the parameter settings used for the experiments for feature selection and the classifier shown in Table 3.

Table 2. Parameters Setting of Feature Selection

Feature Selection	Parameters Setting
Information gain	Weight: 0.01
PCA	Weight: 0.01
Gain Ratio	Weight: 0.1
PSO	Population: 5; Generation: 10
Genetic Algorithm	Population: 5; Generation: 10

3.4 Evaluation Measurement

This study utilizes ten fold cross-validation to validate the training and testing phase. According to [4], cross-validation is efficient in terms of computational time while maintaining the accuracy of the estimation measures. This method divides into ten equal parts and distributes them equally. Then proceed with the learning process ten times.

Table 3. Parameters Setting of Classifiers

Feature Selection	Parameters Setting
DT & RF	Number of tree: 100, gain: 0.01, maximal depth: 500
k-NN	k value: 5
Bagging	Iterarion: 10
Boosting	Iterarion: 10

Evaluation measurements for all experiments used four model evaluations: Accuracy, Precision, Recall, and F-Measure. Where are shown in Eqs. 1, 2, 3, and 4, respectively.

The accuracy value is generated from how accurately the system can classify the data correctly compared to the total data. In other words, the accuracy value is a comparison between the data that is classified correctly with the comprehensive data. Meanwhile, the precision value refers to the number of positive data categories classified correctly divided by the total data classified as positive. Recall shows how much the system correctly classifies positive category data. F-measure weighs precision and recalls equally. In representation, if the F-measure with a good value will represent the classification model both in precision and recall.

$$\text{Accuracy} = \frac{TP + TN}{TP + FN + FP + TN} \tag{1}$$

$$Precision = \frac{TP}{TP + FP} \tag{2}$$

$$\text{Recall} = \frac{TP}{TP + FN} \tag{3}$$

$$F - Measure = \frac{2(Precision \, x \, Recall)}{Precision + Recall} \tag{4}$$

4   Result and Discussion

This study conducted two experiments to predict students' academic performance on three datasets with target classes: multiclass and binary class target. The experiments employed four basic classifiers and four classifiers using the ensemble method. Furthermore, each classifier combines three traditional feature selection techniques: Gain-ratio, PCA, and Info-gain, and feature selection techniques based on metaheuristic algorithms: GA and PSO. In addition, the study used ten fold cross-validation to validate the model and measure the accuracy, precision, recall, and F-Measure value.

Table 4 shows that the prediction model combining feature selection based on genetic algorithms with the DT algorithm on DS3 produced the highest accuracy value of 95.1%, while the results lacked performance on DS1. However, k-NN, Naive Bayes, and ANN

Table 4. Result Base Classifier Performance Measurement

Classifier	Feature Selection	Accuracy Dataset			Precision Dataset			Recall Dataset			F-Measure Dataset		
		1	2	3	1	2	3	1	2	3	1	2	3
Decision Tree	Attributes	68.5	88.9	91.8	70.2	87.4	84.4	71.8	87.4	84.1	71.0	87.4	84.3
	Info - Gain	67.7	90.9	93.2	69.5	89.6	91.3	70.9	89.9	81.3	70.2	89.7	86.0
	PCA	57.3	90.1	91.5	57.4	60.7	85.4	57.3	89.1	80.3	57.4	72.2	82.7
	Gain Ratio	66.9	**93.4**	90.8	67.9	**93.6**	83.0	69.3	91.4	80.2	68.6	**92.5**	81.6
	PSO	72.5	92.4	94.0	73.5	91.0	89.3	74.8	**92.2**	87.0	74.1	91.6	88.2
	GA	**74.8**	91.9	**95.2**	**75.4**	90.7	**92.5**	**75.8**	91.0	**88.6**	**75.6**	90.9	**90.5**
Naive Bayes	Attributes	65.4	86.8	87.8	65.4	85.2	76.9	68.4	84.9	85.4	67.0	85.0	80.9
	Info - Gain	69.0	86.8	92.8	69.1	84.7	84.9	72.2	86.9	89.6	70.6	85.8	87.2
	PCA	62.5	87.1	93.1	62.7	85.0	86.4	66.4	86.9	87.3	64.5	85.9	86.9
	Gain Ratio	67.5	**89.1**	91.4	67.6	**88.0**	82.4	70.2	87.2	88.0	68.9	**87.6**	85.1
	PSO	72.9	88.6	90.8	73.5	87.3	81.3	74.1	86.8	87.6	73.9	87.0	84.3
	GA	**75.4**	88.6	**93.5**	**75.6**	86.8	**86.7**	**77.4**	**87.8**	**89.6**	**76.5**	87.3	**88.1**
k-NN	Attributes	65.0	87.1	90.6	66.1	85.4	86.7	65.7	85.3	74.0	65.9	85.4	79.9
	Info - Gain	59.0	91.4	90.8	59.9	90.0	82.8	59.7	90.7	80.6	59.8	90.3	81.7
	PCA	63.5	89.6	93.4	64.6	88.7	88.6	64.0	87.6	85.0	64.3	88.1	86.8
	Gain Ratio	63.5	90.6	93.2	64.6	89.9	89.0	64.0	88.7	83.7	64.3	89.3	86.3
	PSO	68.3	92.9	**94.8**	69.0	91.7	**93.7**	69.0	**92.4**	85.5	69.0	**92.1**	**89.4**
	GA	**74.6**	92.9	93.5	**75.5**	92.0	89.7	**74.9**	92.0	84.3	**75.2**	92.0	86.9
ANN	Attributes	75.2	89.1	90.9	76.2	87.2	83.2	75.6	89.5	80.7	75.9	88.3	82.0
	Info - Gain	69.2	**92.9**	92.1	70.2	**91.6**	84.1	70.6	92.6	**87.6**	70.4	**92.1**	85.8
	PCA	58.5	89.6	92.1	59.0	88.4	84.6	60.8	88.0	86.0	59.9	88.2	85.3
	Gain Ratio	72.7	89.4	91.2	74.1	87.9	83.0	72.9	88.2	83.8	73.5	88.0	83.4
	PSO	81.0	92.2	**93.2**	81.7	90.5	87.4	**81.5**	92.6	86.2	81.6	91.5	**86.6**
	GA	**81.3**	92.4	93.1	**81.9**	91.2	**87.5**	81.4	91.8	85.3	**81.6**	91.5	86.4

also produced slightly different accuracy results when using a feature selection algorithm based on metaheuristics. On the other hand, the Naive Bayes accuracy value showed the most significant difference of 10% compared to the value without feature selection.

Table 5 shows that Random Forest with the genetic algorithm as its feature selection shows superior and most consistent performance in all performance measures compared to other classifiers in all datasets in this study. Furthermore, all prediction models showed improvement after using feature selection. Therefore, the feature selection technique based on metaheuristics was the most suitable feature selection technique used to predict students' academic performance because it showed the highest accuracy results. The experimental results showed that the combination of feature selection and classifier had significantly improved the performance of the student's academic performance prediction model.

Table 5. Result Ensemble Classifier Performance Measurement

Classifier	Feature Selection	Accuracy			Precision			Recall			F-Measure		
		Dataset											
		1	2	3	1	2	3	1	2	3	1	2	3
Bagging	All Attributes	74.6	92.4	91.9	75.4	91.3	85.6	75.5	91.6	82.6	75.4	91.4	84.1
	Info - Gain	76.1	91.9	93.4	76.7	90.2	93	77.1	92.4	80.5	76.9	91.3	86.3
	PCA	74.2	91.9	93.7	75	90.8	91.4	77.1	90.8	83.2	76.1	90.8	87.1
	Gain Ratio	75.2	93.9	92.8	76.7	93.6	88.5	75.5	92.5	82.2	76.1	93.1	85.2
	PSO	**77.5**	**95.1**	93.5	**78.2**	**94.8**	91.2	78.9	**94.3**	82.7	**78.5**	**94.5**	86.7
	GA	76.7	92.4	**94.6**	77.4	90.9	**91.5**	77.2	92.2	**87**	77.3	91.6	**89.2**
Boosting	All Attributes	72.7	89.1	91.7	73.9	87.1	85.1	73.3	89.5	81.6	73.6	88.3	83.3
	Info - Gain	75.6	91.3	92.1	**76.8**	89.9	84.1	76	91.1	87.6	76.4	90.4	85.8
	PCA	74.2	89.9	91.5	75	88.6	84.9	77.1	88.3	81.1	76.1	88.5	82.9
	Gain Ratio	72.9	92.4	92.9	73.6	91.7	87.5	74.2	91.1	84.4	73.9	91.3	85.9
	PSO	**75.6**	92.9	**94.8**	75.7	91.7	**93.7**	77.5	92.4	85.5	76.6	92.1	**89.4**
	GA	75.4	**93.7**	93.8	76	**92.5**	87.7	**77.7**	93.3	88.6	76.8	92.9	88.3
Voting	All Attributes	75.4	90.9	93.1	76.7	89.4	89.5	75.5	90.3	82.4	76.1	89.8	85.8
	Info - Gain	70.6	90.4	92.6	72.7	88.5	88.3	70.4	90.8	81.7	71.5	89.7	84.9
	PCA	75.4	89.6	91.2	76.5	88.1	87.6	**76.4**	88.7	75.9	76.4	88.4	81.4
	Gain Ratio	73.8	91.6	93.5	74.9	90.5	89.4	74.2	90.6	84.7	74.5	90.6	86.9
	PSO	75.9	93.2	93.9	77.1	92.1	93.1	75.9	92.6	82.9	76.5	92.3	87.7
	GA	**76.1**	**93.9**	**94.9**	**78.2**	92.9	94.3	75.7	**93.5**	85.5	76.9	**93.2**	**89.7**
Random Forest	All Attributes	78.3	91.4	92.9	78.9	90.1	89.3	79.1	90.4	81.9	78.9	90.3	85.5
	Info - Gain	76.1	90.6	92.6	76.6	89.2	88.6	76.9	89.7	81.3	76.8	89.5	84.8
	PCA	75.2	89.9	91.7	76.2	88.3	89.1	76.1	88.9	76.7	76.2	88.6	82.4
	Gain Ratio	73.1	90.9	93.5	74.7	89.6	89.4	73.2	89.9	84.7	73.9	89.7	86.9
	PSO	81.5	92.9	94.3	82.3	91.8	91.9	81.7	92.1	85.2	82.1	92.1	88.5
	GA	**82.5**	**94.7**	**94.9**	**83.2**	**93.9**	**92.2**	**82.9**	**94.1**	**87.6**	**83.1**	**94.1**	**89.8**

5   Conclusion

Predicting students' academic performance is a form of early intervention by educators while monitoring their academic performance. Therefore, maintaining a predictive model with the best accuracy is significantly essential. This comparative work combined classifiers with various feature selection methods to evaluate the predictive accuracy of students' academic performance. This study discussed empirical results demonstrating that feature selection enhanced performance accuracy in predicting students' academic performance. The experimental work on the eight classifiers across the three benchmark datasets showed impressive refinements when combined with feature selection techniques. Based on the comparative studies, the genetic algorithm-based feature selection technique and the Random Forest ensemble classifier showed superior accuracy and consistency across all performance measures compared to other classifiers across all datasets in this study. Furthermore, Table 6 compares the accuracy results on the student benchmark dataset of the proposed work with existing work predicting student academic performance.

The author intends to work on different feature selection techniques and methods to design the optimal feature selection mechanism for future work. It is also important to select several public benchmarks of student academic performance datasets to analyze with optimal combined parameters tuning feature selection techniques and classifier. Furthermore, performance evaluation metrics are suitable for varied target-class domains such as MSE, AUC, Error Rate, and other measures.

Table 6. Accuracy Performances of Proposed Work & Existing Work

Dataset	Proposed Work	Existing Work (%)					
		[8]	[9]	[10]	[11]	[12]	[13]
DS1	**82.5**	–	–	–	80.8	–	75.0
DS2	**94.7**	93.5	81.2	92.2	91.4	–	92.9
DS3	**94.9**	94.8	–	–	–	92.7	90.9

References

1. Sokkhey, P., Okazaki, T.: Hybrid machine learning algorithms for predicting academic performance. Int. J. Adv. Comput. Sci. Appl. **11**, 32–41 (2020)
2. Bujang, S.D.A., et al.: Multiclass prediction model for student grade prediction using machine learning. IEEE Access **9**, 95608–95621 (2021)
3. Son, L.H., Fujita, H.: Neural-fuzzy with representative sets for prediction of student performance. Appl. Intell. **49**(1), 172–187 (2018). https://doi.org/10.1007/s10489-018-1262-7
4. Zhang, Y., Liu, Y.: The research of predicting student's academic performance based on educational data. In: 2021 5th International Conference on Computer Science and Artificial Intelligence, pp. 193–201 (2021)
5. Hassan, H., Anuar, S., Ahmad, N.B.: Students' performance prediction model using meta-classifier approach. In: Macintyre, J., Iliadis, L., Maglogiannis, I., Jayne, C. (eds.) EANN 2019. CCIS, vol. 1000, pp. 221–231. Springer, Cham (2019). https://doi.org/10.1007/978-3-030-20257-6_19
6. Ghosh, M., Guha, R., Alam, I., Lohariwal, P., Jalan, D., Sarkar, R.: Binary genetic swarm optimization: a combination of GA and PSO for feature selection. J. Intell. Syst. **29**, 1598–1610 (2019)
7. Injadat, M., Moubayed, A., Nassif, A.B., Shami, A.: Systematic ensemble model selection approach for educational data mining. Knowl. Based Syst. **200**, 105992 (2020)
8. Rohani, Y., Torabi, Z., Kianian, S.: A novel hybrid genetic algorithm to predict students' academic performance. J. Electr. Comput. Eng. Innov. (JECEI) **8**, 219–232 (2020)
9. Ghorbani, R., Ghousi, R.: Comparing different resampling methods in predicting students' performance using machine learning techniques. IEEE Access **8**, 67899–67911 (2020)
10. Bologna, G.: A rule extraction technique applied to ensembles of neural networks, random forests, and gradient-boosted trees. Algorithms **14** (2021)
11. Ragab, M., Abdel Aal, A.M.K., Jifri, A.O., Omran, N.F., A Saeed, R.: Enhancement of predicting students performance model using ensemble approaches and educational data mining techniques. Wirel. Commun. Mob. Comput. **2021**, 1–9 (2021)
12. Al Duhayyim, M., Marzouk, R., Al-Wesabi, F.N., Alrajhi, M., Hamza, M.A., Zamani, A.S.: An improved evolutionary algorithm for data mining and knowledge discovery. CMC Comput. Mater. Continua **71**, 1233–1247 (2022)
13. Zaffar, M., et al.: A hybrid feature selection framework for predicting students performance. Comput. Mater. Continua **70**, 1893–1920 (2022)
14. Cortez, P., Silva, A.M.G.: Using data mining to predict secondary school student performance (2008)

15. Amrieh, E.A., Hamtini, T., Aljarah, I.: Preprocessing and analyzing educational data set using X-API for improving student's performance. In: 2015 IEEE Jordan Conference on Applied Electrical Engineering and Computing Technologies (AEECT), pp. 1–5. IEEE (2015)
16. Injadat, M., Moubayed, A., Nassif, A.B., Shami, A.: Multi-split optimized bagging ensemble model selection for multi-class educational data mining. Appl. Intell. **50**, 4506–4528 (2020)

A Deep Learning Model for Human Blood Cells Classification

M. Pramodha[1] , S. Ansith[2] , J. V. Bibal Benifa[3], Mohammed Al-Sarem[4,5] ,
J. Hanumanthappa[1] , A. A. Bini[2], Emmanuel Ndagijimana[1] , Faisal Saeed[5,6] ,
Md. Belal Bin Heyat[7] , Abdulrahman Alqarafi[5], Abdullah Y. Muaad[1,8(✉)] ,
and Channabasava Chola[1,3(✉)]

[1] Department of Studies in Computer Science, University of Mysore, Manasagangothri,
Mysore 570006, India
abdullahmuaad9@gmail.com, channabasavac7@gmail.com
[2] Department of Electronics and Communication Engineering, Indian Institute of Information
Technology Kottayam, Kottayam, Kerala, India
[3] Department of Computer Science and Engineering, Indian Institute of Information
Technology Kottayam, Kottayam, Kerala, India
[4] Department of Computer Science, Saba'a Region University, Mareb, Yemen
[5] College of Computer Science and Engineering, Taibah University, Medina 42353,
Saudi Arabia
[6] DAAI Research Group, Department of Computing and Data Science, School of Computing
and Digital Technology, Birmingham City University, Birmingham B4 7XG, UK
[7] IoT Research Center, College of Computer Science and Software Engineering,
Shenzhen University, Shenzhen 518060, Guangdong, China
[8] IT Department, Sana'a Community College, Sana'a 5695, Yemen

Abstract. Microscopic imaging is gaining focus in recent days, especially in
the part of histopathological image analysis. Blood cells plays a critical role in
assessment of health status of patients, especially given the rising frequency of
infectious diseases. Automated blood analysis can aid in detecting early stages of
diseases. In this work, we present study classification of Blood cell using different
transfer learning approaches, MobileNetV2 based model designed for the accurate
multi classification of blood cells. Deep Learning (DL) models require more time
when training on big data sets, to overcome the computation complexity with light
weight model MobileNetV2 is considered. In this paper we present the comparison
among different transfer learning models such as VGG16, VGG19, Resnet50 with
MobileNetV2. The performance evaluated with Accuracy, Precision, recall and
F-Score. MobileNetV2 outperform all other model with accuracy of 97.89%. The
proposed model has improved accuracy for classification blood cell for eight class.

Keywords: Blood Cell Classification · Microscopic Imaging · Convolutional
Neural Network · Deep Learning · Transfer Learning

1 Introduction

Medical imaging is the process of offering a graphic representation of the human body
to help the doctors and radiologists in an effective medical diagnosis and care. There
are certain imaging techniques like CT, X-ray, MRI, microscopic blood smear images,

F. Saeed et al. (Eds.): ICACIn 2022, LNDECT 179, pp. 410–418, 2023.
https://doi.org/10.1007/978-3-031-36258-3_36

PET, and ultrasound upon which experts and doctors can rely for diagnosing diseases and prescribing treatment [1–6]. Different computer-aided diagnosis (CAD) systems can automatically detect numerous haematological disorders, such as AIDS and blood cancer (Leukaemia), by analysing the microscopic images of blood cells with the help of DL and ML algorithms. White Blood Cells (leukocytes), Red Blood Cells (erythrocytes), and platelets (thrombocytes) are major classifications of blood cell [7–9]. Among these, leukocytes are further broken down to five subgroups: monocyte, lymphocyte, neutrophil, basophil, and eosinophil.

Classification and segmentation of White Blood Cells can be done by traditional techniques using manual analysis of WBCs in blood smear images. This conventional method consumes more time and also a challenging task as well as error prone. Different ML and DL techniques have been deployed in the WBC classification over the past two decades [10]. In this proposed work we use CNN architecture for the classification purpose. The use of DL models to classify WBC which can reduce the burden for haematologists and deliver prompt, effective, and accurate results to support medical professionals in the diagnostic procedure at the same time DL can be used in different domain such as text, image and video [8, 11–13].

The following are some of this work's significant contributions:

- A deep learning model is proposed to classify blood cells to multiclass of blood cell.
- Comparison of experiments for different model with latest work have been explored.

This paper is organized as we describe in this section. The introduction for related work is presented in Sect. 2. In Sect. 3 the explanation of proposed models. The Sects. 4 and 5 discuss and explain the result and conclusion of this paper with future work are summarized in Sect. 6.

2 Related Work

There were various works addressed the blood classification. In this section we are going to explore some of them starting by authors in [14]. They proposed a system to descry WBCs. Their model detected WBCs from microscopic images using a simple relationship of R, G, B colours and some morphological operations. They applied SVM and CNN for the detection and they got good accuracy equal to 92.80%. In [15] the authors have used a convolutional neuronal network to detect eight groups of peripheral blood cells. They have designed CNN-based transfer learning model and they have got 96.20% in term of accuracy. The authors in [16] designed a new DL based on a two stage which can detect and classify infected WBCs and RBCs. The result was 72% in term of accuracy. In [17] authors introduced a CNN method for classifying white blood cells into their corresponding five groups, which are monocyte, lymphocyte, neutrophil, basophil, and eosinophil. The validation set indicated an overall accuracy of 91%, while the training set recorded an accuracy of 99%. The proposed system's sensitivity and specificity were 91 and 97%, respectively. The authors in [18] proposed R-CNN technique to detect WBCs and RBCs. They got an average accuracy of 83.25% for RBCs while 99% for eosinophil and 66% for lymphocytes. The lowest testing accuracy was always above 66% for this method. In [19] The authors explored the identification of WBCs in three steps utilizing image processing and machine learning techniques (segmentation, feature extraction and classification). When the number of samples in a dataset is small, the accuracy obtained is rather low. As a result, a deep learning method is used instead of the traditional machine learning method. Capsule Networks (encoder-decoder) approaches obtained 96.86% accuracy using LISC datasets and Deep Learning methods. When it comes to segmentation and classification of blood cells, deep learning methods outperform traditional machine learning methods. With small datasets, Caps Net has achieved a good accuracy rate. In [20], authors introduced the procedure of white blood cell identification, categorization, and quantitative analysis can be done both manually and automatically, according to the authors in [21]. The classification of WBCs manually takes a long time and is highly reliant on the hematologist's knowledge. CapsNet and other convolutional neural networks like AlexNet, VGG16, ResNet-18, and InceptionV3 are used in clinical practise to classify WBCs utilising an automated leukocyte analyzer. All IDB2, BCCD, and CellaVision from the Hospital Clinic of Barcelona are included in the samples, which are from a public dataset made available by Anna et al. [22]. The accuracy rates are consistent with BloodCaps (99.3%), AlexNet (81.5%), VGG16 (97.8%), ResNet-18 (95.9%), and Incep-tionV3 (98.4%). Even with small datasets, BloodCaps, a proposed model capsule network for the categorization of human peripheral blood cells, gives precise classification. In [11] the authors investigated the automatic classification of white blood cells using rapid, accurate and cost-effective data enhancement approaches using DNNs as an alternative to the laboratory environment. Image processing methods and GAN image generation are the approaches used in data augmentation. Utilizing cutting-edge DNNs like VGG, ResNet, and DenseNet that have been trained on the CIFAR-100 datasets are used to examine the classification of leukocytes into their respective classes namely neutrophils, eosinophils, lymphocytes, monocytes, and basophils. On the CIFAR-100 dataset, they applied pre-trained weights. This approach uses the acquired images immediately, in contrast to other approaches that require complex image pre-processing and manual feature extraction prior to classification. It has been demonstrated through extensive testing that the suggested method can accurately classify white blood cells. With a validation accuracy of 98.8%, DenseNet-169, the best DNN model, is ranked first. The proposed approach performs better than existing approaches, which rely on manual feature engineering and sophisticated image processing. In [23] according to authors the accuracy of the measurement of White Blood Cells (WBC) in the blood is 95.89% for categorization and 97.40% for WBC detection. The ability to extract is a crucial sign of pathological conditions, and the suggested method also can detect WBC in raw microscopic pictures. Computer vision-based methods for differential counting of WBC are becoming more prominent due to their benefits over conventional methods. However, many of these techniques aren't applicable to raw microscopic images with multiple WBC because they are made for single WBC images that have already been processed. In addition, the techniques fail to recognise WBC in images. This study provides a K-Means clustering-based image processing strategy for identifying and localising WBC in raw microscopic images, as well as a VGG-16 classifier to categorise those cells. The method may remove dye particles, nucleated RBC, and overlapping cells from multiple-cell pictures in addition to extracting WBC from them. The algorithm can be immediately merged with the initial microscopic images produced by an automated microscope platform to perform WBC differential counting. The suggested algorithms may also be easily calibrated for use with human samples, and they can take the place of expensive haematology analyzers and inefficient manual counting techniques. The authors in [24] presented a computer vision-based strategy for developing a malaria parasite detection algorithm. In order to classify blood cells that have been infected by parasites, the deep learning capsule network is updated, and a supervised pixel-based blood cell segmentation model is developed. They used ANN and CapsNet models to segment and classify blood cells using the Local (DB1) and CDC (DB2) datasets, respectively. Over ANN and CapsNet, the suggested architecture obtained an overall accuracy of 99.1% and 98.7%, respectively. In [25], the authors found that it is difficult to extract discriminate features from unprocessed microscopic images, therefore the commonly used SVM classifier simply requires minor adjustments. These techniques, however, are unable to handle fine-grained classification situations involving up to 40 different types of white blood cells. In addition to a leukocyte classifier, the author has proposed a model based on deep residual learning theory that employs RNN and CNN techniques (Call3Net). A standard dataset of about 100,000 leukocytes from 40 different groups was developed for training and testing. The accuracy gained is 76.84%. Multiple training strategies are chosen and integrated to improve the classifier's generalization performance. This method can also easily handle fine-grained classification situations with up to 40 kinds of white blood cells. The authors in [26] introduced a new automated peripheral blood smear analysis method. It includes software for processing images and an automated microscope for collecting images of a blood sample at the microscopic level. The CNN (U-net) method utilising microscopic images as a dataset yielded a 99.5% sensitivity for WBC extraction. In the last applied segmentation for blood cell which will be future work [27].

3   Proposed Methodology

The proposed approach can be separated into three sections: data preprocessing, feature extraction and classification (Fig. 1).

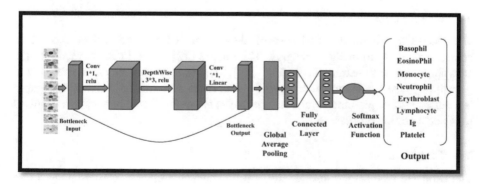

Fig. 1. Architecture of blood cell classification

3.1 A Transfer Learning

Transfer Learning (TL) is described as the knowledge transmission, which was acquired in one area, and then passed down to other for the purpose of classifying and extracting features. TL has been utilized in numerous deep learning applications because refining a pre-formed CNN model is usually a lot quicker and easier than forming a CNN model with randomly initialized weights from scratch. Here MobileNetV2 has been utilized for feature extraction which is one the deep CNN based model [15].

3.2 MobileNetV2

Here MobileNetV2 [28], has been utilized for feature extraction which is one the deep CNN based model. The MobileNetV2 features 53 layers of depth and one Avg pool. It has two primary components, inverted residual block and bottleneck residual block. The conventional VGG architecture, which involves creating a network by adding convolution layers on top of each other to enhance accuracy, is referenced in the design of MobileNetV1. However, when too many convolutional layers are added, the issue of gradient vanishing arises. Residual block in ResNet facilitates information propagation, allowing for feature reuse in forward propagation and preventing gradient vanishing in back propagation. As a result, MobileNetV2 additionally makes use of ResNet's residual structure in addition to continuing to use MobileNetV1's depth separable convolution. The implementation of a linear bottleneck and an inverted residual block in the network are MobileNetV2's two primary enhancements over MobileNetV1 in comparison.

3.3 Data Preprocessing

The most crucial component of the model is the preprocessing method. In this work, we used raw data that has been resized to 224 * 224, the predefined size of our model. To reduce over fitting, we implemented the data augmentation and class-balancing technique, which accelerated the learning process. Later, we divide the dataset into 70% for training, 10% for validation, and 20% for testing. For every class, the dataset was chosen randomly [29].

3.4 Feature Extraction

The Deep Convolutional neural network demonstrated improved performance in all areas, particularly in medical imaging. The images of the blood cells are fed into the convolution layer to extract the features [29]. The result of the convolution layer will pass through the relu layer that converts the data into a non-linear form, then this output is inserted into the pooling layer which removes any redundant data which is acquired in the convolution process.

3.5 Classification

The extracted features are inserted into a fully connected layer to classify images into their respective classes. Dropout layer is utilized to decrease overfitting. These layers cut the part of the data flowing between the fully connected layers, making a model for accurate data classification. The output is computed using the SoftMax function.

3.6 Dataset

Table 1. Cell type and a number of images in each group

Class Name	Number of Images Per class
Basophil	1218
Eosinophil	3117
Erythroblast	1551
Ig	2895
Lymphocyte	1214
Monocyte	1420
Neutrophil	3329
Platelet	2348
Total	17,092

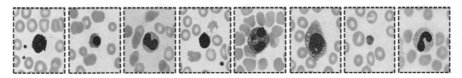

Fig. 2. Data Set Sample Example

The images utilized in this work are part of a public dataset that was made available by Anna et al. [15, 22] from the hospital clinic in Barcelona. In the dataset, there are a total of 17,092 images of human peripheral blood cells from healthy individuals. There are 8 other subtypes of these cells: eosinophil, erythroblast, basophil, Ig, lymphocyte, monocyte, neutrophil, and platelet. Table 1 presents the dataset's statistics in detail. The dataset contains 360 * 363 pixel JPG photos that are all in the RGB color space and format. Clinical pathologists from the Barcelona hospital clinic, annotated each image (Fig. 2).

3.7 Evaluation Metrics

$$\text{Accuracy} = \frac{TP + TN}{TP + FN + TN + FP},$$

$$\text{Precision} = \frac{TN}{TN + FP},$$

$$\text{F1} - \text{score} = \frac{2 \cdot TP}{2 \cdot TP + FP + FN},$$

$$\text{Recall} = \frac{TP}{TP + FN},$$

where TP is true positive, TN is true negative, FP is false positive and FN is false negative for the overall test samples N.

4   Results and Discussion

We performed experiments on human peripheral blood cell dataset [22] with eight class multiclass scenario with state-of-the-art deep learning models namely VGG16, VGG19, ResNet-50 and MobileNetV2 to perform classification. In this section with 4 DL models were validated the eight-class classification.

As we can see in Table 2, the comparison of different model with four different metrics have done.

Table 2. Results on blood cell classification with state-of-the-art deep learning models.

Models	Accuracy	Precision	Recall	F-Measure
VGG16	90.53	92	91	91
VGG19	89.01	90	89	89
ResNet50	96.34	97	97	97
MobileNetV2	97.89	98	98	98

Table 2 present the result of different classification models to classify blood cell were evaluated on testing dataset. The authors in [22] collected data set. The training dataset, as described in Sect. 2.6. The result of the proposed models is 97.89%, 98%, 98% and 98% terms of accuracy, F1 score, precision, and recall. From the Table 1, it is evident that the proposed model has got good result using MobileNetV2 model which was outperform all other models for all four metrics.

5   Conclusions

Classification of Blood Cells has emerged as one of the major research focusing on histopathological images which is subfield of the medical imaging. Blood cell classification is particularly plays an important role in CAD based diagnosis system with the rise in infectious diseases and due lack of facilities or knowledge led to difficulties in early detection. In this work, we suggest to investigate categorization of blood cells using various transfer learning methods, such as modified MobileNetV2. In parallel, we compare proposed model with different transfer learning models. The proposed MobileNetV2 model is more efficient and has less parameters, requires less computation burden compared to other transfer learning models. We achieved improved accuracy, which is equivalent to 97.89%, outperforming the compared models on the data. With eight classes, the categorization of blood cells using our proposed technique yields good performance.

References

1. Al-masni, M.A., Al-antari, M.A., Choi, M.T., Han, S.M., Kim, T.S.: Skin lesion segmentation in dermoscopy images via deep full resolution convolutional networks. Comput. Methods Programs Biomed. **162**, 221–231 (2018). https://doi.org/10.1016/j.cmpb.2018.05.027
2. Al-antari, M.A., Han, S.M., Kim, T.S.: Evaluation of deep learning detection and classification towards computer-aided diagnosis of breast lesions in digital X-ray mammograms. Comput. Methods Programs Biomed. **196**, 105584 (2020). https://doi.org/10.1016/j.cmpb.2020.105584
3. Chola, C., Benifa, J.V.B.: Detection and classification of sunspots via deep convolutional neural network. Glob. Transit. Proc., 0–7 (2022). https://doi.org/10.1016/j.gltp.2022.03.006
4. Al-antari, M.A., Al-masni, M.A., Choi, M.T., Han, S.M., Kim, T.S.: A fully integrated computer-aided diagnosis system for digital X-ray mammograms via deep learning detection, segmentation, and classification. Int. J. Med. Inform. **117**, 44–54 (2018). https://doi.org/10.1016/j.ijmedinf.2018.06.003
5. Al-masni, M.A., Al-antari, M.A., Min, H., Hyeon, N., Kim, T.: 2nd IEEE Eurasia Conference on Biomedical Engineering, Healthcare and Sustainability 2020, ECBIOS 2020, pp. 95–98 (2020)
6. Al-masni, M.A., Kim, W.R., Kim, E.Y., Noh, Y., Kim, D.H.: Automated detection of cerebral microbleeds in MR images: a two-stage deep learning approach. NeuroImage Clin. **28**, 102464 (2020). https://doi.org/10.1016/j.nicl.2020.102464
7. Li, X., Li, W., Xu, X., Hu, W.: Cell classification using convolutional neural networks in medical hyperspectral imagery, pp. 501–504 (2017)
8. Chola, C., et al.: Gender identification and classification of Drosophila melanogaster flies using machine learning techniques, vol. 2022 (2022)
9. Mestetskiy, L.M., Guru, D.S., Benifa, J.V.B., Nagendraswamy, H.S., Chola, C.: Gender identification of Drosophila melanogaster based on morphological analysis of microscopic images. Vis. Comput. (2022). https://doi.org/10.1007/s00371-022-02447-9
10. Baydilli, Y.Y., Atila, Ü.: Classification of white blood cells using capsule networks. Comput. Med. Imaging Graph. **80**, 101699 (2020). https://doi.org/10.1016/J.COMPMEDIMAG.2020.101699
11. Muaad, A.Y., Hanumanthappa, J., Al-antari, M.A., Bibal Benifa, J.V., Chola, C.: AI-based misogyny detection from Arabic Levantine Twitter tweets. In: Proceedings of the 1st Online Conference on Algorithms, 27 September–October 2021, pp. 4–11. MDPI, Basel, Switzerland (2021). https://doi.org/10.3390/IOCA2021-10880
12. Muaad, A.Y., Davanagere, H.J., Al-antari, M.A., Benifa, J.V.B., Chola, C. : AI-based misogyny detection from Arabic Levantine Twitter tweets. Comput. Sci. Math. Forum **2**(1), 15 (2021)
13. Muaad, A.Y., et al.: An effective approach for Arabic document classification using machine learning. Glob. Transit. Proc., 0–5 (2022). https://doi.org/10.1016/j.gltp.2022.03.003
14. Zhao, J., Zhang, M., Zhou, Z., Chu, J., Cao, F.: Automatic detection and classification of leukocytes using convolutional neural networks. Med. Biol. Eng. Comput. **55**(8), 1287–1301 (2016). https://doi.org/10.1007/s11517-016-1590-x
15. Acevedo, A., Alférez, S., Merino, A., Puigví, L., Rodellar, J.: Recognition of peripheral blood cell images using convolutional neural networks. Comput. Methods Programs Biomed. **180**, 105020 (2019). https://doi.org/10.1016/j.cmpb.2019.105020
16. Hung, J., et al.: Applying faster R-CNN for object detection on malaria images, pp. 1–7 (2018). http://arxiv.org/abs/1804.09548
17. Bani-Hani, D., Khan, N., Alsultan, F., Karanjkar, S., Nagarur, N.: Classification of leucocytes using convolutional neural network optimized through genetic algorithm, November, pp. 1–7 (2018)

18. Tobias, R.R., et al.: Faster R-CNN model with momentum optimizer for RBC and WBC variants classification. In: LifeTech 2020 - 2020 IEEE 2nd Global Conference on Life Sciences and Technologies, January 2021, pp. 235–239 (2020). https://doi.org/10.1109/LifeTech48969.2020.1570619208

19. Sajjad, M., et al.: Leukocytes classification and segmentation in microscopic blood smear: a resource-aware healthcare service in smart cities. IEEE Access **5**, 3475–3489 (2017). https://doi.org/10.1109/ACCESS.2016.2636218

20. Long, F., Peng, J., Song, W., Xia, X., Sang, J.: Computer methods and programs in biomedicine BloodCaps: a capsule network based model for the multiclassification of human peripheral blood cells, vol. 202 (2021). https://doi.org/10.1016/j.cmpb.2021.105972

21. Zheng, X., Wang, Y., Wang, G., Liu, J.: Fast and robust segmentation of white blood cell images by self-supervised learning. Micron **107**, 55–71 (2018). https://doi.org/10.1016/j.micron.2018.01.010

22. Acevedo, A., Merino, A., Alférez, S., Molina, Á., Boldú, L., Rodellar, J.: A dataset of microscopic peripheral blood cell images for development of automatic recognition systems. Data Br. **30**, 105474 (2020). https://doi.org/10.1016/j.dib.2020.105474

23. Wijesinghe, C.B., Wickramarachchi, D.N., Kalupahana, I.N., De Seram, L.R., Silva, I.D., Nanayakkara, N.D.: Fully automated detection and classification of white blood cells. In: Proceedings of the Annual International Conference of the IEEE Engineering in Medicine & Biology Society EMBS, vol. 2020, pp. 1816–1819 (2020). https://doi.org/10.1109/EMBC44109.2020.9175961

24. Maity, M., Jaiswal, A., Gantait, K., Chatterjee, J., Mukherjee, A.: Quantification of malaria parasitaemia using trainable semantic segmentation and CapsNet. Pattern Recognit. Lett. **138**, 88–94 (2020). https://doi.org/10.1016/j.patrec.2020.07.002

25. Qin, F., Gao, N., Peng, Y., Wu, Z., Shen, S., Grudtsin, A.: Fine-grained leukocyte classification with deep residual learning for microscopic images. Comput. Methods Programs Biomed. **162**, 243–252 (2018). https://doi.org/10.1016/j.cmpb.2018.05.024

26. Mundhra, D., Cheluvaraju, B., Rampure, J., Rai Dastidar, T.: Analyzing microscopic images of peripheral blood smear using deep learning. In: Cardoso, M.J., et al. (eds.) DLMIA/ML-CDS -2017. LNCS, vol. 10553, pp. 178–185. Springer, Cham (2017). https://doi.org/10.1007/978-3-319-67558-9_21

27. Tavakoli, E., Ghaffari, A., Kouzehkanan, Z.M., Hosseini, R.: New segmentation and feature extraction algorithm for classification of white blood cells in peripheral smear images (2021)

28. Xiang, Q., Zhang, G., Wang, X., Lai, J., Li, R., Hu, Q.: Fruit image classification based on MobileNetV2 with transfer learning technique. In: ACM International Conference Proceeding Series (2019). https://doi.org/10.1145/3331453.3361658

29. Pramodha, M., Muaad, A.Y., Bibal Benifa, J.V., Hanumanthappa, J., Chola, C., Mugahed, A.: A hybrid deep learning approach for COVID-19 diagnosis via CT and X - R ay medical images, pp. 1–10 (2021)

A Transfer Learning Based Approach for Sunspot Detection

Channabasava Chola[1(✉)] 📧, J. V. Bibal Benifa[1(✉)] 📧, Abdullah Y. Muaad[2,6] 📧,
Md. Belal Bin Heyat[3] 📧, J. Hanumanthappa[2] 📧, Mohammed Al-Sarem[4,5] 📧,
Abdulrahman Alqarafi[5], and Bouchaib Cherradi[7,8] 📧

[1] Department of Computer Science and Engineering, Indian Institute of Information
Technology Kottayam, Kottayam, Kerla, India
channabasavac7@gmail.com, Benifa.jhon@gmail.com

[2] Department of Studies in Computer Science, University of Mysore, Manasagangothri,
Mysore 570006, India

[3] IoT Research Center, College of Computer Science and Software Engineering,
Shenzhen University, Shenzhen 518060, Guangdong, China

[4] Department of Computer Science, Saba'a Region University, Mareb, Yemen

[5] College of Computer Science and Engineering, Taibah University, Medina 42353,
Saudi Arabia

[6] IT Department, Sana'a Community College, Sana'a 5695, Yemen

[7] EEIS Laboratory, ENSET of Mohammedia, Hassan II University of Casablanca,
28820 Mohammedia, Morocco

[8] STIE Team, CRMEF Casablanca-Settat, Provincial Section of El Jadida, 24000 El Jadida,
Morocco

Abstract. Realtime space weather activity tracking has improved over the years
due to recent advancements in astronomical instrumentation. Sunspots are known
as important phenomenon of sun and can be addressed on photosphere of sun surface. The occurrences of sunspots determine overall solar activities, sunspots are
being observed from early eighteenth century. In this study, we have implemented
a DL model which automatically detects sunspots from HMI image datasets. A DL
based VGG16 model is used for deep hierarchical features extraction and passed
to softmax layer for classification. The proposed DL approach achieved improved
classification results and model has shown the improved results with HMI data
set which is equal to 97.8%, 96.25% 100%, 98%, and 93.37% for accuracy, precision, recall, F-score and specificity respectively. The proposed DL based model
has achieved improved results with more robust solar spot recognition system to
monitor solar activities.

Keywords: Sunspot · Deep Learning · CNN · Transfer Learning

F. Saeed et al. (Eds.): ICACIn 2022, LNDECT 179, pp. 419–428, 2023.
https://doi.org/10.1007/978-3-031-36258-3_37

1 Introduction

Sun considered to be the core research objective in astrophysics from 17th century [1], conventionally Sunspot observation was carried out by drawing and also included location, sunspot number, and area of the sunspots [2]. In recent days solar physics in conjunction with machine learning, Computer vision and deep learning techniques has showed various developments like event detection like coronal hole, sunspot, prominence, solar flares [3, 4]. In the recent days DNN algorithms outperformed the tasks of classification and detection [5]. There are various studies multidisciplinary approaches addressed in the filed of Microscopic imaging [6, 7], Medical imaging [8–11], CAD based systems [12, 13] and anomaly detection [8, 14, 15]. Solar activates plays an important role in determining the efficient observation of space weather. Sunspots are visualized as dark patches on solar photosphere due variation in surface temperature, sunspots [16] appearance is dynamic in nature and turns out be a challenging task for differentiating among sunspots and group of sunspots manual visualization requires experience human expertise and deep learning based approaches could impact to enhance the decision making systems for better understanding the solar.

In recent days there has been improvement in studies of solar exploration with increased number of space missions and improved instrumentation technologies which lead researchers with large amount of solar activity data [17]. Many researchers with the help of deep learning techniques has introduced various methods to detect the sunspots [18–20], filament recognition [21], flare detection [22–24], and sunspot groups [25, 26] various object of interests for observing the solar activities.

In recent works Deep Convolution Neural Networks (DCNN) has gained focus in solar physics for sun activity tracking, DCNN algorithms for the task of image classification and detection in the field of computer vision gained popularity and input is images with labels. The proposed work consists of SDO data sources. Sunspot and quiet sun images can be seen in Fig. 1 provided by NASA SDO[1] mission database repository.

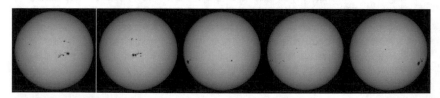

Fig. 1. Sunspot images

The paper organized as follows. In Sect. 2. We discuss the previous works with deep learning and machine learning methods applied to understand the solar activity tracking and classification. In Sect. 3. We address the Proposed model for addressing sunspot detection. In Sect. 4. We present experimental analysis and discussion and finally, in Sect. 5. Conclusion of the proposed model.

---

[1] https://sdo.gsfc.nasa.gov/data/.

2   Related Work

The improved instrumentation led to exponential generation of data such as image data, spectral data and time series data in the field of Astronomy [27]. In recent days the large quantum of data generated by space and ground based observatories, existing data quantity has attracted and initiated multidisciplinary research.

The image processing terminology was introduced for sunspot detection by Colak & Qahwaji [28], Magnetogram and MDI intensitygram images these images are the indicators of magnetic fields visible in the photosphere of sun, image processing technique has used to detect the sunspot and sunspot groups with the accuracy of 99% and 92% respectively [28, 29]. In [30], detection of sunspot with morphological approaches with adoptive threshold based methods with 95% recognition rate on Huairou Solar Observing Station (HSOS) full-disk vector magnetic field images. Next, In [31], Ruben du Toit et al. addressed the task of sunspot detection and tracking with OpenCV library with edge detection and scale-invariant features for localization of sunspot, tracking is followed by Discriminative Correlation Filter with Channel and Spatial Reliability method and Kernelized Correlation Filters were employed for tracking with Michelson Doppler images as input for the proposed approach.

Deep Convolution Neural Networks(DCNN) methods are most widely used in recent days for tracking and sun activities such as sunspot detection [20], Solar flare Prediction [24, 32], ribbons [33], coronal holes [4] and Prominences [4]. In continuation various approaches for solar event detection and classification. Pandey et al. in [24], deep learning based solar flare prediction with DCNN based approach with Full-disk magnetogram images from Helioseismic and Magnetic Imager (HMI) onboard Solar Dynamics Observatory (SDO) were used for flare prediction as binary and multiclass classification with AlexNet, VGG16 and ResNet34 as base architecture for feature extraction. Evaluation performance is done by True Skill Statistics (TSS) with 0.47 and 0.55 for binary classification, for multi-class classification as 0.36, and Heidke Skill Score (HSS) is noted as for Binary classification 0.46 and 0.43 and Multiclass scenario 0.31 scores with alexnet employed feature extraction. In [32], they proposed solar flare prediction is done with Transfer Learning Alexnet based architecture with adaptive average pooling and log softmax is used for classification of flares. The performance of the proposed model is evaluated with TSS and HSS with datatype as augmentation, oversampling, and normal datasets, noted that TSS is 0.47 with oversampling, 0.44 with data augmentation, 0.63 with normal dataset, and HSS is 0.35 for oversampling, 0.37 with data augmentation, 0.62 with normal dataset. Love et al. in [33], Solar flare observations with CNN based approach with AiA 1,600° images were used to detect flare ribbon observations and were able to achieve overall accuracy of 94% with the K-fold validation technique. He et al. in [25], addressed the sunspot group classification with HMI and MDI Magnetogram with DCNN based CornerNet-Sac-cade method is employed for the detection of sunspot group with Mount Wilson Magnetic Classification. The proposed work noted that results with performance metrics as Accuracy (94%), Recall (93%), AP (90%) and Precision (94%). Solar activity monitoring tasks can be eased and can be improved with computer vision techniques.

3 Proposed Methodology

The proposed work deals with the classification of sunspot images into sunspot or quiet sun. Transfer learning methods are adopted with VGG16 as base architecture. The input images were RGB channel with resized to standard size of VGG16 (Fig. 2).

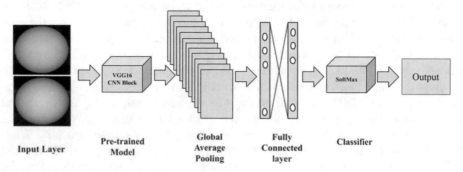

Fig. 2. DCNN based model for classification of sunspot;

3.1 Dataset Preparation

Solar Dynamics Observatory (SDO) Helioseismic and Magnetic Imager (HMI) images were considered. The images used for experimentation are publicly available and retrieved from NASA SDO [34, 20]. All images in this database are resized a fixed size of 224 × 224 pixels. And dataset split into Training, Testing and validation.

3.2 Proposed Architecture

DCNN approach employed to tackle task of sunspot classification with Transfer Learning based model with VGG16 [35] as base architecture. Deep features extracted by the input SDO images passed to softmax for prediction.

In this study we present deep learning based on CNN is used for the classification of sun images into quiet sun or sunspot. Features are generated by raw input images hierarchically based on the depth of the model.

Convolution Neural Networks are imitation of the cerebral cortex of human brain large training amount datasets requires for training a complex model. Features are extracted with operations like filtering, normalization and nonlinear activation operation and learning of algorithm is carried out with backpropagation algorithm and gradient descent optimization algorithms.

Deep Convolution neural network is stack of Input layer, convolution layer, Pooling layer, fully connected layer and output layers. These layers are building block of any deep convolution architecture. In context of image processing feed forward process is adopted for feature extraction, Convolution layer consists of multiple convolution filters of similar size to extract feature. Every filter is matrix of two-dimension with corresponding weights, the value of every neuron for present convolution layer is the result obtained by multiplication of data of previous layer with the convolution filter, and the addition of corresponding offset. The feature extraction involves the sequential scan of filter on input data of upper layer according to feature stride factors. Here Input image of size a * b passed through input layer to convolution block, convoluting with given kernel size x * y gives output known as *feature map*, later non linearity added to the network like Relu, ELU, sigmoid, tanh and LeakyRelu most commonly used functions.

$$C_{ab} = \sigma \left(\sum_{xy} w_{xy} v_{(a+x)(b+y)} + B \right) \tag{1}$$

where C_{ab} referred as unit in feature map at location $(x; y)$, σ represents the activation function. w_{xy} is weights of kernel $v_{(a+x)(b+y)}$ designates an input unit at location $(a + x, b + y)$, and B is bias of feature map.

Activation function considered is ReLU in general form its shown as equation (**) here any negative values is mapped to 0 and rest values no change.

$$f(r) = \max(o, r) \tag{2}$$

The ReLU hyperparameter saturates for negative inputs.

Pooling is most important operation in terms of reducing feature map dimension and overfitting of model, Pooling operation also called as down sampling. There are few pooling operations namely Max pooling and average pooling layers. Here the key concept is to reduce the size by mapping the values based on application or object pixel distribution. Stride factor can be used as pixel shifts over the input matrix. Padding can be used to add pixels which helps to keep width and height of previous layer. When we consider max pooling, it covers the most active feature from the pooling region which implies that it could be generalized for collecting texture information and Average pooling is responsible for background information preservation due to consideration of all features of the pooling region. Hyper parameters in pooling layers are Filter size and stride. Pooling layer represented as $w \times h \times d$ which is width, height and depth and with kernel size as f and stride as s the pooling computed as $w_p \times h_p \times d_p$ represents width, height and depth after the pooling.

$$w_p = (w - f)/s + 1 \tag{3}$$

$$h_p = (h - f)/s + 1 \tag{4}$$

$$d_p = d \tag{5}$$

After the extraction of local features from convolution layer. Fully connected layer applied to enhance the nonlinear mapping, perceiving of global information and aggregation local features to perform classification. Each neuron in the l layer is connected to $l + 1$ layer. The formula for fully connected layer is

$$y_i^{(l)} = f \left(\sum_{i=1}^{n} x_i^{(l-1)} * w_{ji}^{(l)} + b^{(l)} \right) \tag{6}$$

where n is no of neurons in the previous layer, l is the present layer, $w_{ji}^{(l)}$ is the connecting weight of neurons j in present layer and neurons i in the previous layer, $b^{(l)}$ is the bias of j neuron and f as activation function.

The final output layers derived from fully connected layers which are located at the end of convolution blocks, output layer fed input from previous these layers take the output from the hidden layers and process it such that for each data file a pre-defined class is predicted by the network. Classification probability. The classification probability of the image is calculated by the CNN's softmax layer, which is formulated as follows:

$$p(x_i) = \frac{e^{x_i}}{\sum_{K=1}^{K} e^{x_k}} \tag{7}$$

where x_i represents the output of the fully connected layer for class i, K is the total number of classification categories, and p represents the classification probability the outputs from multiple neurons mapped to (0, 1) interval. SoftMax outcome will be classification probability for each category and assigns the maximum classification probability value and corresponding category as the final outputs.

3.3 Performance Evaluation

The proposed sunspot prediction model with binary classification strategy to classify input data as Sunspot or Quiet sun via DCNN model. Performance evaluation is carried out with Accuracy, Precision, Recall, F-measure and Specificity are expressed in following Eqs. 8–12.

$$\text{Precision} = \frac{TN}{TN + FP} \tag{8}$$

$$AZ = \frac{TP + TN}{TP + FN + TN + FP} \tag{9}$$

$$F1 - \text{score} = \frac{2 \cdot TP}{2 \cdot TP + FP + FN} \tag{10}$$

$$\text{Recall} = \frac{TP}{TP + FP} \tag{11}$$

$$SE = \frac{TP}{TP + FN} \tag{12}$$

where TP, TN, FP, and FN were defined to represent the number of true positive, true negative, false positive, and false negative respectively.

4 Results and Discussion

The proposed work deals with the classification of sunspot images into sunspot or quiet sun with the help of 5fold cross validation technique is used to identify the sunspot.

Table 1. Results.

	Accuracy	Precision	Recall	F-measure	Specificity
1st fold	100	100	100	100	100
2nd fold	97.83	95.84	100	97.87	95.6
3rd fold	100	100	100	100	100
4th fold	99.71	100	100	100	88.4
5th fold	91.46	85.43	100	92.14	82.89
Average	97.80	96.254	100	98.00	93.37

Table 1 present classification models to classify sunspot were evaluated on a completely new unseen testing dataset. The data has been collected from different resources. The training dataset, as described in Sect. 1. The performances of the proposed models in terms of accuracy, F1 score, precision, and recall. From the Table 1, we can see that the proposed model has got good result using VGG16 model achieved good results with respect to all evaluation metrics (Fig. 3).

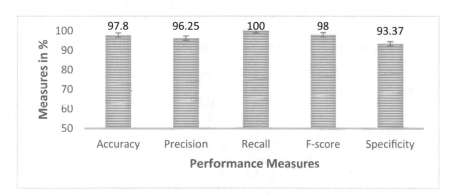

Fig. 3. Performance measures for sunspot classification

The proposed DL approach achieved improved classification results and model has shown the improved results with HMI data set which is equal to 97.8%, 96.25% 100%, 98%, and 93.37% for accuracy, precision, recall, F-score and specificity respectively.

5 Conclusions

Sunspots are known as key object of astrophysics and sunspots are the most prominent feature for assessing space weather and are located in solar photosphere. In this work, we have implemented a deep learning model which automatically detects sunspots from HMI image datasets. Data divided into training, testing and validation subset and passed to classification pipeline. The proposed work focus classification of sunspot via DCNN based approaches has shown average of 97.8% of overall accuracy K-fold cross validation method. The performance is evaluated with Accuracy, Recall, Precision, F-measure and specificity. Based on the experimental setup we conclude that proposed DCNN based model efficient to classify the HMI images into sunspot or quiet sun. Realtime detection and tracking could be addressed as future work with the help of various object tracking algorithms.

Acknowledgments. We are thankful to Dr. Sushree S Nayak, for the useful discussion. The data used for this work is a courtesy of NASA/SDO. We are thankful to Dr Naveen Mourya and HPC Lab, DoS in Computer Science University of Mysore, India.

References

1. Mathieu, S., Lefèvre, L., von Sachs, R., Delouille, V., Ritter, C., Clette, F.: Nonparametric monitoring of sunspot number observations. J. Qual. Technol., 1–15 (2022). https://doi.org/10.1080/00224065.2022.2041376
2. Carvalho, S., Gomes, S., Barata, T., Lourenço, A., Peixinho, N.: Comparison of automatic methods to detect sunspots in the Coimbra observatory spectroheliograms. Astron. Comput. **32**, 1–43 (2020). https://doi.org/10.1016/j.ascom.2020.100385
3. Yi, K., Moon, Y.-J., Lim, D., Park, E., Lee, H.: Visual explanation of a deep learning solar flare forecast model and its relationship to physical parameters. Astrophys. J. **910**(1), 8 (2021). https://doi.org/10.3847/1538-4357/abdebe
4. Baek, J.-H., et al.: Solar event detection using deep-learning-based object detection methods. Sol. Phys. **296**(11), 1–15 (2021). https://doi.org/10.1007/s11207-021-01902-5
5. Tan, M., Le, Q.V.: EfficientNet: rethinking model scaling for convolutional neural networks (2019)
6. Chola, C., et al.: Gender identification and classification of drosophila melanogaster flies using machine learning techniques. Comput. Math. Methods Med. **2022** (2022). https://doi.org/10.1155/2022/4593330
7. Mestetskiy, L.M., Guru, D.S., Benifa, J.V.B., Nagendraswamy, H.S., Chola, C.: Gender identification of Drosophila melanogaster based on morphological analysis of microscopic images. Vis. Comput. (2022). https://doi.org/10.1007/s00371-022-02447-9
8. Al-antari, M.A., Han, S.M., Kim, T.S.: Evaluation of deep learning detection and classification towards computer-aided diagnosis of breast lesions in digital X-ray mammograms. Comput. Methods Programs Biomed. **196**, 105584 (2020). https://doi.org/10.1016/j.cmpb.2020.105584
9. Chola, C., Mallikarjuna, P., Muaad, A.Y., Bibal Benifa, J.V., Hanumanthappa, J., Al-antari, M.A.: A hybrid deep learning approach for COVID-19 diagnosis via CT and X-ray medical images. Comput. Sci. Math. Forum **2**(1), 13 (2021)
10. Mehrrotraa, R., et al.: Ensembling of efficient deep convolutional networks and machine learning algorithms for resource effective detection of tuberculosis using thoracic (chest) radiography. IEEE Access **10**, 85442–85458 (2022). https://doi.org/10.1109/ACCESS.2022.3194152
11. Bin Heyat, M.B., et al.: Wearable flexible electronics based cardiac electrode for researcher mental stress detection system using machine learning models on single lead electrocardiogram signal. Biosensors **12**(6) (2022). https://doi.org/10.3390/bios12060427
12. Al-masni, M.A., et al.: Simultaneous detection and classification of breast masses in digital mammograms via a deep learning YOLO-based CAD system. Comput. Methods Programs Biomed. **157**, 85–94 (2018). https://doi.org/10.1016/j.cmpb.2018.01.017

13. Al-antari, M.A., Hua, C.-H., Bang, J., Lee, S.: Fast deep learning computer-aided diagnosis of COVID-19 based on digital chest x-ray images. Appl. Intell. **51**(5), 2890–2907 (2020). https://doi.org/10.1007/s10489-020-02076-6

14. Chola, C., et al.: IoT based intelligent computer-aided diagnosis and decision making system for health care. In: 2021 International Conference on Information Technology ICIT 2021 - Proceedings, pp. 184–189, July 2021. https://doi.org/10.1109/ICIT52682.2021.9491707

15. Hanumanthappa, J., Muaad, A.Y., Bibal Benifa, J.V., Chola, C., Hiremath, V., Pramodha, M.: IoT-based smart diagnosis system for healthcare. In: Karrupusamy, P., Balas, V.E., Shi, Y. (eds.) Sustainable Communication Networks and Application. LNDECT, vol. 93, pp. 461–469. Springer, Singapore (2022). https://doi.org/10.1007/978-981-16-6605-6_34

16. Yu, L., Deng, L., Feng, S.: Automated sunspot detection using morphological reconstruction and adaptive region growing techniques. In: Proceedings of the 33rd Chinese Control Conference, CCC 2014, pp. 7168–7172 (2014). https://doi.org/10.1109/ChiCC.2014.6896184

17. Tang, R., et al.: Multiple CNN variants and ensemble learning for sunspot group classification by magnetic type. Astrophys. J. Suppl. Ser. **257**(2), 38 (2021). https://doi.org/10.3847/1538-4365/ac249f

18. Ling, L.I., Yan-mei, C.U.I., Si-qing, L.I.U., Lei, L.E.I.: Automatic detection of sunspots and extraction of their feature parameters. Chin. Astron. Astrophys. **44**(4), 462–473 (2020). https://doi.org/10.1016/j.chinastron.2020.11.003

19. Armstrong, J.A., Fletcher, L.: Fast solar image classification using deep learning and its importance for automation in solar physics. Sol. Phys. **294**(6), 1–23 (2019). https://doi.org/10.1007/s11207-019-1473-z

20. Chola, C., Benifa, J.V.B.: Detection and classification of sunspots via deep convolutional neural network. Glob. Transit. Proc., 0–7 (2022). https://doi.org/10.1016/j.gltp.2022.03.006

21. Zhu, G., Lin, G., Wang, D., Liu, S., Yang, X.: Solar filament recognition based on deep learning. Sol. Phys. **294**(9), 1–13 (2019). https://doi.org/10.1007/s11207-019-1517-4

22. Ribeiro, F., Gradvohl, A.L.S.: Machine learning techniques applied to solar flares forecasting. Astron. Comput. **35**, 100468 (2021). https://doi.org/10.1016/j.ascom.2021.100468

23. Nishizuka, N., Sugiura, K., Kubo, Y., Den, M., Ishii, M.: Deep flare net (DeFN) model for solar flare prediction. Astrophys. J. **858**(2), 113 (2018). https://doi.org/10.3847/1538-4357/aab9a7

24. Pandey, C., Angryk, R.A., Aydin, B.: Deep neural networks based solar flare prediction using compressed full-disk line-of-sight magnetograms. Commun. Comput. Inf. Sci. CCIS **1577**, 380–396 (2022). https://doi.org/10.1007/978-3-031-04447-2_26

25. He, Y., Yang, Y., Bai, X., Feng, S., Liang, B., Dai, W.: Research on mount wilson magnetic classification based on deep learning. Adv. Astron. **2021**, 1–15 (2021). https://doi.org/10.1155/2021/5529383

26. Fang, Y., Cui, Y., Ao, X.: Deep learning for automatic recognition of magnetic type in sunspot groups. Adv. Astron. **2019** (2019). https://doi.org/10.1155/2019/9196234

27. Meher, S.K., Panda, G.: Deep learning in astronomy: a tutorial perspective. Eur. Phys. J. Spec. Top. **230**(10), 2285–2317 (2021). https://doi.org/10.1140/epjs/s11734-021-00207-9

28. Colak, T., Qahwaji, R.: Automatic sunspot classification for real-time forecasting of solar activities. In: Proceedings of the 3rd International Conference on Recent Advances in Space Technologies, RAST 2007, pp. 733–738, July 2007. https://doi.org/10.1109/RAST.2007.4284089

29. Colak, T., Qahwaji, R.: Automated McIntosh-based classification of sunspot groups using MDI images. Sol. Phys. **248**(2), 277–296 (2008). https://doi.org/10.1007/s11207-007-9094-3

30. Zhao, C., Lin, G., Deng, Y., Yang, X.: Automatic recognition of sunspots in HSOS full-disk solar images. Publ. Astron. Soc. Aust. **33**(2016), 1–8 (2016). https://doi.org/10.1017/pasa.2016.17

31. Du Toit, R., Drevin, G., Maree, N., Strauss, D.T.: Sunspot identification and tracking with OpenCV. In: 2020 International SAUPEC/RobMech/PRASA Conference SAUPEC/RobMech/PRASA 2020, pp. 1–6 (2020). https://doi.org/10.1109/SAUPEC/RobMech/PRASA48453.2020.9040971

32. Pandey, C., Angryk, R.A., Aydin, B.: Solar flare forecasting with deep neural networks using compressed full-disk HMI magnetograms. In: Proceedings of the 2021 IEEE International Conference on Big Data 2021, February 2022, pp. 1725–1730 (2021). https://doi.org/10.1109/BigData52589.2021.9671322

33. Love, T., Neukirch, T., Parnell, C.E.: Analyzing AIA flare observations using convolutional neural networks. Front. Astron. Sp. Sci. **7**, 1–8 (2020). https://doi.org/10.3389/fspas.2020.00034

34. Baranyi, T., Győri, L., Ludmány, A.: On-line tools for solar data compiled at the debrecen observatory and their extensions with the Greenwich sunspot data. Sol. Phys. **291**(9–10), 3081–3102 (2016). https://doi.org/10.1007/s11207-016-0930-1

35. Simonyan, K., Zisserman, A.: Very deep convolutional networks for large-scale image recognition. In: 3rd International Conference on Learning Representations ICLR 2015 - Conference Track Proceedings, pp. 1–14 (2015)

Arabic Hate Speech Detection Using Different Machine Learning Approach

Abdullah Y. Muaad[1,2](✉) [iD], J. (Jayappa Davanagere) Hanumanthappa[1](✉) [iD],
S. P. Shiva Prakash[3] [iD], Mohammed Al-Sarem[4,5] [iD], Fahad Ghabban[5] [iD],
J. V. Bibal Benifa[6] [iD], and Channabasava Chola[1,6] [iD]

[1] Department of Studies in Computer Science, University of Mysore, Manasagangothri, Mysore 570006, India
abdullahmuaad9@gmail.com
[2] IT Department, Sana'a Community College, Sana'a 5695, Yemen
[3] Department of Information Science and Engineering, JSS Science and Technology University, Mysuru 570006, Karnataka, India
[4] Department of Computer Science, Saba'a Region University, Mareb, Yemen
[5] College of Computer Science and Engineering, Taibah University, Medina 42353, Saudi Arabia
Fghaban@taibahu.edu.sa
[6] Department of Computer Science and Engineering, Indian Institute of Information Technology Kottayam, Kottayam, Kerala, India

Abstract. Hate speech is defined as an expression that targets an individual or community on the aspects like religion, sexual orientation, race, political opinion, and origin. Recently, hate speech on social media especially in the Arabic language has been exponentially increased and led to severe causes. Various studies had been conducted on social media platforms adopted by people to broadcast their opinions. This work aims to develop a model that is able to handle detection and classification of Arabic hate speech and offensive language. The experiments are carried out in using various machine learning (ML) and deep learning (DL) models. In this work, Arabic Hate Speech Detection (AHSD) model is proposed which composed of pre-processing, feature extraction, detection, and classification to identify hate speech on the Arabic benchmark dataset. The proposed model shows improved results. The transfer learning approach model exhibits superior performance compared to all other ML models in terms of accuracy, precision, recall, and F1 scores, achieving improvements of 84%, 79%, 80%, and 79%, respectively.

Keywords: Arabic Text Detection · Deep Learning · Hate speech Detection · Natural Language Processing (NLP) · AraBERT

1 Introduction

Over 1.57 billion people worldwide and most of them live in Arab and Islamic states speak Arabic as their first or second language. According to the Pew Research Center's Forum on Religion demographic survey, these Muslim people live in more than 200

F. Saeed et al. (Eds.): ICACIn 2022, LNDECT 179, pp. 429–438, 2023.
https://doi.org/10.1007/978-3-031-36258-3_38

countries[1] [1]. Meanwhile, in 2009, it was estimated, that Muslims were 23% of the world's population. Approximately 447 million people are native speakers who speak Arabic as their first language and it is also the sixth most widely spoken language in the world [2, 3]. Arabic is written from right to left, also it has a rich morphology and a complex orthography [4, 5]. Simultaneously there are the different domains of Arabic text detection (ATD) such as hate speech detection [6], fake news [7], misogyny [8], Arabic COVID-19 rumors [9] and, so on.

One of the advantages of social media is that it allows individuals and groups with diverse interests to express themselves openly. Because of the openness of social media, issues such as fake news, hate speech, harsh language, rumors, and misogyny can be propagated by a small group of individuals [10, 11]. Detecting vulnerable posts like hate speech or offensive text due to their speed outreach on social media platforms is a challenging task. Although, this problem is well addressed for the English language by many researchers, when it comes to Arabic there is a need for robust and reliable models [12].

The increased usage of different social media platforms among youth for communication as well as for social interaction encourages them to share their opinion. It is noticed that people convey their sentiments, and emotions in online mode through posts and tweets on different social media networks as we mentioned above [13, 14]. The characterization of texts plays an important role due to the increase in online assaults on common people. And there is a rapid rise of users adopting social media as a medium for sharing their thoughts feelings and opinions [15].

In recent days, the increase in Arabic text content on major social networks like Twitter, Facebook, LinkedIn, and YouTube led to a rise in abusive and offensive content which is nearly impossible to handle manually due to the quantum of data being generated on daily basis. There have been various text detection studies on various languages such as English [13], Bengali [16], Tamil [17], Malayalam [17], Urdu [18], and a few studies on Arabic as [19]. One of the challenging tasks in Arabic text data is to identify hate speech or offensive language at the early stages. In addition, Arabic text detection is lack of datasets when compared to well-established studies.

In this work, we have implemented different ML, ensemble, and DL models to detect and classify Arabic hate speech. The structure of this work is as follows: Sect. 2 presents the related work. Section 3 gives an overview of the proposed methodology. After that, the experimental results and discussion have been discussed in Sect. 4. Finally, the conclusion of this work is presented in Sect. 5.

2 Related Work

Arabic hate speech affects the social media space in different forms, such as offensive text, racism, misogyny, sarcasm, sexism, and so on. Hereby, this section presents an overview of various studies that addressed detection of Arabic offensive text and Hate speech [20].

Alshalan et al. in [21] conducted a new model for hate speech using a Twitter dataset with deep learning approaches. In addition, in [22] Alshalan and H. Al-Khalifa also

---

[1] https://www.pewresearch.org/religion/2009/10/07/

investigated several ML-based methods. They used several pre-processing techniques, n-gram as a feature representation. Then, the SVM classifier and Linear Regression (LR) were applied. The highest achieved precision was of 74% and 75% respectively. In continuation of DL-based methods, the word-to-vector (w2V) model is used as a feature extraction technique and then fed into the CNN, GRU, and BERT methods. The results show that DL-based methods yields the good performance by 89% of AUROC, 83% of accuracy, 81% as precision, and F1-Score as 79%. Boucherit and Abainia in [23] proposed a model to detect abusive and offensive language detection in Algerian dialectal Arabic from Facebook comments with new corpus called DziriOFN with manual annotation. The best result by SVM was 74% for binary classification and 66% for multi-classification. They applied several machine-learning approaches for classification the comments using TF-IDF features. Beside the conventional ML classifiers (SVM, Multinomial naïve byes and Gaussian naïve Bayes), they also investigated the impact of several deep learning approaches such as CNN, BiLSTM, and Fast Text methods.

Aldjanabi et al., in [24] proposed a hate speech and offensive speech recognition model. They used different datasets called OSACT-HS, OSACT-OFF, L-HSAB, and T-HSAB. The SVM classifier is used for the experimentation followed by Transformer models namely AraBERT and MarBERT with Multi-Task Learning (MTL) with a task-specific layer. The best performance of the proposed method with OSACT-HS yielded an accuracy of 97.90% and F1-score of 88.73%, OSACT-OFF yielded the an accuracy of 95.20% and F1-score of 92.34, L-HSAB dataset achieved accuracy of 94.20% and F1-score of 87.18% and finally the T-HSAB dataset achieved accuracy of 80.86% and F1-score of 80.50%.

Hadj Ameur et al. [25] proposed a transformer-based model to classify the hate speech and multilabel fake news on the AraCOVID19-MFH dataset with Transformer-based models such as AraBERT, mBERT, AraBERT COV19, mBERT COV19, and Distilbert Multilingual BERT. The performance of the proposed transfer learning-based approach has shown a 92% of F-score for the tasks. Authors in [26] created a new dataset for hate speech and Abusive Language based Levantine dialects. They used Cohen's Kappa for agreement metrics and Krippendorff's alpha for consistency of the annotation. The authors in [27] mentioned that there is lack of models in the early detection of hate speech. Mohamed Aziz Ben Nessir et al. in [28] proposed a model for the detection of hate speech using Quasi-recurrent neural networks and Transformers.

Aly Mostafa et al. in [29] proposed a model which can solve the problem of long-tailed data distribution (Imbalance Data) using loss functions instead of traditional weighted cross-entropy loss. Based on all these works we understood there are different challenges still not covered which we are planning to cover in the future.

3 Proposed Methodology

3.1 The Proposed Architecture for Hate Speech Detection on Arabic Texts

In this paper, a new model for detecting Arabic hate speech in short text using ML and DL is presented. The steps of this model are depicted in Fig. 1. Pre-processing comes first, then representation with feature extraction, and ultimately detection then classification, as well as evaluation in the last step.

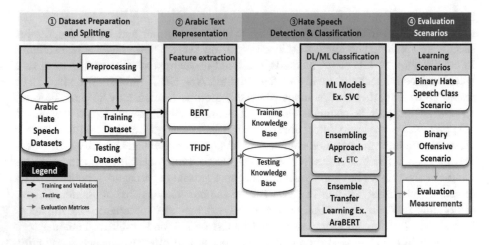

Fig. 1. The Framework Architecture of Hate Speech Detection Model

3.2   Pre-processing

The pre-processing phase can be used to prepare the text for further processing. After that, the procedures of Arabic text representation and feature extraction must be completed. Finally, for both scenarios, categorization is carried out. The text must be pre-processed in order to be ready for further processing and learning tasks. There are different pre-processing techniques such as removing punctuation, slang, stop-words, stemming, lemmatization, and so on. Convert text as unstructured data to structure to be acceptable for further processing by ML models [30, 31].

3.3   Arabic Text Representation and Feature Engineering

After text pre-processing, the next step is text representation and feature engineering for text. The most significant aspect of the requested domain expertise is connected to the problem is feature engineering. There are various varieties of it, including representation, feature extraction [32] and feature selection [33], among others. We are focusing on feature extraction (transformation) in this experimentation more than feature selection. To address issues such as semantic and syntactic information, performance which must be improved. In the result and discussion section, we will go through this in greater detail [34].

3.4   Arabic Text Detection and Classification

The process of identifying text is known as the detection task, and the task of categorizing or classifying it to the right class is called classification task [33, 35, 36]. Various techniques have been conducted in this work, as we will explain the result part in Sect. 5. Two ensembling techniques were adopted with the proposed model. For this purpose, various ML-based classifiers are used. The completed list of these classifiers are presented in Sect. 4.

3.5 Dataset

The dataset has been proposed by merging two datasets Hamdy et al. in 2022 [6] with another data by same author in 2020. The data set name called Fine-Grained Hate Speech detection on Arabic Twitter[2].

Table 1. Distribution of Arabic dataset for binary Hate Speech

#	Category	نوع الصنف	No. of Arabic Documents	Train	Test
1.	Non-hate speech	لا توجد كراهية	17113	13678	3435
2.	Hate speech	كراهية توجد	542	446	96
	Total		17655	14124	3531

3.6 Evaluation Metrics

We have used four evaluation metrics as follows:

$$\text{Recall} = \frac{TP}{TP + FN}, \tag{1}$$

$$\text{Precision} = \frac{TN}{TN + FP}, \tag{2}$$

$$\text{F1} - \text{score} = \frac{2 \cdot TP}{2 \cdot TP + FP + FN}, \tag{3}$$

$$\text{Accuracy} = \frac{TP + TN}{TP + FN + TN + FP}, \tag{4}$$

where TP, TN, FP, and FN were defined to represent the number of true positive, true negative, false positive, and false negative respectively.

4 Results and Discussion

The section presents the discussion of various experimental setup with standalone ML, ensemble learning, and transfer learning models. In addition, the hyper-tuning parameters of these classifiers are addressed as shown in Table 1. We expressed the results for hate speech detection with different models namely as follows: Passive-aggressive

---

[2] https://sites.google.com/view/arabichate2022/home.

434   A. Y. Muaad et al.

(PA), Logistic Regression (LR), Random Forest (RF), Decision Tree (DT), K-nearest Neighbors (K-NN), Linear Support Vector Classifier (Linear SVC), Support Vector Classifier (SVC), Naïve Bayes classifier (NB), Bernoulli Naïve Bayes classifier (BNB), Extra Tree Classifier (ET), Ensemble Bagging Classifier, Ensemble AdaBoost, Ensemble Gradient Boosting Classifier (GB), and Arabic Bidirectional Encoder Representations (Arab-BiER) as a Transformer.

The results of hate speech detection for the binary classification have shown in Table 2. In addition, As we have seen in the Fig. 1, although the task is binary classification but the result is more unsatisfactory, especially for f-score precision, and recall. The main reason is that the data is unbalanced. So, we have tried different ensembling models to solve this problem. However, the results still have the same problem. After a deep investigation, we noted that TFIDF representation was not sufficient. Hence, to avoid this issue, we have decided to implement one transfer learning approaches called AraBERT which were more efficient than all other models (Fig. 2).

Table 2. Evaluation Results for Hate speech Arabic text detection in Binary classification task

Type of Approach	Classifier Model	Accuracy %	Precision %	Recall %	F1-score %
Machine Learning Approach	PA	97	61	55	57
	LR	97	99	51	50
	DT	95	58	58	58
	K-NN	97	49	50	49
	Linear SVM	97	61	54	55
	SVM	97	86	51	52
	BNB	97	65	50	50
	ET	95	57	57	57
	NB	97	49	50	49
Ensemble Learning Approach	Bagging	96	63	59	61
	AdaBoost	97	68	57	60
	GB	97	57	53	54
	RF	97	58	53	54
Transfer Learning Approach	AraBERT	**84**	**79**	**80**	**79**

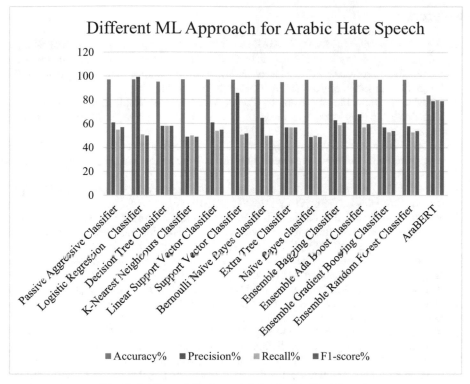

Fig. 2. Different ML Approach for Arabic Hate Speech

5 Conclusions

Detection of Arabic Hate speech, exponential rise in online space. Social media used different tools and platforms to express thoughts and sentiments. This leads to challenging tasks in ATD for filtering and monitoring the content. This article proposed a model to detect hate speech from user comments based on Arabic text. We have carried out our work merging two Arabic data sets about Arabic hate speech detection [7]. This proposed model has got very good results in terms of accuracy, precision, recall, and f-score which were equal to 84%, 79%, 80%, and 79% for hate speech binary classification using transfer learning. The comparison between ML and DL have been done. There are various gaps and limitations which could be considered for future work such as the creation a benchmark dataset with a large size and creation lexicons specific to hate speech detection. One more challenge is finding a model with the capability of working with mixed language problems. Finally, design a new model to cover imbalanced datasets using transfer learning techniques with Zero-shot learning and deep active learning to enhance the performance of the proposed model.

References

1. Albadi, N., Kurdi, M., Mishra, S.: Are they our brothers? Analysis and detection of religious hate speech in the Arabic Twittersphere. In: Proceedings of the 2018 IEEE/ACM International Conference on Advances in Social Networks Analysis and Mining, ASONAM 2018, pp. 69–76 (2018). https://doi.org/10.1109/ASONAM.2018.8508247
2. . Boukil, S., Biniz, M., El Adnani, F., Cherrat, L., El Moutaouakkil, A.E.: Arabic text classification using deep learning technics. Int. J. Grid Distrib. Comput. **11**(9), 103–114 (2018). https://doi.org/10.14257/ijgdc.2018.11.9.09
3. Lee, S., Muaad, A.Y., Jayappa, H., Al-antari, M.A.: ArCAR: a novel deep learning computer-aided recognition for character-level Arabic text representation and recognition. Algorithms (2021)
4. Habash, N.Y.: Introduction to Arabic natural language processing, vol. 3, no. 1 (2010). https://doi.org/10.2200/S00277ED1V01Y201008HLT010
5. Al-Sarem, M., Saeed, F., Alkhammash, E.H., Alghamdi, N.S.: An aggregated mutual information based feature selection with machine learning methods for enhancing IoT botnet attack detection. Sensors **22**(1) (2022). https://doi.org/10.3390/s22010185
6. Mubarak, H., Hassan, S., Chowdhury, S.A.: Emojis as anchors to detect Arabic offensive language and hate speech, pp. 1–21 (2022). 10.1017/xxxxx
7. Nagoudi, E.M.B., Elmadany, A., Abdul-Mageed, M., Alhindi, T., Cavusoglu, H.: Machine generation and detection of Arabic manipulated and fake news, pp. 1–15 (2020). http://arxiv.org/abs/2011.03092
8. Muaad, A.Y., Davanagere, H.J., Al-antari, M.A., Benifa, J.V.B., Chola, C.: AI-based misogyny detection from Arabic Levantine Twitter tweets. Comput. Sci. Math. Forum **2**(1), 15 (2021)
9. Al-Sarem, M., Alsaeedi, A., Saeed, F., Boulila, W., Ameerbakhsh, O.: A novel hybrid deep learning model for detecting Covid-19-related rumors on social media based on LSTM and concatenated parallel CNNs. Appl. Sci. **11**(17) (2021). https://doi.org/10.3390/APP11177940
10. Alkhamissi, B., Diab, M., Ai, R.: Meta AI at Arabic hate speech 2022: multitask learning with self-correction for hate speech classification, no. 2 (2022)
11. Abu Farha, I., Magdy, W.: Multitask learning for Arabic offensive language and hate-speech detection. In: Proceedings of the 4th Workshop on Open-Source Arabic Corpora and Processing Tools, with a Shared Task on Offensive Language Detection, pp. 86–90, May 2020. https://www.aclweb.org/anthology/2020.osact-1.14
12. Alkhair, M., Meftouh, K., Smaïli, K., Othman, N.: An Arabic corpus of fake news: collection, analysis and classification. In: Smaïli, K. (eds.) ICALP 2019. CCIS, vol. 1108, pp. 292–302. Springer, Cham (2019). https://doi.org/10.1007/978-3-030-32959-4_21. To cite this version : HAL Id : hal-02314246
13. Rodriguez, A., Chen, Y.L., Argueta, C.: FADOHS: framework for detection and integration of unstructured data of hate speech on Facebook using sentiment and emotion analysis. IEEE Access **10**, 22400–22419 (2022). https://doi.org/10.1109/ACCESS.2022.3151098
14. Liang, G., He, W., Xu, C., Chen, L., Zeng, J.: Rumor identification in microblogging systems based on users' behavior. IEEE Trans. Comput. Soc. Syst. **2**(3), 99–108 (2015). https://doi.org/10.1109/TCSS.2016.2517458
15. Chiril, P.: Détection automatique des messages haineux sur les réseaux sociaux (Doctoral dissertation, Université Toulouse 3) (2021)
16. Mridha, M.F., Wadud, M.A.H., Hamid, M.A., Monowar, M.M., Abdullah-Al-Wadud, M., Alamri, A.: L-Boost: identifying offensive texts from social media post in Bengali. IEEE Access **9**, 164681–164699 (2021). https://doi.org/10.1109/ACCESS.2021.3134154
17. Roy, P.K., Bhawal, S., Subalalitha, C.N.: Hate speech and offensive language detection in Dravidian languages using deep ensemble framework. Comput. Speech Lang. **75**, 101386 (2022). https://doi.org/10.1016/j.csl.2022.101386

18. Ali, M.Z., Ehsan-Ul-Haq, Rauf, S., Javed, K., Hussain, S.: Improving hate speech detection of Urdu tweets using sentiment analysis. IEEE Access **9**, 84296–84305 (2021). https://doi.org/10.1109/ACCESS.2021.3087827

19. Al-Sarem, M., Saeed, F., Alsaeedi, A., Boulila, W., Al-Hadhrami, T.: Ensemble methods for instance-based Arabic language authorship attribution. IEEE Access **8**, 17331–17345 (2020). https://doi.org/10.1109/ACCESS.2020.2964952

20. Djandji, M., Baly, F., Antoun, W., Hajj, H.: Multi-task learning using AraBert for offensive language detection. In: Proceedings of the 4th Workshop on Open-Source Arabic Corpora and Processing Tools, with a Shared Task on Offensive Language Detection, pp. 97–101, May, 2020. https://www.aclweb.org/anthology/2020.osact-1.16

21. Alshalan, R., Al-Khalifa, H., Alsaeed, D., Al-Baity, H., Alshalan, S.: Detection of hate speech in COVID-19-related tweets in the Arab region: deep learning and topic modeling approach. J. Med. Internet Res. **22**(12) (2020). https://doi.org/10.2196/22609

22. Alshalan, R., Al-Khalifa, H.: A deep learning approach for automatic hate speech detection in the Saudi Twittersphere. Appl. Sci. **10**(23), 1–16 (2020). https://doi.org/10.3390/app1023 8614

23. Boucherit, O., Abainia, K.: Offensive language detection in under-resourced Algerian dialectal Arabic language, pp. 1–9, March 2022

24. Aldjanabi, W., Dahou, A., Al-Qaness, M.A.A., Elaziz, M.A., Helmi, A.M., Damaševičius, R.: Arabic offensive and hate speech detection using a cross-corpora multi-task learning model. Informatics **8**(4), 1–13 (2021). https://doi.org/10.3390/informatics8040069

25. Ameur, M.S.H., Aliane, H.: AraCOVID19-MFH: Arabic COVID-19 multi-label fake news and hate speech detection dataset, May 2021. http://arxiv.org/abs/2105.03143

26. Mulki, H., Haddad, H., Bechikh Ali, C., Alshabani, H.: L-HSAB: a Levantine Twitter dataset for hate speech and abusive language, pp. 111–118 (2019). https://doi.org/10.18653/v1/w19-3512

27. Salminen, J., et al.: Developing an online hate classifier for multiple social media platforms. Hum.-Centric Comput. Inf. Sci. **10**(1), 1–34 (2020). https://doi.org/10.1186/s13673-019-0205-6

28. Aziz, M., Nessir, B., Rhouma, M., Haddad, H., Fourati, C.: iCompass at Arabic hate speech 2022: detect hate speech using QRNN and transformers, pp. 176–180, June 2022

29. Mostafa, A., Mohamed, O., Ashraf, A.: GOF at Arabic hate speech 2022: breaking the loss function convention for data-imbalanced Arabic offensive text detection, pp. 167–175, June 2022

30. Naseem, U., Razzak, I., Eklund, P.W.: A survey of pre-processing techniques to improve short-text quality: a case study on hate speech detection on Twitter. Multimedia Tools Appl. **80**(28–29), 35239–35266 (2020). https://doi.org/10.1007/s11042-020-10082-6

31. Rong, X.: word2vec parameter learning explained, pp. 1–21 (2014). http://arxiv.org/abs/1411.2738

32. Muaad, A.Y., et al.: Arabic document classification: performance investigation of preprocessing and representation techniques, vol. 2022 (2022)

33. Bahassine, S., Madani, A., Al-Sarem, M., Kissi, M.: Feature selection using an improved Chisquare for Arabic text classification. J. King Saud Univ. Comput. Inf. Sci. **32**(2), 225–231 (2020). https://doi.org/10.1016/j.jksuci.2018.05.010

34. Muaad, A.Y., et al.: Artificial intelligence-based approach for misogyny and sarcasm detection from Arabic texts. Comput. Intell. Neurosci. **2022**, 9 (2022). 7937667. https://doi.org/10.1155/2022/7937667

35. Dito, F.M., Alqadhi, H.A., Alasaadi, A.: Detecting medical rumors on Twitter using machine learning. In: 2020 International Conference on Innovation and Intelligence for Informatics, Computing and Technologies, 3ICT 2020, vol. 11, no. 8, pp. 324–332 (2020). https://doi.org/10.1109/3ICT51146.2020.9311957
36. Muaad, A.Y., et al.: An effective approach for Arabic document classification using machine learning. Glob. Transit. Proc., 0–5 (2022). https://doi.org/10.1016/j.gltp.2022.03.003

Modelling COViD-19 Daily New Cases Using GSTAR-ARIMA Forecasting Method: Case Study on Five Malaysian States

Siti Nabilah Syuhada Abdullah[1]($\boxtimes$), Ani Shabri[1], Faisal Saeed[2], Ruhaidah Samsudin[3], and Shadi Basurra[2]

[1] Department of Mathematics, Faculty Science, Universiti Teknologi Malaysia, 81300 Johor Bahru, Johor, Malaysia
nabilah1991@graduate.utm.my, ani@utm.my

[2] DAAI Research Group, Department of Computing and Data Science, School of Computing and Digital Technology, Birmingham City University, Birmingham B4 7XG, UK
faisal.saeed@bcu.ac.uk

[3] School of Computing, Faculty of Engineering, Universiti Teknologi Malaysia, 81300 Johor Bahru, Johor, Malaysia

Abstract. On March 11, 2020, the World Health Organization declared COVID-19 to be in a pandemic status after the number of confirmed cases had surpassed 118,000 cases in more than 110 countries worldwide. To aid decision-makers in battling the epidemic, accurate modelling and forecasting of the spread of confirmed and recovered COVID-19 cases is essential. The non-linear patterns that are frequently seen in these situations have inspired us to create a system that can record such alterations. A hybrid method was approached. Using hybrid models or combining several models has been a common practice to increase forecasting accuracy since the combination of forecasts from many models typically results in improved forecasting performance. Here, an error dataset was obtained from the GSTAR model previously and the error data for each location was modelled using ARIMA model. The final goal of this project is to develop a technique for predicting new COVID 19 cases using a hybrid GSTAR-ARIMA model. From March 16, 2020, to July 23, 2021, a case study was conducted on the number of daily confirmed COVID-19 cases in five Malaysian states. Global Change Data Lab at Oxford University furnished the dataset. GTAR-ARIMA with Uniform weights proves to be a viable forecasting option, ultimately proving to be the best model for forecasting daily new confirmed cases of COViD. GSTAR (1,10) was identified and ARIMA was used as the method of Error Modelling for all 5 states.

Keywords: GSTAR · ARIMA · Hybrid Model · Forecasting

1 Introduction

A deadly strain known as Corona-virus Disease 2019 (COVID-19) was discovered in Wuhan, China, towards the tail end of the year. The World Health Organization proclaimed a pandemic on March 11th, 2020, following the reporting of 118,000 cases

F. Saeed et al. (Eds.): ICACIn 2022, LNDECT 179, pp. 439–448, 2023.
https://doi.org/10.1007/978-3-031-36258-3_39

across 110 countries. There were more patient flows, which led to a lack of hospital beds nationwide and highly stressful situations. It is essential to comprehend the trend and dissemination of this virus to support decision-makers [1–3].

Many modelling, estimation, and forecasting techniques are used in an effort to understand and control this epidemic. Other studies used time-series methods including Auto-Regressive Integrated Moving Average (ARIMA) and Exponential Smoothing to analyze and forecast patterns in the COVID-19 epidemic across many countries, including China, India, and Italy [4–9].

Numerous everyday occurrences were connected not only to past occurrences but also to the places or the area in which they occurred. Time series analysis studies and spatial analysis studies were previously considered separately. When there are fixed sites spread throughout a number of locations, the time series analysis is used. Contrarily, the spatial analysis is used when a large number of locations are unknown but the time is. That said, the space-time analysis gained traction as science and technology developed.

The Generalized Space Time Autoregressive (GSTAR) model was initially brought by Ruchjana in 2002. Prior to this, [10] used the term GSTAR for a separate project in 1995; this project concerned the STAR model with a spatial correlation that happens at the same time and the same parameters in each location. Ruchjana described GSTAR as a STAR model in heterogeneous locations with different parameter values at each location instead [11]. To avoid ambiguous implications, this study uses Ruchjana's interpretation of GSTAR.

One of the most popular time series models is the autoregressive integrated moving average (ARIMA) model. The success of the ARIMA model can be attributed to its statistical properties and the well-known Box-Jenkins model-building process [12]. A number of exponential smoothing methods may be created using ARIMA models [13]. Whilst ARIMA models may explain a wide variety of time series, their fundamental disadvantage is the assumed linear structure of the model. Examples of these time series include pure moving average (MA), pure autoregressive (AR), and combination AR and MA (ARMA) series. The ARIMA model is unable to identify any nonlinear patterns since it is assumed that the time series values have a linear correlation structure.

Using hybrid models or combining several models has become a customary technique to boost the forecasting accuracy ever since the well-known M-competition [14], in which the integration of forecasts from more than one model frequently results in increased forecasting performance. The volume of literature on this topic has greatly expanded since the early work of Reid [15] and Bates and Granger [16]. Clemen [17] provided a detailed analysis and detailed literature in this topic. Utilizing each model's unique feature to find diverse patterns in the data is the basic idea underlying model combinations in forecasting. According to theoretical and empirical studies [18–21], combining several approaches can be an effective and beneficial strategy to improve forecasts.

The goal of this research is to develop a forecasting method for COVID 19 cases. A hybrid method was approached. This is where an error dataset was obtained from the GSTAR model previously and the error data for each location was modelled using ARIMA model. In this research, a case study was conducted, about the daily reported cases of COViD-19 in Malaysia's five most populous states from 16 March 2020 to 23 July 2021. A training set makes up 80% of the data set, while a test set makes up 20%.

The remainder of the essay is structured as follows. We evaluate the experimental setup in the next section. In Sect. 3, we probe deeper into the case study by describing the data. After that, Sect. 4 presents the case's empirical findings. The final reflections are included in Sect. 5.

2 Framework of Study

The basic structure of the suggested methods for forecasting is as seen in Fig. 1. A hybrid method was approached. Here, an error dataset was obtained from the GSTAR model previously and the error data for each state was modelled using ARIMA model. A rough representation of the procedure carried out is as follows:

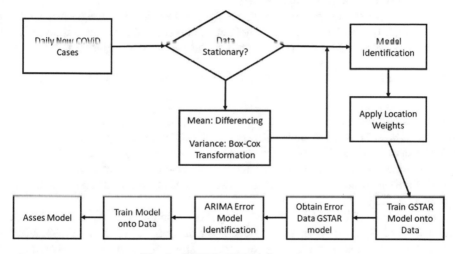

Fig. 1. GSTAR-ARIMA Procedure

The Augmented Dickey Fuller (ADF) test or examination of the MACF and MPACF cross correlation matrix schemes is used to initially determine whether or not the data are stationary. Establishing the temporal and the spatial order comes next when the data has reached a steady state. The determination of location weights is a challenge that frequently arises in GSTAR modelling [22]. There are 3 main weights commonly used. In this study however, uniform location weights are used. Uniform location weight is defined as [22, 23]:

$$w_{ij} = \frac{1}{n_i} \tag{1}$$

where n_i is the state's count of nearby sites to location i in the spatial lag 1. The attributes of the weight in this model are

$$W_{ij} > 0, \ W_{ii} = 0, \ \sum_{j=1}^{N} W_{ij} = 1, \forall_i, \ \sum_{i=1}^{N} \sum_{j=1}^{N} W_{ij} = N \tag{2}$$

The weight value provided here is equally assigned for each location. As a result, this location weight is frequently applied to data that is homogeneous or that uses the same distance for each place [10, 22]. The weight of W_{ij} in lag 1 is expressed by W in the form of $n \times n$ matrix as follows:

$$
W = \begin{bmatrix} 0 & W_{12} & \dots & W_{1N} \\ W_{21} & 0 & \dots & W_{2N} \\ \vdots & \vdots & \ddots & \vdots \\ W_{N1} & W_{N2} & \dots & 0 \end{bmatrix} \tag{3}
$$

Once the GSTAR model is fully developed, a dataset of error values obtained from comparing the GSTAR model output with the original data is obtained. This error dataset is used for error modelling using ARIMA.

An autoregressive integrated moving average, or ARIMA, was designed by Box and Jenkins [24, 25]. The $ARIMA(p, d, q)$ method can be determined using a time series of the actual value, y_t, where t is the time period, and the process is given by:

$$
y_t = c + \varphi_1 x_{t-1} + \varphi_2 x_{t-2} + \dots + \varphi_p x_{t-p} + \varepsilon_t - \theta_1 \varepsilon_{t-1} - \theta_2 \varepsilon_{t-2} - \dots - \theta_q \varepsilon_{t-q} \tag{4}
$$

where c is a constant, $\varphi_1, \varphi_2, \dots, \varphi_p$ are parameters of autoregressive (AR), whereas $\theta_1, \theta_2, \dots, \theta_q$ are moving average (MA) parameters. The random errors, ε_t are assumed to be independently and identically distributed with zero mean and constant variance, σ^2. Theoretically, ARIMA models are the most diverse category of forecasting models for the time series that may be transformed to become stationary using techniques like differencing [25].

The order of ARIMA is determined using PACF and the ACF. All potential models are mentioned for Ljung-Box test diagnostic verification. All of the large p-values for the Ljung-Box statistics indicate a good model. The fact that there are no patterns in the residues suggest that all of the information has been retrieved [25]. The accuracy of the created model would then be assessed using the Root Mean Square Error (RMSE). RMSE will be used to assess the precision of the model developed:

$$
RMSE = \sqrt{\frac{1}{n} \sum_{t=1}^{n} \left(y_t - \widehat{y_t} \right)^2} \tag{5}
$$

This study will compare the forecasting accuracy obtained from modelling COVID-19 confirmed cases using the GSTAR-ARIMA model with those obtained from GSTAR.

3   Case Study: Five Malaysian States

The key goal of this work is to forecast COVID 19 and predict the infection spread. This analysis is based on routine data from confirmed cases received in 5 most populated states in Malaysia. The states and population number reported by The Department of Statistics Malaysia in their official website as of 30th June 2021 are Selangor (6,573,862), Sabah (3,812,391), Johor (3,806,270), Sarawak (2,827,624) and Perak (2,509,587) [25]. The data of the daily confirmed cases used are in between 16th March 2020 and 23rd July 2021. The COVID-19 data in Malaysia are now made readily available online through The COVIDNOW initiative website (https://covidnow.moh.gov.my). The Ministry of Health (MoH) and the COVID-19 Immunization Task Force's open data efforts have been a novel and pleasant experience for the MoH and for everyone in Malaysia. Public and commercial stakeholders may now assess regulations, exchange new perspectives and analyses of the issue, and uphold better standards, thanks to open data. To assess the model's competitiveness, the COVID-19 data set was divided into an 80:20 training and testing set to measure the model effectiveness.

The graphic display of the data is as shown in Fig. 2. This disease has remained relatively under control up until September 2022. Until now, it has risen dramatically since its emergence, hitting a sizeable number of confirmed cases of 2525 on 31 December 2020. Table 1 includes a description of the data used in this analysis. The values for Minimum (Min.), Maximum (Max.), Mean, Median, 1st and 3rd Quartile (1st Qu., 2nd Qu.), are rounded up as they represent New Cases of COVID in each state. Here, we can infer that Selangor had the highest number of new cases daily maximum value. Selangor also had the largest variance meaning that the data fluctuated greatly from day to day. In the 1st quartile, Perak had a value of 0 indicating that it was the last state out of the 5 to have COVID cases.

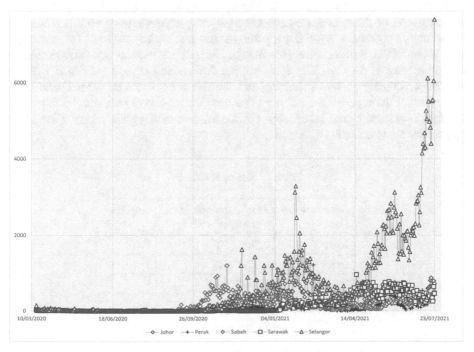

Fig. 2. Daily Confirmed new Cases of COVID-19 in Malaysia

Table 1. Summary of Dataset

	Johor	Perak	Sabah	Sarawak	Selangor
Min.	0	0	0	0	0
1st Qu.	1	0	2	1	6
Median	26	17	91	8	221
Mean	167	66	158	148	710
3rd Qu.	310	83	263	244	958
Max.	1103	1215	1199	960	7672
Variance	53 093.75	12 398.57	33 829.72	47 586.27	1 259 441.00
Std. Dev.	230.42	111.35	183.93	218.14	1 122.25

4 Results and Findings

4.1 Stationary Check

When there is no consistent change in the mean or variance values, the data is considered to be stationary in a time series. According to Markidakis et al. (1992), the visual representation of a time series data plot is frequently adequate to determine if the data is stationary or otherwise. Additionally, the Augmented Dickey Fuller (ADF) test or the MACF and MPACF cross correlation matrix methods are used to explicitly determine the stationarity of the data. The data requires differencing if the MACF and MPACF plots show a steady decline, indicating that the data is not stationary to the mean [22]. In contrast, if the superior and inferior bounds of the lambda (λ) are smaller than zero, the data is not stable and resistant to variations, necessitating the use of a Box Cox transformation to make the data stationary.

Table 2. Stationary Check

	Stationarity of Data	ADF p-value	ndiffs (data)	Lambda
1	Johor	0.5839	1	1
2	Perak	0.0513	0	1
3	Sabah	0.6436	1	1
4	Sarawak	0.5952	1	1
5	Selangor	0.9900	2	1

Table 2 summarizes the stationarity of data. As we can see, P-values below 0.05 demonstrate stationarity, whereas p-values upwards of 0.05 imply non-stationarity. In this study, only Perak's data was readily stationary. All other states' data and differencing were added to these data. Johor, Sabah and Sarawak went through 1 differencing while Selangor had to undergo differencing twice to reach stationarity.

4.2 Construction of GSTAR Model

The very next stage is to establish the temporal and the spatial order after the data is stationary. The order spatial employed here is order spatial 1, as a higher order spatial results in a more complicated model and less interpretability. When it comes to order time, the Vector Autoregressive (VAR) technique is used, which involves examining the least Akaike's Information Criterion (AIC) value (Table 3).

Table 3. Time Order Selection using VAR

	AIC(n)	HQ(n)	SC(n)	FPE(n)
1	49.797	49.899	50.057	4.23E+21
2	48.657	48.844	49.133	1.35E+21
3	48.124	48.396	48.816	7.94E+20
4	47.755	48.113	48.664	5.50E+20
5	47.423	47.865	48.548	3.94E+20
6	46.927	47.454	48.268	2.40E+20
7	46.693	47.305	48.250	1.90E+20
8	46.623	47.321	48.397	1.77E+20
9	46.495	47.277	48.485	1.56E+20
10	46.374	47.242	48.581	1.39E+20
P Selection	10	10	7	10

Based on the table, p = 10 is AR (10), hence the GSTAR model formed is GSTAR (1; 10). Next, the study tackles the task of selecting the weight to be used in the GSTAR Model. The uniform location weight used in the study is shown below:

$$w_{ij} = \frac{1}{n_i} = \frac{1}{5} = 0.2 \tag{6}$$

where $n_i = 5$ declares the 5 states (Johor, Perak, Sabah, Sarawak, Selangor) in the spatial lag 1.

The weight value provided by this location weight is equally assigned to each location. As a result, this location weight is frequently applied to data that is homogeneous or that uses the same distance for each place. The weight of W_{ij} in lag 1 is expressed by W in the form of 5 × 5 matrix as follows:

$$W = \begin{bmatrix} 0 & \cdots & 0.2 \\ \vdots & \ddots & \vdots \\ 0.2 & \cdots & 0 \end{bmatrix} \tag{7}$$

4.3 GSTAR-ARIMA

To further improve the forecasting model, a hybrid method was approached. Here, an error dataset was obtained from the GSTAR model previously and the error data for each state was modelled using ARIMA model (Table 4).

Table 4. ARIMA Error Modeling

State	ARIMA Model
Johor	ARIMA(4,0,0)
Perak	ARIMA(4,1,1)
Sabah	ARIMA(2,1,3)
Sarawak	ARIMA(4,1,0)
Selangor	ARIMA(2,1,2)

Table 5. Models' Performance Comparison

State	Training Set		Testing Set	
	GSTAR	GSTAR-ARIMA	GSTAR	GSTAR-ARIMA
Johor	92.367	86.559*	248.807	116.607*
Perak	72.812	72.048*	106.895	58.690*
Sabah	80.349	68.340*	119.358	49.188*
Sarawak	92.714	80.932*	331.158	135.286*
Selangor	203.709*	206.471	591.977	450.528*

4.4 Performance Model

The table above illustrates the RMSE values for Johor, Perak, Sabah, Sarawak and Selangor for both GSTAR and GSTAR-ARIMA models during the training and testing stages. Models were identified using the Training portion of the data. In the training stages, the GSTAR-ARIMA method outperformed the basic GSTAR model for all states except in Selangor. However, the testing stage has demonstrated astounding capabilities of the GSTAR-ARIMA model where all states performed better using the hybrid model compared to basic GSTAR. It was found that the hybrid model improves the performance significantly compared to the GSTAR models (Table 5).

5 Conclusion and Recommendations

It is safe to conclude that GSTAR-ARIMA is suitable to be used in forecasting daily new confirmed cases of COVID. In the case study, using data from Johor, Perak, Sabah, Sarawak and Selangor, this paper found that GSTAR (1,10) was the best Error Modeling conducted using ARIMA for all 5 states. GTAR-ARIMA with uniform weights proves to be the best model. Despite the fact that the hybrid model has performed well, it is unfortunate that the spread is increasing. Meanwhile, the incidence of infections is rising exponentially, and the number of infections is increasing. If Malaysians wholly take on the responsibility, the incidence of new cases will swiftly begin to decline. The accuracy of these predictions is dependent on a number of external factors. Further studies incorporating more Malaysian States should be carried out to get a more holistic view of the spreading trend. Furthermore, this model could also be adopted to studies of other epidemics such as HIV-AIDS, Polio, etc.

References

1. Velásquez, R.M.A., Lara, J.V.M.: Forecast and evaluation of COVID-19 spreading in USA with reduced-space Gaussian process regression. Chaos Solitons Fractals, 109924 (2020)
2. Yousaf, M., Zahir, S., Riaz, M., Hussain, S.M., Shah, K.: Statistical analysis of fore- casting COVID-19 for upcoming month in Pakistan. Chaos Solitons Fractals, 109926 (2020)
3. Ribeiro, M.H.D.M., da Silva, R.G., Mariani, V.C., dos Santos Coelho, L.: Short-term forecasting COVID-19 cumulative confirmed cases: perspectives for Brazil. Chaos Solitons Solitons Fractals, 109853 (2020)
4. Dehesh, T., Mardani-Fard, H., Dehesh, P.: Forecasting of COVID-19 confirmed cases in different countries with ARIMA models. medRxiv (2020). https://doi.org/10.1101/2020.03.13.20035345
5. Gupta, R., Pal, S.K.: Trend analysis and forecasting of COVID-19 outbreak in India. medRxiv (2020)
6. Chintalapudi, N., Battineni, G., Amenta, F.: COVID-19 disease outbreak forecasting of registered and recovered cases after sixty day lockdown in Italy: a data driven model approach. J. Microbiol. Immunol. Infect. (2020)
7. Kucharski, A.J., et al.: Early dynamics of transmission and control of COVID-19: a mathematical modelling study. Lancet Infect. Dis. (2020)
8. Wu, J.T., Leung, K., Leung, G.M.: Nowcasting and forecasting the potential domestic and international spread of the 2019-nCoV outbreak originating in Wuhan, China: a modelling study. Lancet **395**(10225), 689–697 (2020)
9. Zhuang, Z., et al.: Preliminary estimation of the novel coronavirus disease (COVID-19) cases in Iran: a modelling analysis based on overseas cases and air travel data. Int. J. Infect. Dis. **94**, 29–31 (2020)
10. Terzi, S.: Maximum likelihood estimation of a generalized STAR(p;1p) model. J. Ital. Stat. Soc. **3**, 377–393 (1995)
11. Ruchjana, B.N.: Suatu model generalisasi space-time autoregresi dan penerapanny pada produksi minyak bumi Disertation of Doctoral Program Institut Teknologi Bandung (2002)
12. Box, G.E.P., Jenkins, G.: Time Series Analysis, Forecasting and Control, Holden-Day, San Francisco, CA (1970)
13. McKenzie, E.D.: General exponential smoothing and the equivalent ARMA process. J. Forecast. **3**, 333–344 (1984)
14. Makridakis, S., et al.: The accuracy of extrapolation (time series) methods: results of a forecasting competition. J. Forecast. **1**, 111–153 (1982)
15. Reid, D.J.: Combining three estimates of gross domestic product. Economica **35**, 431–444 (1968)
16. Bates, J.M., Granger, C.W.J.: The combination of forecasts. Oper. Res. Q. **20**, 451–468 (1969)

17. Clemen, R.: Combining forecasts: a review and annotated bibliography with discussion. Int. J. Forecast. **5**, 559–608 (1989)
18. Makridakis, S.: Why combining works? Int. J. Forecast. **5**, 601–603 (1989)
19. Newbold, P., Granger, C.W.J.: Experience with forecasting univariate time series and the combination of forecasts (with discussion). J. R. Stat.. Soc. Ser. A **137**, 131–164 (1974)
20. Palm, F.C., Zellner, A.: To combine or not to combine? Issues of combining forecasts. J. Forecast. **11**, 687–701 (1992)
21. Winkler, R.: Combining forecasts: a philosophical basis and some current issues. Int. J. Forecast. **5**, 605–609 (1989)
22. Fadlurrohman, A.: Integration of GSTAR-X and uniform location weights methods for forecasting inflation survey of living costs in central Java. J. Intell. Comput. Health Inform. (JICHI) **1**(1), 20–25 (2020)
23. Suhartono, Subanar: The optimal determination of space weight in GSTAR model by using cross-correlation inference. J. Quant. Methods J. Devoted Math. Stat. Appl. Var. Field **2**(2), 45–53 (2006)
24. Box, G.E.P., Jenkins, G.M.: Time Series Analysis: Forecasting and Control. Holden-Day, San Fransisco (1976)
25. Abadan, S., Shabri, A.: Hybrid empirical mode decomposition-ARIMA for forecasting price of rice. Appl. Math. Sci. **8**(63), 3133–3143 (2014)
26. Population Clock by State. Department of Statistics Malaysia Official Portal. (n.d.). https://www.dosm.gov.my/v1/index.php?r=columnnew%2Fpopulationclock. Accessed 26 Mar 2022

Information Security

A Comparison of Ensemble Learning for Intrusion Detection in Telemetry Data

Naila Naz[1](✉), Muazzam A. Khan[1], Muhammad Asad Khan[2],
Muhammad Almas Khan[1], Sana Ullah Jan[3], Syed Aziz Shah[4], Arshad[5],
Qammer H. Abbasi[6], and Jawad Ahmad[3]

[1] Department of Computer Science, Quaid-I-Azam University, Islamabad, Pakistan
naznaila470@gmail.com
[2] Hazara University, Mansehra, Pakistan
[3] School of Computing, Edinburgh Napier University, Edinburgh, UK
J.Ahmad@napier.ac.uk
[4] Research Centre for Intelligent Healthcare, Coventry University, Coventry, UK
[5] School of Engineering and Built Environment, Glasgow Caledonian University, Glasgow, UK
arshad.arshad@gcu.ac.uk
[6] James Watt School of Engineering, University of Glasgow, Glasgow, UK

Abstract. The Internet of Things (IoT) is a grid of interconnected pre-programmed electronic devices to provide intelligent services for daily life tasks. However, the security of such networks is a considerable obstacle to successful implementation. Therefore, developing intelligent security systems for IoT is the need of the hour. This study investigates the performances of different Ensemble Learning (EL) approaches applied for intrusion detection in the IoT sensors' telemetry data. We compare the accuracy of various EL approaches in homogeneous and heterogeneous combinations using bagging, boosting, and stacking strategies. These EL approaches apply well-known Machine Learning (ML) models such as Decision Tree (DT), Naive Bayes (NB), Random Forest (RF), Logistic Regression (LR), Linear Discriminant Analysis (LDA) and linear Support Vector Machine (SVM). We evaluate and compare EL approaches for binary and multi-class classification tasks on the ToN-IoT Telemetry dataset for intrusion detection. The results show that stacking EL outperform stand-alone ML algorithms-based classifiers as well as bagging and boosting.

Keywords: Bagging · Ensemble Learning · Intrusion detection · IoT · Stacking Telemetry · ToN-IoT

1 Introduction

Internet of things (IoT) refers to a network of interconnected devices that can communicate with each other without human interference. IoT offers a variety of benefits in every walk of life, ranging from smart homes, smart healthcare, smart cities, smart transportation and smart retail [1]. The use of IoT devices has altered the lifestyle of people and has had a significant positive impact on their lives. In a typical setting, a large

© The Author(s), under exclusive license to Springer Nature Switzerland AG 2023
F. Saeed et al. (Eds.): ICACIn 2022, LNDECT 179, pp. 451–462, 2023.
https://doi.org/10.1007/978-3-031-36258-3_40

number of interconnected IoT devices collect data from the surrounding environment, exchange this telemetry data in a network to perform a task. These devices communicate with other devices using different types of messaging protocols such as Constrained Application Protocol CoAP, MQ Telemetry Transport (MQTT), and Data Distribution Service (DDS) at a specific IoT layer, as shown in Fig. 1. By 2025, it is expected that 41.5 billion IoT devices will be connected and may be vulnerable to numerous types of attacks [2]. As a result, protecting sensitive telemetry data of IoT devices and building intrusion detection systems (IDS) is the need of the day.

IDS is a monitoring device used to track network traffic, detect malicious activity, assess the severity and raise the alarm to the administrator. There are different types of IDS based on various categories such as information source, analysis procedure, detection method, learning method, and knowledge base [3]. IDSs are categorized into three main approaches: Host-based IDS (HIDS), Network- based IDS (NIDS) and Hybrid IDS. HIDS is installed on a local host and keeps track of attacks on the machine where it is installed. It monitors and detects intrusion on the host's file system, system calls, and events and alerts the system of a potential intrusion attempt. NIDS detects attacks by checking network-related data such as network packets, broadcast ranges of network protocols, traffic volume, and IP addresses [4]. The combination of HIDS and NIDS called hybrid IDS is also proposed in the literature to achieve real-time intrusion detection on both the host and network.

Based on detection methodology, IDS are categorized into Signature-based IDS (SIDS) and Anomaly-based IDS (AIDS). In SIDS (also known as misuse detection), the signatures, i.e., patterns of previously known attacks, are stored in a database. An intrusion signature is created by extracting the features from the intruded packet. When an activity is recorded, IDS observes its signature and compares it with stored patterns in the database. If the captured network packet pattern matches a stored attack's signature, an alarm is triggered. SIDS is effective for detecting known attacks, but it does not detect zero-day (unknown) attacks [5]. Thus, continuous identification of new signatures with experts' inspection is required to secure the system. Network intrusion detection and prevention system (Snort) [6] is an example of SIDS. An AIDS (also called behaviour-based detection) mimics the behaviour of the network, users, and computer systems under normal circumstances. Using various techniques, the IDS establishes a baseline from normal behaviour patterns. Any significant deviation of inspected pattern from the baseline gets flagged as an anomaly. A significant deviation in the behaviour of devices from their normal behaviour is identified as intrusion [7].

In literature, multiple IDSs were developed from single ML classifiers or an ensemble of classifiers for different problems. In intrusion detection, researchers mostly adopted EL techniques to enhance the performance of a single ML algorithm. EL is an ML paradigm that combines different ML algorithms to get a model that allows the overall performance to be effectively enhanced. Accumulating the benefits of multiple models, it transforms the weak learners into strong learners [8].

This paper compares the performances of EL techniques (Bagging, Boosting, and stacking) for telemetry data-based IDS. We evaluate these techniques on the ToN-IoT dataset to verify that the proposed stacking model is suitable for IDS design for telemetry data-based IoT networks.

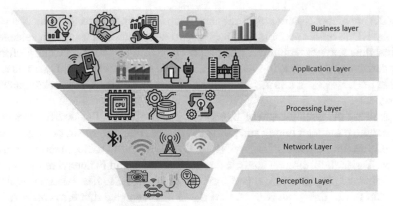

Fig. 1. Essential layers of generic IoT architecture

The remainder of the paper is structured as follows. Section 2 continues the study of existing EL methods used for IDS, while Sect. 3 gives an overview of different EL techniques. Section 4 discusses the description of the dataset and experimental setup, whereas the results are discussed in Sect. 5. The paper is concluded in Sect. 6 with a few closing points.

2 Related Work

Internet of things (IoT) refers to a network of interconnected devices that can communicate with each other without human interference. Machine learning and data analytics on IoT data offers a variety of applications in a variety of domains including smart healthcare [9] smart homes and buildings [10], smart cities [11], smart mobility and transportation [12], smart retail and marketing [13, 14], Energy generation and load forecasting [15, 16] and Smart Energy Management [17].

However, there are several security challenges and privacy concerns in providing these services [2]. There has been significant research efforts in addressing these challenges in the context of IoT security protocols [18], blockchain applications [19, 20], cryptographic methods [21], and machine learning methods for detecting intrusions [22].

In intrusion detection, ensemble methods have received much interest in recent years. This section reviews prevailing IDS research that has supported EL techniques for IoT networks. We present a summary of the literature in Table 1.

In [5], an ensemble-based IDS using XGBoost was discussed with a focus on specific botnet attacks against different protocols, i.e. DNS, HTTP, and MQTT. The proposed IDS was evaluated on the KDDCup99 dataset using five common evaluation metrics. The achieved accuracy of the proposed method was 99.95%. However, the KDDCup99 dataset is derived from the traditional networks, so it does not contain the telemetry data of sensors. Thus it could not be used as an adequate IoT benchmark dataset. Moreover, the proposed IDS generates significant overhead that degrades its performance. An Ada Boost EL via ML methods was implemented in [23] to create an effective and adaptable NIDS. They adopted DT, NB, and artificial neural network (ANN) as base methods.

The model prevents suspicious activities against IoT network specific protocols. UNSW-NB15 and NIMS botnet datasets were used to extract the features using the presented new statistical flow features technique. The proposed technique yields superior performance in comparison to three existing techniques: SVM, Markov chain (MC), and Bayesian network (BN). However, this method may not perform well against attacks against other IoT protocols.

To increase the effectiveness of anomaly-based IDS in IoT, [24] used a random forest classifier with parameter tuning and the different sizes of ensemble trees. They evaluated their AIDS on three data sets, UNSW-NB15, GPRS and NSL-KDD, in terms of FAR and accuracy. The authors employ statistical tests: Friedman and Nemenyi tests. The results revealed that the proposed model performs better for the NSL-KDD dataset (which does not contain sensor data) and worst for GPRS. It also depicted that the proposed method does not provide satisfactory results for the sensors-related data. A SNIDS architecture, ELNIDS, is proposed in [25]. This architecture is based on an ensemble of tree classifiers: boosting, bagging, subspace discriminant, and RUSBoosted tree. The authors created another dataset, RPL-NIDDS17, using the NetSim tool. The results depicted that the ensemble of the Boosted tree gets an accuracy of 94.5%. An ensemble model based on the Bayesian tree and RF as base classifiers and vote as meta classifier is presented in [26]. The suggested IDS model was tested using the KDDcup99 dataset and 10-fold cross-validation. They evaluate their model with accuracy and receiver operating characteristic (ROC) curves. A three-tier paradigm for anomaly-based IDS was developed in [27], consisting of feature selection, modelling of classifier, and validation. The first tier was feature selection, in which the appropriate feature set was selected using a hybrid strategy based on three evolutionary search strategies. The RT and bagging classifiers were ensembled at the second tier and tested on the UNSW-NB15 and NSL-KDD datasets. In the tier of validation, 10fcv was used to validate the model using four performance measures, precision, accuracy, FPR, and recall. The suggested classifiers were tested on the NSL-KDD dataset and scored 85.8%, 86.8%, and 88.0% for accuracy, sensitivity (recall), and detection rate, respectively. In comparison to state-of-the-art approaches, the author reported that the proposed classifiers increase the results on UNSW-NB15.

Both homogeneous methods (using RF) and Heterogeneous ensembles using NB, KNN, RIPPER, DT classifiers and LSTM were proposed in [28]. Classification of binary and multi-class attacks based on mentioned technique was evaluated on the UNSW NB15 dataset. According to the results, the accuracy of LSTM for binary and multi-classification is 80% and 72%, respectively. However, random forest achieved 98% for binary and 87.4% accuracy for multiclassification. The heterogeneous model attained the 0.97% accuracy for binary classification and 85.23% for multi-classification.

A two-phase anomaly detection model for the IIOT network was introduced in [29] study. The model ensemble the SVM and NB classifiers in the first phase. The train and test datasets were created using the Kfc-validation approach. In the second step, the findings of the ANN and RF were used as input for the model's classification. The results of RF and ANN were compared to select the best classifier. The three most commonly available data sets, WUSTL IIOT-2018, N BaIoT and Bot-IoT, were used to test the models. The evaluation metrics precision, recall, and accuracy was greater than 98% for all datasets.

In [8], the authors examined the ensemble approaches of bagging, boosting, and stacking on the UNSW-NB15 dataset for NIDS. They used NB, ANN, J48 and REPTree in the ensemble approaches. The stacking model with REPTree as a base learner and an ANN as a meta learner obtain the highest accuracy of 87.92%. [30] tested several ML algorithms on the ToN-IoT dataset. Random forest is selected to test the performance of EL for telemetry data of IoT devices.

3   Methods of Ensemble Learning

The notion of EL was first introduced in 1979 by Dasarathy [31] to increase the performance of a single ML algorithm. EL is an ML paradigm combining different ML algorithms to enhance the model's overall performance; thus, it transforms the weak learners into a strong learner. There are two main types of EL: homogeneous and heterogeneous EL. This section presents details of EL methods used for intrusion detection.

3.1   Homogeneous EL

Homogeneous EL is an ML technique where a single base model builds upon different subsets of data. Bagging and boosting are examples of homogeneous EL. The term "bagging" refers to "bootstrap aggregating". It is the most intuitive way of EL with a parallel ensemble process. One of the advantages of bagging is that it can also speed up the training process by using parallel computers. It also reduces the variance by averaging many trees. Bagging is more robust than boosting in the case of noisy settings. Random Forest (RF) is an example of the bagging method. In the first step of bootstrap aggregating method, model take N random bootstrap samples with the replacement of the training dataset. Secondly, for each sample, generate a base estimator (predictor) in parallel. Finally, the results from each predictor are assembled to obtain the final output by voting technique for classification problems. Boosting is one of the most significant recent developments in classification methodology. It augments the previous data model's existing results and forces the next data model to put effort into not correctly learned instances. It can be used to solve classification and regression problems. The most common types of boosting are Adaptive boosting (AdaBoost), Gradient, and Extreme Gradient Boosting (XGBoost). For prediction, the boosting algorithms employ the concept of the weak learner and strong learner, utilizing weighted average values. The goal is to reduce the model's bias. The authors [32] explains the working of boosting algorithm in steps.

3.2   Heterogeneous EL

A heterogeneous ensemble is built upon a collection of different classifiers on the same data. Heterogeneous EL overcomes the shortcoming of single ML algorithms. The most common type of heterogeneous EL is stacking. Stacking is a classification paradigm in EL and is known as stacked generalization. As opposed to homogeneous EL, stacking uses a meta-model to combine the predictions of base models, which yields better results. In this study, we perform predictions at two levels. In the first level, all the weak classifiers (base learners) are used to train and predict the output based on the dataset samples. We select Naive Bayes (NB), Linear Discriminant Analysis (LDA), Random Forest (RF), and Linear SVC as base learners. At the second level, we use Logistic Regression (LR) as a meta-model that accepts all the weak learners' predictions as input and generates a final prediction. Figure 2 illustrates stacking model.

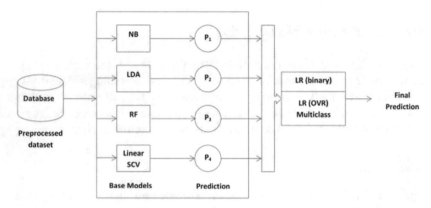

Fig. 2. Essential layers of generic IoT architecture

4   Dataset and Experimental Setup

In this section, we provide detail about the selected dataset, pre-processing, evaluation metrics and proposed methods.

4.1   Dataset

We note that the available datasets that are used in the literature for intrusion detection, such as KDD-CUP99, UNSWNB(2015), CIC-IDS (2017), BOT IOT (2018) and CIC-IDS (2017) etc., do not have heterogeneous IoT sensors in testbeds that may cause a lake of telemetry data [30]. Therefore, we select the TON IoT dataset to evaluate EL methods. TON IoT Telemetry dataset, released in 2020, is collected at UNSW Canberra (Australia). The TON-IoT dataset has telemetry data, operating system data, and network data. It is available on the UNSW web page and comprises seven CSV files. Variety of attacks are available in this dataset. The total number of instances in the data are 401119, out of which 245000 records are normal, and the remaining 156119 represent attacks. The dataset has a significant balance, with normal records accounting for 61% of the dataset and attacked records accounting for only 38%.

4.2 Pre-processing of Dataset

We pre-process the TON IoT dataset prior to using it for evaluation. We perform cleaning, encoding and scaling of the dataset. In cleaning, we remove all missing values and convert nominal values to numeric ones using label encoding and one-hot-encoding. Using min-max scaling, we scale all feature values to the range [0 − 1]. We integrate the seven files into one CSV file, with 22 features for each instance. We removed the date, time, and timestamp features from the data and retained the remaining 19 features for experiments. Alsaedi [30] suggested that these three features could cause the overfitting of ML algorithms during training. All the missing values in the dataset were filled with the median value of that feature. We transform nominal features into numeric data types using the Label encoding method. The "type" feature are encoded with numbers from 1

Table 1. Ensemble Learning Methods for IDS in existing literature

Ref	Year	Study Type	Ensemble Models	ML Classifiers	Ensemble Type		Dataset	Evaluation Metrics	IoT Telemetry Data of diverse sensors	Diverse new Attack Scenarios
					1	2				
[26]	2016	1	Bayesian tree, Random Forest, Voting	—	No	Yes	D1	ACC, ROC	No	No
[24]	2017	1	Random Forest	DT	Yes	No	D2, D4, D5	FAR,ACC	No	Yes
[8]	2017	1	Bagging, Boosting, Stacking	NB, ANN, J48, REPTree	Yes	Yes	D2	ACC	No	Yes
[23]	2018	2	Ada Boost	DT, NB, ANN	Yes	No	D2, D3	ACC, DR, FPR	No	Yes
[25]	2019	1	Boosting, Bagging	Subspace discriminant RUSBoosted tree	Yes	No	D6	ACC, AUC	No	No
[27]	2019	2	Rotation forest, Bagging	—	No	Yes	D2, D5	PRE, ACC, FPR, REC	No	Yes
[28]	2019	1	Bagged RF, voting	NB, KNN, RIPPER DT, LSTM	Yes	No	D2	AUC, TP, FP, TN, FN DR	No	Yes
[5]	2020	1	XGBoost	—	Yes	No	D1	PRE,AC, REC, F-1 S, Sup	No	No
[29]	2021	1	Random Forest	SVM, NB, ANN	No	Yes	D7,D8, D9	PRE, REC, ACC	No	Yes

(continued)

Table 1. (*continued*)

Ref	Year	Study Type	Ensemble Models	ML Classifiers	Ensemble Type		Dataset	Evaluation Metrics	IoT Telemetry Data of diverse sensors	Diverse new Attack Scenarios
					1	2				
[30]	2020	2	Random Forest	LSTM, DT, RF, NB	Yes	No	D10	ACC, PRE, REC, F-1 S	Yes	Yes
This study	2022	1	Stacking Bagging	SVM,NB,LR, RF, LDA, DT	Yes	Yes	D10	ACC, PRE, REC, F-1 S	Yes	Yes

Type 1: Homogeneous Type2: Heterogeneous
Study Type: 1- Hypothetical 2- Empirical
ML models: SVM = Support Vector Machine LDA = Linear Discriminant Analysis DT = Decision tree
LR = Logistic Regression
Dataset: D1 = KDDCup99 D2 = UNSW-NB15 D3 = NIMS botnet D4 = GPRS,D5 = NSL-KDD D6 = RPL-NIDDS17 D7 = WUSTL IIOT-2018 D8 = N BaIoT D9 = BoT-IoT D10 = ToN-IoT
Evaluation ACC = Accuracy PRE = Precision REC = Recall F -1 S = F-1 Score TP = True Positive
Metrics: FP = False Positive TN = True Negative DR = Detection Rate FN = False Negative
Sup = Support
AUC = Area Under Curve ROC = Receiver operating characteristic curves

to 7. The features with large values are scaled down using the Min-Max scalar method before modelling. We use an 80:20 train-test split in our experiments, as suggested by [33] to overcome the overfitting problem. The train-test-split method of the sklearn library is used for splitting data. To compare the performance of EL methods, we select precision, recall, f1-score and accuracy as evaluation metrics.

4.3   Experimental Setup

In order to evaluate the effectiveness of the EL models, the experiments were carried out on an Intel (R) Core (TM) i5 CPU running at 3.20 GHz, 16.0 GB of RAM, and Windows 10 Pro. We implemented our code in Python version 3.7.3 using Jupyter Notebook (Anaconda3). For pre-processing, we used Pandas and sklearn libraries.

5   Results and Discussion

We have integrated seven files (containing different sensor data) into 1 CSV file to get heterogeneous data in one file. We have applied different ML algorithms in EL techniques to find the best method for detecting intrusion based on accuracy, precision, recall and F1-score. Six ML algorithms: Random Forest, Naïve Bayes, Linear SVM,

Table 2. Binary classification results

Results of EL techniques for Binary classification						
Study	Model	Evaluation Matrices				
[30]		Accuracy	Precision	Re-call	F1-Score	
	LR	0.61	0.37	0.61	0.46	
	LDA	0.68	0.74	0.68	0.62	
	RF	0.85	0.87	0.85	0.85	
	NB	0.62	0.63	0.62	0.51	
	SVM	0.61	0.37	0.61	0.46	
	LSTM	0.81	0.83	0.81	0.80	
Homogeneous EL	Bagging with DT	0.82	0.88	0.87	0.86	
	AdaBoost	0.80	0.82	0.80	0.89	
	XGboost	0.82	0.85	0.83	0.82	
Heterogeneous EL	Stacking	0.86	0.87	0.85	0.86	

Table 3. Multi classification results

Results of EL techniques for multi-classification						
Study	Model	Evaluation Matrices				
[30]		Accuracy	Precision	Re-call	F1-Score	
	LR	0.61	0.38	0.62	0.47	
	LDA	0.62	0.46	0.63	0.51	
	RF	0.71	0.69	0.72	0.67	
	NB	0.54	0.59	0.51	0.52	
	SVM	0.60	0.37	0.61	0.46	
	LSTM	0.68	0.64	0.68	0.63	
Homogeneous EL	Bagging with DT	0.76	0.87	0.79	0.79	
	AdaBoost	0.71	0.71	0.72	0.69	
	XGboost	0.76	0.86	0.75	0.73	
Heterogeneous EL	Stacking	0.77	0.80	0.75	0.74	

LDA, Logistic Regression and Decision Tree, are used in bagging, boosting and stacking techniques. [30] tested the performance of single ML algorithms: LR, LDA, RF, NB, SVM and LSTM etc. on ToN-IoT dataset. The results achieved by each classifier on binary and multi-class classification problems are given in Table 2 and Table 3.

We have also applied three EL techniques: bagging, boosting and stacking. Different combination of traditional ML models have been used in stacking technique and the optimum combination was selected in this study. Therefore, in the stacking technique, we integrate RF, NB, Linear SVC and LDA in the base model and LR as the meta-model. Experimental results shows that the best accuracy is 86% in binary classification Table 2 and 77% in multi-classification Table 3, which is achieved by stacking model. The main reason behind the best results is that random forest boosts up the accuracy of other classifiers in ensemble approach and reduce the over-fitting problem of the model. It is evident from the results, that stand alone traditional ML classifiers performance is poor as compared to proposed model ensemble techniques. NB, SVM, LR achieve only 62% accuracy and LSTM (a deep learning model) attain 81% accuracy on the binary classification of ToNIoT dataset. Moreover, Bagging and XGboost both achieve 82% accuracy for binary classification and 76% in multi-classification. However, Adaboost performance is not good in binary as well as multi-classification as compared to other EL techniques. Table 3 also depict that ML models and deep learning model (LSTM) attains upto 68% accuracy only for mluticlass classification.

In contrast, Bagging performance is better than stacking and boosting based on precision in both classifications. Bagging classifier achieves 82% and 87% precision in binary and multi-classification, respectively. The results show that a single ML classifier does not perform better for intrusion detection. So, the performance of these classifiers for IDS on telemetry data could be boosted by using the ensemble technique.

6 Conclusion

This paper compares the performances of bagging-based, boosting-based and stacking-based EL techniques applied for intrusion detection in telemetry data. For experiments, we used the ToN-IoT dataset. The results show that a heterogeneous stacking-based detection model achieves the highest accuracy's of 86% and 77% for binary and multi-class classification problems, respectively. The anomaly-based intrusion detection model comprised an NB, an RF, an LDA and a Linear SVM as base learners and an LR as meta learner. However, a bagging-based framework with DT performs better than other EL techniques of boosting and stacking in terms of precision.

References

1. Raghuvanshi, A., Singh, U.K., Joshi, C.: A review of various security and privacy innovations for IoT applications in healthcare. In: Advanced Healthcare Systems: Empowering Physicians with IoT-Enabled Technologies, pp. 43–58 (2022)
2. Isaac Abiodun, O., et al.: A review on the security of the internet of things: challenges and solutions. Wireless Personal Commun. 1–35 (2021)

3. Al-A'araji, N.H., Al-Mamory, S.O., Al-Shakarchi, A.H.: Classification and clustering based ensemble techniques for intrusion detection systems: a survey. J. Phys. Conf. Ser. **1818**, 012106 (2021)
4. Zarpelao, B.B., Miani, R.S., Kawakani, C.T., Carlisto de Alvarenga, S.: A survey of intrusion detection in internet of things. J. Netw. Comput. Appl. **84**, 25–37 (2017)
5. Bhati, B.S., Chugh, G., Al-Turjman, F., Bhati, N.S.: An improved ensemble based intrusion detection technique using XGboost. Trans. Emerg. Telecommun. Technol. **32**(6), e4076 (2021)
6. Roesch, M., et al.: Snort: lightweight intrusion detection for networks. Lisa **99**, 229–238 (1999)
7. Singh, J., Nene, M.J.: A survey on machine learning techniques for intrusion detection systems. Int. J. Adv. Res. Comput. Commun. Eng. **2**(11), 4349–4355 (2013)
8. Belouch, M., El hadaj, S.: Comparison of ensemble learning methods applied to network intrusion detection. In: Proceedings of the Second International Conference on Internet of things, Data and Cloud Computing, pp. 1–4 (2017)
9. Catarinucci, L., et al.: An IoT-aware architecture for smart healthcare systems. IEEE Internet of Things Journal, **2**(6), 515–526 (2015)
10. Yassine, A., Singh, S., Hossain, M.S., Muhammad, G.: IoT big data analytics for smart homes with fog and cloud computing. Future Generation Comput. Syst. **91**, 563–573 (2019)
11. Caragliu, A., Bo, C.D., Nijkamp, P.: Smart cities in europe. J. Urban Technol. **18**(2), 65–82 (2011)
12. Saarika, P.S., Sandhya, K., Sudha, T.: Smart transportation system using IoT. In: 2017 International Conference On Smart Technologies For Smart Nation (Smart- TechCon), pp. 1104–1107. IEEE (2017)
13. Jayaram, A.: Smart retail 4.0 IoT consumer retailer model for retail intelligence and strategic marketing of in-store products. In: Proceedings of the 17th International Business Horizon-INBUSH ERA-2017, Noida, India, 9 (2017)
14. Ali, S., Shakeel, M.H., Khan, I., Faizullah, S., Khan, M.S.: Predicting attributes of nodes using network structure. ACM Trans. Intell. Syst. Technol. **12**(2) (2021)
15. Mansoor, H., Ali, S., Khan, I., Arshad, N., Khan, M.A., Faizullah, S.: Short-term load forecasting using ami data. ArXiv preprint (2022)
16. Ali, S., Mansoor, H., Arshad, N., Khan, I.: Short term load forecasting using smart meter data. In: Proceedings of the Tenth ACM International Conference on Future Energy Systems, e-Energy 2019, pp. 419–421. ACM (2019)
17. Ali, S., Mansoor, H., Khan, I., Arshad, N., Faizullah, S., Khan, M.A.: Fair allocation based soft load shedding. In: Intelligent Systems and Applications, pp. 407–424. Springer (2020)
18. Granjal, J., Monteiro, E., Silva, J.S.: Security for the internet of things: a survey of existing protocols and open research issues. IEEE Commun. Surv. Tutorials, **17**(3), 1294–1312 (2015)
19. Khan, M.A., Salah, K.: Lotsecurity: review, blockchain solutions, and open challenges. Future Generation Comput. Syst. **82**, 395–411 (2018)
20. Faizullah, S., Khan, M.A., Alzahrani, A., Khan, I.: Permissioned blockchain-based security for SDN in IoT cloud networks. In: 2019 International Conference on Advances in the Emerging Computing Technologies (AECT), pp. 1–6 (2020)
21. Zhou, J., Cao, Z., Dong, X., Vasilakos, A.V.: Security and privacy for cloud-based IoT: challenges. IEEE Commun. Mag. **55**(1), 26–33 (2017)
22. Ali, S., et al.: Detecting DDOS attack on SDN due to vulnerabilities in openflow. In: Proceedings of the International Conference on Advances in the Emerging Computing Technologies (AECT), pp. 1–6. IEEE (2020)
23. Moustafa, N., Turnbull, B., Raymond Choo, K.K.: An ensemble intrusion detection technique based on proposed statistical flow features for protecting network traffic of internet of things. IEEE Internet of Things J. **6**(3), 4815–4830 (2018)

24. Primartha, R., Tama, B.A.: Anomaly detection using random forest: a performance revisited. In: 2017 International Conference on Data and Software Engineering (ICoDSE), pp. 1–6. IEEE (2017)

25. Verma, A., Ranga, V.: Elnids: ensemble learning based network intrusion detection system for RPL based internet of things. In: 2019 4th International Conference on Internet of Things: Smart Innovation and Usages (IoT-SIU), pp. 1–6. IEEE (2019)

26. Wang, Y., Shen, Y., Zhang, G.: Research on intrusion detection model using ensemble learning methods. In: 2016 7th IEEE International Conference on Software Engineering and Service Science (ICSESS), pp. 422–425. IEEE (2016)

27. Tama, B.A., Comuzzi, M., Rhee, K.-H.: TSE-IDS: a two-stage classifier ensemble for intelligent anomaly-based intrusion detection system. IEEE Access, **7**, 94497–94507 (2019)

28. Elijah, A.V., Abdullah, A., Jhanjhi, N., Supramaniam, M., Abdullateef, B.: Ensemble and deep-learning methods for two-class and multi-attack anomaly intrusion detection: an empirical study. Int. J. Adv. Comput. Sci. Appl **10**(9), 520–528 (2019)

29. Priya, V., Sumaiya Thaseen, I., Gadekallu, T.R., Aboudaif, M.K., Nasr, E.A.: Robust attack detection approach for IIoT using ensemble classifier. arXiv preprint arXiv:2102.01515 (2021)

30. Alsaedi, A., Moustafa, N., Tari, Z., Mahmood, A., Anwar, A.: Ton IoT telemetry dataset: a new generation dataset of IoT and IIoT for data- driven intrusion detection systems. IEEE Access, **8**, 165130–165150 (2020)

31. Dasarathy, B.V., Sheela, B.V.: A composite classifier system design: concepts and methodology. Proc. IEEE, **67**(5), 708–713 (1979)

32. Alyasiri, H.: Developing Efficient and Effective Intrusion Detection System using Evolutionary Computation. PhD thesis, University of York (2018)

33. G´eron, A.: Hands-on machine learning with Scikit-Learn, Keras, and TensorFlow: Concepts, tools, and techniques to build intelligent systems. O'Reilly Media (2019)

Performance and Security Evaluation of OpenDaylight and ONOS

Siqi Liu$^{(\boxtimes)}$ and Shao Ying Zhu

School of Computing and Digital Technology, Birmingham City University, Birmingham, UK
Siqi.Liu@mail.bcu.ac.uk, Peggy.Zhu@bcu.ac.uk

Abstract. Software-defined networking (SDN) is a network architecture that separates the network control and data forwarding function. An SDN controller is one of the most important components of SDN, providing a programmable, dynamic and efficient network configuration to help network administrators improve network performance. However, the security challenges in SDN networks are highly similar to those of traditional networks, with controllers in the control layer being particularly vulnerable to security attacks. A denial of service (DoS) attack is one of the most common attacks that threaten the security of SDN controllers. Therefore, the selection of a suitable, high-performance and secure controller for the network is essential. In this paper, two of today's most popular controllers—OpenDaylight (ODL) and the open networking operating system (ONOS)—were selected to conduct a series of tests and be evaluated. CPU utilisation, jitter and throughput of the two controllers were evaluated using sFlow and Iperf under normal and attack traffic. The results show that under normal traffic, ONOS has higher throughput and jitter and lower CPU usage than ODL as the number of switches increases. Under DoS attack, ONOS still has higher throughput than ODL and takes longer to reach 100% CPU usage than ODL. In contrast, under a high-traffic attack such as DDoS, the CPU usage of both controllers almost reaches saturation simultaneously, implying that both controllers perform similarly under a DDOS attack.

Keywords: Software-defined networking (SDN) · ONOS · OpenDaylight · DOS · DDOS

1 Introduction

Software-defined networking (SDN) was first proposed in 2010 [1]. Compared to traditional networks, SDN simplifies network configuration and management, improves network efficiency and provides network programmability, scalability, flexibility, agility and virtualisation that better meets the growing demand [2]. An SDN controller is an essential component that can maintain a global view of the network. However, such controllers are highly vulnerable to attacks, particularly those involving denial of service (DoS), which is one of the most common attack methods used by intruders.

The security and performance of the controller are essential to SDN. Hence, significant changes to the controller design have been made to provide better security and

© The Author(s), under exclusive license to Springer Nature Switzerland AG 2023
F. Saeed et al. (Eds.): ICACIn 2022, LNDECT 179, pp. 463–473, 2023.
https://doi.org/10.1007/978-3-031-36258-3_41

performance. After years of development, a wide variety of controllers have been developed in industry and academia, including ONOS, OpenDaylight (ODL), Floodlight, Ryu, Beacon, NOX and POX. Therefore, a key issue for network administrators to consider is how to select the most appropriate controller for a network. Among the abovementioned open-source controllers, ONOS and ODL are currently the two most widely used [8]. Although SDN controllers and their feature comparisons are an important research area for many researchers, by reviewing the relevant literature, it was found that few studies have tested and compared the performance of these two controllers under normal traffic, and even less research has been done related to the performance and security evaluation of both during stress testing. Hence, it is necessary to conduct a series of tests to evaluate these two controllers to test their performance and security in the event of DoS attacks. The results of this study can provide guidance for others when selecting between these two controllers for their application.

2   Related Work

Among all the open-source distributed controllers, ONOS and ODL are the frontrunners for deployment in enterprise SDNs because both can run cross platform and support various functions [3]. Researchers have conducted a series of experiments to compare both controllers using different testing tools. Tables 1 and 2 provide a summary of the relevant literature comparing the performance evaluation of ONOS and ODL under normal traffic and DoS attacks, respectively.

   As shown in Table 1, Cbench, Ping and Iperf are popular tools used by researchers to study the performance of ODL and ONOS in terms of latency, jitter and throughput. Some discrepancies among these studies could be due to the different testing environments and controller versions. Most of the studies showed that ONOS outperforms ODL in terms of throughput, latency and jitter. Furthermore, few studies have investigated the performance variation of ONOS and ODL under stress testing in the event of a DoS attack, as shown in Table 2. In this research, throughput and CPU usage were selected as performance metrics. Because throughput is a widely used metric for evaluating controller performance, the results of this study's tests can be easily compared to the findings of other studies. The performance of these two metrics for ODL and ONOS under DoS attacks will be closely observed. The results can provide a reference to other researchers who are conducting similar performance evaluations for these two controllers under DoS attacks.

3   Experimental Setup

All implementations and tests were conducted in a virtual environment, using two virtual machines with Ubuntu 21.10 (64-bit) as the operating system running on the Intel Core i5- 10210U CPU @ 1.60 GHz (1 core) and a 4GB RAM laptop. ONOS (X-Wing (LTS) 2.7.0) and ODL controllers (0.6.0-Carbon) were installed on one of the two virtual machines, respectively. Additional software used for testing and evaluation is shown in Table 3.

Table 1. Performance evaluation of ONOS and ODL under normal traffic

Reference	Controllers Version	Testing Tools	Winner Controller
[3]	ONOS (Goldeneye) ODL (Beryllium-SR2)	Cbench	ONOS (throughput, CPU utilisation and latency)
[4]	ONOS (1.12) ODL (Carbon)	Cbench Iperf	ONOS (topology update time) ODL (topology discovery time)
[5]	ONOS (1.5) ODL (0.5)	N/A	ONOS (latency, jitter and throughput)
[6]	ONOS (1.13.2) ODL (0.3.4-Lithium)	Ping	ONOS (throughput) ODL (delay and successful transmission rate)
[7]	ONOS (N/A) ODL (N/A)	Iperf	ONOS (TCP/UDP throughput, latency and jitter)
[8]	ONOS (Nightingale) ODL (Nitrogen)	Iperf Ping	ONOS (jitter, throughput and latency)
[2]	ONOS (N/A) ODL (N/A)	Wireshark Iperf Ping	ODL (TCP/UDP bandwidth and jitter)
[12]	ONOS (N/A) ODL (N/A)	N/A	ONOS (jitter, delay and packets loss rate)
[13]	ONOS (N/A) ODL (N/A)	Iperf Ping	ONOS (jitter, latency and throughput)

Table 2. Performance evaluation of ONOS and ODL under DoS attacks

Reference	Controllers	DoS attack Tools	Winner Controller
[9]	ONOS (N/A) ODL (N/A)	Lmeter Scapy	ONOS (throughput)
[10]	ONOS (Peacock) ODL (Beryllium)	Hping3 Nping Xerxes Tor Hammer LOIC	ODL (packet loss) ONOS (Round trip time (RTT))

In this study, the performance of single node ONOS and ODL controllers was evaluated using Iperf and Sflow-rt. Figure 1 depicts a series of tests carried out on ONOS and ODL. In the test under normal traffic, the throughput and CPU usage changed as the number of switches increased, whereas in the DoS stress tests, the advanced SYN flood attack was selected for the DOS stress tests using hping3 with spoofed IP.

The attack on the linear topology is from Hosts 1 to 2, as shown in Fig. 2. In contrast, in the tree topology, the SYN flood attack will be launched simultaneously from Hosts 1, 3 and 4 to Host 2 to simulate a distributed DoS attack scenario (see Fig. 3).

Table 3. Other tools used for this study

Name	Version
Mininet	2.3.0
Sflow-rt	3.0 -1662
Iperf	2.1.2
Hping3	$3.00 - alpha - 2$

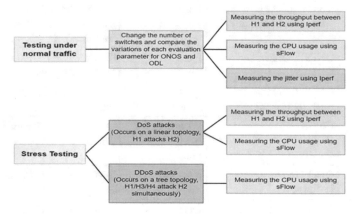

Fig. 1. Testing plan

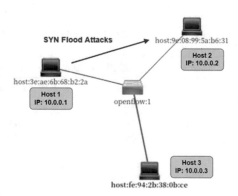

Fig. 2. DoS attack on linear topology

4 Performance Evaluation

This section investigates and summarises the results after running the abovementioned experiment. The test results for both controllers' throughput, CPU usage and jitter are discussed in Subsects. 4.1, 4.2 and 4.3, respectively.

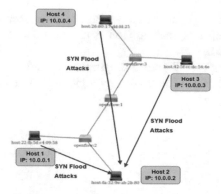

Fig. 3. DDoS attack on tree topology

4.1 Throughput Test

In this section, the changes in throughput for the two controllers using different numbers of switches under both normal traffic and DoS attacks are discussed.

Under Normal Traffic Test. Figure 4 shows the variation in throughput for ONOS and ODL with different numbers of switches (1, 3, 7, 15, 25 and 31) under normal traffic. In our study, ONOS always had a higher throughput than ODL despite the in-creasing number of switches. This finding is consistent with [3 , 5 , 7 , 8 , 11] and [13].

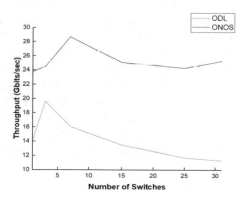

Fig. 4. Throughput variation of ODL and ONOS with different number of switches

The throughput of ONOS peaked when the number of switches was approximately 7, whereas the maximum throughput of ODL was at a switch number of 3, as shown in Fig. 4. After reaching the peak throughput, both ONOS and ODL decreased as the number of switches increased. The throughput of ODL fluctuated more, whereas the ONOS throughput was not affected as much by the increased number of switches. This observation was due to having a larger number of switches that can cause contention at the data layer, with ONOS having more processing power than ODL [11]. In summary, combining our findings with other researchers shows that ONOS performs much better than ODL in terms of throughput. This advantage becomes increasingly apparent as the number of switches increases, making ONOS more suitable for deployment in larger SDN networks than ODL.

Under DoS Attacks Test. Figure 5 illustrates the change in throughput for ONOS and ODL before and after a DoS attack on topologies with different numbers of switches (1, 3, 7).

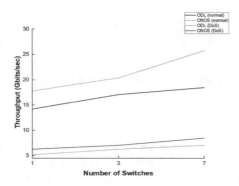

Fig. 5. Throughput variation of ODL and ONOS under DoS attacks

Although the throughput of ONOS and ODL both reduced significantly after DoS attacks, as depicted in Fig. 5, the ONOS throughput was still higher than ODL. Therefore, we believe that ONOS was more reliable under the attack traffic. With further analysis, it was found that the throughput of ONOS decreased more than ODL under DoS attack. However, this finding differs to the one found in [9]. Their test results showed that ONOS throughput did not experience a significant decrease under a DoS attack, whereas the ODL throughput did. The researchers in [9] used Scapy to generate appropriate DoS traffic on the underlying network, as well as using the JMeter Plugin to perform load and performance tests. Therefore, we assumed that the difference in the tools used to generate the attacks is a possible reason for the varying results. Overall, ONOS has higher throughput than ODL under both normal traffic and DoS attack in the throughput test.

4.2   CPU Utilisation Test

In this section, the changes in CPU usage for ONOS and ODL using different topologies under both normal traffic and DoS attacks are investigated.

Under Normal Traffic Test. The test results for ONOS and ODL in each topology under normal traffic are shown in Figs. 6 and 7.

By comparing Figs. 6 and 7, it is clear that the CPU usage of ODL is substantially higher and fluctuates more significantly than ONOS for the same topology, suggesting that ONOS has better performance and is more stable than ODL. Because ONOS and ODL are both Java-based controllers and use the same NIO libraries (Netty), their performance should be identical [3]. However, the CPU usage of ODL is surprisingly

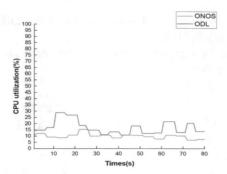

Fig. 6. CPU usage of ONOS and ODL in linear topology

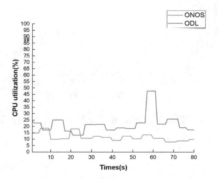

Fig. 7. CPU usage of ONOS and ODL in tree topology

high, even with only one connected switch, and the increase in the number of switches exacerbates the issue. As stated in [3], the reason for this phenomenon could be due to an error in the OpenFlow plug-in for OpenDaylight; in addition, it could also be caused by the switch partitioning algorithm or packet batching. The switch partitioning technique allocates and distributes linked OpenFlow switches to controller worker threads. Packet batching is a mechanism used by OpenFlow controllers to combine several bytes when reading or writing to an underlying network. Its goal is to reduce the number of system calls and increase the overall throughput. These mechanisms do not seem to be well optimised, which seriously affects the CPU utilisation of the controller and reduces the throughput [3]. For this phenomenon, another argument is that the better throughput provided by ONOS is due to the application of a collection of hashes to manage the MAC addresses of SDN devices [11]. Overall, despite the increasing number of switches, ODL's throughput is always lower than ONOS's, whereas the CPU usage is always higher than ONOS's.

Furthermore, combining the throughput of ONOS and ODL measured in Sect. 4.1 with the CPU usage in this section, it can be found that as the number of switches increases, ODL's throughput is always lower than ONOS's, whereas the CPU usage is always higher than ONOS's. Therefore, we propose that the reason ODL performed poorly in the throughput test may be related to its higher CPU utilisation. This conjecture was also confirmed by [3] and [12].

Under DoS Attacks Test. Figures 8 and 9 show the change in CPU usage for ONOS and ODL under stress testing.

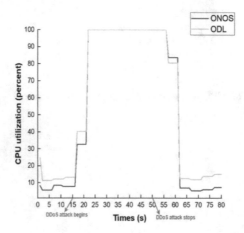

Fig. 8. Change in CPU utilisation under DoS attack on a linear topology

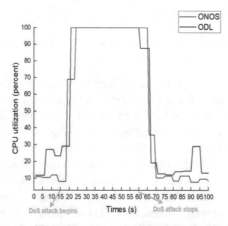

Fig. 9. Change in CPU utilisation under DDoS attack on a tree topology

As depicted in Fig. 8, ODL's CPU usage took less time to reach saturation point than ONOS after the DoS attack began. After the attack stopped, ODL recovered to normal levels faster than ONOS. With the rapid increase in attack traffic over a short period of time, as shown in Fig. 9, the difference in the time to CPU saturation between ODL and ONOS became significantly smaller. Therefore, we believe that the two controllers have similar performance and can be substituted for each other under a high-traffic attack, such as DDoS.

4.3 Jitter Test

The performance of ONOS and ODL was evaluated in terms of jitter. The results of these two controllers under different network topologies using the ping command are shown in Figs. 10 and 11.

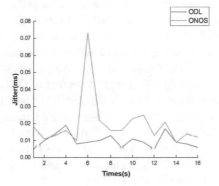

Fig. 10. Jitter of ONOS and ODL in linear topology

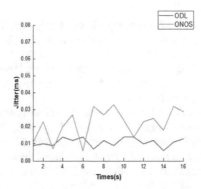

Fig. 11. Jitter of ONOS and ODL in tree topology

Figures 10 and 11 show that the average jitter value for ODL is smaller than ONOS in both topologies but in a small margin, with approximately 0.06 ms maximum differences. This finding is consistent with [2] but contradicts the results of [5, 7, 8, 12] and [13]. In [2], the authors found that ODL had smaller packet jitter than ONOS, but the difference between the two values was not particularly large. In contrast, in [7], by varying the number of switches, the authors found that ONOS showed lower jitter, and the jitter value for ODL fluctuated more significantly than ONOS. Furthermore, in [8], after testing under different network topologies, the authors found that although most ONOS-connected topologies presented lower jitter, the ODL outperformed the ONOS in Torus and Ring topology. The authors in [8] believe that the main reason for this phenomenon is that the ONOS subsystem was developed with a focus on telecommunications, whereas the ODL is more focused on data centres. In summary, in our research, ODL performed better in terms of jitter.

5 Conclusion

The performance of two OpenFlow controllers—ODL and ONOS—was investigated and evaluated. Three evaluation metrics—CPU utilisation, jitter and throughput—were identified and used under two experimental scenarios to test ODL and ONOS performance under normal and attack traffic. The results showed that ONOS performs better in terms of throughput and CPU usage under normal traffic. In the jitter test, the average jitter value of ONOS was slightly higher than that of ODL by a small margin. As the number of switches increased, ONOS demonstrated better performance in terms of CPU usage and throughput than ODL under normal traffic, making ONOS more suitable for networks that need to deploy a large number of network devices. Under DoS attacks, ONOS still has higher throughput than ODL. Under DDoS attacks, the CPU usages were similar between ODL and ONOS, indicating that both controllers are equally secure under heavy DDOS attack traffic.

References

1. Alhapony, F.N., Bkkar, A.M., Almogassipi, K.G.: Study and Simulation for SDN's three layers. In: 7th International Conference on Engineering & MIS 2021 (ICEMIS 2021), pp. 1–6. Association for Computing Machinery, New York, USA (2021)
2. Badotra, S., Panda, S.N.: Evaluation and comparison of OpenDayLight and open networking operating system in software-defined networking. Clust. Comput. 23(2), 1281–1291 (2019). https://doi.org/10.1007/s10586-019-02996-0
3. Darianian, M., Williamson, C., Haque, I.: Experimental evaluation of two OpenFlow controllers. In: IEEE 25th International Conference on Network Protocols (ICNP), pp. 1–6. Institute of Electrical and Electronics Engineers, New York, USA (2017)
4. BAH, M.T., Azzouni, A., Nguyen, M.T., Pujolle G.: Topology discovery performance evaluation of opendaylight and ONOS controllers. In: 22nd Conference on Innovation in Clouds, Internet and Networks and Workshops (ICIN), pp. 285–291. Institute of Electrical and Electronics Engineers, New York, USA (2019)
5. Bedhief, I., Kassar, M., Aguili, T., Foschini, L., Bellavista, P.: Self-adaptive management of SDN distributed controllers for highly dynamic IoT networks. In: 15th International Wireless Communications & Mobile Computing Conference (IWCMC), pp. 2098–2104. Institute of Electrical and Electronics Engineers, New York, USA (2019)
6. Chaipet, S., Putthividhya, W.: On studying of scalability in single-controller software-defined networks. In: 11th International Conference on Knowledge and Smart Technology (KST), pp. 158–163. Institute of Electrical and Electronics Engineers, New York, USA (2019)
7. Lunagariya, D., Goswami, B.: A comparative performance analysis of stellar SDN controllers using emulators. In: International Conference on Advances in Electrical, Computing, Communication and Sustainable Technologies (ICAECT), pp. 1–9. Institute of Electrical and Electronics Engineers, New York (2021)
8. Dissanayake, M., Kunmari, V., Udunuwara, A.: Performance comparison of ONOS and ODL controllers in software defined networks under different network typologies. J. Res. Technol. Eng. 2(3), 94–105 (2021)

9. Latah, M., Toker, L.: Load and stress testing for SDN's northbound API. SN Appl. Sci. **2**(1), 1–8 (2019). https://doi.org/10.1007/s42452-019-1917-y

10. Badotra, S., Panda, S.N.: SNORT based early DDoS detection system using Opendaylight and open networking operating system in software defined networking. Clust. Comput. **24**(1), 501–513 (2020). https://doi.org/10.1007/s10586-020-03133-y

11. Mamushiane, L., Lysko, A., Dlamini, S.: A comparative evaluation of the performance of popular SDN controllers. In: 2018 Wireless Days (WD), pp. 54–59. Institute of Electrical and Electronics Engineers, New York, USA (2018)

12. Mamushiane, L., Shozi, T.: A QoS-based evaluation of SDN controllers: ONOS and Open-DayLight. In: 2021 IST-Africa Conference (IST-Africa), pp. 1–10. Institute of Electrical and Electronics Engineers, New York, USA (2021)

13. Rodriguez, A., Quiñones, J., Iano, Y., Barra, M.A.Q.: A comparative evaluation of ODL and ONOS controllers in software-defined network environments. In: 2022 IEEE XXIX International Conference on Electronics, Electrical Engineering and Computing (INTERCON), pp. 1–4. Institute of Electrical and Electronics Engineers, New York, USA (2022)

Enhancing DDoS Attacks Detection in SOCs by ML Algorithms

Omar Lamrabti[✉], Abdellatif Mezrioui, and Abdelhamid Belmekki

STRS LAB, RAISS TEAM, National Institute of Posts and Telecommunications, RABAT, Morocco
{lamrabti.omar,mezrioui,belmekki}@inpt.ac.ma

Abstract. Security Operation Centers (SOC) are the fundamental pillars for modern cybersecurity. Their critical importance is proven by monitoring, detecting, alerting, and reacting to cyber-attacks in real-time. However, SOC implementations suffer from many limitations that reduce their efficiency. One of the critical attacks faced by SOC teams is Distributed Denial of Service (DDoS) therefore eliminating false positive attacks is an important step undertaken. It is useful to propose an efficient method to detect DDoS attacks based on Machine Learning (ML). We demonstrate how to build an efficient supervised ML system with a low false positive rate, capable of detecting DDoS attacks during dataset processing. We chose the CICDDoS2019 Dataset to achieve this. The experimental results after the use of different ML models shows that XGBoost and Random Forest achieve the highest accuracy with the lowest false positive rate.

Keywords: CNN · DDoS · Decision Tree · Deep learning · KNN · Logistic Regression · Machine learning · Multi-layer Perceptron (MLP) · Naïve Bayes · PCA · Random Forest · RFR · Supervised ML · SOC · XGBoost

1 Introduction

Machine learning is an artificial intelligence demonstrated by machines and is present in every aspect of our daily life, aiming to equip systems with the ability to process and classify colossal data to learn and self-improve via various algorithms. In an active environment like the SOC, adopting machine learning provides an intelligent system capable of classifying different cyber-attacks. In cybersecurity jargon, a Security Operation Center is represented by teams of security specialists, technologies and processes to ensure monitoring, detection and reaction against cyber-attacks that rapidly escalate. The elimination of false positives alerts and extraction of meaningful data from the huge number of cybersecurity data impacts the workload of the SOC. It is in this context that artificial intelligence techniques, more specifically ML and DL, have provided significant assistance to the SOC by increasing its efficiency in detecting real intrusions and decreasing analysis time. False positive alerts occur when an activity is flagged as an alert but the alert is not actually malicious, which systematically wastes resources to address an irrelevant alert. In this paper, we focus on DDoS attacks [1] that are ranked

© The Author(s), under exclusive license to Springer Nature Switzerland AG 2023
F. Saeed et al. (Eds.): ICACIn 2022, LNDECT 179, pp. 474–485, 2023.
https://doi.org/10.1007/978-3-031-36258-3_42

in the top 20 of cyberattacks in 2021 according to Fortinet report [2]. We will leverage ML to detect DDoS attacks and to reduce false positive alerts and demonstrate how to leverage and process a dataset to build an efficient ML system that offers DDoS attack detection with a reduced false positive rate. To this end, we chose the CICDDoS2019 dataset [3] which includes different types of DDoS attacks such as PortMap, NetBIOS, LDAP, MSSQL, etc. We discuss every step regarding data collection, label creation, feature development, ML algorithm selection, model performance evaluation and accuracy score generation.

The main contributions of this work are:

- We use the CRISP-DM process in the phases of the Dataset analysis and the machine learning models applications.
- We handle the problem of unbalanced data in the CICDDoS2019 dataset using Synthetic Minority Sampling (SMOTE) [24] and Random Oversampling (ROS) [40] techniques to balance the minority class in the DataSet.
- We rank the most important features after testing well-known methods regarding features selection as: Principal Component Analysis (PCA) [28] and Random Forest Regressor (RFR) [29].
- We train and test machine learning using our approach in different ML classifiers with optimized ML hyperparameters.
- We discuss the performance metrics results, including Precision, Recall, F1-Score, and Accuracy.

This paper is organized as follows: Section 2 provides background on DoS and DDoS attacks, detection methods, and an overview of ML. In Sect. 3, we review related work on DDoS detection methods and ML in DDoS detection. Section 4 presents our methodology for data processing, feature selection, and ML classifiers. In Sect. 5, we discuss the details of the experiment and give the results and analysis of the experiment with a comparison of the results of other research. Finally, the conclusion and future work are provided in Sect. 6.

2   Background

2.1   Dos and DDoS Attacks

It should be noted that DDoS attack techniques have become not only more frequent but also more sophisticated than ever before. "DoS" and "DDoS" attacks are very similar, designed to flood a target (host, service, networks, etc.) with a huge amount of traffic. The only difference between them is that "DDoS" is an amplified "DoS" attack.

2.2   DDoS Attacks Detection Methods

The detection methods of DDoS attacks can be categorized as statistical, soft-computing, clustering, knowledge-based and hybrid [4–6].

2.3 Overview of Machine Learning

ML algorithms can be categorized as Supervised Learning [11] or Unsupervised Learning [12]. There are many classification techniques or classifiers used in supervised learning, but some widely used ones include Logistic Regression, Linear Discriminant Analysis, K-Nearest Neighbors, Trees, Neural Networks, and Support Vector Machines [13]. On the other hand, in unsupervised learning, the ML algorithms learn to find interesting associations or hidden patterns in unlabeled datasets to classify a data without human intervention.

3 Related Works

Several researches have focused on solutions to detect intrusions, namely DDoS attacks, while other researches have been interested in eliminating false positive alerts. The first finding indicates the existence of many solutions that are either algorithmic or statistical to process extremely large data. Authors [14] describe how to generate labels from SOC investigation notes, to correlate IP, host, and users to generate user-centric features, to select ML algorithms and to evaluate performances of an ML system in a SOC production environment. Authors [37] demonstrate a novel technique which achieves high accuracy scores in comparison with well-known ML classifications models using neural networks for detection and classification of nine intrusion attacks including DoS and DDoS on the ToN_IoT dataset. Hasan T et al. [38] proposes a hybrid intelligent Deep Learning model to secure IIoT by identifying accurately two bots attacks (Gafgyt, Mirai), their model achieve 99.94% in detection rate. Krivchenkov et al. [39] reduce features of used Dataset (UNSW-NB15 and NSL-KDD) using correlation analysis and implementing k-nn ML model to classify traffic and detect DoS attacks. Ajeetha et al. [9] propose a method for detection of DDoS attacks through the traces in the traffic flow. Authors [15] use data mining and combine support vector machines (SVM), decision trees, Naïve Bayes to reduce false positives via a novel model. Authors [16] suggest a DDoS detection model based on data mining and ML techniques. Alamri et al. [17] evaluate the performance of a set of classification algorithms that are widely used in ML-based DDoS attack detection in the Software-Defined Networking (SDN) environment. The existing studies are interested in detecting DDoS attacks, but the elimination of false positive alerts in the context of SOC operations still requires further work. Also, we have investigated, tested and compared height ML models where the others use five models or fewer.

4 Our Approach

4.1 CICDDoS2019 Dataset

CICDDoS2019,created by the University of New Brunswick Canadian Institute for Cybersecurity, contains the most up to date common DDoS attacks, represented as PCAPs and CSVs files concerning 12 DDoS types attacks with 86 features and 13 class labels and 50,063,112 instances as described in Table 1.

Table 1. Class label and their number of instances in the dataset

Class Label	Number of Instances
Benign	56,863
DNS	5,071,011
LDAP	2,179,930
MSSQL	4,522,492
NetBIOS	4,093,279
NTP	1,202,642
SNMP	5,159,870
SSDP	2,610,611
SYN	1,582,289
TFTP	20,082,580
UDP	3,134,645
UDP_Lag	366,461
WEBDDOS	439

4.2   Machine Learning and Dataset Processing Workflow

The CRISP-DM (CRoss Industry Standard Process for Data Mining) [18] is the process model used in this research that involves machine learning and data analytics. The process model consists of six phases, namely: business understanding, data understanding, data preparation, modeling, evaluation and deployment. In our research, the SOC represents the business understanding phase and DDoS attacks and benign events are the content of the Dataset. Figure 1 describes all the phases concerning our proposed approach combined with the use of CRISP-DM. We first involve data discovery and model creation that combine the analysis of historical data, identification of new data sources, collection, correlation and analyzing data across multiple data sources. We look at how to improve model performance for best accuracy by bypassing unbalanced dataset, treating missing and outlier values, applying feature selection to enhance training and testing performance of the model. The objective of the proposed ML workflow is to detect DDoS attacks and reduce the rate of false positives. The features were analyzed to obtain the best features set to detect DDoS attacks. Arriving at this stage, we applied the right kind of ML algorithms. Then we move on to the training and testing phase to evaluate the results of ML Models. We use seven common ML algorithms to predict DDoS attacks, Naïve Bayes, XGBoost, Random Forest, Logistic Regression, KNeighbors Classifier, Multi-layer Perceptron (MLP), Decision Tree Classifier and a Deep Learning using Convoluted Neural Network [19–22].

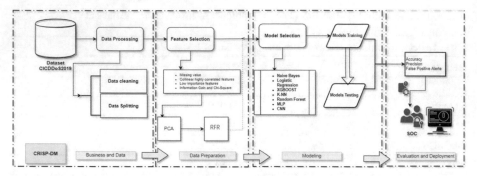

Fig. 1. The proposed ML Working Flow

4.3   Metrics of Evaluation

A weighted average of the four evaluation metrics precision, recall, F1-score and accuracy are used:

$$\text{TPR} = \frac{\sqrt{TP}}{\sqrt{TP + FN}} \tag{1}$$

$$\text{FPR} = \frac{\sqrt{FP}}{\sqrt{FP + TN}} \tag{2}$$

$$\text{Precision} = \frac{\sqrt{TP}}{\sqrt{TP + FP}} \tag{3}$$

$$\text{Recall} = \frac{\sqrt{TP}}{\sqrt{TP + FN}} \tag{4}$$

$$\text{F1 - score} = \frac{2 * Precision * Recall}{Precision + Recall} \tag{5}$$

$$Accuracy = \frac{TP + TN}{TP + TN + FP + FN} \tag{6}$$

where True Positive (TP) represents the number of DDoS attacks correctly recognized as attacks. True Negative (TN) represents the legitimate traffic recognized as legitimate. False Positive (FP) is the classification of legitimate traffic as attacks. False Negative (FN) is the classification of benign traffic as legitimate. Also, Precision is the proportion or ratio of correct positive classifications of the total cases predicted as positive. Recall is the proportion of correct positive classifications of the total number of cases actually positive. Then, the F1 score is the harmonic mean between precision and recall giving a view on the efficiency of the classifier.

4.4   Dataset Processing

The chosen dataset CICDDoS2019 produces a lot of noise that impacts the performance and accuracy of ML. The Data cleaning is the preprocessing step in the Dataset processing; we start by eliminating not number and infinite values in the DataSet and missing values. The analysis of the DataSet shows the presence of 13 class labels used in the system for attack detection. The following class labels are employed: SYN, TFTP, WebDDoS, DNS, Benign, MSSQL, LDAP, NETBIOS, NTP, SSDP, SNMP, UDP, and UDP_Lag. The total number of features extracted from the DataSet is 86. One of the main approaches to solving the problem of unbalanced data is the use of various sampling algorithms. As the machine learning algorithms are directly impacted by the minority class in an imbalanced dataset, we opted to over-sample the CICDDoS2019 minority class using SMOTE algorithm. The application of SMOTE synthetically generates new instances of the minority class (Benign, UDP_LAG and WEBDDOS). Then, we are able to use ML classifiers that act very well with numeric values, the dataset needs normalization, we change the value of the categorical features to numeric value, and we standardize the Dataset given all features value with a defined scale between -1 and 1.

4.5 Features Extractions

To make an important ML model in terms of accuracy and precision, the extraction of a valid and sufficient subset of features that can be used to build efficient models to identify a DDoS attack is an important step. Extracting the relevant features from the dataset is crucial for modeling to differentiate normal behavior from attack behavior. Ranking the most important features is conducted using some existing models for features selection based on:

- Features missing value; Single unique value and low importance features [25].
- Collinear-highly correlated features: This method finds pairs of collinear features based on the Pearson correlation coefficient [26]. For each pair above the specified threshold, it identifies one of the variables to be removed.
- Information Gain and Chi-Square statistics [27]: These methods are applied to measure the importance of each feature and to determine the relationship between two categorical variables. We use this method for feature selection by evaluating the Information gain of each variable in the context of the chosen target variable. In our research, the identification of effective network features for DDoS attack detection was combined by using the PCA method that requires a standardized dataset. We tolerated a threshold of 90% of the variance that was explained by reduction of features from 86 to 37 features. We also use the RFR to determine an optimal feature set.

Crossing all these cited methods results, we are able to select the useful features. The final list of features collected after the application of all the described previous methods is fixed at 37 features and it is applied for the rest of this study.

4.6 Machine Learning Models

The analysis of CICDDoS2019 shows that the dataset contains benign and DDoS attacks and we also find the presence of the following 13 class labels SYN, TFTP, WebDDoS, DNS, Benign, MSSQL, LDAP, NETBIOS, NTP, SSDP, SNMP, UDP, and UDP-Lag. These class labels had been respectively encoded as DDoS or Benign. We use a clustering for the DDoS attack based on the type of protocol used in the attack; we observe that the most of these attacks are carried from both UDP and TCP protocols, which are successively represented by 17 and 6 as shown in e Fig. 2. We split the dataset into 70% training and 30% testing subset. The training data is the portion of the datset on which the model trains to discover and learn patterns, while the testing data is used to evaluate the performance of the model. The next step is to start training and testing the seven chosen ML algorithms, namely, Naïve Bayes, XGBoost, Random Forest, Logistic Regression, KNeighborsClassifier, Multi-layer Perceptron (MLP), Decision Tree Classifier and Convoluted Neural Network. The use of these models requires a tuning of the hyperparameters in order to gain the best performance. The hyperparameters were used appropriately as follows (n_estimators = 50) for RandomForestClassifier, (n_neighbors = 7) for KNeighborsClassifier, and for Multilayer Perceptron classifier we used (alpha = 0.005). For the CNN we used activation: Relu, Pool size: 2, Conv Layer (3), Dense layer: 3, Optimizer: Adam, epochs = 100 and batch_size = 32.

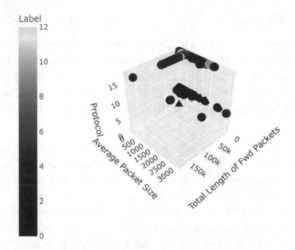

Fig. 2. DDoS clustering by protocol.

4.7 Experimental Section

This experiment is performed on a Windows machine with an Intel(R) Core(TM) i7-10700K CPU @ 3.80 GHz (16 CPUs), and 32768 MB RAM. We use the "Anaconda" platform for deployment of all machine learning models and Python language for development. Scikit-Learn, Pandas, TensorFlow, Keras and Numpy 1.8.2 are the libraries used most in this experiment.

5 Discussion

The classification of Ddos attacks in the CICDDoS2019 dataset where almost all of its data is unbalanced, is a challenge that must be taken seriously and the strong need to improve the accuracy of attack classification has led us to use sampling algorithms such as ROS and SMOTE. Certainly, the results of other research point out the importance and efficiency of using sampling and feature selection algorithms. To this end, the work presented in this paper has contributed to these predecessors while reinforcing the same hypothesis that using sampling methods such as ROS and SMOTE and the selection of features via the principal component analysis (PCA) and the Random Forest Regressor (RFR) have achieved a very high classification accuracy. Table 2 presents the evaluation metrics results, including accuracy, precision, recall, and f-measure.

Table 2. Classifiers results using our approach.

Classifiers	Precision	Recall	F1score	Accuracy
GaussianNB	0.97	0.94	0.96	0,94
XGB Classifier	0.99	0.98	0,99	0,99924
Random Forest Classifier	0,99	0.98	0,98	0,99916
Logistic Regression	0.98	0,99	0,98	0,98684
KNeighbors Classifier	0.99	0.98	0,99	0,99448

We can clearly see that the two models, XGBClassifier and Random Forest Classifier, have achieved the best results. Our analysis of the results obtained allows us to rank the XGBClassifier model at the top of the models. Analyzing further, the XGBClassifier model allowed an accuracy score of 0.99924, with precision, recall, and Fscore 0.99, 0.98 and 0.99 as the final results for the 13 classes tested. The Random Forest Classifier had a slight difference in accuracy with a score of 0.99916 and a precision, recall, and f1score of 0.99, 0.98, and 0.98 for all the 13 classes. Figure 3 represents the accuracy score in detecting DDoS attacks for each tested classifier in detection of the DDoS attacks at CICDDoS19 DataSet.

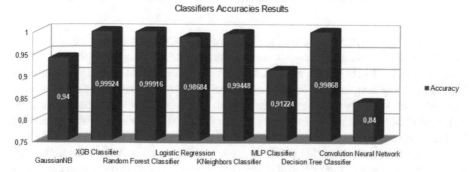

Fig. 3. Classifiers Accuracies scores for DDoS attack Detection in CICDDoS2019 DataSet based on our approach.

Table 3 summarizes the confusion matrix results for the tested dataset using the XGBClassifier and Random Forest models. The confusion matrix summarize the correct and false predictions and determine the performance of the two models achieved the best accuracy results.

Table 3. Confusion matrix on the four distinct events (TP, FP, TN, and FN) obtained using our proposed approach.

	Attack	Normal
Attack	0.97	0.94
Normal	0.99	0.98

Table 4. Results Comparison of related researchs.

Datasets	Classifier	Feature Selection	Accuracy
UNB-ISCX	Semi-supervised ML algorithm	No feature selection	96,28%
CICDDoS2019	DDoSNet using autoencoder and RNN	Feature selection RFR	99%
KDD Cup'99	SVM and DNN	No feature selection	%92,30
CIC DoS dataset	RT, J48, REP Tree, SVM, RF, MLP	No feature selection	95%
CICDDoS2019	SVM, K-NN, DT, NB, RF and LR	Feature selection RFR	99%
UNSW-NB15	Extra-Trees	FSD features	99.88%& 93.71%
CICDDoS2019 (Our Approach)	XGB, RF	PCA & RFR	99,9%

The results' comparison to the related research shows that the most effective features should be selected when creating ML models. From Table 4 we observe that the performance of ML models in studies using feature selection algorithms is better than the other ML without any feature selection, which is confirmed by the results in other studies. The results of our study only continue to confirm this hypothesis by the results obtained in detecting DDoS attacks. The accuracy results obtained for both XGB and RF models show that the prediction of DDoS attacks from the CICDDoS2019 dataset performs best not only with an accuracy rate of 99% but also F-measure with 99%. Concretely, both models correctly predict the illustrated threats via the "Accuracy" index and capture all relevant instances of malicious traffic via the "Recall" index with a 99% success rate.

Applying this work in a real SOC environment will be an opportunity to tackle DDoS attacks more efficiently and with a near-real absent scale of error for false positive alerts.

6   Conclusion and Future Work

In this article, we present our approach to optimize the work time for the SOC team and improve the reaction time detecting serious DDoS attacks using ML. For this purpose, we chose the latest cybersecurity dataset CICDDoS2019, containing several DDoS attacks. CRISP-DM was the main method used during the major phases of our study. In this regard, our approach consisted of processing the dataset, firstly by cleaning noise, normalizing, and managing unbalanced data that impact the learning of machine learning models. Secondly, we took advantage of several feature selection methods used in research to achieve the best results. Afterwards, we examined seven diverse ML algorithms: SVM, K-NN, DT, NB, RF, LR and CNN. The following measurements' accuracy, precision, recall, true-positive ratio, false-positive ratio, and F-score were used in the evaluation. The results of the experiment show that the best accuracy is achieved using XGBClassifier and RF ML algorithms. The model evaluation under the pretext of our approach achieves the highest evaluation metrics not only regarding recall, precision, F-score, and accuracy compared to the existing well-known classical ML techniques but also regarding the rate of false positive alerts, which is positively reduced. The exploitation of these results will be a real significant time saver for SOC teams.

In future work, we are interested in testing the performance of our approach using the two models XGBClassifier and RF on a real dataset and correlating the results with other alerts. The objective is to measure alert accuracy and true positive rate in an SOC environment. Afterwards, we will tackle other types of attacks; we predict using ML in detection of other kinds of cybersecurity attacks.

References

1. Mirkovic, J., Reiher, P.: A taxonomy of DDoS attack and DDoS Defense mechanisms. ACM SIGCOMM Comput. Commun. Rev. **34**(2004). https://doi.org/10.1145/997150.997156
2. Fortinet Cyber Threat Ranking By types, https://www.fortinet.com/resources/cyberglossary/types-of-cyber-attacks
3. Sharafaldin, I., Lashkari, A.H., Hakak, S., Ghorbani, A.: Developing Realistic Distributed Denial of Service (DDoS) attack dataset and taxonomy. In: 2019 International Carnahan Conference on Security Technology (ICCST) (2019). https://doi.org/10.1109/CCST.2019.8888419
4. Kumar, R., Kumar, P., Tripathi, R., Gupta, G.P., Kumar, N., Hassan, M.M.: A privacy-preserving-based secure framework using blockchain-enabled deep-learning in cooperative intelligent transport system. IEEE Trans. Intell. Transp. Syst. 1–12 (2021). https://doi.org/10.1109/TITS.2021.3098636
5. Keshk, M., Turnbull, B., Moustafa, N., Vatsalan, D., Choo, K.-K.R.: A privacy-preserving-framework-based blockchain and deep learning for protecting smart power networks. IEEE Trans. Industr. Inf. **16**(8), 5110–5118 (2020). https://doi.org/10.1109/TII.2019.2957140
6. Raza, S., Wallgren, L., Voigt, T.: Real-time intrusion detection in the Internet of Things. Ad Hoc Netw. **11**(8), 2661–2674 (2013). https://doi.org/10.1016/j.adhoc.2013.04.014
7. Mugunthan, R.: Soft computing based autonomous low rate DDoS attack detection and security for cloud computing. J. Soft Comput. Paradigm **2019**, 80–90 (2019). https://doi.org/10.36548/jscp.2019.2.003

484   O. Lamrabti et al.

8. A review on statistical approaches for anomaly detection in DDoS attacks. Inf. Secur. J. Global Perspective, **29**(3). Accessed 27 May 2022
9. Ajeetha, G., Madhu, P.G.: Machine learning based DDoS attack detection. In: 2019 Innovations in Power and Advanced Computing Technologies (i-PACT)
10. Dong, S., Sarem, M.: DDoS attack detection method based on improved KNN with the degree of DDoS attack in software-defined networks. IEEE Access **8**, 5039–5048 (2020). https://doi.org/10.1109/ACCESS.2019.2963077
11. Chu, J.L., Krzyżak, A.: Analysis of feature maps selection in supervised learning using convolutional neural networks. In: Sokolova, M., van Beek, P. (eds.) AI 2014. LNCS (LNAI), vol. 8436, pp. 59–70. Springer, Cham (2014). https://doi.org/10.1007/978-3-319-06483-3_6
12. Karimipour, H., Dehghantanha, A., Parizi, R.M., Choo, K.-K.R., Leung, H.: A deep and scalable unsupervised machine learning system for cyber-attack detection in large-scale smart grids. IEEE Access **7**, 80778–80788 (2019). https://doi.org/10.1109/ACCESS.2019.2920326
13. Talabis, M., McPherson, R., Miyamoto, I., Martin, J.: Information Security Analytics: Finding Security Insights, Patterns, and Anomalies in Big Data. Syngress, Waltham, MA (2014)
14. Feng, C., Wu, S., Liu, N.: A user-centric machine learning framework for cyber security operations center. In: 2017 IEEE International Conference on Intelligence and Security Informatics (ISI), pp. 173–175 (2017)
15. Goeschel, K.: Reducing false positives in intrusion detection systems using data-mining techniques utilizing support vector machines, decision trees, and naive Bayes for off-line analysis, pp. 1–6 (2016)
16. Seifousadati, A., Ghasemshirazi, S., Fathian, M.: A Machine Learning Approach for DDoS Detection on IoT Devices (2021)
17. Alamri, H.A., Thayananthan, V.: Analysis of machine learning for securing software-defined networking. Procedia Comput. Sci. **194**, 229–236 (2021). https://doi.org/10.1016/j.procs.2021.10.078
18. Chapman, P., Clinton, J., Kerber, R., Khabaza, T., Reinartz, T., Shearer, C., Wirth, R.: CRISP-DM 1.0: Step-by-Step Data Mining Guide. CRISP-DM consortium: NCR Systems Engineering Copenhagen (USA and Denmark) DaimlerChrysler AG (Germany) (2000)
19. Webb, G.I.: Naïve Bayes. In: Sammut, C., Webb, G.I. (eds.) Encyclopedia of Machine Learning, pp. 713–714. Springer, US, Boston, MA (2010)
20. Torlay, L., Perrone-Bertolotti, M., Thomas, E., Baciu, M.: Machine learning–XGBoost analysis of language networks to classify patients with epilepsy. Brain Inform. **4**(3), 159–169 (2017). https://doi.org/10.1007/s40708-017-0065-7
21. Liu, Y., Wang, Y., Zhang, J.: New Machine Learning Algorithm: Random Forest. In: Liu, B., Ma, M., Chang, J. (eds.) ICICA 2012. LNCS, vol. 7473, pp. 246–252. Springer, Heidelberg (2012). https://doi.org/10.1007/978-3-642-34062-8_32
22. -Rymarczyk, T., Kozłowski, E., Kłosowski, G., Niderla, K.: Logistic regression for machine learning in process tomography. Sensors **19**, 3400 (2019). https://doi.org/10.3390/s19153400
23. Hayaty, M., Muthmainah, S., Ghufran, S.M.: Random and synthetic over-sampling approach to resolve data imbalance in classification. Int. J. Artif. Intell. Res. **4**(2), 86–94 (2020). https://doi.org/10.29099/ijair.v4i2.152
24. Wang, J., Xu, M., Wang, H., Zhang, J.: Classification of imbalanced data by using the SMOTE algorithm and locally linear embedding. In: 2006 8th International Conference on Signal Processing (2006)
25. Acuña, E., Rodriguez, C.: The treatment of missing values and its effect on classifier accuracy. In: Banks, D., McMorris, F.R., Arabie, P., Gaul, W. (eds.) Classification, Clustering, and Data Mining Applications. Springer, Berlin, Heidelberg (2004)
26. Liu, Y., Mu, Y., Chen, K., Li, Y., Guo, J.: Daily activity feature selection in smart homes based on pearson correlation coefficient. Neural Process. Lett. **51**(2), 1771–1787 (2020). https://doi.org/10.1007/s11063-019-10185-8

27. Kumar, S., Singh, V.B., Muttoo, S.K.: Bug report classification by selecting relevant features using chi square, information gain and latent semantic analysis. In: 2021 9th International Conference on Reliability, Infocom Technologies and Optimization (Trends and Future Directions) (ICRITO), pp. 1–5 (2021)

28. Guo, Q., Wu, W., Massart, D.L., Boucon, C., de Jong, S.: Feature selection in principal component analysis of analytical data. Chemom. Intell. Lab. Syst. **61**, 123–132 (2002). https://doi.org/10.1016/S0169-7439(01)00203-9

29. Shreyas, R., Akshata, D.M., Mahanand, B.S., Abhishek, C.M.: Predicting popularity of online articles using Random Forest regression. In: 2016 Second International Conference on Cognitive Computing and Information Processing (CCIP), pp. 1–5 (2016)

30. Ravi, N., Shalinie, S.M.: Learning-driven detection and mitigation of DDoS attack in IoT via SDN-cloud architecture. IEEE Internet Things J. (2020)

31. Elsayed, M.S., Le-Khac, N-A., Dev, S., Jurcut, A.D.: DDoSNet: a deep-learning model for detecting network attacks. IEEE Comput. Soc. 391–396 (2020)

32. Karan, B.V., Narayan, D.G., Hiremath, P.S.: Detection of DDoS attacks in software defined networks. In: Proceedings of the 2018 3rd International Conference on Computational Systems and Information Technology for Sustainable Solutions, CSITSS (2018)

33. Bengaluru, India, pp. 265–270, 20–22 December 2018

34. Perez-Diaz, J.A., Valdovinos, I.A., Choo, K.-K.R., Zhu, D.: A flexible SDN-based architecture for identifying and mitigating low-rate DDoS attacks using machine learning. IEEE Access **8**, 155859–155872 (2020)

35. Alzahrani, R.J., Alzahrani, A.: Security analysis of DDoS attacks using machine learning algorithms in networks traffic. Electronics **10**, 2919 (2021)

36. Idhammad, M., Afdel, K., Belouch, M.: Semi-supervised machine learning approach for DDoS detection. Appl. Intell. **48**, 3193–3208(2018). https://doi.org/10.1007/s10489-018-1141-2

37. Latif, H., et al.: Intrusion detection framework for the Internet of Things using a dense random neural network. IEEE Trans. Industrial Inform. **18**(9), 6435–6444 (2022).https://doi.org/10.1109/TII.2021.3130248

38. Hasan, T., et al.: Securing industrial internet of things against botnet attacks using hybrid deep learning approach. IEEE Trans. Netw. Sci. Eng. (2022). https://doi.org/10.1109/TNSE.2022.3168533

39. Krivchenkov, A., Misnevs, B., Grakovski, A.: Using Machine Learning for DoS Attacks Diagnostics. In: Kabashkin, I., Yatskiv, I., Prentkovskis, O. (eds.) RelStat 2020. LNNS, vol. 195, pp. 45–53. Springer, Cham (2021). https://doi.org/10.1007/978-3-030-68476-1_4

40. Pereira, J., Saraiva, F.: Convolutional neural network applied to detect electricity theft: a comparative study on unbalanced data handling techniques. Int. J. Electr. Power Energy Syst. **131**, 107085 (2021)

Enhance E2E Online Shopping System Using Secure Fingerprint Parcel Delivering Method

Dineish Duke A/L Narayanasamy[1], Omar Ismael Al-Sanjary[1(✉)],
Ahmed Saifullah Sami[2], and Manar Y. Kashmola[3]

[1] Faculty of Information Sciences and Engineering, Management and Science University, 40100 Shah Alam, Selangor, Malaysia
omar_ismael@msu.edu.my
[2] Information Technology Management, Duhok Technical Administration College Duhok, Polytechnic University, Kurdistan Region of Iraq, Duhok, Iraq
[3] College of Information Technology, Ninevah University, Mosul, Iraq

Abstract. Nowadays, with the rapid growth of online marketing or online purchasing is increasing day by day as customers tend to purchase or buy goods online rather than physical shopping. This was due to the recent Covid-19 outbreak, which lead to a massive or vast of online shopping businesses or franchises to have emerged. However, with this vast growth of online businesses with massive online purchases being done every day, questions are being raised by customers as were these online businesses secure enough when comes to parcel delivery. With regards to that, a survey was conducted to gather customer's feedback on the current online shopping method, whether was it secure enough or not and this gathered data was the main reason or input that provided purpose to initiate this research, where surprisingly, majority of the customers agreed that the current online shopping is not secured in terms of parcel delivery, in which there is no customer authentication process being done upon parcel delivered. Besides that, no designated parcel lockers are being built for unattended customers and even though there are parcel lockers, these lockers have low-security levels [1]. Therefore, the customer's concerns and the issues highlighted by previous researchers, explain the purpose of this research which was conducted to help curb the issues by implementing an alphanumeric sequence for customer verification and fingerprint sensing for customer confirmation during parcel delivery. Finally, this method will also be implemented for parcel lockers to enhance the security levels upon parcel collection.

Keywords: alphanumeric sequence · fingerprint sensing · secure parcel locker

1 Introduction

This research is aimed at developing and implementing an enhanced End-to-End (E2E) online shopping system using a secure parcel delivery method, which is capable of verifying and confirming customers during parcel delivery and upon parcel collection from system that can generate the alphanumerical sequence for customer verification along with the customer's confirmation by sensing the customer's fingerprint upon item delivery. The following objective is to have a parcel locker with enhanced security level by implementing the alphanumeric sequence verification and the fingerprint confirmation of the customer to the locker upon parcel collection.

The scope of this research is divided into two sections, where the first section is the input scope, and the second section is the output scope. What this means is that the input scope is where the data is coming from, whereas the output scope is where the framework will be implemented. Therefore, in this research, the input data is coming from the customers who will be using this system, as the customers are required to provide the alphanumeric sequence as part of the verification and provide the fingerprint as part of the confirmation. Nonetheless, the output of this research will be implementing these methods in the online shopping systems in Malaysia, which are inclusive of ordering and delivering parcels.

2 Related Work

The main idea of this research paper is to expand upon the context and background of the study, to further define the problem, and provide empirical and theoretical bases for the research. Thus, this paper compares eight different current systems that are like the one that is being proposed to be implemented. As for this paper, those eight systems that were chosen to be compared are the systems of the Online Food Ordering System, Smart Locker System, Online Smart Ration Card System with Multifactor Authentication, On the Regulatory Framework for Last Mile Delivery Robots, Drones for Parcel and Passenger Transportation: A Literature Review, Sustainable Solutions for "Last Mile" Deliveries in the Parcel Industry: A Qualitative Analysis using Insights from Third Party Logistics Service Providers and Public Mobility Experts, Scope of Using Autonomous Trucks and Lorries for Parcel Deliveries in Urban Settings, Innovation in Last Mile Delivery: Meeting Evolving Customer Demands – The case of In-Car Delivery [29, 30].

2.1 Online Food Ordering System

This research paper dealt mainly with the intended use mechanism in the food supply industry. This system allows hotels and restaurants to boost the volume of online food orders for this sort of business. Buyers can choose food menu options for just a few minutes. In advanced food producers, it is possible to deliver efficiently and simply on the customer's premises. Restaurant workers then use these deliveries through easy delivery upon this customer's premises to find out easily how to operate the graphical interface for efficient operation [2].

2.2 Smart Locker System

Numerous sites, such as departments, malls, archives, research centres, and so on. Safety is a major concern for the protection of user's data so that no unauthorized individual will have access to it. These days, seeking a shield for significant details and perhaps even money, all the time searching for security plans. This research suggests an open window mechanism that offers protection in just such a way that many banks, organizations, as well as institutions, will make use of it. There are also several methods to authenticate using a password or perhaps an RFID, but still, the form used for this paper is quite powerful and effective. This guarantees high security for bank manufacturers as well as simplifies the operation. Unauthorized disclosure refused by key design can restrict printing to registered users. Fingerprint has obtained a fingerprint reader and has been submitted for verification. When the fingerprint is functioning well, the door will open immediately, otherwise, a buzzer attached to the device will also be used to give a message [3, 23–25].

2.3   Online Smart Ration Card System with Multifactor Authentication

In this paper, an internet-based smart ration card framework with biometrics authentication mechanisms has been introduced to detect ration falsification. Various factors such as fingerprint biometric approach, Radio Frequency Identification (RFID), colour code combinations as well as SMS input are often used to prevent ration fraud. Throughout this framework, the RFID tag is used that transmits relative key points, as well as the client, needs to demonstrate this tag at just the local store or the user's place before another ration is received. The user will also need to consider giving a thumbprint on the biometric machine to check whether the customer is genuine. In contrast, until making the purchase, the user must also provide opinions on the colour code patterns requested by the system during registration. As a result, the legitimacy of the customer is authenticated using numerous aspects to increase the total network security. Other than the conventional practice of attending and afterwards picking up the ration, a new approach was introduced here where the authorized person can request and receive the supply online at their address. Within that, the authors present a brief overview of the Ration Distribution Model using Smart Card with creates an enabling environment as well as total system protection [4, 28].

2.4   On the Regulatory Framework for Last Mile Delivery Robots

Autonomous delivery robots are now being built all over the globe, and the very first designs are currently being tested over the last-mile delivery of parcels. Estonia plays an important role in this area with all its start-up Starship Technology, which functions not just in Estonia as well as in foreign countries such as Germany, the United Kingdom, and the United States of America (USA), which it seems to have a potential solution to the last-mile challenge. However, the relatively regular presence of logistics robots in public transport exposes flaws in the legislative system for the use of these autonomous vehicles, despite the sophistication of the technological innovations. The associated legal concerns emerge from data privacy over responsibility for wrongdoing in such mundane areas as traffic law, which must be taken into consideration by the logistic service provider. This paper analyses, as well as further, explore the legislative structure for autonomous package distribution robots by discussing the legal consequences. Even though delivery robots can indeed be interpreted as cyber-physical systems in the sense of Industry 4.0, the research adds to the relevant regulatory system of Industry 4.0 at the international level. Finally, the paper addresses possible perspectives as well as suggests concrete ways of compliance [5, 8, 21, 22].

2.5 Drones for Parcel and Passenger Transportation

Delivery drones including 'air taxis' were currently perhaps the most widely debated new developments likely to extend accessibility to the 'third dimension of low-level airspace. This article provides a systematic analysis of 111 interdisciplinary findings (2013–03/2019). The analysis focuses on integrating the ongoing socio-technical debate on civil drones for transport purposes, enabling a (critical) interim evaluation. Four aspects of the study were postponed directing the evaluation process. A sum of 2581 related quotations was segmented into prospective obstacles (426), future difficulties (1037), possible alternatives (737) and projected benefits (381).

Researchers have observed that the controversy is marked primarily by technological and regulatory issues as well as obstacles that are perceived to discourage or hinder the use of drones for parcel and passenger transport. Around the same moment, simple economic aspirations are juxtaposed with nuanced and distinct questions about the social and environmental impacts. Reviewing the most prevalent transport-related claims of traffic avoidance, travel time savings and environmental relief, researchers have observed that there is still a clear need to provide scientific proof of the pledge of using drones for transport [9]. Researchers conclude that the debate on drones for transportation needs further qualification, emphasizing societal benefits and public involvement more strongly [6, 7, 26, 27].

3 Proposed Method

Based on Fig. 1 below, the flowchart diagram explains how the overall system or framework functions or in other words works. First, is where the transmission starts or the overall framework starts, which then the admin will manage the webpage by adding, updating, or deleting the items in the admin portal. Once the Admin has managed the webpage, the customers will then be able to proceed with the online purchasing at the customer's portal. After the customers have done selecting the online items, the customers will then be directed to the checkout or payment page, where here the customers will need to state the customer's credit card or debit card details for the payment process to take place. Once the payment is completed, the system will then generate an alphanumeric sequence or in other words, an invoice number, which the customers are required to present this alphanumeric sequence to the delivery person during the parcel delivery. Therefore, once the items are purchased, the parcel delivery person will then deliver the parcel to the customer's location. The parcel delivery process is divided into two sections, one is the parcel delivery for attended customers and the second is the parcel delivery for the unattended customer.

During the parcel delivery for the attended customer, the delivery person will ask the buyer or customer to provide the alphanumeric sequence to verify if the customer is genuine or not. After the customer has successfully been verified, the delivery person

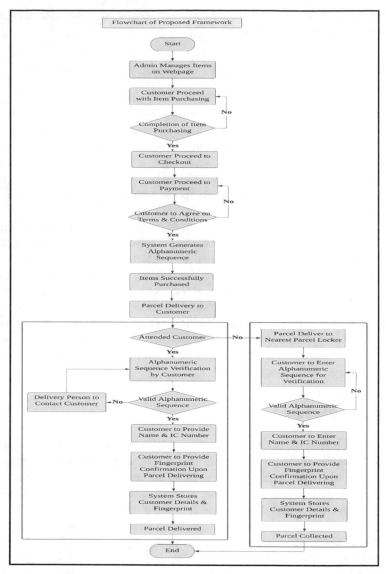

Fig. 1. Flowchart Diagram of Enhanced E2E Online Shopping System using Secure Parcel Delivering

will ask the customer to place the customer's finger on the fingerprint sensor to scan the fingerprint image as part of the customer's confirmation.

Once, both the verification and confirmation processes have been completed, only then the delivery person will hand over the parcel to the customer. As for the parcel delivery for unattended customers, the parcel delivery guy will contact the buyer and customer that, the parcel will be placed in the nearest parcel locker, where the customers will then need to collect the customer's parcel from the nearest parcel looker, in which, the customer is required to key in the alphanumeric sequence to the locker and once it has been verified, the locker will then prompt the customer to provide the customer's fingerprint image by scanning on the fingerprint sensor as part of the customer's confirmation. After the process of verification and confirmation is completed, only then the parcel will be released by the locker for collection.

Table 1. List of Hardware and Software

No.	HARDWARE/SOFTWARE
1.	NodeMCU WIFI Module
2.	Breadboard
3.	Jumper Wires
4.	Fingerprint Sensor
5.	Arduino IDE
6.	Google Chrome Browsing Application
7.	phpMyAdmin
8.	XAMPP Control Panel
9.	IBM SPSS Statistics
10.	Google Forms

Table 1 shows the list of hardware and software used to develop this system, where the NodeMCU WIFI Module is the microcontroller that acts as a server [10], and the breadboard is an electronic board [11]. Jumper wires are used to complete the connections of the NodeMCU WIFI Module to the fingerprint sensor on the breadboard. The fingerprint sensor whereby is a device used to scan the fingerprint image of a particular person [12]. Besides that, the Arduino IDE is used to program the fingerprint sensor [13]. The Google Chrome browsing application is used to open the web pages at the same time to open the phpMyAdmin portal, where this phpMyAdmin portal is a database portal in which all the data will be stored here [14]. Moreover, the XAMPP control panel is used to access the phpMyAdmin portal [15], whereas, Google Forms are used to construct the questionaries or surveys [16], in which the collected or gathered data from the surveys are later then analyzed and sorted using the IBM SPSS Statistics tool, where this tool is used to generate or populate the bar graph, pie chart, percentages, Means, Standard Deviations and many more [17].

There are programming languages used to develop the online webpages or online portals, where the SQL language was used to create the database tables for this system which includes the admin login table, product table, customers table, orders table and order items table [18]. The PHP language used was to develop the online webpages or online portals and is the main language to link the webpages to the database tables. Examples of webpages from this system are the admin page, customer purchasing page, delivery page and so on [19]. Finally, the C++ language, where this language was used to configure the fingerprint sensor, to sense or store the customer's fingerprint as part of the customer's confirmation process [20].

4 Results

Figure 2 below demonstrates the admin's webpage, where here, the admin will manage the items by adding new items, updating the added items and deleting unnecessary items from the webpage. Added, updated, and deleted items will be reflected on the Customer's webpage. Besides that, the admin can view the status of the purchased items.

Fig. 2. Admin's Webpage

Figure 3 demonstrate, the Customer's webpage, where here the customer will start with the purchasing of the items. Moreover, the customer will be able to update the quantity of the item from the shopping cart and on the shopping cart page, the customer can either select the 'Continue Shopping' button to continue shopping or the customer can proceed with the checkout section by clicking on the 'Checkout' button.

Fig. 3. Customer's Webpage

Figure 4 shows the order status webpage or the invoice page, where this will be the final phase of the customer's online purchasing process. Before the final customer phase, the customer must first provide the customer contact details and payment details and must agree to the Terms and Conditions of this system to complete the purchases.

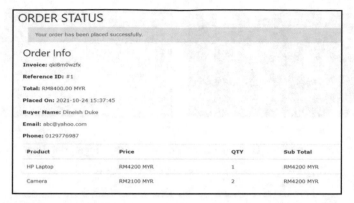

Fig. 4. Order Status Webpage

Figure 5 shows the list of orders that will be visible to the courier, where upon delivery, the attended customer is required to provide the alphanumeric sequence or the invoice sequence to the courier for the verification process. In case the provided alphanumeric sequence is invalid, the courier will need to contact the customer via the customer's phone number for further clarification or confirmation.

Fig. 5. List of Orders Webpage for Delivery Person

Figure 6 (a) displays the design or the interface of the system's prototype, where there are two open areas, which consists of the open area for the power supply to the NodeMCU and an open area for the customers to place the fingerprint on top of the fingerprint sensor. Figure 6 (b) demonstrate the action, where, the customer is required to scan the fingerprint using the fingerprint scanner provided as shown in Fig. 6 (a) by the courier.

After the customer has successfully scanned the fingerprint, the parcel will then be delivered to the customer and the status of the parcel will be updated from 'Verified' to 'Completed' as shown in Fig. 7 (a) and Fig. 7 (b).

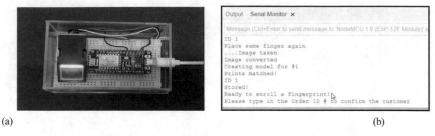

(a)   (b)

Fig. 6. (a) Inner View of the Prototype Interface Design, (b) Fingerprint Scan Completed and Stored

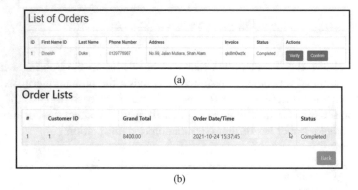

(a)

(b)

Fig. 7. (a) Parcel Status Set to Completed in List of Orders Webpage for Delivery Person, (b) Parcel Status Set to Completed in Order Lists tab of admin's Webpage

5 Conclusion

In a nutshell, from the research paper that has already been done, it can be concluded that this system shows how it functions or works with a newly designed flowchart that is acceptable and beneficial to the customers. Hence, the objectives of developing this system will solve the current problems or issues raised by the customers. This was proven by performing multiple testing which includes Unit Testing, System Testing, Beta Testing as well as Acceptance Testing. Lastly, some works cannot gratify all the validation operations because the technique or method used has its weaknesses that prevented the completion of the validation process. Therefore, we conclude that this area of research has room for improvement by validating the domain engineering directly instead of validating software products during the configuration process. As part of the recommendations for the further development of this "Enhanced E2E Online Shopping System using Secure Parcel Delivering Method" are that developers can enhance this system by including the parcel tracking feature using the Google Map Live Location function that can help customers track the real-time or current location of the parcel. This will also help customers to plan their everyday activities so that the customers will be available to attend or collect the parcel upon delivery. In addition, this system can be enhanced by implementing the "Store and Forward" feature, where in the event of power loss or connectivity interruption, Gateways can store the data internally and then forward or store the data to the server when the device is re-connected or when the power is re-established. This feature is a lossless storing function.

References

1. Huong, T.T., Thiet, B.N.: Smart locker-a sustainable urban last-mile delivery solution: benefits and challenges in implementing in Vietnam (2020)
2. Deepa, T., Selvamani, P.: Online food ordering system. Int. J. Emerging Technol. Innov. Res. ISSN, pp. 2349–5162 (2018)
3. Sharma, A., Jain, A., Bagora, A., Namdeo, K., Punde, A.: Smart locker system (2020)
4. Kamble, B., Dambe, N., Kulkarni, S., Virkar, G.: Online Smart Ration Card System with Multi-factor Authentication. In: Nutan College of Engineering & Research, International Conference on Communication and Information Processing (ICCIP) (2019)
5. Hoffmann, T., Prause, G.: On the regulatory framework for last-mile delivery robots. Machines **6**, 33 (2018)
6. Kellermann, R., Biehle, T., Fischer, L.: Drones for parcel and passenger transportation: a literature review. Transp. Res. Interdisciplinary Perspectives **4**, 100088 (2020)
7. Ducarme, D., Agrell, P.J.: Sustainable solutions for "last mile" deliveries in the parcel industry: a qualitative analysis using insights from third-party logistics service providers and public mobility experts (2019)
8. Kassai, E.T., Azmat, M., Kummer, S.: Scope of using autonomous trucks and lorries for parcel deliveries in urban settings. Logistics **4**, 17 (2020)
9. Hepp, S.B.: Innovation in last mile delivery: meeting evolving customer demands: the case of In-Car Delivery (2018)
10. Parihar, Y.S.: Internet of Things and Nodemcu A review of the use of Nodemcu ESP8266 in IoT products, vol. 6, p. 1085 (2019)
11. Bell, C.: Electronics for Beginners. In: Windows 10 for the Internet of Things, Springer, pp. 307–345 (2021)
12. Gehlot, K.S., Jain, D.: Biometric fingerprint based voting machine using ATmega328P microcontroller, Materials Today: Proceedings (2020)
13. Bell, C.: Introducing the Arduino, ed, pp. 31–70 (2021)
14. Terrell, B.: Creating Data-Driven Web Sites: An Introduction to HTML, CSS, PHP, and MySQL: Momentum Press (2019)
15. Friends, A.: XAMPP installers and downloads for Apache Friends, vol. 14, p. 2020 (2017)
16. La Counte, S.: The ridiculously simple guide to google apps (G Suite): A practical guide to google drive google docs, google sheets, google slides, and google forms: Ridiculously Simple Books (2019)
17. George, D., Mallery, P.: IBM SPSS statistics 26 step by step: A simple guide and reference: Routledge (2019)
18. Weinberg, P.N., Groff, J.R., Oppel, A.J.: SQL, the complete reference: McGraw-Hill (2020)
19. Tatroe, K., MacIntyre, P.: Programming PHP: Creating Dynamic Web Pages: O'Reilly Media (2020)
20. BinUzayr, S.: Mastering C++ Programming Language: A Beginner's Guide: CRC Press (2022)
21. Abushahma, R.I.H., Ali, M.A., Al-Sanjary, O.I., Tahir, N.M.: A region-based convolutional neural network as object detection in images. In: 2019 IEEE 7th Conference on Systems, Process and Control (ICSPC), pp. 264–268. IEEE, December 2019

22. Al-Sanjary, O.I., Ibrahim, O.A., Sathasivem, K.: A new approach to optimum steganographic algorithm for secure image. In: 2020 IEEE International Conference on Automatic Control and Intelligent Systems (I2CACIS), pp. 97–102. IEEE, June 2020
23. Ahmed, A.A., Al-Sanjary, O.I., Kaeswaren, S.: Reserve parking and authentication of guest using QR code. In: 2020 IEEE International Conference on Automatic Control and Intelligent Systems (I2CACIS), pp. 103–106. IEEE, June 2020
24. Al-Sanjary, O.I., Vasuthevan, S., Omer, H.K., Mohammed, M.N., Abdullah, M.I.: An intelligent recycling bin using wireless sensor network technology. In: 2019 IEEE International Conference on Automatic Control and Intelligent Systems (I2CACIS), pp. 30–33, June 2019
25. Ishak, Z., Rajendran, N., Al-Sanjary, O.I., Razali, N.A.M.: Secure biometric lock system for files and applications: a review. In: 2020 16th IEEE International Colloquium on Signal Processing & Its Applications (CSPA), pp. 23–28. IEEE, February 2020
26. Magendran, N., Al–Sanjary, O.I., Aik, K.L.T.: Real-time monitoring and tracking objects for pandemic. In: 2021 IEEE 9th Conference on Systems, Process and Control (ICSPC 2021), pp. 135–140. IEEE, December 2021
27. Alkawaz, M.H., Veeran, M.T., Al-Sanjary, O.I.: Vehicle tracking and reporting system using robust algorithm. In: 2019 IEEE 7th Conference on Systems, Process and Control (ICSPC), pp. 206–209. IEEE, December 2019
28. Mohammed, M.N., Radzuan, W.M.A.W., Al-Zubaidi, S., Ali, M.A., Al-Sanjary, O.I., Raya, L.: Study on RFID-based book tracking and library information system. In: 2019 IEEE 15th International Colloquium on Signal Processing & Its Applications (CSPA), pp. 235–238. IEEE, March 2019
29. Reed, S., Campbell, A.M., Thomas, B.W.: The value of autonomous vehicles for last-mile deliveries in urban environments. Manage. Sci. **68**, 280–299 (2022)
30. Fehling, C., Saraceni, A.: Feasibility of Drones & Agvs in the Last Mile Delivery: Lessons from Germany, Available at SSRN 4065011

Cybersecurity Assessment Construction of Artificial Intelligence

Rachid Batess[1,2](✉) , Younes El Fellah[1] , Reda Errais[1] , Ghizlane Bouskri[3] , and El Houssain Baali[1]

[1] Energy and Agroequipment Department, Institute of Agronomy and Veterinary Hassan II, Rabat, Morocco
Rachidbatess@gmail.com
[2] Continental AG, Hannover, Germany
[3] Volkswagen AG, Wolfsburg, Germany

Abstract. The development of artificial intelligence (AI) in the automotive sector is a major contributor to the increasing number of intelligent transportation systems. Especially, with the rapid emergence and evolution of machine learning and artificial intelligence (AI) capabilities the number of AI models used in the automotive industry has been developed. Within this context; as an automotive company; we have provided efficient and secure AI models that can be used in the autonomous driving industry, and we have regularly investigated and provided the necessary security attributes to ensure that the AI models used in the autonomous driving industry are equipped with the necessary security attributes. From this perspective, the aim of this paper is to provide an overview of the various aspects of the intellectual property eco-system, including the current status of standardization and academic perspectives, also highlights the potential of this area. For this purpose we have performed a D-patching attack on the Yolo v3 model (Mujahid, 2021) to compromise its integrity. The results show that these methods are effective in protecting the model, but they require additional costs to be considered. Finally, To prevent this attack, we proposed a set of countermeasures that can be used to detect the input testing dataset.

Keywords: Artificial Intelligence · Cyber Security · Artificial Neural Networks

1 Introduction

Artificial Intelligence (AI) has become more prevalent due to the increasing number of data storage and the technological advancements that have occurred in its field. These innovations have led to the development of various AI technologies such as voice recognition, natural language processing, and image recognition. However, they have also raised concerns about their potential to affect the security of computers. On the one hand, these technologies can be used to develop effective security systems; while on the other hand, they can be used to launch attacks.

Due to the nature of AI, it is important that the systems are designed to be immune from external interference. This ensures that they can perform their functions properly.

F. Saeed et al. (Eds.): ICACIn 2022, LNDECT 179, pp. 497–507, 2023.
https://doi.org/10.1007/978-3-031-36258-3_44

This paper provide a comprehensive overview of the security concerns related to AI. It aims to explore the various aspects of this technology and how it can be used to protect the confidentiality and integrity of data.

Unlike traditional systems, which are prone to security vulnerabilities, machine learning systems are not designed to explain themselves. This lack of explainability can allow attackers to perform various attacks, such as poisoning and evasion [1].

These types of attacks are very effective and can be used to launch a wide range of attacks against different machine learning models. For instance, attackers can take advantage of the data collected in the training stage to perform various attacks. They can also modify the results of an inference by adding a small parameter to the input samples. As a result, attackers can also perform targeted attacks against different machine learning models. They can also steal the data collected during the training stage.

To address the increasing concerns about the security of AI, the Automotive Research Center is focused on developing effective security solutions that can be used to protect the various aspects of this technology. In this context, this paper also provides a comprehensive overview of the three layers of defense that can be used to protect the data collected and processed by AI systems.

Attack mitigation: Design defense mechanisms for known attacks.

Model security: Enhance model robustness by various mechanisms such as model verification.

Architecture security: Build a secure architecture with multiple security mechanisms to ensure business security [2].

Despite the progress that the industry has made in developing AI systems, there is still a long way to go before they can be fully secure. In order to achieve this, it is important that the various aspects of the technology are thoroughly studied and built on a sound foundation.

2 Introduction of Product Life Cycle and Threat Landscape of Machine Learning

2.1 Machine Learning in Product Life Cycle Using DevOps

Metrics need to be used in the various phases of the MLOps process to ensure that the deep neural networks are safe to operate. This article explains how to use KPls in the MLOps process. The metrics are typically produced in a machine learning cycle, which can take the form of a single or a series of steps. The quality gates and the various steps are also annotated with the names and IDs of the metric catalog. Most of the metrics are generated during the development phase of the image. Data analysis is a critical component of the machine learning process, as it is reflected in the various metrics that are generated in the "Data Analysis" step. In addition to this, field testing is also a part of the life cycle of the machine learning process [3]. Several metrics that are related to uncertainty and detection are also produced at these steps.

2.2 Potential Attacks and Protection Along Machine Learning Product Life Cycle

The AI product life cycle is divided into five phases: planning, data, evaluation, training, and operation. This approach is ideal for developers who are looking to use these phases in a highly agile manner.

The current model does not include all of the details about the data that is collected and stored in the AI model. This will be explained in a case study that will help machine learning developers understand the security of their data.

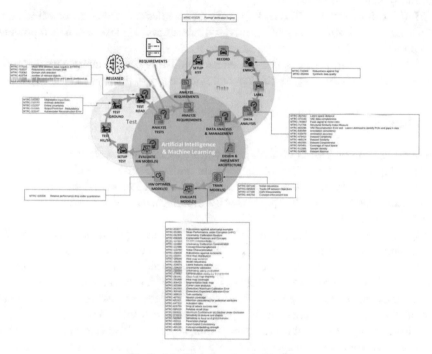

Fig.1. Machine learning in DevOps

3 Five Challenges to AI Security

Despite the potential of AI to build a better world, it still faces many security risks. One of these is the lack of security consideration when it comes to the algorithms' inference results. This vulnerability can allow attackers to perform arbitrary and misleading actions, which can result in the loss of personal safety or property. In critical areas such as healthcare and transportation, such as surveillance, attacks on AI systems can lead to property loss and endanger the safety of people [4].

To mitigate these AI security risks, AI system design must overcome five security challenges:

Software and hardware security: The software and hardware security of various platforms and applications is prone to being exploited by attackers. For instance, attackers can install backdoors in models to perform advanced attacks. Due to the in-built models' in-ability to explain the details of the backdoors, it is difficult to identify them.

Data integrity: In the training stage, attackers can introduce malicious data to the AI models' training environment, which can affect their performance. They can also add a small perturbation to the results of the inference stage to improve their performance.

Model confidentiality: Service providers typically only provide query services to their customers. However, attackers can create a clone model of these services by performing multiple queries.

Model robustness [5]: The lack of training samples that cover all possible corner cases leads to the insufficiency of the model's robustness. This issue can also prevent it from providing correct inference on adversarial cases.

Data privacy [6]: In most cases, attackers would only need to query a trained model to get the details of their users. However, in some scenarios, this can be done repeatedly.

4 Attacks

The latest version of YOLO v3 features a variety of features, such as upsampling and residual skip connections. One of these is its ability to make detections at multiple scales. In order to generate the output, the system uses a 1x1 kernel on a feature map. The resulting detections are then processed by applying a variety of 1x1 detection kernels to various feature maps.

Figure 2 illustrates the attacking of AR-ECU on the input camera by introducing the training patch of noise on the input camera. The adversary who did pre-training on the model has assumed that the model is generic model implementation and open source (i.e., Yolov3 or Yolov5). Therefore, the information about the hardware, software architectures, and code configuration is not necessary for the adversary to successfully generate the noise. As illustrated in Fig. 2, our main object to attack is a person performing the following steps. First, collect the person images dataset. The dataset that we used in this project is from Open Images dataset V6. Subsequently, we use the same detector model, which is the yolov3 model, to train the patch and generate noise. The patch will be added to the source class array in the input image [7]. During the train time, the algorithm will make the source class label vector match the target class label. The patch will be created. In this case, the source class is a person, while the target class is a chair. The algorithm will calculate the loss function to make the patchable fool the detector and output the label as a chair instead of a person. Figure 3 shows the attacking result on the object person to be a chair. The steps can be summarised in three steps as follows. Step 1 retrain the model use your dataset to obtain the patch (noise). Step 2 add the patch into the input image. Step 3 Predict the adversarial image as the input image.

In real case scenario, the generated noise can be manually printed and past it to the object or to camera. In other words, some invasive steps are involved in this attack to fool the model. The result of this attack will be displayed at the AR HUD for the driver to take further action [8].

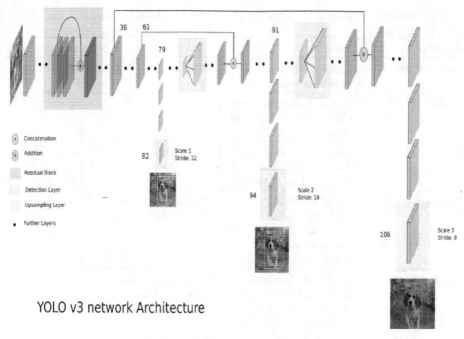

Fig. 2. YOLO v3 network Architecture

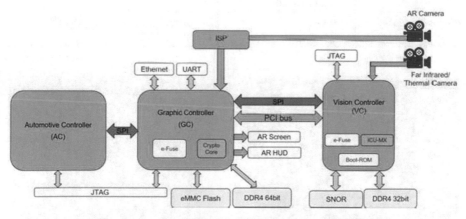

Fig. 3. Concept of **D-patch based attacks on AR-ECU**

5 AI Security Layered Defense

As illustrated in Fig. 5, three layers of defense are needed for deploying AI systems in service scenarios:

- Attack mitigation: Design defense mechanisms for known attacks.
- Model security: Enhance model robustness by various mechanisms such as model verification.

Table 1. Comparison of the state of the art works with our proposed model.

Reference	Model	Dataset	Accuracy (%)
Chen et al. [9]	Labeling Algorithm Producing Palm Mask	1300 images	93%
Nyirarugiraet al. [10]	Particle Swarm Movement (PSO), Longest Common Subsequence (LCS)	Gesture vocabulary	66%
Albawi et al. [11]	Random Forest (RF) and Boosting Algorithms, Decision Tree Algorithm	7805 gestures frames	63%
Fong et al. [12]	Model Induction Algorithm, K-star Algorithm, Updated Naïve Bayes Algorithm, Decision Tree Algorithm	50 different attributes total of 9000 data instance 7 videos	76%
Yan et al. [13]	AdaBoost Algorithm, SAMME Algorithm, SGD Algorithm, Edgebox Algorithm	5500 images for the testing and 5500 for the training set	81.25%
Ren et al. [14]	FEMD Algorithm, Finger Detection Algorithm, Skeleton-Based	Matching 1000 cases	93.20%
Proposed model	YOLOv3	216 images (Train) 15 images(Test)	97.68%

Original Input Image   **Image with noice attached**

Fig. 4. D-patch based attacks on the person object to be deceived as a chair

- Architecture security: Build a secure architecture with multiple security mechanisms to ensure business security [9].

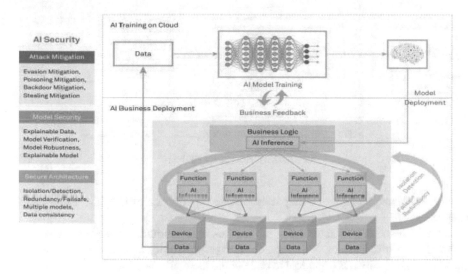

Fig. 5. AI security defense architecture

6 AI Model Security

Due to the high risks associated with AI models, it is important that they are protected from attacks. This can be done through the enhancement of their security [10]. Some of the considerations that can be considered include: preventing damage caused by other attacks, and defending against known ones.

Model detectability: In addition to traditional program analysis, AI models could also be tested with adversarial detection methods to ensure that they are secure.

Currently, the only tools that can effectively test AI models are open data sets. This makes them prone to failure in real deployments. Adapting to adversarial training techniques would cause high performance issues.

In addition to traditional program analysis, AI models could also be tested with adversarial detection methods to ensure that they are secure [11]. This can be done through a variety of security tests, such as pre-processing and post-processing units. These tools could help improve the overall performance of AI systems by detecting and preventing false positives.

Model verifiability: Compared to traditional methods, deep learning models (DNNs) [12] perform better than their counterparts in terms of classification rates and false positive rates. They are widely used in voice recognition and image recognition. However, they should be cautioned about their applications in security and safety sensitive areas. To ensure security, a certified version of a deep learning model should be used. This

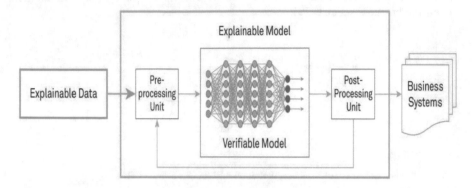

Fig. 6. Model security analysis

process involves mapping the output space and input space of the model to a certain range.

Due to the nature of statistical optimization-based methods, they are not able to tra verse all data distributions. This leaves open the possibility of developing adversarial samples. This is why it is important to understand the working principle of deep learning models [13] before implementing them in real applications.

Model explainability: Currently, most AI models are not considered to be complex black-box systems that can fully understand their decisions. In some applications, like machine translation and chess, we need to understand how these systems make decisions [14].

Despite the inscrutable nature of AI systems, they do not pose a big problem in translation applications. As long as they provide good translation results, they can continue to be considered a complete black-box system.

In some cases, inscrutableness can lead to legal or business risks. For instance, in loan analysis systems, an AI system might be criticized if it cannot provide the necessary basis for its analysis results. In healthcare systems, the basis of AI inference is also important to ensure that the system can perform properly.

One of the most important factors that an AI system should consider when it comes to analyzing a patient's cancer is its working principle. Without a clear understanding of its working principle, it is not possible to design a secure model. This is why it is important that the system's explainability is improved. This can help improve the security of the system [15].

Besides maintaining its own security, an AI model should also be enhanced to prevent it from being damaged by other attacks. This paper shows some of the considerations that need to be considered when implementing these security measures [16].

In the literature, various researchers are working on improving the explainability of AI models by developing new methods (H. Strobelt, 2018) that can perform various tasks such as searching for hidden activation functions. For instance, Morcos et al. proposed a method that can perform statistical analysis to find semantic neurons (A. S. Morcos, 2018).

7   Security Architecture of AI Services

To minimize the risks associated with the development and deployment of AI systems, we must first consider their security risks. This includes analyzing and mapping the various characteristics of their services and their architecture. We also need to design a robust security framework that will allow us to minimize the risks of using.

AI models. In autonomous driving, an AI system should be able to correctly decide on certain actions such as braking, turning, and acceleration. However, if it does not perform these actions correctly, it could seriously endanger the safety of the users. To ensure that the system is secure, various security tests are performed on it. However, in most cases, finding an AI system that can perform these actions consistently will be very challenging. The security design of a system is more important when it comes to dealing with uncertainty. For instance, if a system cannot provide a definite answer regarding the required medicine or detects a possible attack, it is better to consult the doctor instead of providing an inaccurate prediction [17]. The following security measures should be taken to ensure that an AI system is secure. For optimal AI business security, it is important that the various security mechanisms are properly implemented [18]. This can be done through the use of predefined security measures.

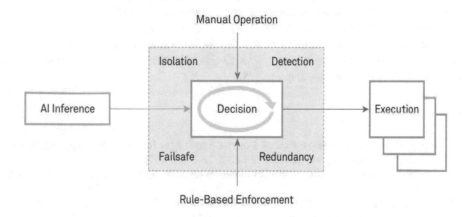

Fig. 7.   Security architecture for AI with business decision making

Isolation: The isolation of AI models from the control system can reduce the attack surface for their use in AI inference [19]. On the other hand, the isolation of the decision module can prevent attacks. When an AI system makes a decision, it can import its output into the integrated decision module, which is an auxiliary suggestion [20].

Detection: With the help of an attack-detection model and continuous monitoring, it is possible to estimate the current risk level and provide a comprehensive analysis of the network security status [21]. When the situation gets too high, the system can intervene and prevent attacks [22].

Failsafe: A security architecture that includes a multi-level framework [23] is required to ensure the safety of an AI-based system when it's needed to perform various operations, such as medical surgery or autonomous driving. The level of certainty that the system provides when it comes to performing its operations must be analyzed. If the result is lower than the threshold, the system reverts back to its traditional manual processing techniques [24].

Redundancy: One of the most important factors that businesses consider when it comes to developing AI models is the security of their data. A method that can be used to analyze the data and make the models more secure is by implementing a multi-model architecture [25]. This allows them to prevent the errors in one model from affecting the system's decisions. It also helps minimize the risk of the system being completely compromised by a single attack.

8 Conclusions

This paper aims to provide a comprehensive overview of the security concerns that are related to the development and use of Artificial intelligence in the automotive industry. We have identified the various potential attacks that could affect the operations and data of machine learning. To prevent these attacks, we have also identified the multiple factors that could affect the availability and confidentiality of the data. The goal is to provide a framework for the development and implementation of cybersecurity assessments. In order to demonstrate the effectiveness of the assessment, we perform a TARA analysis [26] on the existing use case of the Yo- lov3 machine model.

The result of the analysis revealed that the model was able to perform a CIA attack. To prove our work, we also implement an O-patching attack and a countermeasure to prevent the data from being compromised.

References

1. Stoica, D., et al.: A Berkeley View of Systems Challenges for AI, University of California, Berkeley, Technical Report No. UCB/EECS-2017–159 (2017)
2. Szegedy, C., et al.: Intriguing properties of neural networks, arXiv preprint arXiv:1312.6199 (2013)
3. Eykholt, K., et al.: Robust physical world attacks on deep learning models. In: Conference on Computer Vision and Pattern Recognition (CVPR) (2018)
4. Papernot, N., McDaniel, P., Goodfellow, I.: Transferability in machine learning: from phenomena to black-box attacks using adversarial samples, arXiv preprint arXiv:1605.07277 (2016)
5. Jagielski, M., Oprea, A., Biggio, B., Liu, C., Nita-Rotaru, C., Li, B.: Manipulating machine learning: Poisoning attacks and countermeasures for regression learning. In: IEEE Symposium on Security and Privacy (S&P) (2018)
6. Gu, T., Dolan-Gavitt, B., Garg, S.: Badnets: identifying vulnerabilities in the machine learning model supply chain. In: NIPS MLSec Workshop (2017)
7. Tramèr, F., Zhang, F., Juels, A., Reiter, M.K., Ristenpart, T.: Stealing Machine Learning Models via Prediction APIs. In: USENIX Security Symposium (2016)
8. Papernot, N., McDaniel, P., Wu, X., Jha, S., Swami, A.: Distillation as a defense to adversarial perturbations against deep neural networks. In: IEEE Symposium on Security and Privacy (S&P), 2016 L., Chen, J. Fu, Y., Wu, H. Li. And B. Zhen. Hand Gesture Recognition Using Compact CNN via Surface Electromyography Signals. Sensors 2020, 20, 672 (2020)

9. Nyirarugira, C., Choi, H.R., Kim, J., Hayes, M., Kim, T.: Modified levenshtein distance for real-time gesture recognition. In: Proceedings of the 6th International Congress on Image and Signal Processing (CISP) (2013)

10. Albawi, S., Bayat, O., Al-Azawi, S., Ucan, O.N.: Social touch gesture recognition using convolutional neural network. Comput. Intell. Neurosci. (2018)

11. Fong, S., Liang, J., Fister, I., Mohammed, S.: Gesture recognition from data streams of human motion sensor using accelerated PSO swarm search feature selection algorithm. J. Sens. **2015**, 205707 (2015)

12. Yan, S., Xia, Y., Smith, J.S., Lu, W., Zhang, B.: Multiscale Convolutional Neural Networks for Hand Detection. Appl. Comput. Intell. Soft Comput. (2017)

13. Ren, Z., Yuan, J., Meng, J., Zhang, Z.: Robust part-based hand gesture recognition using kinect sensor. IEEE Trans. Multimed. **15**, 1110–1120 (2013)

14. Gu, S., Rigazio, L.: Towards deep neural network architectures robust to adversarial examples. In: International Conference on Learning Representations (ICLR) (2015)

15. Laishram, R., Phoha, V.: Curie: a method for protecting SVM classifier from poisoning attack, arXiv preprint arXiv:1606.01584, 2016. Ai security white paper (2018)

16. Donell, S.: Terminology. In: Akgun, U., Karahan, M., Randelli, P.S., Espregueira-Mendes, J. (eds.) Knots in Orthopedic Surgery, pp. 3–10. Springer, Heidelberg (2018). https://doi.org/10.1007/978-3-662-56108-9_1

17. DIN SPEC 92001–2 artificial intelligence - lifecycle processes and quality requirements -part 2: Robustness (2020)

18. IEEE 7010–2020- IEEE recommended practice for assessing the impact of autonomous and intelligent systems an human well-being (2020)

19. Securing artificial intelligence (sai); security testing of ai. 2020. 15 Machine learning process documentation - devopsvsmlops (2021)

20. Behpour, S.: Are Adversaria robust cuts for semi-supervised and multi-label classification. In: Proceedings of the IEEE Conference on Computer Vision and Pattern Recognition Workshops, pp. 1905--1907 (2018)

21. Chan, A., Tay, Y., Ong, Y.: What it thinks is important is important: Robustness transfers through input gradients. In: Proceedings of the IEEE/CVF Conference on Computer Vision and Pattern Recognition, pp. 332–341 (2020)

22. Chernikova, A., Oprea, A., Nita-Rotaru, C., Kirn, B.: Are self-driving cars secure? evasion attacks against deep neural networks for steering angle prediction. IEEE Security and Privacy Workshops (SPW), pp. 132- 137 (2019)

23. Eykholt, K., et al.: Robust physical-world attacks on deep learning visual classification. In: Proceedings of the IEEE Conference on Computer Vision and Pattern Recognition, pp. 1625–1634 (2018)

24. Ganin, Y., et al.: Domain- adversarial training of neural networks. J. Mach. Learn. Res. **17**(1), 2096--2030 (2016)

25. Huang, Y., Kong, A.W., Lam, K.: Adversaria! signboard against object detector. In: BMVC, p. 231 (2019)

26. Li, J., Zhang, H., Han, Z., Rong, Y., Cheng, H., Huang, J.: Adversaria! attack on community detection by hiding individuals. In Proceedings of The Web Conference **2020**, 917–927 (2020)

27. Lu, S., Duan, L., Deng, D.: Quantum adversarial machine learning. Phys. Rev. Res. **2**(3), 033212 (2020)

28. Morcos, A.S., Barrett, D.G., Rabinowitz, N.C., Botvinick, M.: On the importance of single directions for generalization, arXive preprint arXiv:1803.06959 (2018)

29. Abadi, M., et al.: Deep learning with differential privacy, ACM SIGSAC Conference on Computer and Communications Security (2016)

New 3-Layer Low Complexity Cryptosystem for Color Images Based on Chao, Algebraic Diffusion and ECBC

Faiq Gmira[1,2]([⊠]) and Said Hraoui[3]

[1] Innovative Technologies Laboratory (LTI), University Sidi Mohamed Ben Abdellah, Fez, Morocco
faiq.gmira@usmba.ac.ma

[2] Computer Science & Smart Systems (C3S), Hassan II University of Casablanca, Casablanca, Morocco

[3] Artificial Intelligence and Data Science and Emerging Emerging Systems Laboratory (LIASSE), University Sidi Mohamed Ben Abdellah, Fez, Morocco
said.hraoui@usmba.ac.ma

Abstract. In this paper, based on chaotic dynamical systems coupled with pseudo-random permutations and built using a three-layer architecture, a new color image cryptosystem is proposed. The first layer process performs a confusion using a bit-wise XOR of the original image and a Piece Wise Linear Chaotic Map (PWLCM). The second layer process further enhances the consequent confusion using a one-to-one permutation function over the ring Z/nZ. The third and final layer process performs a diffusion using an enhanced cipher block-chaining mode (ECBC). Evaluation of the proposed cryptosystem proves that with low complexity the diffusion chaotically and sequentially changes the pixels of the image and proves that the avalanche effect is fully ensured. It follows that a high robustness against different attacks is guaranteed.

Keywords: Image encryption · chaos · CBC mode operator · RGB image · ring$\mathbb{Z}/n\mathbb{Z}$ · permutation · confusion · diffusion · PWLCM · confidentiality · integrity

1 Introduction

In order to secure image type data against cryptanalytic attacks, various image encryption methods have been proposed among which, in a non-exhaustive manner, those labeled classical encryption techniques, namely data encryption standard (DES), advanced encryption standard (AES), and elliptic encryption and others based on number theory and algebra concepts [1, 2]. Recently, chaotic cryptography has become widely adopted to secure transmission of multimedia data in communication networks [3, 4]. Its principle is to superimpose the image with a signal generated by a deterministic chaotic system. Besides, Chaotic systems are recognized by their pseudo-random characteristics, sensitivity to initial conditions and ergodicity property, chaotic systems also offer a good compromise between the encryption speed and the high safety performance [5].

© The Author(s), under exclusive license to Springer Nature Switzerland AG 2023
F. Saeed et al. (Eds.): ICACIn 2022, LNDECT 179, pp. 508–519, 2023.
https://doi.org/10.1007/978-3-031-36258-3_45

In general, most of the discrete chaotic cryptography approaches [5–7] are based on a confusion-diffusion architecture [8, 9], in which the confusion shifts the positions of image pixels by using a chaotic map and the diffusion changes the mixed-pixels values through a chaotic sequences. This architecture has yielded results, but some image encryption algorithms were successfully broken by cryptanalysts in various attacks [10, 11].

In order to get a typical cryptosystem complying with Shannon's principles [12], it should consist of two phases namely confusion and diffusion. A plus is the avalanche effect property which becomes is a sought-after property in cryptographic methods, this property being in that a change in any pixel of the original or the encrypted image must be reflected strictly in all the pixels or at least as much as possible with a probability greater than 0.5.

In this context and in this research work all these robustness constraints will be satisfied by proposing a new cryptosystem built according to a 3-layer architecture.

The rest of the paper is organized as follows: the architecture of the new developed cryptosystem is presented in the first section. In the second section, the proposed cryptosystem process is presented. The analysis of the proposed image cryptosystem performance is discussed in the third section. Finally, the fourth section concludes the paper.

2   Cryptosystem Architecture

The architecture of the proposed color image cryptosystem is built on a chaotic three-layer as shown in Fig. 1.

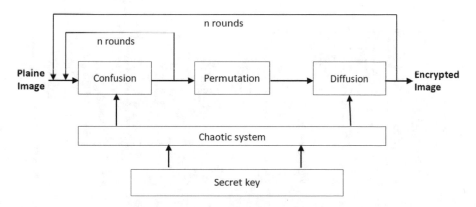

Fig. 1. Cryptosystem architecture

This layer's juxtaposition is made in such a way that it conforms to the Shannon's principles and according to the constraints of robustness. The layers topologies are illustrated in Fig. 2.

This feature will increase security and sensitivity of the proposed cryptosystem.

Layer 1	**Low complexity confusion process layer:** is performed by applying the bitwise operator between Piece Wise Linear Chaotic Map (PWLCM) and the plain-image. This operation has the advantage of having low algorithmic complexity.
Layer 2	**Enhanced diffusion process layer:** is performed by using respectively three one-to-one functions in the ring Z/nZ to make a pseudo-random permutation of pixels of each RGB component.
Layer 3	**Avalanche effect process layer:** is performed by using the ECBC mode to create sequential and chaotic modification of pixels of each color image component obtained from the second layer through.

Fig. 2. Cryptosystem layers

2.1 Low Complexity Confusion Process Layer

This first layer is based on the Piece Wise Linear Chaotic Map (PWLCM) witch defined by the following expression:

$$x(n) = F(x(n-1)) = \begin{cases} x(n-1) \times \frac{1}{p} & \text{if } 0 \leq x(n-1) < p \\ [x(n-1) - p] \times \frac{1}{0.5-p} & \text{if } p \leq x(n-1) < 0.5 \\ F([1 - x(n-1)]) & \text{if } 0.5 \leq x(n-1) < 1 \end{cases} \quad (1)$$

where $x(n) \in [0, 1]$; $n \geq 0$; with $x(0)$ is the initial condition and p the control parameter. For $p \in]0, 0.5[$ the map x(n) is chaotic and has no window in its bifurcation diagram.

PWLCM presents a chaotic aspect effective and high sensitive to initial conditions, an example is shown in Fig. 3.

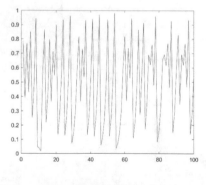 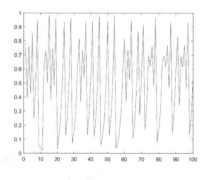

p=0.0312, x0=0.785522   p=0.0312, x0=0.785523

Fig. 3. The states sequences of PWLCM ((p = 0.0312, x0 = 0.785522) and (p = 0.0312, x0 = 0.785523))

Li, Chen and all [13], have proved the cryptographic performances, including ergodicity, random behavior, Lyapunov exponents, uniform invariant density function and an exponential autocorrelation attenuation [13].

2.2 Enhanced Diffusion Process Layer

A permutation of certain objects is a deterministic reorganization of these objects. To use the concept of permutations, for the following a definition, a theorem and a corollary are given respectively:

Definition:
The ring $\mathbb{Z}/n\mathbb{Z}$ is the set formed by the congruence classes of integers prime to n, i.e.: $\mathbb{Z}/n\mathbb{Z} = \{k \in \mathbb{Z}, 1 \le k \le n - 1 \text{ and } \gcd(k, n) = 1\}$.

Theorem:
Any one-to-one function in the ring $\mathbb{Z}/n\mathbb{Z}$ is considered as a permutation in the set of integers $\{1,\ldots, n\}$.

Corollary:
An application is one-to-one in the ring $\mathbb{Z}/n\mathbb{Z}$ if and only if the multiplier coefficient m is invertible i.e. m is prime with n (Euler theorem [14]).

For the proposed scheme, the adopted one-to-one function defined by:

$$f(x) = (m * x + w) mod(n) \tag{2}$$

from corollary, $f(x)$ is a one-to-one function if and only if m is invertible in the ring $\mathbb{Z}/n\mathbb{Z}$.

To implement this theoretical part, the parameters m, w and n are defined as follows: m = 9, w = 3 and n = 16. Analytically a resultant of the permutation, by the defined function, for a sample data is presented in Table 1:

Table 1. Example of one-to-one permutation on the ring $\mathbb{Z}/16\mathbb{Z}$.

Initial Position	Gray level	Final Position	Gray level
0	233	3	79
1	45	8	39
2	106	13	101
3	79	2	106
4	189	7	201
5	57	12	244
6	247	1	45
7	201	6	247
8	39	11	135
9	77	0	233
10	36	5	57
11	135	10	36
12	244	15	7
13	101	4	189
14	211	9	77
15	7	14	211

To visualize graphically the effect of the positions permutation of a series by the defined function, it was applied to a sample test data and the corresponding graph is plotted below in Fig. 4.

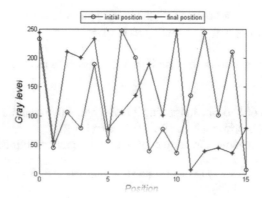

Fig. 4. Effect of permutation by the one-to-one function

This state diagram shows that for each pixel the gray level change by the bijection modulo 16, this defines a permutation in the ring $\mathbb{Z}/16\mathbb{Z}$... Therefore, any bijection modulo [n] (n is the size of the image) defines well a linear congruence generator in the ring $\mathbb{Z}/16\mathbb{Z}$.

2.3 Avalanche Effect Process Layer

This layer is based on an Enhanced Cipher Block-chaining (ECBC) mode. The CBC (Cipher Block-chaining) mode [15] shown in the Fig. 5, applies an Exclusive-OR (XOR) to each block with the encryption of the previous block before it is itself encrypted in the same process. The first plain-image block is coded with a XOR with a random initialization vector (referred to as IV). The vector has the same size as a plain-image block.

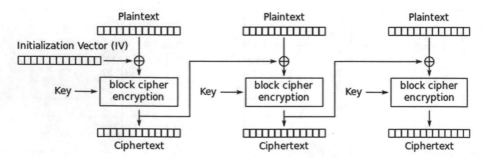

Fig. 5. Cipher Block-chaining (CBC) mode encryption

Among the advantages of CBC has is that it's initialization vector introduce a random parameter for each block, and so CBC gives for identical blocks in a different positions a different ciphers. Yet, the biggest advantage remains that the avalanche effect property is a default property in this mode since the decryption of blocks depends on their previous cipher blocks. Thus, if a block is modified, the decryption of all the following blocks would not be possible.

However, this mode has several drawbacks. The first drawback, is that any error on DC(i) encrypted pixel will affect only the clear pixels MC(i) and MC(i + 1). The second drawback is that for an error that affects the last block, all previous blocks are not affected. A third drawback is that the initialization vector is random but remains unique; consequently reuse of the same vector will lead to brutal attacks. Finally, the initialization vector is independently defined relative to the encrypted data.

To overcome these drawbacks, a corrective improvement is proposed and illustrated in Fig. 6, the pseudo random values of VI are generated from a chaotic map and data obtained from the second layer. Thus, the initialization value for VI depends respectively on the encryption key and the original image context.

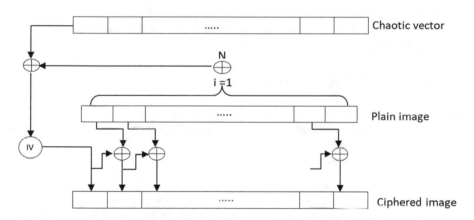

Fig. 6. The Enhanced CBC mode operator (ECBC)

The new initialization vector VI is proposed as follow:

$$\begin{cases} k = (\sum_{i=1}^{n} Vc(i)) mod(n) \\ \rho = Vc(k) \\ n \\ VI = \rho \;\oplus\; MC(i) \\ i = 1 \end{cases} \qquad (3)$$

where MC is the clear data of n blocks and the Vc the chaotic vector obtained by the Eq. (1) where $\oplus$ is bitwise operator.

514   F. Gmira and S. Hraoui

By using the ECBC encryption mode, the encrypted DC data are obtained according to the equation:

$$\begin{cases} DC(1) = VI \ pour\ i = 1 \\ DC(i) = DC(i-1) \oplus MC(i-1) \ pour\ i > 1 \end{cases} \qquad (4)$$

Further, changing the block to the position q, Eq. (3) becomes:

$$VI = \rho \oplus MC(1) \oplus MC(2) \oplus \cdots \oplus MC(q\prime) \cdots \oplus MC(n) \qquad (5)$$

Furthermore performing the XOR operation between Eqs. (3) and (5) gives:

$$MC(q) \oplus MC(q\prime) = 0 \qquad (6)$$

Which is false as MC(q) $\neq$ MC(q').

Hence, a change of MC(1) result in total change in DC(i), and since the Eq. (3) is non-linear then a less disturbance on the input image or on the key will cause a great disturbance on the output image.

Indeed, the former confusion is implemented in an ECBC mode; this allows adding a retroaction mechanism for encryption, creating an avalanche effect as a result, which generates a significant change in the encrypted picture if a single bit of a pixel is modified in the original image.

3 The 3 Layer's Juxtaposition

The proposed cryptosystem process diagram is as shown in Fig. 7.

The first encrypted block takes the value of the initialization vector and it is further processed with all data blocks to achieve diffusion.

The encryption process is as follows:

1. *Read the 24 bits color image;*
2. *Extract the three matrix color channels, MR, MV and MB;*
3. *Construct the chaotic vector PW under constraint size(PW) $> 3 \times L \times M$ (1) by using the Eq. (1).*
4. *Put the three matrices MR, MV and MB in the form of three vectors V_r, V_g and V_b of size $L \times M$;*
5. *Construct the chaotic vectors PW_r, PW_g and PW_b of size $L \times M$, from vector PW generated in step 3;*
6. *Realize the confusion vectors V_r, V_g and V_b with vectors, PW_r, PW_g and PW_b according to formula (7) below:*

$$\begin{cases} V_\alpha x(i) = V_\alpha(i) \oplus PW_\alpha(i) \\ \alpha = r, g \ ou\ b \end{cases} \qquad (7)$$

7. *Construct a linear bijection in Z/nZ (n = L $\times$ M) according to Eq. (2):*

$$f_\alpha(V_\alpha x) = (m_\alpha . V_\alpha x + w_\alpha) mod\,(n) \qquad (8)$$

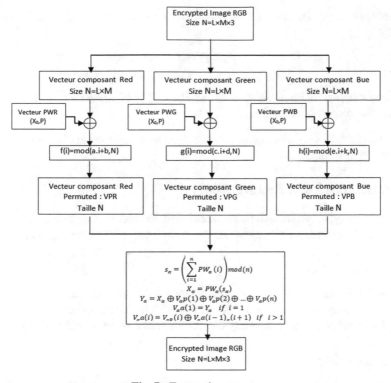

Fig. 7. Encryption process

with m_α first with n (the number of integers m_α is equal to $\varphi(n)$ with φ is the indicator function of Euler.

8. *Permuting the position of pixels in vector $V_\alpha x$ by three f_α functions generated by using the Eq. (8);*

9. *Construction the chaotic initialization vector IV_α by the following steps.*
   a. *Calculate the sum of the chaotic vector values in $\mathbb{Z}/\mathbf{n}\mathbb{Z}$ by the formula:*

$$(a) \ S_\alpha = (\textstyle\sum_{i=1}^{n} PW_\alpha(i)) mod(n) \tag{9}$$

   b. *Assign the value of chaotic vector to position S_α to X_α by the formula:*

$$(b) \ X_\alpha = PW_\alpha(S_\alpha) \tag{10}$$

   c. *Applying a bitwise operator on all values of vector $V_\alpha p$ with X_α according to the equation:*

$$(c) \ IV_\alpha = X_\alpha \ \overset{n}{\underset{i=1}{\oplus}} \ V_\alpha p(i) \tag{11}$$

10. *Apply the encryption mode ECBC presented in section III.3 by applying the formula (12):*

$$(d) \begin{cases} V_\alpha a(1) = IV_\alpha \ pouri = 1 \\ V_\alpha a(i) = V_\alpha p(i-1) \oplus V_\alpha a(i-1) \ pouri > 1 \end{cases} \tag{12}$$

11. *Transform vectors obtained in three matrices RC, VC and BV of size L × M;*
12. *View the encrypted color image.*

To reconstruct the original image, the decryption process is simply the reverse order of the encryption steps.

4   Results and Performance Analysis

The proposed cryptosystem will be evaluated with several tests including the space key analysis, the key sensitivity analysis and the complexity of the proposed cryptosystem.

4.1   Space of the Key

It is found that the key larger size significantly increases the calculation time. The proposed cryptosystem uses a size 128-bit key which eliminates the possibility of any exhaustive attack, because practically that requires $2128 \approx 3.4 \times 1038$ attempts.

4.2   Key Sensitivity

For the key sensitivity analysis, in Fig. 8, the LENA original image is encrypted with keys slightly modified as illustrated below:

Original Image

Image A:
Ciphered image with:
x0 = 0.786321456798120
p =0.033

Image B:
Ciphered image with:
x0 = 0.786321456798121
p=0.033

Image C:
Ciphered image with:
x_0 = 0.786321456798120
p=0.033000000000001

Fig. 8. Images encrypted with slightly modified keys

The number of changing pixel rate (NPCR) and the unified averaged changed intensity (UACI) are the standard metrics used to measure the robustness of a images cryptosystem against differential attacks. The analysis in Fig. 9, it appears that values the NPCR and UACI for all the test cases remain in the range of expected values ($NPCR_{expected} = 99.61\%$, $UACI_{expected} = 33.46\%$) [16–18].

Therefore, the confusion-permutation-diffusion architecture is a good solution since the proposed algorithm offers extreme sensitivity related to the key, which endows the algorithm with a high-level resistance against exhaustive attacks.

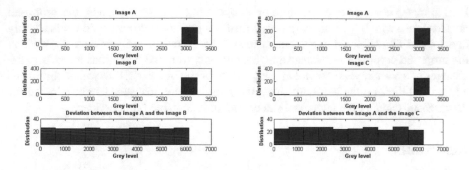

Fig. 9. Sensitivity to the key error

4.3 Complexity of the Proposed Algorithm

Theoretically, the calculation of the complexity of the proposed approach leads to a value of $\theta(n2)$. Practically, the execution time of an image encryption scheme depends on many factors such as, the CPU, the memory size, the operating system, the language of programming, the compiler and code optimization.

In the implementation phase, we used an i5 computer with the following configuration: computer Intel (R) i5 CPU 2.53 Ghz with 4 GB of RAM on Windows 10 Professional and Java eclipse-compiler.

To measure the speed of the algorithm of images with a significant accuracy, we retain the average of a series of tests have been run, and the time encryption and decryption on the test images of different sizes is shown in Table 2.

Table 2. Speed encryption/decryption in second (s) of the proposed algorithm.

size Images							
128 × 128		256 × 256		512 × 512		1024 × 1024	
Encryption time(s)	Decryption time(s)	Encryption time(s)	Decryption time(s)	Encryption time(s)	Decryption time(s)	Encryption time(s)	Decryption time(s)
0.047	0.031	0.096	0.093	0.141	0.156	0.047	0.376

Table 2 gives the speed of execution is reported of the proposed algorithm and other algorithms for different sizes of images. As it can be seen, the proposed scheme has an acceptable speed.

5 Conclusion

In order to ensure data integrity and confidentiality, a based chaotic image cryptosystem has been proposed. This cryptosystem built on 3 layers based on a robust chaotic map, on an algebraic permutation and finally on a correction and improvement of CBC method.

Unlike many other encryption algorithms operating in several rounds, the proposed cryptosystem with low complexity gave high-level performance encryption with an only one round. The obtained experimental results and security analysis have shown that the proposed cryptosystem donate a high sensitivity to initial conditions towards to the changes of secret keys. Further, has a key space of the order of 2128, which eliminates any brutal attack on the key. Ultimately, the proposed cryptosystem can be adopted for data transmission with a high-level safety and an acceptable running time.

References

1. Deamen, J., Rijimen, V.: The design of Rijindal. AES-Advanced Encyption Standard. Information Security and cryptography. Springer, Heidelberg (2002). https://doi.org/10.1007/978-3-662-04722-4
2. Li, C.: Cracking a hierarchical chaotic image encryption algorithm based on permutation. Signal Process. **118**, 203–210 (2016). https://doi.org/10.1016/j.sigpro.2015.07.008
3. Yen, J.-C., Guo, J.-I.: Efficient hierarchical chaotic image encryption algorithm and its VLSI realization. IEE Proc. Vis. Image Signal Process **147**(2), 167–175 (2000). https://doi.org/10.1049/ip-vis:20000208
4. Kanso, A., Ghebleh, M.: A novel image encryption algorithm based on a 3D chaotic map. Commun. Nonlinear Sci. Numer. Simul. **17**(7), 2943–2959 (2012). https://doi.org/10.1016/j.cnsns.2011.11.030
5. Hraoui, S., Gmira, F., Jarar, A.O., Satori, K., Saaidi, A.: Benchmarking AES and chaos based logistic map for image encryption. In: 2013 ACS International Conference on Computer Systems and Applications (AICCSA) (2013)
6. Pareek, N.K., Patidar, V., Sun, K.K.: Image encryption using chaotic logistic map. Image Vision Comput. **24**(9), 926–934 (2006)
7. Kwok, H.S., Tang, K.S.: A fast image encryption system based on chaotic maps with finite representation. Chaos Soliton Fract. **32**(4), 1518–1529 (2007)
8. Patidar, V., Pareek, N., Purohit, G., Sud, K.: A robust and secure chaotic standard map based pseudorandom permutation–substitution scheme for image encryption. Opt. Commun. **284**, 4331–4339 (2011)
9. Gao, T., Chen, Z.: Image encryption based on a new total shuffling algorithm. Chaos Solitons Fractals **38**(1), 213–220 (2008)
10. Ahmad, M.: Cryptanalysis of chaos based secure satellite imagery cryptosystem. In: Aluru, S., Bandyopadhyay, Sanghamitra, Catalyurek, Umit V., Dubhashi, Devdatt P., Jones, Phillip H., Parashar, Manish, Schmidt, Bertil (eds.) IC3 2011. CCIS, vol. 168, pp. 81–91. Springer, Heidelberg (2011). https://doi.org/10.1007/978-3-642-22606-9_12
11. Ahmad, M., Ahmad, F.: Cryptanalysis of image encryption based on permutation-substitution using chaotic map and latin square image cipher. In: Satapathy, S.C., Biswal, B.N., Udgata, S.K., Mandal, J.K. (eds.) Proceedings of the 3rd International Conference on Frontiers of Intelligent Computing: Theory and Applications (FICTA) 2014. AISC, vol. 327, pp. 481–488. Springer, Cham (2015). https://doi.org/10.1007/978-3-319-11933-5_53
12. Alvarez, G., Li, S.: Some basic cryptographic requirements for chaos-based cryptosystems. Int. J. Bifur. Chaos **16**(8), 2129–2151 (2006)
13. Li, S.J., Chen, G.R., Mou, X.Q.: On the dynamical degradation of digital piecewise linear chaotic maps. Int. J. Bifur. Chaos **15**(10), 3119–3151 (2005)
14. Euler, L.: Theorems on residues obtained by the division of powers. Novi Comment. acad. sc. Petrop. **7**, 49–82 (1761)

15. Menezes, A., van Oorschot, P., Vanstone, S.: Handbook of Applied Cryptography. CRC Press, Boca Raton (1996)
16. Wu, Y., Noonan, J.P., Agaian, S.: NPCR and UACI randomness tests for image encryption. Cyber J. Multidisc. J. Sci. Technol. J. Sel. Areas Telecommun. (JSAT) **1**, 31–38 (2011)
17. Maleki, F., Mohades, A., Hashemi, S.M., Shiri, M.E.: An image encryption system by cellular automata with memory. In: Third International Conference on Availability, Reliability and Security, ARES 2008, pp. 1266–1271. IEEE (2008)
18. Gmira, F., Hraoui, S., Saaidi, A., Abderrahmane, J.O., Satori, K.: An optimized dynamically-random chaos based cryptosystem for secure images. Appl. Math. Sci. **8**(4), 173–191 (2014)

Computational Informatics

Crop Recommendation in the Context of Precision Agriculture

Khadija Lechqar[✉] and Mohammed Errais

Computer Science and Systems Laboratory, Faculty of Sciences Ain Chock,
University Hassan II, Casablanca, Morocco
khadija.lechqar@gmail.com

Abstract. Precision agriculture has shown its effectiveness for sustainable agriculture. It is a transversal concept and therefore covers the whole agricultural cycle. Precision agriculture principally aims to manage the agricultural process and resources. These resources are decreasing especially with the global population growth and the damaging effects of climate change. This paper focuses on the first step in implementing precision agriculture, namely the choice of the most appropriate crop. This choice has a direct impact on productivity and later, on the cost of agricultural treatments such as the fertilization process. To build the crop recommendation system, we primarily used the soil properties (pH, humidity, macronutrients NPK, rainfall and temperature). Then, we implemented several models to choose the best one. The implemented models are Fuzzy Logic, Random Forest, Gaussian Naïve Bayes, XGBoost, Logistic Regression and Artificial Neural Network. XGBoost was chosen, since it was the model that achieved the top accuracy with 99.31818%.

Keywords: Precision agriculture · Crop recommendation · Fuzzy Logic · Machine Learning · Deep Learning

1 Introduction

Agriculture is facing several unprecedented challenges at the global level including: population explosion [1], lack of resources and climatic changes. This makes it more difficult to ensure food sufficiency, and has an important negative economic impact, as most countries' economies depend on agriculture [2]. Hence, there is a need to find sustainable solutions that satisfy these challenges while limiting environmental damages. Precision agriculture, which began in the 1980s with the authorization of the use of GPS by civilians in the US [3], is one of those solutions that have yielded satisfactory results. Since then, precision agriculture has been developed by the progressive use of different technologies (Unmanned Aerial Vehicle, robotic systems etc.) [4].

In fact, precision agriculture is a holistic management concept. It is a process consisting first of data collection, then analysis and evaluations of the collected data and finally implementation of the necessary treatment for each agriculture unit to ensure efficiency and precision. This may include, but is not limited to irrigation, crop protection and

F. Saeed et al. (Eds.): ICACIn 2022, LNDECT 179, pp. 523–532, 2023.
https://doi.org/10.1007/978-3-031-36258-3_46

fertilization. Thus, the information technologies used in this process are divided into three principal categories: guidance technologies, recording technologies and reacting technologies [5].

The implementation of this concept is interesting since the rate of adoption of precision agriculture is much lower than forecasts even in developed countries [3, 6, 7]. In this paper, we are interested in the crop recommendation system. In fact, the choice of the appropriate crop must be made as one of the first decisive steps in the agricultural cycle. The correct choice is essential as it directly affects the productivity and therefore the yield.

The paper is organized as follows. First, we present an overview of related works. After that, followed by short theoretical presentations and then implementation of the crop recommendation systems using Fuzzy Logic, Machine Leaning and Deep Learning through Artificial Neural Network. Finally, we will decide which approach is the best.

2 Related Work

2.1 Defining Features

The strength of precision agriculture comes from continuous monitoring of several indicators related mainly to soil's properties and climatic conditions. This enables the most optimal real-time and accurate processing throughout the agricultural cycle.

In fact, soil nutrients have a direct impact on the development and productivity of crops. The most important are macronutrients, namely NPK (Nitrogen, Phosphorous and Potassium) [8]. The level of acidity (soil pH) affects the rate of nutrient absorption for crops. Meanwhile, the temperature and the humidity of soil has an impact on water absorption [9] and rainfall has an impact on plant growth.

2.2 Hardware Tools

In most cases, data collection is based on sensors for measuring and microcontrollers for processing [9–11] (see Fig. 1). It should be noted that the adoption of Variable Rate Technologies for such aims is optimal. In fact, they can also be used in data collection and processing. The cost of establishment of precision agriculture is then lower.

2.3 Used Approaches for Crop Recommendation

The papers related to crop recommendation can be divided into three main categories, depending on the approach utilized:

- Papers implementing one method.
- Papers comparing multiple methods.
- Papers implementing ensemble learning.

Papers Implementing One Method. [12] aimed to determine the best crop to adopt and the best seeding season. The initial dataset was collected from different sources: satellite images, sensor recorded data, irrigation reports, crop data and weather data.

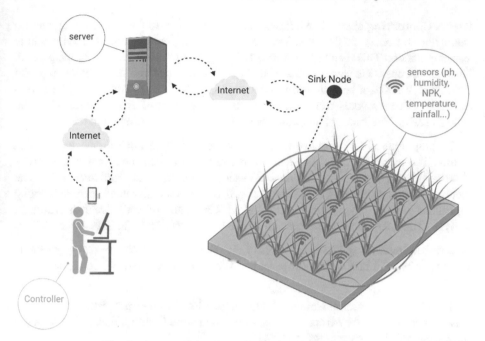

Fig. 1. An overview of crop recommendation architecture

Four inputs were used, namely, rainfall, soil moisture, temperature and atmospheric pressure. Four crops were involved in this study: chilli, rice, cotton and maize. To define the crop to adopt, Naïve Bayes was implemented by exploiting the parallel processing of MapReduce. The history of recorded data over five years was used to decide the best season for sowing.

[9] focused on both hardware and software designs. For the hardware part, three sensors were used to measure successively pH, humidity and temperature. These sensors were connected to a microcontroller to feed the inputs of the processing step. Then, the result, ie the most appropriate crop which consists of eight different cultures, was displayed on the LCD. The software part was based on the fuzzy logic, the whole process took 1–3 min and the accuracy was 77.78%.

[13] proposed a crop recommendation system based on fuzzy logic. The inputs of the dataset mainly concerned soil parameters, elevation and rainfall. The prediction covered eight types of crops. The trapezoidal membership and centroid method were used respectively, for the fuzzification and defuzzification steps. Finally, the accuracy was 92.14%.

A fuzzy multi-layer system was implemented in paper [14]. The idea behind this approach is to avoid the explosion of rules of inference, especially as that study included twelve inputs. The final system was then composed of six fuzzy layers, which allowed to go from 439 to 152 rules. The triangular membership and centroid method were used for the fuzzification and the defuzzification.

Papers Comparing Multiple Methods. [11] proposed an IoT system for crop recommendation. It interfaced the sensor network to a cloud implementing trained machine learning models. This result is used later for irrigation purposes and optimizing yield. The dataset concerned seven inputs related to soil properties and 22 different crops for the output. WEKA, a java platform, was used as the technical platform for the implementation. Three models were selected in this study: multilayer perceptron, decision table and JRip. Their accuracies were respectively: 98.23%, 88.59% and 96%.

[15] highlighted neural networks as a response to the question of choosing the best culture. The initial dataset concerned seven soil properties in input and sixteen crops in output. Keras and Google Collab were used for the technical implementation. The accuracy of this model was 87%. The paper also implemented other machine learning algorithms namely KNN, decision tree, SVM, Gaussian NB and Linear discriminant analysis. Their accuracies were respectively: 64.7%, 70%, 71.56%, 72% and 60.9%.

Papers Implementing Ensemble Learning. The consulted papers using ensemble learning were all based on majority voting. The algorithms used in each article were as follows.

For [16], the dataset contained 15 attributes related to soil properties and sowing season. It considered four crops. The classifiers used were Random Forest, Naïve Bayes and Linear SVM. The accuracy was 99.1%.

In [17], the dataset covered nine inputs and ten crops. Random Forest, CHAID, KNN and Naïve Bayes were used as classifiers. The final proposed system was composed of (IF...Then) rules. The accuracy was of 88%.

2.4 Discussion

We cannot establish a general conclusion, as the datasets used are not the same for all studied papers. However, since we are working on crop recommendation, which is not a complex problem, the approach comparing multiple methods is preferable. In fact, the difference between the aforementioned approaches, in terms of response time and complexity of their implementations, can be ignored. However, comparing several algorithms and choosing the one achieving the best performance, can be a discriminating criteria (see Table 1).

Table 1. Comparison of methods used in related works

	One method	Multiple methods	ensemble learning
Response time	high	medium	medium
Facility	high	medium	medium
Performance	low	high	medium

3 Proposed Approach

The main purpose of this work is to find the best crop according to soil and weather indicators. The crop is a qualitative variable that can take more than two categories, so it is about a multinomial logistic regression. In this section, we present the general theoretical lines of implemented approaches.

3.1 Fuzzy Logic

Fuzzy Logic was introduced by Zadeh in 1965. It considers that truth values are real numbers between 0 and 1, rather than 0 or 1. A Fuzzy controller is based on processing unit, in which three phases can be distinguished: Fuzzification, Inference Engine and Defuzzification. Both the fuzzification and defuzzification phases are based, respectively, on membership functions, to convert crisp values to fuzzified values and conversely. The inference engine uses the Rule Base to decide which rules to keep for the input field. A rule is nothing more than an if-then condition [18].

3.2 Machine Learning Algorithms

Machine learning algorithms are considered as 'soft coded'. In fact, they use repetition to adapt their architecture for better results. Here, we focus on supervised learning as the data are labeled [19].

Random Forest. Decision Trees are limited by being highly sensitive to the original training data. Thus, they can fail to be generalized. Hence, we use Random Forest, which are collections of multiple random decision trees.

To implement Random Forest, we randomly build a new dataset from the initial one, known as bootstrapped dataset. Then, in order to train trees, we randomly select a subset of features for each tree. This reduces the correlation between trees. To make a prediction, majority voting is used for categorical outputs and average for continuous ones. This process of combining results from multiple models is called aggregation [20].

Gaussian Naïve Bayes (GNB) is based on Bayes Theorem and supposes that the variables are independent and distributed according to a normal distribution. This probabilistic classifier calculates the probability of features occurring for each class. The formula for Bayes' Theorem is: [21]

$$P(A|B) = \frac{P(B|A)P(A)}{P(B)}, P(B) \neq 0 \tag{1}$$

XGBoost. eXtreme Gradient Boosting is a sequential ensemble method based on Gradient Boosting Decision Trees, for which, we use, for each iteration, the previous model error residuals. The difference is that in XGBoost, trees are built in parallel. Due to this, XGBoost is a fast model and has good performance [22].

Logistic Regression. This model is used when the dependent variables are categorical. Binary logistic regression occurs when the output can take two possible values. Otherwise, it is multinomial logistic regression. Logistic regression is based on logistic function to build the relationship between inputs and outputs variables [23].

3.3 Artificial Neural Networks

Artificial Neural Networks (ANN) are the key of Deep Leaning. They are inspired by the human nervous system, and so they are composed of node layers. Each node is a logistic regression model. ANN use training data to learn and improve their performance [10].

4 Implementation and Results

4.1 Dataset

The Dataset is a csv file. To build it, we used quantitative features mentioned in part 2.1, namely: N,P,K, rainfall, temperature, humidity and pH. The dependent variable, which is simply the crop to predict, can take 22 different values.

Data Preprocessing consists of data cleaning, data transformation and data reduction. Given our dataset, we proceed in this step to just data cleaning by handling the missing data by replacing them with the mean of each value. In addition, no outliers are detected.

Splitting Data. In order to train the implemented models and test them, we split the dataset into two subsets: 80% for training and 20% for testing.

4.2 Implemented Models

Fuzzy Logic. We relied on two papers to define the membership functions which are based on linguistic values: [9, 13] (see Fig. 2). Then we used the training dataset to define the 180 rules of the inference engine. It is worth mentioning that both membership functions and rules of inference have a direct impact on the performance of a fuzzy logic system.

For the technical implementation, we used JFuzzyLogic, which is a java library implementing fuzzy logic. It uses the Fuzzy Control Language to define all inputs, outputs, memberships' functions and rules. We used Eclipse as an IDE.

Machine Learning. We implemented the following machine learning algorithms: Random Forest, Gaussian Naïve Bayes, Logistic Regression and XGBoost.

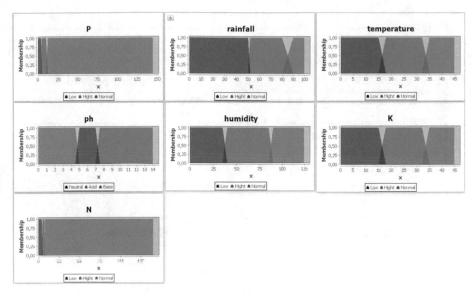

Fig. 2. Membership functions for input features.

For the technical implementation, we used scikit-learn, a Python library used mainly for machine learning. We used Spyder which is a Python IDE.

Artificial Neural Networks. We used Keras, which is a Python library integrating ANN. In addition, we used Google Collab platform for execution. The built model was run with 200 epochs and batches of 40.

4.3 Accuracies and Interpretation

To compare models' performance, we used the accuracies and the runtime. In our case, we proceed to runtime comparison, even if the used hardware is not the same for all implementations. In fact, ANN use Google Collab, which has 12 GB as RAM capacity, while other methods use 8GB as RAM.

According to Fig. 3 and Fig. 4, the XGBoost model is the best model in our case. The type of problem and the size of the dataset can explain this result. Therefore, a hybrid approach would be effective for generalizing. The method to be used will change depending on the size of agricultural plot and thus according to the dataset.

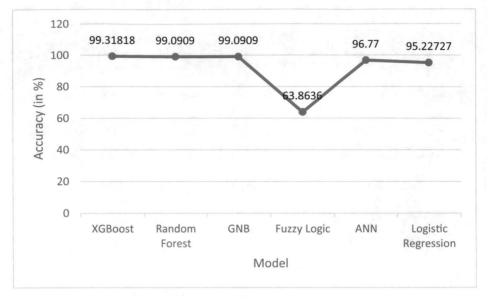

Fig. 3. Accuracies of implemented models

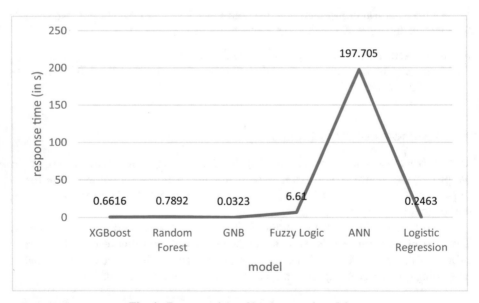

Fig. 4. Response time of implemented models

5 Conclusion and Perspectives

One of the powerful advantages of precision agriculture is the integration of information technologies. This makes it possible to exploit big data and therefore capitalize knowledge. In this paper, we built a crop recommendation system based on soil properties.

Fuzzy logic system, Machine Learning Classifiers and ANN were compared. XGBoost was selected as the best model for predicting the best culture to adopt.

The next step is to define a hybrid solution. It will choose the appropriate method to predict the appropriate crop depending on type of data and farm sizes.

References

1. Thakur, M., Wang, B., Verma, M.L.: Development and applications of nanobiosensors for sustainable agricultural and food industries: recent developments, challenges and perspectives. Environ. Technol. Innov. **26**, 102371 (2022). https://doi.org/10.1016/j.eti.2022.102371
2. Gómez-Godínez, L.J., Martínez-Romero, E., Banuelos, J., Arteaga-Garibay, R.I.: Tools and challenges to exploit microbial communities in agriculture. Curr. Res. Microb. Sci. **2**, 100062 (2021). https://doi.org/10.1016/j.crmicr.2021.100062
3. Lowenberg-Deboer, J., Erickson, B.: Setting the record straight on precision agriculture adoption. Agron. J. **111**(4), 1552–1569 (2019). https://doi.org/10.2134/agronj2018.12.0779
4. Pedersen, S.M., Lind, K.M.: Precision Agriculture – From Mapping to Site-Specific Application. In: Precision Agriculture: Technology and Economic Perspectives, pp. 1–20 (2017)
5. Balafoutis, A., et al.: Precision agriculture technologies positively contributing to ghg emissions mitigation, farm productivity and economics. Sustainability **9**(8), 1–28 (2017). https://doi.org/10.3390/su9081339
6. Cisternas, I., Velásquez, I., Caro, A., Rodríguez, A.: Systematic literature review of implementations of precision agriculture. Comput. Electron. Agric. **176**, 10562 (2020). https://doi.org/10.1016/j.compag.2020.105626
7. Nowak, B.: Precision agriculture: where do we stand? a review of the adoption of precision agriculture technologies on field crops farms in developed countries. Agric. Res. **10**(4), 515–522 (2021). https://doi.org/10.1007/s40003-021-00539-x
8. Jayaraman, V., Parthasarathy, S., Lakshminarayanan, A.R., Sridevi, S.: Crop recommendation by analysing the soil nutrients using machine learning techniques: a study. In: Krishnamurthy, V., Jaganathan, S., Rajaram, K., Shunmuganathan, S. (eds.) Computational Intelligence in Data Science: 4th IFIP TC 12 International Conference, ICCIDS 2021, Chennai, India, March 18–20, 2021, Revised Selected Papers, pp. 15–26. Springer International Publishing, Cham (2021). https://doi.org/10.1007/978-3-030-92600-7_2
9. Gustina, M., Salamah, I., Lindawati, L.: Design and construction of crop suitability prediction system using fuzzy logic classifier method. Log. J. Ranc. Bangun dan Teknol. **21**(3), 139–148 (2021). https://doi.org/10.31940/logic.v21i3.139-148
10. Banavlikar, S.D.T., Mahir, A., Budukh, M.: Crop recommendation system using neural Networks. Int. Res. J. Eng. Technol. **05**, 1475–1480 (2018)
11. Bakthavatchalam, K., et al.: IoT framework for measurement and precision agriculture: predicting the crop using machine learning algorithms. Technologies **10**(1), 13 (2022). https://doi.org/10.3390/technologies10010013
12. Priya, R., Ramesh, D., Khosla, E.: Crop prediction on the region belts of india: a naïve bayes mapreduce precision agricultural model. In: 2018 International Conference on Advanced Computer and Communication Informatics, ICACCI 2018, pp. 99–104 (2018). https://doi.org/10.1109/ICACCI.2018.8554948
13. Banerjee, G., Sarkar, U., Ghosh, I.: A fuzzy logic-based crop recommendation system. In: Bhattacharjee, D., Kole, D.K., Dey, N., Basu, S., Plewczynski, D. (eds.) Proceedings of International Conference on Frontiers in Computing and Systems. AISC, vol. 1255, pp. 57–69. Springer, Singapore (2021). https://doi.org/10.1007/978-981-15-7834-2_6

14. Aarthi, R., Sivakumar, D.: Modeling the hierarchical fuzzy system for suitable crop recommendation. In: Mallick, P.K., Meher, P., Majumder, A., Das, S.K. (eds.) Electronic Systems and Intelligent Computing. LNEE, vol. 686, pp. 199–209. Springer, Singapore (2020). https://doi.org/10.1007/978-981-15-7031-5_19

15. Varshitha, D.N., Choudhary, S.: An artificial intelligence solution for crop recommendation. Indon. J. Electr. Eng. Comput. Sci. **25**(3), 1688–1695 (2022). https://doi.org/10.11591/ijeecs.v25.i3.pp1688-1695

16. Kulkarni, N.H., Srinivasan, G.N., Sagar, B.M., Cauvery, N.K.: Improving crop productivity through a crop recommendation system using ensembling technique. In: Proceedings of 2018 3rd International Conference on Computer System and Information Technology Sustainability Solution, CSITSS 2018, pp. 114–119 (2018). https://doi.org/10.1109/CSITSS.2018.8768790

17. Pudumalar, S., Ramanujam, E., Rajashree, R.H., Kavya, C., Kiruthika, T., Nisha, J.: Crop recommendation system for precision agriculture. In: 2016 8th International Conference on Advanced Computer, ICoAC 2016, pp. 32–36 (2017). https://doi.org/10.1109/ICoAC.2017.7951740

18. Trillas, E., Eciolaza, L.: Fuzzy Logic: An Introductory Course for Engineering Students. Springer, Cham (2015). https://doi.org/10.1007/978-3-319-14203-6

19. El Naqa, I., Li, R., Murphy, M.J.. (eds.): Machine Learning in Radiation Oncology. Springer, Cham (2015). https://doi.org/10.1007/978-3-319-18305-3

20. Oshiro, T.M., Perez, P.S., Baranauskas, J.A.: How many trees in a random forest? In: Perner, P. (ed.) MLDM 2012. LNCS (LNAI), vol. 7376, pp. 154–168. Springer, Heidelberg (2012). https://doi.org/10.1007/978-3-642-31537-4_13

21. Tzanos, G., Kachris, C., Soudris, D.: Hardware acceleration on gaussian naive bayes machine learning algorithm. In: 2019 8th International Conference on Modern Circuits and System Technology, MOCAST 2019, pp. 1–5 (2019). https://doi.org/10.1109/MOCAST.2019.8741875

22. Mitchell, R., Frank, E.: Accelerating the XGBoost algorithm using GPU computing. PeerJ Comput. Sci. **7**, 2017 (2017). https://doi.org/10.7717/peerj-cs.127

23. Kumar, A., Sarkar, S., Pradhan, C.: Recommendation system for crop identification and pest control technique in agriculture. In: Proceedings of 2019 IEEE International Conference on Communication Signal Processing, ICCSP 2019, pp. 185–189 (2019). https://doi.org/10.1109/ICCSP.2019.8698099

Aligning Cultural Heritage Preservation and Knowledge Transfer in Digital Era Between Champions and Agencies

Razifah Othman[1,2(✉)], Othman Ibrahim[1], Fadilah Zaini[3], and Siti Nuur-Ila Mat Kamal[4]

[1] Azman Hashim International Business School, Universiti Teknologi Malaysia, Skudai, Malaysia
`razifah@graduate.utm.my`, `razif879@uitm.edu.my`, `othmanibrahim@utm.my`

[2] Faculty of Information Management, Universiti Teknologi MARA Cawangan Negeri Sembilan, Kampus Rembau, Malaysia

[3] School of Human Resource Development and Psychology, Faculy of Social Science and Humanities, Universiti Teknologi Malaysia, Skudai, Malaysia
`fadilahz@utm.my`

[4] Faculty of Information Management, Universiti Teknologi MARA Cawangan Johor, Kampus Segamat, Malaysia
`sitin509@uitm.edu.my`

Abstract. Culture is the bedrock upon which every country and civilization is built. It cannot be developed in a single week, month, or year. Culture takes a long time to be established in a society. Understanding our cultural background can help us develop a stronger sense of personal identity and foster better communal collaboration. It may play an important role in society by connecting the past, present, and future generations, thus providing a foundation for developing a strong nation and national identity. Furthermore, it may help in alleviating the problems of poverty, unemployment, and inequality in our society. Cultural Heritage (CH) champions educate the community or apprentices ancestors' traditions, beliefs, skills, and knowledge. However, CH agencies play the most important role in preserving and promoting it. This is considered as a huge concern and responsibility for both parties. Therefore, this study aims to establish a new course of action for aligning CH and Knowledge Transfer (KT) with the perspective of CH-related agencies and champions who practice the heritage. The study also offers key implications for future practice and decision-making of heritage agencies, so as to understand and adopt KT in managing and preserving cultural heritage. As a starting point, this study explains issues or implications without considering the existence of traditional/CH knowledge. An evidence-based empirical approach is employed in examining the evolution and trends of CH researchers. This research also provides researchers and practitioners with ideas that may be used to solve or improve upon the mentioned issues and challenges. As a result, the findings support a shift towards identifying several successful projects and their outcomes, as well as determining the possible solutions that can be tailored to preserve and sustain our cultural heritage.

Keywords: Knowledge transfer · cultural heritage · issues · challenges · solutions

1   Introduction

The concept of knowledge is complex and has been widely discussed in academic literature because it can be approached and understood from multiple perspectives. Nowadays, significant Knowledge Transfer (KT) occurs within various organizations, either nations, public or private domains. [1] studied the concept of knowledge from multiple vantage points, including as a mental state, an object, a procedure, a precondition of acquiring new knowledge, and a competency. Learning theories emphasize the significance of active knowledge construction and how it develops over time and with experience [2]. In terms of KT, this collective cognition entails shifting away from peculiar, subjective mental representations of the world and moving towards the concept by which agents with the comparable capacity to act might possibly perceive similar action options in reality [3].

This means that transfer differs from simple information exchange in that it aims to capture, manage, or even create and distribute knowledge, ensuring its availability and applicability, and providing inputs for problem-solving in one or more fields [4]. In the perspective of KT, the field of heritage management is not an exception. The contemporary understanding of CH emphasizes that tangible and intangible elements are indispensable symbols for creating and strengthening personal or social, local, regional, national, and even transnational identity [5]. The European policy framework (2017–2018) on CH is one of the most well-known KT best practices in moving towards a people-centered and holistic approach, exhuming groups of national CH experts to investigate skills and training, and reassigning their knowledge in the heritage professions. In Malaysia, on the other hand, fiscal incentives are used to create a well-established link between conservation programs and sustainable communities. Overall, acceptance of the incentives program has been identified as a critical requirement for encouraging best practices and ensuring conservation program success. However, in this regard, the financial incentives tool was discovered to be insufficiently focused on conforming to the effectiveness of the conservation program. Nonetheless, the participation of the local people should be actively encouraged through education and knowledge sharing and transfer with all stakeholders. As a result, any efforts to preserve CH should aim to protect its architectural and natural forms, and preserve living communities and their intangible heritage and knowledge [6].

2   Evolution and Trends in Cultural Heritage Research

The table below has generated numerous insights for over eight years regarding the diversity of cultural heritage research topics. The outcome is increasingly illuminating regarding the origin and development of numerous significant themes across cultural heritage disciplines. Although the majority of studies have analyzed the relationships between CH and computer technology innovation from a variety of perspectives, there are very few studies on the global factors influencing knowledge and CH evolution. In addition, numerous researchers use abstract and general knowledge as the research object without grouping them into explicit and tacit knowledge and by disregarding the transformation influence mechanism of explicit and tacit knowledge during the knowledge transfer process (Table 1).

Table 1. Thematic Evolution for Cultural Heritage research extracted from Scopus database

From Year duration	To year	Topic area	Inclusion Index	Occurrences	Stability Index
character recognition--2015–2018	crowdsourcing--2019–2020	learning systems	0.50	2	0.10
digital cultural heritages--2015–2018	cultural heritages--2019–2020	artificial intelligence	0.20	2	0.05
digital cultural heritages--2015–2018	digital libraries--2019–2020	interoperability	0.20	2	0.10
digital cultural heritages--2015–2018	historic preservation--2019–2020	digital cultural heritages	0.20	8	0.05
digital humanities--2015–2018	cultural heritages--2019–2020	digital humanities; cultural heritages; linked open datum	0.06	46	0.03
digital humanities--2015–2018	digital libraries--2019–2020	digital libraries	0.17	8	0.05
digital humanities--2015–2018	historic preservation--2019–2020	cultural heritage collections	0.06	3	0.03
historic preservation--2015–2018	crowdsourcing--2019–2020	human computer interaction	0.11	4	0.05
historic preservation--2015–2018	cultural heritages--2019–2020	history; metadata	0.08	9	0.04
historic preservation--2015–2018	historic preservation--2019–2020	historic preservation	0.08	11	0.03
information systems--2015–2018	cultural heritages--2019–2020	visualization	0.14	4	0.05
semantics--2015–2018	crowdsourcing--2019–2020	digital technologies	0.11	2	0.06
semantics--2015–2018	cultural heritages--2019–2020	semantics; data handling; semantic web; ontology; linked data	0.10	7	0.04

(continued)

Table 1. (*continued*)

From Year duration	To year	Topic area	Inclusion Index	Occurrences	Stability Index
crowdsourcing--2019–2020	digital technologies--2021–2022	digital technologies; intangible cultural heritages	0.20	2	0.08
cultural heritages--2019–2020	digital humanities--2021–2022	cultural heritages; digital humanities; ontology	0.25	26	0.05
cultural heritages--2019–2020	digital technologies--2021–2022	knowledge graphs	0.20	2	0.05
cultural heritages--2019–2020	open data--2021–2022	linked data; linked open datum; open data	0.14	5	0.05

Simultaneously, CH related research has gradually evolved into the use of computer technology and innovation such as digital heritage, digital humanities (DH), learning systems, artificial intelligence (AI), visualization, semantic and metadata. Considering the scientific literature, this bibliometric table demonstrates that most academics had conducted studies on digital heritage, digital humanities, and open data themes between 2015 and 2022 (a total of 134 occurrences). In the Knowledge Management/Transfer research sector, only two (2) studies were conducted between 2019 and 2020. This is a nascent field with an emphasis on evidence-based practice in relation to CH. Through spatial representation, the study reveals the relationship between disciplines, fields, and specialists in research creation. The inclusion index is a measurable indicator which show that the research community uses consolidated or integrated terminology. It monitors a field of study and identifies the structure and evolution of newly found research fields, presenting their conceptual, intellectual, and social elements. Occurrences refer to the frequency whereby strongly interconnected concepts or words appear in research-related publications. With 18 and 72 publications, historic preservation and digital humanities are the most often used terms across the years (2015–2022).

3 Issues and Challenges

According to relevant data, one cultural relic and one piece of folk art are lost every minute [5, 7]. Despite the absence of scientific evidence, the disappearance and discontinuation of traditional knowledge should not be overlooked [8]. The following matter is inextricably linked to the current viral phenomenon where there are still minimal and insufficient research and empirical data for mapping KT models or frameworks with elements towards preserving CH for the artistic learning process while creating a meaningful public engagement [6, 9, 10, 11]. Most of the past CH researches had focused solely on digitization and conservation, and disregarded the important aspects of developing and using a technology which depend on humans and their ability to disseminate, absorb and use the learned knowledge. If our heritage resources are not properly explored, handled, studied and promoted, there will be an impact on the availability and accessibility of our unique culture-related information and knowledge. This in turn will prevent us from identifying and claiming the authenticity and integrity of the respective cultural information and knowledge (Table 2).

Table 2. Summary of challenges facing knowledge transfer and cultural heritage protection

PRIORITY	ISSUES & CHALLENGES	RISKS
Urgent priority	• Poor cultural and heritage institutions management and guidelines • Disagreement over CH priorities, specialties, preservation and conservation • Lack of coordination • Lack of awareness about the importance of CH Knowledge • Hazard/natural disaster threats to heritage sites	• Disparity in the services provided in terms of quality and availability • Inefficient use of the various resources available to all governorates • Loss of heritage
Significant priority	• Lack of financial allocations to protect heritage • Lack of digital information about heritage • Lack of knowledge capture, transfer and translation technology and know how	• Financial allocations are not proportional to the size of heritage sites which require conservation • Inadequate international/technological advances, thus affecting the ability of the heritage to attract tourists • Difficulty of transferring, strengthening and mastering CH knowledge
Lower priority	• Vandalism	• Damage of heritage

Experts and "owners" of traditional knowledge and skills may be found worldwide. Those from less developed nations or regions, in particular, are only known to their local community and sometimes by coincidence. Others are widely recognised regionally or globally. Intangible CH is just as important as tangible ones. Elements like songs, music, folk tales, dances, drama, skills, cuisines, crafts, martial arts, and ceremonies are barely preserved. Such matter requires complex KT to access and combine data from in-depth artistry knowledge, capturing visuals as information from accurate observation and displaying them in a natural environment. This is especially important because the cycle in mastering knowledge or skills covers a wide range of topics. Apprentices must go through extensive learning preparation and rigorous hands-on training to gain a good experience covering both theory and practice. Furthermore, it must comprise the exchange and understanding of information, ideas, tasks, procedures, tools, documents, data and many other things [12, 13]. Consequently, it is clear that there is a shortage of high-level experts in traditional customs, heritage skills masters, and the mechanism and strategy choice for passing down the knowledge and skills to the apprentices. Therefore, it is critical to look into quick fixes for future permanency and sustainable cycle of knowledge delivery and heritage skilled workers.

Nevertheless, the lack of concern from higher officials has resulted in a deficiency of CH copyright, intellectual property right, standard policy or regulations, guidelines, and techniques at the national level which are not clearly defined. Also, part of the essential measures in safeguarding the ICH available in its terrain is bestowed to the *Convention for the Safeguarding of the Intangible Cultural Heritage, art.11(a)* by UNESCO [13, 14]. By having this legislative commandment, the legal protection of CH in any level can be identified. It helps the authorities to trace and produce an inventory list of CH available in every state of the country. Furthermore, it expands the idea and proposes evolution initiatives for safeguarding it. Still, it is critical to achieve consistency. Professionals and the public are left perplexed, ultimately resulting in the inability to sustainably preserve the CH assets [15–17]. The lack of archaeological impact assessment and the type of protection and framework offered to the listed items or objects in the Register are also noted as issues.

4   Previous Studies/Projects on KT and CH

Based on the listed projects above, the UNESCO, European and Japan cultural heritage policy has undergone a radical shift in the past several years, shifting from an object-based and object-centric approach to a more holistic one that considers both physical and intangible aspects of cultural heritage. As it views cultural legacy as a common resource, it emphasizes the fact that everyone has a stake in ensuring that it is passed down to the next generation. A more integrated strategy to conservation and management is needed across several policy sectors in order to maximize the advantages for the economy, culture, environment, and society as a whole. New technologies may be used to safeguard cultural heritage and promote public participation at museums and heritage sites, according to the report (Table 3).

In addition to facilitating the transmission of knowledge and access to information, networks are significant sources of KT value. They highlight the significance of mutual responsibility networks and the significant degree of interconnection with other countries, and are characterised by reciprocity and required belonging. Thus, this study emphasises the importance of connections in fostering learning and the nation's growth as a whole. As a result, all three programmes drew on precedents, lessons gained, and appropriate practices when formulating policies to increase public awareness and engagement in preserving traditional knowledge. Malaysia through the AHC Pillars project developed a mentorship program with two companies in order to establish a bespoke experience, quality craftmanship, niche marketing, and public relations activities. Aside from that, new talents are groomed to continue the CH legacy.

Table 3. Previous studies/projects on KT and CH worldwide

Author/ Title	Purpose	Framework	Sample	Results	Implications for practical research theory
1. Alma Mater Studiorum & UNESCO, 2019 / Sustainable Historic Environments Modelling Reconstruction through Technological Enhancement and Community-based Resilience (SHELTER)	To develop a data-driven knowledge framework based on data used by scientists and heritage managers.	SHELTER Framework	9 countries (UK, Netherlands, Belgium, Chechzia, Austria, France, Croatia, Italy, Spain & Turkey), 3 Urban Cities (Ravenna, Italy; Seferihisar, Turkey; Dordrecht, Netherlands) 2 Cross-border (Baixa Limia-Serra Do Xurés Natural Park, Galicia – Spain; Sava River Basin, South East Europe)	1. 1,210 cultural heritage assets collected and ongoing 2. Developed 3 Urban Open Labs (in Ravenna, Seferihisar and Dordrecht) and 2 Cross-border Open Labs (in Sava River Basin and Baixa Limia-Serra)	The ability to link concepts from disaster risk management and climate change adaptation with cultural heritage management in order to provide inclusive and well-informed decision-making on the effects of environmental and climate change on historic sites and buildings.
2. Heritage Consortium, 2019/ Voices of Culture process, a Structured Dialogue	To provide a channel for the voice of the cultural sector in Europe to be heard by EU policymakers	Faro Framework & European Qualification Framework (EQF)	30 cultural heritage stakeholders	Focusing on capacity building for shared stewardship, the group proposed that society is composed of 4 groups of stakeholders: • Public - communities/participation - both public and private • Policymakers/Policy making • Heritage Mediation • Heritage Expertise	The Power of Cultural Heritage Mapping Missions in the Cultural Heritage Sector Transversal Competencies and Methods for Capacity Building Current Challenges & Solutions in the Transmission of Traditional Knowledge

(*continued*)

Table 3. (*continued*)

3. World Bank, 2020/Japan Mainstreaming Disaster Risk Management in Developing Countries, the Resilient Cultural Heritage and Tourism (RCHT) Program	To explore the theoretical underpinnings of knowledge transfer within Japanese multinationals and to enhance the disaster resilience of cultural heritage, sustainable tourism, and communities in their countries	Institutional framework	CH Experts from Japan, Uzbekistan & Bhutan	Concentrates on practical approaches and specific examples, and on lessons learned from previous experiences and disasters. It includes relevant practices and measures for specific key hazards such as fire, earthquakes, floods, and landslides	Identifying and designating its Cultural Properties (CPs) by classifying them into six different categories: Tangible CP, Intangible CP, Folk CP, Monuments, Cultural Landscapes, and Groups of Traditional Building
4. Yayasan Hasanah, 2018/ Hasanah's Arts, Heritage and Culture (AHC) pillars	To involve craftspeople from a variety of fields in teaching selected protégés about their craft.	Evidence-based policymaking framework and social interventions	35 silver craftsmen, 6 woodwork carvers, and 3 fine artists	Sustainability of traditional craft heritage to ensure continuous transmission of craft skills and knowledge from generation to generation via mentorship education	Two MoUs were signed between the craftsmen with Lords Collection and EMASTERS during the exhibition to ensure sustainability of the craft.

5 Recommendations

For the purpose of this study, an evaluation of relevant past researches has been conducted to determine their applicability in investigating the elements that influence Knowledge Transfer acceptance. This analysis was conducted to gain a grasp of the theory that is most frequently utilized and matched to influence the organizational discourse on KT adoption. The investigation indicated 7 elements to be undertaken towards forming a perfect integration in applying KT in any related CH project (Tables 4 and 5).

Table 4. Analysis of constructs used in previous studies/projects

Study/ Project	Elements						
	Data, Information and Knowledge	Resources Mapping	Awareness & Action	Planning & Management	Integrate and prioritize	Partnership	Knowledge Transfer Mentorship and education empowerment
Alma Mater Studiorum & UNESCO (2019)	✓	✓	✓	✓	✓	✓	✓
Heritage Consortium, Europe (2019)	✓	✓	✓	✓		✓	✓
World Bank and Japan, (2020)	✓		✓	✓	✓	✓	✓
Yayasan Hasanah, (2018)	✓		✓	✓	✓	✓	✓

Table 5. Possible solutions for KT and CH Preservation, adopted and amended from [20]

Elements	Description
Data, Information and Knowledge	– Identify factors relating to the understanding of data and information about cultural heritage and adaptive reuse, and their context as knowledge production and skillsets
Resources Mapping	– Mapping types of information, knowledge, sources, skills, types of heritage (tangible, intangible, and natural heritage), tools and technology suitable for the education, preservation and conservation process
Awareness & Action	– Emphasize, cultivate and acknowledge the need to preserve CH. Demand for participation and implementation of heritage management
Planning & Management	– Conduct good data collection, evaluation and comprehensive analysis of CH, finance, values, and etc. – Outline the key policies, legislation and standards relating to CH – Outline actions and measures necessary for the effective management of risks and impacts towards decision making and actions

(continued)

Table 5. (*continued*)

Elements	Description
Integrate and prioritize	– Ascertaining what values and characteristics should be protected. Determining the degree to which the chosen values and characteristics are susceptible to shifts in the environment
Partnership	– Setting up local and international partnerships and management structures for each CH project consisting of countries, CH experts, vendors, private entities, and the public or society
Knowledge Transfer Mentorship and education empowerment	– A journey through school, college, and university is a never-ending process of acquiring new ideas, skills, information, and values while also developing their own set of core beliefs, theories, and habits of thought. T-vet or Heritage Mentorship programs in society

6   Conclusion

Those who seek to improve performance via the use of digital technologies frequently have a specific instrument in mind, but digital transformation should likely be driven by a broader strategy that begins with knowledge. One of the most important topics in educating the public and organization about the need for preserving and disseminating cultural heritage information and knowledge is raising awareness about the importance of doing so. Knowledge transfer had been proven to support the process of transferring expertise, knowledge, skills, and capacities from one knowledge base to another in multiple research perspectives. In relation to heritage, it is also about sharing experiences and best practices, and learning for a society to evolve. It aims to close the gap between what people know and what they can do in relation to their inheritance, culture and societal legacy. The role of knowledge transmission in cultural heritage should be promoted and grown. Cultural heritage organizations work extensively to reduce expenses in managing, promoting, and doing preservation and conservation works. However, educating the society about the importance of traditional knowledge/CH seems to be more efficient and meaningful.

References

1. Alavi, M., Leidner, D.E.: Review: knowledge management and knowledge management systems: conceptual foundations and research issues. MIS Q. **25**(1), 107 (2001)
2. Lee, L., Lajoie, S.P., Poitras, E.G., Nkangu, M., Doleck, T.: Co-regulation and knowledge construction in an online synchronous problem based learning setting. Educ. Inf. Technol. **22**(4), 1623–1650 (2016). https://doi.org/10.1007/s10639-016-9509-6
3. Hammond, J.W., Moss, P.A., Huynh, M.Q., Lagoze, C.: Research synthesis infrastructures: shaping knowledge in education. Rev. Res. Educ. **44**(1), 1–35 (2020)

4. Farnese, M.L., Barbieri, B., Chirumbolo, A., Patriotta, G.: Managing knowledge in organizations: a nonaka's SECI model operationalization. Front. Psychol. **10**, 2730 (2019)
5. Nilson, T., Thorell, K. (eds.): Cultural Heritage Preservation : The Past, the Present and the Future (1:1) (2018). http://urn.kb.se/resolve?urn=urn:nbn:se:hh:diva-37317
6. Sun, C., Chen, H., Liao, R.: Research on incentive mechanism and strategy choice for passing on intangible cultural heritage from masters to apprentices. Sustainability **13**(9), 5245 (2021)
7. Song, X., Yang, Y., Yang, R., Shafi, M.: Keeping watch on intangible cultural heritage: live transmission and sustainable development of Chinese lacquer art. Sustainability **11**(14), 3868 (2019)
8. Radzuan, A.W. (ed.): E-Proceedings of Extended Abstracts The 1st International Symposium on Cultural Heritage (ISyCH) 2021, vol. 1 (2021). http://hdl.handle.net/123456789/3050
9. Radzuan, I.S.M., Ahmad, Y., Zainal, R., Shamsudin, Z., Wee, S.T., Mohamed, S.: Conservation of a cultural heritage incentives programme in a Malay village: assessing its effectiveness. J. Herit. Manag. **4**(1), 7–21 (2019)
10. Amin, R., Faizah, N., Deraman, A., Baker, O.F.: Transforming model to meta model for knowledge repository of malay intangible culture heritage of Malaysia. Int. J. Electr. Comput. Eng. (IJECE) **2**(2), 231–238 (2012). ISSN 2088–8708
11. Khan, M.P., Hussin, A.A., Daud, K.A.M.: Safeguarding intangible cultural heritage through documentation strategy at cultural heritage institutions: Mak Yong's theater performing art. Adv. Sci. Lett. **23**(8), 7890–7894 (2017)
12. Maria, B.A., Tommaso, F.: Knowledge Sharing systems in a historical heritage context: an exploratory study. In: MCIS 2008 Proceedings, pp. 1–15 (2008). https://aisel.aisnet.org/mcis2008/19
13. Azmi, I.M.A.G., Ismail, S.F., Jalil, J.A., Hamzah, H., Daud, M.: Misappropriation and dilution of indigenous people's cultural expression through the sale of their arts and crafts: should more be done? Pertanika J. Soc. Sci. Human. **23**, 165–178 (2015)
14. Zaky, S.Z.N., Azmi, I.M.A.G.: Protection for intangible cultural heritage as a viable tourist product: malaysia as a case study. Malay. J. Cons. Family Econ. **20**, 59–70 (2016)
15. Corradi, G., Feyter, D.K., Desmet, E., Vanhees, K.: Critical Indigenous Rights Studies (Routledge Research in Human Rights Law), 1st edn. Routledge, Abingdon (2020)
16. Bjork, L.: How reproductive is a reproduction? Digital transmission of text based documents. Swedish School of Library and Information Science, University of Borås (2015). https://www.diva-portal.org/smash/get/diva2:860844/INSIDE01.pdf
17. Fry, H.S.K.: Handbook For Teaching And Learning In Higher Education: Enhancing Academic Practice, 4Th edn. T&F INDIA (2020)
18. Straub, E.T.: Understanding technology adoption: theory and future directions for informal learning. Rev. Educ. Res. **79**(2), 625–649 (2009)
19. Jisr, R.E., Maamari, B.E.: Effectuation: exploring a third dimension to tacit knowledge: effectuation: a third dimention. Knowl. Proc. Manag. **24**(1), 72–78 (2017)
20. Pintossi, N., Kaya, D.I., Roders, A.P.: Assessing cultural heritage adaptive reuse practices: multi-scale challenges and solutions in Rijeka. Sustainability **13**(7), 3603 (2021)

A New Landscape of Political Engagement Through Social Media: How Can We Map It?

Norman Sapar[1], Ab Razak Che Hussin[1(✉)], and Nadhmi A. Gazem[2]

[1] Azman Hashim International Business School (AHIBS), Universiti Teknologi Malaysia,
Johor Bahru, Malaysia
norman77@graduate.utm.my, abrazak@utm.my
[2] Department of Information Systems, College of Business Administration-Yanbu,
Taibah University, Medina 42353, Saudi Arabia
nalqub@taibahu.edu.sa

Abstract. The changing dynamics of the mass media world, especially social media, demand a specific and systematic approach to managing and leveraging it. Social media are now widely used without having time to wait for the launch event or the appropriate time frame to be introduced. The benefits are of course wide and very fast access to all over the world and arguably without cost burden. In many countries, political dynamism occurs as times change, and the mechanisms of communication and the approach to relationships are also different. Politics and social media complement each other, politicians who do not use social media consistently are bound to drop out or lose influence and popularity. Popularity on social media is already an individual's priority, in any profession, meaning politics is directly linked to influence and strongly connected with popularity. Young people are a group that is vulnerable to these latest technological trends and makes up a large percentage of the total electorate. In addition, the expenses and focus given to social media are a necessity. The evolution of its development is moving fast as each group pursues popularity to gain influence. Good and bad something is no longer an advantage otherwise popularity can dictate everything. A study of the model of user relations with politics on social media will help a party or politician to gain more confidence and strategize in the challenges of a new political world. The study will be conducted using the positivism approach, a research design based on the quantitative research of the Hypothetico-Deductive Method.

Keywords: Political Engagement · Social Media · Influence · Popularity

1 Introduction

The new age, the new media, and the new generation necessarily the passage of time that takes place requires an adaptation and structural change to many things. The trend of young people is certainly in the media of the generation of his time. Adaptations and approaches are constantly changing sometimes without having time to breathe a sigh of relief, having done something new comes up. Politics and social media are something very synonymous now. Both are very dynamic and need to be harnessed together. The correlation between the two needs to be studied for optimum practice and strategy.

Social media are outlets representing an internet-led online community that shares ideas, opinions, and perceptions with the goal of making an impact [1] and has grown rapidly in recent decades [2]. The basic idea of social media surveys is to generate people's ideas and opinions specifically about a particular political movement. Developments in social media have proved to be important and successful for election campaigns and community participation in keeping up with political developments. Such political activism can be effective because it stimulates the direct involvement of citizens.

In the 2018 14th Malaysian General Election (GE), voters between the ages of 21 and 40 were 41% of the total electorate. That percentage is growing when a voting age of 18 and automatic voter rolls are introduced and voters from this group are expected to make up 58% of voters aged 18 to 40 in the next 15th GE. This age group is a generation that is not only IT-savvy but also IT, smartphones and mass media are already part of their life. However, with the development of technology and innovation that connects any invention with internet access, the reality is there is now no specific age range that can be said to be active in using smartphones and social media. and even people over 60 years old are skilled at leveraging existing technology suggesting that the percentage of voters who can be approached through it is very large above 80%. Surely one day the whole human population will be literate on social.

Social media engagement is a type of engagement that occurs in a specific context and reflects positive or negative attitudes toward a particular influence. In addition, the causes and consequences of social media engagement have been identified to better understand why people interact on social media and the potential outcomes such as loyalty, satisfaction, trust, and commitment [3].

Politicians are passionate about seeking to become popular champions on social media, but do they know how this can be achieved or how to prove their effectiveness? The study will propose a model that politicians can best benefit from to increase their popularity and influence. This study is very important, because the results of past studies from journal sources and indexed articles have not well explore role of social media for political engagement. However, the relevant studies can be used as a reference to strengthen the model to be proposed, especially those related to political engagement in social media.

The purpose of the study is to identify the problem of user engagement in politics on social media as well as the factors that influence it, and then to validate those factors. This study will use quantitative methods based on questionnaires as an approach to confirm the proposed engagement models.

2 Background

Social media is already a key field for the spread of influence and the main war of nerves between political parties and politicians around the world. If earlier the agenda was delivered through television, radio, pamphlets, and talks now the measure of the popularity of all such mechanisms is no longer relevant. The main measure now is the popularity on social media which can be seen from "like", "comment" and "share" on the "post". Through blogs, social media posts, comment sections, "likes" and so on, individuals can find like-minded individuals, convey their messages and engage directly and actively in their areas of interest, should they wish [4].

The evolution of politics on social media is now very fast. Various new applications appear and the need to change the platform of use is necessary sometimes without having time to think. The main social medium, Facebook, has always been the ultimate medium as the adaptations that have always been done have made Facebook so far relevant throughout the social media tech boom. Interestingly, the two-way communication mechanism of all social media is the same as "like", "comment' and "share". So, this paper will look at the correlation in one model.

The defeat of the Barisan Nasional (BN) party in the 14th General Election was undeniably due to the weakness of the machinery and party leaders using social media consistently and strategically. The Pakatan Harapan (PH) team has leveraged this simple and fast-medium with the same large-scale narrative that sparked a wave that managed to give a negative image to BN. PH doesn't use the narrative that they're capable of doing anything, let alone highlighting that they have a brilliant plan and idea but consistently they sow hatred on social media to influence several young and urban voters. As a result, all the services and development and good planning of BN are buried in the war of nerves. We know young and urban voters are people who live side by side with social media and PH has managed to benefit from it.

Social media is just a medium of delivery and gaining popularity, what promises when it fails to be fulfilled becomes material and BN uses it to bring down PH. BN after losing consciousness and being aware of the defeat and learning from failure. The role of social media is best used in an orderly and strategic manner to make up for the defeat and return BN to the national government. As a result, BN managed to win a series of by-elections (PRK) and State Elections (PRN), where social media was best used. At the new PRN Johor, each candidate was provided with a cameraman, videographer, and social media operator to campaign on social media to the maximum. This provided a was systematic and consistent approach to produce a meaningful victory for the BN.

The same is a clear example for the former Prime Minister who was the subject of attacks before and in GE 14 again became a phenomenon and attacked those who attacked him before 100% through social media. As a result, his popularity and influence turned to the opposite in a short period. The trend of increasing the use of social media for political campaigns will continue to increase and it is expected that in the 15th general election it will trigger an extraordinary boom to influence voters. There must be a secret, there must be a way and there must be a strategic relationship of political influence on social media that can be best exploited in a certifiable manner. The study conducted by [5, 6, 7] states Facebook's effectiveness as a political communication tool can best be measured through user reactions and engagement.

Information through multilateral communication between users, followers, and potential agencies makes it easier to organize actions collectively that can have both positive as well as negative consequences. What is spread on social media, such as protest activity, can trigger offline action for real political change [8]. Politicians and their teams need to know to control the information that is spread on social media. There are few guidelines and established approaches for managing social media for political stability or victory and require considerable scrutiny. The introduction of social networks has fostered these require new forms of online political activity and subsequently influenced offline activities, including grassroots, anti-strengthening movements, which in turn are institutionalized and reinserted into the mainstream electoral process. The popularity of individuals who obtain more votes is more important than the ability and competence of politicians as leaders. This situation will certainly affect the progress of a country. A quality leader with integrity who is not popular on social media will not be elected. This must be a situation that will affect the country.

The role of the mass media in politics doesn't need to be explained anymore. Its importance is already a political package of today. It has been revealed that many politicians equip their forces with media officers who specifically play a role on social media. Even the big budgets are being allocated as if they are racing to get skilled and influential to gain popularity. On social media, time plays a critical role. There's no term, or room for delay and it is 'now' for someone to win or conquer something. However, it is also necessary with an informative and strategic approach that is of course through a comprehensive and targeted study.

The rapid change and evolution of social media have made it difficult for researchers to do a study and this field is still new. Although new, the maturity period is fast enough that every day there are changes and developments in related technologies that occur. There is still a lot of space and opportunities to explore so there is also a lot of research that needs to be done and ongoing due to the relevance of this field for some scopes there is a time limit.

3   Social Media Engagement for Political Influence: Challenges

In the age of Facebook, Twitter, and Instagram, political engagement through social media is easier than ever. For members of the millennial generation, where interconnectedness is driven by likes, shares, and retweets on social media sites, politicized discourse often goes viral, perhaps at the cost of meaningful engagement with the topic at hand. The first election in which young citizens participate can be described as their official gateway into politics, and research has found that political events are most formative at the age of 18–19 [9]. Therefore, social mediating with young people for political purposes plays an important role in the development of their future political engagement.

Political socialization is the process through which citizens 'crystalize political identities, values, and behaviors that remain relatively persistent throughout later life' [10]. Although there is an ongoing debate about how stable patterns of political orientations and behavior are throughout the life cycle [11], it is generally believed that early-life experiences form the basis for attitudes towards politics, engagement, and participation patterns in later adulthood [12]. A successful way to mobilize young voters is achieved through peers and parents [11], out-of-voting campaigns that often focus on civic education, political events, and media exposure [13].

Increasingly, political campaigners from different parties are investing in young voters as a specific target group and aim at making political content and issues attractive to this group [14]. Information communicated through mass media can reach a larger audience and contribute to citizens' political socialization through several political campaigns [15]. While we know that the use of media, in general, can positively affect

political engagement and participation [16], young citizens increasingly receive political information through social media [17]; accordingly, political campaigns turned to
online platforms to reach this part of the population [18]. Most research in this field
has shown that receiving political information via social media mobilizes the political
engagement of young citizens in various ways [19–21].

4   Some User Engagement Models from Past Studies

Studies conducted by [22], entitled "Pathways to political (dis-)engagement: motivations
behind social media use and the role of incidental and intentional exposure modes in
adolescents' political engagement", examined the relationship between the four keys of
motivation, political information, entertainment, social exchange, and self-expression,
when using social media networks among adolescents [22].

They found that political information and self-motivation were positively related to
"political engagement" through "intentional mode" [22]. On the other hand, the negative
motivation of entertainment is related to offline, but not to online relationships through
"incidental mode" [22]. But many more attitudes and motivations can influence the user's
relationship to politics on social media that can form a cyclical relationship that needs
to be elaborated to identify specific attitudes.

Here is a model based on the results of the study (Fig. 1).

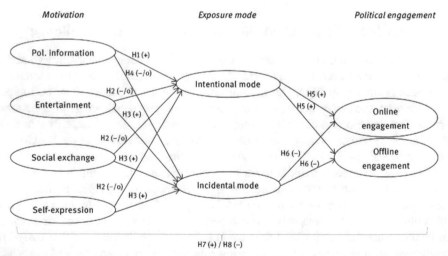

Fig. 1. "The hypothesized indirect relationship between user motivations and political engagement via a mode of exposure. Notes: (+) hypothesized positive effect, (–/o) hypothesized negative
or no effect, (–) hypothesized negative effect."

A review was carried out by [23], with the title "A Framework for The Use of Social
Media for Political Marketing: An Exploratory Study". According to the study, the use
of marketing concepts to attract voters and accumulate participation in political affairs is
critical, in the world of politics, depending on the competition [23]. The authors suggest

that use of social media for political campaigns and making connections has given rise to a new paradigm that offers a new approach in bringing political participation and making connections. The study found voters widely use the internet, mobile phone technology, and social media. Voters were also found to want to get involved in political issues through social media. The proposed model of the study [23] focuses on strategy and can be expanded to detail the scope of how the strategy is used to gain trust, benefit, and build relationships so that the scope of politics can be made more specialized (Fig. 2).

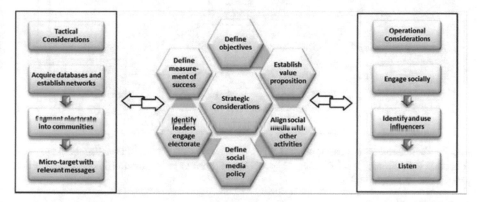

Fig. 2. The concept of the framework used in the study to gain political influence.

Another study on the relationship between Social Network Sites and Political Engagement explores the impact of Facebook connections and uses on Political Protest...", states *past studies have shown a positive correlation between the use of social media networks and political connections, but the understanding of its policy mechanisms is limited because studies do not take into account the dynamism of social media that can influence through multiple channels* [24]. This model of the study was found to be streamlined and connected with attitudes to form a more comprehensive and clearer model (Fig. 3 and Table 1).

When it comes to engagement, it will certainly involve several elements and the environment factors. Based on the factors from previous studies several new factors can be added to be more practical and easier to understand. In addition, segregation and arrangement in order are important for certain segments. The three main segments of the political side that need to be seen are the politician's personality, confidence in his ability to do things, and what will make users make political engagement on social media. Then we will look at the appropriateness of using the factors from the literature study and add to what are the factors that can influence the user for each factor on the political side that will of course look at the motivational and attitudinal factors. Political information, for example, is it beneficial? Is there a narrative? Is it also consistent? Attitudes and motivations towards it will be compiled, questionnaires among experts and users, and then updated in the form of practical and more comprehensive models. Of course, the factors that will be proposed will follow the current situation as an example in terms of strategic, tactical, and operational can attract people. What attitudes and motivations young people will see, perhaps certain posts, this study will elaborate on?

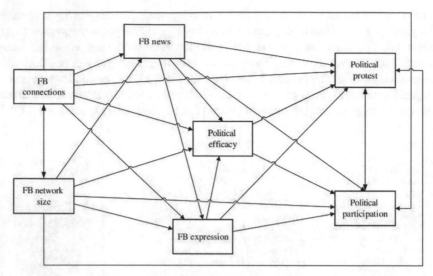

Fig. 3. Theoretical model based on the O-S-R-O-R framework. Note. "FB = Facebook; O = FB size and connections; S = FB news; R = FB expression; O = political efficacy; R = political protest and participation."

Table 1. Summary of factors for political engagement from relevant past studies.

	Researcher	Factors Listed
1	Heiss, Knoll, and Matthes - Pathways to political (dis-) engagement: motivations behind social media use and the role of incidental and intentional exposure modes in adolescents' political engagement	- political information - entertainment - social exchange - self-expression
2	Ayankoya, Calitz, and Cullen - A Framework for the use of social media for Political Marketing: An Exploratory Study	- tactical - strategic - operational
3	Chan - Social Network Sites and Political Engagement: Exploring the Impact of Facebook Connections and Uses on Political Protest and Participation	- news - connection - size

5 Recommendations and Conclusions

The maturity of political parties in Malaysia continues to increase according to current changes. A new political landscape may change through digital technology and digital lifestyle of voters. A party that had previously been a champion or a conventional influence it would not be able to defend its position if it failed to comply with the current situation. Therefore, an in-depth and continuous study is necessary for the preparation and support in the next political strategy. The future political champion is a champion of knowledge according to the current technological developments.

Through the observation of the relevant past studies, it is evident that there is a lot of room to explore and there is scope for model of a relationship that can include aspects of the politician's personality, and competency as well as what is highlighted on social media that will lead users make political connections based on the effect of his/her popularity. Attitudes and motivational aspects will be linked because in past study studies this relationship has not been specifically shown for political connections on social media. This study reveals the need for a model on user engagement in politics through social media by providing knowledge and guidance to political organizations in sustaining the power in digital modern word. Thus, this study will be beneficial in helping political organizations keep up with the current trends of the world today.

References

1. Loader, B.D., Vromen, A., Xenos, M.A.: Performing for the young networked citizen? Celebrity politics, social networking and the political engagement of young people. Media Cult. Soc. **38**(3), 400–419 (2016)
2. Wukich, C., Mergel, I.: Reusing social media information in government. Gov. Inf. Q. **33**(2), 305–312 (2016)
3. van Doorn, J., et al.: Customer engagement behavior: theoretical foundations and research directions. J. Serv. Res. **13**(3), 253–266 (2010)
4. Ekström, M., Östman, J.: Information, interaction, and creative production: the effects of three forms of internet use on youth democratic engagement. Commun. Res. **42**(6), 796–818 (2015)
5. Mohamed, S., Manan, K.A.: Facebook and political communication: a study of online campaigning during the 14th Malaysian general election. IIUM J. Hum. Sci. **2**(1), 1–13 (2020)
6. Khan, A.J.M.A., Rokis, R.: Reviewing the literature on working dual jobs among workers with specific discussion on Malaysian women. IIUM J. Hum. Sci. **2**(1), 14–24 (2020)
7. Sern, T.J., Wenn, B.O.W., Yee, L.L.: Media agenda in politics: how Malaysian RTM Radio Stations cover 14th general election. IIUM J. Hum. Sci. **2**(1), 25–38 (2020)
8. Gladwell, M.: Small change. The New Yorker **4**(2010), 42–49 (2010)
9. Erikson, R.S., MacKuen, M.B., Stimson, J.A.: The Macro Polity. Cambridge University Press (2002)
10. Neundorf, A., Smets, K.: Political socialization and the making of citizens (2017)
11. Bhatti, Y., Hansen, K.M., Wass, H.: The relationship between age and turnout: a roller-coaster ride. Elect. Stud. **31**(3), 588–593 (2012)
12. Krosnick, J.A.: The stability of political preferences: comparisons of symbolic and nonsymbolic attitudes. Am. J. Polit. Sci. **35**, 547–576, 1991
13. Moeller, J., de Vreese, C., Esser, F., Kunz, R.: Pathway to political participation: the influence of online and offline news media on internal efficacy and turnout of first-time voters. Am. Behav. Sci. **58**(5), 689–700 (2014)
14. Sweetser Trammell, K.D.: Candidate campaign blogs: directly reaching out to the youth vote. Am. Behav. Sci. **50**(9), 1255–1263 (2007)
15. Lee, N.-J., Shah, D.V., McLeod, J.M.: Processes of political socialization: a communication mediation approach to youth civic engagement. Commun. Res. **40**(5), 669–697 (2013)
16. McLeod, J.M., Scheufele, D.A., Moy, P.: Community, communication, and participation: the role of mass media and interpersonal discussion in local political participation. Polit. Commun. **16**(3), 315–336 (1999)

17. Schrøder, K.C., Blach-Ørsten, M., Eberholst, M.K.: Is there a Nordic news media system?: a descriptive comparative analysis of Nordic news audiences. Nordic J. Media Stud. **2**(1), 23–35 (2020)
18. Denollet, J.: Interpersonal sensitivity, social inhibition, and Type D personality: how and when are they associated with health? Comment on Marin and Miller (2013). Psychol. Bull. **139**(5), 991–997 (2013)
19. Ekström, M., Shehata, A.: Social media, porous boundaries, and the development of online political engagement among young citizens. New Media Soc. **20**(2), 740–759 (2018)
20. Kahne, J., Lee, N.-J., Feezell, J.T.: The civic and political significance of online participatory cultures among youth transitioning to adulthood. J. Inform. Tech. Polit. **10**(1), 1–20 (2013)
21. Xenos, M., Vromen, A., Loader, B.D.: The great equalizer? Patterns of social media use and youth political engagement in three advanced democracies. Inf. Commun. Soc. **17**(2), 151–167 (2014)
22. Heiss, R., Knoll, J., Matthes, J.: Pathways to political (dis-) engagement: motivations behind social media use and the role of incidental and intentional exposure modes in adolescents' political engagement. Communications **45**(s1), 671–693 (2020)
23. Ayankoya, K., Calitz, A.P., Cullen, M.: A framework for the use of social media for political marketing: an exploratory study, vol. 12, p. 2018. Retrieved, December 2015
24. Chan, M.: Social network sites and political engagement: exploring the impact of facebook connections and uses on political protest and participation. Mass Commun. Soc. **19**(4), 430–451 (2016). https://doi.org/10.1080/15205436.2016.1161803

User-Generated Content (UGC) for Products Reviews Video Factors Derivation Through Weight Criteria Calculation

Siti Zubaidah Mohd Zain[1]([✉]), Ab. Razak Che Hussin[1], and Amri Ab. Rahman[2]

[1] Azman Hashim International Business School, Universiti Teknologi Malaysia (UTM),
81310 Skudai, Johor, Malaysia
zubaidahmohdsiti@graduate.utm.my, abrazak@utm.my
[2] Faculty of Computer and Mathematical Sciences, Universiti Teknologi MARA (UiTM),
18500 Machang, Kelantan, Malaysia
amri@uitm.edu.my

Abstract. The number of user-generated content (UGC) on social media keeps increasing and leads to product review videos by users. UGC video reviews are produced originally by users and managed to reach millions of views on social media sites. UGC has also sparked a massive community and rendered traditional marketing less effective. This situation motivated a study to identify the UGC factors in the product review video. The identified factors are expected to enhance the effectiveness of UGC production in the product review area. Previous studies and related theories have been examined through literature reviews. Some factors from the UGC area are selected and calculated through the weight criteria technique. This process is to figure out the relevant factors for UGC product review. This paper aims to identify factors that can boost UGC for social commerce (s-commerce) through the weight criteria calculation technique. The weight criteria calculation technique involves three steps highlighted in this paper. After completing all three steps, the UGC products review factors were decided and compared to the product review video study domain. These proposed factors will likely encourage s-commerce users to create more effective video-based UGC for product reviews. Later, the proposed factors will be used to further the study toward the research model of this study. The future research model will be developed based on implementation with the selected theory from the information system area. This shows the significant reason for the UGC products reviews video should be derived.

Keywords: User Generated Content · Factor Derivation · Weight Criteria Calculation

1 Introduction

Current businesses place a lot of importance on customer feedback. Even big business such as hotels also rely on third-party booking platforms, which dominated prior research on online customer reviews. Customers' online reviews of a product or service's primary

F. Saeed et al. (Eds.): ICACIn 2022, LNDECT 179, pp. 553–561, 2023.
https://doi.org/10.1007/978-3-031-36258-3_49

supplier on products posted on business-to-consumer (B2C) or consumer-to-consumer (C2C) e-commerce platforms have also received significant attention. These platforms act as traditional middlemen, assisting companies in widening their distribution networks in order to reach more clients.

UGC can be utilized as an alternative to traditional ways to understand client preferences and needs, according to studies from various industries. Thousands of customers can provide instant, educated, and passionate comments online, resulting in "crowd wisdom." Customer reviews on the internet are a prominent type of UGC. Online reviews make it possible for businesses and consumers to exchange content and knowledge about goods and consumption-related activities [7]. This shows how powerfully online reviews influence consumer decision-making and gives marketing managers access to vital market data resources [7].

The UGC complements the aspects of the product review video since it includes user moods, interests, and opinion mining based on user personalization [4], this situation matches with the elements of the product review video. The popularity of using videos as a medium for product reviews is evidenced by the number of channels used to provide product reviews on multiple platforms [2].

Business organizations worldwide are starting to encourage customers to upload online reviews in video formats. A recent study on product-review videos can be a promising marketing method [2]. This brought the relations towards the video-sharing platform for product review videos. The viewer's reaction to the advertising is an excessive increase in ads and effects viewership [1].

This research represents a substantial and timely contribution to the UGC body of knowledge. A number of features that cover the components of product review videos that are appropriate based on the social media platform must support an influential UGC. Those elements must originate from earlier research in the UGC field. This is done to make sure the UGC context is appropriate and aligns with the learning environment.

The purpose of this paper is to describe the weight criterion procedure and UGC product review considerations. Introduction, theoretical background, derivation of factors by weight criteria calculation, comparative procedure of UGC factors selection, and conclusion make up the five sections of this paper.

2 Theoretical Background

As the UGC product review video to increase the interest rate of viewing interest from the users of social media sites involved. Social Influence Theory (SIT) and Theory of Interpersonal Behavior (TIB) are the theories. The core idea of SIT is to refer to an individual's influence on a person's attitudes, beliefs, and actions or behaviors through three processes: compliance, identification, and internalization.

SIT is closely linked to social influences that cause changes in attitudes and behaviors, and it can occur at many "levels." This is why SIT has been chosen as the leading theory in this study. This study involves the results of a survey of the desire to watch from the users' attitude and create trust from the users themselves on the featured UGC product review video. Based on SIT, the study model that will be produced ends with "Watching Intention" as a dependent variable.

TIB is the second theory implemented in this study. TIB involves interpersonal behavior in which behavior is a function of intention, habit strength, and various facilitator variables in any setting. This is in line with this study's dependent variable, which is based on the purpose of "Watching Intention." Furthermore, social factors, as well as intellectual and passionate ideas about behavior, all contribute to behavioral intent. Social factors in the model include norms, roles, and self-concepts. Unlike rational-instrumental assessment of consequences, the concept of effects encompasses positive and negative emotional responses to a decision. Here, the three elements of TIB, perceived consequences, social factors, and affect, are adapted into the future study model.

The foundation for this study is a line of scientific inquiry that starts with problem identification. Finding the issues that need to be addressed and fixed in the context of this UGC research is known as problem identification. UGC has become brittle and less effective as a result of UGC criteria that simply take into account the opinions and comments of users. Opportunities for UGC that have the potential to affect millions of viewers should be thoroughly utilized. After identifying the problem, the method of reviewing the literature has evolved. The control of the factor selection process has also been addressed using information system theory. The theoretical underpinning for this study's factor selection procedure is shown in Fig. 1.

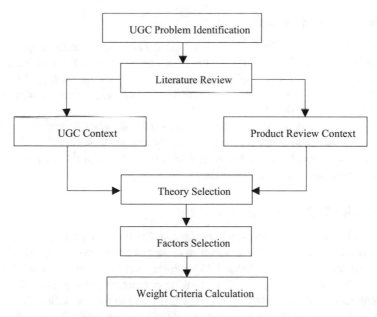

Fig. 1. The theoretical framework.

On the first step of this study, a collection of issues was identified. Understanding the current state of UGC, which people have been using as a platform for product reviews, has led to the concerns. This prompted the next step, which was to analyze earlier related studies conducted by other researchers. The systematic literature review (SLR) method was employed in the literature review process for this investigation [8]. The SLR is focus on UGC and product review context. Those included the models from previous studies and among all the related factors which involved.

The associated IS theories have also been investigated in this work, which is being applied to the field of information systems (IS). It is for continuing and furthering the body of knowledge. The theories connected to human behaviours, such as the theory of TIB, should be explored in accordance with the study's components and SIT. The elements chosen from the UGC and product review contexts should be in line with the research theory.

After the SLR process and theory selection processes were finished, the UGC for product reviews elements were derived. The weight criteria calculation is used in this paper as the method of derivation for the study factors. The next section would explain the steps of through weight criteria calculation technique.

3 Derivation Factors Through Weight Criteria Calculation

The weighted criteria methodology is a useful method for weighing the value of several criteria against programme alternatives. Alternatives' values can be determined by comparing how well they perform against specific metrics.

3.1 Step 1: Relevant Factors Listed

There are several factors of UGC that can influence watching intention in previous study were identified through literature review process. The factors will be selected based on it is significant or not. The most frequently factors are identified that empirically examined two times in previous study. All the factors of UGC for product review and the relevant value of each factor are listed in the Table 1.

The weighting calculation does not end here. This is only the first step. After this, the Total value would be revised in Step 2. The following sub-section has discussed the revision of the relevant factors process. It is because there is to find any factors which brought the same meaning or similarities.

3.2 Step 2: Revised All the Relevant Factors

There are several factors that have similar meaning. The factors that have similar meaning need to be revised and be merged as one. The factors of PU + PEU are grouped together under a single factor called Use Value (UV), which is based on the elements that exist in the factors. Meanwhile, T + C + TIS factors are put under C factor where it is referred to the credibility value that involved in the UGC success factors. Other factors which are I + AI + SI have similarity where it is referring to interactivity factor and put under, I factor. The entire revised factors are listed in the Table 2.

This step clarifies how this technique is used to evaluate the specific criteria based on the weighted importance. The significant value can be identified by evaluating alternatives based on their performance concerning individual criteria. This weighted technique also exposes its importance by treating each factor independently. The next step would be measuring the factor's reliability. The following sub-section presents the step.

Table 1. The factors and its relevant value.

No.	Factors	Previous Studies						Total
		1	2	3	4	5	6	
1	Perceived usefulness (PU)	/						1
2	Perceived ease of use (PEU)	/						1
3	Aesthetics (A)	/						1
4	Trust (T)	/		/				2
5	Interactivity (I)	/		/		/		3
6	Homophily (H)	/						1
7	Sender presence facial expression (SPFE)		/					1
8	Credibility (C)		/					1
9	Price benefits (PB)			/				1
10	Service benefits (SB)			/				1
11	Trust in service (TIS)	/		/				2
12	Perceived risk (PR)			/				1
13	Application interaction (AI)	/		/				2
14	Staff interaction (SI)	/		/	/			3
15	Co-creation (CC)				/			1
16	Empowerment (E)				/			1
17	Community (CM)	/			/			2
18	Self-concept (SC)				/			1
19	Normative (N)					/	/	2
20	Informational (IF)					/		1
21	Social (SO)	/		/		/		3
22	Awareness (AW)					/		1
23	Consideration (CD)					/		1
24	Purchase Intent (PI)					/		1
25	Satisfaction (S)					/	/	2

** 1–6 refers to the main sources of the UGC product review factors study

3.3 Step 3: Factors Reliability Calculation

Based on the Table 2, there are few factors that have been examined two times or more in previous studies. Next step is to calculate the weight of each selected factors using the formula. This following formula is adopted [3] to increase factor reliability.

$$C = B/A$$

Where,

Table 2. The revised factors for weight calculation.

Factors	Revised Factor	Total Factors	Revised Factor Calculation	Total Revised Factor	Total Relationship
Perceived usefulness (PU)	Perceived Ease of Usefulness	1	PU + PEU	1 + 1	2
Perceived ease of use (PEU)		1			
Price benefits (PB)	Product Benefits	1	PB + SB	1 + 1	2
Service benefits (SB)		1			
Community (CM)	Society	2	CM + SO	2 + 3	5
Social (SO)		3			
Aesthetics (A)		1	A	1	1
Trust (T)		2	-	-	-
Interactivity (I)		3	I + AI + SI	3 + 2 + 3	8
Homophily (H)		1	H	1	1
Sender presence facial expression (SPFE)		1	SPE	1	1
Credibility (C)		1	T + C + TIS	2 + 1 + 2	5
Trust in service (TIS)		2	-	-	-
Perceived risk (PR)		1	PR	1	1
Application interaction (AI)		2	-	-	-
Staff interaction (SI)		3	-	-	-
Co-creation (CC)		1	CC	1	1
Empowerment (E)		1	E	1	1
Self-concept (SC)		1	SC	1	1
Normative (N)		2	N	2	2
Informational (IF)		1	IF	1	1
Awareness (AW)		1	AW	1	1
Consideration (CD)		1	CD	1	1
Purchase Intent (PI)		1	PI	1	1
Satisfaction (S)		2	S	2	2

A. The frequency of factor is examined (total relationship)
B. The frequency of factor is significant (significant relationship)
C. The total weight of the factor

Table 3 below shows the factors that have been calculated using the formula to identify the weight of factor. All the selected factors are good weight where it is achieved weight of factor 0.5 and above.

Table 3. The significant relationship calculation.

Factor	Total relationship (A)	Significant relationship (B)	Weight (C)
Perceived Ease of Usefulness	2	2	1
Product Benefits	2	1	0.5
Society	5	4	0.8
Aesthetics	1	1	1
Interactivity	8	4	0.5
Homophily	1	1	1
Sender presence facial expression	1	0	0
Credibility	5	3	0.6
Perceived risk	1	0	0
Co-creation	1	1	1
Empowerment	1	1	1
Self-concept	1	1	1
Normative	2	0	0
Informational	1	0	0
Awareness	1	1	1
Consideration	1	0	0
Purchase Intent	1	1	1
Satisfaction	2	1	0.5

Based on the calculation weight of factor, there are five factors which is sender presence facial expression, perceived risk, normative, informational, and consideration are eliminated because those factors are not achieve weight criteria of factor which is 0.5 or above. Therefore, only thirteen factors are selected due to achieved weight criteria of factor more than 0.5 which is used value, benefits, society, aesthetics, interactivity, homophily, credibility, co-creation, empowerment, self-concept, awareness, purchase intention, and satisfaction.

Then, all thirteen selected factors will be mapped toward the domain of this study. Only those parallel to the study domain will be selected as UGC product review video factors. The selected factors will then be compiled as a preliminary model for this study. After completing this last step of the weight calculation process, all the accepted factors will be compared to the research domain. The factors which correspond to the research domain only would be selected as the UGC products re-views factors. The following section discusses the comparison process.

4 Comparison Process of UGC Factors Selection

This circumstance is similar to the elements of product review video since UGC comprises user moods, interests, and opinion mining based on user personalization [4]. The number of channels used to deliver product reviews on numerous platforms demonstrates the popularity of using videos as a medium for product reviews [2]. Vendor websites, independent websites, and social media pages are all featuring product review videos. Product reviews are cited by consumers as having the biggest influence on their purchasing decisions, according to earlier studies [6].

Customers are being encouraged to contribute online reviews in video formats by businesses all around the world. As a result, according to a new study, product-review videos can be a promising marketing approach [2]. In order to fulfill the criteria, the comparison process should be going through. Table 4 presents the UGC factors which successfully correspond to the product review concept in order to continue this research.

Table 4. The UGC factors toward product review.

UGC Factors	Product Review Concept
- Perceived Ease of Usefulness - Product Benefits - Self-concept - Product Awareness	As the UGC contains user sentiments, interests, and opinion mining based on user personalization [4], this situation is match with the elements of product review video
- Interactivity - Empowerment - Satisfaction	An emerging trend for communication is the use of video as a medium for product reviews. In the coming years, online reviews with videos are expected to gain in popularity. After watching a video regarding a product, 64% of consumers are satisfied and more inclined to purchase it [2]
- Credibility	The recent study stated on the product-review videos can be a promising marketing method [2, 5]

There are eight factors have been identified as being worth further investigation in this study as mention in Table 4. As compared to the Table 3 in sub-Sect. 3.3, there are several factors which not correspond to this research domain. There are society, aesthetics, co-creation, homophily, and purchase intent.

5   Conclusion

The weighted calculation technique is one of the alternatives for decision-making used in this study. The method successfully measures thirteen factors from UGC in the product review area. Even though only eight factors correspond to this study domain which would focus on video types of product reviews format. Five factors do not correspond to the UGC product re-view video domain of study after the comparison process is accomplished. Those factors are society, aesthetics, co-creation, homophily, and purchase intent. Consequently, the following is the list of factors proposed as UGC for product reviews video factors:

- Perceived ease of usefulness
- Product benefits
- Self-concept
- Product awareness
- Interactivity
- Empowerment
- Satisfaction
- Credibility

The factors of UGC for product review video are selected based on the achieved weight of criteria greater than 0.5 and match with the UGC context. The following progress would be a survey preparation toward confirmation of all the factors based on the current situation among social media users. This is important because social media technology keeps upgrading based on technology growth.

References

1. Chakraborty, S., et al.: Advertisement revenue management: determining the optimal mix of skippable and non-skippable ads for online video sharing platforms. Eur. J. Oper. Res. **292**(1), 213–229 (2021). https://doi.org/10.1016/j.ejor.2020.10.012
2. Fitriani, W.R., et al.: Reviewer's communication style in YouTube product-review videos: does it affect channel loyalty? Heliyon **6**(9), e04880 (2020). https://doi.org/10.1016/j.heliyon.2020. e04880
3. Jeyaraj, A., et al.: A review of the predictors, linkages, and biases in IT innovation adoption research. J. Inf. Technol. **21**(1), 1–23 (2006). https://doi.org/10.1057/palgrave.jit.2000056
4. Liu, B., et al.: A reliable cross-site user generated content modeling method based on topic model. Knowl. Based Syst. **209**, 106435 (2020). https://doi.org/10.1016/j.knosys.2020.106435
5. Nanne, A.J., et al.: The role of facial expression and tie strength in sender presence effects on consumers' brand responses towards visual brand-related user generated content. Comput. Hum. Behav. **117**, 106628 (2020). https://doi.org/10.1016/j.chb.2020.106628
6. Sarkar, A.R., Ahmad, S.: A new approach to expert reviewer detection and product rating derivation from online experiential product reviews. Heliyon **7**, e07409 (2021). https://doi.org/ 10.1016/j.heliyon.2021.e07409
7. Tran, L.T.T.: Online reviews and purchase intention: a cosmopolitanism perspective. Tour. Manag. Perspect. **35**, 100722 (2020). https://doi.org/10.1016/j.tmp.2020.100722
8. Wu, J., Chen, D.T.V.: A systematic review of educational digital storytelling. Comput. Educ. **147**, 103786 (2020). https://doi.org/10.1016/j.compedu.2019.103786

A Taxonomy of Dataset Search

Abdullah Hamed Almuntashiri[1,2(✉)], Luis-Daniel Ibáñez[1], and Adriane Chapman[1]

[1] University of Southampton, Southampton, UK
aa1r21@soton.ac.uk
[2] Applied College, Najran University, Najran, Kingdom of Saudi Arabia

Abstract. The demand for and use of data have increased in all life science domains, particularly in scientific communities. Data is organised into datasets which are used in many tasks, e.g. training machine learning (ML) models. Those datasets are stored either privately or publicly in repositories or data portals that can be published on the Web. Due to the need to find and reuse datasets, a new research field has appeared that focuses on the process of searching datasets to meet users' needs. Therefore, the purpose of this paper is to explore the dataset search literature in order to identify the used methods, algorithms, systems and benchmarks and then classify them. We performed a complete search of the dataset search literature on various search engines, scientific sites and digital libraries. We discovered more than 100 dataset search articles, and then we narrowed those articles to 31 after applying the exclusion criteria. As a result, a new dataset search taxonomy has been designed based on the search style that is used by users to retrieve datasets.

Keywords: Dataset retrieval · Dataset search · Structured Data · Data Discovery

1 Introduction

With the rapid development of technology over the last decade, the use of data has become integral in all domains, particularly computer science. Computer and data scientists have a significant need for data in order to implement many tasks, such as training/developing ML models, weather prediction and product recommendation [3]. Developing or training these models requires two basic components: input data and an estimator method [3]. In this case, datasets are the input data. These datasets have become commonly available via open data portals and dataset repositories on the Web [6]. For instance, in 2022, OpenDataSoft[1] contains more than 26,000 datasets and 219 million downloads for these datasets.

Due to the large number of datasets that exist on the Web, specifically on some open data portals, the utilisation of systems/methods to search datasets has become necessary, especially for researchers and data scientists who need to use datasets continuously. [5] mentions that there is a great need for dataset search techniques that can be used by data scientists to search for and integrate datasets. [6] emphasises that the significance of

---

[1] https://data.opendatasoft.com/pages/home/

© The Author(s), under exclusive license to Springer Nature Switzerland AG 2023
F. Saeed et al. (Eds.): ICACIn 2022, LNDECT 179, pp. 562–573, 2023.
https://doi.org/10.1007/978-3-031-36258-3_50

access to datasets has increased among researchers. Furthermore, there is still a dearth of understanding of how users search for datasets online [18].

In the present paper, we aim at answering the following research questions: (a) What are the existing methods of searching datasets? (b) How to classify those methods? We conducted a comprehensive scan of the dataset search literature on some search engines, scientific sites and digital libraries. We looked at more than 100 papers on dataset search and related works. Then, we focused on the algorithms, systems and benchmarks of dataset search, which comprised 31 papers, by using exclusion criteria. Thereafter, we built a new dataset search taxonomy based on the used search styles that could help dataset search users to understand those different search styles. The rest of the paper is organised as follows. Section 2 presents the dataset search definitions. Section 3 illustrates the methodology of constructing our taxonomy. Then, the result of building the taxonomy is described in Sect. 4. Section 5 covers the result discussion. In Sect. 6, some related works are provided. Finally, we conclude the paper in Sect. 7.

2 Dataset Search Definitions

With a rise in the number of datasets as well as the users who need to retrieve these datasets, the importance of data search and discovery has appeared in a domain of complementary disciplines [7]. In addition, [2] points out that data scientists need dataset search techniques in data lakes to search and integrate datasets. Data lakes are great repositories of government, enterprise and web data [2]. Therefore, various researchers and scientists have made a massive effort to construct the fundamental pillars of this new area, including defining the term "dataset search". [9] defined the term "dataset search" as the process of exploration and discovery resulting in providing a user with datasets. Dataset search is the task of providing a query by a user to a system which results in a ranked list of datasets [12]. From the perspective of a data user, [7] defines dataset search as the searching procedure for datasets as well as for a specific point of data in a dataset. There is no comprehensive classification for dataset search that can be followed to explain the prior efforts in this field. [7] built a taxonomy of data-centric tasks but it may not be adequate to demonstrate the dimensions of the entire dataset search domain since it only focuses on the users' activities with data.

3 Methodology

In the present paper, a new taxonomy of dataset search has been built. During the search process to construct this taxonomy, we used four main sources: Google Scholar as a search engine and Science Direct as a scientific platform as well as IEEEXplore Digital Library and ACM Digital Library as digital libraries, and did an initial selection based on title, year and accessibility. In addition, we used the following keywords to find the articles: "Dataset Search", "Data Search", "Data Retrieval", "Structured Data Queries", "Dataset Search Queries", "Table Queries", "How to Search for Datasets" etc. We selected 108 papers by conducting a comprehensive scanning of the title, abstract, introduction and conclusion. Thereafter, the following exclusion criteria were applied in this study: a) Studies that do not focus on searching dataset/structured data directly.

b) Studies that do not clearly explain the search style used c) Studies that lack a full-text version at the source. d) Studies that do not include the used methodology to develop their contributions. This led to approximately 31 papers of algorithms and benchmarks that focus on the dataset search directly. We sought to discover the main differences between those papers.

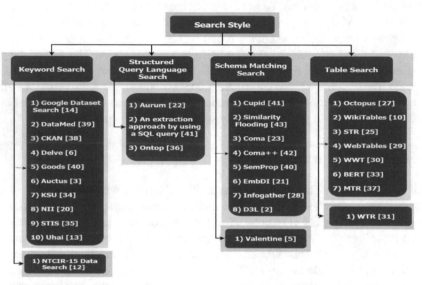

Fig. 1. The Taxonomy of Dataset Search. It includes three levels: the blue represents search style types, the orange is the systems/approaches and the green is the benchmarks.

4   Result

By reading the selected papers to build this taxonomy, we discovered that one of the main differences between the dataset search algorithms is the search style, which is the input form used when searching for a dataset. There are many basic components of searching for datasets, such as the input query and the result presentation. Because the input query enables users to express their requirements for datasets, we decided to choose the search style to be the main pillar of our taxonomy. Figure 1 illustrates the design of the taxonomy of dataset search as well as the algorithms, systems and benchmarks under each search style. Four main search styles were discovered, which are keyword search, search by structured query language, search by using schema matching and search by using tables. These four styles are the ones used to search for datasets/structured data in the dataset search papers. In this section, the search styles will be described, and some systems, methods and techniques will also be explained. Due to the page constraints, we only mentioned the most popular algorithms, systems and a benchmark for each taxonomic branch.

4.1 Keyword Search

Keyword search has become common in a number of communities concerned with dataset search. [11] points out that keyword search is the strategy used for standard information retrieval and web search. Keyword search can also be used to search for datasets using the published metadata for the datasets [1]. Furthermore, [7] mentions that data search on the Web or on data portals allows users to use keyword search for searching published metadata, which is considered as a characteristic. [7] refers to the common use of keyword search while searching for datasets. Most search engines as well as open data portals and repositories use keyword search techniques [4].

There is considerable debate on the efficiency of keyword search while searching a dataset. Many restrictions have been mentioned in various studies, which can be a disadvantage in this type of search. [16] reports that keyword search might not include an excellent selection of words to meet the users' information needs. According to [8], keyword based matching becomes less efficient when a dataset involves structured and semi-structured data. [15] mentions that the existing search engines depend on the keywords entered by users, making it difficult to find datasets that are perfectly matched.

Google Dataset Search. In 2018, Google developed a search engine that is one of the vertical web search engines. The main objective of this system is to discover datasets over metadata. Google Dataset Search supports many essential formats for characterising metadata on the Web, including Schema.org and W3C DCAT markup [14]. This system depends on the Google web crawl to collect metadata. The implementation of the crawl is for all datasets on the Web that utilise the schema.org Dataset class as well as datasets that use DCAT in order to gather the metadata of the dataset [9]. After that, an index of enriched metadata is constructed for all the provided datasets [14]. The Google knowledge graph is used to reconcile the metadata. To facilitate the process of discovering datasets, Google Dataset Search allows the user to interact by entering a search query with information needs. This query is processed via the use of keywords and CQL expressions [9].

CKAN. CKAN is an open-source data management system that enables users to search through the metadata of datasets. It aims to publish, discover, share and use data. One of its strengths is that it is used by many open governmental platforms and portals, such as the UK and Canada [17], and fuels other open portals, including publicdata.eu. CKAN is constructed using Apache Solr that employs Lucene to index the documents [9]. The search process in Solr involves two fundamental tasks: Finding the documents and then ranking the documents. The CKAN platform uses DCAT to characterise the datasets in the catalogues of data [17].

Auctus. Auctus is an open-source dataset search engine that was developed to support the discovery and augmentation of data. The main aim of this engine is to discover and rank a collection of datasets, whether the input query is a keyword or a dataset [3]. Moreover, it provides sophisticated query features such as posing spatial and temporal queries. [3] states that Auctus does not rely on the annotations of schema.org; it can connect to many APIs to capture datasets from various repositories. The metadata for datasets are also inferred automatically from datasets by profiling the ingested datasets, including type detection and statistics computation, which is type-dependent.

DataMed. DataMed is an open-source biomedical data discovery system. It aims to provide a Data Discovery Index (DDI) system that assists users in finding and accessing datasets. As of March 2017, it facilitated search across more than 60 biological data sources and repositories [24]. DataMed includes two main components; the first component is the ingestion and indexing pipeline, and the developers of the system built a search engine as the second component. This search engine uses many sophisticated techniques, including natural language processing (NLP), terminology services and Elasticsearch [29]. This system is considered to be one of the keyword search techniques by using a Google-like search box. Moreover, other advanced search options employ Boolean operators which are provided for professionals to determine the search fields as well as construct specialised queries [29].

NTCIR-15 Data Search. NTCIR-15 Data Search is a test collection for ad-hoc dataset retrieval. Although it is called a "data search", it is more convenient for dataset search tasks [12]. The data collection focuses on certain data published by the Japanese government (e-Stat) as well as the US government (Data.gov) [19]. The aim of this test collection is to process three research problems in the dataset search domain: query understanding, interpret dataset contents and retrieve models for dataset search [12]. Thus, the main task involved in this data search was receiving a query and a ranked list of datasets with a relevance score which should be returned by the system [20]. Overall, 192 questions were collected by using crowdsourcing services that are Amazon Mechanical Turk and Lancers. Afterwards, the entropy between the queries and the language models was calculated to identify the most representative queries for each topic [12]. The developers crawled around 1.5 million pages of e-Stat and Data.gov. Then, a dataset was defined as a pair of metadata and a collection of data files.

4.2 Structured Query Language Search

One of the ways that assists users in searching for datasets is submitting a structured query rather than a keyword query. A structured query can be expressed using query languages, including CQL (Contextual Query Language) and SQL (Structured Query Language). The above-mentioned keywords or CQL are most commonly used for dataset search queries [9]. Although no formal query languages have been devised for dataset retrieval, specific query interfaces in some systems have been developed. These interfaces support several query languages, such as SPARQL queries, SQL and CQL [7].

Aurum. Aurum is a data discovery system that enables users to discover relevant data in a flexible way using the characteristics of the datasets or the syntactic relations among them. This system relies on using an enterprise knowledge graph (EKG) to resolve issues with data discovery. This system aims to implement three fundamental tasks. The first task is building an EKG that aims to decrease the amount of time spent on accessing data sources. The second task is to maintain the EKG, which involves updating the EKG when changing the data without recomputing the relationships between all the data. The final task involves querying the EKG which can be achieved by using the Source Retrieval Query Language (SRQL). The SRQL Language makes use of several concepts that are discoverable elements (DE), an index (G-Index) and discovery primitives (DP) [22].

4.3 Schema Matching

Schema matching is the technique of receiving a dataset as input and returning the related datasets [5]. The main purpose of schema matching is to augment a dataset with the needed data. This technique is based on capturing the relations between the elements of various schemata. In addition, this technique can be used with tabular data or a tabular search to specify if at least two columns have joinable or unionable relations between them [5]. There are several types of matchers based on the information that users need, such as the attribute overlap matcher, semantic overlap matcher, data type matcher and embeddings matcher.

COMA. The COMA system involves a combination of schema-based matchers. This system is considered to be a generic match system that is appropriate for diverse schema types, including relational schemas or XML. The system concentrates on 1:1 matching relationships. The system uses rooted directed cyclic graphs to represent the schema elements. The match results give a collection of pairs which include the elements and score for corresponding similarity. Each matching iteration comprises three stages: user feedback, utilisation of diverse matchers and integration of the single match results [23]. In 2007, [42] extended the COMA system by including two instance-based matchers, which focus on using some constraints and linguistic approaches.

D3L. Dataset Discovery in Data Lakes (known as D3L) involves an approach proposed to contribute to solving the data discovery problem. The approach is based on a novel distance-based framework that identifies the relatedness between a target's features and the datasets in a data lake. This approach aims to extract several types of features [2]. The first type is "attribute name similarity", which focuses on the availability of schema–level information. Secondly, "attribute extent overlap" searches for attributes that share the same values. The third type is "word–embedding similarity", which concentrates on similar attributes in terms of semantics. The next type is "format representation similarity" that uses certain patterns of regular representation and requires the attribute values to follow these patterns. The final type is the similarity of domain distribution employed for numerical attributes. The proposed approach depends on locality-sensitive hashing (LSH), which is used to search for the "nearest–neighbours" in the spaces [2].

Valentine. As a proposed benchmark, Valentine aims to evaluate schema matching techniques for the purposes of dataset discovery. This work concentrates on evaluating 1:1 matching methods. The purpose of Valentine is to facilitate the development of new dataset discovery techniques via several steps: automating the components of the schema matching, adjusting the present techniques and suggesting new matching evaluation metrics. In addition, a ground truth is used to evaluate how well the methods rank matches. A classification of dataset relatedness has been developed to evaluate the methods based on two essential categories; the first category involves unionable relations and the second one includes joinable relations, which stock complimentary data for the conceptual entity [5]. To evaluate the schema matching methods, a set of datasets was selected. The datasets were collected and grouped into two categories: The first category included datasets which had fabricated dataset pairs and the second category involved real-world datasets, including WikiData and Magellan Data.

4.4 Table Search

A table search (or table retrieval) is a method used to search and retrieve datasets. A table search is the process of responding to a search query and providing a list of tables in which the input query is a collection of keywords or a table [26]. This type of search is classified into two types: a keyword-based search, where the query is a keyword, and a table-based search, where a table is an input query [26]. A table search can be used by professionals or experts because the input query can be an existing table, thereby indicating that they have sufficient understanding and knowledge of the required dataset/tables and they need to augment them. The table retrieval process is used in many fundamental tasks, including searching for tables, extending tables such as RDF triples, question answering on tables and augmenting tables [26]. There are a number of table retrieval techniques that use the typical document retrieval methods to rank retrieved tables, in which tables are considered as documents [15].

WebTables System. [29] developed a Web Table system that aims to search a corpus **of** tables on the Web by using keyword search. The implementation of it is based on the attribute correlation statistics database (ACSDb). This system concentrates on schemas that include a single relation. The search system allows users to input a query to receive a list of tables. To rank the results, several ranking functions of increasing complexity were developed, such as filterRank. The WebTables includes three techniques to address some schema-related problems. Firstly, the system helps the beginner database designers to create a relational schema by performing schema auto-complete. Secondly, WebTables allows synonyms to be found automatically through a synonym discovery algorithm. Finally, join-graph traversal is applied in this system to help users to navigate the massive number of schemas in the ACSDb [29].

STR. STR is an abbreviation for Semantic-Table-Retrieval, a modern table retrieval approach. The purpose of this approach is to retrieve a ranked list of tables based on a given keyword query by using novel methods of semantic matching. The fundamental idea of this approach is to overtake lexical matching by describing tables and queries in semantic space, and then measuring the semantic similarity of the outcome [25]. The proposed approach involves three main steps. The first step is content extraction that seeks to represent the raw content, whether a query or a table, as a collection of terms. The second step is semantic representations, in which the terms are embedded in the semantic space. There are two essential types of semantic space. The first type is bag-of-concepts and the second one is embeddings and also there are two alternative types. The final step of the proposed approach is the use of similarity measures that work to measure the similarity between a pair of query-table [25].

WTR. A test collection for Web Table Retrieval has been developed by [31], which is called WTR. The fundamental purpose of this collection is to evolve and improve the WTR domain in several aspects, such as the diversity, a rich context, concentration on the labels for multi-fields and reproducibility, which assist users in reproducing the baseline methods for a comparison of the methods. Several phases for the construction of this benchmark were implemented. The test collection was built by using the WDC Table Corpus 2015. For queries, 60 queries are used from [27] and [32]. In this test collection, some unsupervised methods were used to retrieve the top 20 tables for the whole set of queries that were given, including BM25. Therefore, 6,949 query-table pairs were retrieved to examine the methods. An annotator from Amazon's Mechanical Turk was asked to judge the relevance of the extracted tables in relation to the queries.

5 Discussion

Due to the spread of distribution of existing datasets and their common usage in many different domains, several search styles were developed to facilitate the process of searching for and retrieving datasets. In this work, we present the different styles observed in existing systems, as well as example implementations of each style. However, we note that each style makes fundamental assumptions about the particular domains or user groups. In this section, we expand on the factors that influence the choice of search style.

Discipline. Firstly, the search styles can be as different as the difference in disciplines and the extent of the users' needs to search for datasets. For example, DataMed, a biomedical data discovery system, is developed for users of the biomedical community; therefore, it uses some techniques for enabling expert users to define special characteristics and uses Boolean operators for specialised queries. Domain specific search is often a reflection of other factors for search type, as described below. Each domain has a different amount of user knowledge, data generation, ownership and release restrictions, end-task complexity and types of data generated. For instance, in the biomedical data discovery system, the users are computationally literate, but not database power users who are familiar with complex query writing. The data is generated via scientific research, which requires anonymized and redacted information to be published, but does not incentivize curation of that data for others to use. The tasks have a range of complexities from simple "find data from a specific experiment type" to very complex analysis and integration tasks. Data types are typically heterogeneous, containing semi-structured information, images, etc.

User Knowledge. Secondly, the used styles can largely depend on the level of user experience of the dataset search process. For instance, consider the difference in communities that use keyword versus schema matching search styles. The biomedical community described above is data literate, but not database, information retrieval or dataset search power users. As a community in general, they utilise keyword searches. Meanwhile, the database community, power-users, are comfortable using schema matching to augment existing databases. The search styles diverge depending on the targeted community or population for whom the tools were designed for.

Data Generation, Ownership and Release Restrictions. Who generates the dataset, the rules and restrictions they are operating under and how it is generated will also affect the final search style that can be employed. Consider the difference between the biomedical researchers above, who must publish their data to be compliant with a grant, are not incentivized through any of the personal success criteria to publish data. Effectively, publishing data becomes a thankless chore that takes time from activities that do count towards personal success criteria like publishing papers. On the other hand, many European governments have created a success metric around openness of government data and thus have created a reason to publish. Thus, how and by whom a dataset has been generated will change the search style needed, as some of them are more publicly available, have better resourcing and are more accessible.

Task Complexity. The search style used may also differ according to the extent of task complexity that is required to use a dataset, and the type of needed dataset, such as sort of tables, structured data or unstructured data. This relates to user knowledge, and the tools that are readily available and marketed to their specific community.

6 Related Work

A dataset search survey was published by Chapman et al. [9]. This survey is divided into several sections. One of the sections provides an overview of dataset search as well as dataset search definitions. Several types of searching, such as entity-centric search, centralised search and tabular search, are discussed. Moreover, many current dataset search implementations are mentioned, such as some scientific data systems and data marketplace systems. However, it focused on the systems and techniques in general without considering a taxonomy of those methods.

Another existing related work focuses on information access tasks, especially web tables. This work divided the information access tasks into several major categories: table search, table interpretation, table augmentation, table extraction, question answering and knowledge base augmentation, each of which included at least seven existing works [26]. In the table search tasks section, which is related to the dataset search domain, only two types of search styles were mentioned: keyword-based search and table-based search [26]. However, the focus of this survey is only web tables, and it only addresses two search styles.

[5] defines the matcher as the process of identifying relationships based on knowledge sources and divides the matchers into several categories. The first category of matchers is the attribute overlap matcher that focuses on determining the related columns when the attribute names comprise a syntactic overlap. The second category is the value overlap matcher that refers to the related columns when the corresponding values between them are overlapping. The third category is the semantic overlap matcher that concentrates on deriving the semantics labels of a particular column or the range of its values. The next category is the data type matcher that uses the data type, such as integers, to indicate the relevant or irrelevant columns. The fifth category involves a distribution matcher that utilises value distributions to specify relevant columns. The final category is the embeddings matcher that computes the similarity of identical values depending on the embeddings to determine the related columns [5]. Therefore, our taxonomy can complement their works to contribute to this field.

7 Conclusion and Future Work

Due to the large number of datasets that exist on the Web and the significance of finding them, a novel research field has emerged: "Dataset Search". This research field still needs more development and research to serve all domains. Furthermore, a taxonomy for dataset search based on search styles was constructed. This taxonomy includes four main sub-categories and there are several systems, algorithms and benchmarks under each category. With the rapid development of technology, the use of ML models has become popular in most domains. However, there is a considerable need for dataset search techniques that can be used for ML models by data scientists. In future work, we aim to discover the optimal methods for searching datasets and whether the search style is appropriate for dataset search for implementing a particular task. In addition, we intend to figure out the best search style for searching datasets for ML tasks.

References

1. Ibáñez, L., Simperl, E.: A comparison of dataset search behaviour of internal versus search engine referred sessions. In: ACM SIGIR Conference on Human Information Interaction and Retrieval, pp. 158–168 (2022)
2. Bogatu, A., Fernandes, A.A.A., Paton, N.W., Konstantinou, N.: Dataset discovery in data lakes. In: 2020 IEEE 36th International Conference on Data Engineering (ICDE), pp. 709–720. IEEE (2020)
3. Castelo, S., Rampin, R., Santos, A., Bessa, A., Chirigati, F., Freire, J.: Auctus. Proc. VLDB Endow. **14**, 2791–2794 (2021)
4. Färber, M., Leisinger, A.K.: Datahunter: a system for finding datasets based on scientific problem descriptions. In Fifteenth ACM Conference on Recommender Systems, pp. 749–752 (2021)
5. Koutras, C., et al. Valentine: evaluating matching techniques for dataset discovery. In: 2021 IEEE 37th International Conference on Data Engineering (ICDE), pp. 468–479. IEEE, 2021
6. Akujuobi, U., Zhang, X.: Delve: a dataset-driven scholarly search and analysis system. ACM SIGKDD Explor. Newsl **19**(2), 36–46 (2017)
7. Koesten, L.: A user centred perspective on structured data discovery. In: Companion Proceedings of the Web Conference 2018, pp. 849-853 (2018)
8. Lopez-Veyna, J.I., Sosa-Sosa, V.J., Lopez-Arevalo, I.: KESOSD: keyword search over structured data. In: Proceedings of the Third International Workshop on Keyword Search on Structured Data, pp. 23–31 (2012)
9. Chapman, A., et al.: Dataset search: a survey. VLDB J. **29**(1), 251–272 (2019). https://doi.org/10.1007/s00778-019-00564-x
10. Bhagavatula, C.S., Noraset, T., Downey, D.: Methods for exploring and mining tables on wikipedia. In: Proceedings of the ACM SIGKDD Workshop on Interactive Data Exploration and Analytics, pp. 18–26 (2013)
11. Hearst, M.: Search User Interfaces. Cambridge University Press, Cambridge (2009)
12. Kato, M.P., Ohshima, H., Liu, Y.H., Chen, H.L.: A test collection for ad-hoc dataset retrieval. In: Proceedings of the 44th International ACM SIGIR Conference on Research and Development in Information Retrieval, pp. 2450–2456 (2021)
13. Mibayashi, R., HuuLong, P., Matsumoto, N., Yamamoto, T., Ohshima, H.: Uhai at the ntcir-15 data search task. In: Proceedings of the NTCIR-15 Conference (2020)
14. Brickley, D., Burgess, M., Noy, N.: Google dataset search: building a search engine for datasets in an open Web ecosystem. In: The World Wide Web Conference, pp. 1365–1375 (2019)
15. Agarwal, V., Bhardwaj, A., Rosso, P., Cudre-Mauroux, P.: ConvTab: a context-preserving, convolutional model for ad-hoc table retrieval. 2021 IEEE International Conference on Big Data (Big Data) (2021)
16. Wilson, M., Schraefel, M., White, R.: Evaluating advanced search interfaces using established information-seeking models. J. Am. Soc. Inf. Sci. Technol. **60**, 1407–1422 (2009)

17. Kacprzak, E., Koesten, L., Ibáñez, L., Blount, T., Tennison, J., Simperl, E.: Characterising dataset search—an analysis of search logs and data requests. J. Web Semant. **55**, 37–55 (2019)
18. Kacprzak, E., Koesten, L., Tennison, J., Simperl, E.: Characterising Dataset search queries. In: Companion of the The Web Conference 2018 on The Web Conference 2018 - WWW 2018 (2018)
19. Kato, M.P., Ohshima, H., Liu, Y.H., Chen, H.L.: Overview of the NTCIR-15 data search task. In: Proceedings of the NTCIR-15 Conference (2020)
20. Nguyen, P., et al.: Nii table linker at the ntcir-15 data search task: re-ranking with pre-trained contextualized embeddings data content entity-centric and cluster-based approaches. In: Proceedings of the NTCIR-15 Conference (2020)
21. Cappuzzo, R., Papotti, P., Thirumuruganathan, S.: Creating embeddings of heterogeneous relational datasets for data integration tasks. In: Proceedings of the 2020 ACM SIGMOD International Conference on Management of Data, pp. 15–1349 (2020)
22. Fernandez, R.C., Abedjan, Z., Koko, F., Yuan, G., Madden, S., Stonebraker, M.: Aurum: a data discovery system. In: 2018 IEEE 34th International Conference on Data Engineering (ICDE), pp. 1001–1012. IEEE (2018)
23. Do, H., Rahm, E.: COMA—a system for flexible combination of schema matching approaches. In: VLDB 2002: Proceedings of the 28th International Conference on Very Large Databases, pp. 610–621 (2002)
24. Sansone, S., Gonzalez-Beltran, A., Rocca-Serra, P., Alter, G., Grethe, J., Xu, H., Fore, I., Lyle, J., Gururaj, A., Chen, X., Kim, H., Zong, N., Li, Y., Liu, R., Ozyurt, I., Ohno-Machado, L.: DATS, the data tag suite to enable discoverability of datasets. Scientific Data. 4 (2017)
25. Zhang, S., Balog, K.: Ad hoc table retrieval using semantic similarity. In: Proceedings of the 2018 World Wide Web Conference, pp. 1553–1562 (2018)
26. Zhang, S., Balog, K.: Web table extraction, retrieval, and augmentation: a survey. ACM Trans. Intell. Syst. Technol. **11**, 1–35 (2020)
27. Cafarella, M., Halevy, A., Khoussainova, N.: Data integration for the relational web. Proc. VLDB Endow. **2**, 1090–1101 (2009)
28. Yakout, M., Ganjam, K., Chakrabarti, K., Chaudhuri, S.: Infogather: entity augmentation and attribute discovery by holistic matching with web tables. In: Proceedings of the 2012 ACM SIGMOD International Conference on Management of Data, pp. 97–108. (2012)
29. Cafarella, M., Halevy, A., Wang, D., Wu, E., Zhang, Y.: WebTables. Proc. VLDB Endow. **1**, 538–549 (2008)
30. Pimplikar, R., Sarawagi, S.: Answering table queries on the web using column keywords. Proc. VLDB Endow. **5**, 908–919 (2012)
31. Chen, Z., Zhang, S., Davison, B.D.: WTR: a test collection for web table retrieval. In: Proceedings of the 44th International ACM SIGIR Conference on Research and Development in Information Retrieval, pp. 2514–2520 (2021)
32. Venetis, P., et al.: Recovering semantics of tables on the web (2011)
33. Chen, Z., Trabelsi, M., Heflin, J., Xu, Y., Davison, B.D.: Table search using a deep contextualized language model. In: Proceedings of the 43rd International ACM SIGIR Conference on Research and Development in Information Retrieval, pp. 589–598 (2020)
34. Okamoto, T., Miyamori, H.: Ksu systems at the ntcir-15 data search task. In: Proceedings of the NTCIR-15 Conference (2020)
35. Suadaa, L.H., Maghfiroh, L.R., Fauzi, I.N., Mariyah, S.: Stis at the ntcir-15 data search task: document retrieval re-ranking. In: Proceedings of the NTCIR-15 Conference (2020)
36. Calvanese, D., et al.: Ontop: Answering SPARQL queries over relational databases. Semant. Web **8**(3) (2017)
37. Shraga, R., Roitman, H., Feigenblat, G., Cannim, M.: Web table retrieval using multimodal deep learning. In: Proceedings of the 43rd International ACM SIGIR Conference on Research and Development in Information Retrieval, pp. 1399–1408 (2020)

38. Winn, J.: Open data and the academy: an evaluation of CKAN for research data management (2013)
39. Chen, X., et al.: DataMed – an open source discovery index for finding biomedical datasets. J. Am. Med. Inform. Assoc. **25**, 300–308 (2018)
40. Halevy, A., et al.: Goods: Organizing google's datasets. In: Proceedings of the 2016 International Conference on Management of Data, pp. 795–806 (2016)
41. Jain, A., Doan, A.H., Gravano, L.: SQL queries over unstructured text databases. In: 2007 IEEE 23rd International Conference on Data Engineering, pp. 1255–1257. IEEE (2007)
42. Engmann, D., Massmann, S.: Instance matching with COMA++. In: BTW Workshops, vol. 7, pp. 28–37 (2007)
43. Melnik, S., Garcia-Molina, H., Rahm, E.: Similarity flooding: a versatile graph matching algorithm and its application to schema matching. In: Proceedings 18th International Conference on Data Engineering, pp. 117–128. IEEE (2002)

Adoption Model for Cloud-Based E-Learning in Higher Education

Qasim AlAjmi[✉] [iD]

Department of Education, College of Arts and Humanities, A'Sharqiyah University, Ibra, Oman
Alajmi.qasim@gmail.com

Abstract. Regarding the use of e-learning in higher education institutions, there is a developing trend. A significant upfront infrastructure with several establishments is needed for an e-learning framework. To keep up with the fast change that is essential to globalization, higher education institutions (HEIs) must overcome various obstacles. The classic e-learning approach is therefore insufficient now because of all the difficulties and challenges. The appealing cloud-based e-learning (CBEL) platform offers a scalable and flexible e-learning method that can be accessed from any location, at any time, and using any device. The proposed framework has been built based on various elements: technological evaluation, readiness evaluation, and information culture factors. These were extracted from two well-known technology adoption theories, the Fit-Viability Model (FVM) and the Diffusion of Innovation (DOI) model. Information culture (IC) elements were chosen as a consideration since they have a substantial impact on the adoption of any technology. To investigate its substantial impact on the adoption of CBEL in HEIs in Oman as well as its viability, 14 proposed hypotheses were established. Data was collected from a sample of academics and IT professionals from 32 HEIs in Oman using a structured survey with standardised questions. The Statistical Package for Social Science (SPSS version 25) and Partial Least Squares (SmartPLS version3) were used to evaluate the proposed CBEL model and to examine the relationship between them. Based on the findings, which demonstrated that factors like fit, viability, and information culture strongly influenced HEIs' decisions to adopt CBEL in Oman, the final model was created. The resulting model demonstrated that 68.2% of the critical elements for CBEL adoption were addressed, and that by using this model, the quality of academic services could be improved by 56.1%.

Keywords: Cloud Computing · E-Learning · Cloud-based eLearning · HEIs · Oman

1 Introduction

Cloud computing refers to "sharing computing technology services that provide accessible resources such as storage, computing control, and applications delivered through the Internet as a service" [1]. Cloud computing is also defined as a technology or computing model that relies on the Internet and remote servers to help preserve and maintain data and applications. Various applications of cloud computing, such as Google Apps, G-Drive, Dropbox, Sky Drive, and I-Cloud can be easily integrated with education platforms [2, 3]. It brings new technological advancements where the technology services it provides become computing resources [4, 5]. Cloud computing has been attributed to the use of computing resources, specifically hardware and software, where they are delivered as a service over a network. It is considered highly scalable, with the possibility of creating virtualised resources that can be made available to users. The users do not need any special knowledge on cloud computing to interact with; they can easily communicate via the Internet with remote servers. The servers are able to exchange computing slots by themselves [6]. Cloud computing is one of the emerging technologies that has a significant impact on the learning environment in HEIs [7].

E-learning makes use of the Internet and different digital information on learning and teaching activities as key components in the implementation of modern educational technology. This learning process is implemented by ICT systems inside and outside classroom educational experiences on technology [8]. Terminologies such as Computer-Based Training (CBT), Web-Based Training (WBT), or Internet-Based Training (IBT) are most commonly associated with E-learning. E-learning enables the transfer of skills and knowledge to learners in less time compared to what they spend on traditional learning [9, 10]. E-Learning allows learners to access a broader view of materials in accordance with their personal competence devices with no limitation of time and space. By using any of the E-learning applications such as Web-based learning, virtual education opportunities, computer-based learning, and digital collaboration, all types of information can be transferred in media such as text, images, audio, animation, and video [11, 12]. In conclusion, Cloud Computing (CC) and E-Learning can be combined together to create a Cloud Based E-Learning (CBEL) platform where higher education institutions can benefit from having a modern form of E-Learning.

Cloud computing is a good alternative to running own-managed systems that HEIs should consider due to the relative lack of in-house IT expertise [13], in addition to hardware upgrades and software update issues. Political instability and the potential for irreparable damage to property during a war are other reasons why embarking on cloud computing is a viable option. Given the lower upfront cost of cloud computing services compared to the traditional method on the premise of IT infrastructure models, it has been asserted that the cloud model will not only provide significant development opportunities for the developing world but will go as far as to lessen the development gap with the West [14], as geographical factors will no longer dominate in determining who can and who cannot have access to leading-edge technologies.

1.1 E-Learning

Traditional web-based e-learning is found in most HEIs, which results in problems such as additional investment to maintain the system [15]. The traditional e-learning network consists of six parts, including content creation, Internal Protocol (IP), content management, utilisation and curriculum development, learning management, delivery, and development [16], and these take much time of the HEIs. A cloud computing platform provides an appropriate variety of computing resources with its attractive technology that is more reliable because of its scalability and effective usage in terms of resources. It can also be used under different circumstances with limited resources. The resources used can be applications, network servers, platforms and infrastructure segment, and services. Based on the discussion on the issues of traditional e-learning, cloud computing delivers its services based on the level of demand for sufficient access to the network, effective flexibility, and data resource [17].

1.2 Cloud-Based E-Learning

CBEL provides flexibility in all educational institutions, especially HEIs, and delivers an effective deployment model to fulfil the dynamic demands [18]. It can support a lot of educational institutions and help resolve some of the most common challenges, including quick and effective communication, cost reduction, privacy, security, flexibility, and accessibility. Moreover, CBEL is termed as the next new invention to help sustain the current information technology services and products that are on demand, and this is exactly what HEIs need - moving the processing task from a local device to data-centre facilities- [19]. Many researchers have indicated that cloud computing plays a significant role in changing the nature of learning and business, hence [20], one of these researches have examined the relationship between cloud-based e-learning (CBEL) adoption factors theoretically and empirically, and its influence on the quality of academic services (QAS) in HEIs. Moreover, this study aims to present an extended model of adoption in the context of HEIs in Oman by integrating well-known adoption theories. This study will cover the problem of adopting CBEL.

2 Research Model Constructs and Hypothesis

The proposed model was built based on selected theories, which consist of three dimensions: technology evaluation, readiness evaluation, and information culture factors. Technology evaluation and readiness evaluation were examined by the Fit-Viability Model (FVM) and the DOI model after excluding environmental factors to focus on factors which are essential components of cloud-based e-learning. Figure 1 shows the proposed model with its constructs and hypothesis.

3 Research Questions

The main question of this research is: How can the CBEL adoption model be successfully adopted and validated in HEIs in Oman? To answer this research question, the following sub-questions are to be addressed: Q1: What are the most significant factors that affect the adoption decision of CBEL in higher education institutions (HEIs)? Q2: How to develop an adoption model that can be used for CBEL and QAS in HEIs in Oman? Q3: What is the relationship between these factors, CBEL adoption, and the Quality of Academic Services (QAS) in HEIs in Oman? Q4: How can this model be validated to assess the adoptability of CBEL towards successful implementation for HEIs in Oman?

4 Literature Review

E-learning has found its place on many platforms today, including the higher education sector. Learners are often the main group involved in e-learning systems. Students benefit by using online courses, writing exams, sending feedback, sending projects, and interacting [8]. On the other hand, learners interact with the content, prepare tests, assess the tests, homework, and projects, and communicate the final feedback to the students. The standard e-learning system was developed as a distributed application. The learner in HEIs may use a mobile device or a desktop computer to access an application, which can be a simplified web browser or a dedicated one. However, none of these applications would run without technical requirements in the form of servers and other physical hardware that comprise the whole systems architecture, and these technical requirements are very costly [21]. Any higher educational institution may develop its own e-learning system without having to worry much about the up-front entry-cost that traditionally used to be huge. Adoption of CBEL will handle the e-learning for the HEIs without worrying about any technical requirements.

The impact and effectiveness of e-learning also vary with different types and nature of learning models deployed with technology [22]. Every model contributes differently and uniquely to augmenting the learning experience and learning outcomes of the students. The deployment of e-learning is a means to facilitate and manage learning. It involves the efficient utilisation of all the tools of a virtual learning environment like Moodle, web communication technology, and the Internet for accessing the online learning environment [23]. An electronic framework is also used to support and provide feedback to the students on their projects and assignments. It also facilitates the two-way communication between students and teachers, which is essential for the learning process. The deployment model in which the e-learning platform is used as a medium does not directly contribute to the learning process, but by providing easy and convenient access to the learning material and content, it influences the learning process of students in e-learning platforms.

Cloud computing is a utility computing approach for managing and delivering computer-related functions and services using the Internet [24]. The computing functions are operated and maintained in a data centre, and it extends its resources, capabilities, and services to clients based on subscription basis or size of data use. For example, information-sharing may conveniently occur by connecting people in the network, as cloud computing allows users to store and retrieve digital data on-demand from a remote server, otherwise known as cloud storage. There are cloud computing services (hosting and managing) by third-party providers that HEIs may eventually avail for their own e-learning systems [25]. An e-learning cloud-based architecture is divided into two main layers/parts: infrastructure layer and application layer [26]. Furthermore, four other modules, including the policy module, provision module, the monitoring module, and arbitration module, accompany it. This is discussed further in the next section, which elaborates all these layers and their benefits to e-learning.

5   Research Methodology

The tools for the data collection were divided into the following: document review, in-person interviews, and survey questionnaires. The researcher used a questionnaire as an instrument to collect primary data from higher education institutions and the targeted stakeholders in Oman. The questionnaire was used to collect the quantitative data; to do so, a survey method was applied to identify the population and sampling technique, develop, and validate the instrument.

5.1   The Population

A population in research includes individuals, objects, and events that qualify to be a sample in a study. This research aims to investigate significant factors influencing CBEL adoption in HEIs in Oman. It denotes that the topic under investigation is the use of technology (cloud-based) for a specific purpose (e-learning). For the scope of this study, the whole HEI population in Oman will be the population. Accordingly, 33 HEIs in Oman were targeted, with a total of 1,005 individuals targeted. The targeted population was the top, middle, and low management with IT basic knowledge at least, and cloud computing awareness. The study also includes technicians among the population.

5.2   Sampling Design

This targeted population was selected from different HEIs in the higher education institutions in Oman; those who have experience and relevant information about CBEL. According to [27], it is vital to use a sampling frame when selecting a sample size. The researcher got permission from the respective departments of the institutions to facilitate easier collaboration with the participants. Having identified the sampling method and the respondent's size to be used in the study, the study preceded the data analysis method. However, a pilot study was conducted beforehand. The total participants were N = 312. As long as the sampling method used is census, it means 312 out of 1,005 is valid and representative.

5.3   Validity and Reliability

The author used Cronbach's coefficient alpha to test for reliability of the research instruments as per the literature. The researcher conducted additional trials to determine the consistency of the results and verify the data. The researcher tested the reliability of the data to ensure that the outcome of the research can be used in a larger group and is not limited to only the participants taking part in the research. Carrying out the two tests will ensure that the researcher obtains first-hand information and the data collected will be reliable.

5.4   Pilot Study

With the supporting letter from the Ministry of Higher Education, the author was able to get the list of targeted samples' email IDs in three weeks with personal involvement.

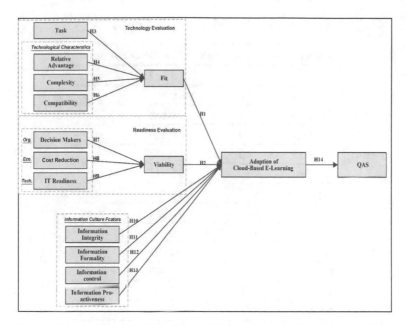

Fig. 1. Proposed Model of Cloud-Based E-Learning Adoption in Omani HEIs

The first group of the invitation was sent to 250 participants; however, 40 responses were received. The reliability test was conducted to assess the internal consistency between each scale measures. Cronbach's Alpha's reliability coefficient test in SPSS was used to test the reliability. According to [28], 0.7 is the satisfactory cut-off for Cronbach's Alpha criterion. Table showed that all the constructs had a satisfactory value (above 0.7) of Cronbach's Alpha, indicating that all the measurements are reliable and confirming the reliability of the instrument measures (Table 1).

5.5 Respondents' Profile

The sample frame for this study was restricted to all computer-knowledgeable staff in Omani HEIs. The results showed that the men are almost twice the women, 65% and 35%, respectively. The results also showed that 113 of the respondents are aged 34–40 (36%), 87 are 41–50 years old (28%), 76 are 26–33 years old (23%), 32 are 51 years or older (10%), and only 6 are in the range of 18–25 years old (2%). Relating to the educational degree of the respondents, Table 5–3 showed that 40.7% of the respondents have a Master of Science degree, 38.5% have a Ph.D. degree, 17% have a bachelor's degree, 8 have a diploma, and only 3 respondents have a high school certification or less. In addition, 113 of the respondents are lecturers, 78 are assistant professors, 24 are associate professors, and 5 are professors, while 49 are technicians. It also showed that most of the respondents (182) are IT, managers, while 24 are deans, 4 are owners, and the rest have different roles (101) under various positions (Faculty, Heads).

580   Q. AlAjmi

Table 1. Reliability Test (Pilot Study)

The Construct	No. Items	Cronbach's Alpha
Fit	5	0.927
Viability	3	0.888
InfoCont	5	0.896
InfoForm	4	0.880
InfoInt	5	0.942
InfoPro	5	0.931
RA	6	0.922
COMX	5	0.914
COMT	6	0.822
Task	4	0.883
Org	6	0.949
IT	6	0.936
Econ	5	0.960
Adoption	5	0.916
AQS	4	0.969

5.6 Descriptive Statistics of Instrument

The statistical results of the measurement items are shown in Table 2. As can be seen, all the items have mean values above 3.5 and Std. Deviation of at least 0.665 (for InfoCont5) and up to 1.018 (for IT5). This indicates that the responses did not deviate from the mean.

Table 2. Results of the measurement items

Item	N	Mean	Std. Deviation	Item	N	Mean	Std. Deviation
Task1	312	3.60	.775	IT1	312	3.96	.842
Task2	312	3.60	.815	IT2	312	3.67	.994
Task3	312	3.66	.786	IT3	312	3.71	.929
Task4	312	3.77	.773	IT4	312	3.67	.905
RA1	312	3.83	.967	IT5	312	3.61	1.018
RA2	312	3.82	.924	IT6	312	3.81	.876
RA3	312	3.91	.869	Vi1	312	3.75	.700
RA4	312	3.90	.849	Vi2	312	3.77	.772

(continued)

Table 2. (*continued*)

Item	N	Mean	Std. Deviation	Item	N	Mean	Std. Deviation
RA5	312	3.91	.875	Vi3	312	3.82	.719
RA6	312	3.90	.843	InfoInt1	312	3.89	.731
ComX1	312	3.73	.784	InfoInt2	312	3.65	.835
ComX2	312	3.68	.904	InfoInt3	312	3.74	.793
ComX3	312	3.65	.872	InfoInt4	312	3.79	.722
ComX4	312	3.79	.883	InfoInt5	312	3.89	.722
ComX5	312	3.74	.915	InfoForm1	312	3.82	.711
ComT1	312	3.58	.826	InfoForm2	312	3.86	.696
ComT2	312	3.69	.736	InfoForm3	312	3.61	.790
ComT3	312	3.68	.802	InfoForm4	312	3.83	.708
ComT4	312	3.77	.734	InfoCont1	312	3.70	.776
ComT5	312	3.73	.758	InfoCont2	312	3.82	.698
ComT6	312	3.74	.731	InfoCont3	312	3.80	.749
Fi1	312	4.02	.732	InfoCont4	312	3.90	.694
Fi2	312	3.71	.762	InfoCont5	312	3.88	.665
Fi3	312	3.84	.725	InfoPro1	312	3.90	.705
Fi4	312	3.68	.790	InfoPro2	312	3.78	.792
Fi5	312	3.85	.697	InfoPro3	312	3.83	.772
Org1	312	3.71	.787	InfoPro4	312	3.79	.796
Org2	312	3.65	.799	InfoPro5	312	3.84	.801
Org3	312	3.55	.840	Adopt1	312	4.15	.804
Org4	312	3.86	.804	Adopt2	312	4.10	.766
Org5	312	3.88	.795	Adopt3	312	3.90	.784
Org6	312	3.78	.809	Adopt4	312	3.78	.842
Eco1	312	3.74	.823	Adopt5	312	3.83	.822
Eco2	312	3.70	.832	ASQ1	312	4.04	.716
Eco3	312	3.75	.770	ASQ2	312	4.03	.743
Eco4	312	3.66	.752	ASQ3	312	3.98	.706
Eco5	312	3.77	.772	ASQ4	312	4.00	.748

6 Result and Conclusion

There are two critical t-values, 1.65 and 2.33. If the t-value is larger than the critical value, the coefficient is significant at a certain level. Accordingly, there are two levels of significance: 5% (t > 1.65) and 1% (t > 2.33). Table 3 presents the results of the

582   Q. AlAjmi

relationships and accordingly, the hypotheses testing. The results showed that hypotheses H5, H10, H11, and H13 were not supported, while the rest were supported.

Table 3. Hypotheses Testing Results

Relations	β	T Statistics	P Values	Result
Adoption → ASQ	0.749	16.967	0.000***	Supported
ComT → Fit	0.446	5.176	0.000***	Supported
Comx → Fit	0.319	4.515	0.000***	Supported
Econ → Viability	0.166	1.904	0.029**	Supported
Fit → Adoption	0.087	1.217	0.112	Not Supported
IT → Viability	0.259	3.262	0.001***	Supported
InfoCont → Adoption	0.243	2.286	0.011**	Supported
InfoForm → Adoption	0.11	1.364	0.086	Not Supported
InfoInt → Adoption	-0.029	0.279	0.390	Not Supported
InfoPro → Adoption	-0.037	0.509	0.305	Not Supported
Org → Viability	0.337	3.749	0.000***	Supported
RA → Fit	0.202	3.203	0.001***	Supported
Task → Fit	0.156	3.282	0.001***	Supported
Viability → Adoption	0.585	7.290	0.000***	Supported

The model was validated by the model fit, which showed an appropriate model validity. The level of the measurement models' quality was at an average commonality index value of 0.777, while the level of structural model quality was at an average redundancy index of 0.173. However, the goodness-of-fit of the model was large at 0.687. This showed that the global model validity was excellent.

The results of this study concluded that (1) there are many factors that should be considered before taking a decision on the adoption of cloud-based e-learning in HEIs in Oman. Most of them are related to infrastructure and cost. In addition, the result showed that (2) factors such as Task's requirements, Relative advantage, Less-Complexity, Compatibility, Cost reduction, Scalability, and Availability can be considered as drivers for the adoption of cloud-based e-learning in HEIs in Oman. This takes us to the point that (3) the higher education institutions in Oman can adopt cloud-based e-learning in order to overcome the challenges of delivering quality academic services (QAS) at the end. Consequently, this will help them for institutional accreditation and programmer accreditation. Therefore, the proposed model is needed to examine these most significant factors that influence the decision on the adoption of cloud-based e-learning in HEIs in Oman. (4) The final model of this study is important and can help decision makers of HEIs in developing countries like Oman decide on adopting CBEL successfully and at the right time to qualify its academic services. (5)The final model showed that 68.2% of the significant factors for the adoption of CBEL were covered, and that using this model will result in a 56.1% improvement in academic service quality.

References

1. Alajmi, Q., Arshah, R.A., Kamaludin, A., Al-Sharafi, M.A.: Current state of cloud-based e-learning adoption: results from gulf cooperation council's higher education institutions. In: 2018 IEEE 9th Annual Information Technology, Electronics and Mobile Communication Conference (IEMCON), pp. 569–575. IEEE (2018)
2. Thomas, P.Y.: Cloud computing a potential paradigm for practising the scholarship of teaching and learning. Electron Libr. **29**(2), 214–224 (2011). (in English). https://doi.org/10.1108/026 40471111125177
3. Yen, T.-F.T.: The performance of online teaching for flipped classroom based on COVID-19 aspect. Asian J. Educ. Soc. Stud., 57–64 (2020)
4. Chengyun, Z.: Cloud security: the security risks of cloud computing, models and strategies, programmer (2010)
5. Al Ajmi, Q., Arshah, R.A., Kamaludin, A., Sadiq, A.S., Al-Sharafi, M.A.: A conceptual model of e-learning based on cloud computing adoption in higher education institutions. In: 2017 International Conference on Electrical and Computing Technologies and Applications (ICECTA), pp. 1–6. IEEE (2017)
6. Alsaadi, H.: Giving voice to the voiceless: learner autonomy as a tool to enhance quality in teaching and learning in higher education. In: Oman National Quality Conference, Research Gate, Muscat (2012)
7. Al-Sharafi, M.A., AlAjmi, Q., Al-Emran, M., Qasem, Y.A., Aldheleai, Y.M.: Cloud computing adoption in higher education: an integrated theoretical model. In: Recent Advances in Technology Acceptance Models and Theories, pp. 191–209 (2021)
8. Almazroi, A.A., Shen, H., Teoh, K.-K., Babar, M.A.: Cloud for e-Learning: determinants of its adoption by university students in a developing country. In: 2016 IEEE 13th International Conference e-Business Engineering (ICEBE), Macau, China, pp. 71–78. IEEE (2016). https://doi.org/10.1109/ICEBE.2016.022. https://ieeexplore.ieee.org/document/7809903/
9. Riahi, G.: E-learning systems based on cloud computing: a review. Procedia Comput. Sci. **62**, 352–359 (2015). https://doi.org/10.1016/j.procs.2015.08.415
10. Abdelazim, A., AlAjmi, Q., AlBusaidi, H.: The effect of Web 2.0 applications on the development of educational communication activities in teaching. Int. J. Inf. Technol. Lang. Stud. **5**(2) (2021)
11. Sai Hemanth, G., Noor Mahammad, S.: An efficient virtualization server infrastructure for e-schools of India. In: Satapathy, S.C., Mandal, J.K., Udgata, S.K., Bhateja, V. (eds.) Information Systems Design and Intelligent Applications. AISC, vol. 434, pp. 89–99. Springer, New Delhi (2016). https://doi.org/10.1007/978-81-322-2752-6_8
12. AlAjmi, Q., Al-Sharafi, M.A., Yassin, A.A.: Behavioral intention of students in higher education institutions towards online learning during COVID-19. In: Arpaci, I., Al-Emran, M., A. Al-Sharafi, M., Marques, G. (eds.) Emerging Technologies During the Era of COVID-19 Pandemic. SSDC, vol. 348, pp. 259–274. Springer, Cham (2021). https://doi.org/10.1007/978-3-030-67716-9_16
13. Sharma, S.K., Al-Badi, A.H., Govindaluri, S.M., A-Kharusi, M.H.: Predicting motivators of cloud computing adoption: a developing country perspective. Comput. Hum. Behav. **62**, 61–69, September 2016. https://doi.org/10.1016/j.chb.2016.03.073
14. Pett, T.L., Wolff, J.A., Perry, J.T.: Information technology competency in SMEs: an examination in the context of firm performance. Int. J. Inf. Technol. Manage. **9**(4), 404–422 (2010). https://doi.org/10.1504/IJITM.2010.035462

15. Lian, J.W.: Critical factors for cloud based e-invoice service adoption in Taiwan: an empirical study. Int. J. Inf. Manag. **35**(1), 98–109 (2015). https://doi.org/10.1016/j.ijinfomgt.2014.10.005

16. Aung, T.N., Khaing, S.S.: Challenges of implementing e-learning in developing countries: a review. In: Zin, T.T., Lin, J.-W., Pan, J.-S., Tin, P., Yokota, M. (eds.) Genetic and Evolutionary Computing, pp. 405–411. Springer, Cham (2016). https://doi.org/10.1007/978-3-319-23207-2_41

17. Elgelany, A., Alghabban, W.G.: Cloud computing: empirical studies in higher education a literature review. Int. J. Adv. Comput. Sci. **8**(10), 121–127 (2017). <Go to ISI>://WOS:000416609500018

18. Subashini, S., Kavitha, V.: A survey on security issues in service delivery models of cloud computing. J. Netw. Comput. Appl. **34**(1), 1–11 (2011). https://doi.org/10.1016/j.jnca.2010.07.006

19. Alkamel, M.A.A., Chouthaiwale, S.S., Yassin, A.A., AlAjmi, Q., Albaadany, H.Y.: Online testing in higher education institutions during the outbreak of COVID-19: challenges and opportunities. In: Arpaci, I., Al-Emran, M., A. Al-Sharafi, M., Marques, G. (eds.) Emerging Technologies During the Era of COVID-19 Pandemic. SSDC, vol. 348, pp. 349–363. Springer, Cham (2021). https://doi.org/10.1007/978-3-030-67716-9_22

20. Priyadarshinee, P., Raut, R.D., Jha, M.K., Gardas, B.B.: Understanding and predicting the determinants of cloud computing adoption: a two staged hybrid SEM - Neural networks approach. Comput. Hum. Behav. **76**, 341–362 (2017). https://doi.org/10.1016/j.chb.2017.07.027

21. Abbas, A., Bilal, K., Zhang, L.M., Khan, S.U.: A cloud based health insurance plan recommendation system: a user centered approach. Future Generation Comput. Syst. Int. J. Escience **43–44**(1), 99–109 (2015). https://doi.org/10.1016/j.future.2014.08.010

22. Al-Samarraie, H., Saeed, N.: A systematic review of cloud computing tools for collaborative learning: Opportunities and challenges to the blended-learning environment. Comput. Educ. **124**, 77–91 (2018). https://doi.org/10.1016/j.compedu.2018.05.016

23. Al Musawi, A.S., Abdelraheem, A.Y.: E-learning at Sultan Qaboos University: status and future. Brit. J. Educ. Technol. **35**(3), 363–367 (2004). https://onlinelibrary.wiley.com/journal/14678535

24. Akande, A.O., Van Belle, J.-P.: Cloud computing in higher education: a snapshot of software as a service. In: 2014 IEEE 6th International Conference Adaptive Science & Technology (ICAST), Covenant University, Ota, Nigeria, 29–31 October 2014, pp. 1–5. IEEE. punumber=1800010

25. Rovai, A.P., Ponton, M.K., Wighting, M.J., Baker, J.D.: A comparative analysis of student motivation in traditional classroom and e-learning courses. Int. J. ELearn. **6**(3), 413 (2007). ERIC Number: EJ763593

26. Pulier, E., Martinez, F., Hill, D.C.: System and method for a cloud computing abstraction layer, Google Patents (2015)

27. Kothari, C.R.: Research methodology: Methods and techniques. New Age International (2004)

28. Hair, J.F., Anderson, R.E., Tatham, R.L., William, C.: Multivariate data analysis. Pearson, New Jersey (2010)

Deep Learning-Based System for Quality Control of Coatings in Recess Punch Manufacturing

Balint Newton Turcsanyi[(✉)], Faisal Saeed, and Emmett Cooper

School of Computing and Digital Technology, Birmingham City University, Birmingham B4 7AP, UK

balint.turcsanyi@mail.bcu.ac.uk, {faisal.saeed, emmett.cooper}@bcu.ac.uk

Abstract. Increasing efficiency of the quality inspection process is an on-going pursuit in all manufacturing-related industries. The research was proposed by Tooling International ltd – a company situated in the UK – in an attempt to solve a decade-long problem faced when undertaking quality inspection of their coated products. The main objective of this research is to develop a model that detects faulty products with unsatisfactory coating. In this research, several convolutional neural network (CNN) architectures were tested in order to find the most suitable one for this particular task. The best performing CNN model delivered 97.68% accuracy which exceeded the company's requirements, providing superior accuracy to when compared to current company methods. This study will be used to develop an automated quality inspection machine, thus enhancing the company's productivity, and will potentially be used as the foundation of further AI based developments in similar manufacturing-related tasks.

Keywords: Computer Vision · Neural Network · Quality Control · Recess Punch Manufacturing

1 Introduction

Artificial intelligence (AI) garners increasing attention throughout various fields from healthcare to manufacturing, due in-part to neural networks achieving human level performance and, in some cases even outperforming humans in specialised tasks [1]. The advantages of applying AI within manufacturing are indisputable, this has been recognised by industry as shown by the rapid growth of research literature on the topic of 'AI in manufacturing' [3].

AI-powered computer vision is a specific application of AI heavily used in myriad ways in manufacturing, from stock management [2], personal hygiene in factories [4], sales analysis and quality control [5]. Using AI for quality control can help factories increase their performance; by using them, human errors can be completely eliminated and the automated system benefits by being highly productive around the clock.

This research has been proposed by Tooling international ltd. Company in the UK, in an attempt to resolve one of their longest ongoing problems with quality control; namely, the detection of an incomplete or damaged coating layer for their products. Their current methods of detecting damaged products is estimated to have 97% accuracy. The company is specialised in the manufacturing of recess punches used by the fastening industry. Tooling international ltd. is one of the over 400 companies in the Germany-based Würth Group.

Figure 1 shows two examples of recess punches, with the critical areas circled. The top portion of the parts with the hexagonal-like shape is the area where any flaws in the coating would significantly reduce the tool's life expectancy, as the coating is not only providing protection from oxidation, but also reduces the friction between the part and the surface of the screw on impact.

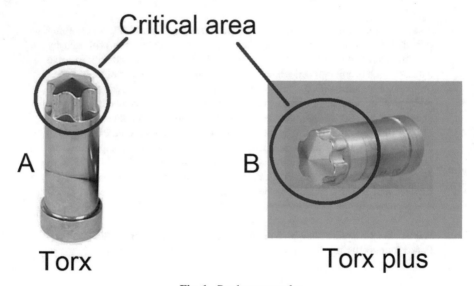

Fig. 1. Product examples

2   Related Works

When conducting a literature review some of the widely available databases have been used such as: Google Scholar, IEEE, ScienceDirect and MDPI, along with the printed resources of the Birmingham City University's library. When searching for related articles online keywords were used such as: AI and quality control, AI in manufacturing, deep learning in quality control, deep learning and computer vision in quality inspection, and articles mainly from the period of 2018 and 2022. The research is heavily based on computer vision-related topics as considering the nature of the problem, this is one of the proposed methods for detecting the incomplete or damaged coating layer of the company's products. In computer vision, the two types of technologies used are the RGB imaging (imaging in the visible light spectrum) and the hyperspectral imaging (imaging in the invisible spectrum). In light of the nature of the defects this research aims to resolve, hyperspectral analysis is not necessary as the colours of the coating (yellow) and the base material (silver) are completely different and easily distinguishable in the visible light spectrum, using hyperspectral imaging would introduce unnecessary extra financial burden to the company.

A study on object detection using DOTA and HRSC2016 datasets proposed the use of oriented R-CNN, creating oriented bounding boxes around objects rather than the old-fashioned horizontal ones [14]. The method used in this research is worth consideration for the research proposed in this study as it has produced good accuracy as a result of the rotated bounding boxes include less of the background regions and have less chance to include multiple objects (in our case coating faults).

[5] conducted a research on a very similar domain and examined different AI models on small part defect detection achieved 98% accuracy in detecting crocked shapes, 99% in the detection of length-, 97.8% in detection of endpoint-size errors and 79.4% at detecting wringing using single short detector network (SSD) and deep learning (DL) algorithm. The research compared YOLO V3, Faster-RCNN, FPN and the SSD algorithms and concluded with the SSD providing the highest accuracy. The data augmentation technique used by the authors was image rotation (0°, 22.5°, 45°, 67.5°, and 90°), this way the authors increased the size of the original dataset by roughly 50%.

A CNN-based research on small scale fault detection achieved remarkable results using for real-time micrometer scale detection of pits (defects on digital display LCD-films occurring during their manufacturing process, mainly caused by dust- and sand contamination) [15]. This research is important for the proposed research not only because the defects our research aims to detect is also on a small scale (borderline microscopic) but also because they developed a model that uses real-time image feed and, in the future, the same method will be examined and potentially used for this project when building the automated quality inspection machine (might not be possible, depending on the performance of the available hardware). Their model predicts boundary boxes around the candidate defects using an image processing algorithm optimised on pits, and after extracting those patches from a high-resolution image feed it applies a binary classification to detect defects in those patches.

Another similar research on this domain was focusing on the quality control in the food industry called suggested the use of hyperspectral imaging as it would increase the range of detectable defects [16]. This can be useful for future development in this particular domain, for example when a similar model is to be built for some slightly different defects such as the detection of cracking, inner holes etc., however it has been qualified as unnecessary by the authors for this project.

A study on deep learning (DL) computer vision system for the quality inspection of printing cylinders [13] achieved a 98.4% accuracy (only accuracy, the study does not mention other measures of the model's performance) and suggest that this result could be further improved by retraining the model using more data on the falsely classified faults.

The main objective of this study was to conduct a review of existing models, once the database is available. This review was focusing on different CNN models as based on several journals in the topic they are the ones most likely to deliver the highest accuracy scores in image classification tasks [9]. Then, as the image-based data is currently not available, the authors created a dataset by using an automated image generator. An explanation of how the image generator works is in the Methods section. The aim is to detect high accuracy rate of the faulty part. During this process, several data augmentation- and feature extraction techniques were tested. The final step was to evaluate and compare the results of all models to choose the one delivering the best accuracy and the least of loss to decide which model to be deployed. The models were evaluated with confusion matrix, measurement of accuracy, precision, recall and F1 score, as the most commonly used accuracy on its own is not enough to justify the model's performance [12].

3   Methods

This section is a detailed description of the methods and machines were be used to conduct this research from data collection to model evaluation.

3.1   Data Collection

As the image-based dataset is currently not available, it was the authors responsibility to generate it. For this task an automated image generator was built seen on Fig. 2. The image generator is consisted of a stepper motor powered by an Arduino and a standard industrial USB camera. The stepper motor with a small magnet attached to it ensured that the parts are held steadily in the same position and rotated with a uniform angle between taking the images. The angle of which the parts were rotated by was set to be 0.7 degrees. All together 3496 images were taken and that was divided into 2806 and 690 images for training and validation.

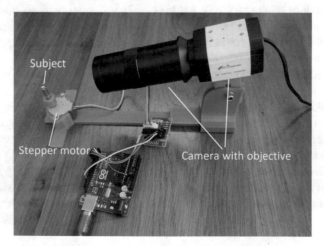

Fig. 2. The automated data gathering machine

3.2 The Dataset

Figure 3 shows an example of a part with a faulty coating from the dataset. The resolution of the raw images was 1280 by 720.

Fig. 3. An example of the generated images from the dataset

3.3 Pre-processing and Data Augmentation

It is clear that the raw images contain a lot of unnecessary information therefore cropping has been applied to the dataset decreasing the resolution from 1280 by 720 to 810 by 538 resulting in a ~53% decrease of the amount of data (pixels) the models needed to process. Figure 4 shows an example of a cropped image compared to a raw one.

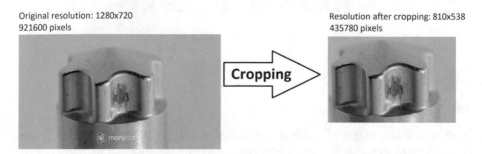

Fig. 4. Visualisation of cropping

During the experiments some data augmentation techniques has been applied to the data, such as rotation and flip of the images, converting them from RGB to grey scale and adding some random noise, this would increase the size of the dataset likely enhancing the overall accuracy of the model [8].

3.4 The Proposed AI Models

The most commonly used AI models for image classification is the convolutional neural network (CNN) [11]. Neural networks consist of an input layer, an output layer and one or more hidden layers. Convolutional neural networks are a type of neural networks that have specifically been designed to process images by having convolution kernel(s) and as a result, these types of models need image inputs. Convolution kernels are two dimensional matrixes with the weighted sum of the pixels of the image as values [7]. Figure 5 shows a diagram of a typical CNN model. CNNs are consisted of several convolutional layers with ReLU activation, pooling layers, a flattening function for feature extraction and this produces the input for the fully connected layer (NN) that performs the classification. The convolution layers pass their values through an activation function to the next layer. Rectifier linear activation function (ReLU) is a function that passes positive values through unchanged but passes 0 if it receives an input below 0. Pooling is a down-sampling step that is used to reduce dimensions and therefore the computational power requirements, by manipulating its value (maxpooling) overfitting can be reduced.

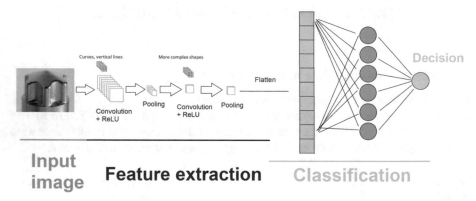

Fig. 5. The CNN Model

3.5 The Model Evaluation

In order to obtain a clear view on the accuracy of the model, a database of images were created of their products with acceptable- and unacceptable coating, this way a labelled database will be available in case should the company wish to test any other machine learning model for the same task in the future. To visualise the performance of the model, a confusion matrix and accuracy/loss graphs were used, and it is expected to achieve 97% and above in success rate in detecting faulty parts.

Accuracy measurement is done by the following equation:

$$\frac{TP + TN}{TP + FP + TN + FN}$$

where TP (true positive): The model successfully detected a faulty product as faulty. TN (true negative): The model successfully detected a not faulty product as not faulty. FP (false positive): The model falsely classified a good part as faulty. FN (false negative): The model failed detecting a faulty part. Out of these possibilities the false negative is the most damaging scenario as it allows faulty parts to proceed without detection, potentially allowing the faulty product to reach the customers.

Precision is the percentage of TP out of all positive findings. Measurement of the precision is done by the equation:

$$\frac{TP}{TP + FP}$$

Recall is the ratio of the true positive instances out of all positive ones, this gives an understanding of how many positive ones the model missed, it is calculated by the equation:

$$\frac{TP}{TP + FN}$$

F1 score is the harmonic mean average of the recall and the precision, it is calculated by the equation:

$$\frac{2 * precision * recall}{precision + recall}$$

F1 score can be useful, however as it gives equal importance to both precision and recall it does not always the best indicator, in some cases weighted F1 score is more suitable [12].

4 Results

Table 1 summarises the results of the 7 tested CNN models. The findings prove the strong viability of CNN in this domain as even the worst performing one achieved over 95%.

The best performing model (model 2 on Table 1) delivered 97.86% accuracy and it consisted of 3 convolutional layers, 7 by 7 kernel size in the first one and 3 by 3 in the rest and 3 fully connected layers with 512, 256, 64 neurons each followed by a maxpooling layer of 2 by 2.

As a last step of evaluation, a set of 215 new images have been created and the model was requested to deliver class prediction on them. Figure 6 shows the resulting confusion matrix where class 0 is the part with defect and class 1 is the good part.

The confusion matrix shows that the model was capable to detect the faulty parts effectively but classified 14 good parts out of 112 to be faulty.

Precision: 1.0
Recall: 0.875.
F1 score: 0.93.

Table 1. The summary of 7 CNN models tested during the study

CNN model ID	Conv layers	Fully connected layers	Input shape	Validation loss	Accuracy
1	3 conv layers with kernel 7x7 in the first 3x3 in the rest	4 layers with 512, 512, 256, 256 neurons	300x300	0.075	97.68%
2	3 conv layers with kernel 3x3 in all	3 layers with 512, 256 ,64 neurons	300x300	0.0362	97.68%
3	3 conv layers with kernel 3x3 in all	2 layers with 32, 32 neurons	300x300	0.2895	95.94%
4	3 conv layers with kernel 5x5 in the first 3x3 in the rest	2 layers with 32, 32 neurons	300x300	0.0462	97.39%
5	3 conv layers with kernel 5x5 in the first 3x3 in the rest	2 layers with 64, 64 neurons	300x300	0.07	97.54%
6	3 conv layers with kernel 5x5 in the first 3x3 in the rest	2 layers with 64, 64 neurons	500x500	0.11	96.23%
7	2 conv layers with kernel 5x5 in the first 3x3 in the second	2 layers with 16, 16 neurons	300x300	0.1053	96.67%

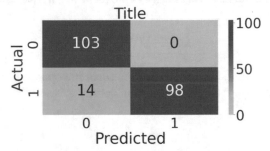

Fig. 6. The model's confusion matrix

5   Conclusion and Future Work

The study concluded that model 2 outperforms the currently used methods by the company and has strong viability to be the foundation of an automated quality inspection machine. As this study aims to provide an algorithm to work in manufacturing, its speed is highly important, it is worth considering applying image segmentation as it is faster and less computationally expensive. To achieve this YOLOv5 model will be tested in future experiments. It is possible that YOLOv5 will deliver even higher accuracy than the current findings.

References

1. Thomsen, M.: Microsoft's deep learning project outperforms humans in image recognition [WWW Document]. Forbes (2015). https://www.forbes.com/sites/michaelthomsen/2015/02/19/microsofts-deep-learning-project-outperforms-humans-in-image-recognition/. Accessed 17 June 2022
2. Mathew, J., Joseph, B.: A study into the use of artificial intelligence in E-commerce stock management and product suggestion generation for end users, sections III.C and IV.D (2021). https://doi.org/10.5281/ZENODO.5100977
3. Zeba, G., Dabić, M., Čičak, M., Daim, T., Yalcin, H.: Technology mining: artificial intelligence in manufacturing. Technol. Forecast. Soc. Change **171**, 120971. https://doi.org/10.1016/j.techfore.2021.120971. Accessed 17 June 2022
4. Fujitsu Develops AI-Video Recognition Technology to Promote Hand Washing Etiquette and Hygiene in the Workplace - Fujitsu Global [WWW Document], 26 May 2020. https://www.fujitsu.com/global/about/resources/news/press-releases/2020/0526-01.html. Accessed 17 June 2022
5. Yang, J., Li, S., Wang, Z., Yang, G.: Real-time tiny part defect detection system in manufacturing using deep learning. IEEE Access **7**, 89278–89291 (2019). https://doi.org/10.1109/ACCESS.2019.2925561
6. Boehm, B.W.: A spiral model of software development and enhancement. Computer **21**, 61–72 (1988). https://doi.org/10.1109/2.59
7. Lei, X., Pan, H., Huang, X.: A dilated CNN model for image classification. IEEE Access **7**, 124087–124095 (2019). https://doi.org/10.1109/ACCESS.2019.2927169
8. Li, W., Chen, C., Zhang, M., Li, H., Du, Q.: Data augmentation for hyperspectral image classification with deep CNN. IEEE Geosci. Remote Sens. Lett. **16**, 593–597 (2019). https://doi.org/10.1109/LGRS.2018.2878773
9. Hasan, H., Shafri, H.Z.M., Habshi, M.: A comparison between support vector machine (SVM) and convolutional neural network (CNN) models for hyperspectral image classification. IOP Conf. Ser. Earth Environ. Sci. **357**, 012035 (2019). https://doi.org/10.1088/1755-1315/357/1/012035
10. Pawar, D.: Improving Performance of Convolutional Neural Network! Medium (2020). https://medium.com/@dipti.rohan.pawar/improving-performance-of-convolutional-neural-network-2ecfe0207de7. Accessed 21 June 2022
11. Le, J.: The 4 convolutional neural network models that can classify your fashion images [WWW Document]. Medium (2018). https://towardsdatascience.com/the-4-convolutional-neural-network-models-that-can-classify-your-fashion-images-9fe7f3e5399d. Accessed 21 June 2022
12. Nighania, K.: Various ways to evaluate a machine learning models performance [WWW Document]. Medium (2019). https://towardsdatascience.com/various-ways-to-evaluate-a-machine-learning-models-performance-230449055f15. Accessed 26 June 2022
13. Villalba-Diez, J., Schmidt, D., Gevers, R., Ordieres-Meré, J., Buchwitz, M., Wellbrock, W.: Deep learning for industrial computer vision quality control in the printing industry 4.0. Sensors **19**, 3987 (2019). https://doi.org/10.3390/s19183987
14. Xie, X., Cheng, G., Wang, J., Yao, X., Han, J.: Oriented R-CNN for object detection. In: 2021 IEEE/CVF International Conference on Computer Vision (ICCV). Presented at the 2021 IEEE/CVF International Conference on Computer Vision (ICCV), IEEE, Montreal, QC, Canada, pp. 3500–3509 (2021). https://doi.org/10.1109/ICCV48922.2021.00350

15. Ban, G., Yoo, J.: RT-SPeeDet: real-time IP–CNN-based small pit defect detection for automatic film manufacturing inspection. Appl. Sci. **11**, 9632 (2021). https://doi.org/10.3390/app11209632
16. De Ketelaere, B., Wouters, N., Kalfas, I., Van Belleghem, R., Saeys, W.: A fresh look at computer vision for industrial quality control. Qual. Eng. **34**, 152–158 (2022). https://doi.org/10.1080/08982112.2021.2001828

Information Systems

FM_STATE: Model-Based Tool for Traceability, Generation, and Prioritization in Software Product Line Engineering

Rabatul Aduni Sulaiman[1](✉) [iD], Dayang Norhayati Abang Jawawi[2] [iD],
and Shahliza Abdul Halim[2] [iD]

[1] Software Engineering Department, Faculty of Computer Science and Information System,
Universiti Tun Hussein Onn Malaysia, Johor, Malaysia
rabatul@uthm.edu.my

[2] Software Engineering Department, School of Computing, Faculty of Engineering,
Universiti Teknologi Malaysia, Johor, Malaysia

Abstract. Software Product Line (SPL) testing requires constant improvement due to current demands. This technique is a challenging task requiring needs of multiple quality methods. Model-Based Testing (MBT) is a method that can reduce testing effort. However, the existing approach emphases on single-phase testing, resulting to only minor improvement in testing result. To handle this issue, this study introduces *FM_STATE*, a tool that comprised three phases of testing which are mapping features through behavior elements, generating test cases, and prioritizing test cases. This method is analyzed by using five evaluation methods based on maximization and minimization. The exclusivity of *FM_STATE* is accepts the feature and behavior elements from.xml files and allows the management of all kinds of SPL model sizes from small to large.

Keywords: Software Product Line · Software testing · Test case generation · Test case prioritization · Hybrid algorithm

1 Introduction

Software Product Line (SPL) focus on findings commonality and variability of related software product to achieve software reusability [1]. SPL engineering is applied in industrial domains for example in automotive or telecommunication domain. Exploitation of commonality and variability components causes testing issues for the product [2]. For example, individual testing requires extra effort due to the development of various new products. Model-Based Testing (MBT) was introduced to tackle this issue [3]. MBT works by discovering faults in test cases and, at the same time tried to reduce effort and cost of testing. In addition, MBT offers faster automatic generation of test cases.

There are two main processes in the scope of SPL which are domain engineering and application engineering [4]. Domain engineering focus on producing reusable core assets based on SPL commonality and variability, while application engineering focus

F. Saeed et al. (Eds.): ICACIn 2022, LNDECT 179, pp. 597–606, 2023.
https://doi.org/10.1007/978-3-031-36258-3_53

on production of SPL products. Domain engineering field requires effective technique in creating reusable assets. Among the challenges, important issue is related with poor variability impressibility caused incorrect clarification in the test model that reuse for generation and prioritization. Another issue is related with test case quality generated from the test model [5], because current industry demand requires testing with minimal cost and high effectiveness [6]. In the generated test case, deficiency of fault detection is measured to prove testing value [7]. These problem require a unique method to handle several kinds of problems and at the same time, increase testing quality.

This study presents *FM_STATE*, which is a Java tool that comprised three phases of testing which are mapping features with behavior elements, test cases generation, and prioritizing. This study starts with handling core assets by using traceability technique. Subsequent, this Java tool process by choosing a appropriate assets to be apply for generation and prioritization. The originality of this method is it combines multiple validation methods to increase testing quality. This tool also capable to handle small or large size SPL product components. The rest of this paper is organized as follows. Section 2 focus on review towards existing studies. Section 3 discusses on the *FM_STATE* tool. Next, Sect. 4 which provides results and discussion of the topic. Lastly, Sect. 5 deliberates of conclusion and the future work.

2 Related Work

The improvement of testing quality needs a trade-off among cost and effectiveness. Such improvement can be obtained from various methods for example traceability method, generation of test case, and prioritization of test case. In [8], automated version of test model tool is offered. This method used hierarchy of Markov Chains. Similar to *FM_STATE*, this paper emphases on generation of test cases process and aim to improve testing cost and effort. [9] has proposed *SMartyTesting* which is one of the tool that apply UML and sequence diagram for test case generation. In *SMartyTesting*, the process begin with a discussion regarding assets in the use case towards the sequence model. In addition, this tool only emphasizes test case generation size and lacks testing coverage.

Furthermore, there are another tool called *SMartyModelling* that used to handle variability using stereotypes relates with UML [10]. This tool offers reusability in SPL core assets of architectural and requirements views. This tool helps to represent variability in UML-based format. [11] proposed *DARWINSPL*, an integrated tool representing spatial variability as FM with constraints. Similar to *SMartyModelling*, this tool is used to support variability representation of SPL. However, *DARWINSPL* uses metamodel to capture features and provision SPL development process. This tool is integrated with *DELTACECORE* to map FM and generate product variants.

[12] proposed *PLEDGE* as a product line editor tool using combinatorial interaction testing using t-wise to minimize the number of products for SPL. Similar to *FM_STATE*, this tool offers generation and prioritization of product configuration. This tool differs from *FM_STATE* because it uses t-wise context for generation and prioritization.

3 FM_STATE TOOL

FM_STATE tool is developed for tracing, generation and prioritization of test model developed from SPL core assets. This tool aims to solve issues of lack of semantics in the test model and to enhance the quality of test cases in SPL. This tool covers three main phases, which are: (i) Traceability between FM and statechart (see Sect. 3.1), (ii) Test case generation (see Sect. 3.2), and (iii) Test case prioritization (see Sect. 3.3).

3.1 Case Study

Educational Robotics case study is adopted from [14, 15] that represents real world educational robotics. This case study is based on SPL. Details of educational robotics based on FM is depicted in Fig. 1. It consists of one root feature called *educational_robotics*. There are eight features which are *Mobile_Robot, Controller, Hardware_features, Communication, Sensory, Micro_Computer, Actuator and Library*. Statechart for educational robots comprises of 31 states and 44 transitions as shown in Fig. 2. This case study embraces the practice of educational robotics to improve students' knowledge through practical with the employment of the robotics technology in learning. This statechart covers two main functionalities which are *MachineController* and *HardwareSettings* as illustrated in Fig. 2.

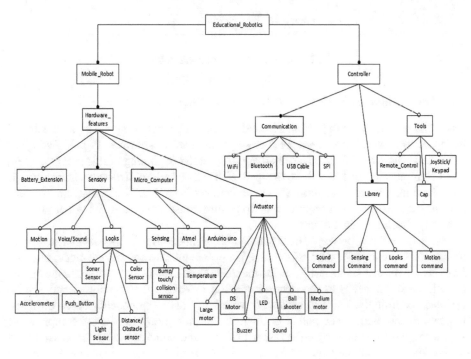

Fig. 1. Feature model of educational robotics

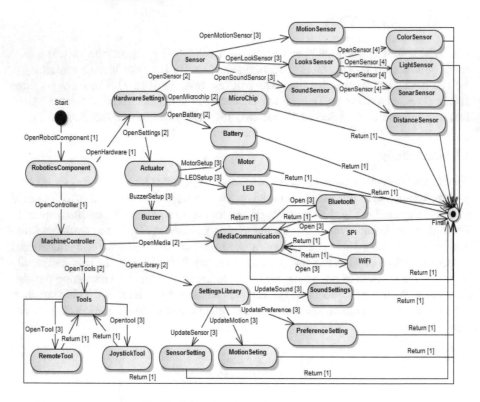

Fig. 2. Statechart for educational robotics

3.2   Traceability for Feature Model and Statechart

FM_STATE starts with importing FM and statechart model components as illustrated in Fig. 3. For traceability, there are four steps involved as shown in Fig. 4. Two types of models which are FM and statechart are used in the beginning of this process. Then, valid configuration of FM will be mapped with the statechart. The statechart model is used for simulation and generation based on features from FM. Directional requirements help to map features with requirements to ensure constraints from FM are considered and fully checked in core assets.

Next, traceability process related to traceability link and query mechanism are implemented. The model-to-text transformation is implemented to convert the model into *.xml* files, which is text-based artifacts. In the *FM_STATE* tool, for every change in the model, the tester needs to regenerate input file to ensure the mapping result can be updated accordingly. The FM represents variability and commonality based on listed features in the model. The nature of documentation tagging in SPLOT comprises of < *feature_model* > tagging to represent FM tree and child of the tree represented by < *feature_tree* > as the name of features. Aside from that, there is also a segment to display existing < constraint >.

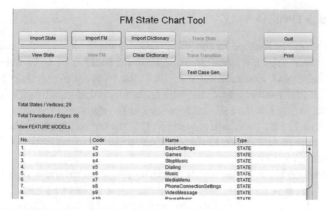

Fig. 3. *FM_STATE* Chart Tool used to import FM and statechart model components.

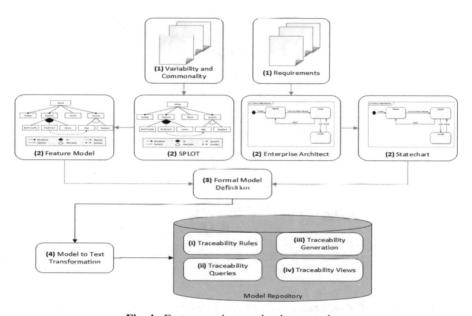

Fig. 4. Features and states development plan

Query mechanism from *XTraQue* standard query is adapted from [7] that represents query based on *XQuery* format. Three primary declarations in this query comprised of *[NAMESPACE]*, *[FUNCTION]*, and *[VARIABLE]*. *NAMESPACE* is required to specify imported document, *FUNCTION* represents mapping function using JAVA language, whereas *VARIABLE* is used to match the text found in records. Output of this process is a set of matching features and states that will be used for the generation process using hybrid algorithm.

3.3   Evaluation Criteria

Evaluation criteria are defined in the *FM_STATE* tool to evaluate the outcome of the approach. Equation 1.1 represents the evaluation for number of test suite n, where x is the test suite generated. Equation 1.2 represents the total execution time T. In terms of coverage criteria, three coverage criteria which are all-transition, all-states and transition-pair are defined in Eq. 1.3. The all-states coverage TS represents the total states covered in the generation. The all-transition coverage TC is the complete transition pairs covered in statechart components. The transition-pair coverage TP is the total transition covered in the test model [16, 17].

APFD defined in Eq. 1.4 is the metric used to evaluate fault detection. In this equation, T is the test suite, n is the test cases, TF is the point of the first test cases ex-posing fault, and m is the number of faults exposed by the test suite. Equation 1.5 represents the average prioritization time P.

$$S(x) = n_x \tag{1.1}$$

$$T = AVG(\Delta time_h) \tag{1.2}$$

$$TS(x) = \frac{ts_n}{ts_t} \tag{1.3}$$

$$TC(x) = \frac{t_n}{t_t}$$

$$TP(x) = \frac{tp_n}{tp_t}$$

$$APFD = 1 - \frac{TF_1 + TF_2 + \ldots. + TF_n}{n \times m} + \frac{1}{2_n} \tag{1.4}$$

$$P = AVG(\Delta Prioritization\_time_h) \tag{1.5}$$

3.4   FM_STATE for Model-based Test Case Generation and Prioritization

FM_STATE implements hybrid algorithm for generation and prioritization from the statechart test model. Advantages of hybrid algorithm include it is able to optimize execution time and produce high quality test case [15]. Overview of the hybrid algorithm based on *Floyd Warshall Algorithm (FWA), Branch and Bound (BBA)* and *Best First Search (BFS) (FWA-BBA-BFS)* has been defined in [15]. This technique can discover all branches in the model. This hybridization is essential to ensure the algorithm can traverse every node to check which path is more promising based on the suggested result from other algorithms.

Even though existing studies proposed hybrid *FWA* and *BBA* as the best technique to generate test cases, results from this hybrid technique is still imprecise due to lack of measurements applied [6]. Nevertheless, the current study indicates that hybrid between *FWA* and *BBA* is able to maximize coverage for all-states since the algorithm procedure will traverse every state [13]. The *FM_STATE* tool based on a hybrid test cases generation algorithm is based on three phases, namely initialization phase, branching phase, and bounding phase. The first step starts with extracting statechart test model artifacts. The model artifacts will be used to assign the nodes, transition, and weight of branches into a matrix *[v,a,w]*, where this information will be called in the branching phase [18].

The weight is calculated starting from the root state until the last node using nodes and transition weight of branches. Variable at the first node is defined as variable ≤ 0. Permutations will be conducted based on the shortest weight values. This process rearranges the ordering of branches from the matrix. To conduct the process, the maximum score *MaxPT*, minimum score *MinPT* and cost *c* are assigned. Maximum and minimum scores help confirm the total cost in the average score defined. The cost will be measured in order to sum the branch values. If the cost is greater than the parent nodes, then the searching process will be terminated. However, if the cost is less than the parent, it will be selected as a node to traverse the next branches. Next, the value of minimum score, maximum score and cost are used to get the total value of score and path weights (cost) in the bounding phase. The branch weight is sorted and test cases are generated based on branch weight until the last node is reached. The total weight will be exposed, and the visited nodes will be computed to stop test cases' redundancy. The process will be repeated until all branches have been traversed.

The objective of prioritization in this tool is to evaluate the faults in the test suite. It is vital to detect faults early to provide faster feedback on the system under test. Compared to other regression techniques, prioritization is an efficient approach to discover faults in the test suite generated as early as possible [14]. In the *FM_STATE* tool, two hybrid string distance techniques, Dice Similarity and Jaro-Winkler, are implemented since these techniques can produce best distance among test cases. To evaluate dissimilarity among test cases, local maximum distance is used to search for maximum distance from the first test case until all test cases have been considered in the prioritized list [19].

For minimization process, Local Maximum Distance acts as a technique that can order a list of test cases faster. Implementation of Local Maximum Distance in a previous study showed that it could cover results more quickly in terms of cover-age, faults, and minimal execution time. The enhanced Local Maximum Distance with Dice Similarity and Jaro-Winkler that have been defined in [7]. The proposed enhancement of Local Maximum Distance improves the selection of the following test cases. Figure 5 shows the process of prioritization that starts with selecting the next test cases into the prioritized list, the last test case in the prioritized list is linked with test cases in the unordered list. The evaluation is based on the minimum distance calculated from the list of unordered test cases. Maximum distance value is use to discover faults as early as possible. The process is repeated until all test cases are inserted into a group of prioritized test case.

R. A. Sulaiman et al.

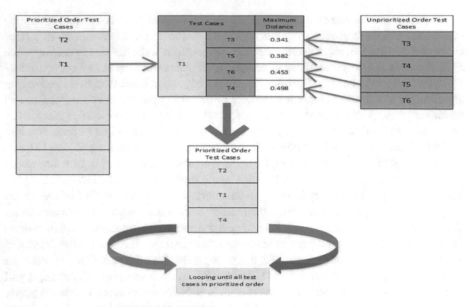

Fig. 5. FM_STATE tool process for prioritization

4 Result and Discussion

An experimental study was conducted on the case study defined in Sect. 3.1 whereas
measurements metrics are defined in Sect. 3.2. The experimental result is compared
with two existing tools that implemented coverage-based [15] and prioritization [18]
approaches. Table 1 shows that for the average size of test suite, *FM_STATE* tool obtained
the best generation time and prioritization time compared to existing approaches.

Table 1. Comparison results of different approaches.

Evaluation Results	FM_STATE tool	Coverage-based with local maximum distance approach
AVG Size of Test Suite	10.32	11.63
AVG Generation Time (*ms*)	0.54	1.00
AVG Execution for Prioritization Time (*ms*)	0.26	0.64
AVG *APFD* (%)	82.58	65.74
AVG Coverage (%)	92.20	69.80

FM_STATE tool obtained the best results in terms of average coverage and *APFD*.
Aside from that, *FM_STATE* also recorded the best maximization and minimization
measures. From this outcome, it is evident that *FM_STATE* is capable of improving
multiple quality measures that are significant in SPL testing since tester demand to have

good quality test case measures. The achieved outcomes indicate the importance of concise mapping with constraint validation, handling multi-objective optimization for cost and effectiveness, and role of faults discovery in test cases. This tool can contribute in handling optimization problem for SPL.

5 Conclusion

This paper presented *FM_STATE*, a tool developed using JAVA that permits features state mapping, generation, and prioritization of products in SPL using hybrid algorithm. This tool handles the challenges in SPL testing in the scope of MBT by evaluating a product by using five measurement criteria. These criteria help to minimize the size of test suite and execution time, in the same time maximize the coverage and rate of fault detected using *APFD* metric. *FM_STATE* tool is able to adapt.*xml* files of FM and statechart. Furthermore, the tool can be extended with additional features including functionality to manage different formats of FM and statechart files.

Acknowledgement. This research was supported by Universiti Tun Hussein Onn Malaysia (UTHM) through Tier 1 Grant (Vot H937).

References

1. Ardimento, P., Boffoli, N., Superbo, G.: Multi software product lines: A systematic mapping study. In: ENASE 2020 – Proceedings of the 15th International Conference on Evaluation of Novel Approaches to Software Engineering, no. Enase, pp. 470–476 (2020). https://doi.org/10.5220/0009469804700476
2. Lee, J., Kang, S., Jung, P.: Test coverage criteria for software product line testing: Systematic literature review. Inf. Softw. Technol.**122**, 106272 (2020). https://doi.org/10.1016/j.infsof.2020.106272
3. Arcaini, P., Inverso, O., Trubiani, C.: Automated model-based performance analysis of software product lines under uncertainty. In: ACM Int. Conf. Proceeding Ser., vol. Part F1716, p. 112 (2021) https://doi.org/10.1145/3461001.3473063
4. Chacón-Luna, A.E., Gutiérrez, A.M., Galindo, J.A., Benavides, D.: Empirical software product line engineering: a systematic literature review. Inf. Softw. Technol. **128**, p. 106389 (2020). https://doi.org/10.1016/j.infsof.2020.106389
5. Sahak, M., Halim, S.A., Jawawi, D.N.A., Isa, M.A.: Evaluation of software product line test case prioritization technique. Int. J. Adv. Sci. Eng. Inf. Technol. **7**(4–2), Special Issue, 1601–1608 (2017). https://doi.org/10.18517/ijaseit.7.4-2.3403
6. Hierons, R.M., Li, M., Liu, X., Parejo, J.A., Segura, S., Yao, X.: Many-objective test suite generation for software product lines. ACM Trans. Softw. Eng. Methodol. **29**(1), 1–46 (2020). https://doi.org/10.1145/3361146
7. Sulaiman, R.A., Jawawi, D.N.A., Halim, S.A.: A dissimilarity with dice-jaro-winkler test case prioritization approach for model-based testing in software product line. KSII Trans. Internet Inf. Syst. **15**(3), 932–951 (2021). https://doi.org/10.3837/tiis.2021.03.007
8. Ergun, B.: Tool support for model based software product line testing, Ozyegin University (2018)

9. Petry, K., OliveiraJr, E., Costa, L., Zanin, A., Zorzo, A.: SMartyTesting: a model-based testing approach for deriving software product line test sequences. ICEIS **2**, 165–172 (2021). https://doi.org/10.5220/0010373601650172

10. Da Silva, L.F., Oliveira, E.: Evaluating usefulness, ease of use and usability of an UML-based Software Product Line Tool. In: ACM International Conference on Proceeding Series, pp. 798–807, (2020). https://doi.org/10.1145/3422392.3422402

11. Nieke, M., Engel, G., Seidl, C.: DarwinSPL: An integrated tool suite for modeling evolving context-aware software product lines. In: ACM International Conference on Proceeding Series, pp. 92–99 (2017). https://doi.org/10.1145/3023956.3023962

12. Henard, C., Papadakis, M., Perrouin, G., Klein, J., Le Traon, Y.: PLEDGE: A product line editor and test generation tool. In: ACM International Conference on Proceeding Series, no. March 2015, pp. 126–129 (2013). https://doi.org/10.1145/2499777.2499778

13. Bardin, S., Kosmatov, N., Marcozzi, M., Delahaye, M.: Specify and measure, cover and reveal: a unified framework for automated test generation. Sci. Comput. Program. **207**, 102641 (2021). https://doi.org/10.1016/j.scico.2021.102641

14. Abdul Manan, M.S., Abang Jawawi, D.N., Ahmad, J.: A Systematic Literature Review on Test Case Prioritization in Combinatorial Testing. In: ACM International Conference Proceeding Series, pp. 55–61, (2021). https://doi.org/10.1145/3490700.3490710

15. Lee, J., Kang, S., Jung, P.: Test coverage criteria for software product line testing: systematic literature review. Inf. Softw. Technol. **122** 106272 (2020) https://doi.org/10.1016/j.infsof.2020.106272

16. Al_Barazanchi, I., Abdulshaheed, H.R., Shibghatullah, A.: The communication technologies in wban. Int. J. Adv. Sci. Technol. **28**(8), 543–549 (2019)

17. Thiab, A.S., Bin Shibghatullah, A.S., Yusoh, Z.I.M.: Internet of things-proactive security approach. J. Eng. Appl. Sci. **13**(9), 2668–2671 (2018)

18. Akhla, A.A.N.: Impact of Real-Time Information for Travellers: A Systematic Review. J. Inform. Knowl. Manage. **21**(4), 2250065–1–2250065–21 (2022)

19. Shibghatullah, A.S., Jalil, A., Wahab, M.H.A., Soon, J.N.P., Subaramaniam, K., Eldabi, T.: Vehicle tracking application based on real time traffic. Int. J. Electr. Electron. Eng. Telecommun. **11**(1), 67–73 (2022)

Learning Programming Difficulties: Toward an Integrated Conceptual Framework

Othman A. Alrusaini[✉]

Department of Engineering and Applied Sciences, Applied College, Umm Al-Qura University,
Makkah 24382, Saudi Arabia
oarusaini@uqu.edu.sa

Abstract. Computer programming languages are becoming more indispensable as our world becomes more computerized. This paper highlights the importance of the factors affecting programming learning difficulties. A survey method was employed to measure the relationships between variables related to programming learning difficulties. A total of 227 students studying for a computer programming diploma at Umm Al-Qura University in Saudi Arabia were surveyed for this purpose. The methodology for the study was structural equation modeling, and AMOS software was used to test the structural and measurement model and study the interlinkages between variables. The theoretical framework employed in this study comprised the technology acceptance model and experiential learning theory. The indirect factors were English language competency, learning materials, educational facilities, and teaching practices. The study concluded that the interactions of these factors explain the programming learning difficulties experienced by students enrolled in programming courses. Moreover, the factors which had the greatest effect on programming learning difficulty were English language competency and educational facilities. Therefore, the factors mentioned above should be considered by educational institutions for enhancement and ongoing development because of their connection with the programming learning difficulties experienced by students.

Keywords: Programming language · learning difficulties · learning environments · teaching · learning materials

1 Introduction

Computer programming is a technical field and profession that is receiving increasing and unprecedented interest. The profession will grow by at least 11% between 2020 and 2030 [1]. Despite this growth in the number of professionals, scholars report that the field is far from being at saturation point because of the many subfields in programming, in particular based on the many languages and technologies in use. In education, there is a similar trajectory, as more students develop an interest in programming and related disciplines. According to [2], the rise in interest in programming skills stems from the fact that they are also a requirement in other fields. For example, when learning statistics

F. Saeed et al. (Eds.): ICACIn 2022, LNDECT 179, pp. 607–622, 2023.
https://doi.org/10.1007/978-3-031-36258-3_54

it can be useful to have knowledge of languages such as R and Python. The researchers found that the failure rate of students on computer programming introductory courses is 28%. Growth in the number of students is also due to the lucrative nature of the profession, with salaries currently standing at around $126,000 per year [1]. However, there are many factors that directly impact students' interest in studying programming.

The body of literature on the effect of English language literacy, learning materials, educational facilities, learning interest, and their effect on programming learning difficulty is yet to be fully established. This paper is guided by five research questions. With regard to the interest of students in learning computer programming: How significant is the effect of English language literacy? How significant is the role played by learning materials? How significant an impact do educational facilities have? How significant is the effect of teaching practices? How significant is learning interest in influencing the learning difficulty experienced by computer programming students?

The paper is organized into six sections. The introduction provides an overview of the aims of the paper. It does so by explaining the basic concepts espoused in the study and setting out the research questions. The literature review provides a scholarly review of the findings of previous studies on this topic and its elements. The theoretical background section presents the theoretical framework, conceptual framework, and for-mulates research hypotheses. The research methodology section explains the methods and procedures that were used for data collection and analysis. The data analysis section provides an in-depth analysis of the findings by assessing the measurement model, the structural model, and model fit indices. The discussion section cross-references the findings of this study with those from previous research studies.

The paper adds value to the current body of literature on the subject because there are currently few studies that discuss all the factors affecting programming learning difficulty as espoused in this study. Specifically, few studies have dealt with the selected variables in combination, probably because of the many factors at play. An investigation examining the direct and indirect effect of the selected factors on learning difficulties through their influence on student interest can open up new possibilities to help the current generation deal with obstacles as they pursue programming courses.

2   Literature Review

Competency in the English language has been closely associated with ease of learning to program. Ruby and Krsmanovic [3] found that English was, at minimum, a desirable skill. The source lamented that this expectation places an extra burden on students whose native language was not English. Similarly, Idris and Ammar [4] reported that 37% of students learning programming claimed that lack of English skills impeded their progress in such courses. The students identified lack of English as one of the major reasons for their underperformance. English knowledge had a moderate correlation with programming ability; a relationship that was also causal.

Learning materials as a factor impacting students' interest in learning programming has been investigated in multiple studies. One of the recommendations in a study by Islam et al. [5] was for management to develop more effective learning materials to make it easier for students to learn programming. This notion is plausible because the quality of these materials can determine whether or not learners find it easy to grasp the content of such technical courses. Similarly, Lahtinen et al. [6] found that the difficulty experienced by beginner programmers was a problem that could be solved by adopting high-quality learning materials, which were tailored to this demographic.

Educational facilities are the physical establishments that encourage and enhance the interest of students in undertaking programming courses. Computer laboratories and libraries are examples of such facilities. Few studies have dedicated attention to this factor. However, Mhash and Alakeel [7] came close to this objective, as part of their purpose was to investigate the role played by educational facilities in promoting the learning of programming among students. Findings from the paper indicated that these facilities were important in the stimulation of understanding and interest among students.

Teaching practices are the strategies and approaches deployed by instructors to capture the interest of students and facilitate their understanding. The effect of teaching practices was the primary focus of a study by Erümit [8]. The paper reported that different teaching practices achieve different levels of success in enhancing students' understanding. The researchers found that students preferred animated classes because they disguised the technical nature of the courses. Islam et al. [5] argued that instructors could inculcate programming skills in students by selecting the best teaching methods.

The role played by learning interest in alleviating learning difficulties in programming is often significant. The study by Wiedenbeck et al. [9] found that learning interest was critical to the success or failure of students enrolled in a computer programming class. More particularly, the findings indicated that those students with a specific interest in computers were less likely to experience problems in programming, which resulted in their higher levels of success. Cheng [4] argued that students with little interest in programming and information technology tended to perform poorly and experienced problems with Visual Programming Environment (VPE) technologies. According to the source, this interest influenced their general perception of programming difficulty, which subsequently impacted their educational journeys.

3   Theoretical Background

3.1   Theoretical Framework

This study has identified the relevant theories as the Technology Acceptance Model (TAM) and experiential learning.

The Technology Acceptance Model (TAM) is a framework that models acceptance of new technology by users. The model asserts that for an individual or a group to accept new technology, they must be assured of its usefulness and ease of use, and social influence must also support such technology [10].

Research on programming learning difficulties shows that the TAM framework is highly relevant. The study by Rafique et al. [11] embraced this model, claiming that it was an integral tool used by analysts to explain the attitudes of programming students towards the languages they were learning. Although the researchers used this model, they also extended it by including additional variables such as intrinsic factors, teaching practices, efficacy problems, and relative usefulness to create a more robust measure of students' interest. Similarly, Cheng [12] extended this framework to measure students' interest in Visual Programming Environment technologies. The additional variables included gender and external encouragement. With regard to gender, male students demonstrated a higher interest in programming compared to their female counterparts. The two studies cited herein seem to affirm the robustness of the TAM framework for analyzing the interest of students in learning programming. For this reason, it is a suitable model for use in our study.

The experiential learning theory posits that students learn more effectively by doing rather than through listening to their instructors. The reasoning behind this theory is that when students engage in practical exercises, concepts are more likely to be retained and the students gain more confidence [13].

The learning cycle in experiential learning occurs in four styles, namely, diverging, assimilating, converging, and accommodating. Those learning in the diverging style prefer to watch rather than do, and they usually have a strong capacity to think [14]. They often work in groups and have a strong desire for social interactions that lead to learning. The assimilating style is apparent in individuals who prefer to learn concepts so as to obtain a clear picture. For this reason, they may not be drawn to group discussions and teamwork, and instead focus on reflective observation and abstract conceptualization. Learners subscribing to the convergence style focus on problem-solving [15]. They prefer to only learn what will ultimately solve their problems. For this reason, they focus on active experimentation rather than learning concepts. Individuals with a strong affinity for the accommodating style prefer practicability. However, they are pragmatic in that they combine active experimentation with abstract conceptualization [13]. In learning programming, experiential learning is widely applicable because of the nature of the technologies, which require intensive practice. Hence, this theory was selected to guide the direction of our study.

3.2 Conceptual Framework

As shown in Fig. 1, the conceptual framework of the study consisted of six factors with five hypotheses. The factors are: English language competency, learning materials, educational facilities, teaching practices, learning interest, and the dependent variable programming learning difficulties.

3.3 Research Hypotheses

By default, programming languages are written in English. It is therefore not surprisingly that a good mastery of the English language can greatly alleviate teething problems when learning to program. Students coming from a background in which English is not their native language often face difficulties [16]. For example, the Gulf region is one such location, where natives speak Arabic as their first language. According to [17], such students may lose interest in programming because of the clear discordance between English and Arabic. It is equally possible that learning programming is a matter of will, implying that regardless of one's first language, one may still excel in programming. Nevertheless, the effect of native language dynamics on the interest of new programming students cannot be ignored, and for this reason. Therefore, we hypothesize the following:

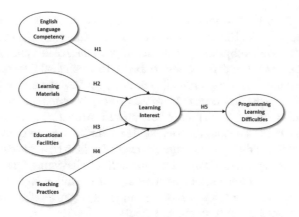

Fig. 1. Conceptual framework of the study.

Hypothesis 1. English language competency has a significant positive effect on the learning interest of programming students.

Because of the technical nature of programming, access to supporting material can be said to be a necessity for students to flourish. This is especially the case for beginner programming students. They need to gain a concrete conceptual understanding of programming tenets before they can proceed to practice and become adept [18]. Learning materials such as books and video tutorials are effective in enabling students to understand the basics of programming languages. Hence, we hypothesize the following:

Hypothesis 2. Learning materials have a significant positive effect on the learning interest of programming students.

From an academic perspective, learning programming needs to be accompanied by supportive educational facilities. Universities and colleges without established computer laboratories tend to receive limited interest from programming students. These labs are helpful to students that may not have their own laptops or do not have paid subscriptions to the software necessary to facilitate their programming endeavors [19]. For this reason, we hypothesize the following:

Hypothesis 3. Educational facilities have a significant positive effect on the learning interest of programming students.

Teaching programming is not the same as teaching other disciplines. Although the objective in both cases is to impart knowledge to students, the strategies and approaches often differ. For example, teaching a history lesson may require an instructor to tell stories about the past. However, this approach is not effective in a programming class, in particular for beginners [20]. There is a need to engage students at a higher level and conventional teaching strategies do not encourage this. Practical lessons, on the other hand, are known to deliver because of the hands-on approach. It is evident that the selection of teaching practices and approaches determines the level of interest among learners. Therefore, we hypothesize the following:

Hypothesis 4. Teaching practices have a significant positive effect on the learning interest of programming students.

In any discipline, the learning interest of students alleviates the difficulties they experience in the process. This interest serves as the intrinsic motivation to enable the students to face difficulties objectively. Additionally, because of this interest, learners will keep trying until they fully understand the concepts. In a technical field such as computer programming, learning interest is critical, especially for beginners. The amount of new terminology and concepts they need to learn may be overwhelming to some. However, it is their interest that will keep them focused on the task. Attracting the interest of students in programming can be challenging. Such students are known to experience few or no problems when undertaking programming courses in their tertiary level education. It is evident that learning interest plays a pivotal role in alleviating learning difficulties among programmers. For this reason, we hypothesize the following:

Hypothesis 5. Learning interest has a significant negative effect on the learning difficulties experienced by programming students.

4   Research Methodology

4.1   Measurement of Constructs and Questionnaire Design

This study used a questionnaire to collect data from respondents. The tool contained close-ended questions that restricted respondents to five answer options. The questions adopted the Likert scale, requiring participants to indicate their level of agreement or disagreement with specific statements. The five levels of agreement were (1) strongly disagree, (2) disagree, (3) neutral, (4) agree, and (5) strongly agree. The actual statements to which respondents reacted are shown in Table 1 below.

Table 1. Questionnaire Items

Variable	Question/Statement	Source
English Language Competency	My written English is excellent	[21]
	My spoken English is excellent	[21]
	I have performed well in English courses/tests	[21]
Learning Materials	I have access to sufficient video and written materials on programming	[18]
	I have the right software, computer programs, and other installations required to learn programming	[18]
	I have a computer/laptop to use in practicing my programming skills	[18]
Educational Facilities	Our institution has a well-equipped computer lab to use for learning programming	[22]
	Our institution's library sufficiently accommodates programming students	[23]

(continued)

Table 1. (*continued*)

Variable	Question/Statement	Source
	Our classrooms are structured in a way that promotes the learning of computer programming	[19]
Teaching Practices	Our lecturers are tolerant of slow-learning students	[24]
	Teaching approaches adopted by our programming lecturers are student-centered	[25]
	Teaching approaches adopted by our programming lecturers encourage intensive practicing	[26]
Learning Interest	I feel very excited to learn new concepts in programming	[27]
	I rarely fail to attend programming classes	[28]
	Whenever we are in class, I feel the urge to ask questions	[29]
Programming Learning Difficulties	I find it hard to follow the lecturer during classes	[30]
	Even when I understand the lecturer, I fail to put the knowledge into practice	[31]
	My scores in assessment tests are poor	[32]

4.2 Sampling and Data Collection

The study investigated diploma students in enrolled in programming courses at the Umm Al-Qura University, Saudi Arabia. The selection of this population was by design because such students are best qualified to explain how the different factors set out in this paper influence their learning interest and consequently their learning difficulties. This population numbered 527, and the sample size computed using the Yamane's formula was 227, as shown in the calculations below:

$$n = \frac{N}{1 + N * (e^2)} = \frac{527}{1 + 527 * (0.05^2)} = 227.4002157 \approx 227$$

where n is the minimum sample size, N is the target population size, and e is the error allowance (0.05).

The sampling technique used was random sampling, which is a probabilistic method of identifying respondents. The method accords everyone in the population an equal likelihood of being selected as a participant in the survey.

The researcher hosted the questionnaire on the Google Forms platform. To make it convenient for respondents, the questionnaire was translated into Arabic. The response rate from the survey was 100% in that all the 227 participants completed the questionnaires. The analysis process began with data preparation, which involved transferring the data into IBM SPSS and coding the responses to reflect the various perspectives expressed therein. The main analytical process employed the structural equation modeling approach. IBM SPSS AMOS version 26 was used for this because it is a software tool specifically designed to analyze data for structural equation models. Finally, the researcher gathered the results of the analysis and considered its implications for the study.

5 Data Analysis

5.1 Analysis of Respondent Demographics

The analysis of demographic characteristics considered the age and gender of respondents involved in the study, as shown in Table 2.

Table 2. Demographic characteristics of respondents

Demographics		Frequency	Percentage (%)
Gender			
	Male	141	62%
	Female	86	38%
Age			
	<20	54	24%
	20 – 25	109	48%
	>25	64	

5.2 Assessment of the Measurement Model

In assessing the measurement model, this section examines the reliability and validity of the study's main construct variables. For reliability, the metrics of interest are indicator reliability and internal consistency, whereas for validity, the metrics of interest are convergent reliability and discriminant reliability.

Indicator reliability is a useful metric in the determination of a variable's reliability because it measures the extent to which constituent variables contribute to the overall variance of the target variable. There are two approaches to computing indicator reliability, namely path-weighting and factorial schemes. This analysis used factor loadings because this method is more reliable. According to [33], an indicator reliability metric of 0.7 is the lowest possible score, a minimum which every variable should score to indicate reliability. In this analysis, the average composite reliability score was 0.88, which is greater than the required 0.7 value. In addition, all the variables met the minimum threshold of reliability by scoring greater than 0.7. Table 3 below summarizes the results.

Table 3. Construct indicators

Construct	Items	Factor Loading	Composite Reliability	Indicators	Cronbach's Alpha	AVE
English Language Competency	EL1	0.942	0.92	3	0.918	0.78
	EL2	0.841				
	EL3	0.870				
Learning Materials	LM1	0.929	0.85	3	0.843	0.66
	LM2	0.780				
	LM3	0.706				
Educational Facilities	EF1	0.952	0.88	3	0.886	0.72
	EF2	0.775				
	EF3	0.802				
Teaching Practices	TP1	0.954	0.88	3	0.878	0.72
	TP2	0.763				
	TP3	0.814				
Learning Interest	LI1	0.799	0.87	3	0.866	0.69
	LI2	0.864				
	LI3	0.822				
Programming Learning Difficulty	PD1	0.837	0.87	3	0.868	0.69
	PD2	0.841				
	PD3	0.814				

Internal consistency is a reliability measure that measures the degree to which a construct's questionnaire items explain the item as a latent variable. Cronbach's alpha is the most suitable metric for measurement of the internal consistency of a research instrument. Here again, Cheah et al. [33] assert that 0.7 should be the minimum score to signal consistency. In this analysis, the average alpha was 0.876, which is within the acceptable range. The most consistent variable was 'English language competency', as it scored 0.918, whereas the least consistent was 'learning materials' scoring an alpha of 0.843. All variables scored well on this metric, affirming their internal consistency. Table 3 summarizes these results.

Variable constructs or questionnaire items need to correlate well with their parent variable so that the parent variable is a fair representation of the underlying exogenous variables. Convergent validity is a measure of how well the parent variable correlates with their exogenous questionnaire items. In measuring this metric, the Average Variance Extracted (AVE) method is the most suitable. According to [34], the AVE should be at least 0.5, and this was met by all variables. The average score was 0.709. English language competency scored highest (0.78), whereas learning materials scored lowest (0.66). Table 3 summarizes the convergent reliability output.

Discriminant validity measures the degree to which the questionnaire items of a construct explain the construct variable to which they belong more than other constructs. In other words, questionnaire items belonging to construct variable A should correlate with that variable more than they correlate with construct variable B [35]. This affirms the distinctiveness of the construct variables, which is critical to the integrity of regression analyses. In this analysis, all construct variables met this requirement as questionnaire items belonging to the same construct variable had a higher correlation with that variable than with other variables. Hence, it is evident that the variables met the discriminant validity requirement. Table 4 summarizes the correlational comparisons that inform the conclusion above.

Table 4. Discriminant validity results

	PD	LI	TP	EF	LM	EL
PD	**0.8811**					
LI	−0.6643	**0.8866**				
TP	−0.6120	0.6349	**0.8878**			
EF	−0.5670	0.6570	0.6083	**0.8943**		
LM	−0.5650	0.6323	0.6064	0.5564	**0.8652**	
EL	−0.7220	0.7110	0.6783	0.6923	0.6563	**0.9186**

PD = Programming learning difficulty; LI = Learning interest; TP = Teaching practices; EF = Educational facilities; LM = Learning materials; EL = English language competency

5.3 Assessment of the Structural Model

The structural equation model for this study shows the interactions between English language competency (EL), learning materials (LM), educational facilities (EF), teaching practices (TP), learning interest (LI), and programming learning difficulties (PD) as shown in Fig. 2 below. It is evident that EL, LM, EF, and TP all significantly and positively influence LI. Similarly, LI significantly influences PD, though this is a negative effect.

5.4 Model Fit Indices

Analysis of the model fit indices indicates that the chi-square value (χ^2) scored by the model is 558.470 (DF = 124, P = 0.000) and that the CMIN/DF score is 4.504 as shown in Table 5. According to [36], a CMIN/DF value of less than 5.0 in a structural equation model indicates a reasonable fit. The implication drawn from these findings is that the structural model applied on the data is significant.

The study also investigated the effect of individual elements in the model on their respective response variables, as shown in Table 6. Firstly, the analysis investigated the collective effect of English language competency, learning materials, educational facilities, and teaching practices on the learning interest of students in the diploma

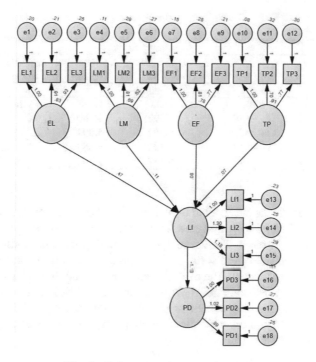

Fig. 2. Full structural equation model.

Table 5. Model fit indices results

Model	NPAR	CMIN	DF	P	CMIN/DF
Default model	47	558.470	124	0.000	4.504
Saturated model	171	0.000	0		
Independence model	18	4310.638	153	0.000	28.174

NPAR = Number of parameters for each model; CMIN = Chi-square value; DF = Degree of Freedom

program at the Umm Al-Qura University, Saudi Arabia. The findings established that the R^2 coefficient denoting this causal relationship was 0.792. This implies that 79.2% of the variation in the learning interest of programming students could be explained by the variation in their English language competency, learning materials, educational facilities, and teaching practices. The F-statistic of the model was 211.273, this meant that at least one of the independent variables was a significant predictor of learning interest. Regarding the constituent independent variables, the educational facilities variable was the most significant, as it scored a beta of 0.293 (t = 5.255, p = 0.000). Nevertheless, all the independent variables were statistically significant.

Table 6. Hypothesis testing results

Summary Stats					
R-Squared		Adj. R-Squared	F Score	Sig/*p*-value	
0.792		0.788	211.273	0.000	
Coefficients					
Hypothesis	Path	Beta	*t*-stat	*p*-value	Decision
H1	EL → LI	0.293	4.260	.000	Supported
H2	LM → LI	0.262	4.891	.000	Supported
H3	EF → LI	0.293	5.255	.000	Supported
H4	TP → LI	0.127	2.301	.022	Supported

Learning interest was observed to have a significant but negative effect on programming learning difficulty, as it scored a beta coefficient of -0.826 (t = -21.974, p = 0.000), as shown in Table 7. The R^2 coefficient denoting the causal effect was 0.682, which means that 68.2% of the variation in programming learning difficulty could be explained by learning interest. The overall implication of the findings was that English language competency, learning materials, educational facilities, and teaching practices significantly affected learning interest, which then significantly predicted the programming learning difficulty among students enrolled in programming courses in the diploma program at Umm Al-Qura University.

Table 7. Hypothesis testing results for the dependent variable

Summary Stats					
R-Squared		Adj. R-Squared	F Score	Sig/*p*-value	
0.682		0.681	482.860	0.000	
Coefficients					
Hypothesis	Path	Beta	*t*-stat	*p*-value	Decision
H5	LI → PD	-.826	-21.974	.000	Supported

6 Discussion, Implications, and Limitations

Our study findings have attested to the model's significance, and affirm the positions held in all the hypotheses. On the issue of the effect of English language competency on learning interest, findings in this investigation have shown that the former is a significant predictor. This agrees with the findings of Idris and Ammar [4], who considered the English language a key requirement to induce interest in learning programming. Learning materials also significantly influenced the learning interest of students in programming. These findings are consistent with Cheah [37], whose paper argued that learning materials can positively impact interest in learning a programming language.

The effect of educational facilities was also investigated in this study. Findings in this regard indicated that these facilities, such as proper classrooms, libraries, and computer laboratories, are a significant predictor of learning interest among programming students. It is for this reason that institutions without these facilities fail to attract a sizeable number of students pursuing related programs. The finding is consistent with López-Pernas et al. [19], who claimed that these facilities tend to ease the progress of students while they are learning programming languages. Teaching practices were also found to be key in enhancing student interest in learning programming. This study found that student-centered approaches, intensive practicing, and lecturers' empathy to slow-learners had a positive impact. The influence of learning interest on the difficulty experienced by programming students was the ultimate test. Our findings established that learning interest significantly alleviated these learning difficulties. This is consistent with the study by Cheng [12], which found that students with little interest in computer technologies tend to find it hard to process and learn programming concepts.

All of the findings of this study imply that the interactions of English language competency, learning materials, educational facilities, teaching practices, and learning interest significantly affect the difficulties experienced by programming students. Such difficulties are significant for beginner students, and need to be alleviated. By working on the learning materials, educational facilities, and teaching practices, institutions can create a conducive environment for students to successfully progress through their programming course. Students may need to work on their understanding and fluency in the English language because this study has established that it is a significant predictor of learning interest. It may be the case that a lack of knowledge of the English language leads to a lack of interest in learning programming. If so, students should consider acquiring a firm grasp of the language before setting out to learn their preferred programming languages.

Scores for the other factors also indicated their significance in determining learning interest. The effectiveness of learning materials in enhancing the learning of programming among students shows that institutions have an opportunity to improve the learning interest of programming students by making available quality materials relevant to programming. Educational facilities also played a significant role in ensuring that programming students developed a sustainable interest in programming. These findings suggest that improvements in the ambience, classroom setup, and library programming inventories can help to increase the students' learning interest in programming. The affirmation that teaching practices are critical to inspire learning interest among programming students indicates that programming learning interest can significantly improve if lecturers adopt a more student-centered approach and teaching methods that encourage intensive practice of concepts. By performing well in English language competency, learning materials, educational facilities, and teaching practices, institutions gain control of programming learning interest, which then alleviates programming learning difficulties.

The primary limitation of this investigation was that it allowed students excessive freedom in their responses to the questionnaire items. This freedom could have prompted them to collude and forward duplicated or almost identical answers. To minimize this limitation, the researcher ensured that the respondents were educated on the effect of such collusion, and the students undertook to refrain from this irregular practice. There was also the risk of incomplete responses to the survey, which would have sabotaged the analysis. For this reason, the researcher adopted brevity in formulating the questions and simplicity in the language used. By mitigating these two limitations the study was able to ensure a high quality of data, which then guaranteed the validity and reliability of the findings.

7 Conclusion and Future Research

In conclusion, this investigation established that the interactions of several factors, namely English language competency, learning materials, educational facilities, teaching practices, and learning interest, explain the learning difficulties experienced by diploma students enrolled in programming courses at the Umm Al-Qura University. Therefore, it is incumbent upon students and their learning institutions to optimize these factors because they have a strong causal association with the learning difficulties experienced in relation to programming courses. Although the investigation attempted to study the factors affecting programming learning interest and difficulty, it did not comprehensively achieve this objective. One area that may need further research concerns approaches students can take in learning English as a requisite language for programming. Another area concerns how institutions can set up their facilities in ways that are responsive to the needs of programming students. Finally, future researchers should investigate the direct effect of English language competency, access to learning materials, availability of educational facilities, and the viability of teaching practices on programming learning difficulty.

References

1. B. o. L. Statistics Computer and Information Research Scientists. Bureau of Labor Statistics (2022). https://www.bls.gov/ooh/computer-and-information-technology/computer-and-information-research-scientists.htm. Accessed 20 Mar 2022
2. Mai, T.T., Bezbradica, M., Crane, M.: Learning behaviours data in programming education: Community analysis and outcome prediction with cleaned data. Futur. Gener. Comput. Syst. **127**(1), 42–55 (2022)
3. Ruby, I., Krsmanovic, B.: Does learning a programming language require learning English? A comparative analysis between English and programming languages. Presented at the EdMedia+ Innovate Learning, June 1 (2017)
4. Idris, M.B., Ammar, H.: The correlation between arabic student's english proficiency and their computer programming ability at the university level. USA Int. J. Manag. Public Sector Inform. Commun. Technol. (IJMPICT) **1**(1), 1–20 (2018)
5. Islam, N., Shafi Sheikh, G., Fatima, R., Alvi, F.: A study of difficulties of students in learning programming. J. Educ. Soc. Sci. **7**(2), 38–46 (2019)
6. Lahtinen, E., Ala-Mutka, K., Järvinen, H.M.: A study of the difficulties of novice programmers. Acm sigcse bulletin **37**(3), 14–18 (2005)
7. Mhashi, M.M., Alakeel, A.L.I.M.: Difficulties facing students in learning computer programming skills at Tabuk University. Recent Adv. Mod. Educ. Technol. **1**(1), 15–24 (2013)
8. Erümit, A.K.: Effects of different teaching approaches on programming skills. Educ. Inf. Technol. **25**(2), 1013–1037 (2019). https://doi.org/10.1007/s10639-019-10010-8
9. Wiedenbeck, S., Sun, X., Chintakovid, T.: Antecedents to end users' success in learning to program in an introductory programming course. In: IEEE Symposium on Visual Languages and Human-Centric Computing (VL/HCC 2007), pp. 69–72 (2007)

10. Al-Emran, M., Mezhuyev, V., Kamaludin, A.: Technology acceptance model in m-learning context: a systematic review. Comput. Educ. **125**(1), 389–412 (2018)
11. Rafique, W., Dou, W., Hussain, K., Ahmed, K.: Factors influencing programming expertise in a web-based e-learning paradigm. Online Learn. **24**(1), 162–181 (2020)
12. Cheng, G.: Exploring factors influencing the acceptance of visual programming environment among boys and girls in primary schools. Comput. Hum. Behav. **92**(1), 361–372 (2019)
13. Kolb, A., Kolb, D.: Eight important things to know about the experiential learning cycle. Australian Educat. Leader **40**(3), 8–14 (2018)
14. Fewster-Thuente, L., Batteson, T.J.: Kolb's experiential learning theory as a theoretical underpinning for interprofessional education. J. Allied Health **47**(1), 3–8 (2018)
15. Bontchev, B., Vassileva, D., Aleksieva-Petrova, A., Petrov, M.: Playing styles based on experiential learning theory. Comput. Hum. Behav. **85**(1), 319–328 (2018)
16. Guo, P. J.: Non-native english speakers learning computer programming: Barriers, desires, and design opportunities. In: Proceedings of the 2018 CHI Conference on Human Factors in Computing Systems, pp. 1–14 (2018)
17. Alaofi, S.: The impact of english language on non-native english speaking students' performance in programming class. In: Proceedings of the 2020 ACM Conference on Innovation and Technology in Computer Science Education, pp. 585–586 (2020)
18. Xie, B,, et al · A theory of instruction for introductory programming skills. Comput. Sci. Educ. **29**(2–3), 205–253 (2019)
19. López-Pernas, S., Gordillo, A., Barra, E., Quemada, J.: Examining the use of an educational escape room for teaching programming in a higher education setting. IEEE Access **7**(1), 31723–31737 (2019)
20. Bers, M.U.: Coding as another language: A pedagogical approach for teaching computer science in early childhood. Journal of Computers in Education **6**(4), 499–528 (2019)
21. Rein, P., Taeumel, M., Hirschfeld, R.: Towards empirical evidence on the comprehensibility of natural language versus programming language. Design Thinking Res. 111–131 (2020)
22. Cline, K., Fasteen, J., Francis, A., Sullivan, E., Wendt, T.: Integrating programming across the undergraduate mathematics curriculum. Primus **30**(7), 735–749 (2020)
23. Zibani, P., Kalusopa, T.: E-resources marketing in African academic libraries: Contexts, challenges and prospects. In: Handbook of Research on Advocacy, Promotion, and Public Programming for Memory Institutions, pp. 261–283 (2019)
24. Priyaadharshini, M., Dakshina, R., Sandhya, S.: Learning analytics: game-based learning for programming course in higher education. Proc. Comput. Sci. **172**(1), 468–472 (2020)
25. Hovey, C.L., Barker, L., Luebs, M.: Frequency of instructor-And student-centered teaching practices in introductory CS courses. In: Proceedings of the 50th ACM Technical Symposium on Computer Science Education, pp. 599–605 (2019)
26. Denny, P., Cukierman, D., Bhaskar, J.: Measuring the effect of inventing practice exercises on learning in an introductory programming course. In: Proceedings of the 15th Koli Calling Conference on Computing Education Research, pp. 13–22 (2015)
27. Babori, A.: Analysis of Discussion Forums of a Programming MOOC. TEM Journal **10**(3), 1442–1446 (2021)
28. Demirkiran, M.C., Tansu Hocanin, F.: An investigation on primary school students' dispositions towards programming with game-based learning. Educ. Inf. Technol. **26**(4), 3871–3892 (2021). https://doi.org/10.1007/s10639-021-10430-5
29. Misra, I., Girshick, R., Fergus, R., Hebert, M., Gupta, A., Van Der Maaten, L.: Learning by asking questions. In: Proceedings of the IEEE Conference on Computer Vision and Pattern Recognition, pp. 11–20 (2018)
30. Medeiros, R.P., Ramalho, G.L., Falcão, T.P.: A systematic literature review on teaching and learning introductory programming in higher education. IEEE Trans. Educ. **62**(2), 77–90 (2018)

31. Tsai, C.Y.: Improving students' understanding of basic programming concepts through visual programming language: The role of self-efficacy. Comput. Hum. Behav. **95**(1), 224–232 (2019)
32. So, M.H., Kim, J.: An analysis of the difficulties of elementary school students in python programming learning. Int. J. Adv. Sci. Eng. Inform. Technol. **8**(4–2), 1507–1512 (2018)
33. Cheah, J.H., Sarstedt, M., Ringle, C.M., Ramayah, T., Ting, H.: Convergent validity assessment of formatively measured constructs in PLS-SEM: On using single-item versus multi-item measures in redundancy analyses. Int. J. Contemp. Hosp. Manag. **1**(1), 25–29 (2018)
34. Sürücü, L., Maslakci, A.: Validity and reliability in quantitative research. Bus. Manag. Stud. An Internat. J. **8**(3), 2694–2726 (2020)
35. Yusoff, A.S.M., Peng, F.S., Abd Razak, F.Z., Mustafa, W.A.: "Discriminant validity assessment of religious teacher acceptance The use of HTMT criterion. J. Phys. Conf. Ser. **1529**(4), 042045 (2020)
36. Marsh, H.W., Hocevar, D.: Application of confirmatory factor analysis to the study of self-concept: First-and higher order factor models and their invariance across groups. Psychol. Bull. **3**(97), 562 (1985)
37. Cheah, C.S.: Factors contributing to the difficulties in teaching and learning of computer programming: A literature review. Contemp. Educ. Technol. **12**(2), 272 (2020)

Users' Perception on Quality Medication Adherence Applications

Madihah Zainal[✉], A. Izuddin Zainal-Abidin, and Suziah Sulaiman

Computer and Information Sciences Department, Universiti Teknologi PETRONAS,
Seri Iskandar, Perak, Malaysia
madihah_19001059@utp.edu.my

Abstract. Mobile application intervention has shown good capability in increasing adherence to medication by providing medication reminders and information. However, current studies showed inconsistent effects and limited evaluation of mobile intervention in improving medication adherence. Identifying the quality medication adherence applications based on the commercial apps available on the current operating platform is crucial to ensure the app's usefulness to the intended consumers. Several quality evaluations tools had been developed to determine the quality of the medication adherence app, however, most of the tools are executed by experts such as researchers, health care professionals, and mobile developers. Therefore, this paper proposes to explore the criteria of quality medication adherence application from the potential users' (patients) perspectives. This study explored potential quality criteria based on the existing literature. Potential users will be given a set of questionnaires based on the Medication Adherence Application Quality (MedAd-AppQ) tool. The finalized criteria of the quality medication adherence app will be determined based on the results from the survey. This study is conducted to provide insights into how the potential patients perceived quality medication adherence application. However, as this research targets to evaluate the medication adherence application developed for the use of patients, thus the app that is created for the health care professionals or clinicians may have different quality measurements based on their specifications.

Keywords: Mobile Application · Quality · Medication Adherence

1 Introduction

With the current technological advancements, mobile health (m-health) applications (apps) have been used to assist users in their medication management. A wide range of medication monitoring platforms were being used in tandem with the growth in mobile phone ownership and subscriptions [1]. Thus, various medication adherence applications have been available for free and paid subscriptions. As of now, there are 704 applications available for medication adherence management across Apple and Android platforms [2], and that number is expected to rise rapidly over time.

F. Saeed et al. (Eds.): ICACIn 2022, LNDECT 179, pp. 623–638, 2023.
https://doi.org/10.1007/978-3-031-36258-3_55

Mobile apps for medication adherence assist users such as chronic disease patients in their medication management through various features such as reminders for medication intake, dosage monitoring, and guidance for meditation instructions [2]. Be-sides, it offers closer communication between the users and their contacts such as family, peers, and co-workers for social support encouraging users to adhere to their prescribed medications [3]. These functions are crucial to reduce non-adherence behaviors, either it is intentionally or unintentionally as medication adherence is multifactorial.

Medication adherence application is one of the most cost-effective interventions to increase adherence as the user only needs to download and subscribe to the application on their current smartphone. Also, mobile technologies support the user's mobility and personalized visualization for the users' convenience [4]. Besides, several applications can provide data and security protection for users' information, which increases the confidence of users to monitor their medication management.

With the abundance of apps in the commercial market, it is important to identify quality medication apps, as medication adherence apps are not regulated or bound by any policy. The app's faulty features or inaccurate measurements may directly affect patients' conditions, especially during the long usage period. For example, a study found that more than 40% of apps for assisting asthma conditions failed to adequately provide suggestions based on the clinical protocols [5]. Therefore, many researchers have studied the quality of medication adherence apps and other digital health applications. As a result, researchers, medical professionals, and app developers developed list of criteria to evaluate quality of the apps.

This paper explores the quality criteria of medication adherence apps based on the potential consumer perspective. A set of questionnaires derived from Medication Adherence App Quality (MedAd-AppQ) [6] with additional items from literature such as password features and rewards features are distributed to the users of medication adherence app. Such items will be scored based on their importance level in the quality evaluation of the medication adherence app. A framework consisting of quality criteria of medication adherence app will be developed based on the questionnaires' responses. This framework could act as guidelines for mobile developers in designing and developing medication apps as it explains the significant criteria valued by the end-users. However, this study is limited to the general chronic disease conditions to get as many as potential users instead of specific disease conditions as different diseases require different measurement specifications.

2   Literature Review

2.1   Medication Adherence

Medication adherence as defined by the World Health Organization (WHO) the degree of an individual's behavior in terms of medication intake with the suggested diet and way of life as prescribed by the medication provider [7]. Medication adherence is critical in improving a person's health as medication has proven to be the most effective intervention and the main approach for improving health conditions [8, 9]. Besides, chronic illness patients need to adhere to their prescribed medications to ensure medication effectiveness in reducing health complications. It is because a medication can effectively control the disease if taken in accordance with the specific frequency, dosage, and timings [8]. Medication adherence is also vital in reducing health care economic costs. Car et al. stated that around US$ 375 billion could be saved annually with good medication adherence [9].

Despite its' effectiveness in controlling and curing chronic illness, most people do not adhere to their medication intake. Several studies stated that medication adherence for chronic disease in developed countries is only 50% [7, 10]. Even in developing countries such as Malaysia, less than 30% of the medication consumer adhere to their prescribed medication [11]. As a result, the numbers will be even lower in developing countries and underdeveloped countries [7] due to the lack of medication facilities and the increased concern about the cost of medication compared to developed and developing countries. A study by Bili and Zha concerning the knowledge level of diabetes patients in South Sudan revealed that around 28% of the patients adhere to their suggested diet [12]. Also, less than 20% of human immunodeficiency virus (HIV) patients in Bangladesh can adhere entirely to their 3.5 years of antiretroviral therapy (ART) [13].

High medication adherence refers to 80% or more medication intake based on the prescriptions given by health care providers [14]. Low and intermediate adherence to medication intake reduces the effectiveness of the medication on the disease, which will increase the risk of hospital re-admission and premature death. Rosen et al. revealed that patients with low and intermediate medication adherence are 2.5 times more likely to be readmissions compared to patients with high medication adherence [15]. Also, more than a thousand premature deaths have been reported in the United States due to poor medication adherence [8]. Furthermore, failure to comply with medication prescriptions increase the risk of health complications for the patient. Still, it leads to medicine wastage which dramatically impacts the overall economic healthcare cost and the environment [16]. According to Berita Harian, Malaysian consumers were detected to waste their medications up to tens of millions of ringgits every year [17].

Medication adherence is influenced by several factors [7]. It includes socioeconomic-related factors, health care system-related factors, condition-related factors, treatment-related factors, and patient-related factors. Examples of socioeconomic factors are patient living conditions, health insurance, limitations in language proficiency, and medication cost. The health care system involves the patient and medicine provider relationship and patient waiting time. Next, condition-related factors are patients' condition regarding their illness, such as lack of symptoms, severity, and depression. The medication regimen's complexity, therapy duration, and side effects are examples of treatment-related factors. Patient-related factor includes motivation and knowledge about the disease. Thus, various technological interventions have been developed to increase medication adherence rates including, mobile interventions. Mobile applications utilize various functions to reduce medication non-adherence behaviors by providing reminders and information for medication intake.

Typically, there are two types of methods used to measure medication adherence: a direct and indirect method [8]. Direct methods refer to the method that directly assesses an individual's medication intake, such as through direct observation of a patient's medication intake or drug detection in the blood test. These methods provide accurate information for measuring medication intake but do not disclose any pattern or factors for non-adherence [18]. Besides, direct methods are more costly than indirect methods [19]. On the other hand, indirect methods can provide a reasonable estimation of medication adherence rate with a lower cost but may heavily depend on user usability [8]. Indirect methods include the patient's self-reporting, patient's questionnaires, pill-counting, assessing pharmacy refill rates, and electronic medication tracking systems. In most studies, patient questionnaires are used to measure the medication adherence rate of the participants, such as the Morisky medication adherence scale (MMAS) [20] and the Basel Assessment of Adherence to Immunosuppressive Medication Scale (BAASIS) [21].

2.2 Existing Quality Evaluation Criteria for Medication Adherence Application

Several existing quality evaluation methods were analyzed to determine the potential significant criteria for the medication adherence app. Most of the evaluation criteria for the medication adherence app are similar in many literatures but were categorized differently based on the researchers' perspectives of the study, such as categorized based on practical and functional features and categorized based on adherence interventions.

Santo et al. evaluated the medication adherence app according to predetermined features and evaluated its quality based on Mobile App Rating (MAR) [1]. The features were divided into two categories, which are practical features and functionality features. Practical features involve the availability of the app on both, the Apple store and Google store, the availability of a free version without third-party advertisement, and the presence of recent updates to solve any software issue. The functionality features consist of 17 features which are beneficial to increase medication adherence, such as history tracking, customizable reminders, drug database, adherence rewards, adherence charts, multiple language options, and flexible reminders audio.

Besides, Dayer et al. evaluated the quality of commercial mobile apps for medication adherence based on five (5) domains; general functions, adherence functions, medication management, connectivity, and health literacy [22]. General functions include the availability of privacy and safety features such as password functions and the Health Insurance Portability and Accountability Act. (HIPPA) statement. Adherence functions refer to alerts, reminders, and tracking functions. Multiple accounts and medication databases are included under the medication management domain. Next, connectivity measures the ability to share data, availability of cloud storage, and ability to create alerts without internet connectivity. Health literacy was measured based on nine (9) domains of the health literacy survey.

Next, the quality of the medication adherence app was evaluated based on the conceptual framework of adherence intervention, including behavioral and educational, in a study by Park et al. [2]. Park et al. distinguished the app based on four (4) important functions: medication reminders, medication tracking, refill reminders, and medication information storage. Also, they included additional features such as backup and export features, reminders without internet connectivity features, and multiple language features, which resulted in 12 features in total. Besides, the authors included the assessment for user-friendliness in the study based on three benchmarks: low, moderate, and high.

In a study by Ahmed et al., the medication adherence app was reviewed based on three (3) categories: evidence base on the development process or app effectiveness, the participation of health care experts during app development, and the adherence intervention strategies [23]. The adherence intervention strategies were divided into three (3) classes: behavioral, educational, and reminder. Behavioral refers to behavior change strategies utilized by the medication adherence app to improve the adherence rate of the user. It includes the availability of alerts, notifications, and reminders through Short Message Service (SMS). The educational strategy refers to features that spread the medication adherence information to the users. Lastly, the reminder strategy refers to alert strategies for medication intake. Examples of reminder strategies are tracking features and gamification features.

Based on the existing assessment of quality medication adherence apps, the potential criteria found were recorded and integrated with the Medication Adherence Application Quality (MedAd-AppQ) tool to be tested for the users' perspective of quality criteria.

2.3 Medication Adherence Application Quality (MedAd-AppQ) Tool

Medication Adherence Application Quality (MedAd-AppQ) tool is chosen for this study as it includes functional criteria to improve medication adherence [6], compared to the other evaluation tools such as Mobile App Rating (MAR) scale. MedAd-AppQ tool was developed by Ali et al. in 2018 specifically to evaluate medication adherence applications on iOS and Android operating systems. The tool consists of 24 items with thee (3) main categories: content reliability, feature usefulness, and feature convenience as shown in Fig. 1.

The content reliability category was developed based on a study by Huckvale et al. and Lewis to evaluate the credibility and reliability of the medical application. This category consists of 7 criteria's and 11 points with 2 sub-categories: the information of app developer and availability of privacy statement and disclaimer statement. 15 criteria's (29 points) in the feature usefulness category were selected based on literature reviews to assess the features of the medication adherence app.

The subcategories of feature usefulness are divided based on features such as scheduling multiple medication regimens, tracking and backup capabilities, availability of disease or medication information, and availability of reminders. Next, feature conveniences include two criteria such as convenience in data entry and the presence of a snooze function. The tool is evaluated using a point system, where one point is given when the app meets statements under the items, with a maximum number of 43 points in total.

3 Methodology

3.1 Study Design

The methodology for this study was divided into three stages: the identification of the quality criteria, the development of questionnaires, and the distribution of the questionnaires.

Stage 1: Identification of the Quality Criteria for Medication Adherence App The first stage is to identify the suitable criteria to evaluate the quality of medication adherence app. Literature reviews are conducted to identify existing methods and tools used to assess the quality of mobile apps used to assist with medication adherence. From the reviews, questionnaires Medication Adherence Application Quality (MedAd-AppQ) tool [6] was chosen to be applied in this study as it is the only tool that was purposely developed to assess medication adherence apps instead of any mobile app such as Mobile App Rating (MAR) scale.

Stage 2: Development of the Questionnaires Next, a set of questionnaires were developed based on the MedAd-AppQ tool and several additional criteria, which are further elaborated in the next section. As MedAd-AppQ is still a new tool and has not been widely applied in academic research, a few experts from the mobile applications and health fields helped to validate and review the questionnaires in this study. Furthermore, a pilot study was conducted to test the reliability of questionnaires. A final questionnaire was developed based on expert validation and the pilot study at the end of this stage.

Stage 3: Distribution of the Questionnaires The questionnaires were distributed to approximately 385 participants for data collection through online platforms such as Facebook, WhatsApp, and Telegram. The criteria of the participants include (1) adults with a minimum age of 18 years old, (2) chronic disease patients or a caregiver and (3) experience in using mobile app.

3.2 Development of Initial Quality Criteria

Based on the existing quality assessment, potential quality criteria have been identified for the medication adherence app. In addition, questionnaires were developed based on the Medication Adherence App Quality (MedAd-AppQ) evaluation tool, which consisted of three main categories (as shown in Fig. 1). The original statements were converted to sentences indicating the importance of the quality criteria in evaluating the medication adherence app, which resulted in different numbers of points for several criteria. For instance, 2 points under the criteria of "availability of the contact details of the app developer," which are "no contact details provided" and "a working email address/contact form is present", is converted to 1 point of "it is important to include app developer's email address, websites, or any contact medium in the app description."

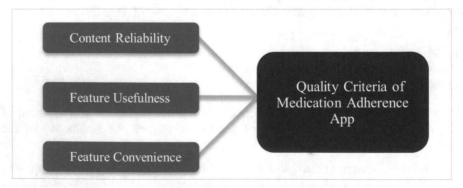

Fig. 1. Main categories of quality criteria medication adherence app for chronic disease

Furthermore, one criterion under the feature usefulness category regarding the feature to schedule medication taper was excluded from the list as the criteria contradicted the app's objective, which was to improve the medication adherence rate. Besides, the application of MedAd-AppQ tool revealed that none of the medication adherence apps consist of the stated feature [6]. Thus, there is no significant reason for the item to be included.

The feature usefulness category was expanded to include adherence rates, which have three criteria. Users of medication adherence apps can monitor their intake patterns by viewing adherence rates. This feature is important to enable the users to reflect on their medication intake and improve their behavior accordingly. This is one of the strategies mentioned in the behavior change technique to improve medication adherence [24]. Two of the criteria of this sub-category were derived from literature [1, 24] and one criterion was derived from existing criteria under tracking and backup capabilities features. In order to make the criteria more specific, one point was moved under the original criterion, and a second point was relocated under the adherence rates criterion.

Furthermore, five new criteria were included in the feature convenience category. The criteria related to the features for flexible alarm sounds [1], compatibility with wearable devices [23], language options [1], password options [1], and customizable time zone [1]. The questionnaires consist of 3 categories: content reliability with 7 criteria and 9 points, feature usefulness with 17 criteria and 25 points, and feature convenience with 7 criteria and 9 points. According to a 5-point Likert scale, the importance of each point will be ranked: 1 strongly disagree, 2 strongly disagree, 3 neutral, 4 agree, and 5 strongly agree. Additionally, each point is accompanied by a comment box that allows participants to suggest new quality criteria.

4 Findings

4.1 Demographic Analysis

This study is still in the process of finalizing the questionnaires and is expected to finish the data collection process by December 2022. The developed questionnaires have been reviewed and improved based on experts' feedbacks. Then, the questionnaires were answered by 39 participants for the preliminary study. This section explains the findings from the preliminary study, which was conducted for nearly a month, from 23rd June 2022 to 15th July 2022. The survey received 82 respondents; however, only 47.6% (39) responses met the criteria for the study after the data filtration process. The detailed result of the demographic section can be refer to Table 1.

Table 1. Demographic percentage.

Demographic	Category	Frequency	Percentage (%)
Gender	Male	12	30.77
	Female	27	69.23
Age	18 - 24 years old	4	10.26
	25 - 34 years old	22	56.41
	35 - 44 years old	2	5.13
	45 - 54 years old	2	5.13
	55 - 64 years old	6	15.38
	65 - 74 years old	3	7.69
Nationality	Malaysia	36	92.31
	Others (Bangladesh, Iraq, Pakistan)	3	7.69
Experience in using medication adherence app	Yes, but in the past	4	10.26
	Yes, currently using the app	1	2.56
	No	34	87.18
Experience in chronic condition	Yes, currently still experience the chronic condition	23	58.97
	Yes, but not anymore	16	41.03
Type of chronic condition(s)	Asthma	3	7.69
	Diabetes	9	23.08
	Hyperlipidemia (High cholesterol)	7	17.95
	Hypertension (High blood pressure)	23	58.97
	Ischemic heart disease	2	5.13
	Stroke	5	12.82
	Others	21	53.85
Period of experiencing the chronic condition(s)	Less than 6 months	2	5.13
	6 to 12 months	7	17.95
	1 to 2 years	3	7.69
	3 to 5 years	9	23.08

(*continued*)

Table 1. (*continued*)

Demographic	Category	Frequency	Percentage (%)
	More than 5 years	18	46.15
Type of medication(s) prescribed	None	1	2.56
	1 type of medication only	4	10.26
	2 - 3 type of medications	13	33.33
	More than 3 type of medications	21	53.85

Among 39 participants, 30.8% (12) were male and 69.2% (27) were female, in range from 18 to 74 years old. Most of the participants were in range of 25–34 years old, which made up 56.4% (22) of the responses, followed by 55–64 years old with 15.3% (6), 18–24 years old with 10.2% (4), 65–74 years old with 7.7% (3), and 5.1% (2) from responses in between 35–44 years old and 45–54 years old. 92.3% (36) of the participants were Malaysian and three person (7.7%) from other countries, such as Bangladesh, Iraq, and Pakistan. Less than 20% of the participants had experienced using app to improve their medication adherence and only 1 (2.6%) of the participants were still using the app while, the rest of them (4) had used it in the past. Example of the apps used are Doctorola, Inpatient care, Wassup and MyUbat. Three (3) apps were used for monitoring family members such as parents and sister, one was used as an aid during doctor's appointments and only one app was used to assist the phone owner with medication management.

4.2   Quality Criteria Categories Analysis (Mean and Standard Deviation)

As part of the statistical analysis of the medication adherence quality criteria, mean and standard deviation were calculated for each statement. The mean of each statement was calculated to show the average of the response for each statement. The standard deviation is calculated to show the dispersed of the results of each statement.

Each statement in content reliability was divided into two categories: (1) details of developer and target audience and (2) privacy policy and disclaimers. For each statement in the content reliability analysis, Table 2 shows the mean and the standard deviation. It can be concluded from the mean value that most participants agreed with the statement since there were no mean values below 3.4. The minimum mean is 3.67, which was "It is important to declare source of funding of the app in the app description". However, the agreement spread is quite high for each statement starting from 0.61 for statement 14 to 1.13 for statement 3. This indicates that statement 3 received the most diverse ratings from 1 to 5 in this category.

Table 2. Mean & Standard Deviation of Content Reliability.

Statement	Mean	Standard deviation
1. It is important to describe the objective of the app in the app description	4.31	0.89
2. It is important to mention the target user of the app in the app description	4.21	0.73
3. It is important to include business analyst's email address, websites, or any contact medium in the app description	3.79	1.13
4. It is important to include developer's email address, websites, or any contact medium in the app description	3.92	0.96
5. It is important to include medical representative's email address, websites, or any contact medium in the app description	4.33	0.66
6. It is important to include developer's name in the app description	3.79	0.98
7. It is important to include medical representative's name in the app description	4.18	0.85
8. It is important to include developer's credentials or qualifications in the app description	3.97	0.93
9. It is important to include medical representative's credentials or qualifications in the app description	4.38	0.63
10. It is important to include conflict of interest in the app description	3.92	0.74
11. It is important to include the privacy statement of user's data in the app or app description	4.28	0.69
12. It is important to declare that the app is not a replacement of health professional' advice in the app or app description	4.38	0.63
13. It is important to declare source of funding of the app in the app description	3.67	0.90
14. It is important to declare approval or certification statement from reliable health institution in the app description	4.31	0.61

For feature usefulness, 28 statements (as seen in Table 3) were needed to explained item under 5 sub-categories, which were (1) capacity to schedule varied dosing regiments, (2) tracking and backup capabilities, (3) drug and disease information, (4) reminders and alerts, and (5) adherence rates. In this category, all statements received above 4.0 for mean value. This indicated that respondents agreed with all the statements for feature usefulness. Furthermore, the standard deviation values for this category are considered low compared to the other categories where the minimum standard deviation is 0.50 and the maximum standard deviation is 0.87. This shows that the participants' rating value or agreement were distributed evenly within the same rating number compared to other categories.

Table 3. Mean & Standard Deviation of Feature Usefulness.

Statement	Mean	Standard deviation
1. It is important for the app to allow users to schedule daily medication	4.46	0.64
2. It is important for the app to allow users to schedule medication based on required day	4.56	0.50
3. It is important for the app to allow users to schedule medication based on required cycle	4.49	0.60
4. It is important for the app to allow users to schedule medications with dosing that vary at different times	4.41	0.68
5. It is important for the app to allow users to record symptoms or side effects of medication	4.56	0.55
6. It is important for the app to allow users to save medication list internally on their phone	4.49	0.68
7. It is important for the app to allow users to save medication list on the cloud	4.51	0.56
8. It is important for the app to allow users to backup medication list internally on their phone	4.33	0.81
9. It is important for the app to allow users to backup medication list on the cloud	4.38	0.71
10. It is important for the app to allow users to manually record doses intake	4.38	0.63
11. It is important for the app to allow users to manually record missed doses	4.41	0.72
12. It is important for the app to automatically record missed doses if users failed to record doses intake after some time	4.26	0.79
13. It is important for the app to allow users to share medication list to other parties such as healthcare professionals or family members	4.28	0.76
14. It is important for the app to provide medication information from the reliable source	4.33	0.62
15. It is important for the app to provide the date of the latest update on the medication information	4.56	0.55
16. It is important for the app to provide disease-related information with the reliable source	4.49	0.60
17. It is important for the app to provide the date of the latest update on the disease-related information	4.36	0.63

(continued)

Table 3. (*continued*)

Statement	Mean	Standard deviation
18. It is important for the app to include medication form (tablet, liquid, cream), shape, colors, icons, or pictures to assist users to identify medication	4.36	0.67
19. It is important for the app to enable users to schedule reminders for the secondary users (such as family members)	4.36	0.74
20. It is important for the app to enable users to receive reminder for medication refill	4.46	0.55
21. It is important for the app to enable users to receive reminder for medical appointment	4.49	0.56
22. It is important for the app to provide warning when user register medication doses higher than maximum doses allowed	4.46	0.64
23. It is important for the app to have feature that able to alert relevant party such as family members or caregivers if users miss their doses	4.44	0.60
24. It is important for the app to enable users to schedule reminders without internet connectivity	4.56	0.50
25. It is important for the app to provide statistics or charts on medication intake trends	4.23	0.87
26. It is important for the app to send congratulatory message when users achieve certain level of adherence rates	4.15	0.74
27. It is important for the app to give badges or points to users who achieve certain level of adherence rates	4.08	0.81
28. It is important for the app to allow users to share medication adherence trends to other parties such as health care professionals or family members	4.26	0.75

Lastly, 9 statements covered the feature convenience category (Table 4). Almost all statements in this category had a mean higher than 4.0 except for statement 4, which mentioned the importance of having snooze or postpone functions. 67% of the statements were strongly agreed by the participants as their mean value are higher than 4.21, while rest of the statements were agreed by the participants. However, the range of the standard deviation for this category is similar to the first category where both categories had quite a high difference between the lowest value standard deviation and the highest value of standard deviation. The standard deviation in feature convenience started from 0.54 to 1.06. The standard deviation values display the spread of participants opinion in valuing each feature for the convenience of using medication adherence app.

In overall, most of the statements in the categories were agreed by the participants as quality criteria for medication adherence app, however the high number in standard deviation should be consider in keeping or removing the item for the finalized framework.

Table 4. Mean & Standard Deviation of Feature Convenience.

Statement	Mean	Standard deviation
1. It is important for the app to provide autocomplete text features during entry of medication list	4.21	0.77
2. It is important for the app to provide text recognition features during entry of medication list	4.38	0.67
3. It is important for the app to provide barcode scanning features during entry of medication list	4.21	0.80
4. It is important for the app to allow users to snooze or postpone reminders functions	3.92	1.06
5. It is important for the app to allow users to choose different notification sounds for reminders or alerts	4.18	0.79
6. It is important for the app to allow users to pair the app with wearable device such as smartwatch	4.38	0.67
7. It is important for the app to provide language options other than English language	4.62	0.54
8. It is important for the app to provide users with password features while accessing the app	4.10	1.02
9. It is important for the app to allow users to change to time zone based on the corresponding time especially when travelling to another country	4.38	0.75

5   Discussions

The purpose of this paper is to identify the quality criteria of medication adherence apps in accordance with end-user perception. The end-users for the medication adherence app refer to the patients or caregivers with the need for medication management assistance. This research emphasizes chronic disease patients as they are more likely to monitor their health than those with no health issues [25]. As part of the initial criteria for participation in this study, participants should have some experience using a mobile app for medication management. However, findings from the pre-liminary study showed the difficulty in finding participants with medication adherence experience. Only 13% out of total participants had the experience in using mobile app and most of them utilized the app to monitor other person medication management such as parent and sister. Therefore, the participants' focus for the data collection process will be broadened to the potential users of medication adherence instead of experienced users.

Furthermore, most of the statements in the survey were agreed by the participants, yet several values of the standard deviation for content reliability and feature convenience were quite high. Further analysis will be needed to decide whether the statement can be kept or remove from the framework as the rating value were very spread in those statements. For example, statement 3 from Content Reliability, which was "3. It is important to include business analyst's email address, websites, or any contact medium in the app description" had 3.79 mean value, which indicated that the statement were agreed by the participants but had 1.13 standard deviation value. Nearly 41% of respondents rated Neutral, Disagree, and Strongly Disagree for this statement, even though the majority rated Strongly Agree or Agree. In short, several improvements and further analyses need to be made to the current questionnaires before being distributed to the targeted participants.

This study aimed to develop a framework of quality criteria for medication adherence app to serve as guidelines to developers or researchers in developing apps utilized for medication management. Although there is abundant medication adherence app, the functions and features usually did not designed to tackle adherence issues from various factors as the app in the current market lacks theories utilization [26, 27]. Studies had revealed positive effects on medication adherence with the used of mobile app developed with framework interventions, such as behavior change techniques (BCT) [28], with the involvement of healthcare professionals (HCP) [29], and utilization of gamification [30] in real-world application [31]. The paper focuses on a general medication adherence app for medication management, not specific to any specific condition. Therefore, the review of the study will not reflect the quality criteria for medication adherence app that target specific diseases condition, such as diabetes patients only and medication adherence app developed for health care professionals or clinicians as the purpose of those apps might differ from each other. Moreover, the criteria are limited to assessing the quality of the mobile app, but it will not indicate the effectiveness or usability evaluation of the mobile app.

6 Conclusion

In conclusion, mobile application intervention had been widely used to increase the medication adherence rate by assisting users in medication management. Despite this, there is a lack of research on patients' perspectives on medication adherence apps. This research aims to provide insight into the quality criteria that are valued by potential users of the medication adherence app. By incorporating this information into the design and development of the app, mobile developers could include key quality features during the development process. Additionally, researchers can use the finalized criteria to develop complex quality criteria that include the perspectives of each stakeholder group, including healthcare professionals, mobile developers, caregivers, and patients. In addition, further research on the effectiveness and usability of identified high-quality medication adherence applications is needed.

References

1. Santo, K., Richtering, S. S., Chalmers, J., Thiagalingam, A., Chow, C.K., Redfern, J.: Mobile phone apps to improve medication adherence: a systematic stepwise process to identify high-quality apps. JMIR mHealth uHealth 4(4), e132 (2016)
2. Park, J.Y.E., Li, J., Howren, A., Tsao, N.W., de Vera, M.: Mobile phone apps targeting medication adherence: quality assessment and content analysis of user reviews. JMIR mHealth uHealth 7(1), 1–12 (2019)

3. Nguyen, H.D., Jiang, X., Poo, C.: Designing a social mobile platform for diabetes self-management: a theory-driven perspective, vol. 1, pp. 89–95 (2015)
4. Rootes-Murdy, K., Glazer, L., Van Wert, M.J., Mondimore, F. M., Zandi, P. P.: Mobile technology for medication adherence in people with mood disorders: a systematic review. J. Affect. Disord. **227**, 613–617 (2018)
5. Nicholas, J., Larsen, M. E., Proudfoot, J., Christensen, H.: Mobile apps for bipolar disorder: a systematic review of features and content quality. J. Med. Internet Res. **17**(8), e198 (2015)
6. Ali, E.E., Teo, A.K.S., Goh, S.X.L., Chew, L., Yap, K.Y.-L.: MedAd-AppQ: A quality assessment tool for medication adherence apps on iOS and android platforms. Res. Soc. Adm. Pharm. **14**(12), 1125–2113 (2018)
7. Kisa, A., Sabaté, E., Nuño-Solinís, R.: Adherence to Long-Term therapies: evidence for action (2003)
8. Aldeer, M., Javanmard, M., Martin, R.: A review of medication adherence monitoring technologies. Appl. Syst. Innov. **1**(2), 14 (2018)
9. Car, J., Tan, W.S., Huang, Z., Sloot, P., Franklin, B.D.: eHealth in the future of medications management: personalisation, monitoring and adherence. BMC Med. **15**(1), 1–9 (2017)
10. Gurumurthy, R.: Importance of medication adherence and factors affecting it. IP Int. J. Compr. Adv. Pharmacol. **3**(2), 69–77 (2018)
11. Azmi H.M, Faha, S.: A national survey on the use of medicines (2016)
12. Bili, A.A.M., Zha, L.: Knowledge of type 2 diabe tes mc llitus and adherence to management guideline s: a cross-sectional study in Juba, South Sudan. South Sudan Med. J. **11**(4), 84–88 (2018)
13. Filimão, D.B.C., Moon, T.D., Senise, J.F., Diaz, R.S., Sidat, M., Castelo, A.: Individual factors associated with time to non-adherence to ART pick-up within HIV care and treatment services in three health facilities of Zambézia Province, Mozambique. PLoS ONE **14**(3), 1–15 (2019)
14. Kleinsinger, F.: The Unmet Challenge of Medication Nonadherence. Perm. J. **22**, 1–3 (2018)
15. Rosen, O.Z., Fridman, R., Rosen, B.T., Shane, R., Pevnick, J.M.: Medication adherence as a predictor of 30-day hospital readmissions. Patient Prefer. Adherence **11**, 801–810 (2017)
16. Azad, A.K., Muhammad, K.R., Hossin, M.M., Robiul, I., Abdullahi, M.M., Islam, M.A.: Medication wastage and its impact on environment: Evidence from Malaysia. Pharmacologyonline **3**, 114–121 (2016)
17. Rahim, R.I.A, Baharudin, N.: Pembaziran Ubat Hospital, Berita Harian (2017)
18. Lam, W. Y., Fresco, P.: Medication adherence measures: an overview. Biomed. Res. Int. **2015** (2015)
19. Anghel, L.A., Farcas, A.M., Oprean, R.N.: An overview of the common methods used to measure treatment adherence. Med. Pharm. Reports **92**(2), 117–122 (2019)
20. Morawski, K., et al.: Association of a smartphone application with medication adherence and blood pressure control: the MedISAFE-BP randomized clinical trial. JAMA Intern. Med. **178**(6), 802–809 (2018)
21. Han, A., et al.: Mobile medication manager application to improve adherence with immunosuppressive therapy in renal transplant recipients: A randomized controlled trial. PLoS ONE **14**(11), 1–18 (2019)
22. Dayer, L.E., et al.: Assessing the medication adherence app marketplace from the health professional and consumer vantage points. JMIR mHealth uHealth **5**(4), e6582 (2017)
23. Ahmed, I., et al.: Medication adherence apps: review and content analysis. JMIR mHealth uHealth **6**(3), e6432 (2018)
24. Carmody, J.K., Denson, L.A., Hommel, K.A.: Content and usability evaluation of medication adherence mobile applications for use in pediatrics. J. Pediatr. Psychol. **44**(3), 333–342 (2019)
25. Singh, K., et al.: Patients' and nephrologists' evaluation of patient-facing smartphone apps for CKD. Clin. J. Am. Soc. Nephrol. **14**(4), 523–529 (2019)

26. Nkhoma, C.D., Soko, J., Banda, K.J., Greenfield, D., Li, Y. C., Iqbal, U.: Impact of DSMES app interventions on medication adherence in type 2 diabetes mellitus: systematic review and meta-analysis. BMJ Heal. Care Inf. **28**(1), 1–9 (2021). https://doi.org/10.1136/bmjhci-2020-100291

27. Chew, S., Lai, P.S.M., Ng, C.J.: Usability and utility of a mobile app to improve medication adherence among ambulatory care patients in Malaysia: Qualitative study. JMIR Mhealth Uhealth **8**(1), e15146 (2020). https://doi.org/10.2196/15146

28. Armitage, L.C., Kassavou, A., Sutton, S.: Do mobile device apps designed to support medication adherence demonstrate efficacy? a systematic review of randomised controlled trials, with meta-analysis. BMJ Open **10**(1), e032045 (2020). https://doi.org/10.1136/bmjopen-2019-032045

29. Al-Arkee, S., et al.: Mobile apps to improve medication adherence in cardiovascular disease: Systematic review and meta-analysis. J. Med. Internet Res. **23**(5), 1–18 (2021). https://doi.org/10.2196/24190

30. Tran, S., Smith, L., El-Den, S., Carter, S.: The use of gamification and incentives in mobile health apps to improve medication adherence: scoping review. JMIR Mhealth Uhealth **10**(2), e30671 (2022). https://doi.org/10.2196/30671

31. Wiecek, E., Torres-Robles, A., Cutler, R.L., Benrimoj, S.I., Garcia-Cardenas, V.: Impact of a multicomponent digital therapeutic mobile app on medication adherence in patients with chronic conditions: Retrospective analysis. J. Med. Internet Res. **22**(8), 1–13 (2020). https://doi.org/10.2196/17834

Compliance Intention for E-learning Continue Usage: Validation of Measurement Items

Ken Ditha Tania[1,2]([✉]), Norris Syed Abdullah[1], Norasnita Ahmad[1], and Samsuryadi Sahmin[2]

[1] Azman Hashim International Business School, Universiti Teknologi Malaysia, Bahru, Johor, Malaysia
kenya.tania@gmail.com, {norris,norasnita}@utm.my
[2] Faculty of Computer Science, Sriwijaya University, Palembang, Indonesia
syamsuryadi@unsri.ac.id

Abstract. As a result of the COVID 19 pandemic, universities are increasingly relying on e-learning. The most important element of e-learning is maintaining usage. The goal of this research is to investigate measure items and validation for compliance intention for e-learning continue usage. This study also looks at the elements that influence compliance intentions, as well as the most critical aspects that influence e-learning usage. This study's research model is focused on the technology acceptance model (TAM), the expectation confirmation model (ECM), the Theory Planned Behavior (TPB), and the Protection Motivation Theory (PMT). A thorough literature review was used to extract the required variables for the compliance intention for e-learning continue usage. The relevant variables were then collected in a testing instrument to hypothesize the relationship between them. The results may be of importance to institutions, academics, administrators, and decision-makers involved in the design, development, and implementation of e-learning.

Keywords: E-Learning · Continue usage intention · compliance intention

1 Introduction

As a result of the COVID-19 pandemic, educational institutions such as universities are rushing to adopt e-learning. The most important element of e-learning is maintaining usage [1–3]. The importance of using e-learning continuously should be emphasized by researchers [4]. Several research on continued usage of e-learning have been conducted. Despite this, no research has been conducted on compliance intention to continue using of e-learning. This study is an extension of work that was first presented at the IRICT Conference [5, 6].

The degree to which users follow organizational IT policies is referred to as compliance intention [7]. IT policies are the regulations and norms that determine the technological options available to users. Compliance is a much broader concept that encompasses both usage and policy-mandated aspects [8]. According to [5], earlier e-learning continuity models did not conduct research from the perspective of compliance. If the student does not intend to follow the standards for using e-learning, it is highly unlikely that e-learning would be utilized indefinitely. Compliance will be applied to assure that a system is utilized indefinitely, such as the Enterprise System [9], so the paper [5] focuses on compliance intention for e-learning continuance use. Recognizing the elements that influence a student's desire to continue using an e-learning system can help not just software engineers, but also instructors and suppliers build the best approaches for maximizing its use. Therefore, this study focuses on measuring items and validation for the compliance intention of e-learning continue usage.

2 Compliance Intention for E-learning Continue Usage

In the field of information technology, compliance is a higher-level concept that encompasses both the use of technology and the mandatory elements that specify how it should be used [8]. Requirements for compliance might come from a variety of sources [10]. A user's compliance intention is their desire to protect information technology systems from potential regulatory harm [11]. On an individual basis, compliance intention is the possibility of a person sticking to the compliance management [12]. Compliance intention has worked in a variety of contexts. The intention of an individual to comply with norms is described as the individual's willingness to engage in voluntary behavior [13]. We concentrate on compliance intention because we believe that in obligatory IT environments, compliance is the fundamental issue. Compliance is a higher-level term that incorporates both policy-mandated usage and elements. Usage is a requirement for compliance, but it is not sufficient. To be compliant, one must utilize technology. Some of these studies have overlooked the reality that e-learning should be examined not only through the lens of IS theories, but also through the lens of other theories, such as compliance intention theory. Moreover, it has been discovered that previous models lack the fundamental constructs for properly proposing a model for the continued usage of e-learning systems. As a result, a thorough review of the literature can provide more light on the issues and lead to the development of more reliable models.

3 Method

To meet the study's objectives, a systematic literature review (SLR) was conducted. The SLR was used to discover and analyze all published papers in a significant research method. The SLR was chosen because of the basis of the investigation. Systematic reviews and meta-analyses are key tools for summarizing data correctly and reliably [17]. The research was conducted in three stages: planning, conducting, and reporting [18]. To be completed, the SLR was processed through several phases. The review protocol was first outlined and illustrated. The criteria for inclusion and exclusion were then discussed. Then there was a discussion about the search approach. The study selection procedure was then defined. The quality assessment was then clarified. The data extraction and synthesis were then discussed in detail.

The methodology of this study encompasses the following steps:

- Construct Extraction
- Measurement Items Derivation
- Validation

4   Constructs Extraction

The researcher extensively evaluated all the primary studies in the SLR. This was done to identify the possible factors that influence e-learning continue usage as well as compliance intention. The previous research yielded several variables connected to continue usage of e-learning. It is, however, impossible to incorporate all variables into a single study model. As a result, the researcher chooses the factors that are most important to the study's goal. Table 1 lists the primary theories and models used in this study to derive the main factors continue usage of e-learning.

Table 1. The Identified Constructs

No.	Theories & Models	Constructs	Studies
1	Theory Acceptance Model	Perceived Usefulness, Perceived Ease of Use, Continuous Intention to Use	[14]
2	Expectation Confirmation Theory	Perceived Usefulness Satisfaction, Continuous Intention to Use	[15]
3	Theory Planned Behaviour	Attitude, Subjective Norm	[16]
4	Protection Motivation Theory	Self-Efficacy, Compliance Intention	[17] [18]

5   Measurement Items Derivation

This research explores in the central ideas for implementing a new scale to assess the continue usage of e-learning based on compliance intention theory. To do so, a reliable and validated measurement equipment was developed for making assumptions the correlation between different variables. There are two portions to the questionnaire that has been constructed. The questions in Section A are intended to aid in the collection of demographic information from respondents. The five-point Likert Scale is used to collect data for examining the constructs in the second section, Section B. Section B's options are specified by a range of options ranging from "Strongly Disagree" to "Strongly Agree." To ensure that the responses were not picked absent mindedly or arbitrarily, an attention-check item was implemented as part of the questionnaire. To ensure that participants have read thoroughly all the questions, they should strongly disagree with this reasonings. Section B contains data that is utilized to validate the research measurement instrument. The Likert Scale responses range from 1 to 5, with 1 indicating Strongly Disagree, 2 indicating Disagree, 3 indicating Neutral, 4 indicating Agree, and 5 indicating Strongly Agree The participants are required to pick a specific response from a list of Likert Scales for each question in this part, which relies on the variables that are significant in affecting e-learning continue usage of e-learning. The study's measures or measurement items are adapted from prior studies. Table 2 shows lists the measurement items and their associated constructs.

Table 2. Constructs and Measurement Items

Constructs	Items	Measuring Items	References
Perceived Usefulness	PU1	Usefulness of e-learning can improve my learning experience	[14]
	PU2	Usefulness of e-learning can increase my learning effectiveness	[14]
	PU3	I believe that usefulness allow me to continuously use of e-learning	[14]
	PU4	I find that usefulness of e-learning will encourage me to continuously comply with available policy in using it	[19, 20]
Perceived Ease of Use	PEU1	Ease of use of e-learning can improve my learning experience	[14]
	PEU2	Ease of use of e-learning can increase my learning effectiveness	[14]
	PEU3	I believe that ease of use allow me to continuously use of e-learning	[14]
	PEU4	I find that ease of use of e-learning will encourage me to continuously comply with available policy in using it	[19, 20]
Satisfaction	SAT1	Rules about the use of e-learning is clear and understandable that motivate me to continuously use it	[21]
	SAT2	I am satisfied with e-learning and will continuously use it	[21]
	SAT3	I believe that being satisfied allows me to use e-learning continuously	[21]
	SAT4	I am satisfied with the policy regarding the use of e-learning and this will encourage me to comply with it	[22]
Attitude	AT1	Using the E-Learning system is not very difficult	[16, 23]
	AT2	Using the E-Learning system is not very complicated	[16, 23]
	AT3	I believe that continuously comply to the e-Learning policy does not create stress for me	[16, 23]
Subjective Norm	SN1	People who influence me think that I should use e-learning continuously	[24]

(continued)

Table 2. (*continued*)

Constructs	Items	Measuring Items	References
	SN2	People whom I respect think that I should use e-learning continuously	[24]
	SN3	People think that it is important to me to continuously comply to e-learning policy while using it	[25]
Self-Efficacy	SE1	I feel comfortable about performing tasks to use e-learning continuously	[26] [27]
	SE2	I have the necessary knowledge to use e-learning continuously	[26] [27]
	SE3	I believe that self-efficacy allow me to continuously use of e-learning	[26] [27]
	SE4	Self-efficacy of using e-Learning will encourage me to continuously comply with e-learning policy	[26] [27]
Compliance intention	CI1	I intend to comply with e-learning usage policy	[12, 28]
	CI2	I intend to continuously comply to the usage policy while using the e-learning	[12, 28]
	CI3	I intend to continuously comply to the usage policy when using the e-learning in the future	[12, 28]
Continue usage of e-learning	CU1	I intend to use e-learning continuously by complying to its usage policy	[15]
	CU2	I plan to use e-learning continuously by complying to its usage policy	[15, 29]
	CU3	I predict I will use e-learning continuously in the futures by complying to its usage policy	[15]

6   Validation

Based on its reflective operational definition of constructs, content validity pinpoints the degree of item acceptance in the assessment [31]. Content validity can be carried out in a variety of methods [32]. The Content Validity Index (CVI) [33] was utilized to validate this research instrument in this study. The CVI technique, according to [32], is adaptable and requires a minimum of three interrater panel members. They also argue that CVI is more realistic in terms of time and cost. To put it another way, CVI is simple and quick to achieve.

As a result, the items were evaluated in terms of content relevance. For each item, the Lynn [33] recommendation for a 4-point ranking system was applied to determine relevance in content. There was also an option for an expert to comment on each item if necessary. Various scale anchors for measuring relevancy were discussed in the literature; however, Davis [30] anchor use was chosen because it is the most used label for content validity by researchers. The labels are: "Not Relevant = 1; Relevant but not important = 2; Relevant but need review = 3; Highly Relevant = 4". The agreement in proportions is 3 and 4. The CVI was calculated using the formula based on their responses [34].

$$I - CVI = \frac{N_c}{n}$$

The number of experts who rate an item as vital is represented by Nc in the preceding formula, while the total number of experts is represented by n. Items with a I-CVI score over a certain threshold, which is determined by the number of experts that rate each item, are kept [34]. Ten experts in the field of information systems were invited to validate the instrument in this study, as shown in Table 3. Only five of them agree to validate the instrument.

Table 3. List of IS Experts invited for Content Validity

Serial	Status of Experts	Numbers of Expert that invited	Numbers of experts that respondent
1	Professors	2	1
2	Associate Prof	2	1
3	Doctors	6	3
Total		10	5

As a result, a total of 5 experts were valid and were used for the ranking, which is higher than the recommended minimum amount [32]. Table 4 shows the relevancy replies of I-CVI evaluators for each item.

$$S - CVI = \frac{\sum (I - CVI)}{p}$$

Many experts agree that an S-CVI of 0.8 or higher is acceptable [31]. However, other academics advise starting at 0.78 [35]. In this study, however, 0.8 is used. As a result, all items were acceptable. Expert replies and recommendations are also modified. As a result, certain adjustments to the items' clarity or language have been made.

Table 4. Experts' content validity evaluation scores

Item	Expert 1	Expert 2	Expert 3	Expert4	Expert 5	Numbers of Relevance	I-CVI	Agreement
PU1	1	4	4	4	3	4	0.8	Acceptable
PU2	4	4	4	4	3	5	1	Acceptable
PU3	3	4	4	4	3	5	1	Acceptable
PU4	-	4	4	4	3	4	0.8	Acceptable
					S-CVI		0.9	
PEU1	4	4	4	4	3	5	1	Acceptable
PEU2	4	4	4	4	3	5	1	Acceptable
PEU3	1	4	4	4	4	4	0.8	Acceptable
PEU4	-	4	4	4	3	4	0.8	Acceptable
					S-CVI		0.9	
SAT1	4	4	4	4	3	5	1	Acceptable
SAT2	1	4	4	4	3	4	0.8	Acceptable
SAT3	4	4	4	4	3	5	1	Acceptable
SAT4	-	4	4	4	4	4	0.8	Acceptable
					S-CVI		0.9	
AT1	4	4	4	4	4	5	1	Acceptable
AT2	4	4	4	4	4	5	1	Acceptable
AT3	4	4	4	4	4	5	1	Acceptable
					S-CVI		1	
SN1	3	4	4	2	3	4	0.8	Acceptable
SN2	3	4	4	2	3	4	0.8	Acceptable
SN3	3	4	4	2	3	4	0.8	Acceptable
					S-CVI		0.8	
SE1	1	4	4	3	3	4	0.8	Acceptable
SE2	3	4	4	4	3	5	1	Acceptable
SE3	1	4	4	4	3	4	0.8	Acceptable
SE4	-	4	4	4	3	4	0.8	Acceptable
					S-CVI		0.85	
CI1	4	4	4	4	3	5	1	Acceptable
CI2	4	4	4	4	3	5	1	Acceptable
CI3	4	4	4	4	3	5	1	Acceptable

(*continued*)

Table 4. (*continued*)

Item	Expert 1	Expert 2	Expert 3	Expert4	Expert 5	Numbers of Relevance	I-CVI	Agreement
						S-CVI	1	
CU1	4	4	4	2	3	4	0.8	Acceptable
CU2	4	4	4	2	3	4	0.8	Acceptable
CU3	3	4	4	2	3	4	0.8	Acceptable
						S-CVI	0.8	
						S-CVI	0.893	

Note: S-CVI = Average scale content validity index of each construct

7   Summary

To examine the e-learning continue usage according to the compliance intention perspective, this study generated reliable measurement items for the constructs retrieved from the SLR. The items have been altered from previous material to fit the study's goals. To be validated, the content of the questionnaire items was analyzed. The Content Validity Index (CVI) was applied, and the items were validated in terms of relevancy and simplicity. Consequently, the number of the items was changed and the items' content was edited and modified in terms of wording. In this ongoing study, the next step that can be a recommendation for further investigation as well, the reliability and validity of the measurement items will be assessed through the pilot and the real survey.

Acknowledgment. This paper is a result of a series of discussions with UNSRI's visiting professor, the PM. Dr. Abdul Razak Che Hussin.

References

1. Ji, Z., Yang, Z., Liu, J., Yu, C.: Investigating users' continued usage intentions of online learning applications. Information **10** (2019)
2. San-Martín, S., Jiménez, N., Rodríguez-Torrico, P., Piñeiro-Ibarra, I.: The determinants of teachers' continuance commitment to e-learning in higher education. Educ. Inf. Technol. **25**(4), 3205–3225 (2020). https://doi.org/10.1007/s10639-020-10117-3
3. Cheng, M., Yuen, A.: Student continuance of learning management system use: a longitudinal exploration. Comput. Educ. **120** (2018)
4. Hayashi, A., Chen, C., Ryan, T., Wu, J.: The role of social presence and moderating role of computer self efficacy in predicting the continuance usage of e-learning systems. J. Inform. Syst. Educ. 15 (2020)
5. Tania, K.D., Abdullah, N.S., Ahmad, N., Sahmin, S.: Student compliance intention for e-learning conitnue use. In IRICT, Malaysia (2020)
6. Tania, K.D., Abdullah, N.S., Ahmad, N., Sahmin, S.: Continued usage of e-learning: a systematic literature review. J. Inform. Technol. Manag. **14** (2022)

7. Hwang, I., Kim, D., Kim, T., Kim, S.: "Why not comply with information security? An empirical approach for the causes of non-compliance. Online Inform. Rev. **41** (2017)
8. Xue, Y., Liang, H., Wu, L.: Punishment, justice, and compliance in mandatory IT Settings. Informat. Syst. Res. **22** (2011)
9. See, B.P., Yap, C.S., Ahmad, R.: Antecedents of continued use and extended use of enterprise systems. Behav. Inform. Technol. (2019)
10. Abdullah, N.S., Indulska, M., Sadiq, S.: Compliance management ontology – a shared conceptualization for research and practice in compliance management. Inf. Syst. Front. (2016)
11. Vance, A., Siponen, M., Pahnila, S.: Motivating IS security compliance: Insights from Habit and Protection Motivation Theory. Informat. Manag. **49** (2012)
12. Hofeditz, M., Ann-Maienaber, N., Dysvik, A., Schewe, G.: Want to' versus 'Have to': Intrinsic and extrinsic motivators as predictors of compliance behavior intention. Human Res. Manag. **56** (2015)
13. Kim, S., Kim, Y.: The effect of compliance knowledge and compliance support systems on information security compliance behaviour. Journal of Knowledge Management **21** (2017)
14. Davis, F.D.: Perceived usefulness, perceived ease of use, and user acceptance of information technology. MIS Q. **22** (1989)
15. Bhattacherjee, A.: Understanding information systems continuance an expectation-confirmation model. MIS Q. **25** (2001)
16. Ajzen, I., Martin, F.: ttitude-behavior relations: A theoretical analysis and review of empirical research. Psychological Bull. **84** (1975)
17. Floyde, A., Lawson, G., Shalloe, S., Eastgate, R., D'Cruz, M.: The design and implementation of knowledge management systems and e-learning for improved occupational health and safety in small to medium sized enterprises. Saf. Sci. **60**, 69–76 (2013)
18. Ifinedo, P.: Understanding information systems security policy compliance: an integration of the theory of planned behavior and the protection motivation theory. Comput. Sec. **31** (2012)
19. Al-Omari, A., El-Gayar, O.F., Deokar, A.: Security policy compliance: user acceptance perspective. In: Hawaii International Conference on System Sciences (2012)
20. Liang, H., Xue, Y., Wu, L.: Ensuring employees' IT compliance: carrot or stick? Inform. Syst. Res. **24** (2013)
21. Al-samarrie, H., Teng, B.K, Alzahrani, A.I..: E-learning continuance satisfaction in higher education: a unified perspective from instructors and students. Stud. Higher Educ. **43** (2017)
22. Nadirov, O., Aliyev, K.D.B., Sharifzada, I., Aliyeva, R.: Life satisfaction and tax morale in azerbaijan: mediating role of institutional trust and financial satisfaction. Sustainability (2021)
23. Wu, B., Zhang, C.: Empirical study on continuance intentions towards E- Learning 2.0 systems. Behav. Informat. Technol. **33** (2014)
24. Lee, M.-C.: Explaining and predicting users' continuance intention toward e-learning: an extension of the expectation–confirmation model. Comput. Educ. (2010)
25. Bin-Nashwan, S.A., Abdul-Jabbar, H., Dziegielewski, S.F., Aziz, S.A.: Moderating effect of perceived behavioral control on islamic tax (Zakah) compliance behavior among businessmen in Yemen. J. Soc. Serv. Res. (2020)
26. Bulgurcu, B., Cavusoglu, H., Benbasat, I.: Information security policy compliance: an empirical study of rationality-based beliefs and information security awareness. MIS Quart. **34**(3) (2010)
27. Ifinedo, P.: Roles of perceived fit and perceived individual learning support in students' weblogs continuance usage intention. Int. J. Educ. Technol. Higher Educ. **15** (2018)
28. Herath, T., Rao, H.R.: Protection motivation and deterrence: a framework for security policy compliance in organisations. Europ. J. Inform. Syst. **18** (2009)
29. Al-Busaidi, K.A., Al-Shihi, H.: Key factors to instructors' satisfaction of learning management systems in blended learnin. J. Comput. High. Educ. **24** (2012)

30. Davis, L.L.: Instrument review: getting the most from a panel of experts. Appli. Nursing Res. (1992)
31. Polit, D.F., Beck, C.T.: The content validity index: are you sure you know what's being reported? critique and recommendations. Res. Nurs. Health **29** (2006)
32. Tojib, D., Sugianto, L.: Content validity of instruments in is research. J. Inform. Technol. Theory Appli. (2006)
33. Lynn, M.R.: Determination and quantification of content validity. Nurs. Res. **35** (1986)
34. Flores, W., Antonsen, E.: The development of an instrument for assessinginformation security in organizations: Examiningthe content validity using quantitative methods. In: International Conference on Information Resources Management (CONF-IRM) (2013)
35. Mikalef, P., Pateli, A.: It flexibility and competitive performance: the mediating role of IT-enabled dynamic capabilities. In: Twenty-Fourth European Conference on Information Systems (ECIS), Istanbul, Turkey (2016)
36. Almaiah, M.A., Al-Khasawneh, A., Althunibat, A.: Exploring the critical challenges and factors influencing the E-learning system usage during COVID-19 pandemic. Educ. Inf. Technol. **25**(6), 5261–5280 (2020). https://doi.org/10.1007/s10639-020-10219-y
37. Bøe, T., Sandvik, K., Gulbrandsen, B.: Continued use of e-learning technology in higher education: a managerial perspective. Stud. High. Educ. (2020)
38. Ashrafi, A., Zareravasan, A., Savoji, S.R., Masoumeh, A.: Exploring factors influencing students' continuance intention to use the learning management system (LMS): a multi-perspective framework. Interact. Learn. Environm. (2020)
39. Vershitskaya, E.R., Mikhaylova, A.V., Gilmanshina, S.I., Dorozhkin, E.M., Epaneshnikov, V.V.: Present-day management of universities in Russia: Prospects and challenges of e-learning. Educ. Inf. Technol. **25**(1), 611–621 (2019). https://doi.org/10.1007/s10639-019-09978-0
40. Wynd, C.A., Schmidt, B., Schaefer, M.A.: Two quantitative approaches for estimating content validity. Western J. Nurs. Res. **25** (2003)

Derivation Citizen Engagement Model for Government Social Media

Ari Wedhasmara[1]([✉]), Samsuryadi[1], and Ab Razak Che Hussin[2]

[1] Faculty of Computer Science, Universitas Sriwijaya, Palembang, Indonesia
{a_wedhasmara,samsuryadi}@unsri.ac.id

[2] Azman Hashim International Business School (AHIBS), Universiti Teknologi Malaysia,
Johor Bahru, Malaysia
abrazak@utm.my

Abstract. Citizen engagement is a terminology that has changed in the current period, starting from citizen involvement until now known as citizen engagement. Citizen engagement is the willingness and ability of regular people to band together, negotiate, and take action on topics they perceive to be important. This paper focused on investigating citizen engagement on government social media. Related to this, the use of government social media is still experiencing difficulties in order to increase the attitude of public involvement towards government social media where the public get all information services and announcements related to the government. To overcome this problem, a citizen engagement model needs to be developed significantly to increase or promote CE on GSM. In summary, the citizen engagement model is able to become a significant guide in presenting all the important factors that need to be possessed in encourage citizen engagement attitudes towards GSM. This essay seeks to develop a model that includes all of these factors. These factors have been identified by the Systematic Literature Review analysis (SLR) method and then applied significantly to increase knowledge about the new contribution in the field of citizen engagement models from the point of view of government social media platforms which become very significant now days.

Keywords: Citizen Engagement · Social Media · GSM · CE Model on GSM

1 Introduction

Web 2.0-based social media applications enable users to produce and trade content based on user-generated content. [1]. The use of social media by the government or Government 2.0, cannot be separated from the development of the use of ICT and Web 1.0 technology by the government or known as e-government, as well as investment in e-government as Government 1.0, where the use of social media by the government is the public face of e-government. The government focuses on sharing, participation, and collaboration of information from government, while e-government provides government services in an open, transparent, and efficient manner, Government 2.0 has the opportunity to realize government openness initiatives through the interaction between government

F. Saeed et al. (Eds.): ICACIn 2022, LNDECT 179, pp. 649–659, 2023.
https://doi.org/10.1007/978-3-031-36258-3_57

and citizens, there are 3 forms of government and citizen interactions, namely from the government to citizens, from citizens to government, and within government [2].

Government social media research has started in 2009 and continues to develop until now. Previous government social media research was generally oriented towards the government side rather than the user side, more about how the government is than how the users are while the content made by the government is still a government image and a political tool rather than increasing government openness. This is what causes the failure to capture the capabilities offered by web 2.0 technology, namely users as main actors in triggering social media interactions [3]. There are very few studies related to user engagement research on government social media [4, 5].

And until now, research in the field of government social media has not made any generalizations regarding the model of citizen and government involvement through social media that fully supports the hypothesis that there is a relationship between citizen involvement factors in government social media, this is the main gap in this study. This study will investigate what factors fully support the hypothesis related to citizen involvement in government social media so that the citizen engagement model for government social media is fully proven.

2   Background of the Problem

Research on the relationship between different constructs that arise from the phenomenon of government social media is still not widely mapped by the research community [3], but this is reflected in the difficulty of finding the determinants of citizen involvement in government social media [6], so that we need a comparison of user engagement models apart from the use of government social media, such as increasing user involvement in crisis management [7], the process of determining the level of engagement with stakeholders [8], the relationship between stakeholders and actors (partnerships) to increase user involvement in project completion [9], determining the level of user engagement through brand posts [10].

As for the weaknesses of these models, there is still no comprehensive explanation regarding citizen involvement in government social media because it is still rarely found in the government sector itself. The authors found three journals related to citizen engagement using social media in the field of crisis management [11–13] and unrelated to the use of government social media. One journal is related to the involvement of citizens with the government but the use of social media is not explained in detail [11].

3   Literature Review

This section reviews all the related literatures regarding user's engagement on government social media accounts through SLR technique. The SLR is chosen because it has already become an important tool [14]. Planning, conducting, and reporting results are the three primary processes as the SLR's procedure. In planning phase, several related questions are highlighted in controlling how the information should be gathered in order to achieve the objectives of this study correctly.

3.1 Citizen Engagement (CE)

Citizens engagement has a broad meaning. We can find references to citizen engagement in citizen participation [15–17], public interaction [18], public engagement [19–22], political participation [23], citizen interactions ([24], online engagement [25], public participation [26, 27], user engagement [28], student participation [29].

To some extent, civic engagement has been defined as an individual or collective behavior that aims to address social problems in society [30, 31]. Although the definition of civic engagement is not clear, the point is the interaction between government and citizenship [32].

3.2 Government Media Social (GSM)

The use of social media by the government has become a concern for researchers; the implementation of social media by the government in Public Relations, Information Systems, and E-Government, is still on the government's point of view, and not government social media [33].

Government social media involve the use of social media platforms as public channels that can be operated with the government while social media users can search for information instantly, share it and create their content on government social media [18]. Government social media specifically focus on citizens both locally and between international relations and citizens living abroad through interactive government social media platforms [34]. Because social media platforms are communication channels between the government and citizens, social media platforms could be used as a new source of information to communicate government agencies in engaging citizen opinion [35]. On the other hand, previous literature on government social media has noted that this implemented platform does provide several advantages for democratic functions for government institutions because government social media can increase transparency and citizen interaction [36].

3.3 CE through GSM

Currently, there has been a change in the use of social media by the government, where previously social media were used as an image and government political media for citizens [3], but now social media are used as a means of government openness to society, as well as a means of citizen engagement [37].

Currently, citizen involvement consists of user-to-user interaction on government social media platforms [38], citizen involvement in government social media [39], government actors, and institutions that influence the external environment of the government and actors. Non-government social media users [40, 41].

4 Methodology

The ultimate goal of this study is to find the factors of citizen engagement in government social media. To fulfill this final goal, several steps of methodology are needed. A search strategy is a scientific approach of a procedure that shows how information source articles about this research domain are organized and carried out. The search strategy applied in this research is through automatic search using several journal search platforms such as Google Scholar, Mendeley Search, and access to available institutional databases such as Science Direct. The automatic search is based on used keywords related to the study domain. Figure 1 below shows the defined keywords that have been studied to investigate this study. Each specified keyword controls the information used in the literature review process.

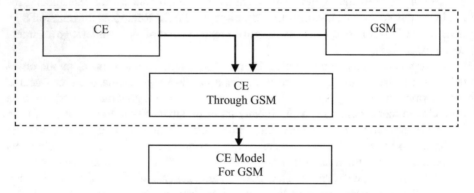

Fig. 1. The literature review searching articles procedure.

Then proceed with providing an overview of the awareness of the problems and suggestions as solutions to existing problems. The result of the process in the SLR method are shown in Fig. 2.

After applying the search strategy and all related articles have been saved, it's time to filter all articles based on the inclusion and exclusion criteria. This action helps the researcher to narrow the study space because there are thousands of articles out there. Since this research focuses on citizen engagement models on government social media, the research concentrates on articles from reputable journals, conferences, and book chapters.

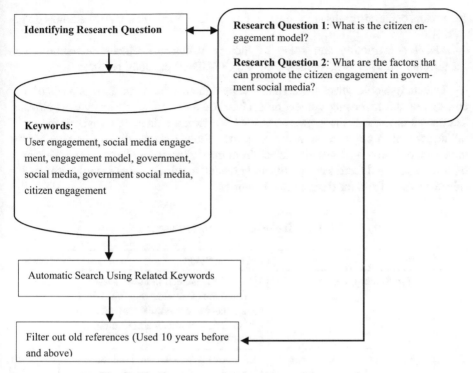

Fig. 2. The literature review searching articles procedure.

5 Model Development

This section is to put forward the proposed research model and to cover the initial model for CE with GSM accounts. As the democratic goals of GSM is always transparent in communication with the citizens, the implementation of GSM in government institutions is a great approach in dealing with any updated news or services from government directly to the citizens through social media platforms [36].

5.1 Derivation of Factors

There are five models of user involvement that serve as the model design theories for this study, as described in the literature review. The following are the five models design theories for developing citizen engagement model towards government social media accounts:

i. Comparing three co-creation efforts, citizens' intentions to participate in governmental co-creation [42].
ii. An integration of STOPS and SMCC shows how the public's active and passive communicative actions influence their responses to tornadoes [11].
iii. How to encourage citizen participation through official social media during the COVID-19 crisis: Opening the mystery [12].

iv. Factors explaining why some citizens engage in E-Participation, while others do not [6].
v. Model construction and study of factors influencing citizen interaction with government TikTok accounts during the COVID-19 pandemic [13].

This study would select and combine several factors from the above models of user engagement as citizen engagement model because the users in this study were citizens. The model is towards the engagement model between citizen and government social media account through social media platforms. Table 1 shows the justification of how the selected factors have been evaluated from the design model theories and the reason of how the selected factors are significantly related to this domain of study based on the selected factors from the design model theories.

Table 1. The selected factors justification.

No.	Factor	Related to Study	Source
1	Trust in Sincere Intentions	This study is needed to find out the expectations of citizens regarding their engagement in government social media will make a significant contribution to the government	[42]
2	Expected Personal Gratification	This study is needed to find out the expectations of citizens regarding their involvement in government social media to achieve personal gain	
3	Dialogic loop	This study is necessary to ensure that there is a means of feedback on government social	12, 13]
4	Situational Motivation	This study is needed to determine the extent to which situational motivation moderates citizen involvement in communication on government social media during a crisis	[11]
5	Crisis Efficacy	This study is needed to find out the extent of the motivation of citizens in responding to crisis information on government social media during times of crisis	
6	Community commitment	This study is necessary to ensure commitment to citizen engagement in government social media	[6]
7	Community ownership	This study is necessary to ensure ownership of government social media while maintaining citizen engagement	

(continued)

Table 1. (*continued*)

No.	Factor	Related to Study	Source
8	Trust in government	This study is necessary to ensure trust in the government by increasing citizen engagement in government social media	
9	Strength of social ties	This research is necessary to increase citizen and government engagement by encouraging interaction between citizens on government social media	
10	Attitude towards the behaviour	This study needs to look at the results of the evaluation of the factors that lead to citizen engagement behaviour in government social media	6, 42]
11	Subjective norm	This study needs to look at environmental factors that influence the behaviour of citizen engagement in government social media	
12	Perceived behavioral control	This study needs to estimate a person's ability to actually demonstrate influence on citizen engagement behaviour in government social media	
13	Behavioural Intentions	This study needs to show the existence of behaviour with citizen involvement in government social media	

Based on the identified factors, an initial model of citizen engagement with government social media is proposed and the model clearly applies the concept of citizen engagement model of government social media from previous researchers to avoid biased concepts in assigning value to citizens engagement on government social media.

From Table 1, all the factors are used to develop the initial model design as proposed in Fig. 3. The CE towards GSM.

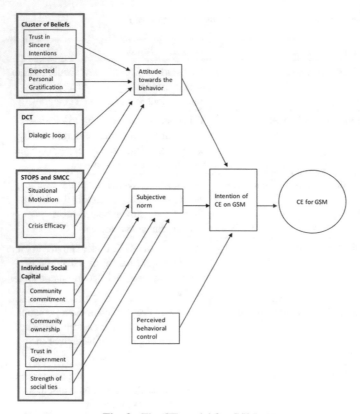

Fig. 3. The CE model for GSM.

6 Conclusion

This study provides an overview of the citizen engagement model in government social media. In order to understand the citizen engagement model, this paper discussed the SLR process. This essay is centered on SLR, which includes a search technique to identify all the elements involving in preserving online reputation while addressing sustainable model ideas. The SLR approach was used to filter out all the unrelated articles and sort out the related matters.

As a result of the SLR, there are thirteen factors which been figured out to generate the CE actions on GSM. The factors are trust in sincere intentions, dialogic loop, situational motivation, crisis efficacy, community commitment, community ownership, trust in government, strength of social ties, attitude towards the behaviour, subjective norm, perceived behavioral control, and intention of CE on GSM. The proposed citizen interaction strategy on GSM is undergoing evaluation for the upcoming related work. In order to validate the model, the next phase entails data collection and analysis.

References

1. Kaplan, A.M., Haenlein, M.: Users of the world, unite! The challenges and opportunities of Social Media. Bus. Horiz. **53**, 59–68 (2010). https://doi.org/10.1016/j.bushor.2009.09.003
2. Khan, G.F.: The Government 2.0 utilization model and implementation scenarios. https://doi.org/10.1177/0266666913502061
3. Medaglia, R., Zheng, L.: Mapping government social media research and moving it forward: a framework and a research agenda. https://doi.org/10.1016/j.giq.2017.06.001
4. Lev-On, A., Steinfeld, N.: Local engagement online: Municipal Facebook pages as hubs of interaction. https://doi.org/10.1016/j.giq.2015.05.007.
5. Reddick, C.G., Chatfield, A.T., Ojo, A.: A social media text analytics framework for double-loop learning for citizen-centric public services: a case study of a local government Facebook use, https://www.sciencedirect.com/science/article/pii/S0740624X16302386 Accessed 10 Feb 2019. https://doi.org/10.1016/J.GIQ.2016.11.001
6. Choi, J.C., Song, C.: Factors explaining why some citizens engage in E-participation, while others do not. Gov. Inf. Q. **37**, 101524 (2020). https://doi.org/10.1016/j.giq.2020.101524
7. Guo, J., Liu, N., Wu, Y., Zhang, C.: Why do citizens participate on government social media accounts during crises? A Civic Voluntarism Perspect. Inf. Manag. (2020). https://doi.org/10.1016/j.im.2020.103286
8. Surucu-Balci, E., Balci, G., Yuen, K.F.: Social media engagement of stakeholders: a decision tree approach in container shipping. Comput. Ind. **115** (2020). https://doi.org/10.1016/j.compind.2019.103152
9. Goodess, C.M., et al.: Advancing climate services for the European renewable energy sector through capacity building and user engagement. Clim. Serv. **16**, 100139 (2019). https://doi.org/10.1016/j.cliser.2019.100139
10. Gutiérrez-Cillán, J., Camarero-Izquierdo, C., San José-Cabezudo, R.: How brand post content contributes to user's Facebook brand-page engagement. The experiential route of active participation. BRQ Bus. Res. Q. **20**, 258–274 (2017). https://doi.org/10.1016/j.brq.2017.06.001
11. Liu, B.F., Xu, S., Rhys Lim, J.K., Egnoto, M.: How publics' active and passive communicative behaviors affect their tornado responses: an integration of STOPS and SMCC. Public Relat. Rev. **45**, 101831 (2019). https://doi.org/10.1016/j.pubrev.2019.101831
12. Chen, Q., Min, C., Zhang, W., Wang, G., Ma, X., Evans, R.: Unpacking the black box: How to promote citizen engagement through government social media during the COVID-19 crisis. Comput. Human Behav. **110**, 106380 (2020). https://doi.org/10.1016/j.chb.2020.106380
13. Chen, Q., Min, C., Zhang, W., Ma, X., Evans, R.: Factors Driving Citizen Engagement With Government TikTok Accounts During the COVID-19 Pandemic : Model Development and Analysis Corresponding Author . J. Med. Internet Res. **23**, 1–13 (2021). https://doi.org/10.2196/21463
14. Nepomuceno, V., Soares, S.: On the need to update systematic literature reviews. Inf. Softw. Technol. **109**, 40–42 (2019). https://doi.org/10.1016/j.infsof.2019.01.005
15. Aladalah, M., Cheung, Y., Lee, V.C.: Towards a model for engaging citizens via Gov2.0 to meet evolving public value. Int. J. Public Adm. Digit. Age. **5**, 1–17 (2017). https://doi.org/10.4018/ijpada.2018010101
16. Ruvalcaba-Gomez, E.A., Renteria, C.: Contrasting the perceptions about citizen participation between organized civil society and government with an open government approach: The case of the state of Jalisco, Mexico. In: 2019 6th International Confernce eDemocracy eGovernment, ICEDEG **2019**. 39–46 (2019). https://doi.org/10.1109/ICEDEG.2019.8734290

17. Caetano, B., Paula, M., De Souza, J.: SoPa: A social media for a participatory society. IEEE Access. **8**, 70627–70639 (2020). https://doi.org/10.1109/ACCESS.2020.2986644
18. Gintova, M.: Understanding government social media users: an analysis of interactions on Immigration, Refugees and Citizenship Canada Twitter and Facebook, https://www.scienc edirect.com/science/article/pii/S0740624X18305756. Accessed 10 Feb 2019. https://doi.org/10.1016/J.GIQ.2019.06.005
19. Cho, M., Schweickart, T., Haase, A.: Public engagement with nonprofit organizations on Facebook. Public Relat. Rev. **40**, 565–567 (2014). https://doi.org/10.1016/j.pubrev.2014.01.008
20. Men, L.R., Tsai, W.H.S.: Infusing social media with humanity: corporate character, public engagement, and relational outcomes. Public Relat. Rev. **41**, 395–403 (2015). https://doi.org/10.1016/j.pubrev.2015.02.005
21. Men, L.R., Tsai, W.H.S.: Public engagement with CEOs on social media: motivations and relational outcomes. Public Relat. Rev. **42**, 932–942 (2016). https://doi.org/10.1016/j.pubrev.2016.08.001
22. Men, L.R., Tsai, W.H.S., Chen, Z.F., Ji, Y.G.: Social presence and digital dialogic communication: engagement lessons from top social CEOs. J. Public Relations Res. **30**, 83–99 (2018). https://doi.org/10.1080/1062726X.2018.1498341
23. Arshad, S., Khurram, S.: Can government's presence on social media stimulate citizens' online political participation? investigating the influence of transparency, trust, and responsiveness. Gov. Inf. Q. **37**, 101486 (2020). https://doi.org/10.1016/j.giq.2020.101486
24. Gálvez-Rodríguez, M. del M., Sáez-Martín, A., García-Tabuyo, M., Caba-Pérez, C.: Exploring dialogic strategies in social media for fostering citizens' interactions with Latin American local governments, https://www.sciencedirect.com/science/article/abs/pii/S0363811117300504. Accessed 10 Feb 2019. https://doi.org/10.1016/J.PUBREV.2018.03.003
25. Wang, X., Liu, Z.: Online engagement in social media: a cross-cultural comparison. Comput. Human Behav. **97**, 137–150 (2019). https://doi.org/10.1016/j.chb.2019.03.014
26. Wagner, S.A., Vogt, S., Kabst, R.: How IT and social change facilitates public participation: a stakeholder-oriented approach. Gov. Inf. Q. **33**, 435–443 (2016). https://doi.org/10.1016/j.giq.2016.07.003
27. Zavattaro, S.M., French, P.E., Mohanty, S.D.: A sentiment analysis of U.S. local government tweets: The connection between tone and citizen involvement, https://doi.org/10.1016/j.giq.2015.03.003.
28. Shugars, S., Beauchamp, N.: Why Keep Arguing? Predicting Engagement in Political Conversations Online. SAGE Open. 9, (2019). https://doi.org/10.1177/2158244019828850
29. Sirait, A.D.S., Fitriani, W.R., Hidayanto, A.N., Purwandari, B., Kosandi, M.: Evaluation of social media preference as e-participation channel for students using fuzzy AHP and TOPSIS. Proc. - 2018 4th Int. Conf. Comput. Eng. Des. ICCED 2018. 158–163 (2019). https://doi.org/10.1109/ICCED.2018.00039
30. Gil de Zúñiga, H.: Social Media Use for News and Individuals' Social Capital, Civic Engagement and Political Participation. J. Comput. Commun. **17**, 319–336 (2012). https://doi.org/10.1111/j.1083-6101.2012.01574.x
31. Warren, A.M., Sulaiman, A., Jaafar, N.I.: Social media effects on fostering online civic engagement and building citizen trust and trust in institutions, https://doi.org/10.1016/j.giq.2013.11.007.
32. Ekman, J., Amnå, E.: Political participation and civic engagement: towards a new typology. Hum. Aff. **22**, 283–300 (2012). https://doi.org/10.2478/s13374-012-0024-1
33. Medaglia, R., Zheng, L.: Extending impact analysis in government social media research: Five illustrative cases. https://doi.org/10.1145/3085228.3085298

34. Silva, P., Tavares, A.F., Silva, T., Lameiras, M.: The good, the bad and the ugly: three faces of social media usage by local governments. Gov. Inf. Q. **36**, 469–479 (2019). https://doi.org/10.1016/j.giq.2019.05.006
35. Belkahla Driss, O., Mellouli, S., Trabelsi, Z.: From citizens to government policy-makers: Social media data analysis. Gov. Inf. Q. **36**, 560–570 (2019). https://doi.org/10.1016/j.giq.2019.05.002
36. DePaula, N., Dincelli, E., Harrison, T.M.: Toward a typology of government social media communication: democratic goals, symbolic acts and self-presentation. Gov. Inf. Q. **35**, 98–108 (2018). https://doi.org/10.1016/j.giq.2017.10.003
37. Bonsón, E., Torres, L., Royo, S., Flores, F.: Local e-government 2.0: Social media and corporate transparency in municipalities, https://www.sciencedirect.com/science/article/pii/S0740624X1200010X. Accessed 10 Feb 2019. https://doi.org/10.1016/J.GIQ.2011.10.001
38. Bao, H., Cao, B., Xiong, Y., Tang, W.: Digital media's role in the COVID-19 pandemic. JMIR Mhealth Uhealth. **8**, e20156 (2020). https://doi.org/10.2196/20156
39. Stone, J.A., Can, S.H.: Linguistic analysis of municipal twitter feeds: factors influencing frequency and engagement. Gov. Inf. Q. **37**, (2020). https://doi.org/10.1016/j.giq.2020.101468
40. Sandoval-Almazan, R., Ramon Gil-Garcia, J.: Towards cyberactivism 2.0? Understanding the use of social media and other information technologies for political activism and social movements, https://doi.org/10.1016/j.giq.2013.10.016.
41. Sumra, K.B., Bing, W.: Crowdsourcing in Local Public Administration. https://doi.org/10.4018/ijpada.2016100103
42. Jong, M.D.T.D., Neulen, S., Jansma, S.R.: Citizens ' intentions to participate in governmental co-creation initiatives : comparing three co-creation configurations. Gov. Inf. Q. **36**, 490–500 (2019). https://doi.org/10.1016/j.giq.2019.04.003

Validation of Measuring Items for Customer Loyalty in E-Commerce Platform

Mira Afrina[1]([✉]), Samsuryadi[1], Ab Razak Che Hussin[2], and Suraya Miskon[2]

[1] Fakultas Ilmu Komputer, Universitas Sriwijaya, Palembang, Indonesia
miraafrina81@gmail.com, samsuryadi@unsri.ac.id
[2] Azman Hashim International Business School (AHIBS), Universiti Teknologi Malaysia, Johor Bahru, Malaysia
{abrazak,suraya}@utm.my

Abstract. The electronic commerce (e-commerce) is a field that involved online enterprises which able to accurately predict the future needs from customers and had influenced on important economic included the social implications. The e-commerce also needed to deal with customer loyalty while dealing with customers' changing consumption habits in brought out for adapting to the new situation while needed to modify their online business activities because of the changing shopping attitudes. Thus, the online shopping must provide trending and satisfied experienced in order to keep the customers be loyal with e-commerce services. Based on this situation, the customer loyalty is compulsory to be maintaining regarding the fast-changing technological evolutions trends which become the reason of customer changing habits. The process of defining customer loyalty evaluation instruments in e-commerce will be discussed in this paper. The definition of this e-commerce customer loyalty evaluation instrument is based on a number of existing theories as well as a variety of other types of literature.

Keywords: Customer Loyalty · Customer Changing Habits · E-commerce

1 Introduction

E-commerce is a field that involves internet businesses that are able to properly predict future client interest pattern and customized customer services [1]. E-commerce has also provided significant benefits in terms of lowering the cost of logistics delivery, increasing the efficiency of logistics delivery, and meeting the diverse and high-quality delivery services. Clients changing consumption habits forces firms to react to the new scenario by modifying their business strategy and marketing. In order to develop customer loyalty, businesses must keep track of their customers' shifting habits.

Customer loyalty is the result of a positive relationship between customers and businesses. A better customer loyalty leads to a higher repurchase intentions in e-commerce businesses. It is becoming a reason on customer loyalty be significant in the marketing community since it is strongly linked to a company's long-term viability. Customer loyalty has already been elevated to a new level of domain known as e-loyalty after

F. Saeed et al. (Eds.): ICACIn 2022, LNDECT 179, pp. 660–670, 2023.
https://doi.org/10.1007/978-3-031-36258-3_58

the revolution of high technology achievement. E-loyalty has become a useful measurement for e-commerce performance. Customer intention to display recurrent purchase behaviour was influenced by e-loyalty, which was driven by a positive attitude toward the company element.

The retail industry and the shopping experience are extremely significant all over the world. Customers prefer to visit a business location at such time to get a sense of how shopping feels. The evolution of shopping experiences via internet technology is transitioning to an online environment and entrepreneurs in e-commerce. Customers nowadays like to purchase from the comfort of their own homes and enjoy having unrestricted shopping experiences everywhere and anywhere, even during working hours.

E-commerce should be aware of trends in consumers' online purchasing patterns in order to earn the loyalty values from customers, as online businesses can gain more share when they can supply all the customers demand. Loyalty values would be established only after the e-commerce site has successfully delivered the factors of customers' satisfaction and trust.

This paper consists of seven section. Sect 1 is for the study's introduction, Sect. 2 is for the study's background, Sect. 3 is for the methodology, Sect. 4 is for the identified constructs, Sect. 5 is for measurements item development, Sect. 6 is for the content validity, and end with the Sect. 7 for this study's conclusion.

2 Background of Study

In the context of e-commerce, several researchers focus on customer life cycle elements. The customer life cycle model has altered as a result of contemporary technological advancements, and it has to be updated. This corresponds to a significant increase in e-commerce participation, which increased from 13 to 35 percent of the internet population between 2007 and 2016 [2]. The development of e-commerce has resulted in a more secure scenario for customers' habits. Customers changing habits refers to customer attitudes and behaviour which are changed by technological services.

Customers began to view e-commerce sites through mobile devices. E-commerce activities are required to achieve customer loyalty to continue using online services. Customers habits are also linked to how often customers have shop online. Then, in order to retain clients committed to e-commerce services, online buying must deliver a trending and satisfying experience.

The next issue is e-commerce research sector is a customer life-styles, needs, and trends, which have resulted in changing client tastes, particularly in relation to their culture. By removing restrictions such as an out-of-date online shopping styles, the evolution of e-commerce has transformed the way businesses may conduct services or systems with clients. Customers nowadays, for example, prefer to shop on the internet at any time and from any location. As a result of this predicament, e-commerce companies have begun to migrate their entire platform to a responsive website that can be seen on mobile devices and makes the purchased via mobile e-commerce apps. This brought to the significant element of measuring the customer loyalty in e-commerce platform. As to go through in details about it, the next section is describing the methodology used for this study.

3 Methodology

A systematic literature review (SLR) was performed to satisfy the study's goals. In a comprehensive research procedure, the SLR was used to locate and summarize all papers published. The nature of the research was the rationale for using the SLR. Systematic reviews and meta-analyses are critical techniques for properly and reliably summarizing data. This study adopted a three-stage approach including planning, conducting, and reporting. The SLR was processed through various stages in order to be accomplished. First, the review protocol was described and depicted. Then, the inclusion and exclusion criteria were explained. After that, the search strategy was discussed. Next, the study selection process was defined. Thereafter, the quality assessment was clarified. Later, the data extraction and synthesis were expounded. The methodology of the present study encompasses the following steps such as identifying constructs, measurements items development, and content validity.

4 Identified Constructs

This study investigates the vital principles to devise a new scale to evaluate the e-learning systems success based on connectivism. To accomplish this, an accurate and validated measurement instrument for evaluating customer loyalty in e-commerce. This section reflects on the constructs that are important in influencing customer loyalty and the respondents are asked to choose one response from a list of Likert Scales for each item. For this study, the Likert scaling method was taken into account for extracting data from the questionnaire.

The Likert scale is a common tool in the field of information systems for presenting a wide range of responses or perspectives ranging from very positive to very negative. The Likert Scale responses range from 1 to 5, with 1 indicating Strongly Disagree, 2 indicating Disagree, 3 indicating Neutral, 4 indicating Agree, and 5 indicating Strongly Agree. Since it will provide the respondents a clearer view of the possible choices, this study uses a five-point Likert Scale [3]. The study's measures or measurement items are adapted from prior studies. Table 1 lists the measurement items and their associated constructs.

5 Measurement Items Development

This study uses a research instrument, namely a validation sheet instrument for expert judgment and a questionnaire instrument for respondents as buyers. In this research, 9 experts are asked to validate the instrument as presented in Table 2. However, only 6 of them accepted to validate the items instrument. The attitudes are the first construct which be as the measurement's items. It is based on customers attitudes on trend, browsing desire, customers interest, products information, and product update. Table 2 is presenting the refining process for attitudes construct.

The second construct is Influenced which refers to the influencing elements such as video content, influencer content, text, photos, trend, and community suggestions. Table 3 is presenting the refining process for the construct of Influenced.

Table 1. Constructs Identification

Construct	Definition	Source
Attitudes	Refers to customers attitudes based on trend, browsing desire, customers interest, products information, and product update	[9]
Influenced	Refers to the influencing elements such as video content, influencer content, text, photos, trend, and community suggestions	[4]
Trend	Refers to latest trend on products and services based on the changing e-commerce environment and interesting content	[4]
Brand Identification	Refers to the crafted name or logo as the products image included the original photos	[10]
Brand Performance	Refers to originality, excellent service, offers, prices, quality, and public perceptions	[5]
Brand Review	Refers to the elements of customers' testimony and comments regarding brand	[6]
Usefulness	Refers to the usefulness of e-commerce site which familiarly known, site benefits, user friendly, and efficiency	[7]
Fulfillment	Refers to easy-to-use e-commerce features and fulfill customers' interest	[11]
Engagement	Refers to sharing customers experience, update notifications based on news, recommendations, and provided reminders	[12]
Loyalty Intention	Refers to repurchase process through the same e-commerce site, having positive image, fulfill customers' needs, and keep updating the new arrival products	[8]
Customer Loyalty in Ecommerce	Refers to be loyal with e-commerce site when it able to control a good value equity, brand equity, and relationship equity with customers	[13]

The third construct is Trend which refers to the latest trend on products and services based on the changing e-commerce environment and interesting content. Table 4 is presenting the refining process for the construct of Trend.

The fourth construct is Brand Identification which refers to the crafted name or logo as the products image included the original photos. Table 5 is presenting the Brand refining process of Identification construct.

Table 2. Refining the Attitudes Items

Section	No.	Item Code	Refined Items
Attitudes	1	A1	I prefer to buy new trendy products that is suitable to my interest based on technological and urbanization influenced
	2	A2	I like to browse through the online shopping websites even when I don't plan to buy anything
	3	A3	I search various online shopping websites that is suitable to my interest just to find out more about the latest styles
	4	A4	I like to see features and descriptions provided in the website that make it easier for me to know the brand and price of the product
	5	A5	I feel happy if there is a new product update from my ecommerce account

Table 3. Refining the Influenced Items

Section	No.	Item Code	Refined Items
Influenced	6	I1	Video content presented by someone popular can attract my shopping interest than text or photos
	7	I2	I like shopping on one of the e-commerce platforms because the influencer icon is trending
	8	I3	The opinions or suggestions of family and close relatives greatly influence me in shopping online
	9	I4	Opinions or suggestions of friends greatly affect me in shopping online
	10	I5	The role of the influencer icon greatly influences me in online shopping

Table 4. Refining the Trend Items

Section	No.	Item Code	Refined Items
Trend	11	T1	I prefer to use e-commerce that always updated with latest products and services
	12	T2	I prefer to use e-commerce site which always adapt with any e-commerce environment change based on current trends
	13	T3	I always prefer to follow trends in keeping updated by using new technology in dealing with excellent e-commerce shopping process
	14	T4	I prefer to follow e-commerce that always updates the information
	15	T5	I prefer to follow e-commerce with interesting content

Table 5. Refining the Brand Identification Items

Section	No.	Item Code	Refined Items
Brand Identification	16	I1	I prefer buying the products which crafted with the brand name or logo
	17	I2	I prefer shopping at prices that match the brand or logo
	18	I3	I prefer to buy products with a public figure icon
	19	I4	I prefer to buy products that look attractive
	20	I5	I prefer to buy products with real photos

The fifth construct in this study is Brand Performance which refers to originality, excellent service, offers, prices, quality, and public perceptions. Table 6 is presenting the refining process of the Brand Performance construct.

Table 6. Refining the Brand Performance Items

Section	No.	Item Code	Refined Items
Brand Performance	21	BP1	I normally buy products or services based on offers and my perceptions toward the brand
	22	BP2	I prefer to buy brand products that people know
	23	BP3	I prefer to buy original products
	24	BP4	I prefer to buy products with excellent service
	25	BP5	I prefer to buy products with appropriate prices and quality

The fifth construct in study involved Brand Review which refers to the elements of customers' testimony and comments regarding brand. Table 7 is presenting the refining process for Brand Review construct.

The sixth construct in this study is Usefulness which refers to the usefulness of e-commerce site which familiarly known, site benefits, user friendly, and efficiency. Table 8 is presenting the refining process for the construct of usefulness.

The seventh construct in this study is Fulfillment which refers to easy-to-use e-commerce features and fulfill customers' interest. Table 9 is presenting the refining process of Fulfillment construct.

The eighth construct of this study is Engagement which refers to sharing customers experience, update notifications based on news, recommendations, and provided reminders. Table 10 is presenting the refining process for Engagement construct.

The ninth construct of this study is Loyalty Intention which refers to repurchase process through the same e-commerce site, having positive image, fulfill customers'

Table 7. Refining the Brand Review Items

Section	No.	Item Code	Refined Items
Brand Review	26	BR1	I totally like the customers' testimony and comments regarding any brand
	27	BR2	I totally like the customers' testimony video and comments regarding any brand
	28	BR3	I am interested in buying products and services because of the many positive comments from customers
	29	BR4	I prefer to buy products with high star rating
	30	BR5	I prefer to buy products that have a high feedback rating

Table 8. Refining the Usefulness Items

Section	No.	Item Code	Refined Items
Usefulness	31	U1	I prefer e-commerce site which able to follow my shopping experience knowledge
	32	U2	I prefer to follow e-commerce that benefits me and other customers
	33	U3	The highlight feature in e-commerce allows me to more quickly select products based on product brands
	34	U4	I prefer to buy products online because it's easier and more practical
	35	U5	Using e-commerce is more efficient than looking for products or services directly in offline stores

Table 9. Refining the Fulfillment Items

Section	No.	Item Code	Refined Items
Fulfillment	36	F1	I prefer easy-to-use ecommerce features
	37	F2	I prefer the clear and easy to understand ecommerce description feature
	38	F3	I am interested in shopping at e-commerce because of the large selection of products
	39	F4	I am interested in shopping at e-commerce because it's simple and practical I am interested in shopping in e-commerce because I can choose products in one service
	40	F5	I prefer easy-to-use ecommerce features

Table 10. Refining the Engagement Items

Section	No.	Item Code	Refined Items
Engagement	41	E1	My need is fulfilled through the service available in e-commerce provider
	42	E2	I like e-commerce website that allow customer share their online shopping experience
	43	E3	I like the e-commerce website that always give notification on news or recommendations on products or services
	44	E4	I like the e-commerce website that always remind me the services their offer
	45	E5	I feel appreciated when communication and interaction with e-commerce providers is always ongoing

needs, and keep updating the new arrival products. Table 11 is presenting the refining process of Loyalty Intention construct.

Table 11. Refining the Loyalty Intention Items

Section	No.	Item Code	Refined Items
Loyalty Intention	46	LI1	I will repurchase at the same e-commerce site or brands when I feel totally satisfied with the previous experiences
	47	LI2	I will have positive image in mind toward e-commerce if I have good experience in previous shopping
	48	LI3	I eager to use e-commerce to fulfill my daily needs
	49	LI4	The products and services offered on the website can influence me to make a repurchase
	50	LI5	I am always looking for new arrival products from e-commerce that make me satisfied in my previous shopping

The last construct for this study is Customer Loyalty in E-commerce which refers to be loyal with e-commerce site when it able to control a good value equity, brand equity, and relationship equity with customers. Table 12 is presenting the refining process for the construct of Customer Loyalty in E-commerce.

The next section content validity is for validating the survey items for this study which already been refined in this section.

Table 12. Refining the Customer Loyalty in E-commerce

Section	No.	Item Code	Refined Items
Customer Loyalty in E-commerce	51	CL1	I choose to be loyal with the e-commerce site when it able to control a good value equity, brand equity, and relationship equity with customers
	52	CL2	I think customer loyalty in e-commerce is good when sellers understand customer needs and behavior
	53	CL3	The experience of buying products and services in e-commerce is the reason for me to be a loyal customer in e-commerce
	54	CL4	I choose to be loyal with the e-commerce site when it able to control a good value equity, brand equity, and relationship equity with customers
	55	CL5	E-commerce is my first choice when I want to do online shopping

6   Content Validity

Content validity pinpoints the degree of item acceptance in the measurement based on its reflective operational definition of constructs. Content Validity Index (CVI) was used to validate this research instrument. CVI approach is flexible and it requires a minimum of 3 experts as panel members of interrater. Therefore, the items were validated in term of its relevancy in content. The 4-point ranking scale were used for each item to measure relevancy in content. In addition, there was a provision for expert to comment in each of the item if need be. The four scale to measure relevancy were used because it is the most frequent label researchers used for content validity. The labels are: "Not Relevant = 1; Relevant but not important = 2; Relevant but need review = 3; Highly Relevant = 4". The proportion agreement is 3 and 4.

In this research, nine experts are asked to validate the instrument as presented in Table 13. However, only six of them accept to validate the instrument. Therefore, a total of five ratters were valid and was used for the ranking which is above the minimum number as S-CVI of 0.8 and above is acceptable and adopted in this study. Therefore, 61 items out of 62 items where acceptable and one item SOC1 (0.4) was rejected because the item score below the threshold.

In addition, experts' responses and suggestions are adapted. Therefore, some revisions regarding the clarity or wording of the items are adjusted. For example, one of the experts suggested that items should be narrow to IT complies in securing classified

Table 13. Refining the Customer Loyalty in E-commerce

Serial	Status of Experts	No of Expert invited	Numbers of experts that respondent
1	Professors	4	3
2	Associate Prof	2	1
3	Doctors	3	2
Total		9	**6**

information than the general perspective of classified information. The expert further observed that the definition of 'social norms' did not march with its items. In other words, those items cannot measure the construct "social norms". Based on the experts' comments, the required improvements were made. In other words, all recommendations are applied, and the revised questionnaire was used for pilot study.

7   Conclusion

In order to examine the success of e-learning systems based on the connectivism learning theory, this study constructed reliable measurement items for the constructs retrieved from the SLR. The items were developed from existing material and tailored to the study's goals. In order to be validated, the content of the questionnaire items was assessed. The elements were verified in terms of relevancy and simplicity using the CVI. As a result, the number of things was adjusted, and the content of the items was edited and changed in terms of phrasing. The reliability and validity of the assessment items will be tested using pilot and real surveys in this ongoing study, which could lead to a suggestion for future exploration. As refers to the findings from this paper, the measurement items are significantly useful towards customer loyalty platform area of research and able to be use by students or any researchers in the same area of study.

References

1. Zheng, K., Zhang, Z., Song, B.: E-commerce logistics distribution mode in big-data context: a case analysis of JD.COM. Ind. Mark. Manag. **86**, 154–162 (2019). https://doi.org/10.1016/j.indmarman.2019.10.009
2. Garín-Muñoz, T., López, R., Pérez-Amaral, T., Herguera, I., Valarezo, A.: Models for individual adoption of eCommerce, eBanking and eGovernment in Spain. Telecomm. Policy **43**(1), 100–111 (2019). https://doi.org/10.1016/j.telpol.2018.01.002
3. Leung, S.O.: A comparison of psychometric properties and normality in 4-,5-,6-, and 11-point likert scales. J. Soc. Serv. Res. **37**(4), 412–421 (2011). https://doi.org/10.1080/01488376.2011.580697
4. Rosqvist, L.S., Hiselius, L.W.: Online shopping habits and the potential for reductions in carbon dioxide emissions from passenger transport. J. Clean. Prod. **131**, 163–169 (2016). https://doi.org/10.1016/j.jclepro.2016.05.054
5. Iyer, P., Davari, A., Zolfagharian, M., Paswan, A.: Market orientation , positioning strategy and brand performance. Ind. Mark. Manag. **81**, 16–29 (2019) https://doi.org/10.1016/j.indmarman.2018.11.004

6. Curina, I., Francioni, B., Hegner, S.M., Cioppi, M.: Journal of retailing and consumer services brand hate and non-repurchase intention : a service context perspective in a cross-channel setting. J. Retail. Consum. Serv. **54**, 102031 (2020) https://doi.org/10.1016/j.jretconser.2019.102031

7. Ergün, H.S., Kuşcu, Z.K.: Innovation orientation, market orientation and e-loyalty: evidence from Turkish e-commerce customers. Procedia-Soc. Behav. Sci. **99**, 509–516 (2013). https://doi.org/10.1016/j.sbspro.2013.10.520

8. Savila, I.D., Wathoni, R.N., Santoso, A.S., Savila, I.D., Wathoni, R.N., Santoso, A.S.: sciencedirect the role of multichannel integration, trust and offline-to-online the role of multichannel integration, trust and offline-to-online customer loyalty towards repurchase intention : an empirical customer loyalty towards repurch. Procedia Comput. Sci. **161**, 859–866 (2019). https://doi.org/10.1016/j.procs.2019.11.193

9. Nel, J., Boshoff, C.: Online customers' habit-inertia nexus as a conditional effect of mobile-service experience: a moderated-mediation and moderated serial-mediation investigation of mobile-service use resistance. J. Retail. consum. Serv. **47**, 282–292 (2019)

10. Merk, M., Michel, G.: (2019) 'The dark side of salesperson brand identification in the luxury sector : when brand orientation generates management issues and negative customer perception.' J. Bus. Res. Elsevier **102**, 339–352 (2018)

11. Yang, Z., Fang, X.: Online service quality dimensions and their relatioships with satisfaction. Int. J. Serv. Ind. Manage. **15**(3), 302–326 (2004)

12. Molinillo, S., Anaya-Sanchez, R., Liébana-Cabanillas, F.: Analyzing the effect of social support and community factors on customer engagement and its impact on loyalty behaviors toward social commerce websites. Comput. Hum. Behav. **108**, 108 (2020)

13. Zhang, R., Li, G., Wang, Z., Wang, H.: Relationship value based on customer equity influences on online group-buying customer loyalty. J. Bus. Res. **69**(9), 3820–3826 (2016)

Derivation Sustainable Model of E-Government Application Usage in Indonesian Local Government

Apriansyah Putra[1]([✉]) [iD], Mahadi Bahari[2], Suraya Miskon[2], and Samsuryadi[1]

[1] Fakultas Ilmu Komputer, Universitas Sriwijaya, Palembang, Indonesia
{apriansyah,samsuryadi}@unsri.ac.id
[2] Azman Hashim International Business School (AHIBS), Universiti Teknologi Malaysia, Johor Bahru, Malaysia
{mahadi,suraya}@utm.my

Abstract. The electronic government (e-government) application is a mechanism which are built for government organization point of used either nationally or locally and become an essential component for the selected population usage. This paper is focused on local government which stated in one selected site of Indonesia. As related, the usage of e-government still is having difficulties in order to keep its sustainability usage among the local citizen. In order to overcome this problem, the sustainable model is significantly needed to controlling the sustainability of e-government application usage in local government organization area. Based on this situation, information technology (IT), e-government applications, and governance policy did offers several factors of influenced in sustaining the usage policy in keep using the e-government application usage. In simple brief, the model which can act as guidelines in dealing with sustaining e-government application usage which containing all related factors to be on track is strongly needed. This paper is the aim on how to develop a model which contains all those factors. Those factors were been figured out by the analysis of systematic literature reviews (SLR) method and later be significantly implement for the enhancement of knowledge regarding a new contribution on sustainable model field for e-government application usage and local government point of view which becomes very important now a days.

Keywords: sustainable · e-government · e-government application · local government

1 Introduction

As the new technology keep been updated, the government organizations did change their strategy in delivering their responsibilities towards citizen such as through the adoption of cloud services which later brought the electronic government or mostly known as e-government applications at the national government level to be used by citizen but unfortunately the achievements of it is still been limited [1]. Electronic government (E-government) applications are government applications which become one-stop service

F. Saeed et al. (Eds.): ICACIn 2022, LNDECT 179, pp. 671–681, 2023.
https://doi.org/10.1007/978-3-031-36258-3_59

platform that integrates big data, artificial intelligence and other new technologies in interacting with citizen [2]. The used of e-government applications is aimed at disseminating information, business handling, the provision of advice and suggestions, and the improvement of government management generally. As of its powerful function-ability, the engagement between citizen and e-government applications should not left behind and be wasted. Even-though, as the new technogical impacts also did risk of maintaining the sustainability on the e-government applications towards the citizen engagement with it. As related, the E-government application is needed to be sustainable in the continuous usage between government organizations and their citizen in order for the national government to support the sharing economy in achieving the sustainable development [3]. To keep the sustainable development through the e-government applications is one of the bigger agenda in the big circle domain of sustainability development towards government strategy rules [4]. This study is aims to investigate the sustainability factors towards e-government application and chooses Indonesia as a case study area in knowing how to sustain the e-government application implementation among Indonesia citizen. Those factors were been figured out by the analysis of systematic literature reviews (SLR) method and later be significantly implement for the enhancement of knowledge regarding a new contribution on sustainable model field for e-government application usage.

2 Background of the Problem

The government applications have been eagerly developed. Unfortunately, there is still a few of studies which been paid attention to the compliance values on government applications use by citizen [2]. This situation is brought the challenge to the national government because the aims of the e-government applications development is in order to make the public institutions or government organizations becomes more transparent and accountable [5]. The lack of studies from researchers regarding this domain will make the consciousness among public towards the important of using e-government applications becomes limited. This is significantly needed to have a study on sustainability of e-government applications usage as it is been developed eagerly then it is needed to be explain well on how it should be keeps using among citizen in order to improve their lifestyles while dealing with government through apps. Nowadays, corruption is widely been discussed on news and all media because of the huge negative impacts which brought by the corruption. It is included the several abuse of public power or office for private benefit by government which included economic corruption and political corruption because of the full monopoly power from government [6]. Those problems occurs because the lack of accountability and transparency which needed to exist between the engagement between government organizations and their own citizen. As solution, the development of e-government application is a great effort which needed in providing the transparent relationship between government and citizen.

Even-though, the sustainability is hard to be achieving in E-government application usage among citizen because of the business relations between government and their citizen in order for sharing economy matters are not always give satisfaction feeling for society and the country environment [3]. This problem is needed to be clear as achieving the aims of e-government application development targeted for. Then, the next issues which be highlighted in continue this study is the significant efforts in finding the sustainability factors in e-government application usage. As needed, a country in Southeast Asia which is Indonesia is selected as the case study subject because of Indonesia is facing a number of independently managed challenges in their country [7]. Furthermore, the implementation of e-government application from local government is not in real consistent situation then this make users do not want to use e-government application in dealing with their problems. This is related with the problem of inconsistent e-government relies platforms such as on different information communication technology (ICT) tools that are critical to delivering government services online [6]. To be highlighted, developing a model of sustainability on e-government usage among citizen is a good approach in showing and guiding elements which able to exist through the model. On the citizen site, the sustainability e-government usage model is able to improve their knowledge in guiding them to keep using and support e-government applications in dealing with government services or matters. On the government site, the sustainability e-government usage model is able to promote the significant engagement relationship between citizen and government. This situation will strength up the relationship between citizen and the current government which able to control the satisfaction feeling from public towards government strategy and development. In simple brief, the e-government sustainability model will enhance the spirits of engagement and concern from government towards their entire citizen widely.

3   Literature Review

The electronic government or most familiar with the modern word as e-government did involved with the ability of implementing the Information Technology (IT) management knowledge and skills as in used with the public organizations possesses services features towards their citizen and been focused on transparency instead of political interference [8]. The International surveys which already been done did indicated that four out of five countries currently having an e-government portal as containing links to their government services [9]. This adaptation becomes one step ahead when a lot of people started to cling over smart phones and tablet. Then, the e-government applications become viral.

3.1   Government

The government is refers to a group of parties which rules the country based on own ideology as either in term of national or local government site which focused on state matters [10]. In ruling strategy, government institutions still were choosing to maintain a push-strategy instead of a pull-strategy. The push-strategy is refers to the none concern on feedback involvement from public and the pull-strategy decided to concern the feedback from public in order to build the virtual and close relationships between government organizations towards publics views [11]. The government obviously always having three main open goals which are the transparency of government activities, the full participation from public towards government services, and the perfect collaboration as much as government can achieved [12]. The three main open goals is followed by other internal growth of goals which included determining the car sales tax, determining the fuel tax, determining the cost of educating the public about environmental awareness, and determining the percentage of government's cooperation in the manufacturer's research and development expenditure [13].

3.2   National Government and Local Government

As mention previously, the government familiarly rules in two main types of government which are national government and local government. The national government is refers to building on the involvement of choosing economic policies [14] while included establish foreign policy, declare war, create and maintain armed forces, regulate interstate foreign trade, make copyright patent laws, establish postal offices, and prepared the coin money. There are also several tasks which be done together with local government power such as raise taxes, provide for the public welfare, criminal justice, borrow money, charter banks, and build roads. Some national governments also concerned the privileged status for rural elites varies for inside and outside of the countries [15]. Another huge problem which always being troublesome in national government area of concern is there's still lack of flexibility in adjusting the plans accordance with concerns that were expressed by the public further raised opposition [16]. This situation make un-calming situation which also make researcher hard to gain needed data and information regarding the internal information from national government side. Currently, the local government is facing international economic crisis which needed the improvement in administration of public sectors and need to cutting the actual costs of local governments spend [17]. It is because of the local governments becomes significant in budgetary constraints. In facts of this situation, the local government did facing troubles in preparing the public services to citizen because of the reduction in their revenues and the limits imposed on their financial capacity. This situation brought the local government to involve with investing in research and development (R&D) sector and it is become more important compared to ever before included conducting planning, execution, and evaluation of R&D because it is crucial for ensuring sustainable competence in local government [18]. There is necessarily important to be sustains in local government competence because of the unbalanced of local government is brought the huge increased on local government debts [19] which later will followed by another problem such as the climate change on public and political awareness which involved the significant local government sustainable energy capacity [20].

3.3   E-Government Application and Indonesia Government

The electronic government is targeted to make public institutions more transparent and accountable while using by e-government programs by citizens in order to access the services and trust of supply-side e-government performance evaluations, per capita income, education, age, and rurality [5]. The main reason of e-government applications promoted transparency is to avoid the corruptions which occur in government domain area and perhaps the e-government could act as one of anti-corruption strategies [6]. This implementation of ICT upgraded trends is offering a great opportunity to citizens in dealing with government sources. Even-though, it is still hard to be avoid regarding the cost factor which did becomes one of the main factors to determine the software development method in government organizations [21]. As familiarly understood by government, latest technology always comes with a great value of cost at first sight and needed some training in the effort to teach citizen in becoming familiar with the e-government applications. Even-though, there is still a few of studies which focused on the compliance values for government applications used by other people [2]. Meanwhile, this situation is brought the challenge to the government because the aims of the e-government applications development is in order to make the public institutions or government organizations becomes more transparent and accountable [5]. Those situations become worst when the elements of documentation crime which is corruption existed in government and going viral on news and all online media that make the issues been spread very fast. It is included the several abuse of public power or office for private benefit by government which included economic corruption and political corruption because of the full monopoly power from government [6]. Those problems occurs because the lack of accountability and transparency which needed to exist between the engagement between government organizations and their own citizen.

Indonesia been selected as a research study area because it is the country which found initiatively on technology evolution and being introduced to promote their citizen performance-related rationalities especially in village area [8]. As the e-government application is one of latest technology and approach on dealing with government services, choosing Indonesia as research area is a good effort in trying to give a new value while promoting e-government applications among Indonesia citizen. Furthermore, Indonesia is eagerly promoting better experience with IT governance and bureaucratic reformation by implementing the e-government development services [21]. Moreover, Indonesia already started the idea of global village through ICT approach in order to follow the global technological development included the e-government application approaches while becoming the prime target in gaining positive impacts through three reasons which are Indonesia's significant information asymmetry, plagued with corruption scandals, and lastly Indonesia situates "information systems governance" research domain [22].

3.4 Sustainability Development Plan

The sustainability is comes from word sustainable which brought the knowledge that be called a model for sustainability and did constitutes an emerging research stream by theoretical or empirical research. Unfortunately, there are still un-sufficient answers as to question on what a sustainable model might involve; then this brought to opinions as an organization's sustainable model should having adaption on surrounding environment while keep being innovative upward through the sustainability [3]. Furthermore, the sustainability also refers to elements which affect human attitudes, selections, and decisions while keeping the important dimensions of sustainability which are economic growth, environmental sustainability, and social responsibility [13]. Those dimensions did pushed private and governments industries to focus on the planet, people, and profits at the same period of time while followed the fundamental role in determining strategies in achieving goals. In developing the initial sustainable model which contains the sustainability development plan regarding the usage of e-government application, there is a sustainable model which been selected as based theory in the modeling design the for this local e-government application domain area. Table 1 is presenting the details about the selected model design theory which been selected as a model based design for the sustainability model of the local e-government application usage among Indonesia citizen. The table consists of model design name, author information, research area, and the specific focus of the model. Then, Table 2 followed afterwards as presenting the selected construct which been figured out from the SLR method of study that be used in developing the proposed sustainable model of e-government application usage in Indonesia local government.

4   Research Method

In order to fulfil all the research objectives, the systematic literature review (SLR) technique is selected as information finding method. In order to follow the procedure established in this method, the previous literature on the investigated issue towards the sustainability of e-government application among government organizations. The area which been selected as the case study area is Indonesia.

4.1   Inclusion and Exclusion Criteria

This section helps the researcher to narrow down the study space as there are thousands of articles out there. Given that this study focuses on e-government platforms, then this study are concentrating on articles from reputable Journals, Conferences Proceedings, and Book Chapters. All these articles must be in English language. Table 1 shows the criteria for this review (Fig. 1).

Table 1. The Inclusion and Exclusion Criteria.

Inclusion Criteria	Exclusion Criteria
Must be in full text	Uncompleted studies
Published in the selected database	Not in English
English version of references	Duplicated studies
Domain of *government, local government, e-government, Indonesia government, sustainability, sustainability IT development, sustainability e-government usage*	Remove un-related domain

In developing the initial sustainable model which contains the sustainability development plan regarding the usage of e-government application, there is a sustainable model which been selected as based theory in the modeling design the for this local e-government application domain area. Table 2 is presenting the details about the selected model design theory which been selected as a model based design for the sustainability model of the local e-government application usage among Indonesia citizen. The table consists of model design name, author information, research area, and the specific

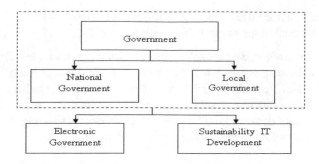

Fig. 1. The defined keywords before study selection.

focus of the model. Then, Table 3 followed afterwards as presenting the selected construct which been figured out from the SLR method of study that be used in developing the proposed sustainable model of e-government application usage in Indonesia local government.

Table 2. The Selected Model Design Theory

Model Design Theories	Author	Research Area	Specific Focus
The sustainable model on competition under government intervention	[23]	Governance	Objectives for social, environmental, and economics

Table 3. The Selected Constructs From SLR

Constructs	Factors
Information technology	ICT development
	Human Computer Interaction (HCI)
	National Telecommunication Infrastructure (TII)
E-Government applications	Government Online Services (OSI)
Governance Policy	Business
	Professional

After completed the process of deciding the dimension for sustainable development plan for this study, another three e-government applications model theories been used in order to figure out the related factors of this study. The three e-government model design theories which been studied are as follows:

- The sustainable model on shared values for e- The E-Government development effects on corruption Model [6].

- The e-government in sustainable public procurement model [4].
- The E-government in the public health sector model [9]

Then, this research study will select and combine several factors from the above theoretical models of e-government applications usage concept toward the research dimensions domain. Next, the model is been developed in order to become guidelines towards the sustainability of e-government applications usage among citizen of Indonesia local government domain area. The following Table 4 is showing the definition on each selected factors for the sustainable model development.

5 Sustainable Model of E-Government Application Usage in Indonesia Local Government

These identified factors will be used in developing the sustainable model of e-government application usage in Indonesia local government. The operational definition for the six factors is as in Table 4.

Table 4. The Definition of Each Factors

No.	Factor	Definition	Source
1	ICT Development	ICT is to attain and maintain the competitiveness while improving the profitability and proven been succeed in current dynamic environment situation	[24]
2	Human Computer Interaction (HCI)	HCI community has increasingly interested in understanding, designing for, and measuring the engagement among user with a host of computer included search applications	[25]
3	National Telecommunication Infrastructure (TII)	TII is the backbone of the fast-growing knowledge economy and involved with the telecommunication services	[26]
4	Government Online Services (OSI)	The OSI involved the users request locally, daily services on online platform such as by using mobile-based apps and then receive the service output	[27]
5	Business	The government involved in supporting the sharing economy in order to achieve the sustainable development	[3]
6	Professional	The professional refers to the ability to listen and communicate effectively which is refers to the demonstration and application of professional behaviours, values and attitudes	[28]

Based on the identified factors, the initial model in sustainable e-government applications usage in Indonesia local government is proposed and the model is clearly implementing the concept of the establish sustainable model from previous researcher in order to avoid any biased concept in delivering the sustainability value of study (Fig. 2).

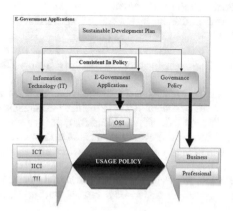

Fig. 2. The Proposed Model

6 Conclusion

This study provides an overview of the sustainable model on e-government application usage for Indonesia local government institutions. In order to understand the sustainable model concept, this paper went through the SLR process. This paper is focused on the SLR which consists of searching strategy to find out all the factors in maintaining the online reputation while dealing with the sustainable model theories. The SLR approach used to filter out all the unrelated articles and sort the related one. As a result of SLR, there are three constructs which been figured out for bring out related factors for sustainable model on e-government application usage in Indonesia local government. The constructs are consisting of IT, e-government application, and governance policy. From those three constructs, there are six factors which been identified such as ICT development, HCI, TII, OSI, business, and professional. For the next related work, the suggested model in sustainable model on e-government application usage in Indonesia local government is going on the validation process. The next stage involved with data collections and analysis in order to validate the model.

References

1. Ali, O., Osmanaj, V.: The role of government regulations in the adoption of cloud computing: a case study of local government. Comput. Law Secur. Rev. **36**, 105396 (2020)
2. Wang, G., Chen, Q., Xu, Z., Leng, X.: Can the use of government Apps shape citizen compliance? The mediating role of different perceptions of government. Comput. Human Behav. **108**, 106335 (2020)

3. Gao, P., Li, J.: Understanding sustainable business model: a framework and a case study of the bike-sharing industry. J. Clean. Prod. **267**, 122229 (2020)
4. Adjei-Bamfo, P., Maloreh-Nyamekye, T., Ahenkan, A.: The role of e-government in sustainable public procurement in developing countries: a systematic literature review. Resour. Conserv. Recycl. **142**, 189–203 (2019)
5. Pérez-Morote, R., Pontones-Rosa, C., Núñez-Chicharro, M.: The effects of e-government evaluation, trust and the digital divide in the levels of e-government use in European countries. Technol. Forecast. Soc. Change **154**, 119973 (2019)
6. Adam, I.O.: Examining e-Government development effects on corruption in Africa: the mediating effects of ICT development and institutional quality. Technol. Soc. **61**, 101245 (2020)
7. Fatimah, Y.A., Govindan, K., Murniningsih, R., Setiawan, A.: A sustainable circular economy approach for smart waste management system to achieve sustainable development goals: case study in Indonesia. J. Clean. Prod. **269**, 122263 (2020)
8. Hapsara, M., Imran, A., Turner, T.: Beyond organizational motives of e-Government adoption: the case of e-voting initiative in indonesian villages. Procedia Comput. Sci. **124**, 362–369 (2017)
9. Tursunbayeva, A., Franco, M., Pagliari, C.: Use of social media for e-Government in the public health sector: a systematic review of published studies. Gov. Inf. Q. **34**(2), 270–282 (2017)
10. Potrafke, N.: General or central government? Empirical evidence on political cycles in budget composition using new data for OECD countries. Eur. J. Polit. Econ. **63**, 101860 (2020)
11. Gruzd, A., Lannigan, J., Quigley, K.: Examining government cross-platform engagement in social media: Instagram vs Twitter and the big lift project. Gov. Inf. Q. **35**(4), 579–587 (2018)
12. DePaula, N., Dincelli, E., Harrison, T.M.: Toward a typology of government social media communication: democratic goals, symbolic acts and self-presentation. Gov. Inf. Q. **35**(1), 98–108 (2018)
13. Rasti-Barzoki, M., Moon, I.: A game theoretic approach for car pricing and its energy efficiency level versus governmental sustainability goals by considering rebound effect: a case study of South Korea. Appl. Energy **271**, 115196 (2020)
14. Steiner, N.D.: Economic globalisation, the perceived room to manoeuvre of national governments, and electoral participation : evidence from the 2001 British general election. Elect. Stud. **41**, 118–128 (2016)
15. Mizuno, N.: Political structure as a legacy of indirect colonial rule : bargaining between national governments and rural elites in Africa. J. Comp. Econ. **44**, 1023–1039 (2016)
16. Van Egmond, S., Hekkert, M.P.: International journal of greenhouse gas control analysis of a prominent carbon storage project failure – the role of the national government as initiator and decision maker in the barendrecht case. Int. J. Greenhouse Gas Control **34**, 1–11 (2015)
17. López, N.R., García, J.M., Toril, J.U., de Pablo Valenciano, J.: Evolution and latest trends of local government ef fi ciency: worldwide research (1928–2019). **261**, 121276 (2020)
18. Lee, H., Choi, Y., Seo, H.: Technological forecasting & social change comparative analysis of the R&D investment performance of Korean local governments. Technol. Forecast. Soc. Chang. **157**, 120073 (2020)
19. Feng, Y., Wu, F., Zhang, F.: Land Use Policy The development of local government financial vehicles in China: a case study of Jiaxing Chengtou. Land Use Policy **112**, 104793 (2020)
20. Kuzemko, C., Britton, J.: Energy Research & Social Science Policy, politics and materiality across scales : a framework for understanding local government sustainable energy capacity applied in England. Energy Res. Soc. Sci. **62**, 101367 (2020)
21. Helingo, M., Purwandari, B., Satria, R., Solichah, I.: The use of analytic hierarchy process for software development method selection: a perspective of e-Government in Indonesia. Procedia Comput. Sci. **124**, 405–414 (2017)

22. Sabani, A., Farah, M.H., Sari Dewi, D.R.: Indonesia in the spotlight: combating corruption through ICT enabled governance. Procedia Comput. Sci. **161**, 324–332 (2019)
23. Hafezalkotob, A.: Competition of domestic manufacturer and foreign supplier under sustainable development objectives of government. Appl. Math. Comput. **292**, 294–308 (2017)
24. Yunis, M., Tarhini, A., Kassar, A.: The role of ICT and innovation in enhancing organizational performance: the catalysing effect of corporate entrepreneurship. J. Bus. Res. **88**, 344–356 (2018)
25. O'Brien, H.L., Cairns, P., Hall, M.: A practical approach to measuring user engagement with the refined user engagement scale (UES) and new UES short form. Int. J. Hum. Comput. Stud. **112**, 28–39 (2018)
26. Alizadeh, T., Farid, R.: Political economy of telecommunication infrastructure: an investigation of the national broadband network early rollout and pork barrel politics in Australia. Telecomm. Policy **41**(4), 242–252 (2017)
27. Zhang, S., Pauwels, K., Peng, C.: ScienceDirect The impact of adding online-to-of fl ine service platform channels on firms' of fl ine and total sales and pro fi ts. J. Interact. Mark. **47**, 115–128 (2019)
28. Brown, T., Yu, M., Etherington, J.. Are listening and interpersonal communication skills predictive of professionalism in undergraduate occupational therapy students? Health. Prof. Educ. **6**(2), 187–200 (2020)

Students' Physical Education Performance Analysis Using Regression Model in Machine Learning

Mohamed Rebbouj[(✉)] [iD] and Lotfi Said

Multidisciplinary Laboratory in Education Science and Training Engineering (LMSEIF),
Sport Science Assessment and Physical Activity Didactic, Normal Higher School (ENS-C),
Hassan II University of Casablanca, BP 50069, Casablanca, Morocco
Mohamed.rebbouj@enscasa.ma

Abstract. Collecting physical tests' results generates a large amount of data to analyse and use its outcomes for intelligent decision making and efficient actions on field [1]. Meanwhile, current known methods are based on transcribing outcomes and use them as a reference for comparison between pre-training and post training results without any constructive project design. However, in order to use data resources toward gaining performance and improving training sessions, a new approach is needed to shape a specific action method based on intelligent analysis system. In this case of study, we analyse data about 6OO high school students in Morocco, aged between 15 and 20 years old (mean:16,21, SD:0,92) in 2021–2022 scholar year, submitted to pass 44 physical and sport tests to collected structured data derived from their performances to be pre-processed, modelled, and analysed using machine learning linear regression approach. This process covers steps of profiling the dataset, defining primary metric as Normalized Root Mean Squared error (0.232) so we can opt for the best model performance. As a result, our model uses Voting Ensemble algorithm with Mean Absolute Error value = 3,607.683, Mean squared error value = 574.27 and R^2 = 0.157. The label prediction is used as a target to reach in the end of the scholar year, where a comparison between predicted and actual values will demonstrate how efficient the training sessions was. The features selection provides an explication of the most impacting variables on students' performance allowing any training planification to relay on their importance respectively based on their density that affects prediction.

Keywords: Physical education · Performance · Machine Learning ·
Algorithms · Regression

1 Introduction

Sport results prediction has always been the interest of tipsters and domain experts as well as those in the betting market [2]. Becoming the most interested field to make prediction using Artificial Intelligence techniques summarized by the Machine Learning and Deep Learning algorithms, where a big amount of data about the teams and players are valuable [3]. Moreover, this modelling approach and results prediction outcomes promotes decision-making toward strategies deployment and training planification [4].

F. Saeed et al. (Eds.): ICACIn 2022, LNDECT 179, pp. 682–692, 2023.
https://doi.org/10.1007/978-3-031-36258-3_60

Regression models' techniques in ML has been used by many researchers in different sectors such as health sector [5, 6], and its cases uses for modelling predictions and times series forecasting for industrial and environmental implementation has shown promising results [7].

For more specific field, physical education and sport activities are designated for students in an age of growth and strength, where performance can be measured continuously in many activities with specific features and performance indicators that are practiced during the scholar year. Early adulthood has been subject of physical activities research about health outcomes [8]; furthermore, psychological and behavioural features' effectiveness in sport activities have been analysed in meta regression approach [9], which means that the regression models are best fit for predicting outcomes of a continuous numeric value.

Therefore, regression approach is commonly used to estimate forecasting models for matches outcomes, or team performance in a match [10], In order to reach such results, many variables have been considered for the modelling process related to the teams score, draw or lose and for some researchers added score difference and other independent variables related to the teams as home or away, distance of travel, weather, position in the league, yellow cards, red cards and motivation before the matches [11]. However, while we are dealing with students' data in a stable environment, we focus our analysis method on how to use the model prediction outcomes for a training planning and monitoring, so we obtain best performance results [12].

For this reason, we set physical and performance features to build a dataset: demographic and descriptive features as sex, age, weight, hight, cardiac frequency and performance features as time recorded in 60 m, 600 m, 1000 m or distance travelled in Cooper, Astrand, Luc-Leger, vertical and horizontal trigger tests. The data collection was carried out during physical education session for students, and weather conditions have not been considered.

Regarding the contribution of regression models in obtaining accurate results; this work describes the use of a machine learning method and analyses the model obtained to predict the outcomes of students' performance, based on 2020/2021 and 2021/2022 years of data.

The structure of this paper is as follows: a machine learning related works overview in Sect. 2, followed by a data collection and analysis method in Sect. 3, and finally a results discussion and conclusion in Sect. 4.

2   Materials and Method

Machine learning techniques were discussed in many works specifying the use of proper algorithms for prediction, where most of the research predict the outcomes of matches results rather than individual performance. [13] predict the distance of a thrown javelin by using regression algorithms, where [14] relay on predicting performance as future success for an athlete-selection process using regression model. In addition, [15] used nonlinear regression models and neural models to predict final swimming performance in distance as a label.

In our paper, we use a regression model with voting ensemble algorithm in Azure ML which is a cloud service for accelerating and managing the machine learning project lifecycle. Authoring ML experiences is done on the ML studio which is a web portal that combines no-code and code-first experiences for a data science platform [16].

2.1 Data Collection

Data has been manually collected on field and transcribed in Microsoft Excel, in a secondary high school by passing the 15 tests and measuring performance results. Data was entered directly to avoid any mistake and to keep transparency. The process of data collection was along with the continuous physical tests during two years for the 600 high school students.

2.2 Data Exploration

We run experiments at scale to train predictive models from the dataset and publish the trained models as services. After creating the ML resource with application insights, key vault and storage account, a compute cluster is shaped in the workspace. After this process, comes the creation of an ML run by uploading the dataset to a defined workspace datastore, the file type is tabular, and the format is configured as delimited, with comma delimiter and UTF-8 encoding without skipping rows; meanwhile, the visualisation of the dataset shows the ID numbers of subjects added, therefore this last feature won't be used. In order to run the ML model, we select the dataset to configure a run under an experiment name by targeting a column (in this case: the Upper_Muscle_Strenght as the label to predict with the condition of being an integer), then we select the ML compute cluster target created before.

In this process and for this type of dataset containing integer and string format in features, the ML tasks offers classification, regression and time series forecasting (to predict values based on time); so in our case we choose the regression since we are looking to predict continuous numeric values with configuration settings that shows the explainability on the best model with the normalized root mean squared error as a primary metric (used to optimize the model).

Concerning the exit criterion, the training job time has been configured to 0.5 h and the metric score threshold is set to 0.085. While the maximum concurrent iterations that should be executed in parallel is set to 2 considering the number of nodes (2) of the compute target (training cluster). The validation type is set to auto so we can enable some data guardrails without requiring a test for the dataset.

3 Results

3.1 Data Transformation

Data preprocessing, feature engineering, scaling techniques and the machine learning algorithm applied to generate the model we use is explained as follows:

- From data source, we have 43 features columns.
- Split into 19 categorical columns, 6 categorical Hash columns and 18 numerical columns.

- For categorical features, we used String cast-Char-Gram Count vectorizer and Mode Cat Imputer-String cast-Label Encoder algorithms. And for the categorical Hash features we used String Cast-Hash One Hot Encoder, then for the numerical features we used Mean Imputer.
- The last used algorithm that finalizes the process was the Prefitted Voting Ensemble.

In this description above, columns are the features describing records in rows; so splitting data is a featurization process, all the encoded and filled columns are fitted by averaging the individual predictions to form a final prediction.

3.2 Model Explanation

Normalized root mean square error (NRSME), also called a scatter index and is a metric that evaluates the model and returns a single value, expressed as absolute value between predicted and observed values in order to explain the best model to use for prediction. NRMSE is defined as follow:

$$NRMSE = \frac{\sum (S_i - O_i)2}{\sum O_i^2}$$

where O_i are observed (actual) values and S_i are simulated (predicted) values.

In our experiment, we have obtained 44 models where Voting Ensemble algorithm shows the lowest NRMSE value (0.23188), as is Stack Ensemble algorithm with 0.23188 and followed by Truncated SVDW rapper, Random Forest with 0.23233. The Table 1 below shows the difference in metrics for the 3 main algorithms.

Table 1. Reported metrics for the first three algorithms.

Metric	Voting Ensemble	Stack Ensemble	Truncated SVDW rapper & Random Forest
Spearman correlation	0.029	−1	0.025
R2 score	−0.051	−0.057	−0.059
Explained Variance	−0.005	2.2204	−0.008
RMSE	25.738	25.738	25.789
NRMSE	0.232	0.232	0.232
Mean absolute error	19.634	20.107	20.06
Mean absolute percentage error	15.301	15.858	15.819
Normalized mean absolute error	0.177	0.181	0.181

The model performance is shown below and explained by the distribution of the predicted values and the values of the model performance metrics, also by exploration of the dataset and the importance of the features (Fig. 1).

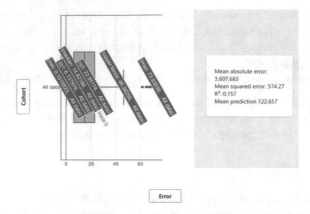

Fig. 1. Model performance chart.

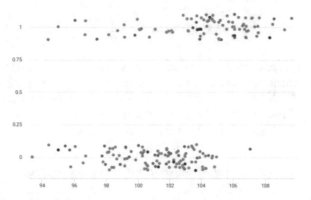

Fig. 2. Dataset explorer chart by gender vs. predicted.

This chart shows how gender category is distributed on the graph; Male are in top around value 1 while the female is below around value 0. The colors designate the age from 15 to 20 years old. We can visualize this chart as aggregate plots to better understand the distribution in respect to a predicted value (Fig. 2).

However, feature importance makes a difference when planning for training session – when taken into consideration -, that's why our regression model explain the importance in a dependence plots chart. We limit the importance in 4 feature that we explore individually as shown below (see Fig. 5).

Important features with high impact are indexed by their density that affects the prediction, the other features are less impactful on the results. Each of these four features can be explored individually in a density plot (Fig. 3).

3.3 Metrics

Metrics are the returned values for the specific type of model supported for use with. For a regression model as, we are deploying in this study, they are designed to estimate

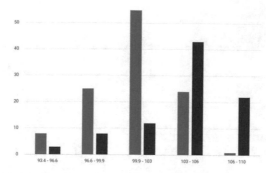

Fig. 3. Aggregate plot for gender vs. predicted label (light blue = F, dark blue = M).

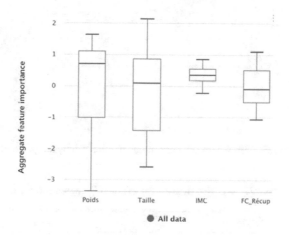

Fig. 4. Top 4 features by their importance compared to all data.

the amount of error; however, the pattern of the residuals (the difference between any one predicted point and its corresponding actual value) can explain potential bias in the model (Fig. 4).

In this case, we have obtained the following metrics shown in the Table 2 below and explained in charts format in the following:

Concerning the relationship between the target feature (true/actual value) and the model's prediction, the predicted vs. true chart plots the model bias toward predicting certain values while the line displays the average prediction, and the shaded area indicates the variance of predictions around that mean. We will put the three charts of the first three accurate algorithms in order to compare their performance.

As noticed, the charts are slightly the same in shape and that because of the approximate values of the NRMSE for the three algorithms. For better explication, the residual chart is a histogram of the prediction errors (residuals) generated for regression and forecasting experiments. This model shows a spread-out residuals distribution with fewer samples around zero, so is considered as not good enough (Fig. 6).

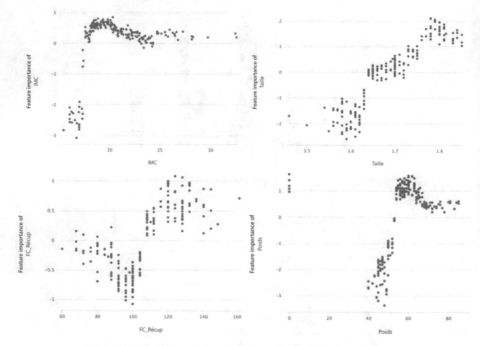

Fig. 5. Top 4 features importance explication by single chart.

Table 2. Reported metrics for evaluating regression model.

Metric	Value
Explained variance	−0.005
Mean absolute error	19.634
Mean absolute percentage error	15.301
Median absolute error	16.374
Normalized mean absolute error	0.177
Normalized median absolute error	0.148
Normalized root mean squared error	0.232
Normalized root mean squared log error	0.238
R^2 score	−0.051
Root mean squared error	25.738
Root mean squared log error	0.191
Spearman correlation	0.029

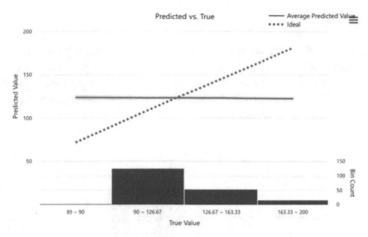

Fig. 6. Predicted vs. true chart for voting ensemble algorithm.

3.4   Prediction

Prediction is made by adding new collected features data in a JSON format. This process take time while we are predicting person by person. However, in order to predict the label (target score = Upper_muscle_strenght) and compare it with the actual results, the Table 3 below contains the 10 first comparison between the actual and predicted values.

Table 3. Actual vs. predicted values using the regression model.

N°	Actual value	Predicted Value
1	100	99
2	130	124
3	106	127
4	108	118
5	98	99
6	150	147
7	86	87
8	138	138
9	87	100
10	110	103

4   Discussion

Training and evaluating a regression model provide a disposal to predict numeric value of a label. And since we are looking to determine variables with impact on students' performance, we can use the prediction as a comparative way to examine the real obtained values after a training session and the predicted values, based on reinforcing the important features of concerned tests (Fig. 7).

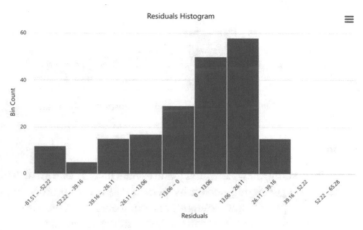

Fig. 7. Residuals chart for the regression model of the voting ensemble algorithm.

The obtained results of analyses contain many metrics allowing to explore the data by gender, age, features and label to reach in the end; regarding the conditions of tests passing. Which should be incorporated in further analyses and may impact the way of dealing with physical education and sport activities by changing the test modalities, dropping some useless test and looking for new ones with higher impact on performance. Considering the comparison between the actual and predicted values, these last approaches the real values and complies in two cases from 10. Which we consider as a good references for planning students' achievement by metrics, even though the model is not as performant as we wished for our first attempt. Therefore, in education environment, the fact of exceeding the expected scores and results is desirable and may strengthen the confidence and the motivation of students in training lessons. Moreover, teachers and sport activities actors may take the predicted values as goals to reach or over pass in the end of a training session.

Different approaches have demonstrated the contribution of neural models in predicting results better than the regression models [17]. Even if the model represents a modest accuracy to make prediction, we can try other algorithms with a developed dataset containing valuable features related to weather, infrastructure and to the students mental and psychological state [18, 19].

5 Conclusion

Predicting students' performance in this case has shown promising results in order to plan training session and for athletes' selection aimed at local and national competitions. Therefore, more features are needed to specify each physical test and activity so the dataset will have a meaningful result for validation and for prediction. This first attempt has changed the way of teaching physical education and sport activities in our school, by incorporating the new AI approach and by involving students in measuring, calculating and data entry. Moreover, the way we input data to predict in a JSON format or in a notebook turns out time consuming and inapprehensible for student to learn quickly, that is why a front-end interface for users could simplify the transcription of the new data and can encourage all the actors around physical activities to use the solution as part for their training program planification to reach the predicted performance value obtained. In this instance, unceasing the dataset feed with the new results may consolidate the prediction process.

References

1. Zhou, Y., Chen, X.: Simulation of sports big data system based on Markov model and IoT system. Microprocess. Microsyst. **80**, 103525 (2020)
2. Hubacek, O., Sourek, G., Zelezny, F.: Exploiting sports-betting market using machine learning. Int. J. Forecast **35**, 783–796 (2019)
3. Paul, R.J., Weinbach, A.P.: Price setting in the NBA GamblingMarket: tests of the Levitt model of sportsbook behavior. Int. J. Sport Finan. **3**(3), 137–145 (2008)
4. Araújo, D., Hristovski, R., Seifert, L., Carvalho, J., Davids, K.: Ecological cognition: expert decision-making behaviour in sport. Int. Rev. Sport Exerc. Psychol. **12**(1), 1–25 (2019)
5. Naji, M.A., El Filali, S., Bouhlal, M., Benlahmar, E.H., Ait Abdelouhahid, R., Debauche, O.: Breast cancer prediction and diagnosis through a new approach based on majority voting ensemble classifier. In: International Workshop on Edge IA-IoT for Smart Agriculture (SA2IOT) August, pp. 9–12, 2021, Leuven, Belgium, Leuven (2021)
6. Ni, Z., et al.: Prediction model and nomogram of early recurrence of hepatocellular carcinoma after radiofrequency ablation based on logistic regression analysis. Ultrasound Med. Biol. **48**(9), 1733–1744 (2022)
7. Kshirsagar, A., Shah, M.: Anatomization of air quality prediction using neural networks, regression and hybrid models. J. Cleaner Prod. **369**, 133383 (2022)
8. Gallant, F., Sylvestre, M.P., O'Loughlin, J., Bélanger, M.: Teenage sport trajectory is associated with physical activity, but not body composition or blood pressure in early adulthood. J. Adoles. Health **71**(1), 119–126 (2022)
9. Barker, J.B., et al.: The effectiveness of psychological skills training and behavioral interventions in sport using single-case designs: a meta regression analysis of the peer-reviewed studies. Psychol. Sport Exerc. **51**, 1469–2292 (2020)
10. Goddard, J.: Regression models for forecasting goals and match results in association football. Int. J. Forecast. **21**(2), 331–340 (2005)
11. Ehmann, P., et al.: 360°-multiple object tracking in team sport athletes: reliability and relationship to visuospatial cognitive functions. Psychol. Sport Exerc. **55**(6), 101952 (2021)
12. Friel, J.: The Triathlete's Training Bible, 2nd edn. VeloPress, Boulder (2004)
13. Maszczyka, A., Gołaś, A., Pietraszewski, P., Roczniok, R., Zając, A., Stanula, A.: Application of neural and regression models in sports results prediction. Procedia. Soc. Behav. Sci. **117**, 482–487 (2014)

14. Maszczyk, A., Zając, A., Ryguła, I.: A neural network model approach to athlete selection. Sports Eng. **13**, 83–93 (2011)
15. Maszczyk, A., Roczniok, R., Waśkiewicz, Z., Czuba, M., Mikołajec, K.: Application of regression and neural models to predict competitive swimming performance. Percept. Mot. Skills **114**(2), 610–626 (2012)
16. Microsoft, What is Azure Machine Learning studio? Microsoft, 04 02 2022. https://docs.mic rosoft.com/en-us/azure/machine-learning/overview-what-is-machine-learning-studio
17. Bartlett, R., Müller, E., Lindinger, S., Brunner, F., Morriss, C.: Three-dimensional evaluation of the kinematic release parameters for javelin throwers of different skill levels. J. Appl. Biomech. **12**, 7–14 (1996)
18. Pezzoli, A., Cristofori, E., Moncalero, M., Giacometto, F., Boscolo, A.: Effect of the environment on the sport performance. In: icSPORTS2013, Villamoura (2013)
19. Menegassi, V.M., et al.: Impact of motivation on anxiety and tactical knowledge of young soccer players. J. Phys. Educ. Sport **18**, 170–174 (2018)

Author Index

F. Saeed et al. (Eds.): ICACIn 2022, LNDECT 179, pp. 693–695, 2023.
https://doi.org/10.1007/978-3-031-36258-3

Printed in the United States
by Baker & Taylor Publisher Services